HANDBOOK OF APPLIED SYSTEM SCIENCE

The *Handbook of Applied System Science* is organized around both methodological approaches in systems science, and the substantive topic to which these approaches have been applied. The volume begins with an essay that introduces three system science methods: agent-based modeling, system dynamics, and network analysis. The remainder of the volume is organized around three broad topics: (1) health and human development, (2) environment and sustainability, and (3) communities and social change. Each part begins with a brief introductory essay, and includes nine chapters that demonstrate the application of system science methods to address research questions in these areas. This *Handbook* will be useful for work in Public Health, Sociology, Criminal Justice, Social Work, Political Science, Environmental Studies, Urban Studies, and Psychology.

Zachary P. Neal is an Associate Professor at Michigan State University, with a joint appointment in the Psychology Department (Ecological–Community Program) and the Global Urban Studies Program. He serves as Editor of the *Journal of Urban Affairs*, and is a Senior Fellow at the Brookings Institution.

HANDBOOK OF APPLIED SYSTEM SCIENCE

Edited by
Zachary P. Neal

Routledge
Taylor & Francis Group
NEW YORK AND LONDON

First published 2016
by Routledge
605 Third Avenue, New York, NY 10017

and by Routledge
2 Park Square, Milton Park, Abingdon, Oxon, OX14 4RN

Routledge is an imprint of the Taylor & Francis Group, an informa business

Library of Congress Cataloging in Publication Data
Handbook of applied system science/[edited] by Zachary Neal.
pages cm
ISBN 978-0-415-84332-4 (hardback: alk. paper)—
ISBN 978-0-415-84334-8 (pbk.: alk. paper)—
ISBN 978-1-315-74877-1 (ebook)
1. Social sciences—Research—Methodology. 2. System theory—Social aspects. 3. Social systems.
I. Neal, Zachary P.
H61.H2788 2015
300.1'1—dc23

ISBN: 978-0-415-84332-4 (hbk)
ISBN: 978-0-415-84334-8 (pbk)
ISBN: 978-1-315-74877-1 (ebk)

DOI: 10.4324/9781315748771

Typeset in Bembo
by Florence Production Ltd, Stoodleigh, Devon, UK

CONTENTS

Detailed contents *ix*

1. **What Is System Science?** **1**
   Z.P. Neal

PART 1
Health and Human Development **21**

2. **The Social Symbiome Framework** **25**
   B.A. Pescosolido et al.

3. **The Impact of Urban Form on Weight Loss** **49**
   A. Ligmann-Zielinska et al.

4. **Promoting Social Contagion of Preventive Behavior during Influenza Epidemics** **68**
   L. Mao

5. **Simulating Syndemic Risk** **84**
   A.W. Batchelder & D. Lounsbury

6. **System Dynamics Modeling and Finding Solutions to the "Wicked" Public Health Problem of Preventing Chronic Diseases** **107**
   N. Unwin et al.

7. **Network Analysis and Psychology** **130**
   M. Vitevitch

8. **Using Cognitive Social Structures to Understand Peer Relations in Childhood and Adolescence** 147
J.W. Neal & M. Kornbluh

9. **Chains of Affection** 164
P.S. Bearman et al.

10. **The Art of Building Dynamic Systems Models** 191
S. Kunnen

PART 2
Environment and Sustainability 207

11. **A System Dynamics Examination of the Willingness of Villagers to Engage in Illegal Logging** 211
R.G. Dudley

12. **Exploring Complexity in a Human–Environment System** 229
L. An et al.

13. **Agent-Based Modeling in Coupled Human and Natural Systems (CHANS)** 267
L. An et al.

14. **Eutropia** 297
D.N. Barton et al.

15. **Using System Dynamics to Model Industry's Developmental Response to Energy Policy** 321
B. Bush et al.

16. **Integrated Agent-Based and System Dynamics Modelling for Simulation of Sustainable Mobility** 341
E. Shafiei et al.

17. **Stakeholder Analysis and Social Network Analysis in Natural Resource Management** 367
C. Prell et al.

18. **The Multilevel Participatory Modelling of Land Use Policies in African Drylands** 384
P. d'Aquino & A. Bah

19. **A Systems Approach to Stakeholder Management** 403
R. Boutilier et al.

PART 3

Communities and Social Change

Communities and Social Change 419

20. Modeling Social Ties and Household Mobility 423
S.S. Metcalf

21. Simulating Sprawl 448
P.M. Torrens

22. Agents of Change 486
M. Ellis et al.

23. The (In)compatibility of Diversity and Sense of Community 512
Z.P. Neal & J.W. Neal

24. Spatializing Social Networks 528
S.M. Radil et al.

25. Social Networks and the Study of Language Variation and Change 553
S.E. Wagner & M.R. Abtahian

26. Community as Method, Community as Net 570
K. Warren & N. Doogan

27. Circles of Association 585
D.S. Connor & R. Levine

28. Using System Dynamics Modeling to Understand the Impact of Social Change Initiatives 599
G.B. Hirsch et al.

About the Contributors 621
Index 624

DETAILED CONTENTS

System Science Method(s)

ABM Net SD

1. **What is System Science?** 1
Zachary P. Neal

PART 1
Health and Human Development 21

2. **The Social Symbiome Framework: Linking Genes-to-Global Cultures in Public Health Using Network Science** 25
Bernice A. Pescosolido, Sigrun Olafsdottir, Olaf Sporns, Brea L. Perry, Eric M. Meslin, Tony H. Grubesic, Jack K. Martin, Laura M. Koehly, William Pridemore, Alessandro Vespignani, Tatiana Foroud, and Anantha Shekhar

Hundreds, if not thousands, of factors have been documented as etiological candidates underlying illness, disease, and treatment. We propose a parsimonious theoretical guide to scientific progress, clinical practice, and improved population health by embracing the interactive, contextual, and dynamic assumptions of systems science, but narrowing its scope through network science's focus on connections as the unifying mechanism of action. Using alcohol dependence as an illustrative case, we introduce the Social Symbiome and consider unique issues in team formation, study design, and analytic tools for rigorous, feasible studies with adequate human protections.

3. **The Impact of Urban Form on Weight Loss: Combining a Spatial Agent-Based Model with a Transtheoretical Model of Health Behavior Change** 49
Arika Ligmann-Zielinska, Sue C. Grady, and Jeremy McWhorter

Since the early 1980s, obesity has become a major health issue in the United States. Here, we develop a GIS-based agent-based model to studying the prevalence of obesity in selected populations and districts in San Diego, California. This model allows for the explicit representation of individual health behavior, the incorporation of the influence of social networks on physical activity, and the spatially heterogeneous depiction of urban form in which the individuals operate. Based on our computational experiments, we conclude that the selected geographic factors play a substantial role in the spatiotemporal variability of individual weight measured with the body mass index.

4. Promoting Social Contagion of Preventive Behavior during Influenza Epidemics: An Agent-based Simulation **68**

Liang Mao

Outbreaks of influenza have drawn unprecedented attention in recent years, and a natural response of people is to adopt flu preventive behaviors, such as getting vaccinated, taking prophylactic medicines, and increasing hand washing. I describe an agent-based epidemic-behavior model developed to simulate the transmission of influenza and the diffusion of flu prophylactic adoption together in a realistic urbanized area of Buffalo, NY, USA. The model is used to evaluate two types of promotional strategies: an incentive strategy for households and a "role-model" strategy for workplaces.

5. Simulating Syndemic Risk: Using System Dynamics Modeling to Understand Psycho-Social Challenges Facing Women Living With and At-Risk for HIV **84**

Abigail W. Batchelder and David Lounsbury

Best practices in system dynamics modeling call for multi-stepped iterative procedures involving problem identification, system conceptualization, model formulation, model simulation, and model evaluation. We describe how we built and validated a system dynamics model of syndemic risk among women living with and at-risk for HIV in a low-resource urban environment, which we use to generate insights and implications for designing clinical, community, and public health interventions for this population. We conclude by reflecting on the system dynamics model-building process, including careful deliberation by the modelers and stakeholders.

6. System Dynamics Modeling and Finding Solutions to the "Wicked" Public Health Problem of Preventing Chronic Diseases **107**

Nigel Unwin, Cornelia Guell, Natasha Sobers-Grannum, and Anders Nielsen

Chronic non-communicable diseases account for over 60% of deaths globally, and share four main behavioral risk factors: smoking, excess alcohol consumption, physical inactivity, and aspects of diet. We review the use of system dynamics modeling to inform the prevention of these diseases, and describe four case studies aimed at smoking cessation, alcohol control, and the prevention of type 2 diabetes and cardiovascular disease. We conclude by making recommendations on how to better realize the potential for improving population health afforded by system dynamics models and similar methodologies.

7. Network Analysis and Psychology **130**
Michael Vitevitch

I highlight examples of network analysis in the domains of Social Psychology (to examine how a group solves simple and complex problems), Biological Psychology (to map out the interconnectivity of the brain), Cognitive Psychology (to understand language use and development), and Clinical Psychology (to redefine psychopathology). I consider the benefits and the hazards of using network analysis, and discuss the gains that network analysis and Psychology can make by collaborating with each other.

8. Using Cognitive Social Structures to Understand Peer Relations in Childhood and Adolescence **147**
Jennifer Watling Neal and Mariah Kornbluh

Focusing on three case studies, we illustrate how three different formats of cognitive social structure network data – locally aggregated structures, slices, and consensus aggregation – can be used to understand different aspects of children's and adolescents' peer relationships. We conclude by examining the benefits and drawbacks of using this unique type of network data.

9. Chains of Affection: The Structure of Adolescent Romantic and Sexual Networks **164**
Peter S. Bearman, James Moody, and Katherine Stovel

We describe the structure of the adolescent romantic and sexual network in a population of over 800 adolescents residing in a midsized town in the Midwestern United States. Using a series of simulation-based tests, we seek to understand the micromechanisms that generate such networks, which have implications for disease transmission dynamics and social policy.

10. The Art of Building Dynamic Systems Models **191**
Saskia Kunnen

Dynamic systems models must be built up in stages, beginning with theoretical notions that are translated into a conceptual model, which in turn is translated into a series of quantitative expressions. I illustrate the steps in this process using the case of adolescent commitment development, and conclude with a discussion of the possibilities and the limitations of this technique.

PART 2
Environment and Sustainability 207

11. A System Dynamics Examination of the Willingness of Villagers to Engage in Illegal Logging 211
Richard G. Dudley

I use a system dynamics simulation to explore Indonesian villagers' motivations to engage in illegal logging, including the need for income, the fact that other villagers and non-villagers are already illegally logging, and the realization of loss of community control over traditional forest areas. The model serves as a tool for understanding the consequences of various proposed strategies to control illegal logging including enforcement of laws, strengthening of community rights, the prevention of outside labor in local forests, and the provision of alternate sources of income.

12. Exploring Complexity in a Human–Environment System: An Agent-Based Spatial Model for Multidisciplinary and Multiscale Integration 229
Li An, Marc Linderman, Jiaguo Qi, Ashton Shortridge, and Jianguo Liu

We use data from the Wolong Nature Reserve for giant pandas in China to simulate the impact of the growing rural population on the forests and panda habitat. By tracking the life history of individual persons and the dynamics of households, this model equips household agents with "knowledge" about themselves, other agents, and the environment and allows individual agents to interact with each other and the environment through their activities in accordance with a set of artificial-intelligence rules. The households and environment co-evolve over time and space, resulting in macroscopic human and habitat dynamics.

13. Agent-Based Modeling in Coupled Human and Natural Systems (CHANS): Lessons from a Comparative Analysis 267
Li An, Alex Zvoleff, Jianguo Liu, and William Axinn

To understand the utility of agent-based models to explore the dynamics of coupled human and natural systems, we compare the features of two such models: one of the Wolong Nature Reserve in China (see Chapter 12) and another of the Chitwan National Park in Nepal. We find that four issues are particularly important for modeling CHANS: representation agents and the landscape, the need for standardized modules, the impacts of scheduling on model outcomes, and using caution in interpreting "surprises" in model outcomes.

14. Eutropia: Integrated Valuation of Lake Eutrophication Abatement Decisions Using a Bayesian Belief Network 297
David N. Barton, Tom Andersen, Olvar Bergland, Alexander Engebretsen, S. Jannicke Moe, Geir I. Orderud, Koji Tominaga, Eirik Romstad, and Rolf D. Vogt

Freshwater eutrophication, the oversupply of nutrients to an ecosystem, is one of the major environmental challenges around the world. By combining systems dynamic, empirical and expert judgment models into a single object-oriented Bayesian network, we aim to assess trade-offs in ecological, social and economic values from improving lake ecological condition through nutrient abatement measures in South-Eastern Norway. We find the integrated systems model is a useful tool for determining where additional data would be most useful, but that uncertainty limits its application to EU policy on eutrophication control.

15. Using System Dynamics to Model Industry's Developmental Response to Energy Policy **321**

Brian Bush, Daniel Inman, Emily Newes, Corey Peck, Steve Peterson, Dana Stright, and Laura Vimmerstedt

Biofuels represent one avenue toward reducing dependence on imported energy as well as reducing the emissions of greenhouse gases. Here, we use the Biomass Scenario Model (BSM), a system dynamics model developed at the National Renewable Energy Laboratory, to explore the potential development of the biofuels industry. This model is designed to analyze the implications of policy on the development of the supply chain for biofuels in the United States by explicitly representing the behavior of decision makers including farmers, investors, fueling station owners, and consumers. We analyze several illustrative case studies that explore a range of policies and discuss how incentives interact with individual parts of the supply chain as well as the industry as a whole.

16. Integrated Agent-Based and System Dynamics Modelling for Simulation of Sustainable Mobility **341**

Ehsan Shafiei, Hlynur Stefansson, Eyjolfur Ingi Asgeirsson, Brynhildur Davidsdottir, and Marco Raberto

Despite rising fuel costs, significant barriers to the adoption of alternative fuel vehicles remain. To understand the diffusion of alternative fuel vehicles in Iceland, we develop a hybrid system model that combines system dynamics and agent-based techniques and takes into account the roles of car manufacturers, car dealers, consumers, energy supply system, fuel stations and government.

17. Stakeholder Analysis and Social Network Analysis in Natural Resource Management **367**

Christina Prell, Klaus Hubacek, and Mark Reed

The increasing use of stakeholder analysis in natural resource management reflects a growing recognition that stakeholders can and should influence environmental decision making, but identifying stakeholders is not always easy. We describe the use of social network analysis to identify the individuals and categories of stakeholders in the Peak District National Park in the United Kingdom, which subsequently guided steps for selecting stakeholders to participate in natural resource management decisions.

18. **The Multilevel Participatory Modelling of Land Use Policies in African Drylands: A Method to Embed Adaptability Skills of Drylands Societies in a Policy Framework** 384
Patrick d'Aquino and Alassane Bah

We describe the use of a participatory agent-based modelling approach called "self-design" in Senegal that lets farmers develop their own conceptual model of environmental management issues. Participants create a qualitative model that reflects the complexity of drylands environmental uncertainty, then use the model to shape uncertainty management principles for policy design. We consider how this unique inclusionary method, which involves and interlinks stakeholders from local to national levels, can lead to a novel multilevel policy design.

19. **A Systems Approach to Stakeholder Management** 403
Robert Boutilier, Witold Henisz, and Bennet Zelner

We introduce a data-driven systems approach for managing the risks and opportunities arising from a focal organization's stakeholder relations. The approach involves measuring the social license to operate (SLO) granted by stakeholders and the strength of the network connections among the stakeholders, then using agent-based modeling to predict which initiatives would do the most to raise the SLO. We demonstrate the approach using data from interviews with over 600 stakeholder organizations at six mines in five African countries.

PART 3
Communities and Social Change 419

20. **Modeling Social Ties and Household Mobility** 423
Sara S. Metcalf

Persistent urban problems like urban sprawl and socioeconomic segregation ultimately emerge from individuals' choices about where to live. To understand this process, I develop an agent-based model to simulate social and economic influences on neighborhood choice in Danville, Illinois. In this model, household agents decide whether and where to relocate within the community based on factors such as neighborhood attractiveness, affordability, and the household's social network. Comparison of simulated and observed migration patterns over a two-year period reveal the significance of components of social network choice such as income, race, and probability of local ties.

21. **Simulating Sprawl** 448
Paul M. Torrens

Suburban sprawl, a relatively recent phenomenon, is among the most important urban policy issues facing contemporary cities, but a well-accepted rationale has not been settled on for explaining and managing the causes of sprawl. To move toward a micro-level explanation of sprawl's origins, I develop a model for simulating the geographic dynamics of suburban sprawl and use it to explore the management of sprawl in three scenarios: a classic central city case, a polycentric urban development case, and the Midwestern megalopolis region centered on Chicago.

22. Agents of Change: Mixed-Race Households and the Dynamics of Neighborhood Segregation in the United States **486**

Mark Ellis, Steven R. Holloway, Richard Wright, and Christopher S. Fowler

We explore the effects of mixed-race household formation on trends in neighborhood-scale racial segregation using an agent-based model. The model reveals that high rates of mixed-race household formation reduce residential segregation considerably, even when preferences for own-group neighbors are high. However, it also reveals that neighborhood-scale segregation of single-race households can remain high even while a growing number of mixed-race households drives down the overall rate of residential segregation, highlighting the importance of parsing neighborhood segregation levels by household type.

23. The (In)compatibility of Diversity and Sense of Community **512**

Zachary P. Neal and Jennifer Watling Neal

Past research has consistently suggested that residentially segregated communities tend to be more socially cohesive, but why? We use an agent-based model of neighborhood social network formation to explore this phenomenon. The model demonstrates that under typical behavioral conditions, where people are more likely to form relationships with those who live nearby and/or are similar to themselves, this pattern will always emerge.

24. Spatializing Social Networks: Using Social Network Analysis to Investigate Geographies of Gang Rivalry, Territoriality, and Violence in Los Angeles **528**

Steven M. Radil, Colin Flint, and George E. Tita

Social processes unfold and are embedded in both geographic space and networks of relations. Highlighting these two kinds of embeddedness, we use spatialized network data to compare the geography of rivalry relations that connect territorially based criminal street gangs in a section of Los Angeles with a geography of the location of gang-related violence.

25. Social Networks and the Study of Language Variation and Change **553**

Suzanne Evans Wagner and Maya Ravindranath Abtahian

From the use of "ain't" to a new pronunciation of "o" in goes, or the adoption of "hoagie" for "sandwich," linguistic features are shown to be disseminated or blocked depending on the characteristics of speakers' networks. Here, we provide an overview of the application of social network theory to the field of variationist sociolinguistics, noting how the general findings hold both in the present and in the past, face-to-face and online, and in communities from Harlem to Hungary.

26. Community as Method, Community as Net: Social Network Analysis as a Tool for Studying Mutual Aid between Therapeutic Community Residents **570**
Keith Warren and Nathan Doogan

Therapeutic communities (TCs) depend on resident peer feedback, in which residents affirm each other for prosocial actions and correct each other for actions that contravene norms, as a primary means of clinical treatment. We examine peer feedback in six different TC units as a dynamic social network to understand why residents participate in this feedback system, and to which peers they are most likely to offer feedback. This approach yields insights into TC processes that are not available through either qualitative research or traditional statistical methods.

27. Circles of Association: The Connections of Community-Based Food Systems **585**
David S. Connor and Ralph Levine

We use a system dynamics framework to illustrate how community-based food systems lie at the center of a number of causal loops which may reinforce a series of potentially positive or negative outcomes including obesity, unemployment, fiscal deficits and farmland loss. The model helps identify levers of (e.g. training for entrepreneurial agriculture) and barriers to (e.g. fast food chains) change and ultimately the most effective pathway toward realizing the benefits of community-based food systems.

28. Using System Dynamics Modeling to Understand the Impact of Social Change Initiatives **599**
Gary B. Hirsch, Ralph Levine, and Robin Lin Miller

Action-oriented social scientists often seek to bring about social change, but anticipating the effects of proposed social changes is also important. We introduce system dynamics modeling as a strategy for examining the consequences of social change, focusing on the particular case of educational reform.

About the Contributors *621*
Index *624*

1

WHAT IS SYSTEM SCIENCE?

Zachary P. Neal

A system is a collection of separate parts that, because they are linked to and affect one another, are interdependent. This interdependence means that understanding how the system works requires understanding not only how each of the parts works individually, but also how they interact with each other. A mechanical clock offers a nice example. Each gear and spring is a separate part, but each of these separate parts interacts with other parts to make the clock run. To understand how the clock runs, it would be useless to study each gear and spring separately. The parts must be studied together, with a view toward understanding how each part interacts with and affects the other parts, and how in combination they produce the outcome that we call "telling time."

Even very complex mechanical clocks with hundreds of parts represent relatively simple, determinist systems. Because each gear interacts with the others in the same way every time, their behavior is highly predictable. This is, after all, what makes clocks useful. But, most of the systems we encounter on a daily basis are much more complicated and their behaviors cannot be so easily predicted. Consider your own neighborhood community. The community is composed of multiple, separate parts – you, your neighbors, the buildings and roads, the businesses and institutions, and so on – that interact with and affect one another. To really understand how the community works, it would be useless to study each part separately. Instead, we need to understand, for example, how you interact with each of your neighbors, or how the roads provide access to different businesses. Still further, we might need to understand how your interaction with your neighbor gave you information about a shortcut road to a business you frequently visit. By understanding how each of these separate parts interacts with the others, we can start to understand how in combination they produce the outcome that we call "the community."

System science methods are a broad collection of methods designed to help understand systems. Although there is no definitive list of such methods, this volume focuses on three that adopt different perspectives on understanding systems and that the Office of Behavioral and Social Sciences Research at the U.S. National Institutes of Health has identified as dominant. *Agent-based models* adopt a bottom-up perspective, seeking to understand systems by examining each interaction among each individual part of the system (e.g. each person's decision about where to live in the neighborhood). *System dynamics* adopts a top–down perspective, seeking to

 DOI: 10.4324/9781315748771-1

understand systems by examining the relationships among key variables that describe the system (e.g. how the number of people and number of businesses depend on each other). Finally, *network analysis* focuses on understanding a system's behavior as a function of the pattern of relationships that exist in-between the system's parts (e.g. which pairs of neighbors are friends and which are not). The common thread that unites these and other system science methods, and distinguishes them from other more conventional research methods, is their focus on understanding systems by looking at all their parts not individually but in combination as interdependent parts of a larger whole. The goal of this chapter is to briefly introduce the logic of these methods; more detailed technical introductions to using these methods can be found in the references below.

Bottom–Up: Agent-Based Models

Agent-based models are a type of computer-based simulation methodology in which agents interact with each other following simple behavioral rules. Because these types of models require the use of computers, and in some cases can be computationally intensive, they have a relatively short history, starting in the early 1970s with Schelling's model of segregation and Conway's "Game of Life" (Gardner 1970). They are often rooted in an epistemological stance known as methodological individualism, which views system-level outcomes as arising or emerging from the micro-level interactions of individual agents. For methodological individualists, a truly complete explanation of a system's outcomes must be framed in terms of the individual agents whose behaviors led to the outcome. This is a kind of reductionist epistemology, but one that innocently asks, if outcomes aren't caused by the actions and interactions of agents and their environments, where else could we possibly look for an explanation? Accordingly, the goal of many agent-based models is to understand the kinds of micro-level interactions that could be responsible for generating a known system-level outcome, or what Epstein (1999) called the generativist's question. To answer this question, Epstein proposed that researchers conduct what he called the generativist's experiment: "Situate an initial population of autonomous heterogeneous agents in a relevant spatial environment; allow them to interact according to simple local rules, and thereby generate–or 'grow'–the macroscopic regularity from the bottom up." This experiment provides a kind of recipe for building and testing agent-based models.

Agents

All agent-based models are composed, at a minimum, of agents and the rules that guide their behavior. These models are highly flexible and can be used to explore systems in many different contexts. In an agent-based model designed to understand a neighborhood community, the agents might be individual people, and the behavioral rules might specify how these people decide with whom to talk or form a friendship. Alternatively, in an agent-based model designed to understand an ecosystem's population dynamics, the agents might be individual animals of various species, and the behavioral rules might specify who eats who, under what conditions, and how often. Thus, what kind of entity the agents represent and what kind of rules they follow depend on the context of the system under investigation.

The agents in an agent-based model, whether they represent people or animals or molecules, are assumed to be autonomous. Each agent has some degree of agency and is not fully controlled by external forces. They are capable of acting on their own, in accordance with the specified behavioral rules. However, agents are not assumed to be completely independent. Indeed, the very essence of system science is the explicit recognition that parts of a system are interdependent. Thus, as agents interact with one another, these interactions can influence their behaviors. In addition to being autonomous yet interdependent, the agents may also be heterogeneous – that is, the agents may have unique characteristics that distinguish them from one another.

Considering the role of agents in a specific agent-based model helps to make these ideas more concrete. Schelling's model of segregation was designed to answer a specific generativist question: what kinds of neighborhood preferences, which guide individuals' decisions about where to live, could be responsible for generating residential segregation? That is, given that we already know residential segregation occurs, what kinds of individual behaviors could be responsible for it? In this model, the agents represent individual people or households. The agents are autonomous because they are free to choose where to live; their decisions are not dictated by an external force such as a law requiring mandatory segregation. But the agents are also interdependent: to the extent that people have preferences about who their neighbors are, one person's decision about where to live depends in part on other peoples' decisions about where to live. Finally, the agents are not all the same. In Schelling's model, people differ from one another on a single, binary characteristic: there are Type A people and Type B people. This characteristic is unspecified and could represent race, ethnicity, religion, or nearly any other observable social characteristic. For Schelling, and indeed for many agent-based models, the focus is on general processes rather than specific variables.

Rules

The agents populate an agent-based model, but the rules tell these agents what to do and how to interact with one another. Although the flexibility of agent-based models means that the rules could specify nearly anything, they nonetheless are expected to be simple, local, and limited. First, the rules that govern agents' behaviors must be simple because people and other entities with agency are not like supercomputers that can consider all possible courses of action and then select the optimal one. Instead, they tend to follow heuristics and rules of thumb that are easy to remember and implement. Second, the rules must be local. People are not omniscient; they do not know what is happening in distant parts of the world, or what is going on inside another person's head. Thus, the information they use to make decisions about what to do comes only from their local, observable environment. Finally, the rules must be limited in number. People generally do not have well-formed plans about what they would do in the event of every conceivable situation (e.g. what would I do if I saw a unicorn?), but rather have just a few behavioral strategies intended to cover the most likely situations.

The stark simplicity of the behavioral rules in Schelling's segregation model offers an excellent example: move to a new house if you are dissatisfied with the composition of your immediately surrounding neighbors. First, it is simple: look at who your neighbors are, and if you don't like them, move. Second, it is local: only take into account the people living immediately adjacent to your house. Finally, it is limited: there is only one rule. Of course, to be implemented, it is also necessary to define what it means for a person to be satisfied or dissatisfied with the composition of the neighborhood. For Schelling, satisfaction depends on a threshold value, X, that can be manipulated by the researcher. People are satisfied with their neighborhood if at least X% of their neighbors is the same type as themselves. Thus, if the researcher sets $X = 75$, then a Type A person would be satisfied if at least 75% of his neighbors were also Type A people, and a Type B person would be satisfied if at least 75% of her neighbors were also Type B people.

Modeling Process

The agent-based modeling process consists of several steps and is frequently iterative. The first step involves building the agent-based model itself, but there are multiple approaches to model building. In some cases, models are built from existing and well-established theories, for example, about the factors that influence when neighborhood residents interact with one another (see Chapter 23). In other cases, models are also informed by empirical data about real people and places, which are used as starting values to simulate realistic populations and environments (see Chapter 3). More recently, agent-based modelers have also explored participatory model-building strategies that directly involve research subjects in the development and testing of the model (see Chapter 18). Finally, as in many fields of research that are incremental, new agent-based models are often developed by adapting and extending earlier models (see Chapter 22). These different model building strategies – theory, data, participatory, and adaptive – are not mutually exclusive, but often are used in various combinations.

After a preliminary agent-based model is built, it is run multiple times not only to debug the model and ensure it is working properly, but also to develop an initial understanding of the types of system-level outcomes the model produces. Running an agent-based model typically consists of two stages: an initialization stage, and an interaction stage. In the initialization stage, the population of agents is created and assigned the various characteristics that might distinguish them from one another. In Schelling's model, the initialization stage is quite simple: a population of N people (the size of N can be set by the researcher) is created; half are Type A and half are Type B. These people are then randomly assigned locations in a grid, where each square in the grid represents a house in the neighborhood. The left box in Figure 1.1 shows a graphic depiction of a neighborhood of 100 people simulated by Schelling's model after the initialization stage. The two types of people are represented by different colors (here, black and gray). Because the people are assigned random locations, there is no segregation present; the Type A and Type B people are evenly mixed throughout the neighborhood.

In the interaction stage, the agents each take a turn following the behavioral rules defined by the model. A single model run, called a "tick," concludes after each agent has taken its turn. Because the focus is often on how system-level outcomes unfold or emerge from agents' repeated interactions, the interaction stage of frequently repeated many times. In Schelling's model, during the interaction stage each person surveys their immediately surrounding neighborhood to determine whether they are satisfied with the composition of their neighbors, given the researcher specified value of X. If the person is satisfied, they do nothing, while if they are dissatisfied, they move to a random unoccupied house. After each person has taken a turn applying this behavioral rule, asking "should I stay or should I go," the interaction stage starts over with each person again taking a turn and applying the same rule. This process repeats until all the people are satisfied with their locations, or until it becomes clear that it is impossible for all people to simultaneously be satisfied.

The three boxes on the right of Figure 1.1 show graphic depictions of neighborhoods after the people have repeatedly followed Schelling's behavioral rule, until all are satisfied. The top box shows the outcome when $X = 75$ – that is, when people are satisfied living in a neighborhood where at least 75% of their neighbors are similar to themselves. When people have such a strong preference about who their neighbors are, the system-level outcome is residential segregation; the gray people live on one side of the neighborhood, while the black people live on the other side. The middle box shows the outcome when $X = 50$ – that is, when people are satisfied living in a neighborhood as long as they are not in the minority (i.e. at least half their neighbors are like them). Interestingly, even this much weaker preference leads to the system-level outcome of residential segregation. Finally, the bottom box shows the

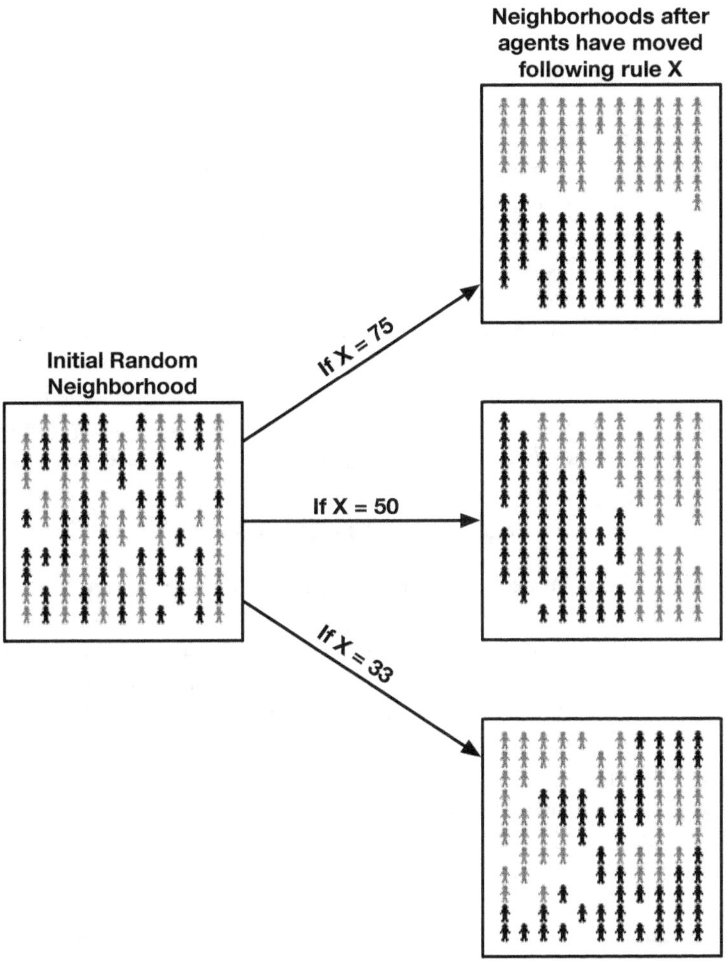

Figure 1.1 Agent-based model of segregation

outcome when $X = 33$ – that is, when people are willing to be in the minority and are satisfied as long as at least one-third of their neighbors are like them. Although every person is willing to be a minority in their local neighborhood, and no person has a preference to live in a segregated neighborhood, the system-level outcome is still residential segregation.

From running an agent-based model just a few times, trying slightly different behavioral rules, some patterns may begin to emerge that provide clues about the relationship between agents' micro-level behaviors and the system-level outcomes they generate. Playing with an agent-based model is an important step in the modeling process because it helps the researcher understand how (and whether) the model works. But, drawing conclusions from the model requires more rigorous experimentation. Although there are many different ways to analyze and experiment with agent-based models, most approaches fall into two broad categories: parameter sweeps and scenario analyses. Conducting a parameter sweep involves systematically examining the system-level outcome for every possible level of a parameter of interest. In Schelling's model, the preliminary results shown in Figure 1.1 seem to suggest that residential segregation will emerge no matter how strong or weak the agents' preferences about their neighbors. Conducting a parameter sweep might involve examining the level of residential segregation that results from running the model at every possible value of *X*, from 0 ("I want all my neighbors to be different from me") to 100 ("I want all my neighbors to be the same as me").

Parameter sweeps are useful for developing theory to bridge the so-called micro-macro gap because they can shed light on how micro-level behaviors give rise to macro-level phenomena. Scenario analyses, in contrast, are useful for considering the potential impact of an intentional or unintentional change to a given system. This type of analysis takes the form of asking a "what if" question: what might happen to this system if *Z* happened? In some cases, *Z* is an intentional intervention designed to solve a system problem – for example, a behavioral intervention designed to reduce the spread of illness (see Chapter 4). In other cases, *Z* is an unintended event that may create a system problem – for example, an environmental disaster that has consequences for the ecosystem's inhabitants. Although Schelling himself did not pursue scenario analyses using his model, many others have adopted his model of segregation as a starting point to explore the potential impact of demographic changes or public policies using scenario analyses.

Strengths and Limitations

Agent-based models have a number of strengths compared to other, more conventional methods of analysis. First, they are highly flexible and adaptable. Because the agents in an agent-based model can represent any kind of entity with the capability of interacting with other entities – a person, an animal, a molecule, a vehicle, etc – these models can be applied in most fields of research. Likewise, because the rules that guide these agents' interactions can specify virtually any kinds of behaviors, these models can be used to explore many different research questions within each field of study.

Second, as a simulation-based method, these models allow researchers to ask and answer questions they might not otherwise be able to consider. Some questions might simply be impossible to study using empirical data because the necessary data would be impossible or impractical to collect. For example, it might be impossible to collect data on the levels of segregation in neighborhoods where residents have different preferences about who their neighbors are. Some preferences simply might not exist anywhere, but conducting a parameter sweep in Schelling's model makes it possible to infer what might happen under every possible preference scenario. In other cases, it might be unethical to collect data. For example, although Schelling's model makes it easy to randomly assign different people to live in different homes throughout a neighborhood, over and over again, this would surely be unethical to do this to real people. Finally, questions about causality can be very difficult to explore using empirical data because researchers rarely have complete control over conditions outside the laboratory. In contrast, because the researcher controls every aspect of an agent-based model, it is possible to conduct true experiments and move toward drawing causal conclusions on issues about which researchers usually can only draw correlational or associational conclusions.

Finally, agent-based models are useful for anticipating unanticipated consequences of both intended and unintended actions. In public health and environmental policy contexts, the goal is often to develop and implement programs or laws that will prevent or remedy problems including epidemics and pollution. However, because health and environmental systems are complex it can be difficult to know precisely what effects such interventions might have, and more importantly, to know whether they might have unintended and undesirable effects. By simulating the system-level outcomes under a variety of possible scenarios, these models can help identify not only the potential benefits but also the potential risks of an intervention.

Despite these strengths, agent-based models are not without their limitations. These models are necessarily simplifications of reality. They cannot (and agent-based modelers should not try to) include every aspect of a system with the goal of mirroring the full complexity of reality. The goal, instead, is to include only the most important or salient features of a system, and to use the resulting simple model to better understand how key processes in the system work. But this requires a careful recognition that, as simplifications, agent-based models leave out a lot of detail. Conclusions drawn from agent-based models are only as robust and generalizable as the models themselves. As a result, while agent-based models can be useful for understanding general processes in abstract systems, they are often not suitable for making predictions about specific processes in specific systems. For example, Schelling's model is useful for understanding how residential segregation might emerge even if no one wants to live in a segregated neighborhood, but may not be useful for predicting the level of segregation in an actual neighborhood.

Top-Down: System Dynamics

Agent-based models are frequently described as a bottom-up approach to understanding systems because they start at the bottom with the actions of individual parts of the system (i.e. the agents), through whose repeated interactions system-level outcomes emerge at the top. In contrast, system dynamics models can be viewed as a top-down approach because they start at the top with a conceptual model of how a system's parts are related, and seek to understand how these system-level relationships govern changes in each individual part of the system at the bottom. System dynamics was created by Jay Forrester in the 1950s and initially applied to understanding systems of corporate decision-making, but later expanded to urban planning (Forrester, 1969) and environmental contexts (Forrester, 1970). Although it developed during the same period as agent-based models, system dynamics has a somewhat longer history in part because system dynamics simulations can be conducted by hand through the solution of systems of equations, and do not necessarily require (but are certainly facilitated by) computer-based simulation. There is no single epistemological perspective that informs system dynamics, but the method is guided by a rejection of unidirectional causality (e.g. *X* causes *Y*). Instead, system dynamics contends that systems are best understood as collections of feedback loops in which *X* causes *Y* but *Y* also causes *X*, and that these feedback loops are the source of the complex and non-linear outcomes observed in systems.

Causal Loops

The construction of a system dynamics model begins with the development of a causal loop diagram, which aims to capture the causal relationships among key variables in the system graphically as arrows linking variables. The causal relationship of one variable, *X*, on another, *Y*, may be positive (denoted in diagrams with a +), indicating that a change in *X* causes a change in the same direction in *Y*. For example, the number of calories one consumes has a positive causal effect on one's weight: an increase in calorie intake causes an increase in weight, while a decrease in calorie intake causes a decrease in weight. Alternatively, the causal relationship may be negative (denoted in diagrams with a –), indicating that a change in *X* causes a change in the opposite direction in *Y*. For example, the amount of exerzcise one gets has a negative causal effect on one's weight: an increase in exercise causes a decrease in weight, while a decrease in exercise causes an increase in weight.

In many conventional non-systems forms of analysis, this is where things stop, at the level of bivariate relationships. However, system dynamics is concerned with systems of causal relationships. These systems are built from bivariate causal relationships in a number of ways. First, a given variable may exert a causal effect on multiple other variables, and likewise a given variable may be caused by multiple other variables. Second, bivariate causal relationships can link together to form longer causal chains in which *X* causes *Y*, which in turn causes *Z*. Finally, these causal chains can form closed loops: *Z*, perhaps after some delay, exerts a causal effect on *X*. Interpretation of causal loop diagrams is focused on examining the effects of these closed loop chains, which can perform two functions in the system. Some causal loops are reinforcing and serve to amplify changes in the variables within the loop; reinforcing loops are denoted in diagrams with an R. Other causal loops are balancing and serve to maintain variables at an equilibrium or status quo; balancing loops are denoted in diagrams with a B. Determining whether a loop is reinforcing or balancing involves examining the nature of the causal relationships composing the chain: chains containing an even number of negative causal effects are reinforcing, while chains containing an odd number of negative effects are balancing.

Schelling used an agent-based model to explore segregation as a system-level outcome, but it is also possible to model a system that produces residential segregation using system dynamics. The top panel of Figure 1.2 shows a very simple causal loop diagram that attempts to explain, in a purposefully caricatured way, segregation. The diagram contains a positive causal relationship from "Segregation" to "Know people like yourself": the more heavily segregated your neighborhood is, the more likely you are to know others like yourself. It also contains a positive causal relationship from "Know people like yourself" to "Segregation": the more you know people like yourself, the more segregated your neighborhood will become as you encourage the people you know to move nearby. These two causal relationships form a single closed loop that, because it contains an even number of negative effects (note: zero is even), is a reinforcing loop. Increases in *X* lead to increases in *Y*, which in turn lead to increases in *X*, which in the long run lead to runaway growth of all variables in the loop. Some might interpret such a reinforcing loop as a problematic vicious cycle that ensures segregated neighborhoods stay segregated, while others might view it as a beneficial social process that allows distinctive neighborhoods to retain their cultural practices.

However, segregation is implicated in two additional causal relationships also shown in Figure 1.2. The diagram contains a positive causal relationship from "Segregation" to "Leave the neighborhood for ethnic food": the more homogeneous your neighborhood, the more you will be inclined to seek new cultural experiences elsewhere. The causal relationship running the other direction, from "Leave the neighborhood for ethnic food" to "Segregation," is negative: the more you leave the neighborhood to satisfy a desire for culture, the less segregated your neighborhood will become as you encourage those you meet to move nearby. These two causal relationships form a single closed loop that, because it contains an odd number of negative effects (i.e. one), is a balancing loop. Increases in *X* lead to increases in *Y*, which in turn lead to decreases in *X*, which in the long run ensures the stability of the variables in the loop. As long as some neighborhood residents have a taste for ethnic foods and other cultural experiences not available locally, levels of segregation in the neighborhood will not grow uncontrolled.

While this causal loop diagram is simplified for the sake of illustration, in practice such diagrams can be much more complex. Causal loops often contain more than just two variables, and in some cases can be composed of extended sequences of variables. Likewise, the diagrams often contain more than just two causal loops, and in some cases can involve many distinct loops that are nested within, or intersect with, one another (see Chapters 5, 11, 27, and 28).

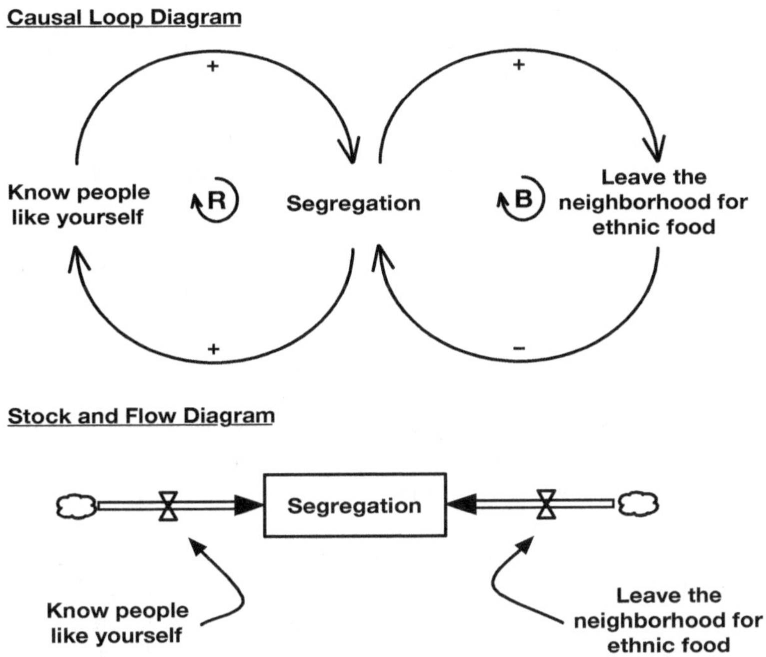

Figure 1.2 System dynamics model of segregation

Stocks and Flows

A causal loop diagram presents a conceptual model of the system that identifies opportunities for key variables to grow, decline, or remain in a stable equilibrium. However, formal analysis of a system dynamics model requires translating these causal loops into stocks and flows. A stock is an accumulation of some resource or quantity, while a flow is a process through which levels of a stock rise or fall. For example, one's weight might represent a stock (i.e. an accumulation of stored energy), while eating and exercise might each represent flows whereby weight can increase or decrease. Stock and flow diagrams can look similar to causal loop diagrams, but have a few key differences. Stocks appear as rectangles to indicate their role as containers of some resource. Flows appear as double-line arrows to indicate their role as pipes or conduits through which resources flow in and out of stocks. Stock and flow diagrams can also include cloud-like icons, which represent stocks that are outside the model's scope. For example, the eating flow fills up the weight stock with additional energy, but the initial source of the food being eaten (e.g. a grocery store, farm, or restaurant) might not be of interest and might simply be represented as a cloud. Within a larger and more comprehensive causal loop diagram, a system dynamics modeler may only translate and diagram selected portions as stocks and flows (see Chapters 5, 6, 11, 16, and 28).

The lower panel of Figure 1.2 shows how the causal loop diagram in the upper panel might be translated into a stock and flow diagram. The neighborhood's level of segregation is a quantity of particular interest, and appears here as a stock represented by a rectangle. The arrow to the left represents the flow of people like you into the neighborhood, where the rate of this flow depends on the extent to which you know people like yourself. The stock from which these residents originate (e.g. another neighborhood, another country) is not of direct interest, and thus appears as a cloud. The arrow to the right represents the flow of people unlike you into the neighborhood, where the rate of this flow depends on the extent to which you leave the neighborhood for ethnic food. Again, the stock from which these residents originate is not of direct interest, and thus appears as a cloud.

The process of developing both causal loop and stock and flow diagrams can help clarify the organization of the system, but the heart of a system dynamics model lies in the formal equations that specify how stocks move through flows. Each flow is represented by a mathematical equation, often in the form of a differential equation, which describes the rate at which the input stock empties into the output stock. It is here that system dynamics models derive their flexibility, because any change relationship that can be described by a mathematical equation can be modeled using system dynamics. Researchers have a wide range of types of equations at their disposal, including equations that specify simple linear or complex nonlinear change patterns, and equations that include parameters that can be experimentally manipulated be the researcher. These equations may be drawn from existing research, including, for example, biological equations that govern the conversion of consumed calories into pounds of weight. They may also be informed by theory – for example, on migration patterns, and built from scratch using a toolbox of basic equation forms (see Chapters 10 and 16).

The analysis of system dynamics models typically involves observing the levels of stocks, which change over time as resources flow in and out. For example, one might observe fluctuations in weight over time as a consequence of energy flows due to eating and exercising. Likewise, one might observe fluctuations in a neighborhood's level of segregation as a consequence of flows (i.e. migration) of different types of people into the neighborhood. Like scenario analysis using agent-based models, system dynamics models can also be used to examine expected system behavior under different scenarios. In a system dynamics model, different scenarios can be simulated by changing the initial levels of the stocks, or the rates of change specified by the flow equations, or both. For example, one might examine how the changes in the segregation stock look different over time if the neighborhood begins with a large stock (i.e. highly segregated) or a small stock (i.e. highly integrated). Likewise, one might examine how changes in the segregation stock look different over time if leaving the neighborhood for ethnic food has a stronger or weaker influence on the rate at which people unlike you move into the neighborhood.

Strengths and Limitations

As a system science method, system dynamics offers many of the same strengths that can be found in agent-based models. System dynamics is highly flexible, capable of modeling nearly any kind of system that can be described in terms of quantities that change over time. It can also be used to explore the expected outcomes of various scenarios, which is helpful for evaluating the effectiveness of proposed interventions and anticipating their potential negative side effects.

However, as a top–down approach, it differs in a number of important ways from agent-based models, and these differences can present both strengths and limitations. A system dynamics model provides a visual overview of the system as a whole and the relationships among its component parts. This is a strength of the method because it can be a helpful conceptual tool to the researcher as ideas about the system are refined, and can serve as an aid in explaining the research to other audiences including the research participants (e.g. patients) and research users (e.g. policy makers). However, this may also be viewed as a limitation, or at least a challenge. Because building a system dynamics model begins with the development of a causal loop diagram and later a stock and flow diagram, the researcher must already have some knowledge of how the system's parts interact with one another both directly and indirectly. This is different from agent-based models, where the researcher discovers these relationships as they emerge endogenously from the model.

System dynamics also differs from agent-based models in the way interactions and relationships must be specified. Although the rules that govern agents' interactions in an agent-based model can be specified using equations, they are more often specified using simple behavioral rules of thumb that mirror the way people actually think and behave. Recall, for example, Schelling's behavioral rule did not involve equations, but rather involved looking at one's neighbors and deciding whether the composition met one's preferences. In contrast, flows in a system dynamics model are always specified using formal mathematical equations, and thus require some familiarity with the types of functional forms that are commonly used in such specifications.

In-Between: Network Analysis

Network analysis differs from both agent-based models and system dynamics in that it is not a simulation-based method, but rather a method for examining the pattern or structure of relationships among the parts of a system. It also has a substantially longer history than the other two system science methods. Its mathematical foundations, known as graph theory, can be traced to the work of Swiss mathematician Leonhard Euler, who used an early version of network analysis to solve the "Bridges of Königsburg" problem in 1735. These tools were initially applied to study people and other social contexts by Jacob Moreno, who in 1934 used an early version of network analysis he called "sociometry" to study the relationships among girls at reform school and explain a series of runaways. Thus, although some have described network analysis as a "new" science (e.g. Barabási, 2002), it is in fact quite old.

All networks are composed of two basic elements: nodes and edges. Nodes are the component parts of a system, while edges are the relationships that link these parts to one another. The fact that the nodes and edges can represent different things depending on the context and type of system under consideration is what gives network analysis its flexibility. For example, when studying a social system, the nodes often represent people, while the edges may represent any of the different types of relationships that might exist between people, including friendship or kinship. When studying a biological system, the nodes might represent proteins, while the edges might represent biochemical reactions that require two different proteins. When studying a neural system, the nodes might represent neurons in an organism's brain, while the edges might represent the synapses that allow information in the form of electrochemical signals to pass from one neuron to another. When studying a physical infrastructure system, the nodes might represent intersections or power stations, while the edges might represent roads or power lines. Because networks can be used to represent nearly any type of system, clarity about what the nodes and edges represent and how they are measured is essential.

A network generally includes all the nodes within a system or population of interest, as well as all the edges that exist among them. Thus, for example, a neighborhood friendship network would include a node for every person in the neighborhood, as well as an edge for every friendship that exists among these neighbors. Analysis of such a network can be conducted at multiple levels that differ in their number of observations, types of research questions, and methods for analysis. Table 1.1 summarizes some key features of network analysis at different levels, using the N^x notation proposed by David Krackhardt, where N represents the number of nodes in the network. Following from the agent-based model example in Figure 1.1, and the system dynamics example in Figure 1.2, the examples in this table highlight how issues of neighborhood segregation and diversity might be examined using network analysis.

Table 1.1 Network analysis of segregation

Level of Analysis	*Number of Observations*	*Typical Research Question*	*Common Metrics*	*Graphic Depiction*
Network	N^0	How does the degree of connectivity of sidewalks in a neighborhood impact residents' ability to meet one another?	• Size • Density • Clustering • Average path length • Degree distribution	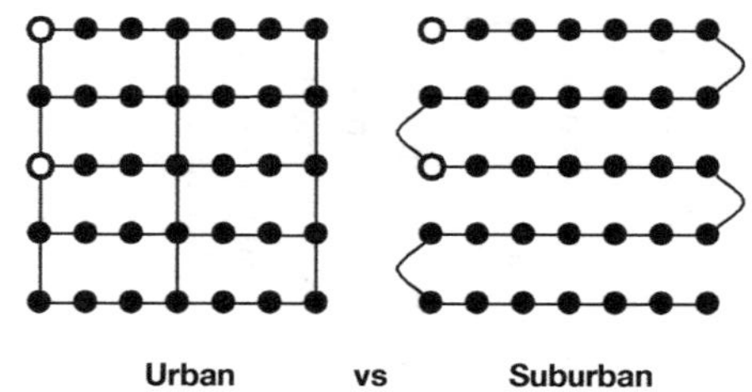
Node	N^1	Do residents with more diverse personal networks have better access to novel information?	• Centrality • Brokerage • Composition	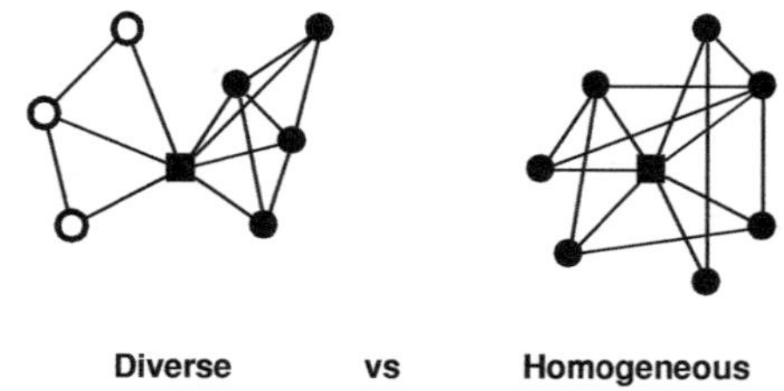
Dyad	N^2	What factors influence the formation of relationships between neighborhood residents?	• Homophily • Multiplexity • Strength	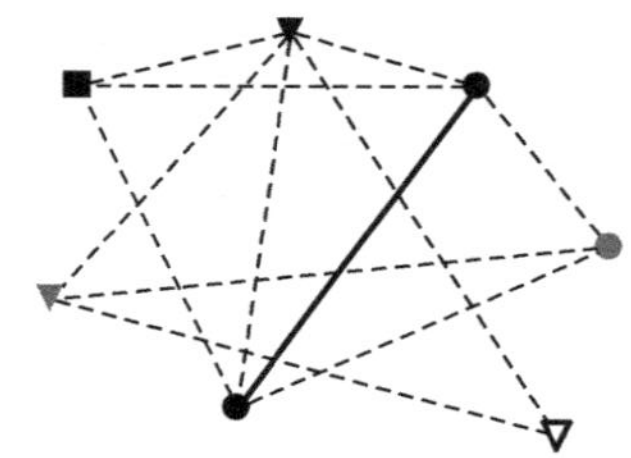
Triad	N^3	Do neighborhood residents really know who is friends with whom?	• Transitivity • Balance • Accuracy	

The Network Level (N^0)

Analysis at the network level focuses on characteristics of a network as a whole, and thus examines N^0 (i.e. 1) unit of observation. The goal is to describe the overall structure of the system, and to understand how this structure impacts the system as a whole or its component parts. In some cases, analysis at the network level will be conducted on a single network, which allows the researcher to describe the network's structure. More often, however, because analysis at the network level involves only one unit of observation per network, researchers will frequently examine and compare multiple entire networks (see Chapter 23).

The simplest metrics used to characterize whole networks are size and density. Size captures the number of nodes in a network, while density captures the number of edges in a network as a percentage of the number of edges possible. In a dense network, most nodes are connected to most other nodes, while in a sparse network, nodes are connected to just a few others or none at all. For example, a rural community might be characterized by a relatively small and dense social network, while an urban community's social network might be large and sparse.

Another pair of metrics – clustering and average path length – is often used in the investigation of small world networks. Clustering captures the extent to which the network is composed of densely connected subgroups, while average path length measures the average distance (i.e. number of edges) between any two nodes in the network. A small world network is a network whose structure has both a high level of clustering and a short average path length. This unique structure has been shown to make the phenomenon of "six degrees of separation" – you have a close circle of friends, but nonetheless are connected to everyone in the world by just a few intermediate links – possible. A final metric, degree distribution, is used in the investigation of scale-free networks and describes the relative proportions of well-connected and poorly connected nodes. Most real-world networks across a wide range of contexts exhibit a very specific degree distribution: a few highly connected nodes, and many poorly connected nodes.

As the examples in Figures 1.1 and 1.2 have shown, agent-based models and system dynamics can both be used to investigate questions about neighborhood segregation and diversity. Such questions can also be explored using network analysis. One might be interested in understanding how the structure of a neighborhood's sidewalk network impacts residents' ability to meet one another, which might be particularly important for breaking down barriers between dissimilar social groups. The figure in the top row of Table 1.1 shows two hypothetical sidewalk networks that mirror patterns commonly observed in urban and suburban settings. In the urban case, each household (node) is only a few sidewalks (edges) away from most other households, which might facilitate meeting neighbors. In contrast, suburban households are separated by long distances in the sidewalk network, which might hinder neighborly encounters. This contrast is particularly visible in the case of the two households represented by the white nodes: only one household lies between them in the urban sidewalk network, while 13 households separate them in the suburban sidewalk network. At the network level of analysis, this difference could be captured by comparing these two networks' average path lengths – that is, the average distance in the sidewalk network between any two households. The urban network has a relatively low average path length suggesting the potential for neighborliness, while the suburban network has a relatively high average path length suggesting the potential for social isolation.

The Node Level (N^1)

Analysis at the node level focuses on the position of individual nodes within a network, and thus can examine N^1 (i.e. N, or the number of nodes) units of observation in a given network. A node's position in a network can be characterized in many different ways, but centrality is among the most widely examined metrics. In a broad sense, centrality aims to capture how "important" a node is in a network based on its position relative to the other nodes. However, there are multiple ways a node might be important, and thus multiple ways of assessing a node's centrality. The simplest form of centrality, degree, counts the number of others to which a node is directly connected. In a friendship network where nodes are people and edges are friendships, a node with high degree is a person with a large number of friends. Another form of centrality, closeness, aims to capture how close in the network a given node is to other nodes by counting the average number of edges between a given node and each of the others. In a friendship network, a node with high closeness might be the first to hear, or the best at rapidly spreading, a juicy piece of gossip. A final common form of centrality, betweenness, focuses on nodes that serve as bridges or brokers between other nodes in the network that otherwise could not reach one another. A node with high betweenness in a friendship network has a privileged position because he or she can decide whether or not to pass along important information (see Chapter 17).

In other cases, the focus of node level analyses may be on characteristics of the network immediately surrounding each node, which is often called an ego network. Within an ego network, the focal node is called ego, while each of the nodes to which ego is directly connected are called alters. As in the case of analyses of the whole network, it can be useful to examine the size and density of an ego network. In the context of a friendship network, looking at the size and density of each node's ego network would indicate who has many friends that all know one another (large and dense), and who has just a few friends that are all strangers (small and sparse). Similarly, it can be useful to examine the composition of ego networks: what kinds of alters is ego connected to?

After characterizing each node's position (for example, using centrality) or ego network (for example, by composition), this information might be used in a node level analysis in at least four different ways. First, the nodes in a single network might be compared to one another, for example, to locate the most and least central people in a social network. Second, a network characteristic might be used as independent variables predicting a non-network outcome. For example, individuals' centrality in a social network might be used to predict their levels of happiness. Third, a network characteristic might be used as an outcome that is predicted by a non-network characteristic. For example, individuals' gender might be used to predict the diversity of their ego network composition, asking whether women have more diverse personal social networks than men.

Finally, one might examine the relationship between two different network characteristics. For example, we might ask whether neighborhood residents with diverse personal social networks have better access to novel information. The figure in the second row of Table 1.1 shows two hypothetical ego networks, where ego is shown as a square and the alters are shown as circles. Focusing on composition, the ego network on the left is diverse because it includes different kinds of alters, while the ego network on the right is homogeneous because all the alters are identical. Focusing on position, the ego on the left is more likely to have access to novel information because she can learn about things from two distinct sets of friends. In contrast, the ego on the right is less likely to have access to novel information because, although he knows the same number of people, they all know each other are likely to all tell him the same thing.

The Dyad Level (N^2)

Analysis at the dyad level focuses on pairs of nodes within a network, and thus can examine N^2 units of observation within a given network – that is, in a network composed of 10 nodes, there are up to 100 (i.e. 100^2) dyads to examine. Some dyad level analyses examine the content or strength of edges that exist between dyads. Edges can vary in their content, for example, with some pairs of people linked by relations of friendship and other pairs linked by relations of kinship. When a single type of edge links a dyad, the dyad is described as having a uniplex relation, while dyads linked by multiple types of edges are described as having a multiplex relation. Likewise, edges can also vary in their strength. For example, some friendships might be stronger than others, and some roads carry more traffic than others. Here, the focus is often on understanding the circumstances under which, or the consequences of, dyads engaging in multiplex (or uniplex) and stronger (or weaker) relations. For example, do individuals linked by strong and multiplex relations have more influence over one another than those linked only by weak and uniplex relations?

Other dyad level analyses seek to understand the factors that influence the formation and dissolution of edges between nodes, and thus more broadly, why the network has the particular structure that it does (see Chapter 9). In some cases, the characteristics of the two nodes in a dyad can influence the likelihood that an edge will form between them. For example, the phenomenon of homophily, captured by the aphorism that "birds of a feather flock together," suggests that a dyad of similar nodes is more likely to interact than a dyad of dissimilar nodes. In other cases, larger structural patterns elsewhere in the network can influence the likelihood of edge formation. For example, the phenomenon of transitivity suggests that two nodes that are both linked to a common third node are likely to be linked to each other. This is the process that underlies the formation of friendships through mutual acquaintances.

Of particular interest in understanding the mechanisms of network evolution is distinguishing processes of selection from processes of influence. Selection processes occur when characteristics of a node cause the formation of an edge. For example, a person who smokes is likely to become friends with other people who smoke. In contrast, influence processes occur when the formation of an edge causes a characteristic of a node. For example, being friends with a smoker may lead a non-smoker to decide to take up smoking. Thus, distinguishing selection from influence involves distinguishing cause from effect in the co-evolution of the nodes' characteristics and the network's structure.

Uncovering the mechanisms responsible for edge formation can be challenging. People often do not know why they interact with some people but not others, so surveying them would not be helpful. And, in cases of networks where the nodes are not people, but rather are animals or traffic intersections, surveys are not possible. Instead, such studies often use simulation-based methods that closely resemble agent-based models. Given an observed network, they ask what kinds of edge-forming processes would need to be at work to produce the observed structure. The figure in the third row of Table 1.1 shows a hypothetical neighborhood social network where strong relationships are shown as solid lines and weak relationships are shown as dashed lines. Here, one might ask, what factors influence the formation of relationships among neighbors? One possibility is proximity: is this the kind of network we would observe if people mostly formed relationships with their nearest neighbors? Another possibility is transitivity: is this the kind of network we would observe if people formed relationships with others who share a mutual acquaintance? Still a third possibility is homophily: is this the kind of network we would observe if people formed relationships with neighbors who are similar to themselves? In this simple example, visual inspection is sufficient to confirm that homophily seems to be the driving force behind this network's formation. Strong ties form between dyads where both nodes have both the same shape and color, while weak ties form between dyads where the nodes have either the same shape or color, and no ties form between dyads where the nodes have neither the same shape nor color.

The Triad Level (N^3)

Analysis at the triad level focuses on sets of three nodes, or triads, and thus can examine N^3 units of observation within a given network. Even in a small network composed of only 10 nodes, there are up to 1,000 (i.e. 10^3) triads to examine. Some triad-level analyses view the network objectively, as a true depiction of the relations that exist among the nodes, and concentrate on identifying the number of different types of triads present in the network. In a network where edges can have a specific direction (e.g. A → B is different from B → A), there are 16 distinctly different ways that edges can be arranged among a set of three nodes. At one extreme, the empty triad, the three nodes have no edges between them. At the other extreme, the complete triad, edges exist in both directions between every pair of nodes. Between these two extremes, may other possibilities exist, including the directed line (A → B → C), the in-star (A → B ← C), and the out-star (A ← B → C). Some of these triadic configurations are associated with specific functions a system can perform, for example, the directed line is functional for moving resources along a supply chain, while an out-star is functional for distributing resources widely. Examining the relative number of each type of triad can provide insight into how the system works.

In networks where edges do not have direction, but do have a positive (e.g. liking) or negative (e.g. hating) valence, there are also a number of different possible arrangements within a triad. For example, A may like B and C, but B and C hate each other. Balance theory contends that triads are balanced when they meet certain conditions that are captured by aphorisms like "a friend of a friend is a friend" and "an enemy of an enemy is a friend." Unbalanced triads are viewed as inherently unstable, and over time are expected to evolve into balanced triads through a combination of edge creation, edge dissolution, and edge valence change. Triad level analyses adopting this approach often focus on examining the extent of balance in a network, and observing a network's evolution toward greater balance over time. Notably, the logic underlying the identification of balanced and unbalanced triads in a network is conceptually identical to the logic underlying the identification of reinforcing and balancing causal loops in a system dynamics model.

Other triad level analyses view the network subjectively, as a representation of an observer's perspective. Each person in a social network has some conception of what the network looks like, and specifically of who is linked to who (see Chapter 8). From this perspective, a triad is composed of an observer node and an observed dyad. The figure in the last row of Table 1.1 shows a graphic depiction of this type of triad: person C has some belief about whether or not (or how strongly, or in what way) person A and person B have a relationship with one another. Some research adopting this approach aims to understand when observers have accurate or inaccurate perceptions of the real network, asking for every observer-observed triad, is the observer right? Other research is less concerned with accuracy per se, and focuses instead on the process of relationship perception, asking for every observer–observed triad, why does the observer think that this dyad has a relationship? Returning once more to the context of neighborhoods and segregation, this type of network analysis may be useful for uncovering imagined us-versus-them dynamics in which residents believe, perhaps erroneously, that neighborhood social relationships are factionalized.

Strengths and Limitations

As with agent-based models and system dynamics, network analysis is a highly flexible method capable of modeling any kind of a system in which the specific pattern of relationships among its component parts is important. But it offers two other important strengths that do not appear in these other methods. First, it provides a way to explicitly reject the assumptions of independence typically imposed by conventional statistical techniques like correlation and regression. Use of such techniques requires the researcher to assume that the units of observation are independent from one another, which except for highly controlled laboratory experiments, is almost always false. Rather than sweeping them under the rug, network analysis brings these patterns of dependence and interaction into the foreground as important phenomena to study in their own right. Second, network analytic techniques are embedded within the extensive, nearly century-long theoretical tradition of structuralism that offers a foundation from which hypotheses can be derived, and that offers a conceptual guide to applying the method.

Perhaps the most significant limitation of network analysis is its stringent data requirements. Unlike agent-based models and system dynamics, which are both simulation-based and can be employed in the absence of empirical data, network analysis is a method for analyzing empirical data which must first be collected. The N^x notation used to distinguish levels of analysis highlights one challenge: the number of observations to be collected increases exponentially with network size. Collecting data on a single network of N nodes requires collecting data on N^2 edges or N^3 perceived edges, making data collection in even fairly small settings resource intensive. A related challenge is that many forms of network analysis have a low tolerance for missing data. The omission of a single node or edge from a network can dramatically alter node-level metrics like betweenness and network-level metrics like average path length. These challenges have led network analysts to search for archival and existing sources of network data, and for innovative ways to collect network data more efficiently.

Linkages

The sections above consider agent-based models, system dynamics, and network analysis as distinct system science methods. Indeed, they each have their own histories, literatures, and researcher communities. However, increasingly researchers have been exploring opportunities for their combined use. The most common combination to date has been agent-based models paired with network analysis (see Chapters 19, 20, and 23). This is a natural marriage because agent-based models provide a framework for simulating the formation of networks from agents' interactions with one another. The combination of agent-based models and system dynamics is also promising because their joint application allows researchers to simultaneously capture the benefits of bottom–up and top–down approaches to understanding systems (see Chapter 16). Despite the visual similarity between causal loop diagrams and network diagrams, there have been relatively few attempts at combining system dynamics and networks (see Chapter 14 for a notable exception). Although these few examples suggest the combination of system science methods holds promise, this is an area of system science research that warrants more attention.

Organization of this Volume

The goal of this handbook is to demonstrate applications of these system science methods to answer research questions in a range of substantive fields. Chapters 2–10 explore applications of system science methods to issues of *health and human development*, ranging from disease prevention to the formation of friendships among children and adolescents. Chapters 11–19 focus on using system science methods to explore questions about the *environment and sustainability*, ranging from the complex interactions between humans and their environments to attempts at influencing local and national environmental policy. Finally, Chapters 20–28 use system science methods to understand *communities and social change*, ranging from urban dynamics like segregation and sprawl to the intentional and unintentional changes communities experience over time. Within each of these three parts, you will find examples of each system science method, as well as examples of the combined application of system science methods in a single study. Although these chapters describe how one or more system science methods were applied to answer a research question, this volume is not designed as a how-to guide or manual to the use of system science methods. In the references below, introductory texts that explain how to use the system science methods are marked with a #, while software programs that have been widely used for system science research are marked with a @.

A header box at the beginning of each chapter summarizes the chapter's primary research question and the system science method(s) it applies, as well as a couple "things to notice." This is not intended to be an exhaustive list of noteworthy features of the chapter, but rather a short list of things the chapter is particularly good at illustrating. After reading a chapter it may be helpful to return to this box and consider a few questions:

- How does this chapter's answer to the research question differ from answers that might be obtained using non-system science methods?
- What other system science methods might be useful for answering this research question?
- For the "things to notice":
    - How does the author do this?
    - How is this different from more traditional approaches to research?
    - How is this helpful for answering research questions and solving problems?

References, Further Reading, and Software

Note: Items marked with a # are recommended as introductions to the use of system science methods, while items marked with an @ are examples of software programs widely used for system science research.

Barabási, A-L. (2002). *Linked: The New Science of Networks*. Cambridge, MA: Perseus.

@ Bastian M., Heymann S., Jacomy M. (2009). Gephi: An Open Source Software for Exploring and Manipulating Networks. International AAAI Conference on Weblogs and Social Media (http://gephi.github.io/)

Borgatti, S. P., Everett, M. G., & Johnson, J. C. (2013). *Analyzing Social Networks*. Thousand Oaks, CA: SAGE.

@ Borgatti, S. P., Everett, M. G. & Freeman, L. C. (2002). Ucinet for Windows: Software for Social Network Analysis. Harvard, MA: Analytic Technologies.

@ Brandes, U., & Wagner, W. (2003). Visone – Analysis and Visualization of Social Networks, pp. 321–340 in Jünger, M. & and Mutzel, P. (Eds.), *Graph Drawing Software*. New York: Springer (http://visone.info/)

Epstein, J. M. (1999). Agent-Based Computational Models and Generative Social Science. *Complexity, 4*, 41–60.

Forrester, J. (1969). *Urban Dynamics*. Cambridge, MA: MIT Press.

Forrester, J. (1970). *World Dynamics*. Cambridge, MA: Wright-Allen Press.

Gardner, M. (1970). The Fantastic Combinations of John Conway's New Solitaire Game "Life." *Scientific American, 223*, 120–123.

@ Isee Systems. (2015). STELLA (www.iseesystems.com/softwares/STELLA-iThink.aspx)

Moreno, J. L. (1934). *Who Shall Survive? A New Approach to the Problem of Human Interrelations.* Washington, DC: Nervous and Mental Disease Publishing Co.

Neal, Z. P. (2013). *The Connected City: How Networks are Shaping the Modern Metropolis.* New York: Routledge.

Newman, M. E. J. (2010). *Networks: An Introduction.* New York: Oxford University Press.

@ North, M.J., Collier, N. T., Ozik, J., Tatara, E., Altaweel, M., Macal, C. M., Bragen, M., & Sydelko, P. (2013). Complex Adaptive Systems Modeling with Repast Simphony. Complex Adaptive Systems Modeling, Heidelberg: Springer (http://repast.sourceforge.net/)

Railsback, S. F., & Grimm, V. (2011). *Agent-Based and Individual-Based Modeling: A Practical Introduction.* Princeton, NJ: Princeton University Press.

Schelling, T. C. (1969). Models of Segregation. *American Economic Review, 59*, 488–493.

@ Smith, M., Milic-Frayling, N., Shneiderman, B., Mendes Rodrigues, E., Leskovec, J., & Dunne, C. (2010). NodeXL: A Free and Open Network Overview, Discovery and Exploration Add-in for Excel 2007/2010 (http://nodexl.codeplex.com/)

Sterman, J. D. (2000). *Business Dynamics: Systems Thinking and Modeling for a Complex World.* Boston, MA: McGraw-Hill.

@ Swarm Development Group. (2009). Swarm (www.swarm.org/)

@ Ventana Systems. (2014). VenSim (http://vensim.com/)

@ Wilensky, U. (1999). NetLogo (http://ccl.northwestern.edu/netlogo/). Center for Connected Learning and Computer-Based Modeling, Northwestern University. Evanston, IL.

PART 1

Health and Human Development

Introduction

The first set of chapters in this volume explores applications of system science methods to issues of health and human development, ranging from disease prevention to the formation of friendships among children and adolescents. The use of system science methods in this area, although not yet widespread, is becoming increasingly common for a couple reasons.

First, finding solutions to pressing public health issues requires incorporating not only technical physiological details (e.g. how does influenza spread and mutate?), but also an understanding of human behaviors (e.g. how often do people wash their hands?), of the local environment (e.g. how close to one another do people live and work?), and of the supply of resources (e.g. how many healthcare professionals are available to administer vaccines?). Thus, research in this area tends to be quite interdisciplinary, necessitating collaborations among physicians, sociologists, geographers, economists, and others. System science methods provide a way for these researchers to speak the same language, and to integrate the unique contributions of their individual fields of expertise.

Second, health-related research has enjoyed relatively greater levels of funding than some other areas of research. One of the largest public funders of health-related research in the United States, the National Institutes of Health (NIH), has explicitly advocated the use of systems science methods. This advocacy has come in the form of the creation of an Institute on Systems Science and Health (ISSH) and System Science e-mail listserv, as well as the announcement of several funding opportunities that either require or encourage the use of system science methods. Thus, there is growing support for the use of system science methods to address issues of public health.

Health

Bernice Pescosolido and colleagues introduce the Social Symbiome framework, which views networks as a way to bridge levels of analysis and areas of expertise to arrive at a more comprehensive understanding of public health issues. Representing a collaboration among 12 scholars from a range of disciplines, they contend that public health issues are embedded in a series of nested networks. At the most microscopic level, adverse health outcomes are the result of breakdowns in molecular networks of individual genes and proteins studied by biochemists. But these are nested within biological networks of organs and tissues studied by physicians.

 DOI: 10.4324/9781315748771-2

But these are nested within social networks of people studied by sociologists. And so on, through a range of units, levels, and disciplines. Rather than argue for a reductionist approach to health, in which solutions are sought at the most microscopic level, they instead advocate a networked approach that explicitly recognizes this nesting and that incorporates the insights of experts studying the unique properties of networks at each level. In this case, the use of network analysis focuses attention not only on the importance of linkages occurring at a specific level (e.g. between genes or between people), but also on the importance of linkages occurring between levels: what do transportation networks between cities have to do with how the seasonal influenza virus mutates from year to year, and how can geographers and biochemists use the tools of network analysis to search for the answer together?

Ligmann-Zielinska and colleagues offer one example of what this might look like in practice. In their investigation of obesity and obesity prevention, they use an agent-based model to incorporate insights from several different disciplines that typically study phenomena at different scales. First, their simulation model incorporates features of the urban landscape often studied by geographers, including the gyms' locations and streets' walkability, which are drawn from real neighborhoods in San Diego, California. Second, it incorporates a mechanism for individuals to change their health behaviors in response to policy interventions, which is typically the domain of public health researchers. Third, using data typically examined by health demographers, the simulated people that populate their model are based on the demographic characteristics of the people actually living in the San Diego area. By combining these related components into a single model, they are able to explore how the success of behavior modification interventions in reducing obesity depends on a range of geographic, behavioral, and demographic factors.

Also using a spatially explicit agent-based model, Liang Mao examines the spread of an influenza epidemic in a simulated version of Buffalo, New York. What makes his model particularly interesting is that it tracks two kinds of contagion simultaneously. Perhaps most obviously, his model simulates the spread of influenza among a population of individuals who, over the course of a day, travel to different locations (e.g. home, work, restaurants) and interact with different people (e.g. family, co-workers, friends). But it also tracks the spread or adoption of flu preventative behaviors like hand washing and vaccination. It thus simulates a kind of public health "race against the clock" in which flu prevention behaviors hopefully spread faster than the flu itself. Here, Mao uses his model to evaluate the expected effectiveness of two different strategies for promoting the adoption of prevention behaviors: a household-based incentive in which vaccines are offered on a buy-one-get-one-free basis, and a workplace role-model program in which some co-workers are recruited to encourage adoption by others. As an evaluative tool, this simulation explores a series of scenarios that can be helpful for identifying the most effective and efficient behavior modification strategies for reducing the spread of infectious diseases.

These two agent-based models are each focused on a specific health problem and on the effectiveness of behavior modifications as preventative tools. However, multiple health problems often occur together and simple behavior modifications are insufficient prevention strategies. Batchelder and Lounsbury use a system dynamics model to investigate the co-occurrence of psycho-social health issues or "syndemic risk" experienced by women with or at risk for HIV. Rather than explore how a specific health problem is transmitted via interactions between people, which might be more commonly investigated using agent-based models or network analysis, their system dynamics model instead focuses on interactions between different aspects of risk in a woman's life. For example, they explore how the relationship between a woman's self-worth, her drug use, and her risk of violence can create a vicious circle, or in the language of system dynamics, a feedback loop. As a woman's self-worth declines, she is more likely to use drugs, which in turn places her at greater risk of violence, which – completing the circle – further depresses her self-worth. Understanding how these components of a woman's life relate to – and in this example, reinforce – one another illuminates for the real scope of issues facing women with HIV, and suggests that strategies for preventing risk of violence, or of drug use, or of depression must be coordinated and systemic.

Each of these chapters offers examples of how system science methods might be used to identify and test possible strategies for the prevention of health problems. But are these methods actually being used to inform public health policy decisions? Unwin and colleagues begin to answer this question in their chapter by reviewing the use of system dynamics models in policy-making contexts. They present case studies drawn from the United States and New Zealand that illustrate national health policies on smoking cessation, alcohol control, type-2 diabetes, and cardiovascular disease being informed by results obtained using these methods. For example, the New Zealand Customs Service commissioned the development of a system dynamics model to understand the feedback loops that exist among national smoking rates, second-hand smoke, and smoking-related deaths. After a series of refinements, and fueled with data from a 2006 tobacco use survey, this model ultimately informed a decision in 2007 to increase funding for smoking cessation programs by NZ$42 million. Unwin and colleagues ultimately conclude that the use of these types of methods in policy decisions remains limited, but their case studies offer a few success stories and highlight the potential of system science methods for health policy-making.

Human Development

System science methods are useful for understanding health problems (i.e. how do people break down?), but they can also be useful for understanding broader processes of human development (i.e. how do people work?). Questions about how people work are often addressed in psychology, but in many different ways. Despite the breadth of the field of psychology, Michael Vitevitch shows how network analysis has found application in almost every branch. In social psychology, networks help us understand how small groups share information and solve problems collaboratively, while in biological psychology, network analysis has been used to map the neural pathways of the human brain. These two cases highlight a single method being applied at two radically different levels, to understand the phenomenon of thinking emerging from interactions *among a group of people*, and from interactions *among a group of brain cells*. Similarly, using networks that link words with similar letters, or that link psychopathologies with similar symptoms, Vitevitch argues that network analysis has been useful to both cognitive psychologists studying language and clinical psychologists studying mental illness by highlighting patterns that might otherwise have gone unnoticed.

Perhaps the most rapid developmental changes occur during childhood and adolescence, and these rapid changes can precipitate a unique set of physical and mental health risks. But children and adolescents can also be challenging to study for a number of reasons. Because their ability to weigh the pros and cons of participating in research is limited, they are afforded a number of special protections as research subjects in many countries. This can limit the feasibility of conducting research on sensitive topics like sex, but can also reduce participation rates in cases where both the child and parent must provide consent to participate. The unique health issues that arise during this developmental stage, together with the unique challenges in studying children and adolescents, make system science methods particularly attractive. Thus, the final chapters in this section focus specifically on how system science methods have been useful for studying human development during this critical period.

Actually measuring the social networks of children and adolescents can be particularly challenging. But Neal and Kornbluh describe an innovative strategy for network data collection – cognitive social structures – that not only overcomes some common practical challenges, but also showcases the breadth of research questions that can be explored using network analysis techniques. Collecting information about social networks ordinarily involves asking research subjects to identify their friends (or enemies, family, colleagues, etc., depending on the study). The cognitive social structures approach adopts a different approach, asking research subjects to identify the friends *of others in the setting*. Each respondent's own perception of the setting's social network can be examined in its own right, but can also be combined or triangulated with others' perceptions to obtain a composite picture of the setting's social network. As this chapter illustrates, by adopting this approach to network analysis, researchers can not only explore how social structure is associated with individual behaviors like depression or sexual initiation, but also how individuals form perceptions of their social environments.

While Neal and Kornbluh were focused on friendships and hanging-out relationships among younger children, other types of relationships emerge as children progress into adolescence, including romantic and sexual relationships. Bearman and colleagues draw on data from the National Longitudinal Study of Adolescent Health, or AddHealth, to study the pattern of these types of relationships in one high school. Their goal is not simply to understand what the romantic network among a group of high school students looks like, but rather to understand how such a network forms. They use a series of simulation-based techniques to identify the unwritten and unconscious rules these students seem to be following when selecting their romantic and sexual partners. For example, they find that adolescents typically avoid dating their "former partner's current partner's partner," or in network terms, avoid creating cycles of length 4. While this behavioral rule is quite abstract, it has important public health implications: it means that adolescent romantic and sexual networks tend to have long, chain-like structures that facilitate the silent spread of disease. This risk, and these behavioral rules, are invisible to the students themselves, but only emerge by adopting a network approach that directly examines larger patterns.

All the various affiliations that form and are explored during adolescence – affiliations to friends and romantic partners, but also affiliations to organizations and interests – play a role in identity development. In the final chapter, Saskia Kunnen builds a dynamic systems model to explore how adolescents' commitments in different life domains shape their identities. Using this as an illustrative example, she highlights how her system model, and indeed all the system science models in this section and the rest of the volume, are informed by theory. The model building process can be viewed as involving multiple steps, beginning first with theoretical constructs, which are then assembled together into a conceptual model, which finally is translated into a series of quantitative expressions. This chapter thus serves as an opportunity to see system science methods as not simply a method, but rather as a tool for expressing theory with precision and transparency.

2

THE SOCIAL SYMBIOME FRAMEWORK[1]

Linking Genes-to-Global Cultures in Public Health Using Network Science

Bernice A. Pescosolido, Sigrun Olafsdottir, Olaf Sporns, Brea L. Perry, Eric M. Meslin, Tony H. Grubesic, Jack K. Martin, Laura M. Koehly, William Pridemore, Alessandro Vespignani, Tatiana Foroud, and Anantha Shekhar

Research Question: How can scholars from different disciplines, working at different levels of analysis, come together to make advances in public health?
System Science Method(s): Networks
Things to Notice:

- Networks exist at multiple scales and can affect one another
- Interdisciplinarity of network analysis

The twenty-first century ushered in a new translational paradigm for understanding the distribution and determinants of human health and treatment outcomes. This orientation arose in part from the success of fully sequencing the human genome, coupled with limited progress toward unraveling chronic disease etiology and health system change. Representations of "helixes to health," "neurons to neighborhoods," "base pairs to bedside," "compound to clinic," or "cells to society" became the metaphorical blueprints for both basic and translational science. However, "translation" came to have different meanings, each of which is key to scientific progress, clinical practice, and improved population health. Here, these meanings are defined and set the principles for a new conceptual framework based on ideas of complexity, transdisciplinarity, and connectedness. Specifically, a genes-to-global cultures frame begins with the interactive, contextual, and dynamic assumptions of Systems Science and draws from Network Science to build one parsimonious variant privileging the explanatory power of network structures, network contents, and network dynamics. A fundamental theoretical plane provides the basic predictive schema with fractal imagery building the extension from the molecular to the geographic levels. Each level of the Social Symbiome (elsewhere called the Network Embedded Symbiome) is constructed and supported from the wealth of classic and contemporary health and health care research across the sciences focused on the influence of networks. While network research is broadly referenced, we use the case of alcohol dependence to focus on how different types of connections, within and across levels, operate to influence risk and outcomes. This vertical integration approach brings unique issues in team formation, study design, and analytic tools to the fore. While these cannot be dealt with in great detail, the basic underpinnings for rigorous, feasible studies with adequate human protections are briefly described.

 DOI: 10.4324/9781315748771-3

Despite unprecedented recent discoveries, health and healthcare research continues to confront etiological, social, and medical challenges in population health. A growing scientific convergence posits that biological and environmental factors operate together through complex pathways to shape health, disease, and treatment outcomes. With growing skepticism toward a genomic-centric approach (Evans, Meslin, Marteau, & Caulfield, 2011), research on prevention, diagnosis, and treatment continues to be central to developing and facilitating intervention strategies with greater public uptake. Further, with diverse disease phenotypes arising from suspiciously similar exposures and processes (e.g., stress, inflammation), physical and mental health problems such as cancer, heart disease, diabetes, and depression are now linked in ways that were vastly underappreciated in the past (Cookson, Liang, & Abecasis, 2009; Eichler, Flint, Gibson, Kong, et al., 2010; Golden, Lazo, Carnethon, Bertoni, et al., 2008; Green, Guyer, & National Human Genome Research Institute, 2011; Hirsch, Iliopoulos, Joshi, & Zhang, 2010; Schadt, 2009). Further, concerns about medical errors, the slow transfer of scientific innovation to clinical practice, and lack of connection between medical care and community conditions have raised similar concerns (Institute of Medicine, 2001; Zerhouni, 2003).

In fact, calls to reconsider research paradigms brought the concept of "translational" science to the fore (Institute of Medicine, 2006; Jones & Duncan, 1995; Shonkoff & Phillips, 2000). Yet, the term "translation" has had different meanings to different scientific audiences. All three of these meanings – moving scientific insights across disciplinary silos (transdisciplinarity), from science to practice (implementation), and from science to the public (dissemination) – are relevant to public health (Pescosolido, 2011).

Taking translational science seriously requires breaking down barriers between biomedical and socio-behavioral science, epidemiology and health services, and research targeted at one level or disease. Further, the challenges in theory, method, analysis, and ethics are daunting. While public health pioneered multi-level models depicting levels of societal influence (Bronfenbrenner, 1979; Brunswick, 2002; Office of Behavioral and Social Sciences Research, 2002; Pescosolido, 2006; Susser & Susser, 1996; Warnecke et al., 2008), the IOM (Institute of Medicine, 2006) identified a fundamental barrier: No organizing theoretical framework has translated the promise of the translational paradigm shift into classes of basic propositions derived from branches of science that have documented important etiological findings for health and healthcare. Even more importantly, no organizing theoretical framework has offered a way to winnow down or organize the plethora of etiological factors with demonstrated promise in unravelling the levers of change to improved health and healthcare.

Thus, the task of rethinking how to push past these challenges remains. Here, we offer one attempt to do so using Network Science as the organizing theoretical path. We suggest that the ability to produce significant scientific breakthroughs depends, in part, on moving the twenty-first-century translational paradigm from seminal idea to a practicable blueprint with theoretically connected data, pioneering analyses, and ethical standards to guide evidence-based research and practice. Creating a multidisciplinary team organized around a shared theoretical perspective, Network Science, holds the potential to minimize disciplinary differences to a manageable level, while expanding analytic tools for dynamic, multi-level data with complex interactions.

Network Science is a variant of Systems Science (Luke & Harris, 2007; Luke & Stamatakis, 2012) that has already suggested new directions for medical and public health research and intervention. As defined, Systems Science focuses on complex systems which "are made up of heterogeneous elements that interact with one another, have emergent properties that are not explained by understanding individual elements of the system, persist over time, and adapt to changing circumstances" (Luke & Harris, 2007:357). Luke and Stamatakis (2012) point to global pandemics, obesity, and tobacco control as three of many critical cases where this approach has broken through research and policy barriers. The Network Science variant provides a specific transformative scientific understanding of complex, interactive systems by shifting focus from *elements (genes, neurons, individuals, nations)* to *interactions (influence, connections, cooperation, links)* among those elements. First and foremost, definition, measurement, and analyses of networks becomes the primary set of factors considered in research.

We begin by providing a brief introduction to the network paradigm, laying out the rationale for why this approach holds particular promise. This is followed by a description of an early health-based prototype, the Network-Episode Model (NEM; Pescosolido, 1991, 2006, 2012), which was developed as a network alternative to dominant individualistic and discipline-based approaches to health care utilization and outcome. This allows a simple demonstration of the utility of a network perspective, a dynamic frame, and cross-level interactions. However, the challenges described above have pushed the boundaries of the NEM to consider epidemiological as well as healthcare problems. In the end, this called for a new, more comprehensive network approach, which we develop here. The Social Symbiome builds on the NEM, centering on Network Science's unifying mechanism of action driving health and healthcare – interactional dynamics shaped by network ties, structures, and processes. We provide a rational for why networks, as opposed to a more linear, single discipline or otherwise traditional approach, is particularly useful. We draw from classic and recent examples of the sometimes hidden power of networks to shape health, disease, and healthcare. To illustrate the operation of networks across levels and time in a more targeted and specific fashion, we focus on alcohol use/dependence/addiction and its treatment.

Systems Science and Complexity – The Network Science Variant

The Human Genome Project made the nature–nurture debate obsolete and called for serious transdisciplinary collaboration. A *genes-to-global cultures* organizing framework provides a response premised on the principle that it is insufficient simply to bring the best scientists across disciplinary or disease-specific expertise together. Rather, it requires scientists to come with a shared vector of theoretical, methodological, and analytic understanding, continuity, and approach. Network Science (Barabasi, 2012; Mabry, Olster, Morgan, & Abrams, 2008; Nunes Amaral & Uzzi, 2007; Vespignani, 2009) cuts across disciplines and has shown great promise in areas of health and health care (Gerberding, 2007; Homer & Hirsch, 2006; Valente, 2010). As Luke and Harris (2007:70) point out, "In public health, much of what we study is inherently relational: disease transmission, diffusion of innovations, coalitions, peer influence on risky behavior, etc."

Network Science shares basic assumptions with Systems Science on the primacy of complexity – that is, how forces across levels interact to shape health, disease, and healthcare outcomes. But embracing the idea of complexity has no single theoretical or methodological expression. At least three methodological variants are commonly cited – agent-based modeling, simulation, and network analysis, each with its particular version of the nature and measurement of complex systems (Luke & Stamatakis 2012). The latter appears to be developing into a unified theoretical version of complexity, and is the one that we employ here.

In 2006, the National Research Council Committee called Network Science "the study of network representations of physical, biological, and social phenomena leading to predictive models of these phenomena" (National Academies Press, 2005). They acknowledged that Network Science may be better defined as a "suite of input attributes and output properties" that "constitute a common core of topics." That list of topics is nearly endless. Network scientists are more likely to agree on what Network Science is not (e.g., not belonging to one discipline, not only about computer networks, not about topology alone, not just an analytic method), even arguing that the lack of precise boundaries and definitions represents a strength (Guerin, 2008).

Network Science is often defined by its unique analytic approach to mapping interactions. Clarifying the distinction between network analysis and a network framework is crucial to separating Network Science from Systems Science, to removing it as an analytic option in the toolbox of Systems Science, and to addressing the "black box" problem in public health. As a theoretical frame, Network Science not only specifies that pieces of society and biology fit together through their interactions or interconnections among large units (as in Systems Science), but that networks are the "active ingredients at each level" (Shonkoff & Phillips, 2000:28). That is, network ties and what flows across them represent the "engine of action" that drives individual behavior, genetic networks turn on or suppresses complex sets of inherited predispositions, and organizational networks explain why the lack of integration in the healthcare system allows individuals to fall through the cracks. In sum, Network Science represents a complex science approach but specifies the nature of complexity by emphasizing (1) interactive ties across units as well as interactions within units, (2) how within and across interactions can reinforce or negate one another, and (3) how the structure, content, function, dynamics, and meanings of these interactions matter.

Social, behavioral, and public health sciences have focused on the influence of network ties for well-being, disease, and healthcare for over a century (Berkman, 1984; Pescosolido & Levy, 2002). A Network Science frame embraces analytic developments from the social sciences in the 1970s (White, Boorman, & Brieger, 1976) as well as those more recently enhanced by natural and physical scientists (Barabasi, 2003). But Network Science, broadly conceived, recognizes and values its theoretical and analytic roots including social support, social capital, egocentric data, and qualitative methods (Erikson, 2013). It embraces classic theoretical underpinnings (e.g., Durkheim, 1951[1897]; Simmel, 1955), its mathematical roots in graph theory (e.g., Koninsberg's "bridge problem"), and recent theoretical and analytic developments (Watts, 2003). Further, some systems scientists who see networks predominately as an analytic approach reject standard design and/or data collection through techniques such as survey methods or randomized clinical trials (RCTs, Brades & Erlebach, 2005; Luke & Stamatakis, 2012). While a Network Science frame requires readjustments in surveys and RCTs, throwing them out only replicates the methodological parochialism of earlier epochs. Surveys have been adapted to collect network data (www.phenxtoolkit.org/) and RCTs may include network-based interventions, whether social or virtual networks (Thorup et al., 2006). All of these theoretical and methodological approaches are necessary to the successful design and specification of a Network Science as translational science.

For example, social support represents one of the most robust influences documented in outcomes from colds (Cohen & Syme, 1985a) to heart disease (Berkman & Syme, 1979); influential at the individual (Valente, 2010), organizational (Ducharme & Martin, 2000), and geographic (Diez Roux, Jacobs, & Kiefe, 2002; Milligan, 2000; Parr, 2003; Pridemore & Grubesic, 2012) levels; and linked to biological processes from heart development to CHD (Berkman & Syme, 1979; Lage, Mollgard, Greenway et al., 2010). A Network Science approach suggests closing the existing bifurcation in health research between social support and social networks studies (Pescosolido, 1991). By understanding the concrete structural connections that create "toxic" workplaces or "protective" homes and communities, as well as seeing how these relate to individuals' reports, a network-based public health approach can unpack and leverage the forces of change in a novel and hopefully effective manner (Koehly & Loscalzo, 2009; Kroenke, Kubzansky, Schernhammer, Holmes, & Kawachi, 2006). Further, medical treatments, even evidence-based practices (EBPs), may be ineffective because they are delivered without support, hope, care, or connection (Pescosolido, 1996). Ineffective delivery can diminish potential efficacy, produce non-compliance, and even magnify problems (Sterman, 2000). Thus, Network Science calls for innovation in RCT research, broadening it to include consideration of network ties among those providing the intervention and/or among the participants' community ties. Both factors now stand outside the traditional RCT, often ignored or seen as confounds to be minimized. Yet, mapping community and treatment-based networks might shed light on the efficacy-effectiveness dilemma, where successful RCT interventions transferred to other settings don't work (e.g., see innovations in the Comprehensive Dynamic Trial, Kadushin, 1983; Rapkin et al., 2012, opposite effects of network support groups in rural and urban areas).

In sum, Network Science both merges and narrows the call for complex and transdisciplinary approaches by drawing on network connectedness within and across units. Together, they build a transformative view of health and healthcare to guide research, based on three principles. *Complexity* signals the growing awareness that understanding most phenomena, including disease, cannot be achieved by focusing on one influence (Ostrom, 2009). *Transdisciplinarity* suggests that the historical division into disciplines and specializations can no longer serve as the only or best road to research progress (Institute of Medicine 2014; Zerhouni, 2003). Network Science addresses the IOM mandate to provide a comprehensive but testable framework (Institute of Medicine, 2001; Shonkoff & Phillips, 2000; Singer & Ryff, 2001). Through a focus on the structure, function, and content of networks, *connectedness* provides a next-generation pathway for the health and healthcare research agenda. Conceptualizing, measuring, and manipulating interconnections among molecular, individual, family, community, and treatment systems allows the discovery of "generalities among systems that, despite their disparate nature, may have similar processes of formation and/or similar forces acting on their architecture" (Bascompte, 2009: 419). Connectedness, once a province of the social sciences, has emerged as the core theme of the landscape of science and public discourse, and has even been proposed as a basic human right (Sporns, 2014).

The Social Symbiome

Background

The proposed theoretical framework represents a natural transdisciplinary extension of the Network-Episode Model (Pescosolido, 1991). This model was developed in response to static and individualistic rational choice approaches to healthcare utilization, and outcomes that had failed to explain how individuals get into the health system (Pescosolido, 1992). The NEM melded insights across social and behavioral disciplines and from past qualitative and quantitative efforts using a social network perspective.

The NEM begins with embedded individuals under the premise that health is a phenomenon given meaning through a social process managed by individuals' social networks in the community and in treatment systems. Since individuals have both agency and *habitus* (i.e., practical consciousness), they improvise and routinize within the possibilities and limits of their social networks. As pragmatists with common-sense knowledge and cultural routines, individuals face changes in their health status by interacting with others who may recognize (or deny) a problem, send them to (or provide) treatment, and support, cajole, or nag them about appointments, medications, or lifestyle. However, as societies also include institutions of social control, individuals may enter the healthcare system with resistance and under coercive requirements (e.g., mandatory school, work, or sports examinations; involuntary commitment). Thus, the NEM challenged the "tyranny of the mean," hypothesizing differential pathways to care (Pescosolido, Brooks-Gardner, & Lubell, 1998) with networks as the mechanism linking influences within and across levels. The NEM has been applied across disease families (Carpentier, Bernard, Grenier, & Guberman, 2010; Carpentier & White, 2002), cross-nationally (Carpentier & Bernard, 2011; Edmonds, Hruschka, Bernard, & Sibley, 2012; Judge, Estroff, Perkins, & Penn, 2008; Rogers, Hassell, & Nicolaas, 1999), and particularly where the focus is on ethnic populations (e.g. Native Americans, Novins, Spicer, Fickenscher, & Pescosolido, 2012; Latinos, Pescosolido, Wright, Alegria, & Vera, 1998; African Americans, Lindsey et al., 2006).

As evidence mounted for its utility, the NEM's vertical dimensions were expanded (Pescosolido, 1996, 2012). Above the individual, the NEM-Phase II refined the notion of social networks into separate contextual levels, including "outside" (community) and "inside" (treatment system) networks (Pescosolido & Boyer, 1999). This distinction preceded the widescale introduction of Systems Science approaches in health and healthcare but reflects its premise of large, interacting units. In fact, the NEM-II defined the *interface of community and treatment systems* as key to understanding initial use and later adherence, since these systems may reinforce or cancel the other's influence. Similarly, the increasing obsolescence of the nature/nurture debate was central to the NEM's expansion of its scope to levels *within* or *below* the individual, expanding its scope to include social epidemiology, and reconceptualizing the illness career. Onset, lay response, diagnosis, and treatment outcomes are intertwined (Pescosolido & Levy, 2002). Social networks may have personality, biology, and genetic underpinnings (Pescosolido, 2006, 2012).

Foundations

At each level, network structures exist, creating a safety net of the sort depicted in Figure 2.1, where there are places of excess and deficiencies in connectedness and content. Poles indicate extremes with underproduction/absence represented by sparse network connections and overproduction/presence by dense network connections.

This safety net conceptualization (Institute of Medicine, 2002; Pescosolido, 2011) allows us to move away from the tendency to think of social and biological structures as fixed or as having only linear effects. Each level of human activity is defined by two essential elements of connectedness. The first, running from right to left, is the notion of integration, generally considered the positive or facilitating aspects of connectedness (e.g., whether individuals have others in their life who provide love, care, and concern). At the cognitive level this would include whether there are expected pathways along which energy is passed in the brain. The second, running from front to back, is regulation, generally considered the negative or suppression aspects of connectedness. For individuals' social networks, this dimension targets issues of norms – that is, whether there are people in the network who provide guidance, establish boundaries, and negatively sanction behavior outside societal expectations (see Huitsing et al., 2012 on the structure of positive and negative social relationships).

According to the initial sociological conceptualization that focused on suicide, having moderate levels of both integration and regulation are best and provide the most protection from suicide (Durkheim 1951[1897]). The effect is not linear. Risk increases exponentially the closer social groups or societies move toward the extreme poles. These poles define two sets of dangerous regions. The first exists where the connectedness in networked social structures are too sparse. Loose social structures, ones in which there is too little integration (referred to as an egoistic group, society, or structure) or too little regulation (referred to as an anomic

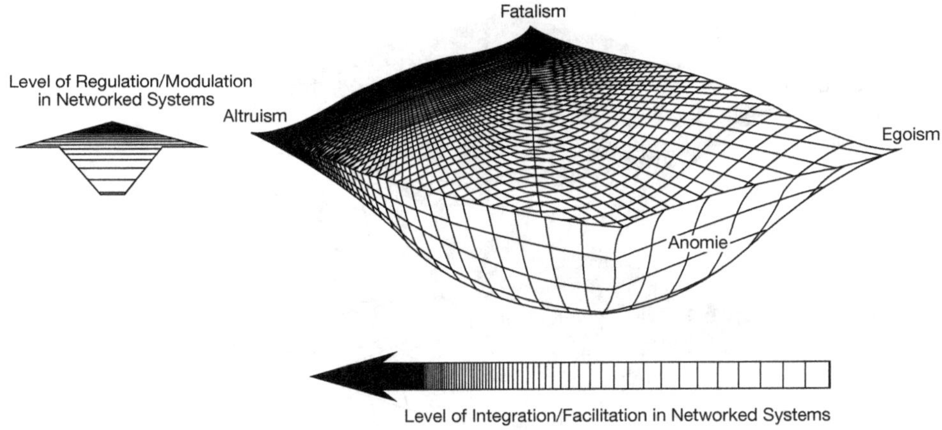

Figure 2.1 The theoretical predictive plane

structure), increase the risk of suicide for individuals who live in them. In the face of crisis, for example, the networks in their life are too sparse, making it hard for them to grasp the safety net. There is no social cushion to absorb the shock of life traumas, including loneliness. The second set of regions does not reference lacunae but surfeits. In the too tightly knit networks, the safety net, in essence, closes up. These networks are unable to provide a soft place to land and become like walls that shatter when individuals hit them during crises. For example, close-knit connections in military units, families, and religious groups engender a sense of responsibility and loyalty, where individuals may sacrifice their lives for others in the network.

While Figure 2.1 suggests a homeostatic view, a number of additional assumptions, at least at the societal level, escape those conservative assumptions. This is not a fixed configuration across societies. Each society establishes the bounds of the normative and this can vary dramatically across time and place. This is also a dynamic conceptualization. Individuals, organizations, and societies can change their location on the net, allowing for contingencies and dynamic processes in the conceptualization of risk for health/illness, recognition/response, and good/bad outcomes (Hoogland et al., 2013). Societies have the characteristic of plasticity, and plasticity and adaption are also core features of brain networks. As different neural elements and circuits become engaged in the course of cognition and behavior, their mutual relations continually adjust and their placement and embedding within the overall network shift in response to each individual's history of behavioral and social engagement.

The view in Figure 2.1, representing the theoretical predictive plane, can exist at any level. To match the notion of complex systems, the major interacting units must be considered, defined as part of the research frame, with propositions developed within and across levels. The Social Symbiome, shown in Figure 2.2, unites levels through network connections using Abbott's (2001) translation of mathematical fractal imagery for social structures and processes. The safety net is a complex but self-similar geometric shape repeated at many levels, with those in Figure 2.2 representing major interacting units (Barabasi, 2009). Other levels are not only possible, but critical. For example, the organizational system level includes staff networks within treatment centers (intraorganizational ties), links of various sorts (shared personnel, contracts) between organizations that make up the healthcare system, including the relevant educational, religious, and voluntary organizations that provide important services (interorganizational ties, Morrissey et al., 2002; Pescosolido, 1996). At the personal network level, individuals' reports of their important matters and health matters networks are neither the same nor equally predictive of

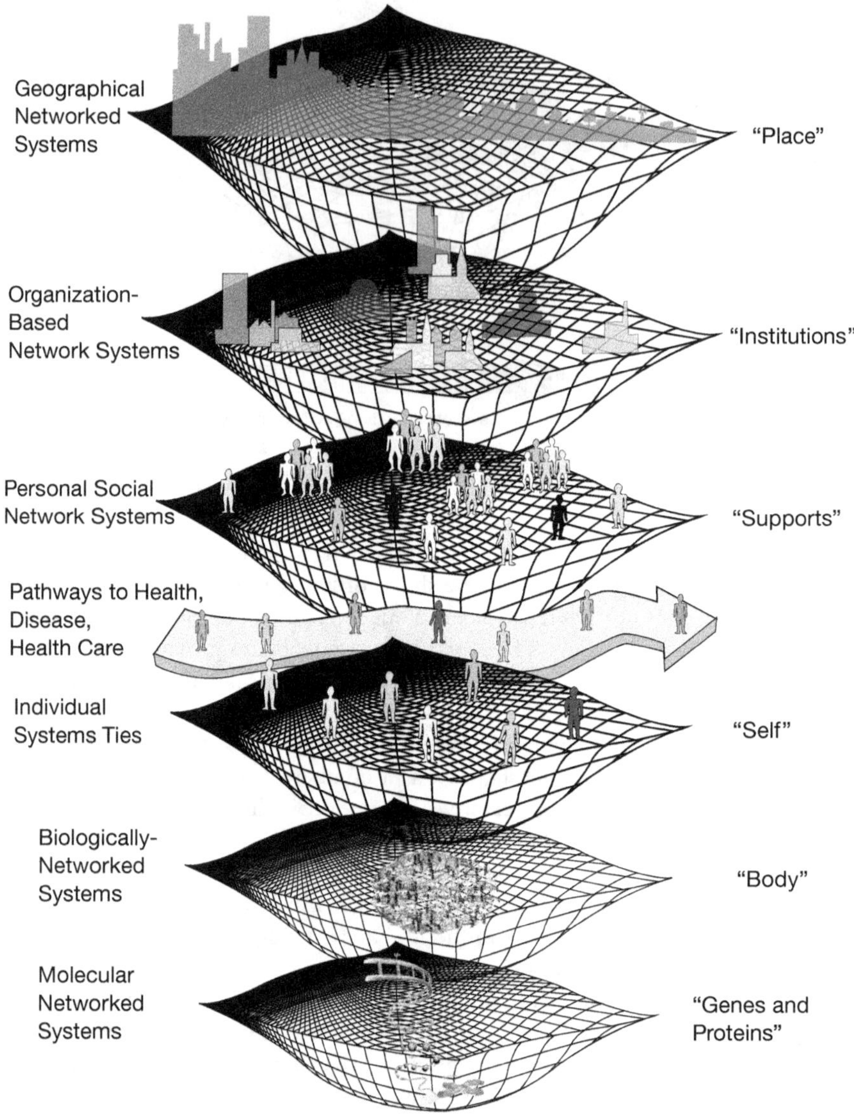

Figure 2.2 The Social Symbiome (elsewhere called the Network Embedded Symbiome)

health and health care outcomes (Perry & Pescosolido, 2010; see also Small, 2013 on important matters). Taken together, the basic levels depicted in Figure 2.2 are designed to represent the classic concept of *social embeddedness*, the degree to which actors (either individuals or molecules) are enmeshed in networks with variable structure and content (Granovetter, 1983; Kawachi, Ichiro, & Berkman, 2001).

Though the concept of a social safety net originated in social science (Berkman et al., 2000), it can be naturally extended to biological levels. For example, as noted above, the multiscale human brain networks range from individual nerve cells to brain systems engaged in sensory, motor, and cognitive function (Sporns, Tononi, & Kotter, 2005). These brain networks are mediating as well as responding to behavioral and cognitive processes at the systems end of the scale, and they result from and modulate molecular signals and genetic regulation at the cellular end of the scale. Because collapsing all potentially interacting biological elements into a single unit has been shown to result in poor models (Butts, 2009), we represent this as a parallel concept of *biological embeddedness*.

Hertzman's term for the mechanism linking social and biological structures is the process of *biological embedding* (Hertzman & Boyce, 2010; Hertzman, 2000) – that is, how social experiences (e.g., traumas) and circumstances (e.g., day-to-day lived experiences) alter biological processes to shape human outcomes (e.g., epigenetic regulation). Environmental experiences, including recurring network-based interactions, get "underneath the skin" (Hyman, 2009; Pescosolido, 2015; Taylor, Repetti, & Seeman, 1997) or "into the mind" (Toyokawa, Uddin, Koenen, & Galea, 2012), conferring advantage or detriment, depending on timing, duration, and other contextual influences. Health disparities arise through biological embedding because low socioeconomic status, documented to shape and be shaped by social network connections (i.e., social embeddedness), result in a "generalized susceptibility" (i.e., biological embeddedness) that degrades well-being in "non-linear and non-specific ways" (Hertzman & Boyce, 2010: 333). However, Hertzman and colleagues do not elaborate on how biological embeddedness also signals biological limits on social interaction. For example, recent research suggests that the brain may impose cognitive limits on social networks by defining the capacity for close relationships (Saramäki et al., 2014).

Illustrating Social Symbiome Levels

The conceptual scaffolding for the Social Symbiome levels is bolstered by three preceding sets of theories. First, general systems or network frames that are directed at understanding society and individuals' behavior in it have been proposed, some of them quite early on (e.g., Boulding, 1956; Castells, 2005; Christaller, 1966[1933]; Giddens, 1984; Monge & Contractor, 2003; Simmel, 1955; Tilly, 1984). Second, systems models that consider networks as an influence, but not the underlying and unifying dynamic, have also been proposed (e.g., Andersen, 1995; Ostrom, 2009). Third, as noted earlier, even more widespread in the area of health and healthcare are contextual or environmental influences for human development, in general, or for specific health problems (Ennett et al., 2008; see Pescosolido, 2006 for a review).

These efforts support, rather than replace, the Network Science approach presented in Figure 2.2. The insights have not been integrated across these three theoretical avenues, in part, because disciplinary foci have often dictated boundaries of interest while the grand theoretical schemes have had either no or many substantive foci. On the former, for example, organization-based network systems, among the most developed theoretically and empirically, has been a focus in public administration, entrepreneurship, of central interest to researchers in public health and public administration departments (Aldrich, 1972; Aldrich & Whetten, 1981; Van de Ven, 1975). More recently, urban geographers' long-term concern of networks of place has produced a network-based framework that mirrors the complexity of the top level in Figure 2.2 (e.g., Neal's 2013 conceptualization of cities as networks *and* networks of cities), as well as reinforces the importance of personal networked systems (i.e., Neal's networks *within* cities). This is exacerbated by the differential timing of interest in network approaches across siloed disciplines. For example, personal networked systems have been the domain of sociology since its origins with similar early importance in geography (noted above). Physics (Barabasi, 2009, or perhaps the earlier development of many-body theory and condensed matter physics in the late 1950s, see Cohen, 2008), psychology in the 1970s (Sarason, 1976), economics in the 1980s (Jackson, 2010), or brain science at the turn of the century (Sporns, Tononi, & Kotter, 2005) have more recently focused on connections rather than units. Similarly, health and illness behavior theories looked to community-based social ties as instigating or legitimating claims of entry into the "sick role" (Parsons, 1951) but have had a muted and undulant prominence in this area until fairly recently (Pescosolido, 2006).

In sum, the combination of (1) multiple level approaches that (2) take a network perspective and (3) are tailored to a particular set of human problems is rare. The innovation lies in synthesis, not in unique discovery. Thus, we draw liberally from the insights in these seminal works, synthesizing what seems to us as fundamental insights. However, we do not go as far as considering how particular previous theoretical schemes, as useful as they may be, mesh exactly with the Social Symbiome.

What seems most novel to us is the incongruous synthesis of generality and specificity necessary to the conceptualization, design, and analyses of network empirical studies. To provide this simultaneous sense of breadth and depth in a limited space, each of the categorical levels depicted in Figure 2.2 is defined, a brief accounting of health research where a network approach has proven useful is provided, and a more in-depth illustration of network studies focused on alcohol use, abuse, and dependence is presented. The area of substance abuse represents an ideal candidate for integrative theorizing, since drinking is seen as highly social – that is, tapping into normative and geographic influences as well as genetic and biological ones (Pescosolido et al., 2008). In pre- or proto-network theories (e.g., Sutherland's Differential Association Theory), alcohol and drug use were seen as initiated by and through delinquent peers (Williams et al., 2007), an idea that contemporary research continues to support (Sznitman, 2013).

Dynamic Network-Shaped Pathways to Health, Disease, and Health Care

Social, technological, and organic systems are all dynamic, mutually contingent and reinforcing or diluting. The early notions of pathways to care (e.g., the "illness career"; Clausen & Yarrow, 1955) and life course approaches ("linked lives"; Elder, 1998) are represented in the Social Symbiome as the undulating arrow. Changes in health, illness, and disease can be studied as patterns or pathways and examined for trajectories and turning points. Individuals' responses to the onset of health problems likewise tend to be more of a "problem response string" than a one-time, either/or decision-making process. The solutions are often driven by formal and informal consultation with others (Pescosolido, 1991) but limited or facilitated by access networks of the health care system. Finally, social networks change in response to the biographical disruption that illness represents, and not always in expected ways or in the same way for different health problems (Perry & Pescosolido, 2012).

The structure and supportive/disruptive content in social networks are considered critical to outcome trajectories in alcohol treatment and prevention (Gordon & Zrull, 1991; Homish & Leonard, 2008; Peirce, Frone, Russell, Cooper, & Mudar, 2000). For individuals socially embedded in low alcohol networks or who form new ones through interventions, substance abuse is likely to be low, to increase the likelihood of successful intervention, or to result in changes in an individual's social ties. Social networks that support heavy drinking present a fundamental challenge to individually oriented therapeutic approaches. In marriage, the association between drinking and the network tie is dynamic and bidirectional throughout the early years. However, because ties are moderately stable over time, they serve to maintain existing drinking patterns (Homish & Leonard, 2008). However, networks do not exert a simple one-way social influence. Heavy drinkers maintain and shape their social networks to support their levels of alcohol use.

Geographic Networked Systems

The top net or level of the Social Symbiome represents both real and virtual links between large geographic areas (i.e., place). Face-to-face contact, though impersonally determined by technological (e.g., airline) or political/economic (e.g., trade) routes, can provide a vector of disease or of health information/resources. Virtual networks that escape specific physical locations, mapped on a local to global scale, can be leveraged to promote healthy or unhealthy behavior (e.g., blogs, message boards, and websites), advance treatments (e.g., telemedicine), and monitor adverse events (e.g., following the impact of a health event through the Twittersphere; the contextualization of the transmission of etheric pathogens, Eisenberg et al., 2012). Thus, the global system (Colizza, Barrat, Barthelemy, & Vespignani, 2007; the macro-urban networks of Neal, 2013), nation states (Pridemore & Grubesic, 2012), communities (Christakis & Fowler, 2009), neighborhoods (Diez Roux, 2011; Entwisle, Faust, Rindfuss, & Kaneda, 2007), and census blocks (Mazumdar, King, Zerubavel, & Bearman, 2010) have been shown to shape coronary vascular disease, cancer, autism, general health status, and mental health (Cummins, Curtis, & Diez-Roux, 2007; Echeverria, Diez-Roux, Shea, Borrell, & Jackson, 2008; Klassen & Platz, 2006).

Network Science uncovered the roots and patterned spread of epidemics such as HIV, SARS, and H1NI as a spatial diffusion phenomenon (Barabasi, 2009; Vespignani, 2009). Obesity researchers have documented associations between BMI, local geography, and food distribution patterns, creating public health solutions beyond a focus on individual behavior. Finally, the WHO contended that social networks carried the policy message of health promotion as the "new public health" globally to collaborators in other sectors, organizations, and all levels of governance (Kickbusch, 2003).

Relatively little research on alcohol-related networked global systems has been conducted, but preliminary research is promising. Jernigan argues that the globalization of the production and marketing networks in the alcohol industry, even among those organizations that aim at education and support for responsible drinking, have become part of "an integrated and global strategy of branding and promotion" (Jernigan, 2008:11). According to Jernigan, a small number of transnational firms create and control networks of producers, importers, advertisers, and distributors. They shape the context and norms of drinking, once only subject to local drinking practices and patterns, by embedding them in global business networks. Without understanding global commodity chains and networked relations within and across links in the chain, public health efforts at the state level are likely to have limited success.

Networked individuals and systems drive the spatial clustering of problem drinking and alcohol-related problems. At lower geographic levels, research relates the density of alcohol outlets to alcohol problems in both cross-sectional and longitudinal studies. For example, problem drinkers tend to be grouped in neighborhoods, with social networks tying individuals' drinking behavior to drinking behavior of others. The presence of alcohol outlets signals both observable drinking behavior and drinking norms in social networks (Scog, 1985; Scribner, Cohen, & Fisher, 2000).

Organization-Based Networked Systems

In the next system level, networks connect individuals to healthcare providers, providers within organizations to each other, providers to service system organizations, and organizations to others in the organizational field (Menchik & Meltzer, 2010; Pescosolido, 1996; West, Barron, Dowsett, & Newton, 1999). Networks have been used widely to describe and mend "cracks in the system," as well as to understand the relationship between organizational strategy, structure, and performance under policy change, especially as recent economic forces have created greater interdependencies (Anderson & McDaniel, 2000; Bazzoli, Shortell, Dubbs, & Chan, 1999; Burns & Wholey, 1993; Krauss, Mueller, & Luke, 2004; Moore, Smith, Simpson, & Minke, 2006; Provan, Fish, & Sydow, 2007; Soda & Usai, 1999; Valente, Chou, & Pentz, 2007).

Both intra-organizational networks and inter-organizational networks determine how culture and climate support or motivate the work of those in various organizations or are served by them. Intra-organizational networks affect how individuals are cared for in a treatment setting and just as importantly, whether discharge planning occurs, whether families are brought into treatment decisions, and whether follow-up services are secured (Ashida et al., 2009; Cross, Parker, & Cross, 2004; Wright, 1997).

Inter- and intra-organizational networks create and reproduce organizational cultures and climates that affect the health of their members (both employees and clients). Alcohol research on the inter-organizational level suggests that innovative private sector substance abuse treatment centers are more likely to scan the organizational network (Knudsen & Roman, 2004). With social embeddedness expectations in the Social Symbiome, too-dense organizational networks stifle the adoption of EBPs because they do not have the "weak ties" or "structural holes" (Burt, 1992) that bridge network groupings, providing efficient pathways for new information to push policy change (Valente et al., 2007). Program efforts have leveraged the network perspective to positive outcomes, implementing a central authority to integrate systems in mental health, substance abuse, and homelessness (Morrissey et al., 2002).

The most detailed inter-organizational research on drinking networks has been conducted in the workplace, and shows that employee drinking patterns are affected by social networks in two ways. Work groups may expect or demand some level of drinking among members through drinking norms (e.g., heavy drinking among hard core journalists) and they may cover for problem drinkers by providing instrumental support (Martin, Roman, & Blum, 1996).

Personal Networked Systems

This level represents individuals' constellations of ties. Personal networks, until recently, represented the most easily understood and well-researched level of the Social Symbiome (Berkman et al., 2000; Friedman & Aral, 2001; Umberson & Montez, 2010). They can be listed in response to specific queries (i.e., name generators) about supporters, regulators, health advisors, or they can be recorded using new RFID technologies. The point is that they do not assume ties based on certain statuses and roles (e.g., marriage equals support). In fact, the absence of expected relationships (e.g., spousal support) from socio-demographic characteristics is crucial for health and health care (Pescosolido & Wright, 2004). Kin, friend, co-worker, or provider composition of personal networks provides different types and levels of support (Wellman & Wortley, 1990).

At the overall level, individuals who report no or few personal ties (i.e., those who are socially isolated) have been shown to have health risks comparable to individuals with high blood pressure, obesity, sedentary life style, and smoking (Cornwell & Waite, 2009; Koehly & Loscalzo, 2009). Network ties buffer the experience of stress, reduce the need for help, provide emotional support, transmit norms and values, encourage screening, influence adoption of health messaging, and offer material aid, services, and information (Carpentier & White, 2002; Koehly & Loscalzo, 2009; Rogers et al., 1999; Valente et al., 2007). Importantly, in line with the Social Symbiome, networks also serve as vectors of health risk, morbidity, and mortality (Colizza et al., 2007; Stickley & Pridemore, 2010).

With regard to alcohol use, population studies suggest that individuals with more diverse personal networks high in social integration engage in more social interaction but drink less (Cohen & Lemay, 2007). Friendship networks depend on the drinking norms held in those networks (Seeman, Seeman, & Budros, 1988). Those that support drinking have an influence on heavy drinking and alcohol related problems (Homish & Leonard, 2008; Leonard, Kearns, & Mudar, 2000). Selection (i.e., which network ties are formed with whom) and socialization (i.e., what network-based norms are learned and/or enforced) processes are at work (Bauman & Ennett, 1996). Likewise, the structure of family networks, even among families with substance abuse issues, matters. Closeness and social monitoring are systematically related to low alcohol misuse among adolescents, parents, and peers (Ennett et al., 2008). In sum, personal networks provide functions and messages, sometimes clear and sometimes confusing, for health issues like alcohol consumption (Gordon & Zrull, 1991). While embedded in these networks, individuals may forge strong or weak ties to them, or include or exclude kin, medical professionals, or other relevant individuals in their personal network systems.

Individual Networked Systems

Individuals are not networks. Yet, individuals have always depended on others, whether this means that they have tight local ties or live under what Rainie and Wellman (2014) call the new *network operating system* (i.e., looser, more fragmented ties that have greater reach because of the Internet and globalization). While networks vary by time and place, in both the historical and geographical sense, identities and resources are thought to be shaped in part by the linked status sets which, in turn, have been linked to health status and health outcomes (Pescosolido and Rubin, 2000; Thoits, 1983). Social interaction processes frame health problems and translate directly into behaviors, including the use of preventative service or low adherence for prescription drugs (Day et al., 2005). Further, socio-demographics (e.g., age, gender), increasingly less predictive in describing individuals' social experiences and exposures, appear to have important interactive (in the statistical sense) effects with network influences (Wenzel et al., 2012).

With respect to alcohol use, interconnected social statuses and personality traits produce different levels of risk, and alcohol problems tend to occur at the confluence of multiple linked and interacting psychosocial variables. For example, high levels of stress (associated with gender, race, and socioeconomic status), low feelings of self-efficacy, an avoidant coping style, and learned alcohol expectancies combine to produce a very strong likelihood of alcohol use and drinking problems, particularly for men (Evans & Dunn, 1995). Moreover, alcohol expectancies are learned in large part through behavior modeling and social interactions in the context of family and peer networks (Brown et al., 1999). While Cohen and Syme (1985b) once suggested that social support may merely proxy personality, their later studies (Cohen & Lemay, 2007) did not appear to support this. In fact, low network engagement and high powerlessness appear to produce the highest levels of drinking (Seeman et al., 1988).

Biological Networked Systems

This level of the Social Symbiome refers to human cellular-level networks. While a relatively new focus, the shift in biological systems (e.g., from the structure and function of brain regions like the frontal cortex or amygdala to the signaling of transport links) represents a renewed enthusiasm for systems biology. The connectivity structure of biological systems, from protein interaction networks (Jeong, Mason, Barabasi, & Oltvai, 2001) to networks of the brain (Bullmore & Sporns, 2009; Sporns, 2011), has been implicated in complex chronic disease

(Lage et al., 2010; Lo, Wang, Chou, Wang, et al., 2010), while specific disturbances of network interactions are associated with various brain and mental disorders (Menon, 2011). Schizophrenia involves severe disruptions of integrative cognitive processes and non-invasive imaging studies show widespread disconnectivity (e.g., reduced densities in long-range cortico-cortical projections, disruptions in functional coupling among remote brain regions, Fornito et al., 2012; Zhou et al., 2012). However, as Kelly et al. (2011) report, connectome (i.e., the networked brain system) research is only now moving in the direction of discovery science on the neural bases of variation in behaviors like alcohol abuse and dependence.

Molecular Networked Systems

Traditional understanding of the link between genes and biological function, rooted in the one gene–one protein hypothesis, has been replaced by a view that eRNAs (non-protein coding) function as endogenous network control molecules facilitating gene-to-gene communication and multi-tasking (Mattick & Gagen, 2001). With technological developments, genetics is now undergoing several shifts, including the movement from gene-centric to network-centric approaches, mapping gene networks in Crohn's disease, cancer, schizophrenia, and in cancer's comorbidity with other chronic conditions (Cookson et al., 2009; Eichler et al., 2010; Hirsch et al., 2010; Schadt, 2009; Wu, Liu, & Jiang, 2008).

According to recent estimates, genetics are thought to account for at least half of the risk for alcohol dependence (Kalsi, Prescott, Kendler, & Riley, 2008). Recent Network Science research documents gene co-expression networks in identifying epigenetic modifications for individuals diagnosed with alcohol dependence (Ponomarev et al., 2012), integrates research on molecular and cellular changes, suggesting a central role for chromatin modification (Kalsi et al., 2008), and uncovers common biological networks that underlie the genetic risk for alcoholism (Kos et al., 2013).

Cross-Level Interconnections

As might be expected, research from a Network Science perspective is emergent; but connections across major units are suggestive. On the geographical level, larger networks facilitate the use of mental health services in NYC while decreasing their use in Puerto Rico, reflecting key differences in cultural contexts (Pescosolido et al., 1998). Kenis and Knoke (2002) posit that organizational, field-level networks shape inter-organizational collaborative ties. On the link between organizational and personal networks, one-third of the reduction of drinking for AA members appears to be due to the program's network components serving as mediators (Kaskutas, Bond, & Humphreys, 2002) in the personal sphere and as the mechanism underlying critical network change in school and neighborhood contexts (Ennett et al., 2008). On the interface of community and treatment systems, individuals' commitment to treatment for alcohol problems was greater when social network members were brought into the treatment setting in the early stages of treatment (Galanter & Pattison, 1984). Personal ties have also been implicated in brain function, structure, and neuroimmunological activity (Glass & McAtee, 2006; Singer & Ryff, 2001). On the biology-society interface, data on alcohol dependence reveal that the typical 4–6% elevated risk for alcohol dependence associated with the candidate gene GABRA2 is virtually eliminated at higher levels of family support, even though many of these families have a substance abuse problem (Pescosolido et al., 2008). Further, networks operating at one level can counter negative ties at another or reinforce them in line with ideas of biological embeddedness and social embeddedness (Alegria et al., 2012; Hertzman & Boyce, 2010;

Pescosolido, 2015). For example, an individual near a dangerous "pole" at the Biological Networked Systems level (e.g., connected cluster of elevated stress markers) of the Social Symbiome may be hypothesized to be offset by having ties of both care and regulation in the Personal Social Network System level. In cancer, the effect of distal factors (e.g., larger social conditions, toxic exposures) can be redressed through social networks (Warnecke et al., 2008). GxE evidence also suggests multiple effects (e.g., low social support, traumatic exposure and PTSD, Amstadter et al., 2009; rural/urban residence, gender and G1287A polymorphism (G/G Geneotype) on major depression, Xu et al., 2009).

A Brief Note on Logistic and Ethical Issues

Space prohibits detailed treatment of design, analysis, and ethical implications of this approach. But ignoring the fact that Network Science requires innovation, or at least alteration, of standard considerations would be misleading. Network scientists come from all disciplines, including physics, medicine, sociology, public health, neuroscience, and genetics, and can bring their expertise to bear, combining existing data sources with primary data collection.

More importantly, sampling constraints and resulting data gaps represent major unresolved issues in Network Science (Handcock & Gile, 2010; Kossinets, 2006; Morris, 2004). Connecting *genes-to-global cultures*, at least at this point, is best served by a research design that emplaces the total population of a geographic area as a living laboratory, an early focus of the Chicago School in sociology (e.g., Park, Burgess, & Janowitz, 1968; see also Sampson, 2012). This avoids biases in assessing comorbidities, for example, since diagnosed health outcomes will not depend on access, diagnosis, or practice differences across health care systems where national public health registries do not exist (Wennberg, 2010).

Network Science incorporates analytic methods from early graph theory to new large-scale algorithms (Barabasi, 2003; Barrat, Barthelemy, & Vespignani, 2008; Jackson, 2010; Lazer et al., 2009; Newman, 2010; Nielsen, 2011; Snijders & Boskers, 2011; Watts, 2004). Innovative analyses integrate data summarized through standard parametric models, relating parameters via higher-order models (i.e., hyperparameters), with standard inference approaches to assess statistically significant interactions. Network Scientists have pioneered clustering and visualization approaches, including community detection algorithms for gene studies (Vazquez, Flammini, Maritan, & Vespignani, 2003) and 3D visual techniques to capture and describe the spatial interrelationships of multi-level data (Green, Hoppa, Young, & Blanchard, 2003; Lee, Li, Shi, et al., 2006; Lee et al., 2008). But without carefully designed qualitative components, the meaning underlying human connectedness will likely be missed. Finally, continuous ethical monitoring must accompany the design and fielding of multi-level research. Relevant ethics guidelines for research, linking biological samples to other kinds of data are only now harmonizing (Caufield, Brown, & Meslin, 2007; Caulfield et al., 2008; Meslin & Garba, 2011; Meslin & Quaid, 2004), determining whether data should be considered minimal risk (Evans et al., 2011).

Conclusions

Luke and Stamatakis (2012:358) see the entry of System Science as "a historical moment for public health science" because its findings have demonstrated utility and impact on a national policy level. Following through on the promise of the twenty-first-century translational paradigm will require strong scientific leadership, tolerance for high-risk projects, and advanced theoretical frameworks, methodological tools, and analytic skills. Yet these characteristics also present opportunities for great rewards and potentially transformative results. A framework lays out the common set of influences thought to have explanatory potential, limiting but demanding rigorous alignment with data collection and analysis. Without a theoretical architecture, the ability to separate the most salient factors can be divisive and overwhelming (Ostrom, 2009; Sterman, 2000). The Social Symbiome, a blueprint for translational science, offers one approach emerging from science and from the broader frame of Systems Science to break through current barriers. Here, we lay out the blueprint as a first step, with more theoretical (i.e., proposition) and methodological (i.e., novel research designs) work to follow.

Network Science meets five criteria essential to merging complexity, transdisciplinarity, and connectedness in the design of a theoretical framework (Pescosolido, 2006, 2012). It (1) considers and articulates all contextual levels documented to have an impact on past empirical research, (2) offers an underlying mechanism (networked interactions) that connects levels, is dynamic, and allows for a way to narrow focal research questions to allow accumulation of scientific knowledge, (3) employs the network metaphor as an analytic language familiar to the biomedical, natural, physical, and socio-behavioral sciences to facilitate synergy, (4) understands the need for and uses methodological tools proven useful across the sciences, and (5) provides a tangible way to improve population health, whether through direct medical intervention, public health prevention, or larger scale social and policy change. The last two points continue to be particularly important for health.

Yet Network Science is still primitive in transdisciplinary integration (Mabry et al., 2008; Shonkoff & Phillips, 2000). Theoretical propositions need to be developed, applied, and tailored to classes of diseases, to specific disease phenotypes, to individuals' health and illness behaviors, and to system functioning. But the movement to a theoretical framework that captures the intricate realities of population health and health care systems is essential. The Social Symbiome is novel and timely. It does not reject the insights or value of other approaches, but absorbs them where relevant to a Network Science approach and, where not in sync, leaves them for other scientists to develop. No single framework can capture the set of heterogeneous influences on human health, yet we have attempted to answer the call for marking new pathways to understanding the etiology of health, illness, and disease, and for integrating scientific advances to improve prevention and treatment efforts.

Following production of this chapter, The Social Symbiome Framework was renamed the Network Embedded Symbiome Framework.

References

Abbott, A. (2001). *Chaos of Disciplines.* Chicago, IL: University of Chicago Press.

Aldrich, H. E. (1972). Organizational boundaries and inter-organizational conflict. *Human Relations, 24*(4), 279–293.

Aldrich, H., & Whetten, D. (1981). Organization-sets, action-sets, and networks: Making the most of simplicity (pp. 385–408). In P. Nystrom & W. Starbuck (Eds.), *Handbook of Organizational Design* (pp. 385–408). New York: Oxford University Press.

Alegria, M., Pescosolido, B., Williams, S., & Canino, G. (2012). Culture, race/ethnicity and disparities: Fleshing out the socio-cultural framework for health services disparities. In B.A. Pescosolido, J.D. McLeod, J.K. Martin & A. Rogers (Eds.), *The Handbook of the Sociology of Health, Illness, and Healing: Blueprint for the 21st Century*, pp. 361–380. New York: Springer.

Amstadter, A.B., Koehen, K.C., Ruggiero, K.J., Acierno, R., Galea, S., Kilpatrick, D.G., & Gelernter, J. (2009). Variant in RGS2 moderates posttraumatic stress symptoms following potentially traumatic event exposure. *Journal of Anxiety Disorders, 23*(3), 369–373.

Andersen, R. (1995). Revisiting the behavioral model and access to medical care: Does it matter? *Journal of Health and Social Behavior, 36*(1), 1–10.

Anderson, R., & McDaniel, R.R., Jr. (2000). Managing health care organizations: Where professionalism meets complexity science. *Health Care Management Review, 25*(1), 83–92.

Ashida, S., Hadley, D.W., Vaughn, B.K., Kuhn, N.R., Jenkins, J.F., & Koehly, L.M. (2009). The impact of familial environment on depression scores after genetic testing for cancer susceptibility. *Clinical Genetics, 75*, 43–49.

Barabasi, A.-L. (2003). *Linked: How Everything is Connected to Everything Else and What It Means*. New York: Plume.

Barabasi, A.-L. (2009). Scale-free networks: A decade and beyond. *Science, 325*(5939), 412–413.

Barabasi, A.-L. (2012). The network takeover. *Nature Physics, 8*, 14–16.

Barrat, A., Barthelemy, M., & Vespignani, A. (2008). *Dynamical Processes on Complex Networks*. Cambridge: Cambridge University Press.

Bascompte, J. (2009). Disentangling the web of life. *Science, 325*, 416–419.

Bauman, K.E., & Ennett, S.T. (1996). On the importance of peer influence for adolescent drug use: Commonly neglected considerations. *Addiction, 91*, 185–198.

Bazzoli, G., Shortell, S., Dubbs, N., & Chan, C. (1999). A taxonomy of health networks and systems: Bringing order out of chaos. *Health Services Research, 33*(6), 1683–1717.

Berkman, L. (1984). Assessing the physical health effects of social networks and social support. *Annual Review of Public Health, 5*, 413–432.

Berkman, L., & Syme, L. (1979). Social networks, host resistance, and mortality: A nine-year follow-up study of Alameda County residents. *American Journal of Epidemiology, 109*, 186–204.

Berkman, L., Glass, T.A., et al. (2000). From social integration to health: Durkheim in the new millennium. *Social Science & Medicine, 51*(6), 843–857.

Boulding, K. (1956). General systems theory. *Management Science, 2*(3), 197–208.

Brades, U., & Erlebach, T. (2005). *Network Analysis: Methodological Foundations*. New York: Springer.

Bronfenbrenner, U. (1979). *The Ecology of Human Development: Experiments by Nature and Design*. Cambridge, MA: Harvard University Press.

Brown, S., Tate, S., Vik, P., Haas, A., & Aarons, G. (1999). Modeling of alcohol use mediates the effect of family history of alcoholism on adolescent alcohol expectancies. *Experimental and Clinical Psychopharmacology*, 7(1), 20–27.

Brunswick, A. (2002). Phenomenological perspectives on natural history research: The Longitudinal Harlem Adolescent Cohort Study. In E. Phelps, J. Furstenberg, F.F. & A. Colby (Eds.), *Looking at Lives: American Longitudinal Studies of the Twentieth Century* (pp. 219–244). New York: Russell Sage Foundation.

Bullmore, E.T., & Sporns, O. (2009). Complex brain networks: Graph-theoretical analysis of structural and functional systems. *Nature Reviews Neuroscience, 10*, 186–198.

Burns, L., & Wholey, D. (1993). Adoption and abandonment of matrix management programs: Effects of organizational characteristics and interorganizational networks. *The Academy of Management Journal, 36*(1), 106–138.

Burt, R. (1992). *Structural Holes: The Social Structure of Competition*. Cambridge, MA: Harvard University Press.

Butts, C. (2009). Revisiting the foundations of network analysis. *Science, 325*, 414–416.

Carpentier, N., & Bernard, P. (2011). The Complexities of Help-Seeking: Exploring Challenges Through a Social Network Perspective. In B.A. Pescosolido, J.D. McLeod, J.K. Martin & A. Rogers (Eds.), *The Handbook of the Sociology of Health, Illness, & Healing: Blueprint for the 21st Century* (pp. 465–481). New York: Springer.

Carpentier, N., Bernard, P., Grenier, A., & Guberman, N. (2010). Using the life course perspective to study the entry into the illness trajectory: The perspective of caregivers of people with Alzheimer's disease. *Social Science & Medicine, 70*, 1501–1508.

Carpentier, N., & White, D. (2002). Cohesion of the primary social network and sustained service use before the first psychiatric hospitalization. *Journal of Behavioral Health Services and Research, 29*(4), 404–418.

Castells, M. (2005). *The Network Society: A Cross-Cultural Perspective*. Cheltenham: Edward Elgar.

Caufield, T., Brown, R., & Meslin, E.M. (2007). Challenging a well established consent norm?: One time consent for biobank research. *Journal of International Biotechnology Law, 4*(2), 69–74.

Caulfield, T., McGuire, A.L., Cho, M., Buchanan, J., Burgess, M.M., Dainilczyk, U., et al. (2008). Research ethics recommendations for whole genome research: Consensus statement. *PLOS Biology, 6* (March).

Christakis, N.A., & Fowler, J.H. (2009). *Connected: The Surprising Power of Our Social Networks and How They Shape Our Lives*. New York: Little Brown.

Christaller, W. (1966) [1933]. *Central Places in Southern Germany*: Upper Saddle River, NJ: Prentice Hall.

Clausen, J. A., and Yarrow, M.R. (1955). Pathways to the mental hospital. *Journal of Social Issues, 11*, 25–32.

Cohen, M.L. (2008). Essay: Fifty years of condensed matter physics. *Physical Review Letters, 101*, 25001.

Cohen, S., & Lemay, E. (2007). Why would social networks be linked to affect and health practices? *Health Psychology, 26*(4), 410–417.

Cohen, S., & Syme, S.L. (1985a). *Social Support and Health*. New York: Academic Press.

Cohen, S., & Syme, S.L. (1985b). The study and application of social support. In S. Cohen & S.L. Syme (Eds.), *Social Support and Health* (pp. 3–22). Orlando, FL: Academic Press.

Colizza, V., Barrat, A., Barthelemy, M., & Vespignani, A. (2007). Predictability and epidemic pathways in global outbreaks of infectious diseases: The SARS case study. *BMC Medicine, 5*(34).

Cookson, W., Liang, L., & Abecasis, G. (2009). Mapping complex disease traits with global gene expression. *Nature Reviews Genetics, 10*(3), 184–194.

Cornwell, E.Y., & Waite, L.J. (2009). Social disconnectedness, perceived isolation, and health among older adults. *Journal of Health and Social Behavior, 50*, 31–48.

Cross, R., Parker, A., & Cross, R. (2004). *The Hidden Power of Social Networks: Understanding How Work Really Gets Done in Organizations*. Cambridge, MA: Harvard Business Review.

Cummins, S., Curtis, S., & Diez-Roux, A.V. (2007). Understanding and representing "place" in health research: A relational approach. *Social Science & Medicine, 65*, 1825–1838.

Day, J., Bentall, R., Roberts, C., Randall, F., Rogers, A., Cattell, D., Power, C. (2005). Attitudes toward antipsychotic medication: The impact of clinical variables and relationships with health professionals. *Archives of General Psychiatry, 62*(7), 717–724.

Diez Roux, A.V. (2011). Moving beyond speculation: Quantifying biases in neighborhood health effects research. *Epidemiology, 22*(1), 40–41.

Diez Roux, A.V., Jacobs, D.R., & Kiefe, C.I. (2002). Coronary Artery Risk Development in Young Adults (CARDIA) Study. *Diabetes Care, 25*(11), 1976–1982.

Ducharme, L., & Martin, J.K. (2000). Unrewarding work, coworker support, and job satisfaction: A test of the "buffering" hypothesis. *Work and Occupations: An International Sociological Journal, 27*(2), 223–243.

Durkheim, E. (1951[1897]). *Suicide*. New York: Free Press.

Echeverria, S., Diez Roux, A.V., Shea, S., Borrell, L.N., & Jackson, S. (2008). Associations of neighborhood problems and neighborhood social cohesion with mental health and health behaviors: The Multi-Ethnic Study of Atherosclerosis. *Health and Place, 14*(4), 853–865.

Edmonds, J., Hruschka, D., Bernard, H.R., & Sibley, Lynn. (2012). Women's social networks and birth attendant decisions: Application of the Network-Episode Model. *Social Science & Medicine, 74*, 452–459.

Eichler, E.E., Flint, J., Gibson, G., Kong, A., et al. (2010). Missing heritability and strategies for finding the underlying causes of complex disease. *Nature Reviews Genetics, 11*, 446–450.

Eisenberg, Joseph N.S., James Trostle, Reed J.D. Sorensen, and Katherine F. Shields. (2012). Toward a systems approach to enteric pathogen transmission: From individual independence to community interdependence. *Annual Review of Public Health, 33*, 239–257.

Elder, Glen H. Jr. (1998). The life course as developmental theory. *Child Development, 69*(1), 1–12.

Ennett, S., Foshee, V., Bauman, K., Jussong, A., Faris, R., Hipp, J., & Reyes, H. (2008). The social ecology of adolescent alcohol misuse. *Child Development, 79*(6), 1777–1791.

Entwisle, B., Faust, K., Rindfuss, R., & Kaneda, T. (2007). Networks and contexts: Variation in the structure of social ties. *American Journal of Sociology, 112*(5), 1495–1533.

Erikson, E. (2013). Formalist and relationalist theory in social network analysis. *Sociological Theory, 31*, 219.

Evans, M., & Dunn, N. (1995). Alcohol expectancies, coping responses and self-efficacy judgments: A replication and extension of Cooper et al.'s 1988 study in a college sample. *Journal of the Study of Alcohol and Drugs, 56*(2), 186–193.

Evans, J.P., Meslin, E.M., Marteau, T.M., & Caulfield, T. (2011). Deflating the genomic bubble. *Science, 331*, 861–862.

Fornito, A., Zalesky, A., Pantelis, C., & Bullmore, E.T. (2012). Schizophrenia, neuroimaging and connectomics. *Neuroimage*.

Friedman, S., & Aral, S. (2001). Social networks, risk-potential networks, health, and disease. *Journal of Urban Health, 78*(3), 411.

Galanter, M.D., & Pattison, M.E. (1984). The use of social networks in office management of the substance abuser. In M.D. Galanter (Ed.), *Advances in the Psychosocial Treatment of Alcoholism* (pp. 98–113). Washington, DC: American Psychiatric Press.

Gerberding, J. (2007). Health protectionomics: A new science of people, policy, and politics. *Public Health Grand Rounds, September 19, 2007.* Washington, DC: George Washington University School of Public Health and Health Services.

Giddens, A. (1984). *The Constitution of Society: Outline of the Theory of Structuration.* Cambridge: Polity Press.

Glass, T.A., & McAtee, M.J. (2006). Behavioral science at the crossroads in public health: Extending horizons, envisioning the future. *Social Science & Medicine, 62*(7), 1650–1671.

Golden, S.H., Lazo, M., Carnethon, M., Bertoni, A.G., et al. (2008). Examining a bidirectional association between depressive symptoms and diabetes. *Journal of the American Medical Association, 299*(23), 2751–2759.

Gordon, A., & Zrull, M. (1991). Social networks and recovery: One year after inpatient treatment. *Journal of Substance Abuse Treatment, 8*, 143–152.

Granovetter, M. (1983). The strength of weak ties: A network theory revisited. *Sociological Theory, 1*, 201–233.

Green, C., Hoppa, R.D., Young, T.-K., & Blanchard, J.F. (2003). Geographic analysis of diabetes prevalence in an urban area. *Social Science & Medicine, 57*(3), 551–560.

Green, E.D., Guyer, M.S., & National Human Genome Research Institute. (2011). Charting a course for genomic medicine from base pairs to bedside. *Nature, 470*, 204–213.

Guerin, R. (2008). All you ever wanted to know and did not dare ask about network science. *What is Network Science? Panel Presentation, ACM CoNEXT 2008.* Madrid, Spain.

Handcock, M., & Gile, K. (2010). Modeling social networks from sampled data. *Annals of Applied Statistics, 4*(1), 5–25.

Hertzman, C. (2000). The biological embedding of early experience and its effects on health in adulthood. *Annals of the New York Academy of Sciences, 896*, 85–95.

Hertzman, C., & Boyce, T. (2010). How experience gets under the skin to create gradients in developmental health. *Annual Review of Public Health, 31*, 329–347.

Hirsch, H.A., Iliopoulos, D., Joshi, A., & Zhang, Y. (2010). A transcriptional signature and common gene networks link cancer with lipid metabolism and diverse human diseases. *Cancer Cell, 17*(4), 348–361.

Homer, J.B., & Hirsch, G.B. (2006). System dynamics modeling for public health: Background and opportunities. *American Journal of Public Health, 96*, 452–458.

Homish, G., & Leonard, K. (2008). The social network and alcohol use. *Journal of the Study of Alcohol and Drugs, 69*(6), 906–914.

Hoogland, E., Patten, I., & Berghmans, S. (2013). The human brain – from cells to story. *Frontiers in Human Neuroscience*, 7(359), 1–7.

Huitsing, G., van Duijn, M.A.J., Snijders, T.A.B., Wang, P., Sainio, M., Salmivalli, C., & Veenstra, R. (2012). Univariate and multivariate models of positive and negative networks: Liking, disliking, and bully-victim relationships. *Social Networks, 34*, 645–657.

Hyman, S.E. (2009). How adversity gets under the skin. *Nature Neuroscience, 12*, 241–243.

Institute of Medicine. (2001). *Crossing the Quality Chasm: A New Health System for the 21st Century.* Washington, DC: National Academies Press.

Institute of Medicine. (2002). *Reducing Suicide: A National Imperative.* Washington, DC: National Academies Press.

Institute of Medicine. (2006). *Genes, Behavior, and the Social Environment.* Washington, DC: The National Academies Press.

Institute of Medicine. (2014). Convergence: Facilitating transdisciplinary integration of life sciences, physical sciences, engineering, and beyond. Washington, DC: National Research Council.

Jackson, M. (2010). *Social and Economic Networks.* Princeton, NJ: Princeton University Press.

Jeong, H., Mason, S.P., Barabasi, A.L., & Oltvai, Z.N. (2001). Lethality and centrality in protein networks. *Nature, 411*, 41–42.

Jernigan, D. (2008). The global alcohol industry: An overview. *Addiction, 104*(Suppl.), 6–12.

Jones, K., & Duncan, C. (1995). Individuals and their ecologies: Analysing the geography of chronic illness within a multilevel modelling framework. *Health and Place, 1*, 27–40.

Judge, A., Estroff, S., Perkins, D., & Penn, D. (2008). Recognizing and responding to early psychosis: A qualitative analysis of individual narratives. *Pyschiatric Services, 59*(1), 96–99.

Kadushin, C. (1983). Mental health and the interpersonal environment: Reexamination of some effects of social structure on mental health. *American Sociological Review, 48*, 188–198.

Kalsi, G., Prescott, C., Kendler, K., & Riley, B. (2008). Unraveling the molecular mechanisms of alcohol dependence. *Trends in Genetics, 25*(1), 49–55.

Kaskutas, L.A., Bond, J., & Humphreys, K. (2002). Social networks as mediators of the effect of Alcoholics Anonymous. *Addiction, 97*, 891–900.

Kawachi, I., & Berkman, L. (2001). Social ties and mental health. *Journal of Urban Health, 78*(3), 458–467.

Kelly, C., et al. (2011). Reduced interhemispheric resting state functional connectivity in cocine addiction. *Biological Psychiatry, 69*, 684–692.

Kenis, P., & Knoke, D. (2002). How organizational field networks shape interorganizational tie-formation rates. *Academy of Management Review, 27(2)*, 275–93.

Kickbusch, I. (2003). The contribution of the World Health Organization to a new public health and health promotion. *American Journal of Public Health, 93*, 383–388.

Klassen, A.C., & Platz, E.A. (2006). What can geography tell us about prostate cancer? *American Journal of Preventative Medicine, 30*(2 Suppl.), S7–17.

Knudsen, H., & Roman, P. (2004). Modeling the use of innovations in private treatment organizations: The role of absorptive capacity. *Journal of Substance Abuse Treatment, 26*(1), 51–59.

Koehly, L.M., & Loscalzo, A.E. (2009). Adolescent obesity and social networks. *Preventing Chronic Disease, 6*(3), A99.

Kos, M.Z., Yan, J., Dick, D.M., Agrawal, A., Bucholz, K.K., Rice, J.P., Johnson, E.O., Schuckit, M.A., Kuperman, Kramer, J., Goate, A.M., Tischfield, J.A., Foroud, T., Nurnberger Jr., J., Hesselbrock, V., Porjesz, B., Bierut, L.J., Edenberg, H.J., & Almasy, L. (2013). Common biological networks underlie genetic risk for alcoholism in African- and European-American populations. *Genes, Brain and Behavior, 12*, 532–542.

Kossinets, G. (2006). Effects of missing data in social networks. *Social Networks, 28*(3), 247–268.

Krauss, M., Mueller, N., & Luke, D. (2004). Interorganizational relationships within state tobacco control networks: A social network analysis. *Preventing Chronic Disease, 1*(4), 1–25.

Kroenke, C., Kubzansky, L., Schernhammer, E., Holmes, M., & Kawachi, I. (2006). Social networks, social support, and survival after breast cancer diagnosis. *Journal of Clinical Oncology, 24*(7), 1105–1111.

Lage, K., Mollgard, K., Greenway, S., et al. (2010). Dissecting spatio-temporal protein networks driving human heart development and related disorders. *Molecular Systems Biology, 6*, 381.

Lazer, D., Pentland, A., Adamic, L., Sinan, A., Barabasi, A.-L., Brewer, D., et al. (2009). Life in the network: The coming age of computational social science. *Science, 323*(5915), 721–723.

Lee, C.S.-L., Li, X., Shi, W., Cheung, S.C.-N., & Thornton, I. (2006). Metal contamination in urban, suburban and country park soils of Hong Kong: A study based on GIS and multivariate statistics. *Science of the Total Environment, 356*(1–3), 45–61.

Lee, D.-S., et al. (2008). The implications of human metabolic network topology for disease comorbidity. *Proceedings of the National Academy of Science, 105*, 9880–9885.

Leonard, K.E., Kearns, J., & Mudar, P. (2000). Peer networks among heavy, regular and infrequent drinkers prior to marriage. *Journal of the Study of Alcohol and Drugs, 61*, 669–673.

Lindsey, M., Korr, W., Broitman, M., Bone, L., Green, A., & Leaf, P. (2006). Help-seeking behaviors and depression among African American adolescent boys. *Social Work, 51*(1), 49–58.

Lo, C.Y., Wang, P.N., Chou, K.H., Wang, J., et al. (2010). Diffusion tensor tractography reveals abnormal topological organization in structural cortical networks in Alzheimer's disease. *Journal of Neuroscience, 30*(50), 16876–16885.

Luke, D.A., & Harris, J.K. (2007). Network analysis in public health: History, methods, and applications. *Annual Review of Public Health, 28*, 69–93.

Luke, D., & Stamatakis, K. (2012). Systems science methods in public health: Dynamics, networks, and agents. *Annual Review of Public Health, 33*, 357–376.

Mabry, P., Olster, Deborah, M., Glen, & Abrams, D. (2008). Interdisciplinarity and systems science to improve population health: A view from the NIH Office of Behavioral and Social Sciences Research. *American Journal of Preventative Medicine, 35*(2 Suppl.), S211–S224.

Martin, J., Roman, P., & Blum, T.C. (1996). Job stress, drinking networks, and social support at work: A comprehensive model of employees' problem drinking behavior. *Sociological Quarterly, 37*(4), 201–221.

Mattick, J., & Gagen, M. (2001). The evolution of controlled multitasked gene networks: The role of introns and other noncoding RNAs in the development of complex organisms. *Molecular Biology Evolution, 18*(9), 1611–1630.

Mazumdar, S., King, M., Zerubavel, N., & Bearman, P. (2010). The spatial structure of autism in California, 1993–2001. *Health and Place, 16*, 539–546.

Menchik, D., & Meltzer, D. (2010). The cultivation of esteem and retrieval of scientific knowledge in physician networks. *Journal of Health and Social Behavior, 51*, 137–152.

Menon, V. (2011). Large-scale brain networks and psychopathology: A unifying triple network model. *Trends in Cogntive. Sciences, 15*, 483–506.

Meslin, E., & Garba, I. (2011). Biobanking and public health: Is a human rights approach the tie that binds? *Human Genetics, 130*(3), 451–463.

Meslin, E., & Quaid, K. (2004). Ethical issues in the collection, storage, and research use of human biological materials. *Journal of Laboratory and Clinicale Medecine, 144*, 229–234.

Milligan, C. (2000). Bearing the burden: Towards a restructured geography of caring *Area, 32*(1), 49–58.

Monge, P., & Contractor, N. (2003). *Theories of Communication Networks*. New York: Oxford University Press.

Moore, S., Smith, C., Simpson, T., & Minke, S.W. (2006). The influence of partnership centrality on organizational perceptions of support: A case study of the AHLN structure. *BMC Health Services Research, 6*, 141.

Morris, M. (Ed.). (2004). *Network Epidemiology: A Handbook for Survey Design and Data Collection*. Oxford: Oxford University Press.

Morrissey, J., Calloway, M., Thakur, N., Cocozza, J., Steadman, H.J., & Dennis, D. (2002). Integration of service systems for homeless persons with serious mental illness through the ACCESS program. Access to Community Care and Effective Services and Supports. *Psychiatric Services, 53*(8), 949–957.

National Academies Press. (2005). *Network Science*: The National Academies Press.

Neal, Z.P. (2013). *The Connected City: How Networks are Shaping the Modern Metropolis*: New York and London: Routledge.

Newman, M. (2010). *Networks: An Introduction*. Oxford: Oxford Oxford University Press.

Nielsen, M. (2011). *Reinventing Discovery: The New Era of Networked Science*. Princeton, NJ: Princeton University Press.

Novins, D., Spicer, P., Fickenscher, A., & Pescosolido, B.A. (2012). Pathways to care: Narratives of American Indian adolescents entering substance abuse treatment. *Social Science & Medicine, 74*, 2037–2045.

Nunes Amaral, L.A., & Uzzi, B. (2007). Complex systems – a new paradigm for the integrative study of management, physical, and technological systems. *Management Science, 53*(7), 1033–1035.

Office of Behavioral and Social Sciences Research. (2002). Progress and promise in research on social and cultural dimensions of health: A research agenda. Bethedsada, MD.

Ostrom, E. (2009). A general framework for analyzing sustainability of social-ecological systems. *Science, 325*, 419–422.

Park, R.E., Burgess, E.W., & Janowitz, M. (1968). *The City*. Chicago: University of Chicago Press.

Parr, H. (2003). Medical geography: Care and caring. *Progress in Human Geography, 27*(2), 212–221.

Parsons, T. (1951). *The Social System: The Major Exposition of the Author's Conceptual Scheme for the Analysis of the Dynamics of the Social System*. New York: Free Press.

Peirce, R., Frone, M., Russell, M., Cooper, L., & Mudar, P. (2000). A longitudinal model of social contact, social support, depression, and alcohol use. *Health Psychology, 19*(1), 28–38.

Perry, B., & Pescosolido, B. (2010). Functional specificity in discussion networks: The influence of general and problem-specific networks on health outcomes. *Social Networks, 32*, 345–357.

Perry, B., & Pescosolido, B. (2012). Social network dynamics in the face of biographical disruption: The case of "first timers" with mental illness. *American Journal of Sociology, 18*(1), 134–175.

Pescosolido, B. (1991). Illness careers and network ties: A conceptual model of utilization and compliance. In G.L. Albrecht & J.A. Levy (Eds.), *Advances in Medical Sociology* (pp. 161–184). CT: JAI Press.

Pescosolido, B. (1992). Beyond rational choice: The social dynamics of how people seek help. *American Journal of Sociology, 97*, 1096–1138.

Pescosolido, B. (1996). Bringing the "community" into utilization models: How social networks link individuals to changing systems of care. In J. Kronenfeld (Ed.), *Research in the Sociology of Health Care* (Vol. 13, pp. 171–198). Greenwich, CT: JAI Press.

Pescosolido, B. (2006). Of pride and prejudice: The role of sociology and social networks in integrating the health sciences. *Journal of Health & Social Behavior, 47*(September), 189–208.

Pescosolido, B. (2011). Taking "the promise" seriously: Medical sociology's role in health, illness, and healing in a time of social change. In B.A. Pescosolido, J.K. Martin, J.D. McLeod & A. Rogers (Eds.), *The Handbook of the Sociology of Health, Illness, and Healing: A Blueprint for the 21st Century* (pp. 3–20). New York: Springer.

Pescosolido, B. (2012). Organizing the sociological landscape for the next decades of health and health care research: The network episode Model III-R as cartographic subfield guide. In B.A. Pescosolido, J.K. Martin, J.D. McLeod & A. Rogers (Eds.), *The Handbook of the Sociology of Health, Illness, and Healing: Blueprint for the 21st Century* (pp. 39–66). New York: Springer.

Pescosolido, B. (2015). Linking the social brain to the social world through network connections. In R.K. Schutt, L.J. Seidman & M.S. Keshavan (Eds.), *Psychosocial Neuroscience: Health and Illness in Context*. Cambridge, MA: Harvard University Press.

Pescosolido, B., & Boyer, C. (1999). How do people come to use mental health services? Current knowledge and changing perspectives. In A.V. Horwitz & T.L. Scheid (Eds.), *A Handbook for the Study of Mental Health: Social Contexts, Theories, and Systems* (pp. 392–411). New York: Cambridge University Press.

Pescosolido, B.A. & Rubin., B.A. (2000). The web of group affiliations revisited: Social life, postmodernism, and sociology. *American Sociological Review 65*, 52–76.

Pescosolido, B., Brooks-Gardner, C., & Lubell, K. (1998). How people get into mental health services: Stories of choice, coercion and "muddling through" from "first-timers." *Social Science and Medicine, 46*(2), 275–286.

Pescosolido, B., & Levy, J. (2002). The role of social networks in health, illness, disease and healing: The accepting present, the forgotten past, and the dangerous potential for a complacent future. *Social Networks & Health, 8*, 3–25.

Pescosolido, B., Perry, B., Long, J., Scott, M.J., Nurnberger, J.I. Jr., & Hesselbrock, V. (2008). Under the influence of genetics: How transdisciplinarity leads us to rethink social pathways to illness. *American Journal of Sociology, 114*(s1), S171-S201. doi: doi:10.1086/592209

Pescosolido, B., & Wright, E. (2004). The view from two worlds: The convergence of social network reports between mental health clients and their ties. *Social Science & Medicine, 58*, 1795–1806.

Pescosolido, B., Wright, E., Alegria, M., & Vera, M. (1998). Social networks and patterns of use among the poor with mental health problems in Puerto Rico. *Medical Care, 36*(7), 1057–1072.

Ponomarev, I., Wang, S., Zhang, L., Harris, A., & Mayfield, D. (2012). Gene coexpression networks in human brain identity epigenetic modifications in alcohol dependence. *The Journal of Neuroscience, 32*(5), 1884–1897.

Pridemore, W., & Grubesic, T.H. (2012). Community organization moderates the effect of alcohol outlet density on violence. *British Journal of Sociology, 63*(4), 680–703.

Provan, K., Fish, A., & Sydow, J. (2007). Interorganizational networks at the network level: A review of the empirical literature on whole networks. *Journal of Management, 33*, 479–516.

Rainie, L. and Wellman, B. (2014). *Networked: The New Social Operating System*: MIT Press.

Rapkin, B., Weiss, E., Lounsbury, D., Thompson, H., Goodman, R., Schechter, C., & Padgett, D. (2012). Using the Interactive Systems Framework to support a quality improvement approach to dissemination of evidence-based strategies to promote early detection of breast cancer: Planning a Comprehensive Dynamic Trial. *American Journal of Community Psychology, 50*, 497–517.

Rogers, A., Hassell, K., & Nicolaas, G. (1999). *Demanding Patients? Analysing the Use of Primary Care*. Philadelphia, PA: Open University Press.

Sampson, R. J. (2012). *The Great American City: Chicago and the Enduring Neighborhood Effect*. Chicago: University of Chicago Press.

Saramäki, J., Leicht, E.A., López, E., Roberts, S.G.B., Reed-Tsochas, F., and Dunbar, R.I.M. (2015). Persistence of social signatures in human communication. *Proceedings of the National Academy of Sciences, 111(3)*, 942–947.

Sarason, S. B. (1976). Community psychology, networks, and Mr. Everyman. *American Psychologist, 31*(5), 317–328.

Schadt, E.E. (2009). Molecular networks as sensors and drivers of common human diseases. *Nature, 461*, 218–223.

Scog, O. (1985). The collectivity of drinking cultures: A theory of the distribution of alcohol consumption. *British Journal of Addiction, 80*, 83–99.

Scribner, R., Cohen, D., & Fisher, W. (2000). Evidence of a structural effect for alcohol outlet density: A multilevel analysis. *Alcoholism: Clinical and Experimental Research, 24*(2), 188–195.

Seeman, M., Seeman, A., & Budros, A. (1988). Powerlessness, work, and community: A longitudinal study of alienation and alcohol use. *Journal of Health and Social Behavior, 29*(3), 185–198.

Shonkoff, J., & Phillips, D. (2000). *From Neurons to Neighborhoods: The Science of Early Childhood Development. Institute of Medicine Committee Report*. Washington, DC: National Academy Press.

Simmel, G. (1955). *Conflict and the Web of Group Affiliations*. New York: Free Press.

Singer, B., & Ryff, C. (2001). *New Horizons in Health: An Integrative Approach*. Washington, DC: National Academy Press.

Small, M. L. (2013). Weak ties and the core discussion network: Why people regularly discuss important matters with unimportant alters. *Social Networks, 35*, 470–483.

Snijders, T.A.B., & Boskers, R.J. (2011). *An Introduction to Basic and Advanced Multilevel Modeling*, 2nd edition. New York: Sage Publications.
Soda, G., & Usai, A. (1999). The dark side of dense networks: From embeddedness to indebtedness. In A. Grandori (Ed.), *Interfirm Networks: Organization and Industrial Competitiveness* (pp. 276–302). London: Routledge.
Sporns, O. (2011). *Networks of the Brain*. Cambridge: MIT Press.
Sporns, O. (2014). Network neuroscience. In G. Marcus & J. Freeman (Eds.), *The Future of the Brain: Essays by the World's Leading Neuroscientists*. Princeton, NJ: Princeton University Press.
Sporns, O., Tononi, G., & Kotter, R. (2005). The human connectome: A structural description of the human brain. *PLoS Computational Biology, 1*, 245–251.
Sterman, J. (2000). *Business Dynamics: Systems Thinking and Modeling for a Complex World*. New York: McGraw-Hill/Irwin.
Stickley, A., & Pridemore, W. (2010). The effects of binge drinking and social capital on violent victimization: Findings from Moscow. *Journal of Epidemiology and Community Health, 64*, 902–907.
Susser, M., & Susser, E. (1996). Choosing a future for epidemiology: II. From black box to Chinese boxes and eco-epidemiology. *American Journal of Public Health, 86*(5), 674–677.
Sznitman, S.R. (2013). Exploring the promise of mandatory random student drug testing by comparing it to other school drug prevention strategies. *Addiction, 108(5)*, 848–850.
Taylor, S., Repetti, R., & Seeman, T. (1997). Health psychology: What is an unhealthy environment and how does it get under the skin? *Annual Review of Psychology, 48*, 411–447.
Thoits, P. (1983). Multiple identities and psychological well-being: A reformulation and test of the social isolation hypothesis. *American Sociological Review, 48*(2), 174–187.
Thorup, A., Petersen, L., Jeppesen, P., Ohlenschlaeger, J., Christensen, T., Krarup, G., & Nordentoft, M. (2006). Social network among young adults with first-episode schizophrenia spectrum disorders. *Social Psychiatry and Psychiatric Epidemiology, 41*, 761–770.
Tilly, C. (1984). *Big Structures, Large Processes, Huge Comparisons*. New York: Russell Sage.
Toyokawa, S., Uddin, M., Koenen, K., & Galea, S. (2012). How does the social environment 'get into the mind'? Epigenetics at the intersection of social and psychiatric epidemiology. *Social Science & Medicine, 74*, 67–74.
Umberson, D., & Montez, J. (2010). Social relationships and health: A flashpoint for health policy. *Journal of Health and Social Behavior, 51*, 554–566.
Valente, T. (2010). *Social Networks and Health: Models, Methods, and Applications*. Oxford: Oxford University Press.
Valente, T., Chou, C., & Pentz, M. (2007). Community coalitions as a system: Effects of network change on adoption of evidence-based substance abuse prevention. *American Journal of Public Health, 97*(5), 880.
Van de Ven, A. H., Emmet, D.C., & Koenig Jr., R. (1975). Frameworks for interorganizational analysis. In *Interorganization Theory*. Kent, OH: Kent State University Press.
Vazquez, A., Flammini, A., Maritan, A., & Vespignani, A. (2003). Global protein function prediction in protein-protein interaction networks. *Nature Biotechnology, 21*(697–700).
Vespignani, A. (2009). Predicting the behavior of techno-social systems. *Science, 325*(5939), 425–428.
Warnecke, R.B., Oh, A., Breen, N., Gehlert, S., Paskett, E., Tucker, K.L., et al. (2008). Approaching health disparities from a population perspective: The National Institutes of Health Centers for Population Health and Health Disparities. *American Journal of Public Health, 98*(9), 1608–1615.
Watts, D. J. (2003). *Six Degrees: The Science of A Connected Age*. New York: W.W. Norton & Co.
Watts, D. (2004). The "new" science of networks. *Annual Review of Sociology, 30*, 243–270.
Wellman, B., & Wortley, S. (1990). Different strokes from different folks: Community ties and social support. *American Journal of Sociology, 96*(3), 558–588.
Wennberg, J. (2010). *Tracking Medicine: A Researcher's Quest to Understand Health Care*. Oxford: Oxford University Press.
Wenzel, S., Hsu, H.-T., Zhou, A., & Tucker, J. (2012). Are social network correlates of heavy drinking similar among black homeless youth and white homeless youth? *Journal of the Study of Alcohol and Drugs, 73*, 885–889.
West, E., Barron, D., Dowsett, J., & Newton, J. (1999). Hierarchies and cliques in the social networks of health care professionals: Implications for the design of dissemination strategies. *Social Science & Medicine, 48*, 633–646.
White, H., Boorman, S., & Brieger, R. (1976). Social structure from multiple networks. I. Blockmodels of roles and positions. *American Journal of Sociology, 88*, 135–160.

Williams, J., VanDorn, R., Ayers, C., Bright, C., Abbott, R., & Hawkins, D. (2007). Understanding race and gender differences in delinquent acts and alcohol and marijuana use: A developmental analysis of initiation. *Social Work Research*, *31*(2), 71–81.

Wright, E. (1997). The impact of organizational factors on mental health professionals' involvement with families. *Psychiatric Services*, *48*(7), 921–927.

Wu, X., Liu, Q., & Jiang, R. (2008). Align human interactome with phenome to identify causative genes and networks underlying disease families. *Systems Biology*, *25*(1), 98–104.

Xu, Y., Li, F., Huang, X., Sun, N., Zhang, F., Liu, P., & Zhang, K. (2009). The norepinephrine transporter gene modulates the relationship between urban/rural residency and major depressive disorder in a Chinese population. *Psychiatry Research*, *168*(3), 213–217.

Zerhouni, E. (2003). The NIH Roadmap. *Science*, *302*(5642), 63–72.

Zhou, J., Gennatas, E.D., Kramer, J.H., Miller, B.L., & Seeley, W.W. (2012). Predicting regional neurodegeneration from the healthy brain functional connectome. *Neuron*, *73*, 1216–1227.

3

THE IMPACT OF URBAN FORM ON WEIGHT LOSS

Combining a Spatial Agent-Based Model with a Transtheoretical Model of Health Behavior Change

Arika Ligmann-Zielinska, Sue C. Grady, and Jeremy McWhorter

Research Question: How might an intervention aimed at changing behavior impact the prevalence of obesity in San Diego?
System Science Method(s): Agent-based models
Things to Notice:

- Simulation of a real spatial environment using empirical data
- Testing the effectiveness of a hypothetical intervention

Since the early 1980s, obesity has become a major health issue in the United States. Although the prevalence of obesity is attributed to a myriad of biosocial, chemical, economic, and cultural factors that constitute a complex hierarchical system, published research suggests that the most important contributors to the obesity epidemic are individual behaviors (diet and physical activity), social interactions, and local environments. Three major improvements to existing obesity analytical approaches have recently been proposed. First, we should extend the methods to include a temporal dimension so that potential explanatory pathway(s) of obesity epidemics can be generated and evaluated. Second, we should make use of the constantly growing individual health databases. Third, we should include spatial heterogeneity and dependence of food and physical activity systems. This chapter reports on a research project that incorporates these three postulates in a spatial empirical agent-based model of obesity dynamics. The model provides a platform for computational experimentation, in which a synthetic population of heterogeneous human agents occupies a GIS-based urban environment. The model allows for the exploration of obesity prevalence by incorporating empirical health and geographic data collected for three selected neighborhoods in San Diego, California, USA, and integrated into a model that simulates weight change measured using the body mass index (BMI) due to the combined impact of health behavior and the built environment. The model is equipped with the transtheoretical sub-model of behavior change that mimics a public health intervention aiming at long-term lifestyle change. Based on the results we conclude that significant differences in obesity dynamics exist between the neighborhoods in San Diego. We also observe that a simple policy intervention can substantially impact population-specific weight loss over a period of five years. Using computational experimentation, we demonstrate how agent-based modeling can augment the conventional obesity analytical frameworks, by providing a means of studying the dynamics of obesity in spatially heterogeneous population and physical activity systems.

 DOI: 10.4324/9781315748771-4

Worldwide, obesity claims more deaths annually than starvation and other underweight-related causes of death combined. Obesity is defined as a body mass index (BMI) of 30.0 or higher calculated by dividing an individual's weight / height^2. Normal weight is defined as a BMI 18.5 to 24.9, with overweight BMI 25.0 to 29.9. Since 1980, the number of obese persons has doubled across the globe (2014). The United States holds the title of *the fattest nation in the world* – a dubious distinction—as evidenced by an obesity prevalence rate of 31.8% (FAO, 2013). Although the prevalence of obesity is attributed to a myriad of biosocial, chemical, political, and cultural factors that constitute a complex system, published research suggests that the most important risk factors for obesity are individual behaviors (diet and physical activity), socioeconomic status, and local residential environments (Alvanides, Townshend, & Lake, 2010; Campbell & Dhand, 2000; Haslam & James, 2005; Hill, Wyatt, Reed, & Peters, 2003; Larkin, 2003; Townshend, Ells, Alvanides, & Lake, 2010). Consequently, there is an urgent need to address the problem of obesity from a variety of perspectives.

The well-known conceptual frameworks of obesity – namely, ANalysis Grid for Environments Linked to Obesity (Swinburn, Egger, & Raza, 1999), and International Obesity Taskforce (Kumanyika, Jeffery, Morabia, Ritenbaugh, & Antipatis, 2002) – are useful tools to represent the complex web of food-activity interrelationships, and theorize the potential 'explanatory pathways' of the obesity epidemic (Pearce & Witten, 2010). These models, however, have limited operational value due to their static, aspatial, and highly abstract nature. Conversely, real world applications to study obesity employ in-depth national surveys such as the Nutritional Health and Nutrition Examination Survey (CDC, 2014) or Behavioral Risk Factor Surveillance System – BRFSS (CDC, 2010), and geographic information systems – GIS (Curtis & Lee, 2010; Drewnowski, Rehm & Solet, 2007; Leslie et al., 2007; Plantinga & Bernell, 2007; Stewart et al., 2011). A major limitation of these studies, however, is the inadequate representation of causality due to the restraining nature of conventional (spatial) statistical methods. Pearce and Witten (2010) propose three major improvements to existing obesity modeling and analysis approaches: (1) introducing dynamics by adding a temporal dimension, (2) making use of additional health databases, and (3) incorporating geographic aspects of food and physical activity systems.

In response to general trends concerning obesity dynamics, agent-based models (ABMs) have received increasing attention as tools to investigate how dynamic processes shape the distribution of health outcomes (Auchincloss & Diez Roux, 2008; Galea, Riddle, & Kaplan, 2010). In their pioneering work Yang et al. (Yang, Diez Roux, Auchincloss, Rodriguez, & Brown, 2011, 2012) demonstrated how ABMs might help to better understand the determinants of walking and identified the most promising interventions to increase physical activity by walking. Evidence-based studies on health have concluded that "who you are" (e.g. age, gender, race, income, social status, routines, and lifestyle habits) is the main predictor of your overall health, but that "where you live" also matters (Brewis, 2010; Pickett & Pearl, 2001). Agent-based modeling provides the opportunity to systematically model how people's interactions within their local environment(s) offer protection from, or contributes to the obesity epidemic.

The objective of this chapter is to address a critical barrier to understanding the micro-level complexities of obesogenic systems (systems that contribute to obesity) by developing and evaluating a prototype of a spatial empirical ABM, which we call oABM (agent-based model of obesity). Spatial agent-based modeling is a method of computational experimentation aimed at modeling dynamic systems, in which individual entities (like humans) operate in a common heterogeneous geographic environment (Brown, Riolo, Robinson, North, & Rand, 2005; Ligmann-Zielinska, 2010). ABM is well suited to address the complex issue of obesity for a number of reasons. First, it directly simulates the interactions between people and those urban structures that affect weight loss and gain. Second, it represents obesity as a system-level property resulting from unbalanced and unhealthy diet reinforced by the lack of physical activity. By operating at an individual-level, the oABM accounts for the variability in sociodemographic, economic, health, and locational characteristics of people, their interactions, feedbacks, and effects on the emergence of obesity. Third, oABM simulates temporal variability and, compared to other modeling approaches, provides a relatively easy procedure to track causality. Tracking obesity is a longitudinal process composed of different stages, including transitional and maintenance lifestyles. Consequently, explicit representation of temporal variability is critical in studies that assess the long-term implications of public health policy interventions.

The oABM presented here allows for exploration of obesity prevalence by incorporating individual, geographic and empirical health data integrated into a model that simulates weight change measured using the BMI. Our oABM is designed to examine different levels of energy expenditure though two distinct forms of physical activity: resistance training and aerobics. The overarching objective is to simulate the joint impact of health behavior and the built environment on BMI dynamics. We choose three distinct spatial factors previously identified as potential drivers of physical activity: accessibility to exercise amenities (Roux et al., 2007; Sallis et al., 1990), neighborhood walkability (Frank, Andresen, & Schmid, 2004; Leslie et al., 2007; Yang et al., 2012), and neighborhood safety (Wilson, Kirtland, Ainsworth, & Addy, 2004). Assuming that a portion of overweight and obese individuals (represented as agents in the model) in a neighborhood decides to lose weight, and the lifestyle change intervention is framed in a selected model of health behavior augmented by social support, we focus on the following research questions: "To what extent does accessibility to physical activity facilities influence BMI change?" "Do different configurations of the built environment affect walkability?" "How does neighborhood safety affect exercise and, as a consequence, BMI dynamics?"

The remainder of this chapter is organized as follows. First we introduce the case study of San Diego, California. We then describe two variants of our oABM, before outlining six computational experiments conducted using the oABM. What follows is a detailed description of datasets and data processing necessary to parameterize the model. Analysis of the results is reported, after which we reflect on the limitations of the study, the challenges faced when developing agent-based models, and the opportunities for improvement. We conclude the chapter by providing insight into future directions of research.

Study Site

For our study, we selected a region characterized by high obesity prevalence, albeit with a diverse population to better model the country as a whole. In 2010, San Diego, California was the 10th most obese (26 per 100 people) large city in the United States (Figure 3.1), and of the other top obese cities, it was one of the most demographically heterogeneous (CDC, 2012). San Diego is also an attractive location for modeling obesity because of its limited weather variability throughout the year, which minimizes confounding in activity due to weather and allows for maintaining the same physical activity routines over longer periods of time.

Given the large size of San Diego and the inability to sample all corridors of the city, we chose three distinct districts to represent the city as a whole. These include *Emerald Hills*, a low- to medium-income area with a large presence of African Americans; *La Jolla*, a predominately white and high-income area; and *Logan Heights*, a predominately Latino and low-income area (Figure 3.1, Table 3.1) (U.S. Census Bureau, 2014). The physical environments of the districts were evaluated using a number of spatial metrics, including compactness, the density of physical activity centers (referred to as *gyms*), and walkability (Figure 3.2, Table 3.1) (Galster et al., 2001).

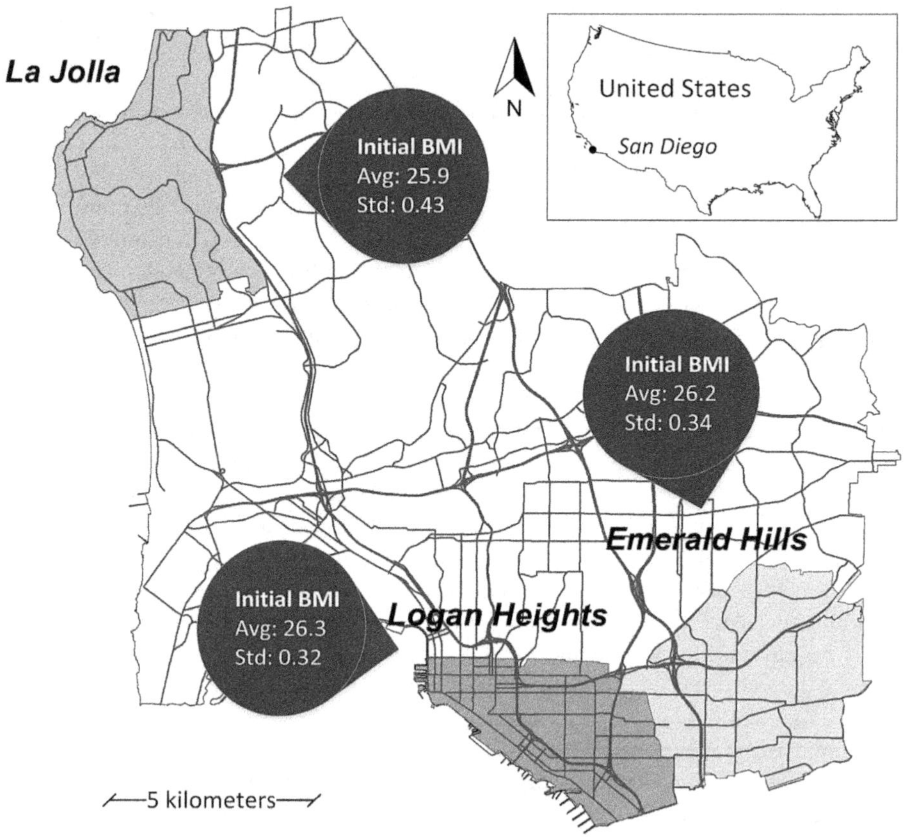

Figure 3.1 Three districts in San Diego, CA, used in this study. BMI: body mass index. Initial BMI represents the start value of all ABM simulations, scaled down from BRFSS data collected in 2009 for the whole of San Diego.

Overall, these three areas are characterized by population and environment heterogeneity, serving as a microcosm of spatial diversity of the San Diego metropolitan area. Finally, previous studies in San Diego suggest that community design and access to exercise amenities have a considerable influence on individual physical activity in the city (Frank et al., 2004; Sallis et al., 1990) making this an ideal city to study the role of physical activity on obesity prevalence in districts with different populations and environmental characteristics.

Model Description

Our agent-based model of obesity (oABM) is a simulation method in which adult individuals (adult ≥ 18 years) are represented by heterogeneous decision-making entities (agents), who follow their daily diets and perform physical activities in a shared spatial environment. The synthetic population of human agents occupies an artificial space composed of parcel lots (places of agents' residence), fitness centers (gyms), walkable roads and crime zones. The fundamental agent behavioral mechanism is the energy balance model (EBM) of energy intake and energy expenditure to imitate weight dynamics (Figure 3.3). Each agent is equipped with weight in kg (w), height in m (h), sex (s), and age in years (y) derived from empirical data. These attributes

Table 3.1 Selected spatial metrics of the tree districts

Metrics	*Description*	*Emerald*	*Logan Hills*	*Lajolla Heights*
Adults	Total number of adults	53213	64299	34471
Female	Female population (% of adults)	50.5	41.1	51.7
White	White population (% of total adults)	28.0	20.0	90.0
Black	Black population (% of total adults)	30.0	10.0	1.0
Latino	Latino population (% of total adults)	42.0	70.0	9.0
Low income	Population with income below poverty level (%)	29.0	43.0	10.0
Agents	Number of agents	263	319	171
Area	District area in square miles	10.5	7.2	10.1
Perimeter	District perimeter in miles	16.6	15.8	15.8
Population density	Average number of adults per square mile	5068	8930	3413
Gym density	Average number of gyms per square mile	0.4	1.4	4.0
Gym proximity	Average number of gyms in 5 mile radius from agent	0.3	0.5	3.3
Crime	Neighborhood crime index (0 lowest, 1 highest)	0.2	0.2	0.7
Walkability	Walkable roads in neighborhood	4.2	5.6	4.5
Agent clustering	Variation coefficient of nearest neighbor distance (%)	56.0	67.0	62.0
Gym accessibility	Agents within 0.5 mile walking distance from gyms (%)	10.0	8.9	29.0
Shape index	Compactness (most compact is square with value 1)	1.3	1.5	1.2

Number of agents in the oABM is proportional to adult population where one agent represents approximately 200 people. Data processing: Network Analyst in ArcGIS (www.esri.com/software/arcgis/extensions/networkanalyst), GeoDa (http://geodacenter.asu.edu/), and FRAGSTATS (www.umass.edu/landeco/research/fragstats/fragstats.html). Census data for 2010. The study concerns only adults due to obesity data availability.

are used to estimate agent's *BMI*, equation 1 (WHO, 2006), and its basal metabolic rate, *BMR*, equation 2 (in calories), which is the energy expended daily at rest (Mifflin et al., 1990). Agents also have a lifestyle attribute called a physical activity level (*PAL*, unitless) which varies from sedentary to vigorous. *PAL* and *BMR* are then used to calculate the total energy expedited by agents in a day (*TEE*, in calories) (FAO, 2004):

$$BMI = w/h^2 \tag{1}$$

$$BMR = 10w + 6.25h + 5y + c \tag{2}$$

$$TEE = BMR \times PAL \tag{3}$$

where $c = 5$ for males and $c = -161$ for females (Mifflin et al., 1990).

The EBM operates as follows (Figure 3.3). For a given day, an agent retrieves its *BMR* and *PAL* to calculate *TEE*. An agent carries a value of calories – *agcal* (estimated based on secondary data), which are consumed daily. An agent is also assigned empirically derived workout time (in minutes) and intensity (calories burned per minute) to calculate total calories burned

Figure 3.2 Spatial configurations of the three districts

(referred to as *burned*) during physical activity like muscle strength-training and aerobic. At the end of the day, agent's weight is updated as follows:

$$w_{t+1} = w_t + (agcal - TEE - burned)/bf \quad (4)$$

where w_t is weight in kg at the beginning of the day, w_{t+1} is weight in kg at the end of the day, and $bf = 7716$ calories/kg is an approximated amount of energy needed to burn 1 kg of body fat (FAO, 2004). What follows is a daily update of *BMI*, *BMR*, and *TEE*. Importantly, after each year, the agent updates its *BMR* and *TEE* to account for its metabolism decrease with increasing age (Mifflin et al., 1990).

The EBM reflects a simple projection of agent's current lifestyle into the future. In order to simulate an increase in exercise and, consequently, weight loss, we introduced a public health policy intervention by adopting the transtheoretical model of behavior change (TTM), Figure 3.4. The TTM, first proposed by Prochaska and colleagues in the 1970s (Prochaska & DiClemente, 1992), provides a temporal framework leading to healthier diet behavior. TTM has been used in a number of interventions including smoking cessation, stress management, and weight change (Dray & Wade, 2012; Johnson et al., 2008). The model is composed of five stages: precontemplation, contemplation, preparation, action, and maintenance. TTM is implemented in our oABM as follows (Figure 3.4). The agent starts from its base lifestyle according

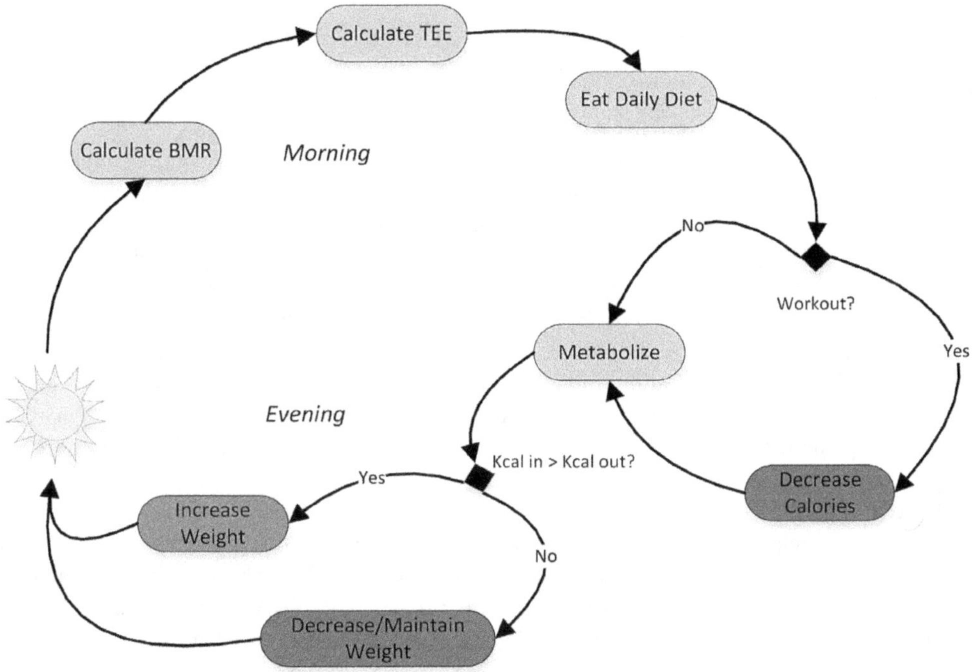

Figure 3.3 The energy balance behavioral rule applied daily to agents. BMR: basal metabolic rate; TEE: total energy expenditure based on agent's physical activity level.

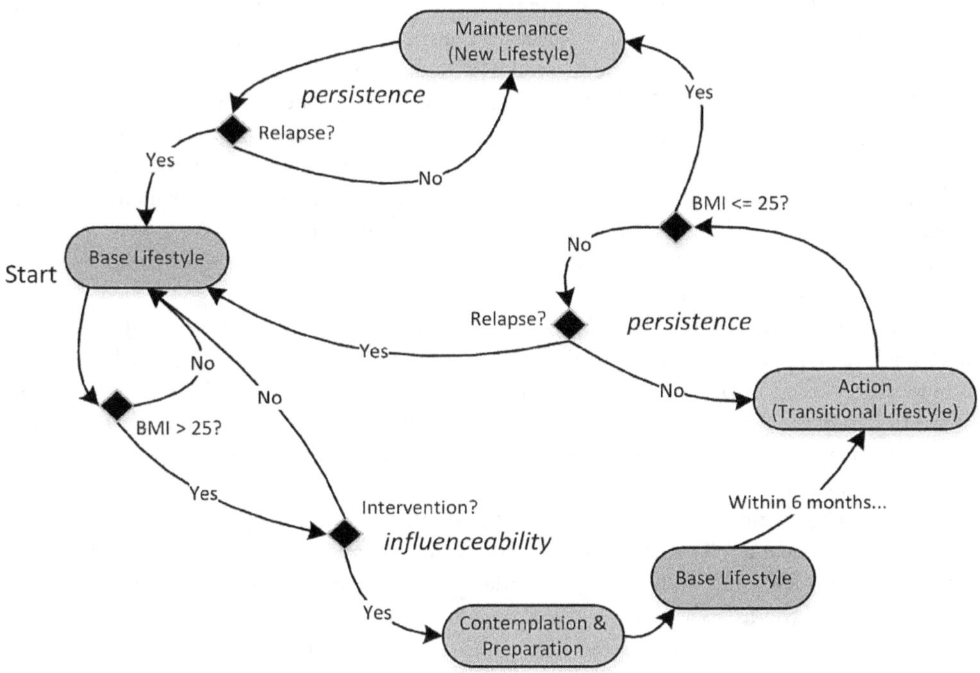

Figure 3.4 Transtheoretical model of health behavior change applied to agents as policy intervention

to EBM (precontemplation). If its BMI > 25.0 (i.e., exceeds the normal range) the agent considers lifestyle change (contemplation), provided that it can be *influenced* by its physician, dietitian, or other specialist. What follows is the preparation stage during which the agent maintains its base lifestyle. Within one day to six months (*readiness*), the agent makes gradual adjustments to behavior by increasing its physical activity (action). For simplicity, we assume that its diet remains unchanged and focus solely on fitness. Moreover, during the transitional lifestyle, the agent may get *support* from its social network. The agent evaluates its BMI on a daily basis, and if it is dissatisfied with the results, it continues the transitional stage, provided that its *persistent*. If a relapse occurs, the agent goes back to its initial lifestyle. If the agent's BMI falls within the normal range, it moves to maintenance, during which its energy input is balanced by energy output. As before, the agent may relapse into its base lifestyle, after which the whole TTM cycle is repeated.

We introduced the spatial component of the model into the TTM by adjusting the agent's burned calories. This modification is dependent on the choice of exercise (a workout in a gym, a walk or a combination of both). Walking can be further reduced due to high crime in agent's neighborhood.

In summary, the EBM requires five attributes: age, weight, height, workout, and calories burned, which are populated based on data obtained from the 2009 BRFSS (CDC, 2010). The TTM needs additional four variables. Due to the lack of empirical observations, influenceability, persistence, and readiness are all populated with simulated values whereas social support is approximated based on previous studies (Kiernan et al., 2012; Leahey, Kumar, Weinberg, & Wing, 2012; Wing & Jeffery, 1999).

Computational Experiments

To account for different combinations of the three spatial factors evaluated in the model, namely, *accessibility to gyms*, *walkability*, and *neighborhood safety*, designed six different computational experiments. Each experiment was executed 500 times, using simple random sampling. The oABM output (called avgBMI in the following sections) is the value of BMI averaged over all agents in a given district and reported for each model run at the end of model execution. The oABM runs for five years with one day increments, which amounts to 1825 execution loops. We selected five years as the simulation time period to assure that the experiments run long enough to render a trend in BMI rather than short-term fluctuations (i.e. the model reached a stable equilibrium after 1825 steps), and short enough so that permanent and major changes in the built environment can be ignored.

The first model (EXP0, Table 3.2) is a lifestyle maintenance model with EBM as the only mechanism driving BMI dynamics. In this version of the oABM, the agents, who emulate individuals recorded in BRFSS, do not change their eating, PAL, and workout habits over the course of model execution. They follow the same lifestyle from day to day, they do not interact, and they do not evaluate their environment in order to select physical activity sites. In this sense, EXP0 produces a benchmark output distribution of avgBMI which is a simple projection of recorded observations into the future. This oABM is therefore conceptually closer to microsimulation than a fully fledged ABM, and is often referred to as a protomodel or a model with protoagents (North & Macal, 2007). Models two through six (EXP1 to EXP5, Table 3.2) are all variants of the lifestyle change TTM, in which access to physical activity is represented using different combinations of selected aspects of the workout-relevant built environment: gym access, walkability, and neighborhood safety. In EXP1 (labeled GYM), the only type of exercise

Table 3.2 Key inputs and differences among the six versions of oABM. Refer to the text for input explanations. *True:* input present in a particular version; *OR:* either one physical activity or the other can be done (but not both), *AND:* both physical activities can be done by agents.

oABM component or input	*EXP0*	*EXP1*	*EXP2*	*EXP3*	*EXP4*	*EXP5*
EBM	True	True	True	True	True	True
Agent group (type)	True	True	True	True	True	True
Weight	True	True	True	True	True	True
Height	True	True	True	True	True	True
Age	True	True	True	True	True	True
Length of workout	True	True	True	True	True	True
Calories burned	True	True	True	True	True	True
TTM	–	True	True	True	True	True
Influenceability	–	True	True	True	True	True
Readiness	–	True	True	True	True	True
Social support	–	True	True	True	True	True
Persistence	–	True	True	True	True	True
Gym access	–	True	–	–	OR	AND
Walkability	–	–	True	True	OR	AND
Safety	–	–	–	True	True	True

allowed is workout in gyms, which predominantly represents muscle strength (resistance) training. The factor influencing calories burned is distance to the nearest gym. The further the gym, the less likely that the agent will do its strength training in a given day. EXP2 (WALK) is the other extreme case, where agents can only walk/run in their neighborhood. It represents mainly aerobic or cardio training. In this case, the lower the surrounding walkability, the less likely it is for the agent to walk or run in a given day. EXP3 is a modification of EXP2, in which walkability is further modified by neighborhood safety (SAFE WALK) – the higher the crime index, the less likely that the agent takes a walk in a given day.

EXP4 and EXP5 represent combinations of resistance training (GYM) and aerobics (SAFE WALK). Specifically, in EXP4 agents can select either gym training or walking/running (chosen randomly with equal probability), whereas EXP5 allows for a combination of GYM and SAFE WALK. The specific rule driving EXP5 is as follows. An agent randomly selects what fraction of workout in a given day is allocated to GYM. SAFE WALK is then calculated as the remainder of the daily workout. What follows is that the GYM fraction is further influenced by accessibility to activity amenities (as in EXP1) whereas the SAFE WALK fraction is influenced by neighborhood walkability and safety (as in EXP3). Note that EXP5 was designed to reflect physical activity guidelines for adults issued by the U.S. Department of Health and Human Services (HHS, 2008). These physical activity guidelines introduce variation to exercise that encourages lifestyle change endurance by combining muscle strength-training and aerobic. This combination of activity has been shown to burn body fat rather than muscle, leading to better physique, much more effective weight reduction, and ensuring such health benefits as decreasing blood pressure and/or glucose levels (Church, Blair, Cocreham, & et al., 2010; HHS, 2008; Sigal et al., 2007; Thorogood et al., 2011). In all five experiments, the decision to actively burn excess calories is independently made every day. The model was implemented in Python (www.python.org/) and executed using the computing resources in the High Performance Computer Center at Michigan State University (http://icer.msu.edu/).

Data Acquisition and Processing

For the oABM, we utilize publicly available data. Except for the TTM parameters, all inputs are derived from empirical observations. Model inputs comprise agent data and spatial data. Below we describe data sources and data processing required to fine-tune the oABM.

Agent Data

The EBM requires demographic and body measurement data collected at an individual level. We used a subset of the 2009 California BRFSS survey data collected from adults only, and aggregated for selected zip codes (n = 408) in the San Diego metropolitan area. The data was pre-processed by Survey Research Group (http://s-r-g.org/).

Research suggests that significant differences in BMI exist among different demographic groups, and variables like sex, race, or income can be used as parameters to differentiate population groups (Sallis et al., 2009). Consequently, based on the sample data, we subdivided the BRFSS variables into four empirical agent groups: male (M); female white (FW); female black/Latino and low income (FBL); and female black/Latino and other income (FBO). For each agent grouping, we generated probability density functions for the following model inputs used in EBM: age, weight, height, and workout time. These probability density functions were later

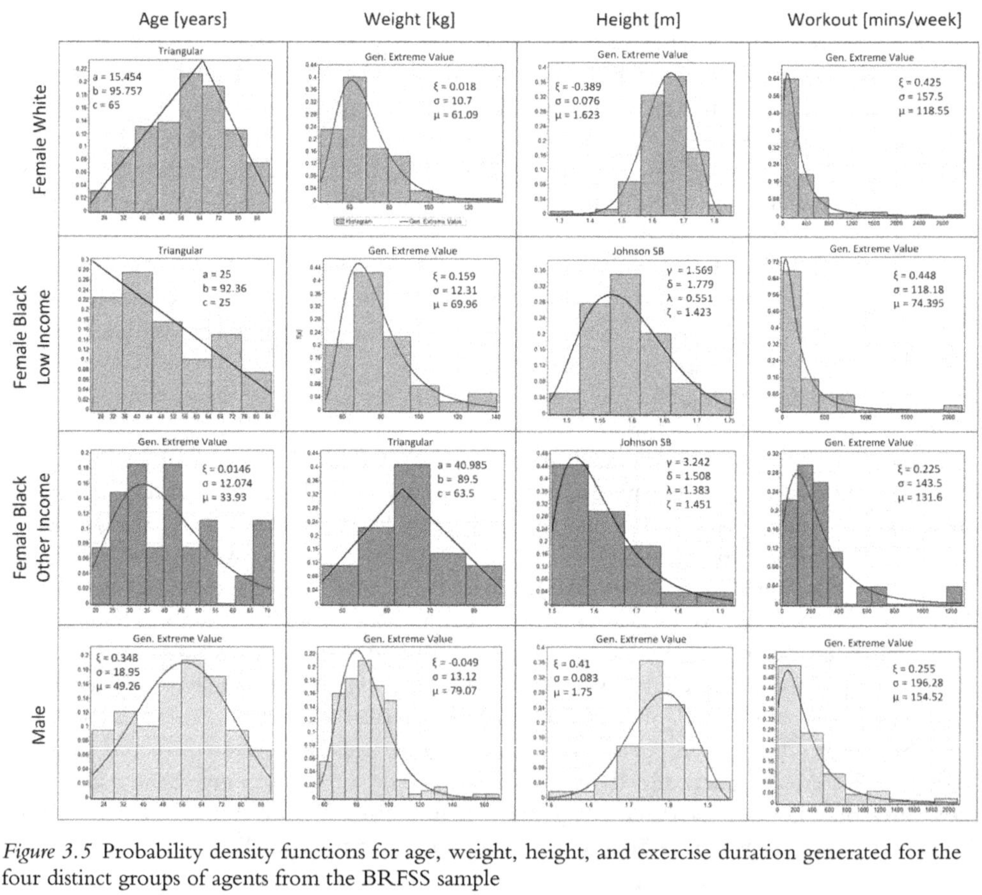

Figure 3.5 Probability density functions for age, weight, height, and exercise duration generated for the four distinct groups of agents from the BRFSS sample

Table 3.3 Parameters for the triangular probability distribution of calories burned per minute for the four distinct groups of agents

Group	*Calories per minute*
FW	a = 2.8, b = 9.1, c = 5.7
FBL	a = 3.3, b = 6.6, c = 10.5
FBO	a = 2.7, b = 8.7, c = 5.4
M	a = 3.6, b = 11.5, c = 7.2

Data source: www.nutristrategy.com/

sampled in order to build diverse parameter sets used in model runs, i.e., when a simulation was populated, each agent's value for every input was randomly drawn from these functions. Figure 3.5 summarizes the distributions.

EBM also requires estimates for PAL, calories consumed, and calories burned to calculate the total energy expenditure and daily weight change (Equations 3 and 4). Since such measurements are not recorded in BRFSS, we used secondary resources to approximate the probability density functions for these variables. Table 3.3 shows the estimates for calories burned per agent group. PAL data by sex, age, and BMI for adults comes from National Academies Press (NAP, 2005) and Roberts and Dallal (2005). Calories consumed were derived from data on mean energy intake among adults by sex and age reported by CDC (Wright & Wang, 2010).

We did not find any data that could be directly utilized to set the TTM parameters (Figure 3.4). Consequently, we had to resort to stylized data. After a number of test simulations, we set the probability density functions of influenceability to a uniform distribution with minimum = 0.001, and maximum = 0.05, and the probability density functions of persistence to a uniform distribution with minimum = 0.95, and maximum = 0.999. Since, by definition, both influenceability and persistence can vary from zero to one, the assumed distributions denote that agents are not easily influenced but, if they decide on lifestyle change, they are fairly committed. We also assumed that, to lose weight, agents have to perform additional workout burning from 250 to 760 calories daily, so that they lose from 0.5 to 1.5 pounds per week.

Spatial Data

The BRFSS data is de-identified and, as a result, it is not geocoded. To allocate agents, we used ancillary information on the number of adults, their sex, race, and income per census tract (U.S. Census Bureau, 2014). These variables were used to subdivide adult population into FBL, FBO, FW, and M. We narrowed down the acceptable locations only to residential parcels (Table 3.4) and then randomly spread a number of points proportional to the adult population in a given tract (Figure 3.2). After attribute initialization, agents were assigned to the points in a manner that reflects the percentage of FBL, FBO, FW, and M in a given tract. AvgBMI at time zero in each of the three study districts in San Diego is shown in Figure 3.1.

Data on recreation and athletic centers (physical activity centers in Figure 3.6) and calculated distance to the nearest workout facility was retrieved for each agent point. Walkability was operationalized based on previous conceptual and empirical studies (Saelens, Sallis, & Frank, 2003). We used walkable roads (roads with pedestrian-oriented design, street connectivity, and sidewalk) as the base spatial dataset (Table 3.4). We first derived a road density surface which was then reclassified into nine equal-interval categories, where high walkability is associated with high density of walkable roads (Table 3.4, Figure 3.6 left). Our rationale for linking

Table 3.4 Spatial data used in oABM

Map	*Date(s)*	*Source(s)*	*Description*
Roads	2012	SANDAG	Polylines for roads
Residential parcels	2009	SANDAG	Polygons of parcels lots with residential land use
Census Tracts	2010	Census	Census tract polygons with data on sex, income, and race
Gyms	2012	SANDAG	Point data for businesses with recreational services
Crime	2009	SDPD	Total crimes per police neighborhood

Sources: United States Bureau of Census (Census), San Diego Association of Governments (SANDAG), San Diego Police Department (SDPD).

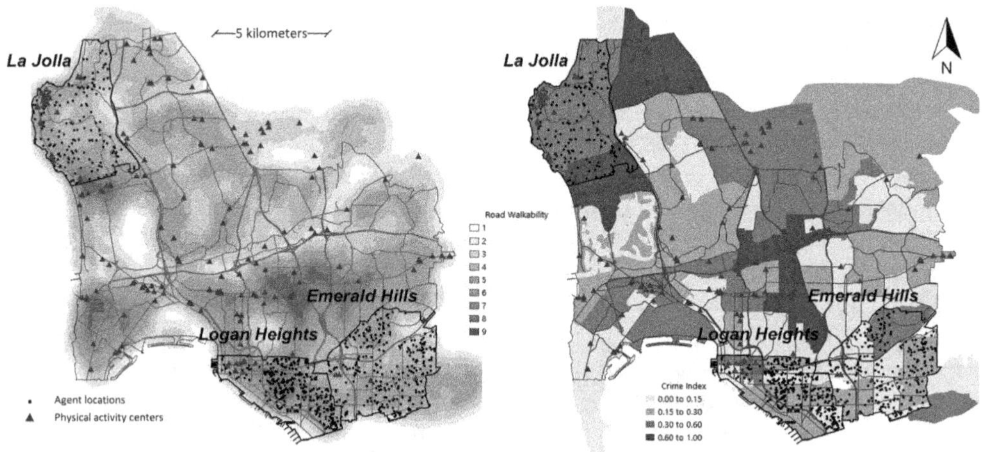

Figure 3.6 Exercise amenities, walkability, and crime in San Diego (see Table 3.4 for data source)

high road density areas with longer exercise is based on a San Diego study by Saelens, Sallis, Black, & Chen (2003). Data from that study showed that, in the high walkability neighborhoods, individuals walk/run on average about 210 minutes per week, as opposed to low walkability communities where the duration of walking/running amounts to about 140 minutes weekly. Finally, agent neighborhood safety was evaluated using total crimes per police neighborhood, normalized by the maximum number of crimes recorded in San Diego (Table 3.4, Figure 3.6 right).

Results and Discussion

The eighteen distributions of output avgBMI for all five experiments (EXP0–EXP5) are rendered in two ways. First, we chart the distributions by experiment (a total of six diagrams with three box plots each) to demonstrate the consequences of spatial heterogeneity on avgBMI with all other processes unchanged (Figure 3.7). Second, we display the results of all experiments per district (three diagrams with six box plots each) to facilitate the comparison between interventions and their influence on avgBMI within a given study site (Figure 3.8).

EXP0 (Figure 3.7) serves as a baseline scenario. At a base lifestyle, after five simulated years, avgBMI is elevated (BMI > 30) demonstrating high obesity prevalence in all three districts. Although the mechanisms driving the null model are spatially independent, significant differences

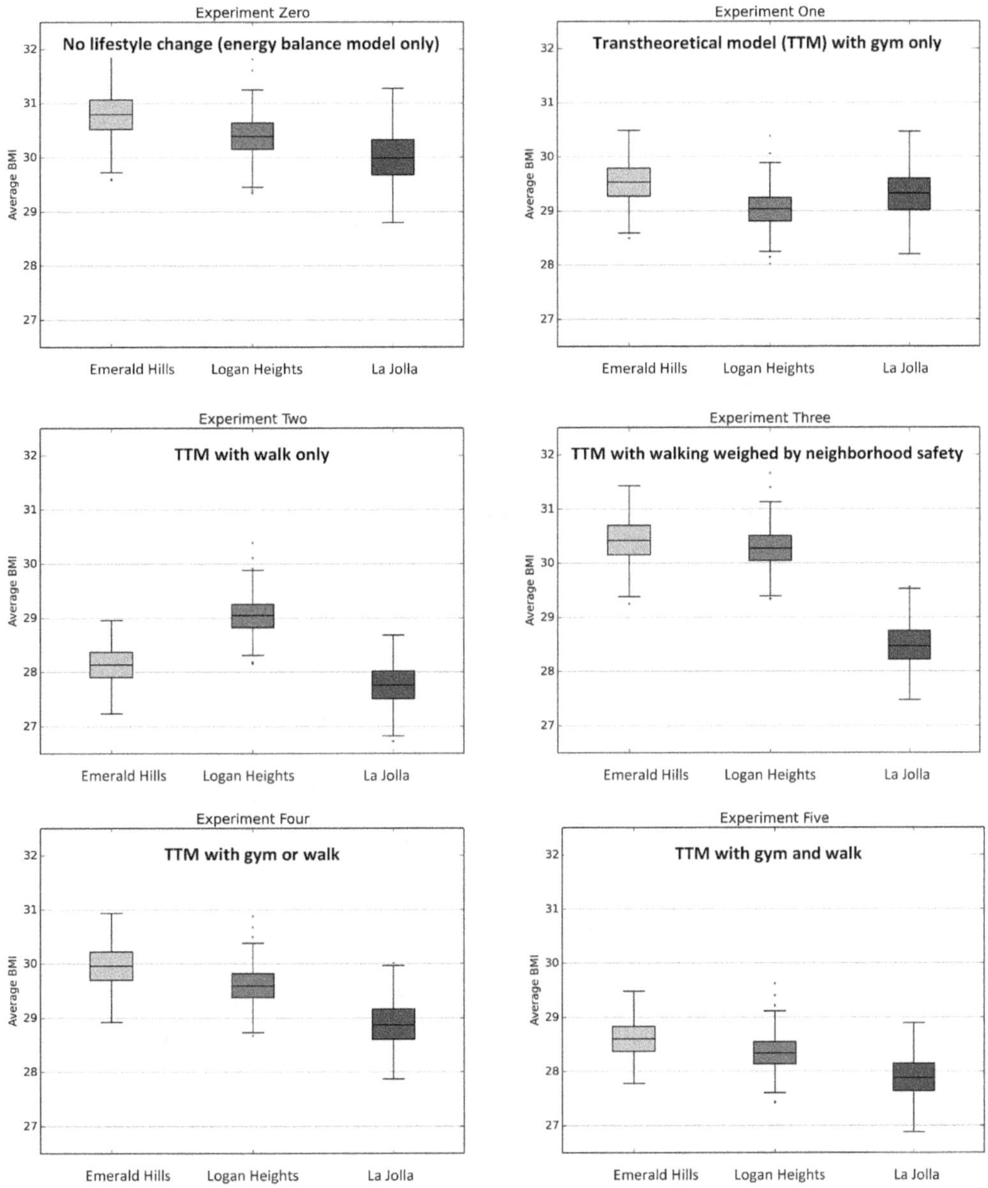

Figure 3.7 Box plots (distributions) of average BMI per all agents in a district at the end of model execution for n = 500 model runs shown by experiment

in avgBMI between districts are observed (F-test (2, 1497) = 45.96, p-value = 0.00) (Figure 3.7, top left). These differences appear to be due to the dominating agent groups within districts (Table 3.1). For example, output avgBMI for *La Jolla* = 30.00 (95% CI, 29.96–30.04) compared to *Emerald Hills* = 30.79 (95% CI 30.76–30.83). This difference is not surprising, since *Emerald Hills* has a relatively high fraction of adult population falling into the FBL group with the highest initial BMI and, at the same time, *La Jolla* has a relatively large FW group with

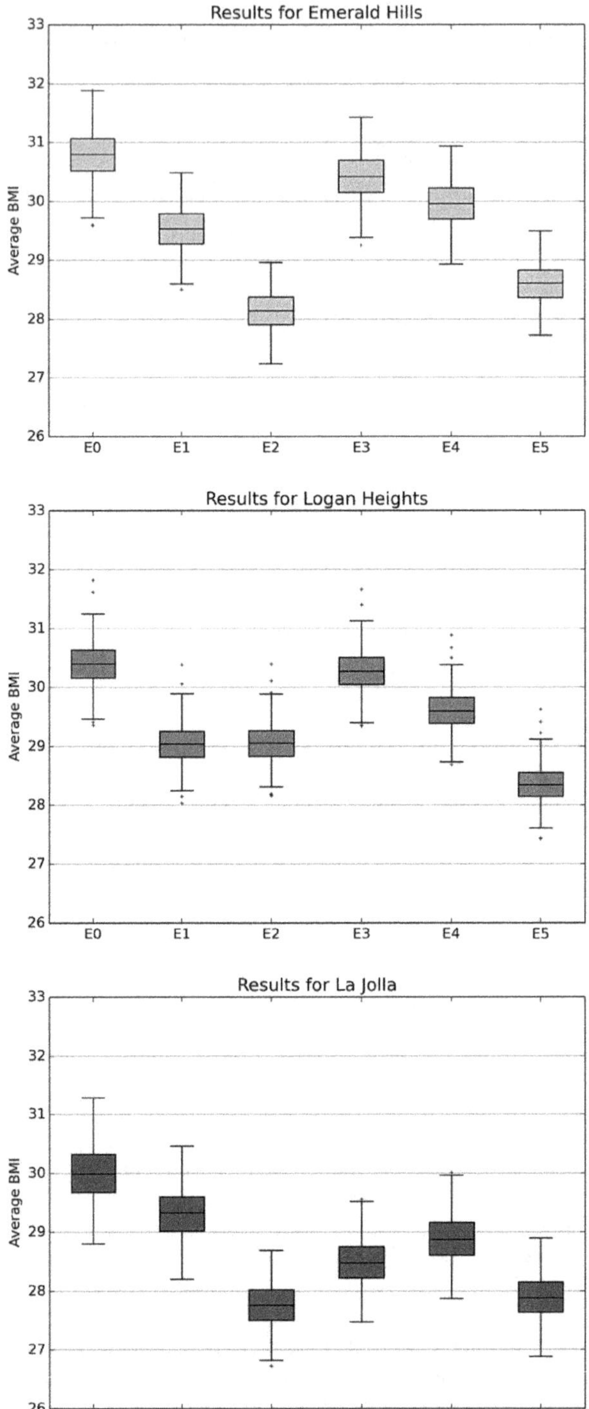

Figure 3.8 Box plots (distributions) of average BMI per all agents in a district at the end of model execution for n = 500 model runs shown by district

the lowest initial BMI. Observe, however, that this association is neither univariate nor linear. To validate the results, we compared the outcomes from the baseline scenario to an independently derived projection. We used adult BMI from the Nutritional Health and Nutrition Examination Survey database (CDC, 2014), recorded every other year from 1999 to 2009, generated a trend line, and extrapolated the trend to 2014 (R^2 = 0.95). The resulting avgBMI = 30.0, which is similar to the avgBMI in EXP0. We can therefore conclude that the base oABM exhibits satisfactory performance.

EXP1 through EXP5 are all equipped with the TTM mechanism. As anticipated, the behavior change model leads to a decrease in avgBMI, even though only a small fraction of overweight and obese agents follows the intervention (due to low agent influenceability). Since we tried to minimize the impact of agent locational distribution on model behavior (the *shape index* and *agent clustering* are relatively similar across the districts—Table 3.1) we hypothesize that between-district differences in the TTM experiments are caused solely by the spatial differentiation of gym location, walkability, and crime. There were significant differences between the three districts for all five experiments (data not shown, p-value 0.000 for EXP0–EXP5). The influence of accessibility to fitness centers (EXP1) on avgBMI reduction is moderate. The results indicate that the impact of GYM on avgBMI is more complex than expected. Based on the three gym metrics in Table 3.1, we expected that the highest drop in avgBMI would be observed in *La Jolla*. While all three districts show avgBMI reduction, it is *Logan Heights* that exhibits the greatest reduction in avgBMI (–1.36), most probably due to the fact that, although *La Jolla* has the greatest number of gyms that are relatively accessible, it also has the smallest population of agents that require lifestyle change to reduce their BMI. EXP2 (WALK only) produces a considerably greater drop in avgBMI from baseline than EXP1 (across all three districts –2.08). Again, the outcomes suggest that the relationship between the fraction of population that employs TTM and the spatial configuration of walkable roads is more complex. While *Logan Heights* has the highest average walkability (Figure 3.6), it is *Emerald Hills* and *La Jolla* that record the greatest BMI reduction (–2.66, –2.25), possibly due to the relatively high population of FBL.

Neighborhood safety (EXP3) significantly undermined weight loss due to walkability. *Logan Heights* and *Emerald Hills* showed only a slight reduction in avgBMI (–0.38, –0.12) when compared to the base scenario. The reduction of avgBMI was greatest for *La Jolla* (–1.51), despite it having the highest crime index of the three districts (Table 3.1), suggestive that this population is walking in locations other than the city (e.g., perhaps walking along the shoreline).

In the last two experiments, we allow agents to choose GYM or WALK (but not both – EXP4) or they can combine walking with resistance training (EXP5). In both cases, we observe that the between-district decrease resembles the output distribution of the EBM model, where change in avgBMI from base lifestyle is the greatest in *La Jolla* (–1.13), followed by *Emerald Hills* (–0.84) and *Logan Heights* (–0.81). Importantly, the downward shift in avgBMI is the most pronounced in the final, most flexible model (GYM and WALK), in which the likelihood of exercise in any given day is the highest (overall reduction in avgBMI = –2.12; *Emerald Hills* = –2.11, *Logan Heights* = –2.05, *La Jolla* = –2.20).

The differences in avgBMI from base for each activity presented were significantly different (data not shown, p-value = 0.00). Recall that the last scenario has the highest potential for long-term maintenance of BMI reduction. These results show the most promise in achieving the goal of 5% reduction in average BMI over the next ten years, which could draw the health care costs down by about $28 billion in the state of California alone (Levi, Segal, Laurent, Lang, & Rayburn, 2012; Robert Wood Johnson Foundation, 2012).

The results from the five experiments by district are reported in Figure 3.8. Beginning with *Emerald Hills*, which has the highest avgBMI of the three districts (30.79) WALK (EXP2) and

GYM and WALK (EXP5) show the greatest reduction in avgBMI (–2.66, –2.20) from the base scenario. BMI was least reduced in the SAFE WALK experiment (–0.38) despite the crime index being lower in *Emerald Hills* compared to *Logan Heights* and *La Jolla* (0.7 vs. 0.2, 0.2) suggestive that the perception of crime is more important in influencing outdoor activity than actual crime. In *Logan Heights* utilizing GYM and WALK (EXP5) reduced avgBMI twice (–2.05) as much as participating in these activities independently (GYM only = –1.39; WALK only = –1.35) again, demonstrating the value of multiple forms of activity. Finally, *La Jolla*, which had the lowest avgBMI of the three districts, showed WALK (–2.25) and GYM and WALK (–2.11) to be the most important activities for reducing BMI. The high reduction by WALK may be associated with walking in locations other then the city, such as the shoreline. All of these avgBMI differences in activities from the base lifestyle are statistically significant (p-value = 0.00).

The Role of Agent-Based Modeling in Studying Obesogenic Systems

The major benefit of utilizing agent-based modeling in obesity research is the ability to utilize cross-sectional datasets to model longitudinal intervention scenarios, and identify place-specific obesity interventions for public health policy and practice. Cross-sectional health datasets however, are limited by their lack of georeferencing individual activity spaces and potential environmental exposures and detailed representation of social networks. This study attempted to address these data limitations by implementing geographic methods to best allocate surveyed individuals (BRFSS) within three districts, define population (sociodemographic) characteristics most at-risk of obesity to represent the "agents" of study and to utilize available information on lifestyle practices to improve our understanding of behavioral-activity interventions. Another advantage of agent-based models over other commonly used models in obesity research is the ability to actively apply theoretical applications of behavior change and develop assumption scenarios to model and assess the complex interaction between obesity, behavior and environment interactions. Behavioral qualitative variables, like the type of lifestyle, persistence during the transitional period, or the underlying factors of the decision to participate in an intervention, are critical in developing rules that guide the actions of agents. Agent-based modeling allows for tracing the results back to consequences. It demonstrates how modeled behavior, rather than probabilities, affects weight gain and weight loss. It provides a means to evaluate the effectiveness of interventions. Finally, it allows for coupling the dominant drivers of weight change with other indirect causes, which together shape the dynamics of obesity.

The ABM presented here, however, has some distinct limitations that are worth mentioning. For simplicity, this study concentrates solely on calories expended during exercise, when in fact individual weight change is a product of diet and exercise, with such entangled factors as social influence, norms affecting food choices, cultural conventions, or conformism. A comprehensive analysis of the identified sociodemographic groups (FBL, FBO, FW, and M) could provide more valuable information used to design a suite of interventions tailored to specific high-risk cohorts. We excluded the influence of proximity to beaches on physical activity – a potentially significant factor in coastal areas (Bauman, Bellew, & Wales, 1996).

Regardless of these drawbacks, ABM is an immensely useful method for policy-relevant analysis of obesity prevalence. This study found that, across districts of distinct population and environmental characteristics, multiple forms of activity, specifically utilizing the gym and walking, are significantly associated with obesity reduction. In *La Jolla*, which had a high index of crime, walking was still an important contributor to obesity reduction possibly because of the availability of safe local amenities such as the waterfront. Importantly, the perception of crime limits walking activity more than the actual crime of an area. Thus crime and physical activity need to be studied in relation to the availability of resources and other environmental characteristics. Black women of low income remain the primary population group to experience obesity and future research should continue to focus on interventions that specifically address this population.

Summary

In this chapter we demonstrate how spatial ABM can be employed to study the prevalence of obesity in selected populations and districts of a major metropolitan area. We develop a prototype model, where individual total energy expenditure is modified by physical activity. We extend the ABM with the transtheoretical model of health behavior change that serves as an intervention aimed at obesity reduction. To parameterize the model we use publicly available (BRFSS) individual and spatial data. Based on the computational experiments, we conclude that the selected geographic factors: accessibility to physical activity centers, walkability, and neighborhood safety, play a substantial role in the spatiotemporal variability of individual weight measured with the body mass index. To a lesser extent, the factors also influence the magnitude of weight loss in the overweight and obese agents.

References

Alvanides, S., Townshend, T. G., & Lake, A. A. (2010). Obesogenic Environments: Challenges and Opportunities. In Lake, A., Townshend, T.G. & Alvanides, S. (Eds.), *Obesogenic Environments Complexities, Perceptions, and Objective Measures* (pp. 215–220). Chichester, UK: Wiley-Blackwell.

Auchincloss, A. H., & Diez Roux, A. V. (2008). A New Tool for Epidemiology: The Usefulness of Dynamic-Agent Models in Understanding Place Effects on Health. *American Journal of Epidemiology, 168*(1), 1–8. doi: 10.1093/aje/kwn118

Bauman, A. E., Bellew, B., & Wales, N. S. (1996). *NSW Health Promotion Survey 1994: Towards Best Practice for the Promotion of Physical Activity in the Areas of NSW.* Washington, DC: National Centre for Health Promotion.

Brewis, A. A. (2010). *Obesity Cultural and Biocultural Perspectives.* New Brunswick, NJ: Rutgers University Press.

Brown, D. G., Riolo, R., Robinson, D. T., North, M., & Rand, W. (2005). Spatial Process and Data Models: Toward Integration of Agent-Based Models and GIS. *Journal of Geographical Systems,* 7(1), 25–47.

Campbell, P., & Dhand, R. (2000). Obesity. *Nature, 404*(6778), 631–631.

CDC. (2010). Behavioral Risk Factor Surveillance System Survey Data. Atlanta, GA: U.S. Department of Health and Human Services, Centers for Disease Control and Prevention.

CDC. (2012). Overweight and Obesity. Retrieved February 22, 2012, from: www.cdc.gov/obesity/data/index.html

CDC. (2014). National Health and Nutrition Examination Survey. Retrieved January 1, 2014, from: www.cdc.gov/nchs/nhanes.htm

Church, T. S., Blair, S. N., Cocreham, S., et al. (2010). Effects of Aerobic and Resistance Training on Hemoglobin a1c Levels in Patients with Type 2 diabetes : A Randomized Controlled Trial. *JAMA, 304*(20), 2253–2262.

Curtis, A., & Lee, W.-A. (2010). Spatial Patterns of Diabetes Related Health Problems for Vulnerable Populations in Los Angeles. *International Journal of Health Geographics, 9*(1), 43.

Dray, J., & Wade, T. D. (2012). Is the Transtheoretical Model and Motivational Interviewing Approach Applicable to the Treatment of eating Disorders? A Review. *Clinical Psychology Review, 32*(6), 558–565. doi: http://dx.doi.org/10.1016/j.cpr.2012.06.005

Drewnowski, A., Rehm, D. C., & Solet, D. (2007). Disparities in Obesity Rates: Analysis by ZIP Code Area. *Social Science & Medicine, 65*(12), 2458–2463.

FAO. (2004). Human Energy Requirements Report of a Joint FAO/WHO/UNU Expert Consultation 17–24 October 2001. *Food and Nutrition Technical Report Series* (pp. 103). Rome: Food and Agriculture Organization of the United Nations.

FAO. (2013). The State of Food and Agriculture (pp. 100). Rome: Food and Agriculture Organization.

Frank, L. D., Andresen, M. A., & Schmid, T. L. (2004). Obesity Relationships with Community Design, Physical Activity, and Time Spent in Cars. *American Journal of Preventive Medicine, 27*(2), 87–96.

Galea, S., Riddle, M., & Kaplan, G. A. (2010). Causal Thinking and Complex System Approaches in Epidemiology. *International Journal of Epidemiology, 39*(1), 97–106.

Galster, G., Hanson, R., Ratcliffe, M. R., Wolman, H., Coleman, S., & Freihage, J. (2001). Wrestling Sprawl to the Ground: Defining and Measuring an Elusive Concept. *Housing Policy Debate, 12*(4), 681–717.

Haslam, D. W., & James, W. P. T. (2005). Obesity. *The Lancet, 366*(9492), 1197–1209.

HHS. (2008). 2008 Physical Activity Guidelines for Americans: U.S. Department of Health and Human Services.

Hill, J. O., Wyatt, H. R., Reed, G. W., & Peters, J. C. (2003). Obesity and the Environment: Where Do We Go from Here? *Science, 299*(5608), 853–855. doi: 10.1126/science.1079857

Johnson, S. S., Paiva, A. L., Cummins, C. O., Johnson, J. L., Dyment, S. J., Wright, J. A. (2008). Transtheoretical Model-based Multiple Behavior Intervention for Weight Management: Effectiveness on a Population Basis. *Preventive Medicine, 46*(3), 238–246. doi: http://dx.doi.org/10.1016/j.ypmed.2007.09.010

Kiernan, M., Moore, S. D., Schoffman, D. E., Lee, K., King, A. C., Taylor, C. B. (2012). Social Support for Healthy Behaviors: Scale Psychometrics and Prediction of Weight Loss Among Women in a Behavioral Program. *Obesity, 20*(4), 756–764. doi: 10.1038/oby.2011.293

Kumanyika, S., Jeffery, R. W., Morabia, A., Ritenbaugh, C., & Antipatis, V. J. (2002). Obesity Prevention: The Case for Action. *International Journal of Obesity, 26*(3), 425–436.

Larkin, M. (2003). Can Cities be Designed to Fight Obesity? *The Lancet, 362*(9389), 1046–1047.

Leahey, T. M., Kumar, R., Weinberg, B. M., & Wing, R. R. (2012). Teammates and Social Influence Affect Weight Loss Outcomes in a Team-Based Weight Loss Competition. *Obesity, 20*(7), 1413–1418. doi: 10.1038/oby.2012.18

Leslie, E., Coffee, N., Frank, L., Owen, N., Bauman, A., & Hugo, G. (2007). Walkability of Local Communities: Using Geographic Information Systems to Objectively Assess Relevant Environmental Attributes. *Health & Place, 13*(1), 111–122.

Levi, J., Segal, L. M., Laurent, R. S., Lang, A., & Rayburn, J. (2012). *F as in Fat: How Obesity Threatens America's Future: 2012*. Washington, DC: Trust for America's Health.

Ligmann-Zielinska, A. (2010). Agent-based Models. In B. Warf (Ed.), *Encyclopedia of Geography*. Thousand Oaks, CA: Sage Publications.

Mifflin, M., St Jeor, S., Hill, L., Scott, B., Daugherty, S., & Koh, Y. (1990). A New Predictive Equation for Resting Energy Expenditure in Healthy Individuals. *The American Journal of Clinical Nutrition, 51*(2), 241–247.

NAP. (2005). *Dietary Reference Intakes for Energy, Carbohydrate, Fiber, Fat, Fatty Acids, Cholesterol, Protein, and Amino Acids (Macronutrients)*. The National Academies Press.

North, M. J., & Macal, C. M. (2007). *Managing Business Complexity Discovering Strategic Solutions with Agent-Based Modeling and Simulation*. Oxford: Oxford University Press.

Pearce, J., & Witten, K. (2010). Conclusions: Common Themes and Emerging Questions. In Pearce J. & Witten K. (Eds.), *Geographies of Obesity Environmental Understandings of the Obesity Epidemic* (pp. 313–322). Farnham, UK: Ashgate.

Pickett, K. E., & Pearl, M. (2001). Multilevel Analyses of Neighbourhood Socioeconomic Context and Health Outcomes: A Critical Review. *Journal of Epidemiology and Community Health, 55*(2), 111–122.

Plantinga, A. J., & Bernell, S. (2007). The Association Between Urban Sprawl and Obesity: Is It a Two-Way Street? *Journal of Regional Science, 47*(5), 857–879. doi: 10.1111/j.1467–9787.2007.00533.x

Prochaska, J. O., & DiClemente, C. C. (1992). The Transtheoretical Approach. *Handbook of Psychotherapy Integration, 2*, 147–171.

Robert Wood Johnson Foundation. (2012). *Bending the Obesity Cost Curve in California*. Washington, DC: Trust for America's Health.

Roberts, S. B., & Dallal, G. E. (2005). Energy Requirements and Aging. *Public Heath Nutrition-CAB International, 8*(7A), 1028.

Roux, A. V. D., Evenson, K. R., McGinn, A. P., Brown, D. G., Moore, L., Brines, S. (2007). Availability of Recreational Resources and Physical Activity in Adults. *Journal Information, 97*(3).

Saelens, B. E., Sallis, J. F., Black, J. B., & Chen, D. (2003). Neighborhood-Based Differences in Physical Activity: An Environment Scale Evaluation. *American Journal of Public Health, 93*(9), 1552–1558.

Saelens, B. E., Sallis, J. F., & Frank, L. D. (2003). Environmental Correlates of Walking and Cycling: Findings from the Transportation, Urban Design, and Planning Literatures. *Annals of Behavioral Medicine, 25*(2), 80–91.

Sallis, J. F., Hovell, M. F., Hofstetter, C. R., Elder, J. P., Hackley, M., Caspersen, C. J. (1990). Distance Between Homes and Exercise Facilities Related to Frequency of Exercise Among San Diego Residents. *Public Health Reports, 105*(2), 179.

Sallis, J. F., Saelens, B. E., Frank, L. D., Conway, T. L., Slymen, D. J., Cain, K. L. (2009). Neighborhood Built Environment and Income: Examining Multiple Health Outcomes. *Social Science & Medicine, 68*(7), 1285–1293.

Sigal, R. J., Kenny, G. P., Boulé, N. G., Wells, G. A., Prud'homme, D., Fortier, M. (2007). Effects of Aerobic Training, Resistance Training, or Both on Glycemic Control in Type 2 DiabetesA Randomized Trial. *Annals of Internal Medicine, 147*(6), 357–369.

Stewart, J., Battersby, S., Lopez-De Fede, A., Remington, K., Hardin, J., & Mayfield-Smith, K. (2011). Diabetes and the Socioeconomic and Built Environment: Geovisualization of Disease Prevalence and Potential Contextual Associations Using Ring Maps. *International Journal of Health Geographics, 10*(1), 18.

Swinburn, B., Egger, G., & Raza, F. (1999). Dissecting Obesogenic Environments: The Development and Application of a Framework for Identifying and Prioritizing Environmental Interventions for Obesity. *Preventive Medicine, 29*(6), 563–570.

Thorogood, A., Mottillo, S., Shimony, A., Filion, K. B., Joseph, L., Genest, J. (2011). Isolated Aerobic Exercise and Weight Loss: A Systematic Review and Meta-Analysis of Randomized Controlled Trials. *The American Journal of Medicine, 124*(8), 747–755.

Townshend, T. G., Ells, L., Alvanides, S., & Lake, A. (2010). Towards Transdisciplinary Approaches to Tackle Obesity. In Lake A., Townshend, T.G. & Alvanides, S. (Eds.), *Obesogenic Environments Complexities, Perceptions, and Objective Measures* (pp. 11–20). Chichester, UK: Wiley-Blackwell.

U.S. Census Bureau. (2014). United States Census 2010. Retrieved from: www.census.gov/

WHO. (2006). BMI Classification. Global Database on Body Mass Index. Retrieved January 10, 2014, from: http://apps.who.int/bmi/index.jsp?introPage=intro_3.html

Wilson, D. K., Kirtland, K. A., Ainsworth, B. E., & Addy, C. L. (2004). Socioeconomic Status and Perceptions of Access and Safety for Physical Activity. *Annals of Behavioral Medicine, 28*(1), 20–28.

Wing, R. R., & Jeffery, R. W. (1999). Benefits of Recruiting Participants with Friends and Increasing Social Support for Weight Loss and Maintenance. *Journal of Consulting and Clinical Psychology, 67*(1), 132–138. doi: http://dx.doi.org/10.1037/0022-006X.67.1.132

Wright, J. D., & Wang, C.-Y. (2010). Trends in Intake of Energy and Macronutrients in Adults from 1999–2000 through 2007–2008. *NCHS data brief* (No. 49, p. 8). Hyattsville, MD: National Center for Health Statistics.

Yang, Y., Diez Roux, A. V., Auchincloss, A. H., Rodriguez, D. A., & Brown, D. G. (2011). A Spatial Agent-Based Model for the Simulation of Adults' Daily Walking Within a City. *American Journal of Preventive Medicine, 40*(3), 353–361.

Yang, Y., Diez Roux, A. V., Auchincloss, A. H., Rodriguez, D. A., & Brown, D. G. (2012). Exploring Walking Differences by Socioeconomic Status Using a Spatial Agent-Based Model. *Health & Place, 18*(1), 96–99. doi: http://dx.doi.org/10.1016/j.healthplace.2011.08.010

4

PROMOTING SOCIAL CONTAGION OF PREVENTIVE BEHAVIOR DURING INFLUENZA EPIDEMICS

An Agent-Based Simulation

Liang Mao

Research Question: What level of compliance is needed for an influenza prevention intervention to be effective?
System Science Method(s): Agent-based models & Networks
Things to Notice:

- Testing the effectiveness of a hypothetical intervention
- Simulation of a real spatial environment using empirical data

Outbreaks of influenza have drawn unprecedented attention in recent years. A natural response of people is to adopt flu preventive behaviors, such as getting vaccinated, taking prophylactic medicines, and increasing hand washing. Health agencies have devised various strategies to promote these behaviors in a population, for example, vaccine campaigns and free hand-sanitizer programs. However, these promotional strategies have been primarily studied in small groups of individuals. Their effectiveness remains unclear for large metropolitan populations due to limited personnel and health resources for such experiments. To address this problem, an agent-based epidemic-behavior model has been developed to simulate the transmission of influenza and the diffusion of flu prophylactic adoption together in a realistic urbanized area of Buffalo, NY, USA. Two types of promotional strategies were designed for evaluation: an incentive strategy for households and a "role-model" strategy for workplaces. Different compliance levels of households and workplaces with these strategies have been investigated as a sensitivity analysis. The simulation results show that an influenza epidemic cannot be controlled if the group compliance level is lower than 10%. The incentive strategy for households is effective to control the influenza outbreak if 50% or above compliance can be achieved. Setting role-models in workplaces fails to significantly improve the adoption rate even for a 90% compliance level. This simulation model offers a tool for health policy makers to experiment promotional strategies over a large population, and the results shed insights into the preventive practices before and during influenza seasons.

After several waves of flu pandemics in the past decade, there has been a rapidly growing interest in understanding human response to threats of infectious diseases (Fenichel, Kuminoff, & Chowell,

DOI: 10.4324/9781315748771-5

2013; Jones & Salathe, 2009; Sadique et al., 2007). Preventive behavior is one type of human responses aiming to reduce the risks of influenza infection through preventive actions, such as getting vaccination, taking prophylactic medicines, and increasing hand washing. Although these behaviors have been widely recommended by health agencies, their adoption rate remains low and differs remarkably over population, due to various socioeconomic status, culture, knowledge, etc. For instance, the seasonal flu vaccination coverage was estimated to be only 45% in the 2012–2013 season in the US population, and its state variability continues to be large (CDC, 2013). To improve, health agencies and scientists have devised promotional strategies to increase the adoption of preventive behaviors – for example, vaccine campaigns, free hand-sanitizer giveaway, and educational programs (Moran, Nelson, Wofford, Velez, & Case, 1996; Talaat et al., 2011).

These behavior promotional strategies have long been researched in the literature to understand their effectiveness in promoting adoptions, but most studies have been confined in small groups of populations (100–1,000 people in size) – for example, healthcare workers, schoolchildren, and elderly people). This is because limited personnel and resources, such as vaccines and hand sanitizers, could not allow such experiments being conducted in a large population. As a result, little is known for the effectiveness of these strategies for large populations in metropolitan areas, where influenza outbreaks are highly possible to occur. The lack of such knowledge may cloud decision makers when facing influenza epidemics in large cities, or pandemics across a country.

With the recent rise of computer-based epidemic models, there have been a small number of studies to simulate the effectiveness of behavior promotional strategies among large populations (Fu, Rosenbloom, Wang, & Nowak, 2011; Galvani, Reluga, & Chapman, 2007; Perisic & Bauch, 2009; Vardavas, Breban, & Blower, 2007). These studies employed the system science methodologies, such as the agent-based modeling and network approach, to simulate discrete individuals as autonomous agents, who decide to adopt a preventive behavior based on their self-evaluation of trade-offs between costs and benefits. These models, then, examined how the population-level properties (e.g., the overall adoption rate) emerge from collective behavior of individuals, as well as their interactions. With the rapid advance in computer technology, these models can simulate complex behavior of large populations and allow the evaluation of behavior promotional strategies in a virtual environment, breaking through previous bottlenecks of epidemiological experiment. No doubt that these studies have brought a leap in understanding human responsive behavior and shed insights into the design of promotional strategies, but they were based on hypothetical populations with few demographic characteristics and simplified social interactions between individuals. Due to their lack of realism, suggestions from these studies are limited as general guidelines and may not be applicable to real urban areas where demographic features and contact patterns of individuals are complicated. So far, little research has been devoted to exploring behavior promotional strategies in realistic metropolitan areas.

To fill this knowledge gap, this research intends to develop a spatially explicit agent-based model to simulate human preventive behavior in response to an influenza outbreak in a population of approximately 1 million in Buffalo, NY. The simulation model is more realistic than hypothetical models by incorporating observed demographic, social, behavioral characteristics of human individuals. Based on such a model, a number of behavior promotional strategies are simulated and evaluated, in terms of their effectiveness in increasing adoptions as well as reducing flu infections. The remainder of this chapter is organized as follows. The method section that follows describes the design of an agent-based epidemic model and the promotional strategies being simulated. The result section presents and compares the effectiveness of simulated strategies. The last section concludes this chapter with a discussion of its implications.

Methodologies

Agent-Based Epidemic-Behavior Simulation Model

Model Assumptions. The design of a spatially explicit agent-based model follows a dual-diffusion framework proposed by Mao and Yang (2011). The model construction was based on five fundamental assumptions:

1. Each individual is a discrete modeling unit that differs in their characteristics (e.g., age, gender, occupation, infection status, and geographic location) and behaviors (e.g., having social contacts, movements between locations, and preventive behavior against diseases).
2. Individuals travel between three time periods and four types of locations in a day, and have contact with different groups of individuals. Specifically, individuals have contact with co-workers during the daytime at workplaces, with family members during the nighttime at homes, with friends during the pastime at service places or neighbor households. These spatiotemporally varying contacts form a dynamic social network.
3. Influenza spreads through physical proximity among individuals. The contacts between individuals at a location transmit the disease virus, while the travel of individuals between locations disperses the disease across the urban area. The development of influenza in an individual follows its natural history.
4. The adoption of preventive behavior reduces the chance of disease transmission between individuals. That is, it lowers not only the chance of contracting a disease, but also the likelihood of infecting others.
5. Individuals make decisions toward an adoption of preventive behavior based on their own characteristics (e.g., knowledge, experience, and personal traits) and interpersonal influence from surrounding individuals (e.g., family supports and role model effects; Glanz, Rimer, & Lewis, 2002). Because individuals vary in their willingness to adopt, the preventive behaviors also diffuse from a few early adopters to the early majority and then over time throughout the social network, known as the 'social contagion' process (Rogers, 1995).

With these five assumptions an epidemic can be conceptualized as two competing diffusion processes through the human social network: the diffusion of influenza and the diffusion of preventive behavior (Mao and Yang, 2011). These two diffusion processes take place simultaneously during an epidemic and interact with one another in opposite directions, producing – that is, the transmission of influenza motivates the diffusion of preventive behavior, which in turn limits the disease transmission. From a perspective of system theory, these two population-level diffusion processes emerge from a massive number of individual-level interactions, leading to a negative feedback loop that stabilizes the human-disease system.

Model Implementation. To build a working model, these five assumptions are implemented in the urbanized area of Buffalo, NY, USA. The study area is located at the western end of New York State, and on the eastern shore of Lake Erie (Figure 4.1). Surrounded by great lakes and rural areas, it is relatively separated from other metropolitan areas, and the nearest urbanized area, Rochester, is about 120 km to the east. A travel survey of 2002 by the Greater Buffalo–Niagara Regional Transportation Council (GBNRTC, 2002) indicates that most people living in this area take their daily activities within it. For these reasons, the study area can be treated as a closed system, where influenza epidemics could develop with little outside influence.

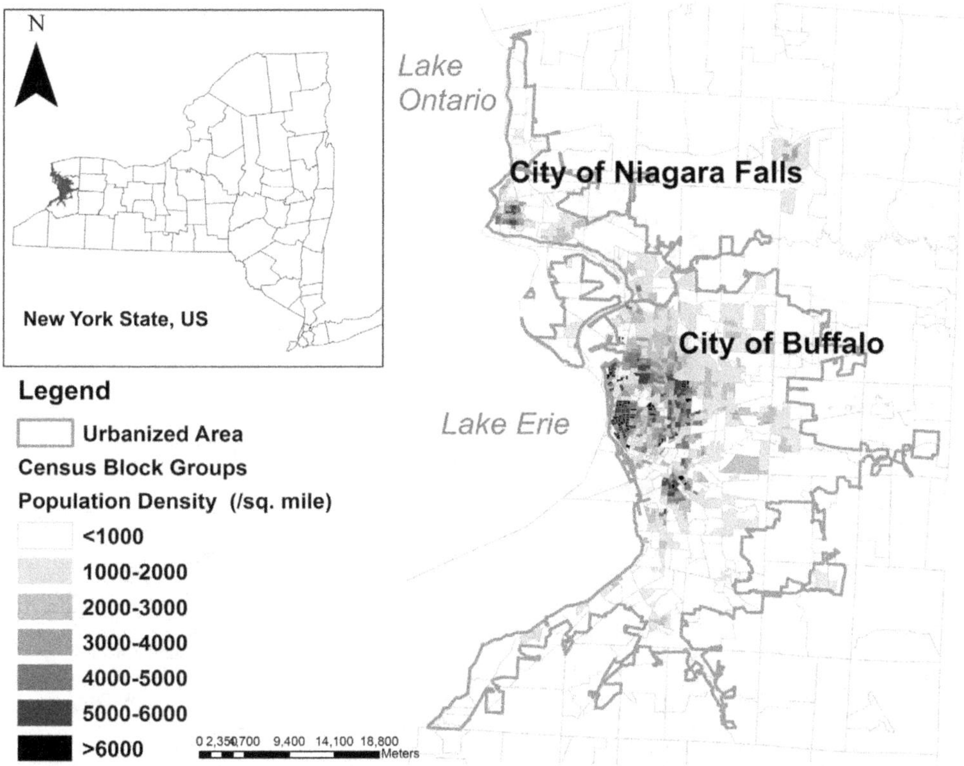

Figure 4.1 The geographic location and spatial scope of study area, and the population density by census block group

Assumptions 1 and 2: The implementation of these two assumptions adopts a previous algorithm developed by Mao and Bian (2010), and thus is not discussed in detail in this chapter. Based on the census data (US Census, 2000), 985,001 individuals in the study area are programmed as modeling units associated with a set of attributes (e.g., age, occupation, infection status, time and location of daily activities) and events (e.g., traveling between locations for activities and having contact with other individuals), as shown in Figure 4.2. For each census block group, table cross-referencing and Monte Carlo simulation are performed to assign demographic traits to individuals, so that the modeled demographic distributions match real distributions. Individuals are then grouped based on simulated household memberships and allocated to 400,870 households according to the land parcel data that indicate spatial location.

These individuals are further modeled to travel between their homes and 36,839 business locations (schools, workplaces, service places) and neighbor households (in the same census block) to carry out their daily activities, such as working, shopping, recreation, and friend visiting, which expose them to the risk of infection. The assignment of individuals to workplaces utilizes a business database purchased from Reference USA (2009). A worker is randomly assigned to a workplace that satisfies two criteria: (1) the employer type of the workplace in the business databases should match the employer type of the worker from the census data; (2) the shortest distance between the home and workplace in the real road system should match the trip distance to work estimated from the census data (trip duration to work × speed of a trip mode). Unlike ordinary workers, school-aged individuals in each household are assigned to a nearest school as

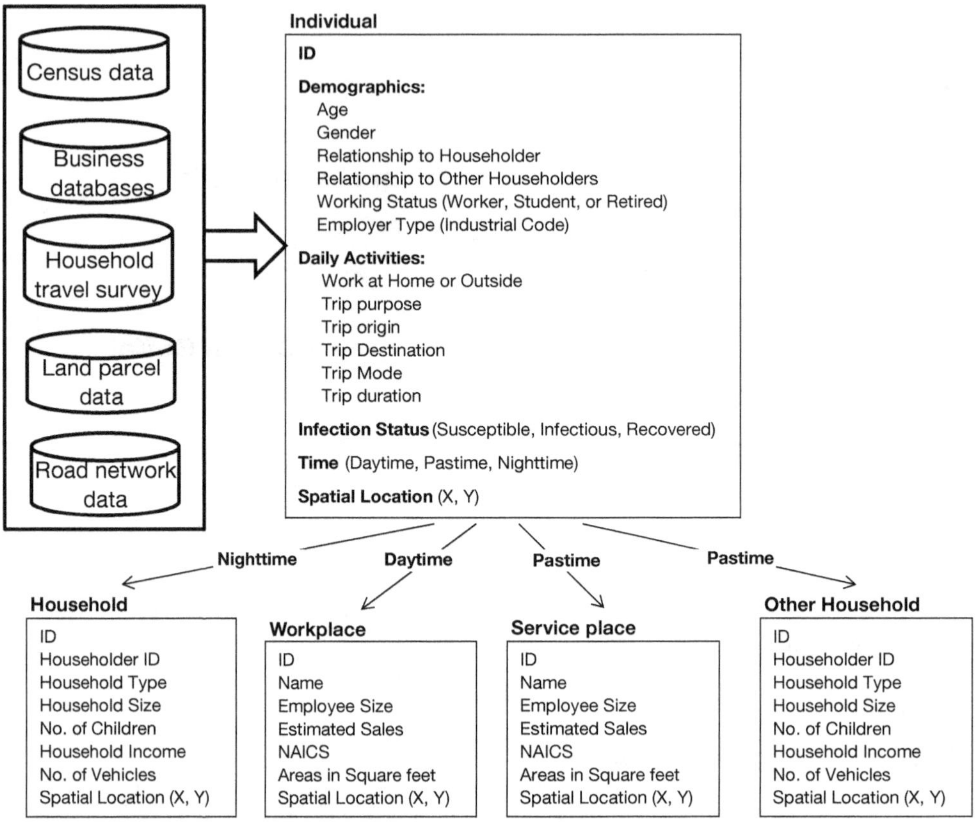

Figure 4.2 The assignment of individuals to households, workplaces, service places and neighbor households based on the attribute and spatial information of individuals (revised based on Mao & Bian, 2010)

their workplace. To differentiate the weekdays and weekends, individuals are not assigned to work at weekends except those who work in service-oriented businesses (Mao & Bian, 2010). The assignment of individuals to service places and neighbor households employed statistical distributions from the regional travel survey report (GBNRTC, 2002), and used similar criteria as the workplaces to identify destinations. The simulated mobility of all 985,001 individuals weave a spatio-temporally varying contact network that allows influenza transmission.

Assumptions 3 and 4: These two assumptions are relevant to the diffusion of influenza, and are formulated by the classic susceptible-infectious-recovered (SIR) model (Anderson & May, 1992). Each individual possesses an infection state (as an attribute) and a series of events, as shown in Figure 4.3.

The progress of influenza starts with an individual in a "susceptible" state who may receive infectious agents if having contact with an *infectious* neighbor in the network. The receipt of disease virus is controlled by a transmission probability P, as specified in Equation 1:

$$P = E_{contact} \times I_{age} \times (1 - E_{Adoption}) \qquad (1)$$

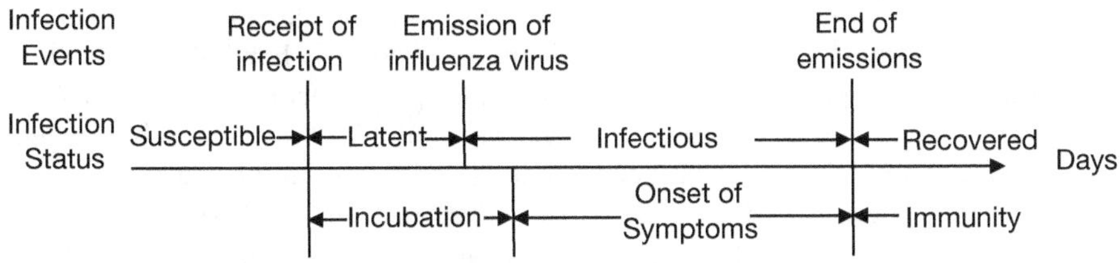

Figure 4.3 The infection events, periods and status in the natural history of influenza.

Table 4.1 Key parameter settings for influenza diffusion

Parameter	*Empirical estimate*	*Literature*
Basic reproduction Number R_0	1.2– 1.3	(Chowell, Miller, & Viboud, 2007; Mills, Robins, & Lipsitch, 2004)
Effectiveness of contacts $E_{contact}$	$E_{close\_contact}$: 1 $E_{ocassional\_contact}$: 0.2	Calibrated based on R_0
Likelihood of symptom manifestation	0.5	(Ferguson et al., 2005; Miller et al., 2010)
Latent period	2 days	(Heymann, 2004)
Incubation period	3 days	(Halloran et al., 2008; Heymann, 2004)
Infectious period	Children: 7 days Youth and Adults: 4 days Senior: 4 days	(CDC, 2008; Heymann, 2004)
Efficacy of flu-prophylaxis ($E_{Adoption}$)	Symptomatic: 40% Susceptible: 70%	(Hayden, 2001; Longini, Halloran, Nizam, & Yang, 2004)

E_{contact} is a real number between [0, 1], indicating the effectiveness of a contact to transmit diseases, dependent on the closeness and duration of the contact. I_{age} is an age-specific infection rate, also ranging between [0, 1]. $E_{Adoption}$, between 0 and 1, indicates the mitigation efficacy from a preventive behavior on infection (*Assumption 4*). The receipt of infection triggers a "*latent*" state, during which the disease agents develop internally in the body and are not emitted. The end of latent period initiates an "infectious" state, in which this individual is able to infect other susceptible contacts and may manifest disease symptoms ("symptomatic"). After the infectious period, this individual either recovers or dies from the disease. In such a manner, the diffusion of influenza among the population is simulated as a collective outcome of changes in infection states of all individuals. As shown in Table 4.1, values of these key parameters are acquired from the current literature to describe the natural history of influenza.

For the $E_{Adoption}$ in Equation 1, the adoption of flu prophylactics (e.g., oseltamivir, also known as Tamiflu) is taken as a typical example of preventive behavior. The flu prophylactic is chosen because its efficacy has been quantified in many clinical studies, and is more conclusive than other preventive behaviors, such as hand washing and facemask wearing. Both symptomatic and susceptible individuals may adopt flu prophylactics, which in turn affects their ability to transmit the influenza virus. For symptomatic individuals, their ability to infect others could be reduced by 40% if using flu prophylactics (Longini et al., 2004). For susceptible individuals who have adopted, the probability of being infected can be reduced by 70% (Hayden, 2001).

Assumption 5: To simulate the diffusion of preventive behavior, individuals are modeled as decision makers with an adoption state: either Adopted or Not Adopted. For individuals who have developed flu symptoms, their likelihood of adopting flu prophylactics is set to be 75%, based on health surveys by Stoller et al. (1993) and McIsaac et al. (1998). For all other individuals, the decision of adoption is simulated based on their own characteristics and inter-personal influence from their social networks (Funk, Gilad, Watkins, & Jansen, 2009; L. Mao & Yang, 2011). This research modifies a threshold model (Granovetter & Soong, 1983) to determine how individuals change their adoption states, as specified in Equation 2:

$$AdoptionState_i(t) = \begin{cases} \text{Adopted, if } \alpha_i(t) \geq T_{p,i} \text{ or } \mu_i(t) \geq T_{r,i} \\ \text{Not adopted,} \qquad \text{Otherwise} \end{cases} \tag{2}$$

Where

$$\alpha_i(t) = \frac{\text{Number of adopters in } i\text{'s network neighbors at } t}{\text{Total number of } i\text{'s network neighbors}}$$

$$\mu_i(t) = \frac{\text{Number of infections in } i\text{'s network neighbors at } t}{\text{Total number of } i\text{'s network neighbors}}$$

For a given time step t, individual i will evaluate the proportion of adopters among his/her social contacts, as the peer pressure of adoption $\alpha_i(t)$. Once the peer pressure reaches a threshold $T_{p,i}$ (called the threshold of peer pressure), individual i will decide to adopt. Meanwhile, individual i will also evaluate the proportion of symptomatic individuals among his/her social contacts, as the perceive risks of infection $\mu_i(t)$. If the perceived risk exceeds another threshold (termed the threshold of infection risk), individual i will also adopt. While the $\alpha_i(t)$ and $\mu_i(t)$ represent the inter-personal influence between individuals, the individualized thresholds ($T_{p,i}$ and $T_{r,i}$) reflect personal characteristics of individuals.

In order to estimate these two individualized thresholds, a health behavior survey was conducted online for one month (March 12–April 12, 2010) to recruit participants (L. Mao & Bian, 2011). A total of 262 voluntary participants living in the study area were invited to answer two questions: (1) "Suppose you have 10 close contacts, including household members, colleagues, and close friends, after how many of them *get influenza* would you consider using flu prophylactics?", and (2) "Suppose you have 10 close contacts, including household members, colleagues, and close friends, after how many of them start to *use flu drugs* would you consider using flu prophylactics?" Their responses were summarized into two frequency distributions, one for the thresholds of perceived risks based on Question 1 (Figure 4.4a) and the other for the thresholds of peer pressure according to Question 2 (Figure 4.4b).

Based on these two distributions, the Monte Carlo parameterization was then performed to assign threshold values to each modeled individual. For example, an adult male would have a 10% chance (a randomly generated number < 0.1) to be assigned a threshold of infection risks at 20% (i.e., adoption when 20% of contacts were infected), and 50% chance (a randomly generated number > 0.5) to be assigned a threshold of peer pressure at infinity (i.e., never adopting). In such a way, the diffusion of preventive behavior among the population is represented as a collective outcome of changes in adoption states of all individuals. The diffusion of disease would increase the perceived risks by individuals *μi(t)* in Equation 2, and stimulates them to adopt flu prophylactics, which in turn impedes the diffusion of disease, as formulated in Equation 1.

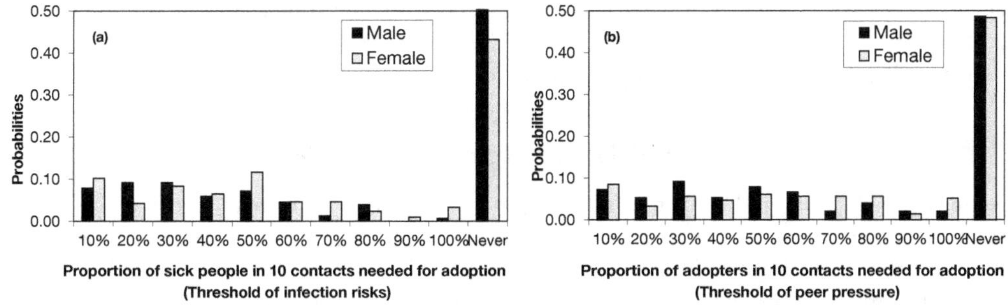

Figure 4.4 Estimated distribution of the threshold of infection risks (a) and threshold of adoption pressure (b) by gender. The *x*-axis indicates the proportion of influenza cases in the contacts of a participant that is needed to convince this participant to adopt. The *y*-axis shows the frequency of such proportion occurring in the survey results.

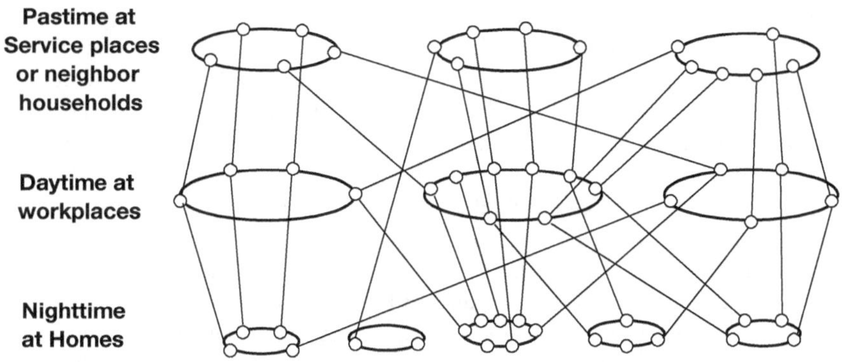

Figure 4.5 An illustration of model simulation flow chart (adopted from Bian et al., 2012). Ovals represent the four types of locations at three time periods. Small circles on an oval represent individuals who interact with one another within a location. Solid-line ovals denote close contacts, while dash-line ovals denote occasional contacts. The straight lines represent the travel trajectory of an individual at different locations.

Model Initialization and Validation. The dual-diffusion model is simulated over 150 days, covering a general flu season (from December to May). At the beginning of simulation, all individuals are assigned susceptible states, and non-adopters of flu-prophylaxis. To initialize the simulation, five infectious individuals are randomly introduced into the study area at the first day. The simulation takes a tri-daily time step (nighttime, daytime, and pastime), and runs the two diffusion processes concurrently in each time step. Figure 4.5 depicts the tri-daily time-step simulation, starting from household contacts during the nighttime at homes, then coworker contacts during the daytime at workplaces, then other social contacts during the pastime at service places or neighbor households, and back to the nighttime at household again. This loop is repeated 150 times to simulate flu spread in 150 days. At each step, the agent's infection and adoption statuses are updated.

The final outcomes are two diffusion curves – namely, the epidemic curve (the daily number of new cases) and the adoption curve (the daily number of new adopters), all averaged from 50 model realizations. In each realization, the first five infectious individuals, their contacts, and the infections of these contacts are randomized. To further validate the model, the simulated epidemic curve is compared to weekly reports of laboratory confirmed specimens in flu season 2004–05 of the study area (NYSDOH, 2005). The season 2004–05 is chosen because available reports indicate a severe influenza epidemic taking place in this area.

Design of Behavior Promotional Strategies

As formulated above, individuals will adopt preventive behavior, if their perceived infection risk or peer pressure exceeds a threshold. Following such concepts, there are two possible pathways to motivate the adoption, or in other words, to assist individuals to overcome their thresholds. One is to lower the thresholds of adoption, for example, reducing the costs of adoption. The other is to elevate individuals' perceived infection risk or peer pressure, for instance, informing individuals about the severity of influenza, or establishing role models among individuals. Corresponding to these two pathways, this chapter explores two widely considered strategies for behavioral promotion – namely, an incentive strategy for households and a role-model strategy for workplaces.

For the incentive strategy, once an individual decides to adopt flu prophylactic, the health agencies will offer free prophylaxis for other household members of this individual. This "buy one and get others free" strategy aims to remove thresholds of other household members through financial incentives, and thus motivate the adoption of the entire household. To account for the compliance of households with this strategy, this chapter investigates four compliance levels: 0%, 10%, 50%, and 90%. The compliance is a group behavior of household members to follow a strategy, while the adoption is an individual behavior toward taking a preventive action. The 0% compliance represents a baseline scenario, in which the strategy is not applied and individuals' adoption are completely voluntary. The 10% compliance models a low compliance scenario in which only 10% of the households being offered free flu prophylaxis will decide to comply.

With respect to the role-model strategy, once a workplace is affected by influenza (i.e., at least one flu case being identified), a role model of adoption (e.g., the employer) will be established in the workplace to motivate other co-workers to adopt. In the simulation, the role models are assumed to be twice as influential as other workers. In other words, the role models are counted twice by other co-workers when evaluating their infection risk and adoption pressure. Similar to the incentive strategy, four compliance levels of workplaces are investigated, including 0%, 10%, 50%, and 90%.

For each promotional strategy, the four compliance levels are simulated respectively by the epidemic-behavior model with 100 realizations, resulting in 400 realizations in total (4 compliance level × 100 realizations). The resultant adoption curves are constructed to reflect how the preventive behavior diffuses over time. In addition, because the ultimate goal is to help control influenza, the epidemic curves are also plotted to indicate the epidemiological effectiveness. The adoption curves and epidemic curves are compared between the four compliance levels to evaluate their effectiveness. An influenza epidemic is assumed to be controlled if the disease attack rate can be reduced under 5%.

Results and Discussion

Baseline Epidemic

During the simulated epidemic, 7.49% of the population would develop influenza symptoms (Figure 4.6a), which is a reasonable estimation as the observed rates vary between 5% and 20% (CDC, 2008). Meanwhile, approximately 40% of the population has adopted flu prophylactics

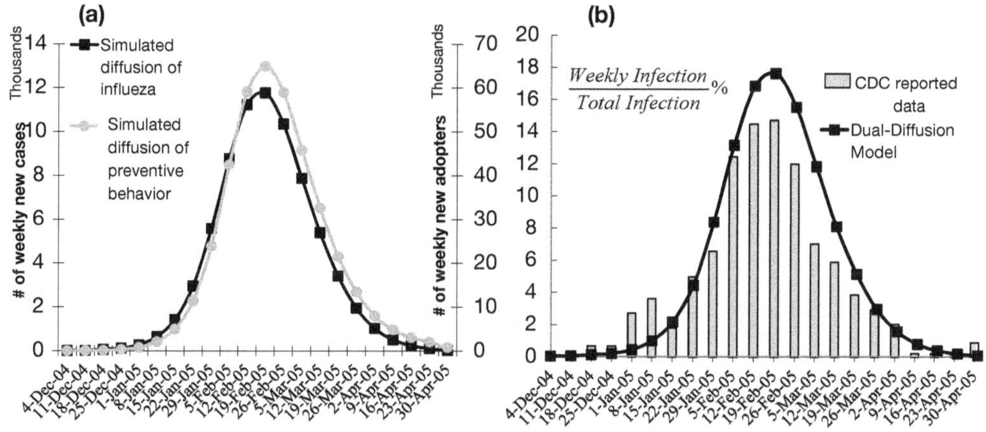

Figure 4.6 (a) Temporal dynamics of dual-diffusion in an influenza epidemic. The *y*-axis on the left shows the weekly number of new cases (aggregated from the daily number), while the *y*-axis on the right shows the weekly number of new adopters. (b) The comparison of simulation results to the weekly laboratory confirmed influenza specimens in 2004–05. The *y*-axis indicates the percentage of weekly infections in total infections of the entire epidemic.

during the course of the epidemic. Although these two diffusion processes differ in magnitude, they follow quite similar temporal trends. Both curves took off in early January, and peaked in the middle of February. This is in line with other empirical studies that reported a 90% correlation between the number of diagnosed influenza cases and flu antiviral drug sales (Das et al., 2005; Magruder, 2003). Figure 4.6b shows a consistency between the simulated and observed flu infections in the study area. The comparisons with empirical studies and disease reports suggest that the model could be a close representation of influenza transmission, as well as a reliable tool to explore behavior promotional strategies in the study area.

Incentive Strategy for Households

Figure 4.7 shows adoption curves resulting from the household incentive strategy, and the associated adoption rates. The periodic sharp drops show the difference between weekdays and weekends. Since people are not assigned to work during weekends, their daily contacts are significantly reduced, leading to decreases in both infections and adoptions. As the compliance level increases from 0% to 90%, the adoption curve tends to be taller and narrower, and reaches the peak faster. At the compliance level of 90%, the adoption curve has the highest peak above 20,000 new adopters per day (Figure 4.7a). The incentive strategy also increases the overall adoption rate from 40% to 46% or above (Figure 4.7b). Interestingly, the overall adoption rate rises dramatically when the compliance is under 10%, but only improve slightly from 46% to 48%, as the compliance level goes beyond 10%.

The epidemic curves and disease attack rates in Figure 4.8 indicate how the incentive strategy limits the diffusion of influenza. The baseline scenario has the highest epidemic peak, with about 2,000 influenza cases at the peak day. The overall attack rate (7.5%) is greater than any other scenario. As the household compliance level increases from 0% to 10%, the shape of the epidemic curve only changes mildly (Figure 4.8a), and the disease attack rate is slightly reduced (Figure 4.8b). A remarkable change can be observed when the compliance level rises to 50%. The resulting epidemic curve is apparently flatter than that of the baseline scenario. The 50% incentive strategy

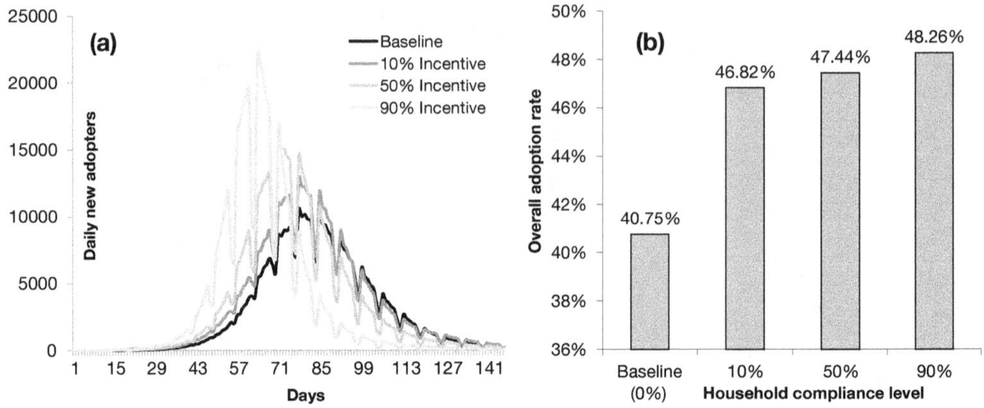

Figure 4.7 (a) The adoption curves and (b) the overall adoption rates under the incentive strategy at household compliance level of 0%, 10%, 50% and 90%, respectively

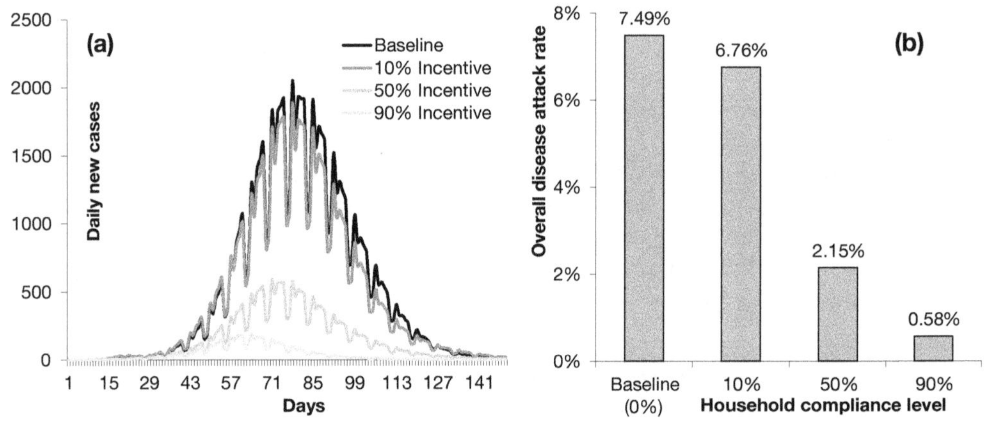

Figure 4.8 (a) The epidemic curves and (b) the overall attack rates of influenza under the incentive strategy at household compliance level of 0%, 10%, 50% and 90%, respectively

considerably lowers the epidemic peak to approximately 500 new cases, and contains the disease attack rate to only 2.15% of population. The 90% incentive strategy produces another remarkable reduction in infection. The epidemic curve is further suppressed with an even smaller peak size (only 188 new cases) and lower attack rate (0.58%).

Results presented in Figures 4.7 and 4.8 suggest three noteworthy points. First, the voluntary adoption of individuals is not adequate to control an influenza epidemic. As shown in the baseline scenario, although 40% of the population may voluntarily adopt the preventive behavior, infection remains above 5% of the population. External forces, such as interventions, are needed to further promote the adoption. Second, applying an incentive strategy in a low-compliance population (below 10%) does not produce significant effects on epidemics. The higher the compliance, the more effective the incentive strategy will be. A compliance level of 50% is an effective threshold to control influenza in the study area. Third, when the compliance level is above 10%, the incentive strategy leads to similar adoption rates but distinctly different disease attack rates. This implies that similar stockpiles of flu prophylactics (adoption rates) may produce different control effectiveness (disease attack rates). For the incentive strategy, a high-compliance population uses up the stockpiles quickly before they were sick. Therefore, only a few people get infected later and the epidemic can be largely mitigated. On the contrary, a low-compliance population depletes the stockpile slowly, possibly after they are infected. Hence, similar expenses of flu prophylactics produce worse results (more infection). These findings above recommend that the incentive strategy could be effective in influenza control and prevention, but its effectiveness is highly dependent on the compliance of the population. A survey of population compliance in advance is essential for success.

Role-Model Strategy for Workplaces

The curves in Figure 4.9 show that the workplace role-model strategy produces slight effect on the diffusion of preventive behavior. For any compliance level, the role-model strategy leads to similar peak time at day 78, and similar peak size around 10,000 new adopters. However, the overall adoption rates can be increased 3–6% from the baseline scenario.

With respect to the disease diffusion, Figure 4.10 indicates that the role-model strategy poses few effects on containing influenza. The role-model strategy with a low compliance (10% level) produces a similar curve and attack rate as those from the baseline scenario. As the compliance level rises to 50%, the epidemic curve starts to be flattened, but the overall attack rate is only reduced 0.5%. Even a considerably high compliance level (at 90%) merely produces a 1% reduction in the attack rate. Therefore, the role-model strategy fails to bring the disease attack rate fewer than 5%.

The simulation results above suggest that the role-model strategy might not be an effective strategy for the study area. A possible reason is that a large proportion of the population may have extremely high thresholds of adoption. Based on the health behavior survey conducted in this research, about half of the respondents would never adopt flu prophylactics unless they were ill (Figure 4.4). This large population is difficult to be convinced to adopt flu prophylactics when they are healthy, even though role models are set around them. Since these people only adopt preventive behavior after they get infected, the number of infections would be fairly high

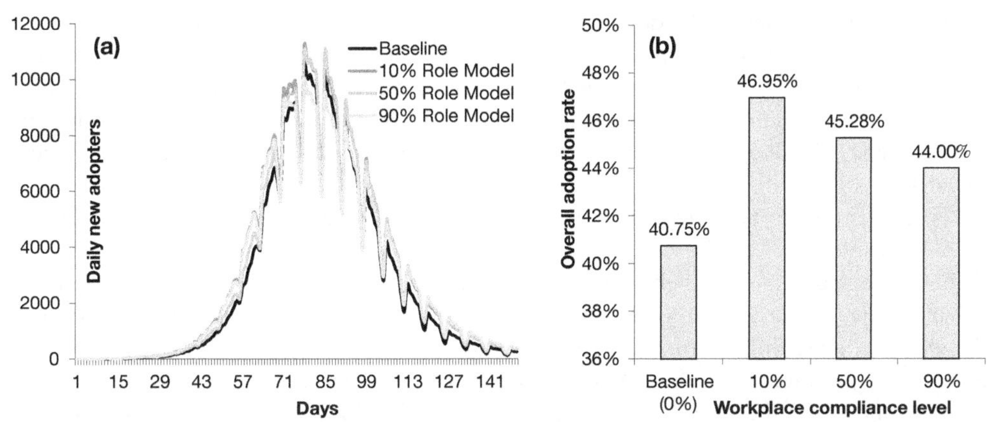

Figure 4.9 (a) The adoption curves and (b) the overall adoption rates under the role-model strategy at workplace compliance level of 0%, 10%, 50% and 90%, respectively

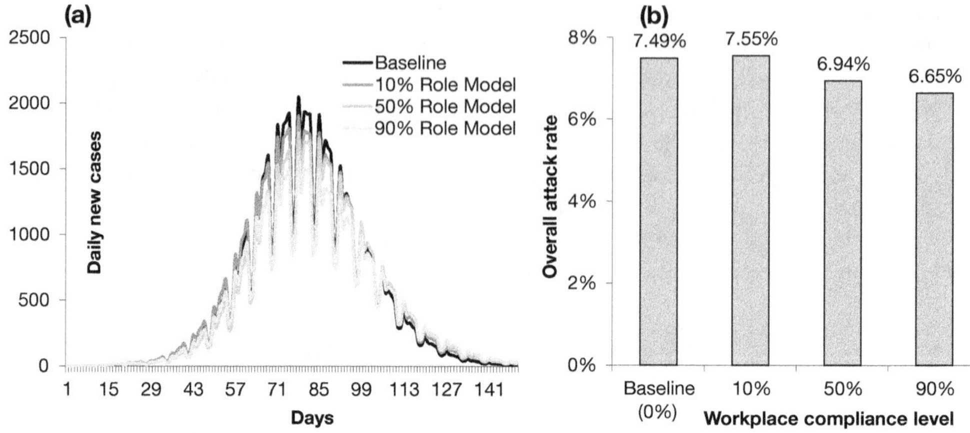

Figure 4.10 (a) The epidemic curves and (b) the overall attack rates of influenza under the role-model strategy at workplace compliance level of 0%, 10%, 50% and 90%, respectively

and the effectiveness of this strategy is greatly weakened. The failure of the role-model strategy suggests another key factor to success, i.e., the adoption-related thresholds of individuals. This role-model strategy might be effective when a vast majority of a population has low thresholds of adoption. A prior survey about adoption thresholds of a population is critical for policy makers. Health behavioral studies are needed to understand why those people have extremely high thresholds of adoption. Appropriate education programs, thus, can be tailored for these people to lower their thresholds and promote adoption among them.

Limitations

There are two limitations in this research that suggest the future directions. First, the current work only considers the effects of person-to-person contacts on human preventive behavior. Other contributing factors, such as the Internet and mass media, have not been taken into account. For example, people who watch TV frequently may be convinced by mass media toward adoption, rather than by surrounding people. Future research needs to incorporate the effects of other media. Second, individuals adopt preventive behavior primarily dependent on their evaluation of surrounding individuals. Other characteristics of individuals themselves have not been considered, such as their own belief, socio-economic status, and habits, all of which can be featured by different lifestyles. The effects of individual lifestyle on adoption is important (Langlie, 1977; Rosenstock, Strecher, & Becker, 1988), and should be added into future research.

To alleviate these two limitations, the PRIZM lifestyle segmentation data (published by Nielsen Claritas Inc.) can be exploited. The PRIZM data defines U.S. households in terms of 66 demographically and behaviorally distinct types, or "segments," to discern consumer behavior: their likes, dislikes, lifestyles, purchase behaviors, and media preferences. With such lifestyle data, a health behavior survey can be conducted to relate the lifestyles to individuals' adoption of preventive behavior. A regression model, or a binary discrete choice model, can be established from the survey. By incorporating the PRIZM data and the regression model, a more sophisticated model can be developed to predict individual adoption of preventive behavior.

Conclusion

The pathways to human health behavior are complex because the causes involve interconnected social, economic, and environmental factors. Computer simulation models can elucidate these pathways and relationships, and can be used to assess benefits and harms of intervention options, particularly for large urbanized populations. For these advantages over classic epidemiology experiments, the Department of Health and Human Services (HHS) has been devoted to advancing the use of predictive and system-based simulation models to understand the health consequences of disease preventive strategies. This research is one of the first attempts to answer this call and develops an agent-based epidemic-behavior model to explore effective strategies that promote health behavior among a realistic urban population. The model conceptualizes the preventive behavior as a practice that spreads among people through inter-personal influence, commonly known as the "social contagion." The diffusion of preventive behavior takes place simultaneously with the diffusion of influenza disease. The interactions between the two form a negative feedback loop that stabilizes the human-disease system.

Based on the working model, two behavior promotional strategies have been simulated and evaluated, including an incentive strategy for households and a role-model strategy for workplaces. The results show that the effectiveness of both strategies is highly dependent on two factors: the compliance of individuals to these strategies and the thresholds of individuals toward adoption. Specifically, the incentive strategy requires a 50% compliance to achieve high adoption rate as well as low influenza attack rate. The success of the role-model strategy needs a population that has a large number of low-threshold individuals. This strategy is thus not suggested for this study area, because half of its population may be reluctant to adopt preventive behavior unless they were sick. Prior knowledge of a population's compliance and adoption thresholds is necessary for health agencies before considering these promotional strategies.

References

Aiello, A. E., Perez, V., Coulborn, R. M., Davis, B. M., Uddin, M., & Monto, A. S. (2012). Facemasks, hand hygiene, and influenza among young adults: a randomized intervention trial. *PloS one,* 7(1), e29744.

Anderson, R. M., & May, R. M. (1992). *Infectious Diseases of Humans: Dynamics and Control.* Oxford: Oxford University Press.

Bian, L., Huang, Y., Mao, L., Lim, E., Lee, G., Yang, Y., . . . Wilson, D. (2012). Modeling individual vulnerability to communicable diseases: A framework and design. *Annals of the Association of American Geographers, 102*(5), 1016–1025.

CDC. (2008). Key facts about seasonal influenza (flu). Retrieved January, 2009, from www.cdc.gov/flu/about/disease/index.htm

CDC. (2013). Flu vaccination coverage, United States, 2012–13 influenza season. Retrieved December 3, 2013, from www.cdc.gov/flu/fluvaxview/coverage-1213estimates.htm

Chowell, G., Miller, M. A., & Viboud, C. (2007). Seasonal influenza in the United States, France, and Australia: Transmission and prospects for control. *Epidemiology and Infection, 136*(6), 852–864.

Das, D., Metzger, K., Heffernan, R., Balter, S., Weiss, D., & Mostashari, F. (2005). Monitoring over-the-counter medication sales for early detection of disease outbreaks – New York City. *Morbility and Mortality Weekly Report, 54*(Suppl.), 41–46.

Fenichel, E. P., Kuminoff, N. V., & Chowell, G. (2013). Skip the trip: Air travelers' behavioral responses to pandemic influenza. *PlOS One, 8*(3), e58249.

Ferguson, N. M., Cummings, D. A. T., Cauchemez, S., Fraser, C., Riley, S., Meeyai, A., . . . Burke, D. S. (2005). Strategies for containing an emerging influenza pandemic in Southeast Asia. *Nature, 437*(7056), 209–214.

Fu, F., Rosenbloom, D. I., Wang, L., & Nowak, M. A. (2011). Imitation dynamics of vaccination behaviour on social networks. *Proceedings of the Royal Society B: Biological Sciences, 278*(1702), 42–49.

Funk, S., Gilad, E., Watkins, C., & Jansen, V. A. A. (2009). The spread of awareness and its impact on epidemic outbreaks. *Proceedings of the National Academy of Sciences, 106*(16), 6872.

Galvani, A. P., Reluga, T. C., & Chapman, G. B. (2007). Long-standing influenza vaccination policy is in accord with individual self-interest but not with the utilitarian optimum. *Proceedings of the National Academy of Sciences, 104*(13), 5692–5697.

GBNRTC. (2002). Greater Buffalo–Niagara Regional Transportation Council 2002 Regional Transportation Survey Results Report. Buffalo, NY.

Glanz, K., Rimer, B. K., & Lewis, F. M. (2002). *Health Behavior and Health Education: Theory, Research, and Practice* (3rd ed.). San Francisco, CA: Jossey-Bass.

Granovetter, M., & Soong, R. (1983). Threshold models of diffusion and collective behavior. *Journal of Mathematical sociology, 9*(3), 165–179.

Halloran, M. E., Ferguson, N. M., Eubank, S., Ira M. Longini, J., Cummings, D. A. T., Lewis, B., . . . Cooley, P. (2008). Modeling targeted layered containment of an influenza pandemic in the United States. *Proceedings of the National Academy of Sciences, 105*(12), 4639–4644.

Hayden, F. G. (2001). Perspectives on antiviral use during pandemic influenza. *Philosophical Transactions of the Royal Society of London. Series B, 356*(1416), 1877–1884.

Heymann, D. L. (2004). *Control of Communicable Diseases Manual* (18th ed.). Washington, DC: American Public Health Association.

Jones, J. H., & Salathe, M. (2009). Early assessment of anxiety and behavioral response to novel swine-origin influenza A (H1N1). *PlOS One, 4*(12), e8032.

Langlie, J. K. (1977). Social networks, health beliefs, and preventive health behavior. *Journal of Health and Social Behavior, 18*(3), 244–260.

Longini, I. M., Halloran, M. E., Nizam, A., & Yang, Y. (2004). Containing pandemic influenza with antiviral agents. *American Journal of Epidemiology, 159*(7), 623–633.

Magruder, S. F. (2003). Evaluation of over-the-counter pharmaceutical sales as a possible early warning indicator of human disease. *Johns Hopkins APL Technical Digest, 24*, 349–353.

Mao, L., & Bian, L. (2010). Spatial-temporal transmission of influenza and its health risks in an urbanized area. *Computers, Environment and Urban Systems, 34*(3), 204–215.

Mao, L., & Bian, L. (2011). Agent-based simulation for a dual-diffusion process of influenza and human preventive behavior. *International Journal of Geographical Information Science, 25*(9), 1371–1388.

Mao, L., & Yang, Y. (2011). Coupling infectious diseases, human preventive behavior, and networks—A conceptual framework for epidemic modeling. *Social Science & Medicine, 74*(2), 167–175.

McIsaac, W. J., Levine, N., & Goel, V. (1998). Visits by adults to family physicians for the common cold. *Journal of Family Practice, 47*(5), 366–369.

Miller, E., Hoschler, K., Hardelid, P., Stanford, E., Andrews, N., & Zambon, M. (2010). Incidence of 2009 pandemic influenza A H1N1 infection in England: A cross-sectional serological study. *The Lancet, 375*(9720), 1100–1108.

Mills, C. E., Robins, J. M., & Lipsitch, M. (2004). Transmissibility of 1918 pandemic influenza. *Nature, 432*(7019), 904–906.

Moran, W. P., Nelson, K., Wofford, J. L., Velez, R., & Case, L. D. (1996). Increasing influenza immunization among high-risk patients: Education or financial incentive? *The American Journal of Medicine, 101*(6), 612–620.

NYSDOH. (2005). 2004-2005 Influenza Season Statewide Summary Report Retrieved 2009, from www.health.state.ny.us/diseases/communicable/influenza/surveillance/

Perisic, A., & Bauch, C. T. (2009). Social contact networks and disease eradicability under voluntary vaccination. *PLOS Computational Biology, 5*(2), e1000280.

Reference USA. (2009). U.S. Business Databases, 2009, from www.referenceusa.com/Static/Home# businessDatabases

Rogers, E. M. (1995). *Diffusion of Innovations* (4th ed.). New York: Free Press.

Rosenstock, I. M., Strecher, V. J., & Becker, M. H. (1988). Social learning theory and the health belief model. *Health Education & Behavior, 15*(2), 175–183.

Sadique, M. Z., Edmunds, W. J., Smith, R. D., Meerding, W. J., De Zwart, O., Brug, J., & Beutels, P. (2007). Precautionary behavior in response to perceived threat of pandemic influenza. *Emerging Infectious Diseases, 13*(9), 1307.

Stoller, E. P., Forster, L. E., & Portugal, S. (1993). Self-care responses to symptoms by older people. A health diary study of illness behavior. *Medical Care, 31*(1), 24–42.

Talaat, M., Afifi, S., Dueger, E., El-Ashry, N., Marfin, A., Kandeel, A., . . . Afifi, S. (2011). Effects of hand hygiene campaigns on incidence of laboratory-confirmed influenza and absenteeism in schoolchildren, Cairo, Egypt. *Emerging Infectious Diseases*, *17*(4), 619.

US Census. (2000). Census 2000 Data Releases, 2010, from www.census.gov/main/www/cen2000.html

Vardavas, R., Breban, R., & Blower, S. (2007). Can influenza epidemics be prevented by voluntary vaccination? *PLOS Computational Biology*, *3*(5), e85.

Vessey, J. A., Sherwood, J. J., Warner, D., & Clark, D. (2007). Comparing hand washing to hand sanitizers in reducing elementary school students' absenteeism. *Pediatric Nursing*, *33*(4), 368.

5

SIMULATING SYNDEMIC RISK

Using System Dynamics Modeling to Understand Psycho-Social Challenges Facing Women Living With and At-Risk for HIV

Abigail W. Batchelder and David W. Lounsbury

Research Question: How do the psycho-social challenges faced by women living with HIV depend on the convergence of multiple risk factors?
System Science Method(s): System dynamics
Things to Notice:

- Model building as a stepwise process
- Triangulation from multiple data sources

Best practices in system dynamics model development call for application of a multi-stepped, iterative procedure involving problem identification, system conceptualization, model formulation, model simulation, and model evaluation. This procedure unfolds differently for each model-building project, with some steps requiring more investment than others depending on the problem of focus, the availability of supportive data and information, and the targeted audience or "users" of the model. In this chapter we describe how we built and validated a system dynamics model of syndemic risk among women living with and at-risk for HIV in a low-resource, urban environment. We explain how we applied three sources of evidence to the model-building process, underscoring key decision-points we encountered along the way. We use the model to illustrate divergent patterns of syndemic risk using simulated profiles. These profiles generated important insights and implications for designing clinical, community, and public health interventions for this vulnerable population, including providing a deeper understanding of the dynamics of syndemic risk. Specifically, our model emphasized the need for individualized multi-aimed psychosocial interventions, prioritizing safety planning and substance abuse treatment, while addressing unmet psychosocial challenges and maximizing resilience. Finally, we reflect on how every system dynamics model-building project is informed by a process of careful deliberation by the modelers and their participating stakeholders, with the desired outcome of a deeper understanding of the problem and ways to effectively address it.

In this chapter, we apply system dynamics modeling to explore core tenets of syndemic theory, a public health framework used to explain how some groups or communities experience a much higher burden of disease than others. System dynamics is applied here to evaluate assumptions about *syndemic risk* that have been predominantly tested on cross-sectional samples using general

DOI: 10.4324/9781315748771-6

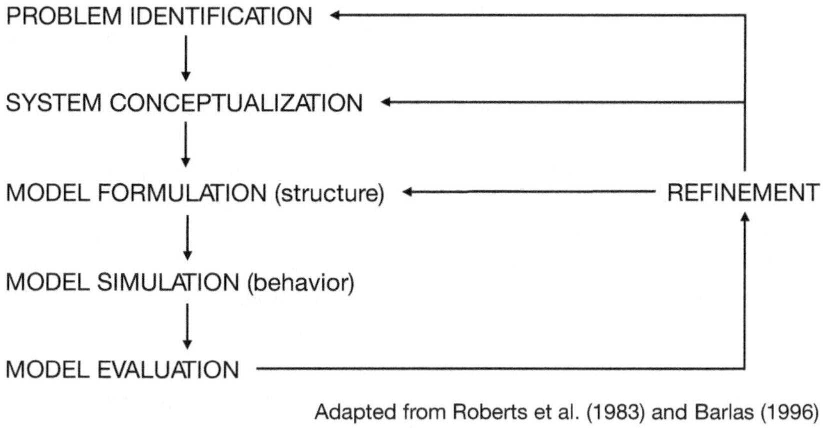

Figure 5.1 The iterative nature of system dynamics model development and validation

linear statistical assumptions (Milstein, 2008; Milstein & Homer, 2006). The Center for Disease Control and Prevention's "Syndemic Prevention Network" identified system dynamics modeling as a powerful tool for assessing the interdependent relationships among co-occurring risk factors and risk regulators (CDC, 2013). In particular, system dynamics modeling can help illustrate how factors and regulators interact with each other over time, and could be used to evaluate the potential utility of a variety of interventions or policies to mitigate syndemic risk in target populations.

Best practices in system dynamics model development call for the application of a multi-stepped, iterative procedure involving *problem identification*, *system conceptualization*, *model formulation*, *model simulation*, and *model evaluation* (see Figure 5.1) (Martinez-Moyana & Richardson, 2013; Roberts, Anderson, Deal, Garet, & Shaffer, 1983). However, as we will show, this procedure unfolds differently for each model-building project, with some steps requiring more investment than others depending on problem of focus, the availability of supportive data and information, and the targeted audience or "users" of the model.

System dynamics model building is inherently "mixed methods" research. Both qualitative and quantitative techniques are used in every project. For example, problem definition often begins by using the modeler's own substantive knowledge to generate a set of elements (i.e., constructs, variables) that are considered to be related to their topic of interest. This approach supports *systems thinking*, which moves away from looking at isolated events and their associated antecedents, to look at how these elements interact as parts of a *system* (Richardson & Pugh, 1981a, 1981b). In system dynamics, qualitative *causal loop diagrams* (CLDs) are used to graphically notate the *causal structure* of elements of a system. CLDs illustrate one or more feedback loops, which are hypothesized to generate particular *behavior patterns* associated with the topic of interest. Behavior patterns can often be plotted over time, to show trends in key elements of the system (i.e., *reference modes* shown as *behavior-over-time graphs*). Together, these qualitative tools are used to inform the design of a formal mathematical model, which is used to simulate the dynamics of a problem (as well as ways to address it), wherein every variable, parameter and relationship is quantified.

The mathematical model is expressed as a *stock-and-flow diagram*, where *stocks* represent the accumulation (i.e., level, prevalence) of a variable at a given moment and *flows* represent rates of change (i.e., the first order derivative or incidence) between two stocks. The final system dynamics model is a set of algebraic and differential equations, each verified to be dimensionally correct (units on the left and right side match) and conceptually meaningful (consistent with available evidence, plausible and face-valid), that can be used to test the utility of particular *policies*, or interventions of interest, hypothetically implemented and assessed over a given period, or *time horizon*.

Model formulation and simulation are performed with the aid of computer simulation software that enables representation of relationships among variables in a wide variety of ways, including: (1) stochastic or deterministic, (2) continuous or discrete, (3) linear or non-linear, and (4) simultaneous or lagged (Meadows & Robinson, 1985). Computer simulation software packages, such as Powersim, iThink/Stella, and Vensim, are among the most widely used packages for creating system dynamics models. These software packages allow the user to work with a graphical interface to build the model, equation by equation, examining preliminary simulation runs in an iterative fashion, to check and recheck the extent to which the model's *behavior* (simulated output) conforms to key assumptions and sources of evidence that are guiding the project. Vensim by Vantana Systems, Inc. was used to develop the system dynamics model featured in this chapter.

Problem Identification

The quality of a system dynamics project ultimately rests upon the clarity of purpose for the modeling exercise (J.W. Forrester, 1971, 1987; Oliva, 2003; Wolstenholme, 1983). In the current project, the identified problem is the disproportionate number of women living with HIV in low-resource urban communities, like the Bronx, and applying principles of *syndemic theory*. Our aims are: (1) to better understand the drivers of psychological and physical volatility among women with and at-risk for HIV, as it exists in their daily lives (including *history of childhood sexual abuse*, *emotional distress*, *substance use*, *perceived financial hardship*, *violence*, and *self-worth*) and (2) to apply our understanding to begin to inform community programs and interventions that can effectively reduce risk levels among women living with and at-risk of HIV.

Epidemiological Burden of HIV in Women

In the current project, the increasing and disproportionate number of women living with HIV in the Bronx, NY, constitutes an important reference mode and the identified problem driving the current modeling project. Women in the United States continue to be affected by HIV/AIDS, mainly through heterosexual contact (CDC, 2013). In New York City, the prevalence of HIV/AIDS among women has grown steadily since 2001. By 2012, there were an estimated 32,500 cases among women in New York City with more than 10,000 cases in the Bronx (see Figure 5.2) (NYCDHMH, 2014). Almost 80% of these women reportedly acquired the virus from heterosexual sex; however, other psychosocial constructs, including history of childhood sexual abuse, emotional distress, substance use, financial hardship, and violence – particularly intimate partner violence – are thought to have played a role in their acquisition of HIV (Brier & Runts, 1993; CDC, 2013; Cook et al., 2007; Somalai et al., 2000; Zierler & Krieger, 1997). These psychosocial constructs are intimately related to low levels of sexual self-care, defined as both the avoidance of high-risk behaviors, including transactional sex and unprotected sex, as well as active engagement in condom use negotiation, sex with only one uninfected partner, or abstinence (Bedimo, Kissinger, & Bessinger, 1997; El-Bassel, Gilbert, Wu, Go, & Hill, 2005; El-Bassel et al., 1997; Lane et al., 2004; Ward, 1993).

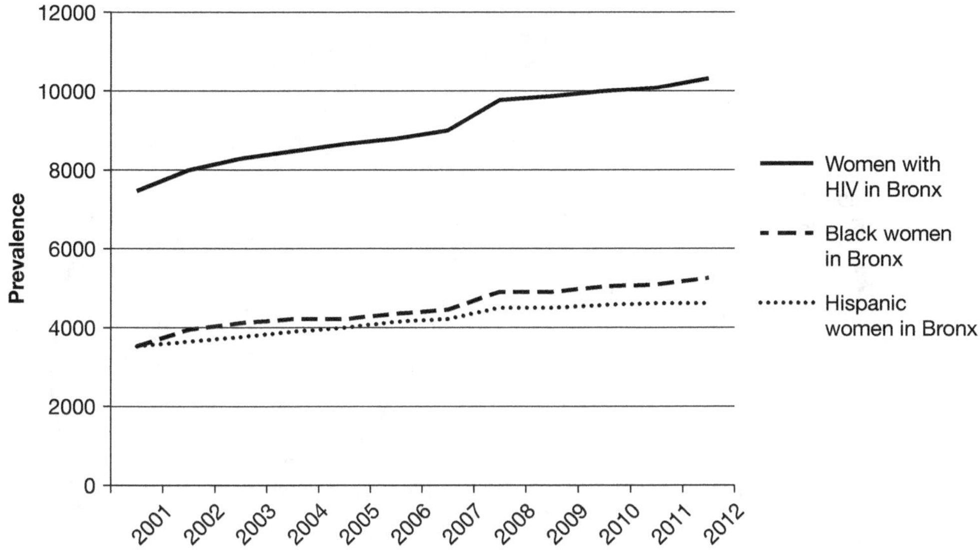

Figure 5.2 HIV prevalance among women in the Bronx, NY (2001–2012)

Syndemic Theory

Singer (1992) first described synergistically interacting afflictions that contribute to excess burden of disease as a "syndemic." Substance abuse, violence, and AIDS (SAVA) were the first described syndemic (Singer, 1992). In 2002, the CDC initiated the Syndemic Prevention Network to promote exploration of syndemics in public health work. Milstein later defined syndemic orientation as a way of conceptualizing the connections between and among health-related problems (Milstein, 2008). Although originally applied as a population health-related concept for research and intervention, Stall (Stall et al., 2003) and others (Kurtz, 2008; Mustanki, Garofalo, Herrick, & Donenberg, 2007) wrote about individual-level syndemics involving the additive relationships among variables such as history of childhood sexual abuse, emotional distress, substance abuse, and adult victimization among men who have sex with men (MSM).

System Conceptualization

Our conceptualization of syndemic risk among women residing in low-resource urban communities is conveyed as a causal loop diagram (CLD) (see Figure 5.3). The feedback structure shown here constitutes our working *dynamic hypothesis* of the major drivers of syndemic risk in this population. Its purpose is to show how *behavior* is generated by an explicit feedback *structure*, which is the hallmark of the system dynamics method (Forrester, 1994; Oliva, 2003; Richardson, 1991; Sterman, 2000).

CLDs comprise construct or variable names, arrows, and positive (+) and negative (–) signs. The arrows show the hypothesized causal pathways between variables. Positive (+) and negative (–) signs are used to indicate the nature of the hypothesized causal association between two variables. Positive signs indicate that two variables change in the same direction, if both X (the cause) and Y (the effect) increase or decrease over time, from t_n to t_{n+1}. Negative signs indicate that the two variables change in opposite directions. For example, as X increases, Y will decrease,

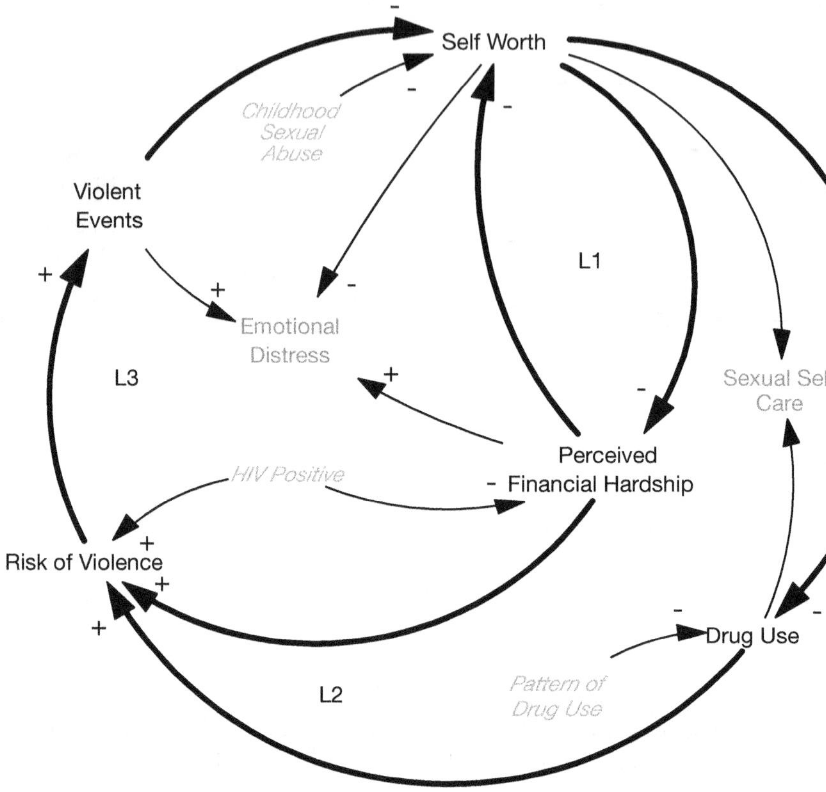

Figure 5.3 Casual loop diagram of syndemic risk

or vice versa, from t_n to t_{n+1}. In addition to the sign of each link, each resultant loop also has a sign. By counting the number of links that have a negative (–) sign, the loop's sign, or *polarity*, can be determined. If a loop has an odd number of negative (–) signs, it is said to be *balancing* (or negative) loop. Balancing loops serve to stabilize the system, bringing variables into steady states (i.e., homeostasis, equilibrium). If the loop has all positive (+) signs, or an even number of negative (–) signs, it is said to be *reinforcing* (or positive) loop. Reinforcing loops indicate either growth or decay, over time. Reinforcing loops can also indicate the presence of a vicious cycle, where a problem becomes worse over time, often at an increasing rate of speed. In summary, the sign of the loop is the algebraic product of the sign of its links (Richardson, 1986; G. P. Richardson & Pugh, 1981).

Our final CLD presents three reinforcing loops (positive loops) and two indicator variables (*emotional distress* and *sexual self-care*). In our CLD, the first reinforcing loop (L1) involves *self-worth* and perceived financial hardship. Incremental changes in *self-worth* lead to changes in the opposite direction in *perceived financial hardship* (e.g., as *self-worth increases*, *perceived financial hardship* also decreases). However, as *perceived financial hardship* decreases, *self-worth* increases. Hence, this loop contains two negative links, resulting in a positive loop, or a reinforcing feedback structure.

The second reinforcing loop (L2) involves four constructs: *self-worth, substance use, and violence*, which we operationalized as *risk of violence* and *violent events*. Given the categorical nature of the type of *substance use* construct, this loop is only activated when the baseline characteristic of frequent substance user or woman prone to relapse are programmed. This cycle demonstrates that incremental changes in *self-worth* over time lead to incremental changes in *substance use*, in the opposite direction (i.e., as *self-worth* increases, *substance use* decreases). As *substance use* increases, *risk of violence* also increases. As *risk of violence* increases, the *rate of violent events* increases. Finally, as *violent encounters* increase, *self-worth* incrementally decreases. This reinforcing loop is an example of a potential vicious cycle: increasing *substance use* and *violence* contributing to a decreasing of *self-worth*.

The third reinforcing loop (L3) represents causal influences of *risk of violence* for women who are not using substances (i.e., non-users and women prone to relapse when they are not using substances). This loop involves four constructs: *self-worth, perceived financial hardship, risk of violence, and violent events*. This cycle demonstrates how an incremental change in *self-worth* leads to an incremental change in the opposite direction in *perceived financial hardship* (e.g., as *self-worth* increases, *perceived financial hardship* decreases). It is of note that this feedback structure is not suggesting that actual financial hardship is caused by *self-worth*; rather, we suggest level of *self-worth* incrementally contributes (albeit minutely) to perception of the severity of financial hardship. An incremental change in per*ceived financial hardship* can then leads to an incremental change in the same direction in *risk of violence* (e.g., as *perceived financial hardship* increases, *risk of violence* increases). An incremental change in *risk of violence* can then lead to an incremental change in the same direction in *violent events* (e.g., as *risk of violence* decreases, the number of *violent events* decreases), which then leads to an incremental change in the opposite direction in *self-worth* (e.g., as the number of *violent events* increases, *self-worth* decreases). As this loop contains two positive and two negative links, the loop has an overall positive sign and is, therefore, reinforcing.

Sources of Evidence for an Applied Project

In the current project, three sources of evidence were used to support the our CLD, our dynamic hypothesis, shown in Figure 5.3, namely: (1) a review of published literature of the psychosocial syndemic among similar populations of women (written databases), (2) qualitative feedback from community stakeholders ("mental models"), and (3) purposeful secondary analyses of data from a health study of women living with and at-risk of HIV in the Bronx (numerical databases). Our use of these multiple sources of evidence is consistent with best practices in system dynamics modeling (Forrester, 1980, 2009). Appropriately synthesizing information from across these data sources is the objective, appreciating that information can be misapplied and/or be fallible.

Review of published literature. We first used peer-reviewed research literature, prioritizing meta-analyses and systematic reviews when available, to identify initial constructs associated with syndemic risk, including *emotional distress, substance use, perceived financial hardship, violent events*, and *HIV-status*. The same review was used to inform plausible patterns of influence among these constructs. Notably, we had not identified the construct of *self-worth* as a construct to be included in our CLD during our initial literature review. However, after consultation with community stakeholders (described below), a return to the literature revealed that *self-worth*, conceptualized as a related concept to *self-esteem* (Rosenberg, 1979), was consistently and independently associated with all of our current elements of syndemic risk (Bagley & Ramsay, 1986; Batchelder et al., 2013; Dietz, 1996; Ethier et al., 2004; Gutierres & Puymbroeck, 2006; Jacobs & Kane, 2011; Lundberg & Kristenson, 2008; Somalai et al., 2000; Sonis & Langer, 2008; Sowislo & Orth, 2012). Similarly, several other related constructs that were identified by our community stakeholders were subsequently confirmed to be associated with syndemic risk by published literature (i.e., risk of violence, frequency of violent events and frequency of substance use) (Dietz, 1996; Golder & Logan, 2011; Logan, Cole, & Leukefeld, 2002; Orth, Robins, &

Widaman, 2012; Selenko, Batinic, & Paul, 2011), and were added to our CLD. To organize this literature, we developed an extensive table of evidence, prioritizing meta-analyses and systematic reviews.

Qualitative Feedback from Community Stakeholders: The Bronx Community Research Review Board (BxCRRB). The BxCRRB is a community-academic partnership made up of approximately ten Bronx community members, demographically similar to the population of interest, who work to increase the potential benefits of research to the Bronx community. We met with the BxCRRB on three occasions and obtained qualitative feedback on the constructs included in our CLD.

At our first meeting, we presented our list of initial syndemic constructs (*history of childhood sexual abuse*, *distress*, *substance use*, *financial hardship*, and *violence*) without conveying our hypothesized relationships. Members were asked if the identified constructs were meaningfully related to sexual self-care, and how they felt the constructs were related to one another. Board members emphasized how a "lack of structure" or being treated poorly early in life can contribute to low levels of "self-worth," "self-esteem," and perceived "validity." Members of the board highlighted how low levels of self-worth drive many of the proposed syndemic variables including *emotional distress*, *substance use*, *risk of violence*, and *perceived financial hardship* (not actual financial hardship, but a "giving up" that fosters a sense of "hopelessness about poverty"). The members affirmed that the identified constructs were important and confirmed our hypothesized interrelationships. By consensus, the members suggested we add *self-worth* and focus on *perceived financial hardship*.

Analyses of Cohort Data. We also conducted secondary analyses of a four-year cohort study that recruited women living with and at-risk for HIV/AIDS in the Bronx (the Natural History of Menopause in HIV Injection Substance Users; PI Ellie Schoenbaum, R01DA13564; $n = 620$). Women in the cohort had a median age of 45 years (range 35–71); 50% were Black and 39% were Hispanic; 52% were HIV+ at baseline; and all endorsed past HIV-related high risk behaviors, such as engagement in intravenous drug use, sex with known or suspected HIV infected male, >4 partners in the last five years, and transactional sex for money or substances. History of childhood sexual abuse, emotional distress, illicit substance use, perceived financial hardship, violent events, and HIV-status were all collected as part of the cohort study's patient surveys at baseline and follow-up time points at six-month intervals. Data from all nine time-points were used to compute indicators, annualized and semi-annualized average frequency counts, and to evaluate bivariate relationships between constructs over time. We also conducted a K-means cluster analysis using the cohort data to identify meaningful patterns of syndemic risk. We used these results to develop clinically relevant simulation profiles or *reference modes*. In system dynamics model-building, a reference mode is a graphical or verbal description of the behavioral process or phenomenon of interest (Forrester, 1961; Oliva, 2003). A description of these references modes is provided below, in Model Simulation.

Model Formulation

Model formulation involves translating a CLD into a working simulation tool. For the current project, we used Vensim software to accomplish this. There are three basic types of structure detail: levels or *stocks*, rates of flow or *flows*, *auxiliary variables*, and *parameters* (constants). Stocks are accumulations of material or information at a given moment. Flows are what increase or decrease a stock, incrementally, over time. Constants are variables that have a fixed value (at least over a specified simulation time period or *time horizon*). Stocks are represented by a rectangle. Flows are represented by a double-lined arrow and a faucet-like icon. Often one end of a flow structure will be attached to a cloud icon. The *cloud* represents a sink, or a source of material of information that accumulates outside of the model boundary. The equations are programmed easily using Vensim's stock-and-flow graphical interface.

Note that the entire model need not be programmed in order to run the model. Preliminary runs are often generated to ensure that model structure generated plausible, valid simulated output. Often, initial runs are not sensible. Here, the formulation was iteratively refined and the model was repeatedly re-run until it generated output that was more realistic and supportable by available data, theory, and other source of evidence.

As equations are built, modelers must ensure that they are choosing the correct structure detail (i.e., Is the variable a stock or a flow? Is it meaningfully named? Are variables dimensionally consistent?). Table functions are often used to ensure that the effect of a causal variable is properly converted into units that ensure dimensional consistency. Most model structures use one or more of these to represent a single construct. In addition to these, there is a wide variety of other pre-programmed functions that can be applied (complete documentation for Vensim is available on-line: www.vensim.com/documentation/index.html?20770.htm).

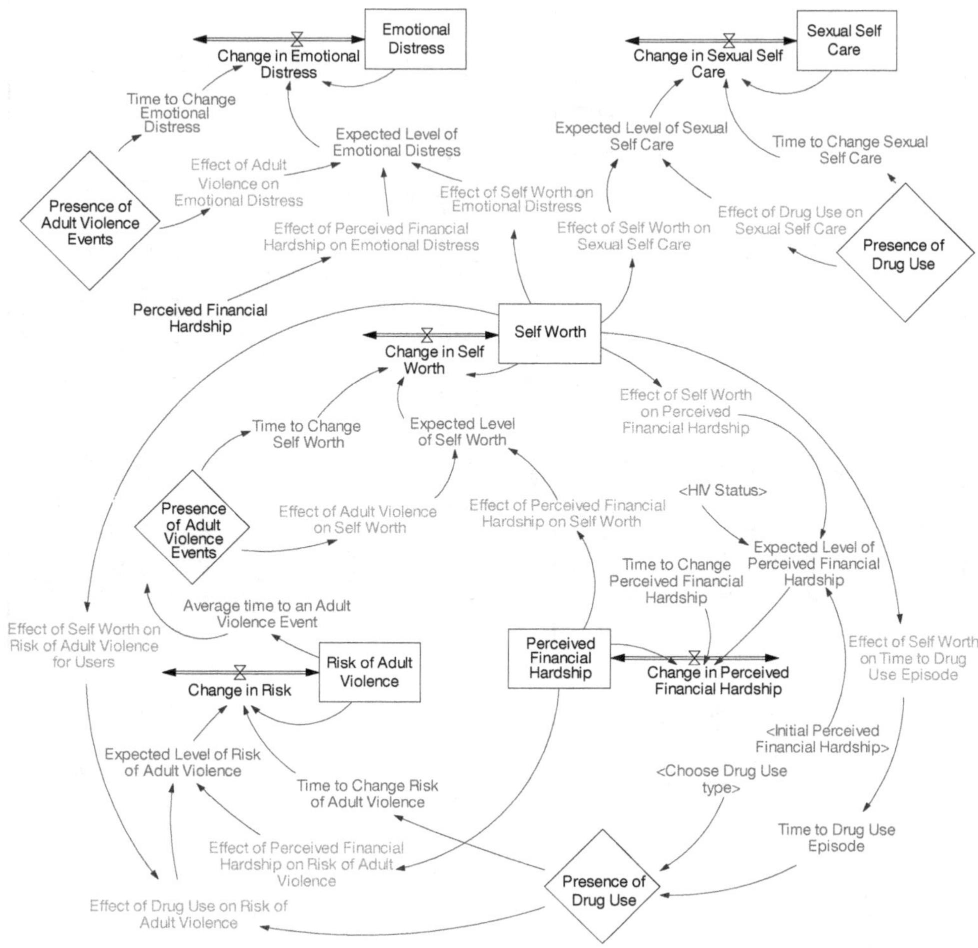

Figure 5.4 Stock-and-flow diagram of syndemic risk

Table 5.1 System dynamics model equations by syndemic risk construct

Construct	*Type of Equation*	*Equation*	*Units*
Sexual Self-Care	Stock	Sexual Self-Care = INTEG (Change in Sexual Self-Care)	SSC Units [Range = 1 to 100; Initial value = 50]
	Flow	Change in Sexual Self-Care = (Expected Level of Sexual Self-Care – Sexual Self-Care)/Time to Change Sexual Self-Care	SSC Units/Week
	Auxiliary	Expected Level of Sexual Self-Care = Effect of Self-Worth on Sexual Self-Care×Effect of Drug Use on Sexual Self-Care	SSC Units
	Auxiliary	Time to Change Sexual Self-Care = IF THEN ELSE (Presence of Drug Use = 1, 0.5, 3)	Weeks
	Auxiliary	Effect of Drug Use on Sexual Self-Care = WITH LOOKUP (Presence of Drug Use, ([(0,0) – (1,2)], (0,1), (1,0.5)])	Dmnl
	Auxiliary	Effect of Self-Worth on Sexual Self-Care = WITH LOOKUP (Self-Worth, ([(0,0) – (100,100)], (0,0), (39,18), (50,50), (57,81), (100,100)])	SSC Units
Emotional Distress	Stock	Emotional Distress = INTEG (Change in Emotional Distress)	Emotional Distress Units [Range = 1 to 100; Initial value = 10]
	Flow	Change in Emotional Distress=(Expected Level of Emotional Distress – Emotional Distress)/Time to Change Emotional Distress	Emotional Distress Units/Week
	Auxiliary	Effect of Adult Violence on Emotional Distress = WITH LOOKUP (Presence of Adult Violence Events, ([(0,0) – (1,2)], (0,0), (1,0.3)])	Dmnl
	Auxiliary	Expected Level of Emotional Distress = (Effect of Self Worth on Emotional Distress + Effect of Perceived Financial Hardship on Emotional Distress) + (Effect of Adult Violence on Emotional Distress×(60 – (Effect of Self Worth on Emotional Distress + Effect of Perceived Financial Hardship on Emotional Distress)))	Emotional Distress Units
	Auxiliary	Effect of Perceived Financial Hardship on Emotional Distress = WITH LOOKUP (Perceived Financial Hardship, ([(0,0) – (100,10)], (0,0), (100,5)])	Emotional Distress Units
	Auxiliary	Time to Change Emotional Distress = IF THEN ELSE (Presence of Adult Violence Events = 1, 0.5, 10)	Weeks
	Auxiliary	Effect of Self-Worth on Emotional Distress = WITH LOOKUP (Self-Worth, ([(0,0) – (100,60)], (0,55), (2,49), (7,40), (11,32), (18,25), (24,18), (32,12), (40,9), (50,6.5), (62,4.2), (76,2.6), (100,1)))	Emotional Distress Units

Construct	*Type of Equation*	*Equation*	*Units*
Self-Worth	Stock	Self-Worth= INTEG (Change in Self-Worth,Initial SW – (Childhood Sexual Abuse×5) – (HIV Status×2))	Self-Worth Units [Range = 1 to 100; Initial value = 25]
	Flow	Change in Self-Worth = (Expected Level of Self-Worth – Self-Worth)/ Time to Change Self-Worth	Self Worth/Weeks
	Auxiliary	Time to Change Self-Worth = IF THEN ELSE (Presence of Adult Violence Events = 1,0.5, 8) IF THEN ELSE (Presence of Adult Violence Events = 1, 1, 8)	Weeks
	Auxiliary	Expected Level of Self-Worth = Effect of Adult Violence on Self-Worth× Effect of Perceived Financial Hardship on Self -Worth	Self-Worth Units
	Auxiliary	Effect of Perceived Financial Hardship on Self-Worth = WITH LOOKUP (Perceived Financial Hardship, ([(0,0) – (100,1)], (0,1), (100,0.5)])	Self-Worth Units
	Auxiliary	Effect of Adult Violence on Self-Worth = WITH LOOKUP (Presence of Adult Violence Events, ([(0,0) – (1,100)], (0,100), (1,0)])	Self-Worth Units
Violence	Stock	Risk of Adult Violence = INTEG (Change in Risk, Initial Risk of Adult Violence + (HIV Status×2))	Risk Units [Range = 1 to 100; Initial value =25]
	Flow	Change in Risk = (Expected Level of Risk of Adult Violence – Risk of Adult Violence)/Time to Change Risk of Adult Violence	Risk Units/Week
	Auxiliary	Time to Change Risk of Adult Violence = IF THEN ELSE (Presence of Drug Use = 1, 1, 14)	Weeks
	Auxiliary	Presence of Adult Violence Events = PULSE TRAIN (Average time to an Adult Violence Event, TIME STEP, Average time to an Adult Violence Event, 1000)	Violent Acts
	Auxiliary	Effect of Perceived Financial Hardship on Risk of Adult Violence = WITH LOOKUP (([(0,0) – (100,100)], (0,0), (100,50)])	Dmnl
	Auxiliary	Expected Level of Risk of Adult Violence = Effect of Drug Use on Risk of Adult Violence + Effect of Perceived Financial Hardship on Risk of Adult Violence	Risk Units
	Auxiliary	Effect of Self-Worth on Risk of Adult Violence for Users = WITH LOOKUP (Self-Worth, ([(0,0) – (100,100)], (0,50), (50,40), (100,30)])	Risk Units
	Auxiliary	Effect of Drug Use on Risk of Adult Violence = Presence of Drug Use× (Effect of Self-Worth on Risk of Adult Violence for Users)	Risk Units

Table 5.1 Continued

Construct	*Type of Equation*	*Equation*	*Units*
	Auxiliary	Average time to an Adult Violence Event = WITH LOOKUP (Risk of Adult Violence, ([(0,0) – (100,100)], (0,100), (20,52), (60,8), (100,4)))	Weeks
Perceived Financial Hardship	Stock	Perceived Financial Hardship = INTEG (Change in Perceived Financial Hardship)	Poverty Units [Range = 1 to 100; Initial values = 25 for low or 75 for high]
	Flow	Change in Perceived Financial Hardship = (Expected Level of Perceived Financial Hardship – Perceived Financial Hardship)/Time to Change Perceived Financial Hardship	Poverty Units/Week
	Auxiliary	Expected Level of Perceived Financial Hardship = Initial Perceived Financial Hardship + (Effect of Self Worth on Perceived Financial Hardship/10) – (HIV Status×2)	Poverty Units
	Auxiliary	Effect of Self Worth on Perceived Financial Hardship = – Self-Worth	Poverty Units
	Parameter	Time to Change Perceived Financial Hardship = 20	Weeks
Drug Use	Auxiliary	Presence of Drug Use = IF THEN ELSE (Choose Drug Use type = 1, 1, IF THEN ELSE (Choose Drug Use type = 2, PULSE TRAIN (Time to Drug Use Episode, 1, Time to Drug Use Episode, 1000), 0])	Dmnl (1=Relapse; 2=Non-user)
	Auxiliary	Time to Drug Use Episode = Effect of Self-Worth on Time to Drug Use Episode	Weeks
	Auxiliary	Effect of Self-Worth on Time to Drug Use Episode = WITH LOOKUP (Self-Worth, ([(0,0) – (100,40)], (0,1), (50,8), (100,26)])	Weeks
HIV Status	Parameter	HIV Status = 0	Dmnl (0 = HIV–; 1 = HIV+)
Childhood Sexual Abuse	Parameter	Childhood Sexual Abuse = 0	Dmnl (0 = No; 1 = Yes)

In the current project, we formulated our system dynamics model to represent the hypothesized behavioral and psychosocial dynamics of an individual woman affected by syndemic risk, over the course of a two-year period (Time horizon = 104 weeks). We applied the *first-order smooth* to represent *information delays* (described in detail below) for all of our major constructs of syndemic risk (i.e., *emotional distress, perceived financial hardship, risk of violence, self-worth* and *sexual self-care*), with the exception of *substance use* and *violent events*, for which we utilized a *pulse function* (also described in detail below). Our final model is a fifth-order model, which is determined by adding up the number of stocks comprising the model.

Our final model is presented in Figure 5.4, as a stock-and-flow diagram. It includes all the constructs featured in our final CLD (see Figure 5.3). Table 5.1 lists each construct in the diagram with detail about the equations used to define it. For example, sexual self-care is defined by a set of six elements, including one stock, one flow, and four auxiliary variables.

Quantification of Model Constructs

We can classify the constructs included in our model of syndemic risk as either *endogenous or primary, exogenous,* or *indicator. Endogenous or primary constructs* are those that form a feedback structure, or loop. These include: *substance use* and *violent events* as well as *risk of violence, perceived financial hardship,* and *self-worth. Exogenous constructs* are those that have a direct effect on one or more constructs in the model, but are not part of a feedback structure. These include *history of childhood sexual abuse*, which has a fixed effect on initial self-worth, and *HIV-status*, which has a fixed effect on *risk of violence* and *perceived financial hardship*, and *pattern of substance use*, which was programmed as either a non-user or an active user whose substance use varied over time depending on *self-worth*. Like exogenous constructs, *indicator constructs* are not part of a feedback structure, but can be computed as an outcome of the effect or influence of one or more constructs in the mode. These include *sexual self-care* and *emotional distress. Emotional distress, perceived financial hardship, risk of violence, self-worth,* and *sexual self-care* were all formulated on a 100-point scale, ranging from 0 to 100 units, allowing for ease of comparison of simulated output.

Representation of Information Delays

System dynamics models can be used to simulate the flow of both materials and information throughout a system. The rate of flow is represented as a delay function. Per Sterman (2000), in a material delay the stock is the quantity of material (e.g., money) in transit and the output of the delay is the flow. In information delays it is the level or intensity of the perceived information that is the stock, which may represent perception of, for example, attitudes, values, emotions, cognitions, behaviors and/or experiences. In our model, all of our primary constructs, that are not discrete events (i.e., *substance use* and *violent events*), are delays in perception, which are most expediently represented using a first order *smooth.* (Note that a modeler could also use the first order material delay to represent a psychological or social problem. See, for example, Levine (2000) for further explanation about when a first order material delay could be used to represent changes in information.) The first order smooth, is a generic structure used to simulate a balancing (negative) feedback process. The smooth structure is defined by an initial value, a goal value, and an average time delay to attain the goal. For example, we present the smooth structure for the construct of *sexual self-care* (Figure 5.5). Applying the smooth structure, *sexual self-care* and our other major constructs are essentially treated as a cumulative variables, or stocks, whose dynamics move toward a specified goal, or steady state, over time. You can think of the

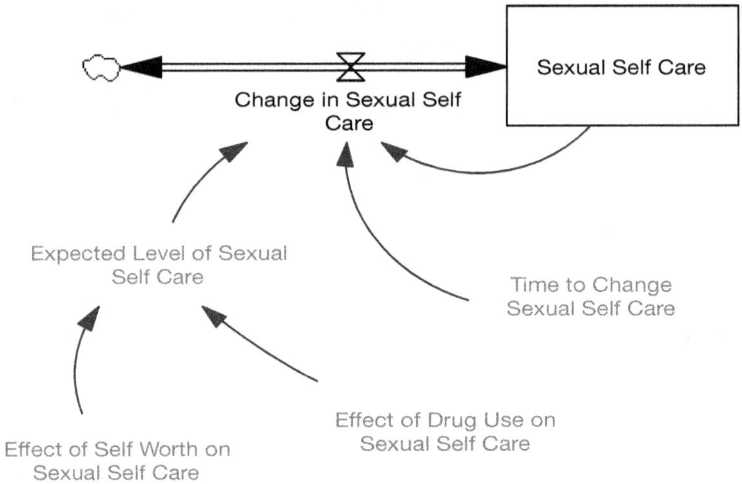

Figure 5.5 Stock-and-flow diagram of sexual self-care (first-order "smooth")

smooth as the dynamic average of the construct, computed across the simulated time horizon. The time delay may be constant or variable.

The simulated output of the smooth structure displays a trajectory over a specified time horizon, starting from its initial value and trending towards its ultimate goal, or expected value. The rate of change is a function of the specified time delay. The rate of change is intended to close the gap, or the difference between the expected value (or goal), and the smooth stock level (or actual value). The stock adjusts toward the goal over time, unless it is impacted by another variable. The gap between the actual value and the expected value (or goal) is closed according to the specified smoothing time. The magnitude of the gap would decline to zero over the smoothing time if the net inflow were held constant, resulting in the actual value approaching the expected value (i.e., "goal").

In our modeling project, the formulation of each smooth information delay and its parameter values was informed by our three sources of evidence: trend analyses of our women's cohort data, published literature, as well as feedback from the BxCRRB. This process included choices of how to integrate constructs (e.g., whether to add or multiple to constructs).

Managing Uncertainty of Discrete Events

Although system dynamics models generally simulate continuous processes, as opposed to discrete events, our representation of syndemic risk is linked to specific events, such as a violent incident in the community. We utilized the Vensim pulse train function to manage uncertainty about the probability of occurrence of a discrete happening. The decision to apply the pulse function was based on evidence for the relationships between violent events, driven by *risk of violence*, linked to *substance use* and *HIV-status* (see Table 5.1 for equations using the pulse function). Using the pulse function, a violent event was defined as a function of the average time to violence and change in risk of violence. Similarly, with the pulse function, an episode of substance use was defined as a function of an assumed time to substance use for substance users. In general, the pulse function was used to associate the probability of experiencing a discrete event to a specified threshold, relative to another variable of interest at a given *time step*. The model's time step is the length of time elapsing between each computational iteration of the simulation run (i.e., in calculus, the *dt*). Based on cohort data and literature on substance abuse and violent events among similar populations, we were able to develop equations to estimate the pulse rate based on large national samples of frequency of violent events based on previous experience of violence.

Model Simulation

Our system dynamics model was designed to simulate syndemic risk, based on the psychosocial dynamics of an individual woman's lived experience, over a hypothetical two years (104 weeks) time horizon. Our K-means cluster analyses of the cohort data were used to identify meaningful patterns of syndemic risk. We used these results to develop clinically relevant simulation profiles. High and low *perceived financial hardship* and *active and inactive substance use* were identified as the most impactful dimensions of syndemic risk (see cluster analysis results, summarized in Table 5.2). We confirmed the validity of these derived syndemic risk profiles by obtaining qualitative feedback from our community stakeholders and by documenting support in the published literature (e.g., Meyer et al., 2011).

Table 5.2 Syndemic risk profiles derived from cluster analyses of cohort data

	Syndemic Risk Profile (N=597)				
	Profile 1	*Profile 2*	*Profile 3*	*Profile 4*	
Selected Variables Assessed in Women's HIV Risk Cohort Study	LOW Perceived Financial Hardship and INACTIVE Substance Use n = 296 (50%)	HIGH Perceived Financial Hardship and INACTIVE Substance Use n = 163 (27%)	LOW Perceived Financial Hardship and ACTIVE Substance Use n = 54 (9%)	HIGH Perceived Financial Hardship and ACTIVE Substance Use n = 84 (14%)	F-test ($df = 3$); $p < .01$
Variable Name	*Proportion of Sample at Baseline*				
Endorsed History of Violent Events	62%	54%	63%	66%	SIG
Self-Reported History of Childhood Sexual Abuse	0.39	0.40	0.30	0.46	NS
HIV Positive	0.60	0.44	0.63	0.44	SIG
Variable Name	*Average Score Across Follow-Up Points*				
Perceived Financial Hardship (1–3)	1.56	2.49	1.32	2.34	SIG
Self-Reported Drug Use	3%	6%	78%	80%	SIG
Self-Reported Violent Events	13%	12%	15%	23%	SIG
Emotional Distress (0–60)	34.39	40.18	34.90	39.33	SIG
Number of Sexual Self-Care Behaviours (0–4)	3.49	3.46	3.89	2.97	SIG

Simulation of Syndemic Risk Profiles

Our system dynamics model was independently calibrated and run to generate simulated output for each of the four syndemic risk profiles, as identified by our K-means cluster analysis (see Figure 5.6). In order to simulate these, profiles 1 and 2 were programmed as non-substance users and profiles 3 and 4 were programmed as active users. In profiles 1 and 3, baseline *perceived financial hardship* was set to an initial value of 25 units profiles 2 and 4 baseline *perceived financial hardship* was set at 75 units. All other constructs were the same at baseline across profiles (i.e., *emotional distress* = 40 units, *risk of violence* = 50 units, *self-worth* = 25 units, and *sexual self-care* = 50 units). Note that all profiles presented here were simulated as if they denied *childhood sexual abuse* and being HIV+.

The outputs are simulated by the integration of the equations in the model at regular intervals, including feedback between equations. Across all four of the syndemic risk profiles, *violent events* were directly followed by decreases in *self-worth, sexual self-care* and an increase in *emotional distress.* In profiles 2 and 4, *substance use* events were closely followed by decreases in *sexual self-care* and increases in *emotional distress.* Additionally, the interrelationships between *violent events, substance use, sexual self-care, emotional distress,* and *self-worth* became increasingly complex when *perceived financial hardship* increased, as evidenced in profiles 2 and 4 compared to profiles 1 and 3, respectively. Further, *self-worth* was a central construct, which decreased in the presence of *violent events* in all four profiles. There was significant variability in the time it took for *self-worth* and *emotional distress* to recover after *violent events* in all profiles.

Given the complex trajectories associated with *substance use*, this method enabled us to include the patterns of *substance use* without modeling the complexity of the biopsychological underpinnings of addiction, which was beyond the scope of this model. Our decision to build the system dynamics model for users and non-users was based primarily on the cohort data, but included our other two sources of data. The cluster analysis revealed two profiles of substance use: those who reported using and those who denied substance use at all time points.

Model Evaluation

Evaluation of the model includes both structural validity tests as well as behavioral validity tests (Barlas, 1996). Per Barlas (1996), we first carried out a *theoretical structure test*, by comparing the original model equations, based on the cohort study, with the generalized knowledge from our three sources of data together. This process included asking members of the BxCRRB about the proposed syndemic constructs in our first meeting. Members of the BxCRRB agreed that the originally proposed syndemic constructs were interrelated. However, they also stated that we were missing a critical psychological construct: self-worth. Members of the BxCRRB explained how self-worth contributes to all the proposed syndemic constructs and should be added to the model in relation to *perceived financial hardship* (as described above) and *substance use*, both of which were described to contribute to increased *risk of violence.* For example, a woman with low self-worth was described as more likely to be at-risk of violence (e.g., intimate partner violence) when she felt financially desperate or needed illicit substances. After this meeting, relationships between self-worth, and the related psychological construct of self-esteem, were researched in the published empirical literature and self-worth was added to the model. Notably, neither self-worth nor self-esteem was evaluated in the cohort data. We were able to include self-worth as a construct based mainly on input by our community stakeholders and further supported by peer reviewed empirical literature.

The *structural validity* of our model of syndemic risk was established by ensuring each equation conformed to basic tests of dimensional consistency and was properly calibrated

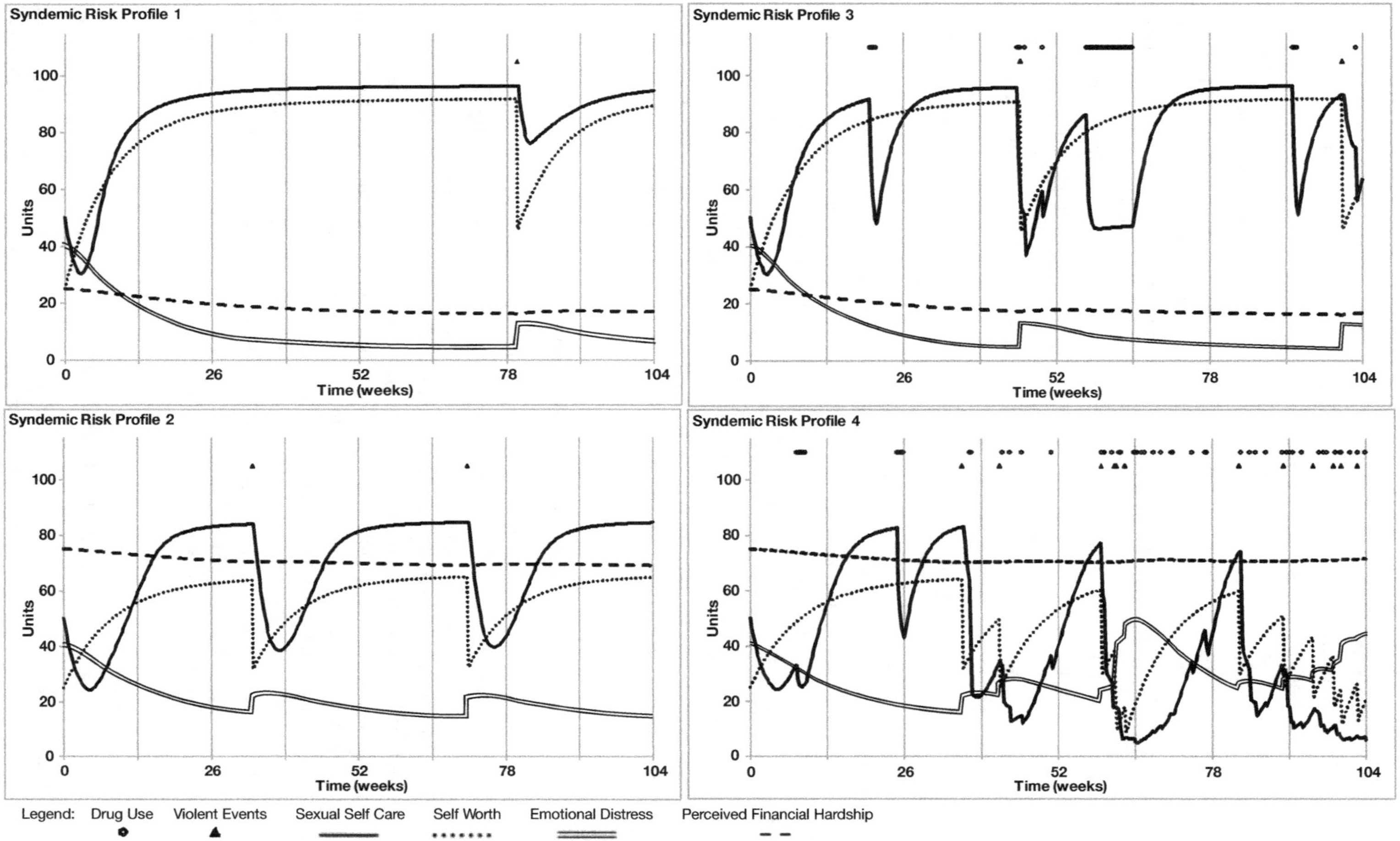

Figure 5.6 Simulate profiles of syndemic risk (1 through 4)

(i.e., the units of measurement or quantification of the constructs or variables on each side of the equation were the same). The model was then calibrated to ensure replication of expected reference modes, as supported by one or more of our sources of evidence. Therefore, after confirming the dimensional consistency and calibration of all of the equations, we compared the output for each pair of constructs with data we had for each bivariate relationship, such as *perceived financial hardship* and *self-worth*. Discrepancies between simulated output and sources of evidence resulted in iteratively revisiting the three sources of evidence and editing the equations.

Next, basic *behavior pattern tests*, including a qualitative features analysis (Carson & Flood, 1990), were conducted. This involves generating and evaluating simulated output of the model. Our simulated results show how the identified syndemic constructs (*emotional distress, substance use, perceived financial hardship, violent events* and *self-worth*) interact synergistically beyond what can be assessed using an additive framework. The four profiles resulted in output that is consistent with published literature (e.g., Catalano et al., 2009; El-Bassel et al., 2005; Jewkes, 2002; Meyer et al., 2011), demonstrate face validity, and depict consistent patterns as well as significant variability.

Violent events were consistently followed by behavioral and psychosocial sequela (e.g., reductions in *sexual self-care* and *self-worth,* and increases in *emotional distress*), indicating that strategies to improve safety, in addition to interventions aimed at addressing psychosocial needs, may be more effective than attempting to improve self-worth or reduce emotional distress independently among this population. The model also represented associations between reductions in *sexual self-care* and increases in active *substance use* and *emotional distress*, consistent with the literature (e.g., (Meyer, Springer, & Altice, 2011)). Further, the interrelationships between *emotional distress, substance use, violent events, self-worth* and *sexual self-care* were more complex when *perceived financial hardship* was greater, also consistent with the literature (Meyer et al., 2011). Ultimately, our finding indicated a need for prioritizing safety and substance abuse treatment in combination with interventions aimed at addressing the related psychosocial needs of this population.

Reflection on Methods

Throughout this project, we encountered several challenges unique to our conceptualization of the problem of syndemic risk and our choice to use system dynamics modeling to explore syndemic theory and simulate the dynamics over time. These challenges ranged from the operationalization of psychological variables using stock and flow equations to the parameterization of complex interrelated constructs, including deciding not to include related biological, psychological and social constructs (e.g., the physiological underpinnings of addiction, the typical alleviation of depressive symptoms over time, and the historical context of oppression). We turned, time and again, to our three sources of evidence: (1) a review of published literature relevant to the defined syndemic (written databases), (2) qualitative feedback from community stakeholders ("mental models"), and (3) purposeful secondary analyses of data from a health study of women living with and at-risk of HIV in the Bronx (numerical databases). Collectively, our use of these sources of evidence proved to be robust in defining and exploring the dynamics of syndemic risk.

Technical Challenges

Throughout the early stages of the literature review, CLD development, and model development, we searched for the best way to mathematically represent and simulate psychosocial variables using stock and flow system dynamics structures. We identified relationships in our three sources of data that indicated how constructs trended toward an expected trajectory (e.g., a simulated woman's emotional distress trending toward an expected trajectory based on related constructs such as *self-worth* and *violent events*). Based on our three sources of evidence, we ultimately identified the first order smooth structure and the pulse train function to simulate the relationships we identified between constructs.

Limiting Model Boundary

Given the current development of syndemic theory and its relatively limited dissemination in public health circles to date, we grappled with the problem of model parsimony. While all the included constructs are clearly multifaceted and involve contextual and historical qualifications, we intentionally excluded constructs we thought to be secondary to conveying the inter-relationships of the proposed syndemic, based on syndemic theory. These exclusions included physiological and psychological processes (e.g., the natural progression of depression and addiction), explicit interpersonal interactions (e.g., the role of the perpetrator in acts of violence), and specifying the types of violence (e.g., intimate partner violence, random acts of violence, violence associated with transactional sex). In addition, we questioned whether to include history of disempowerment as a contributing exogenous variable in the system dynamics model, to address the effects of racial and ethnic disparities evident, similar to the exogenous effect of *HIV status* and *history of childhood sexual abuse* on syndemic risk.

Organizing and Documenting Support for Key Assumptions

The validity of our modeling ultimately rests on the strength and accessibility of the assumptions and decisions we made throughout our model building process. To address this, we created an extensive appendix of evidence to support the inclusion of each construct in our model, its causal influence on other constructs in the model, its plausible range of values, as well as the strength of its interdependencies with the other constructs.

Valuing Qualitative Evidence

Beyond our use of the peer-reviewed literature and the cohort data we had available to inform our modeling, we found qualitative input from our community sources (i.e., BxCRRB) to be of great utility. For example, we were able to incorporate deliberative comments and insights from the stakeholders (e.g., the assertion by BxCRRB members about the central role of *self-worth* in understanding the dynamics of syndemic risk). Notably, the usefulness of quantitative data for system dynamics modeling is often limited by under- or over-reporting, missing or incomplete data, poor reliability, and little or no information about change over time about a given variable or sets of variables.

To overcome these, we need systems methods that strive to understand and reconcile linear and nonlinear, qualitative and quantitative, and reductionist and holist approaches to data analysis and interpretation (Krygiel, 1999; Trochim, Cabera, Milstein, Gallagher, & Leischow, 2006). The epistemology of system dynamics is supportive of a perspectivist and constructivist epistemology and philosophy of science (Barlas, 1989, 1996; Tebes, 2005). System dynamics models embody the modelers' theory about how a system works and aim to represent the system "adequately with respect to a purpose" (Barlas, 1996). By virtue of the constructivist epistemology, system dynamics models are inevitably a simplification of reality, aiming to represent a problem in order to accomplish a purpose. "No model can claim absolute objectivity, for every model carries in it the modeler's worldview. Models are not true or false but lie on a continuum of usefulness," (Barlas, 1996, p. 187). Therefore, unless a model is built exclusively on empirical data, models are not measuring constructs; rather, they quantify and assess the problem and purpose of the model (e.g., to increase understanding about the interrelationships among constructs of interest).

Unique Benefits of System Dynamics Modeling

The use of the system dynamics methodology offers several benefits to researchers. It enables the use of types of evidence inconsistently utilized by researchers, such as qualitative information and the lived experiences of stakeholders. It facilitates extending the time horizon beyond available empirical data, as a way to understand possible future outcomes of a given systems problem (Homer & Hirsch, 2006). However, the longer the time horizon, the more difficult it is to assess the validity of the model's output and ensure potentially unforeseen important constructs and equations are included. Finally, system dynamics modeling compliments interpretation of results derived from general linear methods, as system dynamics modeling enables explicit analyses of causal feedback structures that are rarely testable use statistical data analytic approaches. In addition, system dynamics modeling provides a way to interpolate data points from databases that may have gaps or missing information.

Our results, particularly the system dynamics model, demonstrate how this methodology can be used by clinicians and policy makers to identify points of leverage among variables that are historically challenging to measure: psychological and to a lesser extent behavioral and social variables. This methodology can also be used to simulate the effects of different types of interventions to assess long-term effects and unintended consequences of simulated interventions. Using system dynamics modeling enables the exploration of psychological feedback and unforeseen consequences of the relationships between psychological, behavioral, and social variables over an extended time horizon. System dynamics modeling also facilitates the simulation of data without requiring costly support for a longitudinal comprehensive research study.

By involving the BxCRRB we were able to utilize the insight and lived experience of key stakeholders. Specifically, the BxCRRB members identified *self-worth* as a highly relevant additional construct, which was not only associated with the previously identified syndemic constructs (Dietz, 1996; Ethier et al., 2006; Jacobs & Kane, 2011; Sowislo & Orth, 2012), but may be in the causal pathways linking the constructs together, as indicated in the CLD (Figure 5.2). Further empirical work is needed to evaluate the longitudinal relationships between self-worth and the other syndemic constructs. The involvement of BxCRRB exemplifies how key stakeholders and qualitative methods strengthen the system dynamics modeling process.

Conclusion

We believe this project makes an important, though incremental, contribution to the syndemic literature. Our results hypothesize a particular structure and set of behavioral patterns that emerged out of our identified constructs of syndemic risk. Our model emphasizes the deficit-oriented nature of syndemic theory and indicates a need for individualized multi-aimed psychosocial interventions. In addition, this and other system dynamics modeling projects can be used to estimate cost of developing and sustaining such psychosocial interventions.

The development of this system dynamics model brought attention to the deficit nature of the extant literature on syndemic theory. Not all women with these challenges experience the ramifications of the negative reinforcing interrelationships. Future work should explore balancing feedback structures that explain how some women emerge out of, or are resilient despite of, syndemic risk (e.g., religious or spiritual beliefs, particular kinds of social support, or emotion regulation) (Nyamathi, Keenan, & Bayley, 1998; Thompson, Thomas, & Head, 2012).

Our results emphasize the need for individualized multi-aimed interventions that prioritize safety planning and substance abuse treatment, while addressing unmet related psychosocial challenges (e.g., Smith & Romero, 2010). Given the evidenced relationship between perceived financial hardship and frequency of violent events, working with women to feel economically empowered may improve sexual self-care, as many women engage in high-risk situations due to economic necessity (Prather et al., 2012; Sonis & Langer, 2008). Our modeling suggests that clinicians should consider how co-occurring psychosocial problems might perpetuate one another. For example, a woman may feel she cannot leave an abusive partner because of financial hardship, while the abuse is decreasing her self-worth and resulting in emotional distress, which eventually leads to a substance use relapse. Clinicians must also account for differences between individuals, as the women depicted in Profiles 1–4 would likely benefit from different interventions.

System dynamics modeling is useful for comparative studies of resources required to implement and sustain multi-aimed simulated interventions. While individualized multi-aimed interventions may seem cost-prohibitive, this methodology includes feedback structures that integrate interactions over time, enabling cost-benefit analysis of multi-aimed interventions without the resources and support necessary for multiple empirical longitudinal studies. This and other system dynamics models can be used to simulate the effects and cost of different multi-aimed interventions among specific profiles over time (Ahmad, 2005; Homer, Hirsch, & Milstein, 2007; Tengs, Osgood, & Chen, 2001).

Overall, our model provided a deeper understanding of the complex dynamics of syndemic risk. System dynamicists use the iterative process to gain confidence in their models, which Oliva notes has been underplayed in literature about system dynamics methods (Oliva, 2003). Confidence is gained as the model's structure replicates expected or known behaviors, which is the ultimate objective of the iterative nature of the system dynamics methodology (J. W. Forrester & Senge, 1980). This work also exemplified the value that social scientists, clinicians and community members collectively bring to each step of the system dynamics modeling method, including problem identification, system conceptualization, model formulation, model simulation, and model evaluation. In addition, this work shows how statistical methods, such as K-means cluster analysis, can be used to create a framework or rationale for model formulation. Future work will involve further development of the model to explore the design, implementation, and positive impact of multi-aimed intervention strategies and polices that can effectively address systemic risk for underserved communities.

Acknowledgments

We would like to thank Drs. Jeffrey Gonzalez and Ellie Schoenbaum for the guidance and mentorship they provided and Anton Palma, MPH and Gregory Jo, AB for their assistance in building, testing and validating our syndemics simulation tool. We would also like to thank the Bronx Community Research Review Board (BxCRRB) for their invaluable contribution to our understanding of syndemic risk among affected women. Finally, I would like to thank Gary Hirsch and Jack Homer, senior experts in system dynamics modeling, for their generous feedback and support throughout this project. This work was supported in part by the National Institute of Drug Abuse (R01 DA13564; PI: Ellie Schoenbaum, M.D.).

References

Ahmad, S. (2005). The cost-effectiveness of raising the legal smoking age in California. *Medical Decision Making, 25*(3), 330–340. doi: 10.1177/0272989x05276859

Bagley, C., & Ramsay, R. (1986). Sexual abuse in childhood: Psychosocial outcomes and implications for social work practice. *Journal of Social Work and Human Sexuality, 4* (journal article), 33–47.

Barlas, Y. (1989). Multiple tests for validation of system dynamics type of simulation models. *European Journal of Operational Research, 42*(1), 59–87.

Barlas, Y. (1996). Formal aspects of model validity and validation in system dynamics. *System Dynamics Review, 12*(3), 183–210.

Batchelder, A. W., Brisbane, M., Litwin, A. H., Nahvi, S., Berg, K. M., & Arnsten, J. H. (2013). "Damaging what wasn't damaged already": Psychological tension and antiretroviral adherence among HIV-infected methadone-maintained substance users (journal, electronic), June 25, 2013. doi:0.1080/09540121. 2013.766303

Bedimo, A., Kissinger, P., & Bessinger, R. (1997). History of sexual abuse among HIV-infected women. *International Journal of STDS & AIDS, 8* (journal article), 332–335.

Brier, J., & Runts, M. (1993). Childhood sexual abuse: Long-term sequelae and implications for psychological assessment. *Journal of Interpersonal Violence, 8* (journal article), 312–330.

Carson, E. R. & Flood, R. L. (1990). Model validation: philosophy, methodology, and examples. *Transactions of the Institute of Measurement and Control, 12*, 178–185.

CDC. (2013). HIV among Women. National Center for HIV/AIDS, Viral Hepatitis, STD, and TB Prevention, Division of HIV/AIDS Prevention. Retrieved from: www.cdc.gov/hiv/topics/women/pdf/women.pdf

Cook, J. A., Grey, D. D., Burke-Miller, J. K., Cohen, M. H., Vlahov, D., Kapadia, F., . . . Levine, A. (2007). Illicit substance use, negative affect and their association with highly active antiretroviral therapy in HIV-positive women. *Substance & Alcohol Dependence, 15*(89), 74–81.

Dietz, B. E. (1996). The relationship of aging to self-esteem: the relative effects of maturation and role accumulation. *International Journal of Aging and Human Development, 43*(3), 249–266.

El-Bassel, N., Gilbert, L., Wu, E., Go, H., & Hill, J. (2005). HIV and intimate partner violence among methadone-maintained women in New York City. *Social Science Medicine, 61*(1), 171–183.

El-Bassel, N., Schilling, R. F., Irwin, K. L., Faruque, S., Gilbert, L., Von Bargen, J., . . . Edlin, B. R. (1997). Sex trading and psychological distress among women recruited from the streets of Harlem. *American Journal of Public Health, 87*(1), 66–70.

Ethier, K. A., Kershaw, T. S., Lewis, J. B., Milan, S., Niccolai, L. M., & Ickovics, J. R. (2004). Self-esteem, emotional distress and sexual behavior among adolescent females: Inter-relationships and temporal effects. *Journal of Adolescent Health, 38* (journal article), 268–274.

Ethier, K. A., Kershaw, T. S., Lewis, J. B., Milan, S., Niccolai, L. M., & Ickovics, J. R. (2006). Self-esteem, emotional distress and sexual behavior among adolescent females: inter-relationships and temporal effects. *Journal of Adolescent Health, 38*(3), 268–274.

Forrester, J. W. (1961). *Industrial Dynamics*. Cambridge, MA: MIT Press.

Forrester, J. W. (1971). *Principles of Systems* (2nd ed.). Cambridge: Wright-Allen Press.

Forrester, J. W. (1980). Information sources for modeling the national economy. *Journal of American Statistical Association, 75*(371), 555–574.

Forrester, J. W. (1987). The model versus a modeling process. *System Dynamics Review, 1*(1), 133–134.

Forrester, J. W. (1994). System dynamics, systems thinking, and soft OR. *System Dynamics Review, 10*(2–3), 245–256.

Forrester, J. W. (2009). Some basic concepts in system dynamics (S. S. o. Management, Trans.). Cambridge, MA: Massachusetts Institute of Technology.

Forrester, J. W., & Senge, P. M. (1980). Tests for building confidence in system dynamics models. *TIMS Studies in the Management Science, 14*, 201–228.

Golder, S., & Logan, T. K. (2011). Cumulative victimization, psychological distress, and high-risk behavior among substance-involved women. *Violence and Victims, 26*(4), 477–495.

Gutierres, S. E., & Puymbroeck, V. C. (2006). Childhood and adult violence in the lives of women who misuse substances. *Aggression and Violent Behavior, 11*(5), 497–513.

Homer, J., & Hirsch, G. (2006). System dynamics modeling for public health: background and opportunities *American Journal of Public Health, 96*(3), 452–458.

Homer, J., Hirsch, G., & Milstein, B. (2007). Chronic illness in a complex health economy: The perils and promises of downstream and upstream reforms. *System Dynamics Review, 23*(2–3), 313–343. doi: 10.1002/sdr.379

Jacobs, R. J., & Kane, M. N. (2011). Psychosocial predictors of self-esteem in a multiethnic sample of women over 50 at risk for HIV. *Journal of Women & Aging, 23*(1), 23–39.

Jewkes, R. (2002). Intimate partner violence: causes and prevention. *The Lancet, 359*, 1423–1429.

Krygiel, A. J. (1999). Behind the wizard's curtain: An integration environment for a system of systems. Washington, DC: Institute for National Strategic Studies.

Kurtz, S. P. (2008). Arrest histories of high-risk gay and bisexual men in Miami: Unexpected additional evidence for syndemic theory. *Journal of Psychoactive Substances, 40*(4), 513–521.

Lane, S., Rubinstein, R., Keefe, R., Webster, N., Cibula, D., Rosenthal, A., & Dowdell, J. (2004). Structural violence and racial disparity in HIV transmission. *Journal of Health Care for the Poor and Underserved, 14* (journal article), 319–335.

Levine, R.L. (2000). System dynamics applied to psychological and social problems. Oral presentation at the 18th International Conference of the System Dynamics Society: August 6, 2000, Bergen, Norway.

Logan, T. K., Cole, J., & Leukefeld, C. (2002). Women, sex, and HIV: Social and contexual factors, meta-analysis of published interventions, and implications for practice and research. *Psychological Bulletin, 128*(6), 851–885.

Lundberg, J., & Kristenson, M. (2008). Is subjective status influenced by psychosocial factors? *Social Indicators Research, 89*(3), 375–390.

Martinez-Moyana, I. J., & Richardson, G. P. (2013). Best practices in system dynamics modeling. *System Dynamics Review, 29*(2), 102–123.

Meadows, D. H., & Robinson, J. M. (1985). *The Electronic Oracle: Computer Models and Social Decisions.* New York: John Wiley & Sons.

Meyer, J. P., Springer, S. A., & Altice, F. L. (2011). Substance abuse, violence, and HIV in women: A literature review of the syndemic. *Journal of Women's Health, 20*(7), 991–1006.

Milstein, B. (2008). Hygeia's constellation: Navigating health futures in a dynamic and democratic world. Atlanta, GA: Center for Disease Control, Syndemic Prevention Network.

Milstein, B., & Homer, J. (2006). Background on system dynamics simulation modeling with a summary of major public health studies. Retrieved April 11, 2005, from www.systemswiki.org/images/f/f5/SD_background_for_public_health(4.11.05).pdf

Mustanki, B., Garofalo, R., Herrick, A., & Donenberg, G. (2007). Psychosocial health problems increase risk for HIV among urban young men who have sex with men: Preliminary evidence of a syndemic in need of attention. *Annals of Behavioral Medicine, 34*(1), 37–45.

Nyamathi, A., Keenan, C., & Bayley, L. (1998). Differences in personal, cognitive, psychological, and social factors associated with substance and alcohol use and nonuse by homeless women. *Research in Nursing & Health, 21*(6), 525–532.

NYCDHMH. (2014). New York City HIV/AIDS Annual Surveillance Statistics 2012; HIV epidemiology and field services program: New York City Department of Health and Mental Hygiene.

Oliva, R. (2003). Model calibration as a testing strategy for system dynamics models. *European Journal of Operational Research, 151*, 522–568.

Orth, U., Robins, R., & Widaman, K. F. (2012). Life-span development of self-esteem and its effects on important life outcomes. *Journal of Personality and Social Psychology, 102*(6), 1271–1288.

Prather, C., Marshall, K., Courtenay-Quirk, C., Williams, K., Eke, A., O'Leary, A., & Stratford, D. (2012). Addressing poverty and HIV using microenterprise: Findings from qualitative research to reduce risk among unemployed or underemployed African American women. *Journal of Health Care for the Poor and Underserved, 23* (journal article), 1266–1279.

Richardson, G. (1986). Problems with causal loop diagrams. *System Dynamics Review, 2*(2), 158–170.

Richardson, G. P. (1991). *Feedback Thought in Social Science and Systems Theory.* Waltham, MA: Pegasus Communications.

Richardson, G. P., & Pugh, A. L. (1981a). *Introduction to System Dynamics Modeling with DYNAMO.* Cambridge, MA: MIT Press.

Richardson, G. P., & Pugh III, A. L. (1981b). *Introduction to System Dynamics Modeling.* Portland, OR: Productivity Press.

Roberts, N., Anderson, D., Deal, R., Garet, M., & Shaffer, W. (1983). *Introducation to Computer Simulation: A System Dynamics Modeling Approach.* Reading, MA: Addison-Wiley Publishing Company.

Rosenberg, M. (1979). *Conceiving the Self.* New York: Basic Books.

Selenko, E., Batinic, B., & Paul, K. (2011). Does latent deprivation lead to psychological distress? Investigating Jahoda's model in a four-wave study. *Journal of Occupational and Organizational Psychology, 84*(4), 723–740.

Singer, M. (1992). A dose of substances, a touch of violence, a case of AIDS: Conceptualizing the SAVA Syndemic. *Free Inquiry in Creative Sociol, 24*(2), 99–110.

Smith, L., & Romero, L. (2010). Psychological interventions in the context of poverty: Participatory action research as practice. *American Journal of Orthopsychiatry, 80*(1), 12–25.

Somalai, A. M., Kelly, J. A., Heckman, T. G., Hackl, K., Runge, L., & Wright, C. (2000). Life optimism, substance use, and AIDS-specific attitudes associated with HIV risk behavior among disadvantaged inner-city women. *Journal of Women's Health-Gender Based Medicine, 9*(10), 1101–1111.

Sonis, J., & Langer, M. (2008). Risk and protective factors for recurrent intimate partner violence in a cohort of low-income inner-city women. *Journal of Family Violence, 23*(7), 529–538.

Sowislo, J. F., & Orth, U. (2012). Does low self-esteem predict depression and anxiety? A meta-analysis of longitudinal studies. *Psychological Bulletin, 139*(1), 213–240.

Stall, R., Mills, T., Williamson, J., Hart, T., Greenwood, G., Paul, J., . . . Catania, J. (2003). Association of co-occuring psychological health problems and increased vulnerability to HIV/AIDS among urban men who have sex with men. *American Journal of Public Health, 93*(6), 939–942.

Sterman, J. D. (2000). *Business Dynamics: Systems Thinking and Modeling for a Complex World* (Vol. First). Boston, MA: Irwin/McGraw-Hill.

Tebes, J. K. (2005). Community science, philosophy of science, and the practice of research. *American Journal of Community Psychology, 35*(3), 213–230.

Tengs, T. O., Osgood, N. D., & Chen, L. L. (2001). The cost-effectiveness of intensive national school-based anti-tobacco education: Results from the tobacco policy model. *Preventive Medicine, 33*(6), 558–570. doi: 10.1006/pmed.2001.0922

Thompson, M. S., Thomas, M. E., & Head, R. N. (2012). Race, socioeconomic status, and self-esteem: The impact of religiosity. *Sociological Spectrum, 32*(5), 385–405.

Trochim, W. M., Cabera, D. A., Milstein, B., Gallagher, R. S., & Leischow, S. J. (2006). Practical challenges of systems thinking and modeling in public health. *American Journal of Public Health, 96*(3), 538–546.

Ward, M. (1993). A different disease: HIV/AIDS and health care for women in poverty. *Culture, Medicine, and Psychiatry, 17*, 413–430.

Wolstenholme, E. F. (1983). System dynamics: A system methodology or a system modelling technique. *Dynamica, 9*(II), 84–90.

Zierler, S., & Krieger, N. (1997). Reframing women's risk: Social inequalities and HIV infection. *Annual Reviews of Public Health, 18*, 401–436.

6

SYSTEM DYNAMICS MODELING AND FINDING SOLUTIONS TO THE "WICKED" PUBLIC HEALTH PROBLEM OF PREVENTING CHRONIC DISEASES

Nigel Unwin, Cornelia Guell, Natasha Sobers-Grannum, and Anders Nielsen

Research Question: How have system dynamics models been used to understand chronic disease and inform disease prevention policies?
System Science Method(s): System dynamics
Things to Notice:

- Range of diseases studied using system dynamics
- Role of system science to inform policy

Chronic (non-communicable) diseases (CDs), which include cardiovascular disease, diabetes, cancers and chronic respiratory disease, are responsible for over 60% of deaths globally. In fact, these conditions are responsible for the majority of all deaths in all regions of the world, with the exception of sub-Saharan Africa, where infectious disease deaths still predominate (World Health Organization, 2011a, 2011b). There are four major behavioral risk factors for CDs: smoking, excessive alcohol consumption, unhealthy diet, and physical inactivity (World Health Organization, 2011a, 2011b). These behaviors are related to increased levels of biological risk factors, including overweight and obesity, abnormal lipid levels (fats) in the blood, raised blood pressure, and raised blood glucose (sugar). These risks remain major public health challenges in highly developed countries, where obesity, physical inactivity and diabetes in particular continue to increase. However, it is in developing countries where the greatest rise in these risk factors and associated rates of CDs over the next 20 years will occur (World Health Organization, 2005).

These rising levels of physical inactivity, overweight and obesity, and related CDs, seen in virtually all countries of the world, rich and poor, have to date proved largely intractable to interventions intended to halt never mind reverse them. This, arguably, is the major public health challenge facing the vast majority of countries globally. It is increasingly acknowledged that the underlying determinants of these risk factors are embedded in the social and economic organization of modern societies (Stuckler & Siegel, 2011). In recognition of this, the United Nations High Level Meeting (UNHLM) on CDs in September 2011 highlighted

 DOI: 10.4324/9781315748771-7

the potential importance of multi-sectoral government actions, particularly using legislative and fiscal policies, for reducing CD risk (World Health Organization, 2011a, 2011b). However, with the exception of tobacco control, evidence on effectiveness to guide how and where to intervene is largely lacking.

Theoretical frameworks and models can be helpful in identifying intervention strategies and their potential impact (Davies, Roderick, & Raftery, 2003). In this regard, systems science offers advantages over other multilevel or ecological frameworks (Committee on an Evidence Framework for Obesity Prevention Decision Making Institute of Medicine, 2010). In particular, systems science acknowledges that the behavior of a system producing a particular outcome, such as increasing obesity, is often unpredictable, non-linear, and complex, with feedback loops and time delays, and cannot be understood simply as the sum of its component parts (Kohl et al., 2012; Sterman, 2006). This complex behavior of systems can mean that well-intentioned interventions, even those based on good evidence from controlled studies, can have unintended consequences when applied in the real world. A simple example of this is that an intervention that successfully increases physical activity at school may not lead to any overall increase in activity due to reductions at other times (Kohl et al., 2012).

If systems thinking and modeling is to positively contribute to public health policy it will need to be used by policy makers. While there is no single best approach to supporting policy makers to use research evidence, such as from systems science and modeling, there is a consensus that approaches that enable policy makers and researchers to work together can be some of the most effective in helping to translate evidence into policy (Bowman et al., 2012; Hanney, Gonzalez-Block, Buxton, & Kogan, 2003). Such approaches should facilitate the generation of policy options that are recognized by the policy makers as locally appropriate, implementable, and effective (Bowman et al., 2012).

System dynamics modeling (SDM), which describes the behavior of systems in terms of stocks, flows and feedback loops (Meadows, 2008; Sterman, 2006), is well suited to the task of engaging with policy makers. A recently published study (Martinez-Moyano & Richardson, 2013) highlights the stated importance in the field of SDM to stakeholder involvement. The authors identified six stages in the system dynamics modeling process, and for each stage identified what constitutes best practice. Some key findings of this study are highlighted in Table 6.1. The common theme that runs across all the stages of the modeling process is explicitly involving and addressing the needs of the "problem owners" (i.e. clients, such as policy makers). Indeed, under the 5th stage, "model use, implementation and dissemination," best practice includes making sure that "the entire modeling process revolves around the problems of concern of the audience (problem owner, client)" (from Table 10 in Martinez-Moyano & Richardson, 2013). While this emphasis on involving the client is hardly unique to SDM, it is a recognized strength of SDM that it provides a well-established methodology for doing so (Willis, Mitton, Gordon, & Best, 2012). The natural starting point for engaging with policy makers is working with them to identify and define the problem and to conceptualize the system generating it (stages 1 and 2). Group model building is an established methodology developed within SDM to do this (Vennix, 1999; Willis et al., 2012).

The use of systems thinking, and SDM in particular, to inform public policy goes back many years, with Forester's study on the impact of social housing on urban environments in the United States, published in 1969, being a seminal example (Forrester, 1969). Another example is the application over many decades of SDM to the impact of human activity on the environment, and a detailed review of models applied in this area was written over 30 years ago (Meadows and Robinson, 1985). The application of SDM to public health problems, particularly to CDs, is much more recent and the subject of this chapter. We have two main aims. The first is to

Table 6.1 Stages of the System Dynamics Modelling Process, with selected examples of best practice

Stage of modeling process	*Example of best practice*
Problem identification and definition	Listen carefully and reflectively to clients ('problem's owners'), clarify purpose of modeling
System conceptualization	Generate a dialogue with clients that focused on their mental models and dynamic hypotheses
Model formulation	Develop the structure through a series of simple to more comprehensive models, adding detail as needed to improve realism and show policy impacts
Model testing and evaluation	Compare simulated behavior with real behavior (data) using statistical measures of pattern fit
Model use, implementation and dissemination	Make sure that the entire modeling process revolves around the needs of the client
Design of learning strategy/ infrastructure	Use simplified causal loop diagrams to communicate how the system works rather than relying on the model to tell its own story
Design of learning strategy/ infrastructure	Use simplified causal loop diagrams to communicate how the system works rather than relying on the model to tell its own story

(Adapted from Martinez-Moyanoa and Richardson, 2013)

review applications of SDM to guiding policy interventions designed to lower prevalence of the following chronic disease risk factors: smoking, obesity, unhealthy diet, physical inactivity and excess alcohol consumption. As part of this review we describe in detail four illustrative case studies. Our second aim is to draw lessons from the studies reviewed and to suggest a pragmatic research agenda for the further application of SDM to guiding policy on the prevention of chronic diseases.

Methods

Literature Search Strategy

The literature search for this chapter was performed using the ProQuest Dialog online search engine (as provided by the Royal Society of Medicine Library) which enables simultaneous searching of three databases: DH-DATA (Health Administration, Medical Toxicology & Environmental Health), Embase and MEDLINE. The search was done without time-cut-offs and limited to the languages Danish, Dutch, English, French, German, Icelandic, Norwegian, and Swedish. The search terms and flow is shown in Figure 6.1. In addition to the chronic disease risk factors listed in the aims, we included the term "diabetes" because of the strong association between type 2 diabetes and obesity. In addition, we also searched Web of Science using the same search terms, as potentially relevant journals such as the "System Dynamics Review," and "Systems Research and Behavioral Science" are indexed here but not in ProQuest Dialog. No additional relevant articles were found. Our final check was to search using Google and the string "system dynamics, policy, public health, diabetes." This resulted in about 280,000 hits. A manual assessment of the first 300 listed hits resulted in 157 relevant references of which only 16 were not already included in the main search, none of the additional 16 were relevant to our topic. The final search of the aforementioned databases was performed on March 16, 2014, and it is the results from this that are shown in Figure 6.1.

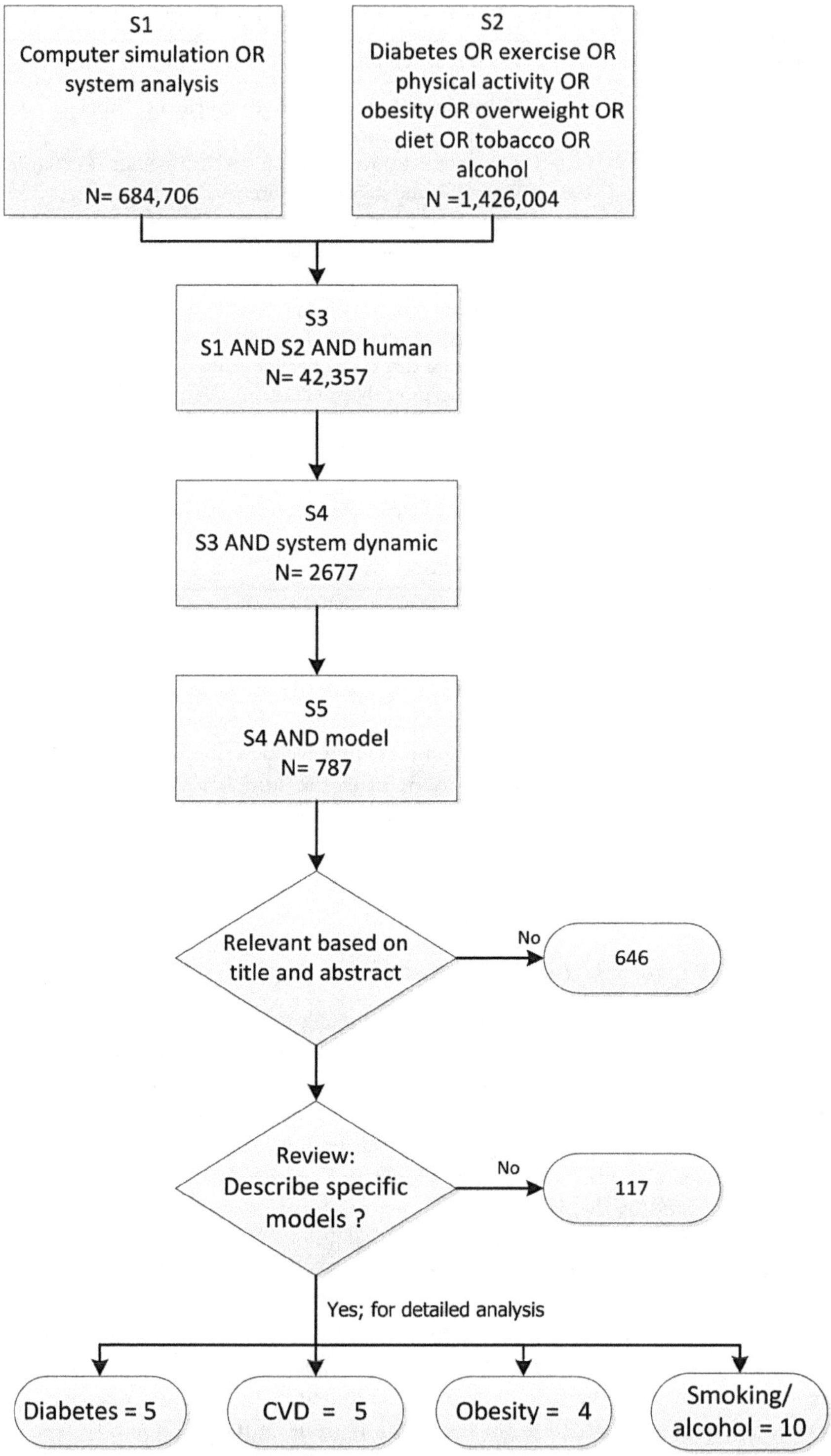

Figure 6.1 Summary of the literature search strategy and findings

Working Definition of System Dynamics Modeling

It was clear after an initial review of the citations identified by the literature search that the term "system dynamics" is not consistently used, and identifies a heterogeneous range of modeling approaches. This partly reflects the fact that within the data bases searched there is no indexing term for "system dynamics modelling." It was essential therefore to examine each article against a set of explicit criteria to determine if it described a "system dynamics model." In drawing up these criteria we aimed to be consistent with the definition used by the Institute of Medicine (Committee on an Evidence Framework for Obesity Prevention Decision Making Institute of Medicine, 2010) that system dynamics is "a methodology for mapping and modeling the forces of change in a complex system in order to better understand their interaction and govern the direction of the system" (p. 74). More specifically, we were guided by the description of the Systems Dynamics Society, who refer to the "system dynamics approach" (System Dynamics Society, 2014). Core to the approach is aiming for an understanding of the characteristics of the system that generates or exacerbates the problem being investigated, and fundamental to this is defining what should be included in the system (i.e. where its boundaries lie).

System boundaries may be defined by addressing the question "What are the smallest number of components that generate the dynamic behavior (problem) being investigated?" A complementary question is, "Are all actions that are essential to the behavior included in the system?" (Richardson, 2011). Thus, system dynamics aims to define a bounded system that of itself, through its structure, generates the behavior of interest. This is referred to as the "endogenous point of view." Exogenous factors, those outside the boundaries of the system, may disturb it but are not required to generate the behavior of interest (System Dynamics Society, 2014).

The bounded system is conceptualized and modeled as being made up of stocks, flows, and causal feedback loops between the stocks, and it is this structure that is responsible for the dynamic, complex, system behavior. "Stocks" represent accumulations within the system, and could be people, information, materials, depending on the dynamic behavior being modeled. "Flows" are movements into and out of stocks. "Feedback loops" are flows out of and back into stocks, causally affecting the levels of those stocks. Feedback loops may be reinforcing (positive) or balancing (negative), and are described in detail elsewhere in this handbook.

In line with the above description of systems dynamics modeling, we needed to see clear descriptions of stocks, flows and feedback loops within the models in order to include them in this review. Applying this last criterion in particular excluded several studies that were described by the authors as "systems models," but were typically Markov models, with transition probabilities from one state (or "stock") to another but no feedback loops within the system.

Criteria Used to Guide our Review of the Models Identified and Choice of Case Studies

This chapter is not intended to be a comprehensive and representative review of all the literature on this topic but rather we aimed to present an overview and highlight concrete examples of system dynamics models developed by researchers and policy makers in collaboration to address real world application in the field of CD prevention. Unfortunately, many of the reviewed papers did not provide the detail necessary to trace the evolution of models, their development in participatory collaboration, or application to shape policy making. The selected case studies, however, do give some indication for the utility of system dynamics in engaging both researchers and stakeholders in systems thinking in order to tackle complex issues of CD prevention.

Findings

As shown in Figure 6.1, we identified 787 citations that were then reviewed independently by two of the authors who read the title and abstract. Of these articles, 141 were found to be potentially relevant and a copy of the full article was retrieved. Figure 6.2 shows the publication of these 141 articles by 5-year periods, illustrating a steep increase in the number of publications since the year 2000. Citations tended to be concentrated in a relatively small number of journals, with 7 journals having 5 or more publications each. The *American Journal of Public Health* led the way, with 13 publications, followed by *Tobacco Control* (8), *Diabetes Care* (7), and *Preventive Medicine* (6). *PLOS One*, *Medical Decision Making* and the *British Medical Journal* each contributed 5 citations.

After full text review of the 141 potentially relevant articles, we identified 24 that describe specific models and thus met our criteria for detailed review. The articles identified were grouped into four categories (Figure 6.1): diabetes, obesity, smoking/alcohol, and cardiovascular disease (CVD). From these studies we picked four case studies to illustrate the development and application of system dynamics modeling in particular settings. The New Zealand Tobacco Control Policy model shows how policy makers from the Ministry of Health were closely consulted to develop a SDM to map the possible dynamic changes to smoking from different policy options. Similarly, a SDM on policy alternatives for alcohol control was developed in the 1980 in the United States (US). The third case study, also from the US, the Center for Disease Control and Prevention (CDC) diabetes SDM, illustrates the way in which a careful mapping of the diabetes system can aid stakeholders in setting realistic targets. Finally, we describe the adaptation of a SDM model from one setting to another: a CVD prevention model developed in the US that was adapted to New Zealand. Key characteristics of these four case studies are summarized in Table 6.2.

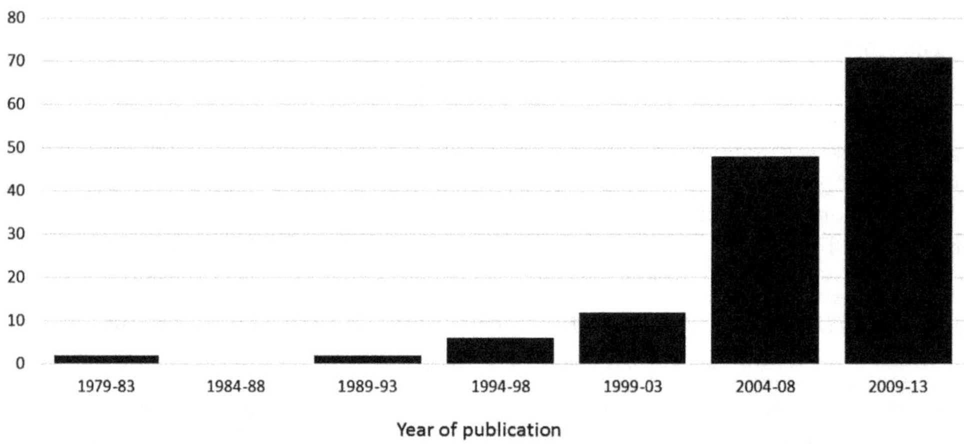

Figure 6.2 Number of potentially relevant citations identified by the search strategy by year of publication

Table 6.2 Summary of the main features of the models and their development described in the four case studies

Model features	*US alcohol control model (Holder & Blose, 1987)*	*New Zealand Tobacco Control Policy model (Cavana & Clifford, 2006)*	*CDC diabetes system modeling project (Jones et al., 2006; Milstein et al., 2007)*	*New Zealand CVD model – application in Counties Manukau (Kenealy et al., 2012)*
Objectives	To develop an integrating structure to explain the interaction between community drinking patterns and various stimulus factors. To use the model to project the potential impact of specific prevention strategies.	To determine the influence of price on the use and consequences of tobacco in New Zealand. To provide a decision support tool when considering strategic policy options in tobacco control.	To understand diabetes population dynamics. To track the rates at which people develop diabetes, are diagnosed with the disease, and die, and assess the effects of various preventive-care interventions.	To determine the impact of interventions from established cardiovascular (CV) risk factors through to CV events and deaths. To assess the usefulness of a national and a local system dynamics model of cardiovascular disease to planning and funding decision makers.
Systems conceptualization and model development	Model building was informed by the theories and results of published research in alcohol epidemiology, prevention and control strategies.	Developed as a consequence of a series of model-building workshops attended by policy makers from Ministry of Health, New Zealand (NZ) and New Zealand Customs Service (NZCS).	Built in iterative, cooperative process by health planners and researchers in learning labs and workshops.	Developed as a national system dynamics model of CVD causation and potential interventions by the NZ Ministry of Health in consultation with Homer, built on the CDC CVD system modelling project (Homer, et al., 2008, 2010). Adapted to local county context in iterative process through individual consultation and workshops with health managers, practitioners and indigenous health representatives.

Table 6.2 Continued

Model features	*US alcohol control model (Holder & Blose, 1987)*	*New Zealand Tobacco Control Policy model (Cavana & Clifford, 2006)*	*CDC diabetes system modeling project (Jones et al., 2006; Milstein et al., 2007)*	*New Zealand CVD model – application in Counties Manukau (Kenealy et al., 2012)*
Types of data	Data from published research, National Alcohol Drinking Surveys (1964–1979).	Main input variables such as smoking initiation and intensity were derived from New Zealand, Tobacco Use Survey, Mortality Hazard ratios were taken from NZ Census-Mortality Study and demographic data from Statistics NZ.	Main input variables such as population growth and death rates, health insurance coverage, obesity, prediabetes and diabetes prevalence, diabetes detection, obesity prevalence, glucose self-monitoring, eye and foot examinations: US Census Bureau, National Health Interview Survey, National Health and Nutrition Examination Survey, Behavioral Risk Factor Surveillance System, and research literature.	In national NZ model, some use of the US estimates of quality and use of primary care services (Behavioral Risk Factor Surveillance Survey). National NZ data substituted by local data, in particular for population stock sizes, prevalence and incidence rates; taken from local evaluations of Interventions (sometimes involving reanalysis of data).
Model testing and evaluation	Model was tested at the national, not community level due to lack of availability of appropriate data. Authors tested for significant differences between mean yielded by the model and the mean based on national consumption patterns and found no significant difference.	The qualitative development of model was subject to several iterations. The final quantitative model was verified by checking that the base case reproduced current prevalence (2001–2004) and validated against five different morbidity and mortality metrics.	Testing and second iteration of model in consultation with stakeholders. Model calibration using historical data for US adult population.	Only reported that all models produced a close visual fit to historical data points for diabetes, obesity, smoking and reproduced the Census population projection.

Table 6.2 Continued

Model features	*US alcohol control model (Holder & Blose, 1987)*	*New Zealand Tobacco Control Policy model (Cavana & Clifford, 2006)*	*CDC diabetes system modeling project (Jones et al., 2006; Milstein et al., 2007)*	*New Zealand CVD model – application in Counties Manukau (Kenealy et al., 2012)*
Model use and implementation	Used in three counties of the US to explore the impact of prevention strategies (public education, increased retail price) on alcohol consumption.	Model was used to predict changes in smoking prevalence; tobacco-attributable mortality; under two main tobacco cessation policy frameworks.	Authors report policy change in diabetes program planning in Vermont, Minnesota, California, Alabama, Tennessee and Florida.	Counties Manukau adaptation assessed for usefulness for local policy makers.
Policy insights and impact	Simulation results were generally similar across counties but there was sufficient variation to suggest prevention efforts may need to be modified by community. Second generation of this model was used in San Diego, California by local decision-makers to plan a drinking and driving prevention program.	Results were used during the 2007 annual budget process to provide supporting evidence for additional investment in tobacco cessation services in New Zealand. Authors opine that the application of this model has served to increase awareness among policymakers of the dynamics of tobacco control.	"Reality check": prevalence reduction targets shown to be unrealistic due to "backing up" phenomenon (improved diabetes treatment leads to higher diabetes prevalence "stock").	Accepted that the main value of the outputs of the model was the ability to explore the relative magnitude of changes that could be brought about by interventions (effect of smoking cessation underestimated and effect of PA and nutritional interventions overestimated). Some use in making case for long term finance – more use on national level predicted.

Guiding Policy on Smoking and Tobacco Control

From as early as the 1970s and 80s, health systems researchers have been using the system dynamics approach to model the effects of policy on various risk factors and health-related outcomes. One of the first uses of system dynamics models we found was in the evaluation and prediction of the impact of tobacco-related policies. We found articles demonstrating the use of system dynamics modeling for smoking-related policies developed in the 1970s and published from as early as 1982 (Roberts, Homer, Kasabian, & Varrell, 1982). The authors reported this model as a simplified and preliminary one which served to test the efficacy of certain policy initiatives in reducing smoking-related mortality and morbidity. The model would not be considered best practice today since much of the data inputs (according to the authors) were developed by reasonable guesses and intuition rather than empirical data. It does, however, provide a significant foundation for the successful use of system dynamics model in tackling chronic diseases in public health.

Subsequently, system dynamics models have been used in examining smoking-related policies in New Zealand (Tobias, Cavana, & Bloomfield, 2010) and the United States (US) (Ahmad & Billimek, 2007; Ahmad & Franz, 2008; Tengs, Ahmad, Moore, & Gage, 2004; Tengs, Osgood, & Chen, 2001). The Tobacco Policy model developed by Tengs et al. (Tengs et al., 2004) has been used to predict changes in smoking prevalence and smoking-related outcomes for a variety of policy scenarios (Tengs, Ahmad, Savage, Moore, & Gage, 2005; Tengs, Osgood, & Lin, 2001). One such scenario involved an answer to the question: "If cigarettes could be made safer, and if policy makers were to mandate their safety, then what might be the public health implications?" The authors were able to use their model to determine that even if a legislative mandate to improve the safety (e.g. by removing certain toxins from the smoke) of cigarettes made cigarette use more attractive (thus increasing smoking prevalence), the reduction in risk from safer cigarettes will still result in net public health gain. The authors ended by inviting policy makers to use these results to facilitate their decision making on federal policy mandating safer cigarette production.

The New Zealand Customs Service worked with the Ministry of Health to develop a system dynamics model that analyzed public policy issues related to the collection of tobacco excise duties (Tobias et al., 2010). The New Zealand Tobacco Policy Model was developed during collaboration between academics and policy advisers, and provided users with reliable estimates of the health effects of smoking cessation intervention programs. These estimates were used in the New Zealand's Ministry of Health annual budget process to provide evidence for investment in cessation services in New Zealand. This model is described in more detail in the case study below.

Other simulation models have also been used in tobacco-related policy development, the most prominent of which is the Simsmoke model developed by Levy et al. (Levy, de Almeida, & Szklo, 2012; Levy, Boyle, & Abrams, 2012; Levy & Friend, 2001), which has been used in several states in the US (Arizona, California, Minnesota and Kentucky) and in countries such as Brazil, Korea and Italy. In addition, Simsmoke has recently been used to evaluate the global impact of adopting the MPOWER (Tobacco Free Initiative) tobacco control policies in different countries and territories from 2007 to 2010. The Simsmoke model provides simulations based largely on Markov processes which describe methods whereby the components of the model (e.g. people) are in a set of states (e.g. diseased or not diseased) and the probability distributions governing their transitioning through these states are used to predict the distribution of their outcomes. The fact that the probability of transition to a state is dependent only on the current state, and is completely independent of previous states is a distinctive feature of Markovian-based simulations. These simulations do not incorporate explicit feedback loops, a feature we have considered essential for the definition of a system dynamics model, but it has proven itself a useful tool in area of tobacco-related policy modeling.

Case Study: New Zealand Tobacco Control Policy

Of the models examined in our review, the Tobacco Control Policy model developed by Cavana and Tobias (Cavana & Clifford 2006; Cavana & Tobias, 2008; Tobias et al., 2010) for use by the Ministry of Health in New Zealand is a good example of the application of system dynamics modeling for one of public health's more vexing problems. The main aim in the development of this model was to assist the Ministry of Health (MOH) in evaluating the dynamic consequences of tobacco control policies in New Zealand (NZ).

The model-building exercise appears to have been commissioned by the New Zealand Customs Service (NZCS) as well as the MOH in New Zealand and these were the main participants in the model-building workshops. These workshops were organized to provide an opportunity for the clients to describe and elucidate the problem; great consideration appears to have been given to listening to and adequately understanding the client's perspective. The main workshops were preceded by training sessions in causal loop diagrams and there were several iterations in the workshops to decide on the organizing question ultimately used in model building.

After much refinement the conceptual framework for the system dynamics simulation model which was used in the tobacco policy model for the MOH is presented in Figure 6.3. The model consists of four sectors: population; smoking prevalence; second-hand smoke; and tobacco-attributable deaths arranged in the characteristic stocks, flows and feedback loops of a systems dynamics model.

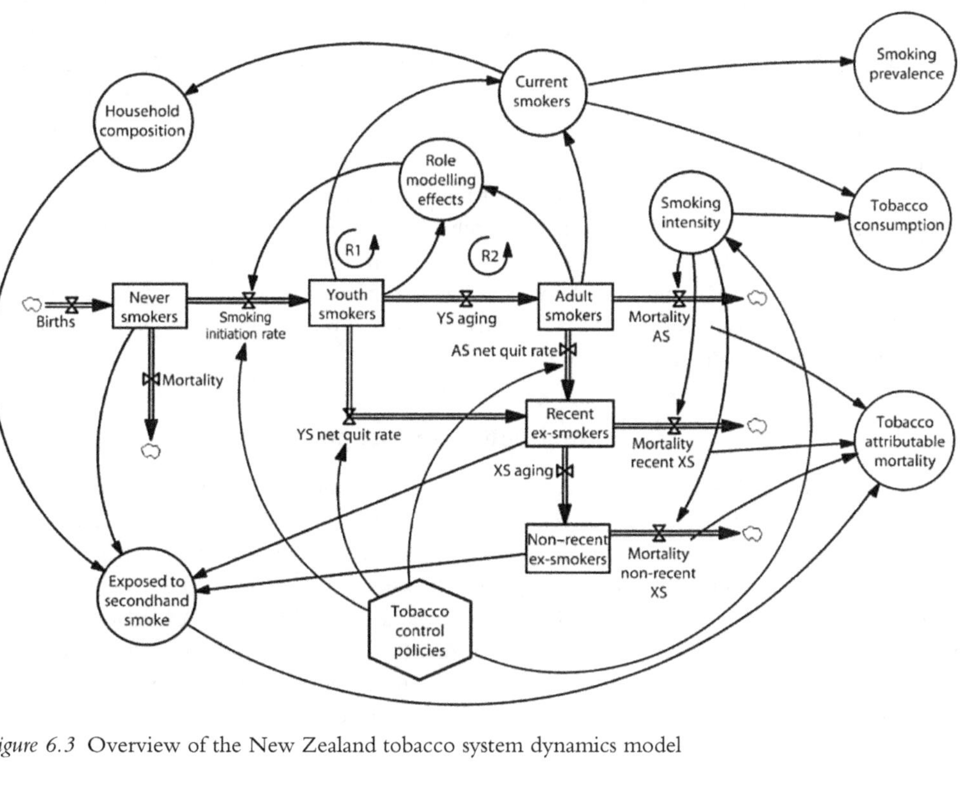

Figure 6.3 Overview of the New Zealand tobacco system dynamics model

Model developers used smoking initiation rates derived from the 2006 New Zealand Tobacco Use Survey (NZTUS) to produce the stock titled "youth smokers." Using the net quit rate provided by the NZTUS for "youth smokers," a batch of this population will flow into the stock of recent ex-smokers, while the remainder will age to become "adult smokers." The stock of "adult smokers" may die in this state, become recent and later "non-recent smokers" before death ensues. The volume of the stock and rate of flow between states is influenced by various factors illustrated in the model such as tobacco control policies which may act to decrease the smoking initiation rates while increasing youth and adult net quit rates, and smoking intensity acting to influence mortality on current and ex-smokers alike. The presence of youth and adult smokers in a household through "role modeling" may act to increase smoking initiation rates and increase smoking prevalence via reinforcing causal feedback loops (R1 and R2 respectively). Having developed the model, its structure was discussed extensively with subject matter experts in New Zealand post-model development, allowing them an opportunity to revise the structure if necessary.

The authors ensured that the model behavior was consistent with available reference data. For example, the model was verified by checking that the base case assumptions matched known prevalence data and that the population projections generated agreed with national population projections published by New Zealand's national statistics department.

Having presented the model in the published literature in its developmental stage, the authors invited policy makers to frame policy questions and assess suitability of the model for exploring them, and to obtain the necessary empirical data to run the model and agree plausible ranges for sensitivity analysis.

To date, the NZ Tobacco Policy model has been used to examine the effects of changes in excise duties on tobacco smoking, and to inform government decision making by estimating the long-term health effects of enhanced smoking cessation interventions used in tobacco control. Using data extracted from the 2006 New Zealand Tobacco Use Survey, the NZ Census-mortality study and demographic data from Statistics New Zealand the authors derived estimates of population cessation rates from smoking behaviors and applied these over a 50-year period under business as usual and enhanced cessation intervention scenarios. The main output variables were smoking prevalence, tobacco consumption and tobacco-attributable mortality. The authors noted in their conclusion that the model generated reliable estimates of the effects of interventions designed to enhance smoking cessation on health and on tobacco use. Most significantly, the results informed a decision announced in May 2007 to increase funding for smoking cessation by NZ$42 million over four years. It is this real-life application of the NZ model that distinguishes it as an exemplary model.

Guiding Policy on Alcohol

As with smoking, SDMs have been used for at least two decades to make predictions about the potential impact of changes in alcohol-related policies. A model developed by Holder and Blose (1987) was used to assess the impact of a set of prevention strategies on the level of drinking-related family disruptions and alcohol-related work problems in three counties of the United States using data from 1970 to 1984. The outcome variables in this model are related to injury prevention which is now classified as a chronic disease. The model follows some of the best practices of system dynamics model development, including what appears to be a group process in system conceptualization, adequate model testing and good use of the model to compare expected outcomes from several policy scenarios.

A little over twenty years later, Scribner et al. (2009) published a more complex system dynamics model which they used to indicate what combination of simulated interventions targeting heavy episodic drinkers at a moderately "dry" campus would extinguish heavy episodic drinkers, replacing them with light and moderate drinkers. The article provides a good example of the steps in the dynamic modeling process with perhaps the only significant omission being the involvement of stakeholders such as persons involved in the development of college campus policy.

Case Study: An Example of System Dynamics Modeling of Alcohol Control from Earlier Years

The model developed by Holder and Blose (1987) provides a good foundational study of the use of system dynamics modeling in public health in earlier years. The authors clearly articulated their understanding of the importance of system thinking in implementation and evaluation of public health policies. They sought to develop "an integrating structure or model" to which they would add mathematical specifications that could represent the interactions and feedback processes occurring in the real world. Their main stated aim was to:

> incorporate as many system factors and relationships as appropriate into the model, to test the model against national consumption data and to use the validated model to make predictions about potential changes in selected alcohol-related problems in three test communities in the United States.
>
> *(Holder & Blose, 1987, p. 125)*

Two of the early first steps in model development today are problem identification and system conceptualization. In this study, they used a comprehensive review of the literature to inform these components (Holder & Blose, 1987). The model was informed by the theories and results of published research in alcohol epidemiology, prevention and control strategies. For example:

- The basic epidemiology (i.e. age and gender distribution) of consumption was informed by results of national surveys.
- The stimuli for changes in consumption were developed based on work by Jessor and Jessor, 1975, and Johnson et al. published in 1977.
- Based on the work of Hoadley et al., Levy and Sheflin and Ornstein, the authors included in the model an equation to model the inverse relationship between price and consumption.
- Maisto and Rachal in 1980 reported that areas with higher minimum purchase ages for alcohol have lower per capita consumption and their findings were utilized in model development.

Using these and other findings the authors identified the main problems in this area of research, and conceptualized a model to inform the planned implementation of alcohol prevention strategies in the United States.

Four types of prevention strategies were modeled: raising the retail price of all alcoholic beverages by 25% once, indexing the price of all beverages to the Consumer Price Index (CPI) each year; reducing high-risk alcohol consumption through public education and raising the minimum drinking age to 21 for all alcoholic beverages. Having developed the graphical and mathematical components of the model, it was tested against the national level data for consumption rates for the time period of interest. They found that the model estimates were not significantly different from survey values obtained for the same period.

The model was then loaded with local data from three counties (Wake County, North Carolina, Washington County, Vermont, Alameda County, California) which differed significantly in their alcoholic beverage control legislation. The impact of the aforementioned prevention strategies was explored via computer simulation for each county model. The developers looked at the impact of these strategies on alcohol-related family disruptions and alcohol-related work problems. They found that in the absence of any intervention, both family disruptions and work problems would increase over the simulation period and that these increases were related to the projected economic growth of each county. The authors clearly described development of the model from identification to conceptualization, model formulation and simulation. The authors did not mention any direct communication with stakeholders in their problem description and they did not mention whether the model impacted actual policies in these counties. Despite these omissions, the model process clearly illustrates system dynamics modeling and its application in prevention of alcohol related harm. The main omission, at least in what is reported, is the direct involvement of policy makers in either developing the model or in evaluating its outputs.

Guiding Policy on Diet, Physical Activity, and Obesity

A range of computer modeling has been used to understand the complexity of diet and nutrition, and to a lesser extent physical activity, but few attempts of system dynamics modeling have yet been published in this area. Conner and Levine (2006) developed a system dynamics conceptual framework with causal loops within the agricultural and food system. Their framework aims to inform strategies such as policy interventions towards more sustainable community-based food systems that can counteract obesity and farmland loss but has not been taken to a full SDM.

With a focus on obesity prevention or reduction, however, several SDMs have been developed. Abdel-Hamid (2003) used SDM to model the impact of energy metabolism on body weight. The model focuses on internal processes such as hormonal regulation, body composition such as fat mass and energy metabolism. By modeling exercise regimes in comparison to nutritional weight-loss interventions they experimented with the varied interactions between different intensity levels of exercise and different nutrient compositions of diets (e.g. high carbohydrate over balanced diet). Although the model could highlight the complexity of these interactions, with mixed results, researchers and practitioners could not reach consensus over a most favored weight-loss strategy.

Flatt (2004) also developed a SDM on the metabolic system in a two-compartment model (one compartment being the energy store of glycogen and the other of fat) that included personal factors such as aspects of lifestyle and genetic makeup and "environmental" factors such as food availability, diversity, and palatability. Similarly to Abdel-Hamid (2003), Flatt found that adiposity can be maintained under a diverse range of conditions and also highlighted that genetic metabolic differences (altering the competition between glucose and fat oxidation) may affect the development of obesity.

Madahian and colleagues' (2012) SDM moved beyond a model of metabolic subsystems and included environmental and behavioral factors such as sweet beverage preference to understand the impact of diet and physical activity on childhood obesity. The model was used to simulate a reduction in body mass index in several intervention strategies and found the same effect of 10 minutes of exercise in combination of 100 calories less energy intake as of 30 minutes exercise a day regime. No stakeholder involvement or group model building was described for any of the above SDMs.

The most complex and comprehensive attempt that we are aware of to map the system of diet, physical activity and environmental, psychological, social and physiological factors that underlie obesity can be found in the Obesity System Map developed by the UK's Government Foresight Programme (Butland et al., 2007). The causal loop diagrams were developed through a detailed process of stakeholder involvement. However, a quantitative system dynamics model was not developed, and so this work does not fit the criteria that we used to select studies for this review. The quantitative model that was developed for this work was a microsimulation model, which simulated levels of obesity and associated health outcomes to the year 2050 under different assumptions of trends in obesity prevalence. It is relevant to note that this approach to modeling was partially justified by the study investigators on the grounds that the "systems mapping work clearly demonstrated that the determinants of obesity were too complex" to be able to reliably model the effect of changing them (McPherson, Marsh, & Brown, 2007, p. 1). A separate exercise in this large program of work undertook "qualitative" modeling, in which stakeholders were encouraged to use the causal loop diagrams to think through and compare the potential impact of different policy options (Chipperfield et al., 2007). However, there is no evidence presented on whether this exercise actually led to any changes in policy.

Guiding Policy on Diabetes

There are several diabetes computer simulation models such as the Cardiff Diabetes Model, the Sheffield Diabetes Model, the U.K. Prospective Diabetes Study (UKPDS) Model, the Economic Assessment of Glycemic Control and Long-term Effects (EAGLE) Model, the Center for Outcomes Research (CORE) Diabetes Model, the Archimedes Model, the Diabetes Decision Analysis and Cost of Type 2 (DiDACT) Model, and the CDC/RTI (Centers for Disease Control/Research Triangle Institute) Type 2 Diabetes Progression Model (Mount Hood 4 Modeling Group, 2007). However, these models do not use system dynamics but other computer simulation approaches, mostly of a Markov model type.

SDMs for diabetes have been developed relatively recently. Applications range from modeling the impact of current and emerging technologies on diabetes and its complications and co-morbidities, in the US (Edwards et al., 2009), and estimating need and cost of eye-care services including diabetic retinopathy and screening frequency of diabetes in Finland (Tuulonen, Salminen, Linna, & Perkola, 2009). An SDM developed by Shabestari and Roudsari (in the UK, 2009) simulates the cost-benefit of web-based diabetes education, accounting for factors such as education and technology refusal. Considering increased incidence the model indicated that web-based patient education and continuous e-learning, although more expensive in its set up than traditional educational approaches, could lead to a marked slowing in the rise of diabetes treatment costs. There is no description of stakeholder involvement in problem definition, model development or model evaluation in any of these publications.

Case Study: The Centre for Disease Control (CDC) Diabetes System Modeling Project (US)

We chose the diabetes model developed within the CDC diabetes system modeling project to illustrate the application of systems dynamics on understanding diabetes population dynamics (Jones et al., 2006; Milstein et al., 2007). Stocks in this model represent various subpopulation groups, moving to various stages of the disease from healthy to prediabetes, to diabetes without and then with complications. Diagnosed and undiagnosed diabetes is also noted. Modifiable influences of diabetes control include clinical management, self-management, lifestyle and

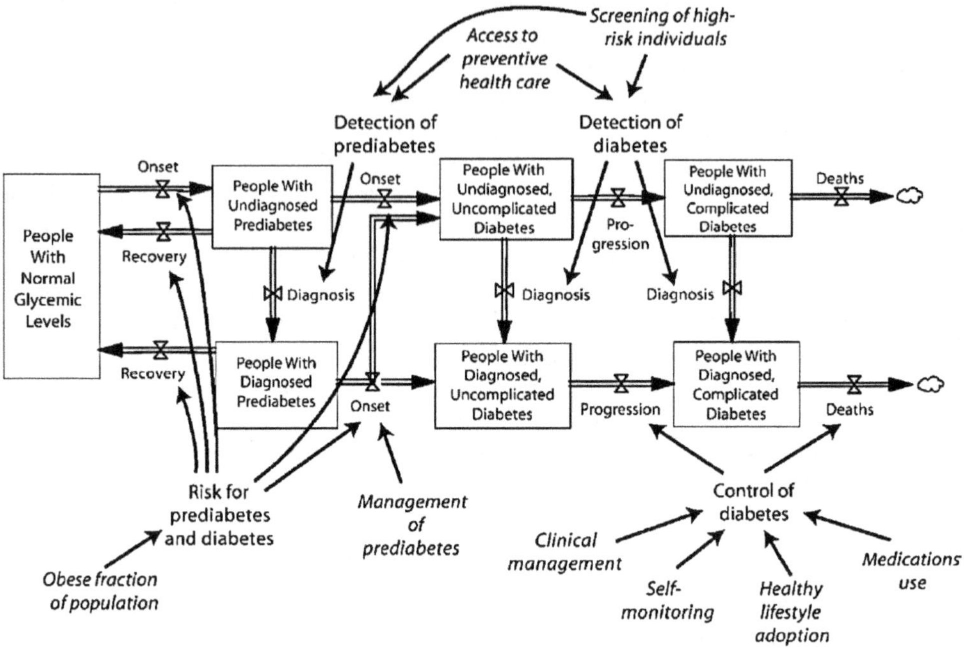

Figure 6.4 Model overview of the CDC (US) diabetes system dynamics model

medication use. Other factors included in the system are obesity prevalence and risk. The model was built in an "iterative, cooperative, continuous process" by health planners and researchers in the Division of Diabetes Translation and others from the CDC National Center for Chronic Disease Prevention and Health Promotion, according to both policy needs and scientific knowledge (Jones et al., 2006). A more detailed description of the model development stage is not published but outlined in conference proceedings. They used "learning labs" and a series of workshops with CDC staff with a specialty in diabetes prevention and control, as well as nutrition and physical activity, smoking, cardiovascular health, adolescent and school health, adult and community health, and chronic disease in general to conceptualise the model and interact with simulation model. Later they engaged with stakeholders to test and further develop a second iteration of the model.

The model was first used to simulate three policy strategies or scenarios: enhanced clinical management of diabetes, increased management of prediabetes, and thus prevention of diabetes, and reduced obesity prevalence (Jones et al., 2006). Figure 6.4 is an overview of the model structure. One of several important insights from this model was the "backing-up" phenomenon. This refers to reduced outflow from a stock that may mean its level changes little or even increases. For example, more successful treatment of diabetes will lead to lower rate of progression to complicated cases and death, thus tending to increase the stock of people with diabetes – i.e. diabetes prevalence. However, while lack of decreased diabetes prevalence may disappoint a policy maker keen to achieve such a target, clearly lower complication rates in people with diabetes is a positive health outcome. Another important insight from this model is that of all the potential interventions examined (i.e. interventions of those areas in italics in Figure 6.4) only reducing the prevalence of obesity would lead in the long term to a reduced prevalence of diabetes and death rates from diabetes. In all the other scenarios examined prevalence and mortality would continue to rise until 2050.

Showing that targets for diagnosed prevalence reduction were highly unrealistic, the model could serve as a "reality check" in policy application (Milstein et al., 2007). Milstein and colleagues (2007) report that subsequently the model has been used to affect policy change by diabetes program planners in Vermont who have worked with members of the CDC Diabetes System Modeling team to set consistent and achievable objectives for diabetes-related outcomes according to estimated diabetes prevalence trajectories; and Milstein also refers to its use by health planners in Minnesota, California, Alabama, Tennessee, and Florida are currently exploring similar uses.

Guiding Policy on Cardiovascular Disease

For Cardiovascular Disease (CVD), system dynamic models have been used in similar ways to their use in diabetes. The El Paso County Model (Texas, US) (Hirsch, Homer, Evans, & Zielinski, 2010; US) was developed to inform planning for CVD prevention, early detection and treatment. Stocks represent healthy to diseased population groups. Strategies that can be evaluated within the model include those related to lifestyle and environment, and medical and mental health care – such as taxation policies, improvements to access to healthy diets and physical activity as well as smoking cessation, weight loss, and improved primary care. The model also captures population changes in smoking, obesity and hypertension. The simulations indicated that lifestyle and environmental interventions alone could significantly reduce first-time CVD events, deaths and consequence costs.

Another US CVD simulation model was developed with CDC and experts at Austin/Travis County, Texas (US) by Homer and colleagues (2008, 2010). As with the CDC diabetes system modeling project, the aim was to map the most significant contributors to the problem and develop a computer simulation model that is able to compare alternative policy scenarios and intervention strategies. Local experts were involved to ascertain features of the local context such as linking stress to varied accessibility to mental health services (Homer et al., 2008). The model is populated with major risk factors such as diet, physical activity, smoking, obesity, high cholesterol, hypertension and diabetes, as well as utilization of primary care, air pollution, anti-smoking social marketing, junk food and tobacco taxes, to name a few. The model is able to indicate which interventions can reduce CVD-related deaths rapidly or have a more gradual impact on costs of related risk factors (Homer et al., 2010). Policy makers in Austin, Texas, have subsequently used the model to affect policy decisions and engage other stakeholders with systems thinking (Loyo et al., 2013).

Case Study: The New Zealand CVD Model – Application in Counties Manukau

We picked a final case study of system dynamics modeling that illustrates in what way an initial model can be adapted to other national and local settings. Building on the Austin/Travis County model (Homer et al., 2008, 2010) and in consultation with Homer, the New Zealand Ministry of Health developed a national system dynamics model of CVD causation, and potential interventions in 2008/2009 (Kenealy et al., 2012). The New Zealand model follows the Austin/Travis County model with population stocks of no previous CVD event (e.g. heart attack or stroke) flowing to post CVD event. The population, however, is divided into the major ethnic groups of Maori, Pacific, and European/Other to be able to evaluate health disparities. Causal links are established with risk factors such as obesity prevalence, smoking prevalence and uncontrolled chronic disease prevalence, as well as factors such as poor diet, physical inactivity, taxes and regulations, health promotion social marketing and quality of care. The New Zealand

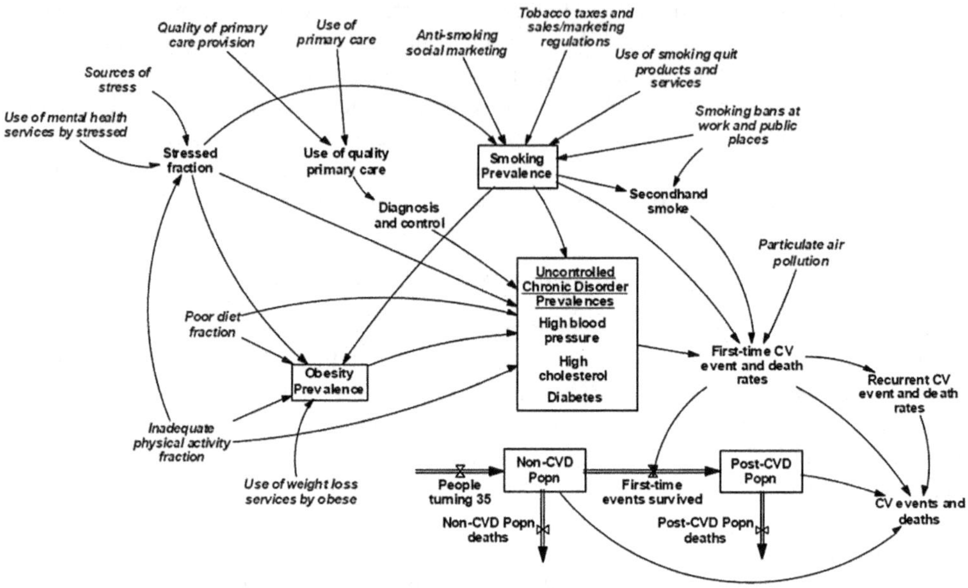

Figure 6.5 Overview of the New Zealand CVD system dynamics model

model does not aim to model the financial cost of medical interventions but focuses on health outcomes.

Kenealy and colleagues (2012) describe the adaption of this New Zealand model for the Counties Manukau population and assessed for its usefulness for local decision makers such as health managers, practitioners and indigenous health representatives. In three phases, relevant stakeholders were identified, then using an iterative process the stakeholders explored and adapted the model (mainly to account for ethnic and gender variations in the population) using local data. Finally, the stakeholders were asked to evaluate the usefulness of the process. The researchers asked if the model enabled the stakeholders to:

- make a difference to decisions;
- confirm current decisions;
- improve confidence with decisions made;
- support advocacy for a decision, or
- facilitate the process of decision making within and across different disciplines or interest groups.

Clinicians and policy makers rated the model as very valuable and saw its future potential but they also highlighted many shortcomings, in particular regarding its local level use. They appreciated the use of local data but suggested further refinements to the model – for example, including "invisible" migrant populations that the current model does not capture. They would have also appreciated greater detail about each intervention relating to their particular clinical or policy work. A health cost component, which was taken out of the original CDC model from which the New Zealand model was adapted, would have also aided their planning and funding decision-making. More generally, stakeholders were concerned about the uncertainties of the model predictions – for example, a lack of empirical data on the interaction effect of variables, and other shortcoming of the modeling such as turning continuous risk factors into binary variable (exercising or not, hypertension or not). Overall, they saw greater potential in the model for national-level funding decisions rather than planning smaller local interventions, and also noted that the complexity of the model compromised its user-friendliness. Rather than using detailed model outputs, stakeholders noted that predictions of the current model could serve them to qualitatively compare decision options. The authors highlighted that the model development was an ongoing process and that refinements have been made since this study with stakeholders had been concluded; however, no further use of the model is published as yet.

Challenges, Strengths and Limitations of Systems Dynamics Modeling to Date in Guiding Policy on Chronic Disease Prevention

In theory at least SDM has some clear strengths in helping to assist policy makers in evaluating the potential impact of different policy scenarios. It has a well-established methodology for group model building, can incorporate broad system boundaries and inclusion of a wide range of variables, model over long time horizons and is relatively accessible, with modern software, to those without a strong mathematical background (Willis et al., 2012).

One of the surprises, therefore, in carrying out the work for this chapter was the small literature that we found on the application of SDM to risk factors for chronic diseases. This is despite high-profile publications, such as in the *American Journal of Public Health* (e.g. J. D. Sterman, 2006), organizations, such as the Institute of Medicine (e.g. Committee on an Evidence Framework for Obesity Prevention Decision Making Institute of Medicine, 2010) and funding bodies, such as National Institutes of Health (e.g. Office of Behavioral and Social Sciences Research), extolling the potential advantages of systems thinking, including specifically SDM, for addressing chronic disease prevention and other health issues. Of the limited number of studies that we did, we found that the vast majority were from the US, with New Zealand being the next best represented. To date, therefore, there is limited published experience with SDM to inform policy making for chronic disease prevention.

The literature we did find, however, provided good insights into the potential uses of SDM for informing policy on the prevention and control of chronic diseases. The work of Holder and Blose (1987) on alcohol-related harm illustrated the use of SDM in first describing the dynamic, worsening, nature of a public health problem and then showing the potential impact of different types of interventions to help alleviate the problem. The CDC diabetes model (Jones et al., 2006) also nicely illustrates modeling to show how without intervention the problem will progress and the relative impact of different policies designed to reduce its impact. In addition, this model can be seen as having provided a "reality check" for policy makers. Polices aimed at improved care of people with prediabetes and diabetes, while clearly having health benefits, were shown not to be capable of reducing the increase in diabetes prevalence or diabetes related mortality over the next 30–40 years. Reduction in diabetes prevalence and mortality could only be achieved by dramatically reducing the prevalence of obesity. In addition to the implications for policy, this exercise also shows how the model can help policy makers to set realistic targets. It is reported by the authors of the CDC diabetes model that it has been used by policy makers in different US states to help set realistic targets, although unfortunately we were not able to find any literature that provided further detail on this.

The clearest examples of directly involving policy makers in model building or use were found in the two case studies from New Zealand. The New Zealand Tobacco policy model (Cavana & Clifford 2006; Cavana & Tobias, 2008; Tobias et al., 2010), for example, was used by policy makers to provide plausible estimates of the effect of smoking cessation policies. These estimates contributed directly to the decision to substantially increase funding for smoking cessation services. The New Zealand CVD model was based on an American model (Kenealy et al., 2012) and illustrates a process of adaptation that actively involved health care workers, patients representatives and decision makers. This is the only case study in which we found a description of stakeholder feedback on the utility of the model and the modeling process. It is noteworthy that stakeholders were concerned about uncertainties in the model predictions, but nonetheless thought that the outputs could help them to qualitatively compare different decision options.

Given the increasing prominence on the potential for government policy measures to assist in chronic disease prevention, we were particularly interested to learn if and how policy makers had been involved in the modelling process, and whether it proved useful in guiding policy formulation. Disappointingly, the vast majority of studies that we reviewed either did not involve policy makers/stakeholders in model development, or failed to report that they did. The two best descriptions of policy-maker involvement in full, quantitative SDMs were in the two case studies from New Zealand, described in the preceding paragraph. Had we also included studies that were limited to qualitative system dynamics modeling (e.g. produced causal loop diagrams but not quantitative models), then we may have found more evidence of stakeholder involvement. The UK Foresight project on obesity (Butland et al., 2007), for example, worked very closely with stakeholders in deriving the obesity system causal loop diagrams. It is argued that involving stakeholders in deriving causal loop diagrams can offer them "policy insight," leading to better informed policy making, without the need to develop a quantitative model (Lane, 2012).

In conclusion, despite the existence of good examples of SDM being applied to inform policy on the prevention of chronic diseases, its use has been limited and its overall impact on improving health unclear. In fact, this is the case for all forms of dynamic modeling and simulation methods and their application to population health problems (whether SDM, agent-based modeling, discrete event modeling, etc.): they are seen as having huge potential to help solve public health problems arising within complex systems, but to date their use has been relatively sparse (Maglio, Sepulveda, & Mabry, 2014). From the perspective of using SDM, and similar techniques, to engage with policy makers we agree with Maglio and colleagues (2014) that a major challenge remains demonstrating their value to decision makers. Doing this would be assisted by studies that better described how decision makers were involved in the modeling process, evaluations of how the process could be made more useful to them, and longer term follow-up of what substantive changes in policy actually resulted. Part of this work will need to address the tension, noted in the New Zealand CVD case study, between producing models that are accessible enough to enable stakeholders to understand and ask questions of them, and detailed enough to adequately capture the behavior of interest. A second challenge is increasing the exposure to modeling and simulation methods in the core training of the broad range of professionals who need to work together to improve population health (Maglio et al., 2014). These include not only public health and other health professionals, but urban planners, government civil servants, and business leaders, among others. Not only will this increase the receptivity of these professionals to the use of dynamic modeling for informing solutions to population health problems, it will also improve their ability to work together. Interdisciplinary collaboration is crucial for effective problem solving in public health.

In this chapter, we have focused on SDM. However, in closing we acknowledge that systems thinking in public health cannot be encompassed by a single discipline or even a single "systems thinking" approach; rather, it requires an interdisciplinary integration of approaches aimed at understanding and reconciling linear and nonlinear, qualitative and quantitative, and reductionist and holistic thinking (National Cancer Institute, 2007).

References

Abdel-Hamid, T. K. (2003). Exercise and diet in obesity treatment: an integrative system dynamics perspective. *Medicine & Science in Sports & Exercise, 35*(3), 400–413.

Ahmad, S., & Billimek, J. (2007). Limiting youth access to tobacco: comparing the long-term health impacts of increasing cigarette excise taxes and raising the legal smoking age to 21 in the United States. *Health Policy, 80*(3), 378–391.

Ahmad, S., & Franz, G. A. (2008). Raising taxes to reduce smoking prevalence in the US: a simulation of the anticipated health and economic impacts. *Public Health, 122*(1), 3–10.

Bowman, S., Unwin, N., Critchley, J., Capewell, S., Husseini, A., Maziak, W., et al. (2012). Use of evidence to support healthy public policy: a policy effectiveness feasibility loop. *Bulletin of the World Health Organization, 90*, 847–853.

Butland, B., Jebb, S., Kopelman, P., McPherson, K., Thomas, SE., Mardell, J., et al. (2007). *Foresight. Tackling Obesities: Future Choices. Project.* London: Goverment Office for Science.

Cavana, R. Y., & Clifford, L. V. (2006). Demonstrating the utility of system dynamics for public policy analysis in New Zealand: the case of excise tax policy on tobacco. *System Dynamics Review, 22*(4), 321–348.

Cavana, R. Y., & Tobias, M. (2008). Integrative system dynamics: analysis of policy options for tobacco control in New Zealand. *Systems Research and Behavioral Science, 25*(5), 675–694.

Chipperfield, T., O'Brian, R., Bolderson, T., Eidinow, E., Shafner, L., & Butland, B. (2007). *Foresight. Tackling Obesities: Future Choices—Qualitative Modelling of Policy Options.* London: Goverment Office for Science.

Committee on an Evidence Framework for Obesity Prevention Decision Making Institute of Medicine. (2010). *Bridging the Evidence Gap in Obesity Prevention: A Framework to Inform Decision Making.* The National Academies Press.

Conner, D. S., & Levine, R. (2006). Circles of association: the connections of community-based food systems. *Journal of Hunger and Environmental Nutrition, 1*(3), 5–25.

Davies, R., Roderick, P., & Raftery, J. (2003). The evaluation of disease prevention and treatment using simulation models. *European Journal of Operational Research, 150*(1), 53–66.

Edwards, R. A., Graham, A., Williams, M., Amati, M., Wright, C., Lee, M., et al. (2009). Quantifying a strategic view of diabetes technology impacts: A system dynamics approach. *Diabetes Technology and Therapeutics, 11*(7), 411–418.

Flatt, J. P. (2004). Carbohydrate-fat interactions and obesity examined by a two-compartment computer model. *Obesity Research, 12*(12), 2013–2022.

Forrester, J. W. (1969). *Urban Dynamics.* Portland, OR: Productivity Press.

Hanney, S., Gonzalez-Block, M., Buxton, M., & Kogan, M. (2003). The utilisation of health research in policy-making: concepts, examples and methods of assessment. *Health Research Policy and Systems, 1*(1), 2.

Hirsch, G., Homer, J., Evans, E., & Zielinski, A. (2010). A system dynamics model for planning cardiovascular disease interventions. *American Journal of Public Health, 100*(4), 616–622.

Holder, H. D., & Blose, J. O. (1987). Reduction of community alcohol problems: computer simulation experiments in three counties. *Journal of Studies on Alcohol, 48*(2), 124–135.

Homer, J., Milstein, B., Wile, K., Pratibhu, P., Farris, R., & Orenstein, D. R. (2008). Modeling the local dynamics of cardiovascular health: risk factors, context, and capacity. *Preventing Chronic Disease, 5*(2), A63.

Homer, J., Milstein, B., Wile, K., Trogdon, J., Huang, P., Labarthe, D., et al. (2010). Simulating and evaluating local interventions to improve cardiovascular health. *Preventing Chronic Disease,* 7(1), A18.

Jones, A. P., Homer, J. B., Murphy, D. L., Essien, J. D., Milstein, B., & Seville, D. A. (2006). Understanding diabetes population dynamics through simulation modeling and experimentation. *American Journal of Public Health, 96*(3), 488–494.

Kenealy, T., Rees, D., Sheridan, N., Moffitt, A., Tibby, S., & Homer, J. (2012). A "whole of system" approach to compare options for CVD interventions in Counties Manukau. *Australian and New Zealand Journal of Public Health, 36*(3), 263–268.

Kohl, H. W., Craig, C. L., Lambert, E. V., Inoue, S., Alkandari, J. R., Leetongin, G., et al. (2012). The pandemic of physical inactivity: global action for public health. *The Lancet, 380*(9838), 294–305.

Lane, D. C. (2012). What is a "policy insight"? *Systems Research and Behavioral Science, 29* (6), 590–595.

Levy, D., de Almeida, L. M., & Szklo, A. (2012). The Brazil SimSmoke Policy Simulation Model: the effect of strong tobacco control policies on smoking prevalence and smoking-attributable deaths in a middle income nation. *PLOS Medicine, 9*(11).

Levy, D. T., Boyle, R. G., & Abrams, D. B. (2012). The role of public policies in reducing smoking: the Minnesota SimSmoke tobacco policy model. *American Journal of Preventive Medicine, 43*(5 Suppl. 3), S179–186.

Levy, D. T., & Friend, K. (2001). A computer simulation model of mass media interventions directed at tobacco use. *Preventive Medicine, 32*(3), 284–294.

Loyo, H. K., Batcher, C., Wile, K., Huang, P., Orenstein, D., & Milstein, B. (2013). From model to action: using a system dynamics model of chronic disease risks to align community action. *Health Promotion Practice, 14*(1), 53–61.

Madahian, B., Klesges Robert, C., Klesges, L., & Homayouni, R. (2012). System dynamics modeling of childhood obesity. *BMC Bioinformatics, 13*(Suppl+12), A13.

Maglio, P. P., Sepulveda, M.-J., & Mabry, P. L. (2014). Mainstreaming modeling and simulation to accelerate public health innovation. *American Journal of Public Health, 104*(7), 1181–1186.

Martinez-Moyano, I. J., & Richardson, G. P. (2013). Best practices in system dynamics modeling. *System Dynamics Review*, 102–123.

McPherson, K., Marsh, T., & Brown, M. (2007). *Foresight. Tackling Obesities: Future Choices – Modelling Future Trends in Obesity and their Impact on Health.* London: Government Office for Science.

Meadows, D. H. (2008). *Thinking in Systems: A Primer.* White River Junction, VT: Chelsea Green Publishing Company.

Meadows, D.H., & Robinson, J.M. (1985). *The Electronic Oracle: Computer Models and Social Decisions.* Chichester: John Wiley & Sons.

Milstein, B., Jones, A., Homer, J. B., Murphy, D., Essien, J., & Seville, D. (2007). Charting plausible futures for diabetes prevalence in the United States: a role for system dynamics simulation modeling. *Preventing Chronic Disease, 4*(3).

Mount Hood 4 Modeling Group. (2007). Computer modeling of diabetes and its complications: a report on the Fourth Mount Hood Challenge Meeting. *Diabetes Care, 30*(6), 1638–1646.

National Cancer Institute. (2007). *Greater Than the Sum: Systems Thinking in Tobacco Control* (Other): National Cancer Institute. Office of Behavioral and Social Sciences Research. Systems Science. Retrieved February 1, 2014, from: http://obssr.od.nih.gov/scientific_areas/methodology/systems_science/index.aspx

Richardson, G. P. (2011). Reflections on the foundations of system dynamics. *System Dynamics Review, 27*(3), 219–243.

Roberts, E. B., Homer, J., Kasabian, A., & Varrell, M. (1982). A systems view of the smoking problem: perspective and limitations of the role of science in decision-making. *International Journal of Bio-Medical Computing, 13*(1), 69–86.

Scribner, R., Ackleh, A. S., Fitzpatrick, B. G., Jacquez, G., Thibodeaux, J. J., Rommel, R., et al. (2009). A systems approach to college drinking: development of a deterministic model for testing alcohol control policies. *Journal of Studies on Alcohol and Drugs, 70*(5), 805–821.

Shabestari, O., & Roudsari, A. (2009). Potential Return on Investment (RoI) on web-based diabetes education in UK. *Studies in Health Technology and Informatics, 143*, 258–263.

Sterman, J. D. (2006). Learning from evidence in a complex world. *American Journal of Public Health, 96*(3), 505–514.

Stuckler, D., & Siegel, K. (Eds.). (2011). *Sick Societies: Responding to the Global Challenge of Chronic Disease.* Oxford: Oxford University Press.

System Dynamics Society. (2014). Introduction to system dynamics. Retrieved January 31, 2014, from: www.systemdynamics.org/what-is-s/

Tengs, T. O., Ahmad, S., Moore, R., & Gage, E. (2004). Federal policy mandating safer cigarettes: a hypothetical simulation of the anticipated population health gains or losses. *Journal of Policy Analysis and Management, 23*(4), 857–872.

Tengs, T. O., Ahmad, S., Savage, J. M., Moore, R., & Gage, E. (2005). The AMA proposal to mandate nicotine reduction in cigarettes: a simulation of the population health impacts. *Preventive Medicine, 40*(2), 170–180.

Tengs, T. O., Osgood, N. D., & Chen, L. L. (2001). The cost-effectiveness of intensive national school-based anti-tobacco education: results from the Tobacco Policy Model. *Preventive Medicine, 33*(6), 558–570.

Tengs, T. O., Osgood, N. D., & Lin, T. H. (2001). Public health impact of changes in smoking behavior: results from the Tobacco Policy Model. *Medical Care, 39*(10), 1131–1141.

Tobacco Free Initiative. MPOWER. Retrieved February 1, 2014, from: www.who.int/tobacco/mpower/en/

Tobias, M. I., Cavana, R. Y., & Bloomfield, A. (2010). Application of a system dynamics model to inform investment in smoking cessation services in New Zealand. *American Journal of Public Health, 100*(7), 1274–1281.

Tuulonen, A., Salminen, H., Linna, M., & Perkola, M. (2009). The need and total cost of Finnish eyecare services: a simulation model for 2005–2040. *Acta Ophthalmologica, 87*(8), 820–829.

Vennix, J. A. M. (1999). Group model-building: tackling messy problems. *System Dynamics Review, 15*(4), 379–401.

Willis, C. D., Mitton, C., Gordon, J., & Best, A. (2012). System tools for system change. *BMJ Quality and Safety, 21*(3), 250–262.

World Health Organization. (2005). Preventing chronic diseases: a vital investment: WHO global report. Geneva: World Health Organization.

World Health Organization. (2011a). Global status report on noncommunicable diseases 2010: description of the global burden of NCDs, their risk factors and determinants. Geneva: World Health Organization.

World Health Organization. (2011b). United Nations high-level meeting on noncommunicable disease prevention and control. Retrieved January 25, 2014, from: www.who.int/nmh/events/un_ncd_summit2011/en/index.html

7

NETWORK ANALYSIS AND PSYCHOLOGY

Michael Vitevitch

> **Research Question:** How have networks been used in the different branches of psychology?
> **System Science Method(s):** Networks
> **Things to Notice:**
> - Interdisciplinarity of network analysis
> - Networks as a common "language" for otherwise disparate research topics

Psychology is the science that studies human behavior in the mind and brain. The broad scale and domain of investigation, and multiple-levels of analysis in Psychology makes the statistical methods and theoretical framework of network analysis well-suited to address various questions in Psychology. Selected examples of network analysis are highlighted in the domains of Social Psychology, Biological Psychology, Cognitive Psychology, and Clinical Psychology. In the area of Social Psychology, network analysis has been used to examine the spread of gossip, fads, and other information. In the highlighted example, a laboratory-based experiment demonstrated how certain network structures affect how quickly a group finds a solution to either a simple or complex problem. In the area of Biological Psychology, network analysis has been used to explore the rich neuroanatomical connectedness of the brain at the level of individual neurons, and—through the use of diffusion spectrum magnetic resonance imaging—at the level of the whole organ. Cognitive Psychology is the branch of Psychology that studies mental processes including perception, learning, memory, thinking and problem solving, and language. A number of Psycholinguists have applied network analysis to various aspects of language, including phonology and semantics, to better understand language development in children, as well as how adults understand spoken language. Network analysis has been used in Clinical Psychology to re-envision how to define psychopathology. Rather than view mental illness as a collection of symptoms caused by a latent variable, a network can be used to model symptom space, with nodes representing specific symptoms and edges connecting symptoms that co-occur, thereby redefining the comorbidity of and diagnostic boundaries of mental disorders. Other areas of Psychology as well as related fields could also benefit from the theoretical and methodological perspective that the network analysis approach offers. There are, however, several important points to bear in mind when considering the adoption of this approach.

Psychology is the science that studies human behavior in the mind and brain. This simple definition hides the numerous ways in which human behavior is studied, and the various levels at which explanations are provided. The behavior of an individual can be examined with a

DOI: 10.4324/9781315748771-8

developmental approach to determine when certain abilities are acquired (and sometimes lost) over the lifespan. Human behaviors can also be compared to behaviors of non-human animals with a comparative approach to determine when such abilities might have first appeared in the phylogenetic timespan. A computational approach can be used to create a computer program or a mathematical formula to simulate certain behaviors with a wide range of parameter values to further determine the limits of human abilities. These and other approaches can provide scientific explanations at a microscopic or biological level, such as how neurotransmitters and nerve cells work, at the level of the mind of an individual person, such as how our senses are fooled by various perceptual illusions, or at the level of larger groups of people, such as how in the *bystander effect* being surrounded by a large number of people makes it less likely for an individual to receive assistance than if the individual in need is instead surrounded by a smaller number of people (Darley & Latané, 1968).

The wide range of topics that are explored in Psychology, as well as the multiple levels at which explanations are provided makes the discipline of Psychology a fertile field for *network analysis*. Many fields besides Mathematics and Sociology began using network analysis to examine the complex systems that they study after the publication of work by Watts and Strogatz (1998) on *small-world networks* (for brief descriptions of how network analysis has been used in other fields see Watts, 2004, and Barabási, 2009). Small-world networks are highly interconnected (as measured by a relatively high clustering coefficient), and have a small average shortest-path length. Interest in small-world networks increased in part because they provided a much better model of many real-world systems than either the lattices or the random networks typically studied by mathematicians (e.g., Erdös & Rényi, 1959).

The recent work in psychology using network analysis should not be confused with earlier work in certain areas of psychology that have appealed to the metaphor of a network, or that have employed networks of a different type, such as artificial neural networks (e.g., Rosenblatt, 1958) or semantic networks (Quillian, 1967). The remainder of the present chapter will consider several examples of network analysis from various subdisciplines of Psychology. The cases presented here are not intended to provide an exhaustive review of the field, but were instead selected to illustrate the ways that network analysis has been employed in different areas of Psychology to understand human behavior at various levels (e.g., biological, individual, group, etc.).

Although there are large differences in scale in the examples described here—ranging from the cell to society—the use of network analysis enables researchers in various subdisciplines of Psychology to examine the same question in different ways. It is often difficult to integrate or relate findings from different levels of analysis; this is especially true in trying to relate the mind to the brain. However, the use of network analysis across different scales provides researchers with a common language to facilitate communication and the dissemination of important findings.

The selected examples also serve to illustrate the novel insights into a field that network analysis can provide, and in some cases, the novel insights that Psychology can provide to network analysis. By examining how various subdisciplines of Psychology have employed network analysis, researchers in other fields may see similar opportunities to employ network analysis to address their own research questions in their own field of inquiry.

Social Psychology: The Spread of Innovations in Groups

Sociologists, Anthropologists, and Social Psychologists have long used network analysis—where it is often referred to as Social Network Analysis (SNA)—not to examine abstract mathematical principles, but to understand how real people in a group, such as a Karate club, interact with each other (Zachary, 1977). Given the rich history of SNA (Wasserman & Faust, 1994), and the continued use of network analysis to examine the spread of diseases (Barthelemy, Barrat, Pastor-Satorras & Vespignani, 2005), fads (Gronlund & Holme, 2005), and gossip (Lind, da Silva, Andrade & Herrmann, 2007) through social groups, the first case we will consider is an experiment by Mason, Jones, and Goldstone (2008) that examined how the structure of a group influences how quickly the solution to a simple or a complex problem spreads through a group.

Mason et al. performed a set of three experiments, but we will focus on Experiment 2, where 150 undergraduate students were divided into small groups ranging in size from 7 to 18 people (median of 10 people per group) to play a number guessing game on a computer that was connected to a server that controlled the experiment. The participants in the study were asked to guess a number between 0 and 100. A guess that was close to the correct answer received a high number of "points," whereas a guess that was not close to the correct answer received a low number of "points." After a participant made their guess, the computer showed the participant what numbers their "neighbors" in the group guessed, and how many points those guesses earned. Because participants were asked to try to accumulate as many points as possible, seeing the guesses made by their "neighbors" in the group (and the points those guesses received) inspired each participant to make a better guess in the next round of the game. Participants were given 15 rounds of guessing to find the number selected by the computer.

Who the "neighbors" were of each participant (i.e., which guesses and points each person saw) was dependent on the type of network structure that the server running the experiment selected for that group of participants: lattice, fully connected, random, or small-world network. In a lattice, a node is only connected to its immediate neighbors on either side. In a fully connected network each node is connected to every other node in the network. In a randomly connected network, the network had the same number of connections as the other networks, but the connections are place between nodes at random. Finally, in a small-world network most nodes can be reached from every other node by a small number of connections (Watts & Strogatz, 1998). In the case of the fully connected network, a participant would have access to the guesses made by everyone else playing the game, whereas in the remaining cases participants would have access to a limited number of guesses made by other players in the game constrained further by the structure of the network (lattice, random, or small-world).

To examine how network structure influenced the ability of the group to find the solution to problems varying in complexity, Mason et al. had either one number be the correct answer that each group was trying to find (i.e., a simple problem), or three numbers as the correct answers that each group was trying to find (i.e., a complex problem). Performance in the number guessing game was measured in several ways, but we will focus on the fairly intuitive measure called *speed of convergence*, which refers to the number of rounds of guesses that the group members needed to find the right answer. For the simple problem (the correct answer was one number), Mason et al. found that participants found the right answer with the fewest rounds of guessing when each participant had access to the guesses of all of the other members of the group (i.e., the fully connected network performed best). For the complex problem (the correct answer was three numbers), Mason et al. found that participants found the right answer with the fewest rounds of guessing when each participant had access to a limited number of guesses by the other members of the group; specifically when the group was connected like a small-world network.

The results of Mason et al. (2008) are interesting for several reasons. First, small-world networks have received much attention in the literature for their unique topological properties and the ramifications of these properties on network dynamics (e.g., the spread of disease, etc.). However, in the case of a simple problem (the correct answer was one number), the solution to the problem spread to other members in the group quickest when the group was structured like a fully connected network, not a small-world network. Despite the attention that small-world networks have received, the small-world network is not a panacea for all problems. For *simple* problems, the fully connected network may instead be optimal (in terms of how quickly the whole group converges on the solution), showing that different network structures may be better suited for different types of problems.

Second, the results of Mason et al. suggest somewhat counter-intuitively that sometimes it is not good to know everything—sometimes it is better to know less. Specifically, when a group had a complex problem to solve, convergence on the solution by everyone in the group occurred more quickly when the information available to each individual in the group was limited (by the constraints imposed by the small-world network structure). In other words, limiting the information that is available to an individual could actually improve the overall performance of the group. If one considers the results of Mason et al. regarding search by a group for a solution at the level of individual cognition—a person searching through their own memories for a particular fact—there might be important lessons to learn regarding how to best search through one's own memory for information to solve a problem.

Finally, the work of Mason et al. (2008) illustrates one way that the research method often employed in Psychology—controlled, laboratory-based experiments—can contribute to the field of network analysis. Although much has been learned from the mathematical analysis of networks (Bollobas, 2013), and from computer simulations (see Lazer & Friedman, 2005 for an agent-based computer analogue of the work by Mason et al., 2008), the experiment is the only research method that enables a researcher to establish a causal relationship for a given phenomenon. A laboratory-based analogue of a real-world system created in the context of a psychology experiment could enable researchers to explore the influence and interaction of variables of interest in a way that cannot be done in the real-world for practical or ethical reasons. The field of Psychology not only stands to benefit from using network analysis, but network scientists might also benefit from collaborating with researchers in Psychology to examine various questions via experiments.

Biological Psychology: Physical and Functional Networks in the Brain

Recall that Psychology is more than the study of human behavior. A growing segment of the field examines the brain and how neural activity is related to the mental activities we commonly refer to as perceptions, thoughts, and emotions. Invaluable knowledge about which parts of the brain are involved in which cognitive activities has been gained by studying individuals with various types of brain damage. Phineas Gage, a railroad worker who suffered damage to his frontal lobe after an iron used to tamp down explosives struck him in the head, and the patient H.M., who suffered amnesia after surgery to cure him of his epilepsy, are among the most well-known and well-studied individuals in this regard (James & MacKay, 2001; Vanderstoep, Fagerlin & Feenstra, 2000). The uniqueness of such cases, however, limits the ability to generalize the findings more broadly, especially to healthy, intact individuals.

Additional understanding of the relationship between the mind and the brain (in patients and healthy individuals) was obtained with the development of various non-invasive tools used to image the brain and neural activity, such as electroencephalography (EEG), magneto-encephalography (MEG), positron emission tomography (PET), computer axial tomography (CAT-scan), magnetic resonance imaging (MRI), functional magnetic resonance imaging (fMRI), and near infrared spectroscopy (NIRS). For more in-depth reviews of these neuro-imaging and electrophysiological techniques see Bandettini (2009), Luck (2005), and Hernandez,

Wager, and Jonides (2002). These techniques have been useful for identifying localized regions of the brain that are involved in a particular cognitive process. However, more than just a single, localized region of the brain is activated when we experience a given perception, thought, or emotion. Rather, networks of brain regions contribute to our cognitive experience, making network analysis an important tool for understanding how the human brain works.

With the establishment of the Human Connectome Project, a consortium of several research centers, *connectomics* has emerged as the field that seeks to understand the human brain by making a detailed mapping of neural pathways, sort of a neuronal wiring diagram of the brain (Lichtman & Sanes, 2008). The neuron-to-neuron connectivity pattern of the nematode *C. elegans* was mapped in the 1980s (White, Southgate, Thomson & Brenner, 1986), producing a network that contained 302 neurons as nodes and over 7,000 synapses as edges. Current efforts are under way to map the neuron-to-neuron connectivity pattern of humans (Seung, 2013). Given that some estimates suggest that the human brain contains about 86 billion neurons (Herculano-Houzel, 2009), a neuronal wiring diagram of the human brain is an ambitious project.

Although the scale of the human brain makes mapping it a daunting enterprise, progress is being made. Indeed, a branch of connectomics known as *projectomics* is mapping not neuron-to-neuron connections, but the connections that exist between regions of the brain to understand how those brain regions interact to give us our cognitive experiences.

In the projectomics approach a specific region of the brain serves as a node, and edges represent connections between those brain regions. To determine which brain regions are connected a relatively new form of magnetic resonance imaging known as Diffusion Tensor Imaging (DTI) is used to trace tracts of white matter in the brain (Bandettini, 2009; Basser, Mattiello & LeBihan, 1994). White matter in the brain consists primarily of a type of nerve cell called glia, and myelinated axons. Because water diffuses along tracts of white matter faster than it does perpendicularly to the tracts, DTI produces images of structures in the brain that could not previously be imaged noninvasively. More important to the current discussion, the images of white matter tracts indicate the directionality of connections between brain regions, enabling researchers to map the networks that connect various regions of the brain.

There are important concerns to keep in mind when deciding what constitutes a node and what constitutes a connection (Butts, 2009). However, a number of studies of brain networks using a variety of definitions of nodes and edges, different imaging hardware and imaging protocols, and a variety of participants have found a common set of network features to describe how the human brain is structured, including hubs and communities. Network *hubs* refer to nodes that are connected to a large number of other nodes in the network. Supposed network hubs include the precuneus, anterior and posterior cingulate cortex, insula, and portions of the superior frontal, temporal and lateral parietal cortex (Power, Schlaggar, Lessov-Schlaggar & Petersen, 2013; Sporns, 2014).

Communities refer to subsets of nodes in a network that tend to be more highly connected to other nodes in the same community than to nodes in other communities or elsewhere in the network (Newman & Girvan, 2004). Communities of nodes in the brain have emerged that correspond to portions of the brain involved in visual processing, and in moving the body and sensing those body movements (Sporns, 2014).

Neuroscientists were aware of such processing regions in the brain prior to the network analyses of the brain, but confirming the existence of such regions via network analyses opens up the nervous system to the principles and tools of network analysis (see also Changizi & Destefano, 2009), which could lead to interesting new discoveries or resolve long-standing debates in the field. One such issue relates to the extent that brain functions are localized to certain regions of the brain, versus distributed throughout the brain (Uttal, 2002). Sporns (2013) suggests that the hubs and communities found in the brain provides a unique structure that allows for both localized and distributed processing to co-exist.

Another long-debated issue in Psychology and related fields like Philosophy relates to the nature of consciousness: can sensations and feelings be reduced simply to neuro-chemical activity in the brain (Churchland, 1981)? Work by Puschmann, Weerda, Klump and Thiel (2013) suggests an interesting alternative. Perhaps it is not simple neuro-chemical activity in the brain, but temporally synchronized neuro-chemical activity in networks in the brain that produce conscious awareness. Puschmann et al. examined a perceptual phenomenon known as change deafness (Vitevitch, 2003), in which listeners fail to detect what might be described as obvious changes to sounds in the environment (see Levin & Simons (1997) for the visual analogue known as change blindness). Puschmann et al. presented to listeners two successively presented complex auditory scenes that consisted of six auditory streams. Listeners were asked to indicate if the two auditory scenes were the same, or if one stream changed between the two scenes.

Changes in the auditory scene—whether listeners noticed the change or not—produced activation in the auditory cortex (a region in the temporal lobe) of the brain. When listeners indicated they heard a change in the auditory scene—even if there wasn't a change—activation was observed in the insula and anterior cingulate cortex. Most interesting, Puschmann et al. observed that only correctly perceived changes in the auditory scenes resulted in the integration of activation in auditory cortex and the insula. The integration of activation in functionally specialized areas of the brain is known as effective connectivity (Friston, 2011). The demonstration of effective connectivity by Puschmann et al. in the successful detection of auditory changes suggests that temporally synchronized networks in the brain may play a role in producing the much-debated construct known as consciousness.

Network analysis of the physical connectivity that exists among regions of the brain, and of the functional connectivity among regions of the brain that are simultaneously activated during a task has the potential to lead to significant advances in our understanding of normal brain functioning, and may serve to bridge the large divide between the mind and the brain. These techniques also hold much promise for increasing our understanding of the brain functioning that accompanies various brain disorders (Menon, 2011; Simpson, DuBois-Bowman & Laurienti, 2013). Indeed, Wang et al. (2010) observed that the functional network in the motor cortex (part of the frontal lobe in the brain, which controls intentional movements of the body) undergoes reorganization during recovery from a stroke, perhaps providing an opportunity for therapeutic intervention.

Cognitive Psychology: Language and Network Analysis

Cognitive Psychology is the branch of Psychology that studies mental processes including perception, learning, memory, thinking, problem solving, and language. Many of these mental processes are studied in human and non-human animals, but it has been argued that humans alone use language to communicate (Scott-Phillips & Blythe, 2013); although other animals do have sophisticated forms of communication, those methods are not considered to be a language. Given the unique status of language in humans, we will focus our discussion of Cognitive Psychology on the mental processes associated with language, including the comprehension, production, and acquisition of language.

The American Speech-Language-Hearing Association (ASHA, 1982) defined language as "a complex and dynamic system of conventional symbols that is used in various modes for thought and communication." Furthermore, language "evolves within specific historical, social, and cultural contexts" (ASHA, 1982). As a complex system, language is amenable to network analysis.

The changes that take place in language—either the long-term evolution of language across generations, or the short-term changes seen in the development of an individual learning a language—also make language suitable for analysis with network analysis techniques, especially those techniques used to examine network growth. Indeed, researchers in a number of disciplines have recognized these characteristics of language, and have used network analysis to examine various aspects of language, as indicated by the works listed on a web-based bibliography of research on linguistic, cognitive, and brain networks: www.lsi.upc.edu/~rferrericancho/linguistic_and_cognitive_networks.html.

Language further "is described by at least five parameters—phonologic, morphologic, syntactic, semantic, and pragmatic" (ASHA, 1982). *Phonology* refers to the sounds that are used in a given language. For example, the African language !Kung uses a variety of click-like sounds—like the /!/[1] in its name—in the words of the language, but English does not employ such sounds to form words. See note 1 at the end of the chapter for the URL of a sound file of a Linguist producing the phoneme /!/.

Morphology examines the smallest grammatical units of a language. For example, the word "untied" contains three grammatical units: the prefix *un-* (meaning *not*), the root word *tie*, and the suffix *-d* (or *-ed*), which is used to form the past tense of the verb. *Syntax* refers to the rules that form acceptable sentences in a language. For example, in the English sentence "He went to the store," the words are in an order that forms a meaningful English sentence. However, "To went the store he," contains the same words, but not in an order than creates a meaningful sentence in English.

Semantics refers to the meanings conveyed by words, phrases and sentences, whereas *pragmatics* refers to the meaning conveyed by context. For example the sentence, "He saw the man with binoculars" could mean: (a) a person looked through a pair of binoculars at a man in the distance, or (b) a person saw a man who had a pair of binoculars in his hand. The true meaning of that sentence depends on the context in which it is spoken, either while someone is looking through an optical device, meaning (a), or while attempting to distinguish one man from other men who are carrying telescopes, microscopes, periscopes, etc., meaning (b).

Looking at the phonological parameter of language, Vitevitch (2008) constructed a network containing approximately 20,000 English words as nodes. A connection was placed between nodes if the addition, deletion, or substitution of a phoneme turned one word into the other. This method of assessing similarity between strings of symbols is more broadly known as *Hamming* or *Levenshtein distance*, and has been widely used in Psycholinguistics (Greenberg & Jenkins, 1966; Landauer & Streeter, 1973). Thus, the node for the word "cat" would be connected to the node for the word "scat" (a one-phoneme addition), the node for the word "_at" (a one-phoneme deletion), and the node for the words "fat," "cot," and "cab" (one-phoneme substitutions indicated by underlining). Figure 7.1 shows a small portion of this network; not all words and connections are shown.

Analysis of the whole network revealed several noteworthy characteristics about the structure of the mental lexicon (Vitevitch, 2008; see also Arbesman, Strogatz & Vitevitch, 2010). First, less than 50% of the nodes in the phonological network were connected to each other in the large, interconnected group of nodes known as the *giant component*. There were also many smaller groups of words that were connected to each other, but not to the giant component (known in the network analysis literature simply as *components*), and a large number of words that were not connected to any other word (nodes that are not connected to anything are known as *isolates*). This finding is striking because in networks in other domains (e.g., social networks) it has been observed that the giant component contains 80–90% of the nodes in the system (Newman, 2001).

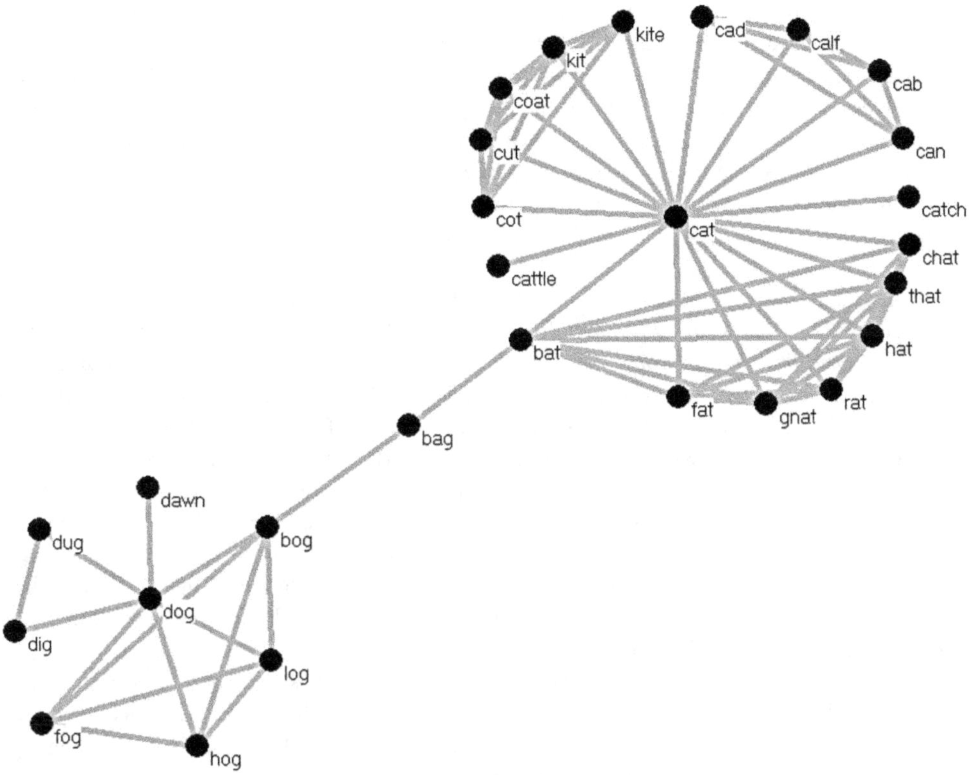

Figure 7.1 A small portion of the network examined in Vitevitch (2008). Each word is, of course, connected to other words in the lexicon, but only a few words are displayed here for the sake of image clarity. Edges are placed between nodes if one sound in a word can be changed to form the other word.

Second, the giant component exhibited small-world characteristics: a "short" average shortest path length and a high clustering coefficient (Watts & Strogatz, 1998). In addition to modeling many real-world systems better than lattices or random networks, computer simulations have shown that the structure of small-world networks facilitates rapid and efficient searches through them (Kleinberg, 2000). Interestingly, rapid and efficient search of the mental lexicon to retrieve the meaning of a word from memory is also a hallmark of language processing.

A third interesting characteristic that was observed in the network of phonological words was *assortative mixing by degree*. The mixing pattern of a network refers to a general tendency for how nodes in a network connect to each other. In a social network, mixing might be defined based on the age of the entities, resulting in the observation that people in the network tend to have as friends people that are of a similar age.

Mixing, however, can be defined on a variety of other characteristics, including the number of connections that a node has (i.e., degree). If nodes with many connections tend to connect to other nodes that also have many connections—there is an overall positive correlation in the degree of connected nodes—the network is said to exhibit assortative mixing by degree. If nodes with many connections tend to connect to other nodes that have few connections—there is an overall negative correlation in the degree of connected nodes—the network is said to exhibit disassortative mixing by degree. If there is no correlation in the degree of connected nodes, then there is no observable mixing pattern (Newman, 2002).

Although assortative mixing by degree has been observed in social networks (demonstrated by computing Pearson's *r* over the whole network for the degree of nodes that are connected) with values of *r* ranging from .1 to .3, the phonological networks examined by Vitevitch (2008) and by Arbesman et al. (2010) exhibited much higher values—as high as.7. Given the role that assortative mixing by degree plays in maintaining connectivity in a network when nodes are removed from the network (Newman, 2002), the higher levels of assortative mixing by degree in the phonological networks may account for the robustness of the language faculty despite the damage to the brain caused by a stroke or other types of trauma (Vitevitch, Chan & Goldstein, 2014).

Finally, the network of English examined by Vitevitch (2008) and the networks of languages examined by Arbesman et al. (2010) exhibited degree distributions that deviated from a power-law. The *degree distribution* represents the number of nodes that have a given number of connections in the network. Networks with degree distributions that follow a power-law are known as *scale-free* networks (Albert & Barabási, 1999; see Fox Keller, 2005 for additional discussion of scale-free networks). Scale-free networks have attracted much attention because of certain structural and dynamic properties, such as remaining relatively intact in the face of random failures in the system, but vulnerability when attacks are targeted at well-connected nodes (Albert, Jeong & Barabási, 2000). Although the phonological networks are not scale-free, they nevertheless are structured in such a way that affords rapid and efficient search, as well as robust connectivity in the face of damage, perhaps hinting to an alterative way of structuring an organization to achieve both rapid and robust processing.

The use of network analysis to examine language (e.g., Arbesman et al., 2010; Vitevitch, 2008) revealed a unique configuration of network structures that has not been observed in other domains (e.g., social, economical, biological, technological). The efficiency of language processing—given how often humans speak, speech errors are relatively rare—hints at alternative network structures that yield efficient and robust dynamics that might prove useful in other domains, highlighting how Psychology might contribute more broadly to the field of network analysis.

Just as research in Psychology might contribute more broadly to the field of network analysis, certain network measures might prove especially useful in increasing our understanding of normal and disordered language processing. Consider the work of Borgatti (2006) on keyplayers, or nodes in a network that, when removed, result in the network fracturing into several smaller components. In Figure 7.1, if the word "bag" (and its connections) were removed from the network, two smaller components would be obtained: "dog" and the words connected to it, and "cat" and the words connected to it, with no way to get from "dog" to "cat."

Vitevitch and Goldstein (2014) extracted a set of 25 words that held such "key" positions in the larger phonological network studied by Vitevitch (2008), and another set of 25 words that were comparable to the "keywords" on a number of psycholinguistic characteristics (e.g., how long the words were, how often the words were used in English, etc.). Vitevitch and Goldstein found that the keywords were responded to more quickly and accurately in a variety of language tests than the words that were similar on a variety of psycholinguistic characteristics, suggesting that the position of words in the phonological network plays an important role in the retrieval of those words from memory.

Keyplayers in a company facilitate the overall flow of information among disparate groups in the organization (Borgatti, 2006). One might, therefore, try to engineer the structure of a group to increase the productivity of the company. Similarly, one might attempt a network-inspired intervention in the domain of psycholinguistics to facilitate language recovery in individuals with acquired language disorders, such as various types of aphasia that can accompany strokes, by focusing on the reacquisition or rehabilitation of keywords in the mental lexicon (Vitevitch & Goldstein, 2014b).

Another example of network analysis shedding light on language research, and language research also pointing to new areas for network analysts to investigate further can be seen in work by Hills et al. (2009), who made networks of the 130 nouns that most children know, based on norms of language development (Dale & Fenson, 1996). Note that the analysis by Hills et al. was of a network in which the connections between words were based on *semantic* similarity, rather than phonological similarity as in most of the psycholinguistic studies described above. Semantic similarity was defined based on norms of semantic association (Nelson, McEvoy & Schreiber, 1999); when presented with the word "cat," most people respond with the word "dog." Thus, in the network constructed by Hills et al. there was a direct connection between "cat" and "dog" (cf., the longer path between "cat" and "dog" in the phonological network in Figure 7.1).

Hills et al. then used the longitudinal norms that indicate when a word is typically learned by a child (Dale & Fenson, 1996) to determine which principle of network growth best accounted for the semantic network. One principle of network growth that Hills et al. examined was *preferential attachment*, which gained much attention for its role in producing scale-free networks (Barabási & Albert, 1999). In preferential attachment, a new node being added to the system is more likely to connect to an existing node that has high degree (many connections) than to an existing node that has low degree (few connections).[2] In the context of language development, a child will more likely learn a new word if it is semantically similar to a known word that has many semantic connections than if it is semantically similar to a known word with few semantic connections (see Steyvers & Tenenbaum, 2005).

Interestingly, the semantic network examined by Hills et al. (2009) exhibited a degree distribution that followed a power-law, suggestive of a scale-free network structure. Although evidence for preferential attachment had been reported in a number of other domains (but see Clauset, Shalizi & Newman (2009) for methodological concerns regarding some of these reports; see also Fox Keller, 2005), preferential attachment did not account very well for the growth of the semantic network examined by Hills et al (2009). Instead, a principle that Hills et al. called *preferential acquisition* seemed to account better for growth in the semantic network. In preferential acquisition, a word is more likely to be learned if it is well-connected to other words in the learning environment (i.e., the more elaborate semantic network of an adult). This differs from preferential attachment, where a word is more likely to be learned if it is related to a word that is well-connected (not that the word itself is well-connected as in preferential acquisition).

While it is possible that Hills et al. (2009) could have discovered the growth principle that best accounted for semantic acquisition without using network analysis, the methodology and theoretical mind-set of network analysis helps researchers view old questions in a different way and to ask often-examined research questions in a new way. One of the fundamental ideas in network analysis is that the structure of the network influences the dynamics that take place in that network (Watts & Strogatz, 1998). For a field like Psycholinguistics—which focuses on the psychological processes involved in the comprehension, production, and acquisition of language—this raises the intriguing possibility that the ability to understand, speak or learn a word may not be solely dependent on the characteristics of the individual word—as is commonly assumed in mainstream Psycholinguistics—but may instead depend on how that word relates to other words that are known and stored in that part of memory called the mental lexicon.

Another intriguing aspect of network analysis is that it provides a common framework for examining changes in a system on multiple time-scales. In the case of language, one can use network analysis to examine the development of language in an individual (Hills et al. 2009), as well as changes on a longer, evolutionary time-scale (Solé, Corominas-Murtra, Valverde & Steels, 2010). In mainstream Psycholinguistics such investigations occupy their own intellectual silos; therefore, adopting the techniques of network analysis might help researchers from disparate fields communicate with each other despite the barriers created by the terminology of traditional academic disciplines.

The work reviewed in this section illustrates that psychologists can make interesting new discoveries in their domains of interest with the tools of network analysis. The work in this section also illustrates that some of the domains of interest to psychologists offer efficient and robust systems with unique network structures for network analysts to examine further. Perhaps other systems that require many of the same characteristics of the language system—rapid and accurate search, robustness to damage, continued growth—could engineer a network structure that achieves these goals using language networks as a model.

Finally, network analysis of language has been done at a number of different scales ranging across the different parameters that define language, and has been extended to examine how interlocutors interact and adapt to each other linguistically (Mehler, Lücking & Weiß, 2010). Network analysis of social interactions has also been combined with other tools from natural language processing, text analysis and computational linguistics to shed light on the attitudes and emotions expressed through written language, a field known as sentiment analysis (Tan et al., 2011). Looking at smaller scales, however, Fedorenko and Thompson-Schill (2014) suggest that there is still much work to been done when considering how language processing is actually carried out in brain networks.

Clinical Psychology: (Re-)defining Psychopathology

Network analysis has been described as equal parts theory and equal parts methodology: "networks offer both a theoretical framework for understanding the world and a methodology for using this framework to collect data, test hypotheses, and draw conclusions" (Neal, 2013, p. 5). The previous sections considered work from various subdisciplines of Psychology in which the methodological tools of network analysis were employed to provide insight into their respective domains of interest. In the present section we will look at one way that networks can be used as a theoretical framework in Psychology by considering the work of Cramer, Waldorp, van der Maas and Boorsboom (2010) in which they reconsider how to define Psychopathology.

The mainstream conceptualization of psychological disorders subscribes to the fundamental assumption that some attribute (or attributes) that cannot be directly observed causes the symptoms associated with a particular disorder (Cramer et al., 2010). These latent attributes must be assessed indirectly by the presence or absence of certain observable variables. For example, the latent construct for the disorder known as *major depressive disorder* must be measured indirectly by the presence of a depressed mood, changes in sleep (e.g., insomnia or hypersomnia), fatigue, etc. (American Psychiatric Association, 2000). The latent construct for major depressive disorder is thought to be the cause of these symptoms.

Cramer et al. (2010) suggest that common and successful therapeutic interventions like *cognitive therapy* are problematic for the latent variable perspective of psychopathology. The central premise of cognitive therapy is that thoughts, feelings, and behavior are influenced by each other (Beck, 2014). By learning new ways to think and respond in certain situations, one can change one's feelings about a situation. For example, if someone makes a mistake at work, they may think to themselves, "I can't do anything right. I'm worthless," which leads to feelings of worthlessness and sadness. With repeated episodes of such thoughts and feelings over time, clinically diagnosed depression might emerge.

However, cognitive therapy trains individuals to respond differently when a mistake is made at work again. Instead of recalling all the past mistakes that one made, you might focus on all the important things that you do that go smoothly, or instead recall the mistakes of others that you caught, etc. Such thoughts could give the individual hope, and boost the esteem of the individual, instead of making the individual feel worthless and sad as before. From the latent variable perspective of psychopathology there is some underlying construct that causes one to feel worthless, sad, and perhaps eventually leading to major depressive disorder; the way that one thinks about or interprets one's own feelings and behaviors should not matter. The success of cognitive therapy in treating major depressive disorder, however, is problematic for the latent variable perspective of psychopathology.

Another problem with the latent variable perspective of psychopathology is the arbitrary manner of drawing a diagnostic cut-off. In this perspective each symptom is considered equal in weight, so that if a certain number of symptoms is observed, one is diagnosed with a clinical condition. Cramer et al. (2010) provide the following example of Alice and Bob to illustrate that equal weighting of symptoms may not accurately capture the likelihood of developing a psychopathology like major depressive disorder (MDD):

> Suppose that Alice displays two MDD symptoms—depressed mood and loss of interest—while Bob displays two other MDD symptoms—psychomotor and weight problems. On an intuitive level, it is plausible that Alice's symptoms are more likely than Bob's to eventually result in a full-fledged depression. In other words, some symptoms appear to be more *central* features of depression than others.
>
> *(Cramer et al., 2010, p. 143; emphasis in original)*

Rather than view psychological disorders such as major depressive disorder through the latent variable perspective, Cramer et al. (2010) constructed a network to model symptom space, with nodes representing specific symptoms and edges connecting symptoms that co-occur. The network view allows for node strength to be represented to capture the relative importance of some symptoms over others, and edge weighting to capture how often symptoms co-occur; see Figure 7.2. (Note that the inclusion of node strength and weighted edges differentiates the network of Cramer et al. from some of the networks described in the other sections of this chapter.)

Furthermore, the network of symptom space offers a novel way to view *comorbidity*, or the simultaneous occurrence of two or more disorders. Consider now the disorder known as *generalized anxiety disorder*. Among the symptoms that are used to diagnose generalized anxiety disorder are changes in sleep (e.g., insomnia or hypersomnia) and fatigue (American Psychiatric Association, 2000)—recall that both of these symptoms are also indicative of major depressive disorder. In the network perspective, symptom space nodes representing symptoms like changes in sleep and fatigue serve as what Cramer et al. call *bridge symptoms* that connect symptom nodes that are unique to major depressive disorder and to generalized anxiety disorder. The network view of comorbid symptoms differs significantly from the latent variable view of psychopathology, which suggests that comorbidity arises because of a direct relationship between two latent variables.

In addition to providing a unique way to view comorbidity of psychological disorders, the network view proposed by Cramer et al. (2010) recognizes that many mental disorders do not have defining features that clearly distinguish disorder X from disorder Y. Instead, many mental

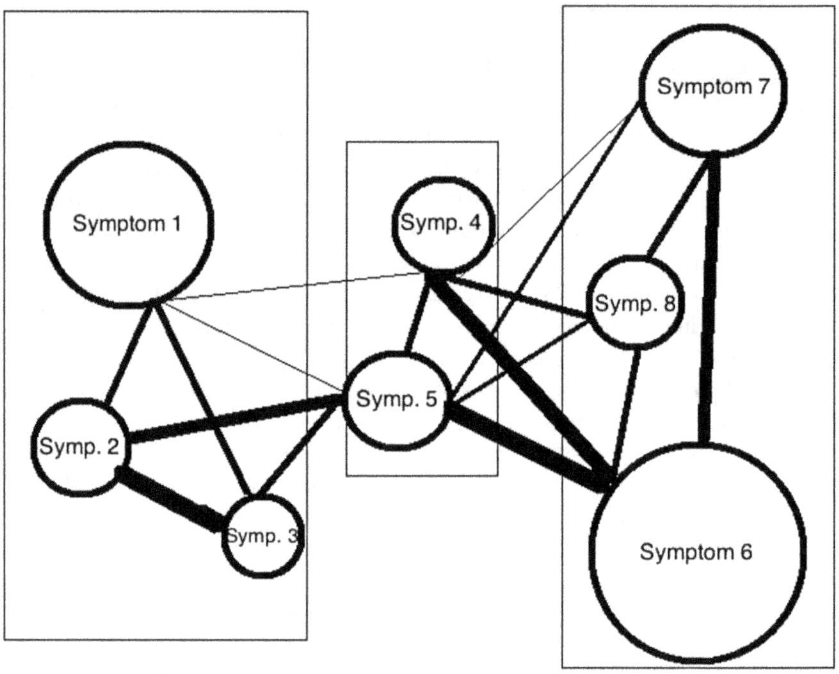

Figure 7.2 A network of symptom space for two mental disorders (adapted from Cramer et al., 2010)

disorders have fuzzy boundaries. These characteristics of mental disorders are not readily captured by the latent variable approach, leading to problems in diagnosis and ultimately treatment.

In the figure, each node represents a particular symptom. Larger nodes indicate more frequent symptoms, and thicker edges represent a higher frequency of two symptoms co- occurring. The box around symptoms 1–3 represents symptoms of a particular mental disorder (Disorder A) that do not overlap with symptoms of another mental disorder. The box around symptoms 6–8 represents non-overlapping symptoms of a different mental disorder (Disorder B). The box around symptoms 4 and 5 represents *bridge symptoms* that are part of both Disorder A and Disorder B.

By visualizing mental disorders as a network of symptom space one recognizes that there may be multiple etiologies (e.g., genetic, environmental, etc.) that interact to produce the observed symptoms, instead of a single latent variable. The network approach also accepts that there may be variation among individuals in terms of which symptoms are exhibited, and recognizes that some symptoms are more likely to lead to a clinical diagnosis than other symptoms.

Most important, the network approach allowed Cramer et al. (2010) to make testable predictions about the comorbidity of major depressive disorder and generalized anxiety disorder that the latent variable approach could not account for. For example, the symptom network demonstrated that there are certain pathways to comorbidity that are more likely than others. For example, exhibiting the core symptoms of depressed mood and loss of interest were more likely to result in both major depressive disorder and generalized anxiety disorder than exhibiting other subsets of symptoms. From the network approach, it was also predicted that comorbidity might be directional in nature; indeed, Cramer et al. (2010) found that it was more likely to observe major depression leading to comorbidity with generalized anxiety than the other way around. It is not clear how the latent variable approach to psychological disorders could account for these observations.

The network model of the comorbidity of major depressive disorder and generalized anxiety disorder proposed by Cramer et al. (2010) is a provocative re-examination of issues in the definition, diagnosis, and treatment of psychopathology. As such, it raises a number of interesting questions for future research to consider (the Open Peer Commentary that accompanies the article by Cramer et al. points to a number of such questions that remain to be answered). It is described here not to illustrate a new model of psychopathology that is now universally accepted (again, see the Open Peer Commentary that accompanies the article by Cramer et al. for statements that suggest this perspective has not been widely accepted), but rather to illustrate how the theoretical and methodological perspective that network analysis offers can benefit various areas of Psychology and related fields by forcing researchers to critically evaluate fundamental assumptions about phenomena of interest. In some cases the fundamental assumptions that are challenged are theoretical in nature, whereas in other cases the fundamental assumptions that are challenged are statistical in nature. Because network analysis is equal parts theory and equal parts methodology, it can offer researchers in a variety of fields a unique set of tools to shed new light on the problems that drive the field.

Conclusion

Network analysis had its origin in Mathematics (where it is known as *graph theory*), and made many advances through Social Network Analysis. However, the discovery over the past decade that many systems—despite differences in their age, function, and size—exhibit a common set of structural characteristics has fueled the emergence of a new field of study known as *Network Science* (Barabási, 2009). The examples from several areas of Psychology considered in this chapter speak to the variety of systems that exhibit similar network characteristics. Groups of people trying to solve a problem (a social network), neurons in the brain and regions of the brain (a biological network), and representations of words in memory (a cognitive network) all display small-world characteristics and heavy-tailed degree distributions despite being from different domains and varying in a number of other ways. The commonality in network structures across domains means that a great many fields might benefit from network analysis techniques.

Network analysis provides a theoretical and methodological framework that researchers can use to examine a system of interest at the *micro-level*, the *meso-level*, and the *macro-level*. At the micro-level one can analyze individual agents in the system. Using macro-level measures one can describe the overall structure of the system. Finally, at the meso-level one can describe the system at scales between the micro- and macro-levels. Most other research approaches consider only one level of analysis.

Although network analysis allows researchers to examine several levels of the same system, there are still challenges to be faced when moving from one network to another network involved in the same problem. It is not clear, for example, how to best map the brain network involved in language processing to the cognitive networks involved in language processing, to social networks of interlocutors communicating via language. Similarly, it is not clear how damage at one layer influences processing at the other layers. Much additional work is required to resolve these issues of multilayer networks (Kivelä et al., 2014).

A foundational assumption of network analysis is that the structure of a network influences the dynamics that take place in that network (Watts & Strogatz, 1998). This assumption enables researchers who study networks in one domain (e.g., language) to make novel predictions based on the observations of networks in another domain (e.g., social networks); consider the work on keywords (Vitevitch & Goldstein, 2014) based on the work on keyplayers (Borgatti, 2006) described in a previous section. However, it is also important to keep in mind that while the similar structure observed in networks from different domains may impose similar constraints on the dynamics of each system, very different mechanisms may have initially generated those systems (cf. Barabási, 2012; Fox Keller, 2005).

Network analysis holds much promise for various areas of psychology (and vice versa), and can be used to address a wide variety of problems and research questions. Adopting this approach can lead to new discoveries, and to an expansion in the range of questions that one investigates. It is important to note, however, that one must think carefully about how well entities and the relationships among those entities in a given domain map onto nodes and connections in a network representation before adopting network analysis (see Butts, 2009 for a similar point regarding network interventions).

Notes

1. In Linguistics, the solidi // are used to indicate that the character in the brackets is a symbol from the International Phonetic Alphabet (IPA), an agreed upon set of symbols to represent the various sounds of the languages of the world. IPA is the preferred method to transcribe spoken language, and is often the only way to document languages that do not employ a writing system. The particular phoneme (a *phoneme* is the basic unit of sound in a language), /!/, is known as an alveolar click, which is produced by pulling the tip of the tongue down sharply from the area just behind the upper teeth known as the alveolar ridge. To hear a linguist produce this sound, go to www.phonetics.ucla.edu/course/chapter1/ipaSOUNDS/Con-60b.AIFF (from Ladefoged, 2005).
2. More precisely (Barabási & Albert, 1999), the probability p_i that a new node is connected to node i is:

$$p_i = \frac{k_i}{\sum_j k_j}$$

where k_i is the degree of node i and the sum is made over all pre-existing nodes j.

References

Albert, R. & Barabási, A.L. (1999). Emergence of scaling in random networks. *Science, 286*, 509–512.

Albert, R., Jeong, H. & Barabási, A.L. (2000). Error and attack tolerance of complex networks. *Nature, 406*, 378–382.

American Psychiatric Association. (2000). *Diagnostic and Statistical Manual of Mental Disorders* (4th ed., text rev.).

American Speech-Language-Hearing Association. (1982). Language (relevant paper). Available from www.asha.org/policy.

Arbesman, S., Strogatz, S.H. & Vitevitch, M.S. (2010). The structure of phonological networks across multiple languages. *International Journal of Bifurcation and Chaos, 20*, 679–685.

Bandettini, P.A. (2009). What's new in neuroimaging methods? *Annals of the New York Academy of Sciences, 1156*, 260–293.

Barabási, A.L. (2009). Scale-free networks: A decade and beyond. *Science, 325*, 412–413.

Barabási, A.L. (2012). Network science: Luck or reason. *Nature, 489*, 507–508.

Barabási, A.L., & Albert, R. (1999). Emergence of scaling in random networks. *Science, 286*, 509–512.

Barthelemy, M., Barrat, A., Pastor-Satorras, R. and Vespignani, A. (2005). Dynamical patterns of epidemic outbreaks in complex heterogeneous networks. *Journal of Theoretical Biology, 235*, 275–288.

Basser, P.J., Mattiello, J., LeBihan, D. (1994). MR diffusion tensor spectroscopy and imaging. *Biophysical Journal, 66*, 259–267.

Beck, A.T. (2014). Advances in cognitive theory and therapy: the generic cognitive model. *Annual Review of Clinical Psychology, 10*, 1–24.

Bollobas, B. (2013). *Modern Graph Theory*. New York: Springer.

Borgatti, S.P. (2006). Identifying sets of key players in a network. *Computational, Mathematical and Organizational Theory. 12*, 21–34.

Butts, C.T. (2009). Revisiting the foundations of network analysis. *Science, 325*, 414–416.

Changizi, M.A. & Destefano, M. (2009). Common scaling laws for highway systems and the mammalian neocortex. *Complexity, 15*, 11–18.

Churchland, P.M. (1981). Eliminative materialism and the propositional attitudes. *The Journal of Philosophy, 78*, 67–90.

Clauset, A., Shalizi, C.R. & Newman, M.E.J. (2009). Power-law distributions in empirical data. *SIAM Review, 51*, 661–703.

Cramer, A.O.J., Waldorp, L.J., van der Maas, H.L.J., and Borsboom, D. (2010). Comorbidity: A network perspective. *Behavioral and Brain Sciences, 33*, 137–193.

Dale, P.S., & Fenson, L. (1996). Lexical development norms for young children. *Behavior Research Methods, Instruments, & Computers, 28*, 125–127.

Darley, J. M. & Latané, B. (1968). Bystander intervention in emergencies: Diffusion of responsibility. *Journal of Personality and Social Psychology, 8*, 377–383.

Erdös, P. & Rényi, A. (1959). On random graphs. *Publicationes Mathematicae, 6*, 290–297.

Fedorenko, E. & Thompson-Schill, S.L. (2014). Reworking the language network. *Trends in Cognitive Sciences, 18*, 120–126.

Fox Keller, E. (2005). Revisiting "scale-free" networks. *BioEssays, 27*, 1060–1068.

Friston, K.J. (2011). Functional and effective connectivity: a review. *Brain Connectivity, 1*, 13–36.

Greenberg, J. H., & Jenkins, J. J. (1966). Studies in the psychological correlates of the sounds system of American English. *Word, 20*, 157–177.

Gronlund, A. & Holme, P. (2005). A network-based threshold model for the spreading of fads in society and markets. *Advances in Complex Systems, 8*, 261–273.

Herculano-Houzel, S. (2009). The human brain in numbers: a linearly scaled-up primate brain. *Frontiers in Human Neuroscience, 3*, 00031.

Hernandez, L., Wager, T.D., & Jonides, J. (2002). Introduction to functional brain imaging. In John Wixted and Hal Pashler (Eds.), *Stevens Handbook of Experimental Psychology, Third Edition, Volume 4: Methodology in Experimental Psychology*. New York: John Wiley & Sons, pp. 175–221.

Hills, T., Maouene, M., Maouene, J., Sheya, A., & Smith, L. (2009). Longitudinal analysis of early semantic networks: preferential attachment or preferential acquisition? *Psychological Science, 20*, 729–739.

James, L.E. & MacKay, D.G. (2001). H.M., Word knowledge and aging: Support for a new theory of long-term retrograde amnesia. *Psychological Science, 12*, 485–492.

Kivelä, M., Arenas, A., Barthelemy, M., Gleeson, J.P., Moreno, Y. & Porter, M.A. (2014). Multilayer networks. *Journal of Complex Networks, 2*, 203–271.

Kleinberg, J. M. (2000). Navigation in a small world, *Nature, 406*, 845.

Ladefoged, P. (2005). *A Course in Phonetics* (5th ed.). Belmont, CA: Wadsworth Publishing.

Landauer, T. K., & Streeter, L. A. (1973). Structural differences between common and rare words: Failure of equivalence assumptions for theories of word recognition. *Journal of Verbal Learning and Verbal Behavior, 12*, 119–131.

Lazer, D. & Friedman, A. (2005). The hare and the tortoise: The network structure of exploration and exploitation. *ACM International Conference Proceeding Series, 89*, 253–254.

Levin, D. T. & Simons, D.J. (1997). Failure to detect changes to attended objects in motion pictures. *Psychonomic Bulletin & Review, 4*, 501–506.

Lichtman, J.W. & Sanes, J.R. (2008). Ome sweet ome: What can the genome tell us about the connectome? *Current Opinion in Neurobiology, 18*, 346–353.

Luck, S.J. (2005). *An Introduction to the Event-related Potential Technique*. Cambridge, MA: MIT Press.

Mason, W. A., Jones, A., & Goldstone, R. L. (2008). Propagation of innovations in networked groups. *Journal of Experimental Psychology: General, 137*, 422–433.

Mehler, A., Lücking, A., & Weiß, P. (2010). A network model of interpersonal alignment. *Entropy, 12*, 1440–1483.

Menon, V. (2011). Large-scale brain networks and psychopathology: A unifying triple network model. *Trends in Cognitive Sciences, 15*, 483–506.

Neal, Z.P. (2013). *The Connected City: How Networks are Shaping the Modern Metropolis*. New York: Routledge.

Nelson, D.L., McEvoy, C.L., & Schreiber, T.A. (1999). *The University of South Florida Word Association, Rhyme, and Word Fragment Norms*. Available at: www.usf.edu/ freeassociation/

Newman, M.E.J. (2001). The structure of scientific collaboration networks. *Proceedings of the National Academy of Science, 98*, 404–409.

Newman, M.E.J. (2002). Assortative mixing in networks. *Physical Review Letters, 89*, 20889701.

Newman, M. E. J. & Girvan, M. (2004). Finding and evaluating community structure in networks. *Physical Review E, 69,* 026113.

Lind, P.G., da Silva, L.R., Andrade, Jr., J.S. & Herrmann, H.J. (2007). Spreading gossip in social networks. *Physical Review E, 76,* 036117.

Power, J.D., Schlaggar, B.L., Lessov-Schlaggar, C.N. & Petersen, S.E. (2013). Evidence for hubs in human functional brain networks. *Neuron, 79,* 798–813.

Puschmann, S., Weerda, R., Klump, G. and Thiel, C.M. (2013). Segregating the neural correlates of physical and perceived change in auditory input using the change deafness effect. *Journal of Cognitive Neuroscience, 25,* 730–742.

Quillian, M. R. (1967). Word concepts: A theory and simulation of some basic semantic capabilities. *Behavioral Science, 12,* 410–430.

Rosenblatt, F. (1958). The perceptron: A probabilistic model for information storage and organization in the brain. *Psychological Review, 65,* 386–408.

Scott-Phillips, T.C. & Blythe, R.A. (2013). Why is combinatorial communication rare in the natural world, and why is language an exception to this trend? *Journal of the Royal Society Interface, 10,* 20130520.

Seung, S. (2013). *Connectome: How the Brain's Wiring Makes Us Who We Are.* New York: Mariner Books.

Simpson, S.L., DuBois Bowman, F. and Laurienti, P.J. (2013). Analyzing complex functional brain networks: Fusing statistics and network science to understand the brain. *Statistics Surveys,* 7, 1–36.

Solé, R. V., Corominas-Murtra, B., Valverde, S. & Steels, L. 2010. Language networks: Their structure, function and evolution. *Complexity, 15,* 20–26.

Sporns, O. (2013). Network attributes for segregation and integration in the human brain. *Current Opinion in Neurobiology, 23,* 162–171.

Sporns, O. (2014). Contributions and challenges for network models in cognitive neuroscience. *Nature Neuroscience, 17,* 652–660.

Steyvers, M., & Tenenbaum, J.B. (2005). The large-scale structure of semantic networks: Statistical analyses and a model of semantic growth. *Cognitive Science, 29,* 41–78.

Tan, C., Lee, L., Tang, J., Jiang, L., Zhou, M., & Li, P. (2011). User-level sentiment analysis incorporating social networks. In Proceedings of the 17th ACM SIGKDD International Conference on Knowledge Discovery and Data Mining (pp. 1397–1405). ACM.

Uttal, W.R. (2002). Précis of the new phrenology: The limits of localizing cognitive processes in the brain. *Brain and Mind, 3,* 221–228.

Vanderstoep, S.W., Fagerlin, A. & Feenstra, J.S. (2000). What do students remember from Introductory Psychology? *Teaching of Psychology, 27,* 89–92.

Vitevitch, M.S. (2003). Change deafness: The inability to detect changes in a talker's voice. *Journal of Experimental Psychology: Human Perception and Performance, 29,* 333–342.

Vitevitch, M.S. & Goldstein, R. (2014). Keywords in the mental lexicon. *Journal of Memory & Language, 73,* 131–147.

Vitevitch, M.S., Chan, K.Y., Goldstein, R. (2014). Insights into failed lexical retrieval from network science. *Cognitive Psychology, 68,* 1–32.

Wang, L., Yu, C., Chen, H. Qin, W., He, Y. Fan, F., Zhang, Y., Wang, M., Li, K., Zang, Y., Woodward, T.S., & Zhu, C. (2010). Dynamic functional reorganization of the motor execution network after stroke. *Brain, 133,* 1224–1238.

Wasserman, S. & Faust, K. (1994). *Social Network Analysis: Methods and Applications.* Cambridge: Cambridge University Press.

Watts, D. J. (2004). The "new" science of networks. *Annual Review of Sociology, 30,* 243–270.

Watts, D.J. & Strogatz, S.H. (1998). Collective dynamics of "small-world" networks. *Nature, 393,* 409–410.

White, J.G., Southgate, E., Thomson, J.N. & Brenner. S. (1986). The structure of the nervous system of the nematode *Caenorhabditis elegans. Philosophical Transactions of the Royal Society of London B, 314,* 1–340.

Zachary, W. W. (1977). An information flow model for conflict and fission in small groups. *Journal of Anthropological Research, 33,* 452–473.

8

USING COGNITIVE SOCIAL STRUCTURES TO UNDERSTAND PEER RELATIONS IN CHILDHOOD AND ADOLESCENCE

Jennifer Watling Neal and Mariah Kornbluh

Research Question: How do children perceive peer relationships, and how are these relationships associated with behavior?
System Science Method(s): Networks
Things to Notice:

- Difference between "real" and perceived networks
- Linkage between research question and type of network data

Research in sociology (e.g., Moody, 2001), developmental psychology (e.g., Gifford-Smith & Brownell, 2003), and education (e.g., Farmer et al., 2002) highlights the utility of social network analysis (SNA) for understanding peer relations in childhood and adolescence. SNA is useful for understanding interracial and cross-gender relationships (e.g., Moody, 2001; Neal, 2010a; Hallinan & Smith, 1989; Shrum, Cheek, & Hunter, 1988). SNA is also useful for exploring how peer network structure is related to a variety of childhood and adolescent behaviors including aggression (e.g., Faris & Felmlee, 2011; Farmer et al., 2002; Neal & Cappella, 2012), peer victimization (e.g., Cappella & Neal, 2012), academic engagement (e.g., Cappella, Kim, Neal, & Jackson, 2013; Ryan, 2001), and substance use (e.g., Ennett & Bauman, 1993; Henry & Kobus, 2007; Kobus & Henry, 2010). Moreover, longitudinal application of SNA allows researchers to understand processes of peer selection and influence for these behaviors (Veenstra, Djikstra, Steglich, & Van Zalk, 2013).

Most peer relations researchers who use SNA have relied on one of three sources of network data: (1) self-report surveys or interviews, (2) behavioral observations, or (3) peer-report surveys or interviews gathered using social cognitive mapping. Surveys or interviews of self-reported relationships are among the most widely used sources of SNA data (Marsden, 1990), but are often problematic in research on peer relations for two reasons. First, to gather accurate whole network data in a setting (e.g., classroom, school), self-report measures require high response rates (e.g., > 80–90%). In the U.S., it is often not feasible to obtain response rates this high in settings of children and adolescents due to difficulties securing active parental consent and absences during data collection. Second, children and adolescents tend to exhibit self-serving

 DOI: 10.4324/9781315748771-9

biases and may overestimate their number of connections (Neal & Cappella, 2014). In contrast, observations generally result in complete network data and rely on objective behavioral indicators of peer relationships. Observational network data are also common in studies of young children who typically are not developmentally able to provide valid self-reported network data (e.g., Hanish, Martin, Fabes, Leonard, & Herzog, 2005; Schaefer, Light, Fabes, Hanish, & Martin, 2010). However, observations are time-consuming, resource intensive, and require regular access to the research setting (Marsden, 1990). In social cognitive mapping, children or adolescents are asked in surveys or interviews to list groups of peers who "hang out together" in a setting (Cairns & Cairns, 1994; Gest, 2008). Because this procedure relies on peer report, it allows researchers to gather complete network data, even in settings where response rates are low. However, social cognitive mapping is generally used for peer group identification (Neal & Neal, 2013), and is somewhat more limited than self-report and observational data in identifying the specific fine-grained structure of peer relationships within a peer group (Neal, 2008). Specifically, unlike self-report and observational methods, social cognitive mapping does not provide information about whether children are connected to many or few peers within these groups.

In addition to the three common sources of network data listed above, there is a growing body of research that uses cognitive social structures (CSS) to examine childhood and adolescent peer relations (e.g., Neal, 2008; Pittinsky & Carolan, 2008). CSS provide an assessment of individuals' perceptions of the entire network structure in a setting. Initially developed by Krackhardt (1987) to explore adults' perceptions of relationships in the workplace, this method holds key advantages over other data sources for researching child peer relations. First, like social cognitive mapping, CSS can be used to gather complete network data from only a subset of child respondents in a setting (Neal, 2008). Second, CSS provide a mechanism for exploring how individual children as well as important adults (e.g., teachers) perceive the entire social network structure between peers in a setting (e.g., classroom). This creates new avenues for research that focuses on individuals' awareness to social relationships in a setting and how their perceptions of social relationships impact their behavior. For example, Neal & Cappella (2014) have explored how children's perceptions of their position in their classroom network are associated with aggressive behaviors.

This chapter provides an overview of the use of CSS to understand peer relations in childhood and adolescence, and is written with a practical eye toward how CSS can be used to answer substantive research questions about children's peer relations. First, we provide a brief overview of CSS and the measurement of cognitive social structures with children and adolescents. Second, drawing on Krackhardt (1987), we describe the three formats of data that CSS provides – locally aggregated structures, slices, and consensus aggregation. Third, we illustrate how these three formats of data have been applied in the recent literature on childhood and adolescent peer relationships to answer three types of research questions. Finally, we end with a consideration of the benefits and drawbacks to using CSS to examine childhood and adolescent peer relations.

Introducing CSS Data

CSS seeks to provide information about respondents' perceptions of the relationships in a setting (Krackhardt, 1987; Neal, 2008). Therefore, to gather CSS data, researchers typically use survey or interview methods that ask respondents to identify the presence or strength of a relationship between each pair of actors in the setting. Peer relations researchers have used survey methods to collect CSS data on classroom peer networks from children (e.g., Cappella, Neal, & Sahu, 2012; Neal, Neal, & Cappella, 2014; Neal, 2010a) and teachers (e.g., Neal, Cappella, Wagner,

Please CIRCLE the names of all of the kids in your classroom that Rachel hangs out with often:

Artie	**Dave**	**Kitty**	**Mike**	**Ryder**	**Tina**
Becky	**Finn**	**Kurt**	**Noah**	**Sam**	**Wade**
Blaine	**Jake**	**Marley**	**Quinn**	**Santana**	
Brittany	**Joe**	**Mercedes**	**Rory**	**Sugar**	

Are there any other kids at your school that Rachel hangs out with often?

First Name and Last Name	Grade

If Rachel does not hang out with ANY of the other children at your school, please check this box: ☐

Figure 8.1 Sample page from a CSS survey for children (adapted from Neal, 2008)

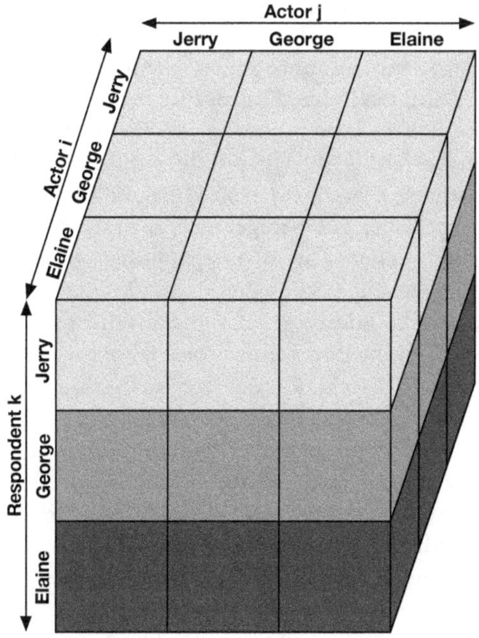

Figure 8.2 Three-dimensional CSS matrix

& Atkins, 2011; Pittinsky & Carolan, 2008). As described in Neal (2008), these surveys typically include a separate page for each child in the setting (e.g., classroom). The survey instructs respondents to circle the names of children on a full class roster that the designated child on the page "hangs out with often." There is space to enter the names of other individuals outside the setting that the designated child hangs out with and respondents are instructed to check a box if the designated child on the page does not hang out with anyone (see Figure 8.1 for an example survey page). These surveys are feasible in settings that include up to 30–40 children and adolescents, and have been used successfully with children as young as seven years old.

As Krackhardt (1987) noted, "the amount of information in a cognitive social structure far exceeds that in a traditional social structure" (p. 113). Most methods of network data collection (i.e., self-reported surveys, observations) result in a two-dimensional matrix, R, where each cell R_{ij} indicates the presence or strength of a relationship between actors i and j. However, CSS is unique in that it provides an additional dimension of information – namely, information about respondents' *perceptions* of the social network structure in a setting. Thus, CSS results in a three-dimensional matrix, R, where each cell R_{ijk} indicates the presence or strength of a relationship between actors i and j as perceived by respondent k (Krackhardt, 1987). CSS data can be collected to yield data on directed relationships like advice (e.g., actor i gives advice to actor j) or undirected relationships like hanging out (e.g., actors i and j hang out with one another). In this chapter, we will focus only on reports of the presence or absence of undirected relationships. Figure 8.2 provides an illustrated example of a three-dimensional CSS matrix that includes three actors: Jerry, Elaine, and George.

Three Types of CSS Data and Accompanying Research Questions

The three-dimensional data structure resulting from CSS is complex, and therefore Krackhardt (1987) described three different methods of aggregating CSS data into more manageable formats. The first method of aggregating CSS data produces *locally aggregated structures*. Here, the three- dimensional matrix with cells R_{ijk} is reduced to a two-dimensional matrix with cells R_{ij} that only reflect self-reported information on relationships (i.e., relationships that the respondent k is a part of). Thus, locally aggregated structures essentially mirror what researchers might get if they used a traditional self-report survey to gather network data. Figure 8.3 illustrates this form of CSS data reduction in a CSS matrix of three individuals: Jerry, George, and Elaine. Here, a locally aggregated structure would be created using only the cells indicating Jerry's report of his own relationships with Elaine and George, George's report of his own relationships with Jerry and Elaine, and Elaine's report of her own relationships with Jerry and George.

The second method of aggregating CSS data produces *slices* (Krackhardt, 1987). Unlike locally aggregated structures, slices take advantage of the breadth of perceptual data on relationships collected using CSS. Here, the three-dimensional matrix with cells R_{ijk} is reduced to a set of two-dimensional matrices with cells R_{ij}, one for each respondent k. Each of these two-dimensional matrices represents respondent k's perception of the entire network structure in a setting. Thus, a slice includes respondent k's perceptions of both his or her own relationships, and other relationships in which he or she is not directly involved. Figure 8.4 illustrates how Jerry, George, and Elaine each have their own slice of the CSS matrix that represents their perceptions of all the relationships that exist in the setting (i.e., their view of the whole network). Slices provide rich information about how individual respondents perceive social networks in settings.

The third method of aggregating CSS data produces *consensus aggregations*. Consensus aggregations triangulate across individual respondents' perceptions of the network structure. Here,

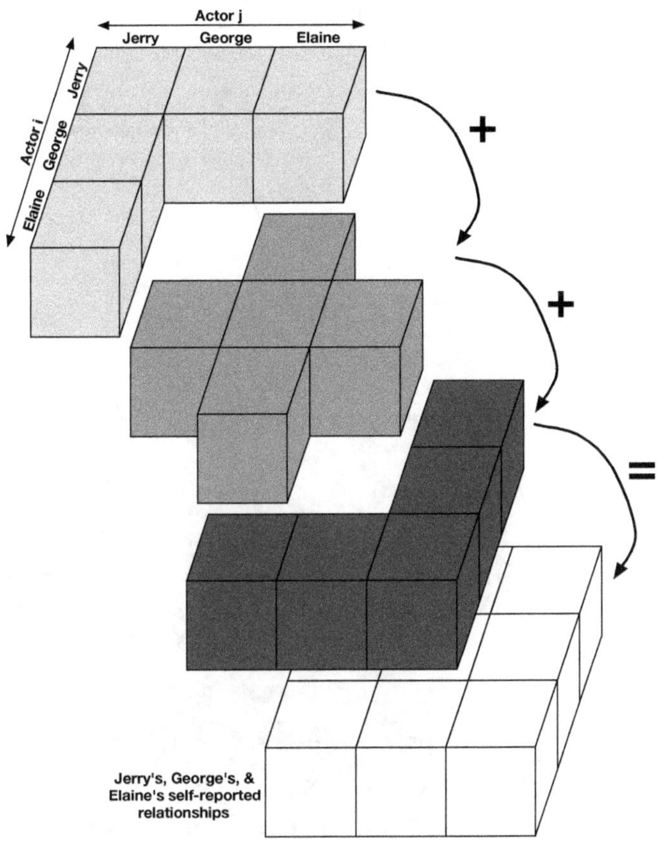

Figure 8.3 Locally aggregated structure

the three- dimensional matrix with cells R_{ijk} is reduced to a two-dimensional matrix with cells R_{ij} where each cell indicates the consensus of respondents' reports on the relationship between actors *i* and *j*. As illustrated in Figure 8.5, this is typically accomplished by summing reports of the relationships between actors *i* and *j* across all respondents *k*. Thus, in a consensus aggregation, the cell indicating a relationship between Jerry and George would indicate the number of respondents (i.e., Jerry, George, and Elaine) that perceive the relationship to exist. It is also common to divide each cell in the consensus aggregation by the total number of respondents yielding the proportion of respondents that perceive the relationship to exist. Proportions can be helpful for the interpretation of each cell in the consensus aggregation because they account for the total number of respondents that could possibly indicate a relationship between each pair of actors (e.g., if the proportion in the cell for Jerry and George is 0.5, then 50% of respondents perceive a relationship between Jerry and George).

Valued consensus aggregation matrices preserve information about the salience of relationships between each pair of actors to others. However, many existing network measures still require simple, dichotomous matrices that only indicate whether a relationship is present or absent. If necessary, Krackhardt (1987) noted that valued consensus aggregation matrices could be dichotomized using a threshold function. For example, using consensus aggregation matrices that include proportions in each cell, a threshold function of 0.5 would only count a relationship as present if at least 50% of respondents perceived it. More recently, Neal (2008) advocated the

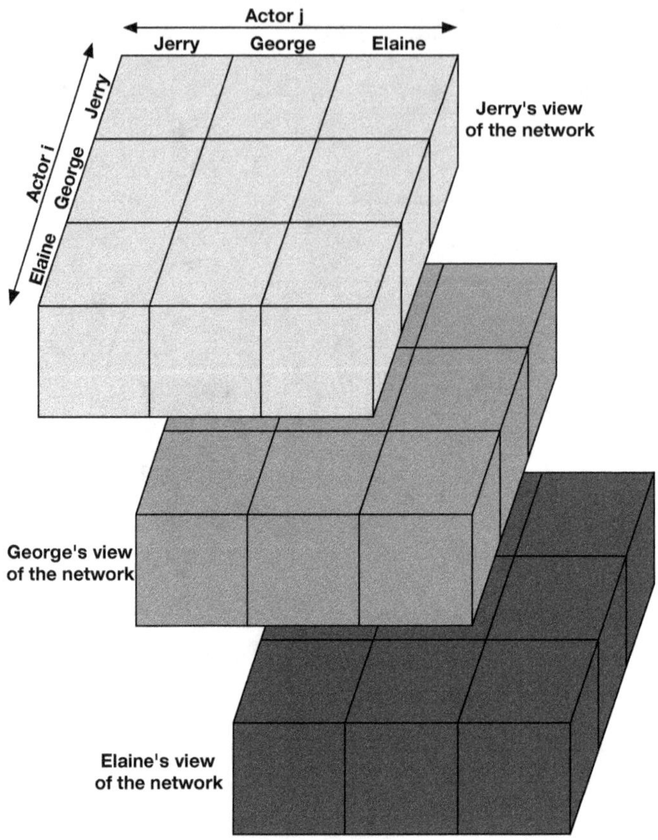

Figure 8.4 Slice

use of a binomial test to dichotomize consensus aggregation matrices based on whether a greater number of respondents perceived a tie than expected by chance (i.e., $p < .05$).

The three types of data formats provided by CSS (i.e., locally aggregated structures, slices, and consensus aggregations) can be mixed and matched in complementary ways to address different research questions about children's peer relations. First, there is a growing interest in understanding children's social perceptions, which are linked to social adjustment (Cillessen & Bellmore, 1999) and theorized to lead to more effective navigation of classroom relationships (Cappella, Neal, & Sahu, 2012). CSS data can address the research question: *What dimensions of spatial (e.g., sitting nearby) and social (e.g., sex similarity) proximity predict children's self-reported relationships (i.e., locally aggregated structure) and peer-inferred relationships about classmates (i.e., consensus aggregation)* (e.g., Neal, Neal, & Cappella, 2014)?

Second, recent research has highlighted the importance of teachers in shaping children's classroom relationships (Farmer, Lines, & Hamm, 2011; Gest & Rodkin, 2011). Moreover, teacher attunement, or the degree to which teachers perceive the same classroom relationships as their students, has been linked to improved peer experiences among students (e.g., Hamm, Farmer, Dadisman, Gravelle, & Murray, 2011). CSS data can address the research question: *What contextual factors (e.g., grade, class size, teaching practices) predict agreement between individual teachers' perceptions of the network (i.e., slices) and the consensus of students (i.e., consensus aggregation)* (e.g., Neal, Cappella, Wagner, & Atkins, 2011)?

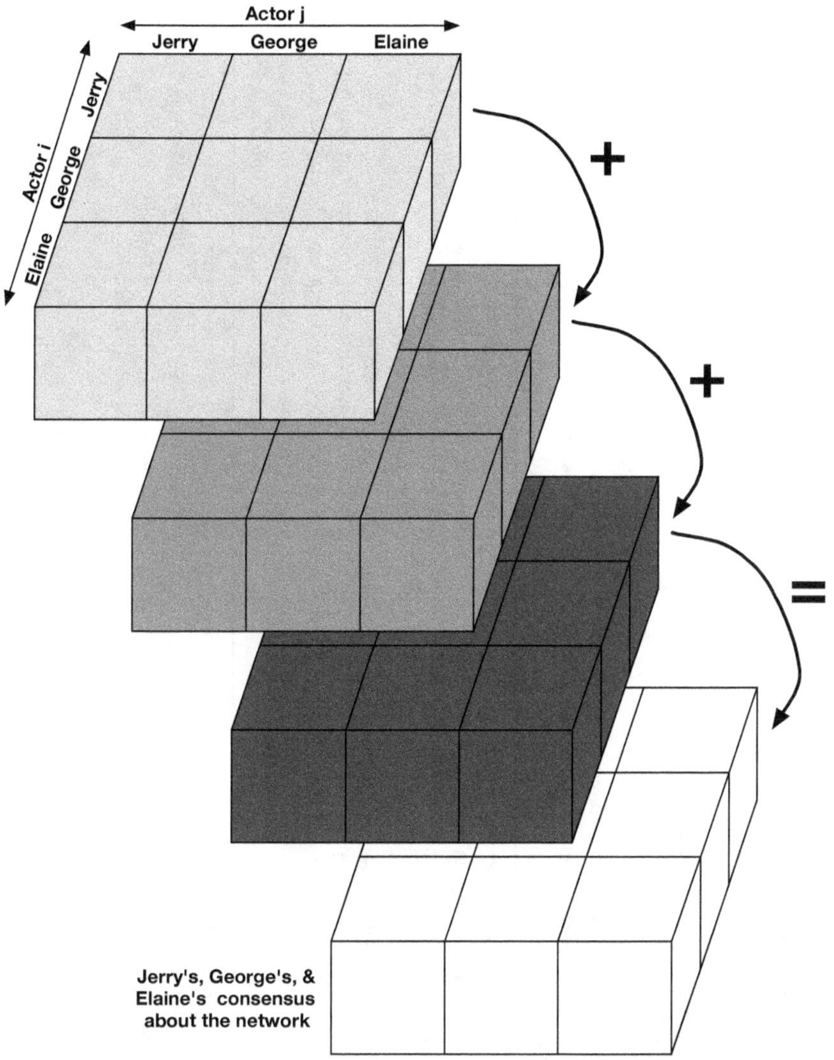

Figure 8.5 Consensus aggregation

Finally, beyond data on perceptions of networks, CSS can be used to explore research questions linking network position and behavior that are often reserved for more traditional forms of network data (e.g., self-report, observation). For example, researchers have been interested in how children's position within their classroom peer networks link to classroom behaviors, including aggression (e.g., Gifford-Smith & Brownell, 2003; Neal, 2007, 2010b). CSS data can address the research question: *How are children's positions in their social networks (i.e., consensus aggregation) associated with behavior and/or contextual factors* (e.g., Neal, 2009)?

The three research questions identified above serve as an outline for the next several sections of this chapter. More specifically, in these sections, we provide examples of how CSS data have been used to answer each of these questions in recent research on peer relations in children and adolescent (see Table 8.1). However, it is important to note that these three research questions only illustrate some of the possibilities that the rich CSS data hold for research in this area.

Table 8.1 Summary of examples

Research question	*Type of CSS data*	*Key findings*
Example 1. What predicts children's self-reported relationships and peer-inferred relationships?	**Locally aggregated structures.** Each participating child provided reports of his/her own relationships to assess self-reported relationships. **Consensus aggregation.** The consensus of peer reports on each classroom relationship was used to assess peer-inferred relationships.	Spatial and social proximities were positive predictors for both self-report and peer-inferred relationships. Peer-inferred measures are not interchangeable with self-report measures, and could lead to inflated estimates of gender segregation within the classroom.
Example 2. What predicts teacher and student agreement on classroom networks?	**Slices.** Teachers provided reports of each of their students' "hanging out" relationships in the class to assess teacher perceptions of the classroom network. **Consensus aggregation.** The consensus of peer reports on each classroom relationship was used to assess students' perceptions of classroom relationships.	Consistency in teacher and student perceptions of the classroom network is dependent on several contextual factors (class size, behavioral norms, and the teacher's classroom organization). Teacher report of classroom networks can be a reliable proxy for aggregated student report under conditions in which class sizes are small, aggressive norms are low, and when there are high levels of classroom productivity.
Example 3. How are children's network positions associated with behavior?	**Consensus aggregation.** The consensus of peer reports on each classroom relationship was used to assess the structure of classroom networks.	Ego network density was a positive predictor of teacher-rated relational aggression. In contrast, there was a significant quadratic effect of centrality on peer-nominated relational aggression. Findings highlight that different aspects of network position are associated with teacher-rated and peer-nominated relational aggression.

Example 1: What Predicts Children's Self-Reported Relationships and Peer-Inferred Relationships?

Neal, Neal, and Cappella (2014) were interested in understanding how predictors of children's self-reported relationships differ from predictors of peer-inferred relationships between classmates. Substantively, answering this research question highlighted factors of spatial (i.e., sitting nearby) and social proximity (i.e., demographic and behavioral similarities) that are associated with children's self-reported relationships and their inferences about classmates' relationships. In particular, it sheds light on factors that are linked to childhood relationship formation as well as factors children use as markers for the existence of peer relationships. This is important because the children's ability to form peer relationships, as well as their ability to make accurate inferences about peer relationships, may encourage both social and school adjustment (Cappella, Neal, & Sahu, 2012). Methodologically, answering this research question provided important information about the degree to which peer-reported relationships are adequate proxy measures for self-reported relationships in situations where response rates are low (Cairns & Cairns, 1994; Neal, 2008).

Setting and Sample

The researchers collected data from 426 African American students in 34 second-through-fourth grade classrooms located in five urban schools in low-income neighborhoods. The sample was 53% female and included roughly equal percentages of second (30%), third (37%), and fourth (32%) grade students.

Data Collection and Measures

To assess self-reported and peer-inferred relationships, Neal, Neal, and Cappella (2014) collected CSS survey data on classmates' "hanging out" relationships from participating children in each classroom (see example in Figure 8.1). These CSS data were then aggregated to produce *locally aggregated structures* that reflected the self-reported relationships of participating children in each classroom. A self-reported relationship was coded as present when both children involved in the relationship mutually reported "hanging out often." The CSS data were also used to create *consensus aggregations* that represented the consensus of peer reports on each classroom relationship, or peer-inferred relationships. More specifically, Neal, Neal, and Cappella (2014) used a. 0.5 threshold function similar to that described by Krackhardt (1987) to create a dichotomous consensus aggregation for each classroom. That is, a peer-inferred relationship was coded as present when more than 50% of classroom peer reports indicated that child *i* and *j* were involved in the relationship "hanging out often." They counted both reports indicating that child *i* hangs out with child *j* and reports that child *j* hangs out with child *i*, yielding undirected data.

In addition to collecting CSS data on self-reported and peer-inferred relationships, Neal, Neal, and Cappella (2014) also collected data on spatial and social proximities for each possible pairwise combination of participating students. Spatial proximity was assessed using children's nominations of classmates that sit next to them. Social proximity was assessed using three indicators: sex similarity (i.e., are both members of the dyad the same sex?), academic behavioral similarity (i.e., are both members of the dyad "good at school"?), and athletic behavioral similarity (i.e., are both members of the dyad "good at sports"?) (Lease, McFall, Treat, & Viken, 2003).

Data Analysis

Data were analyzed using two logistic regression models, one predicting self-reported relationships and one predicting peer-inferred relationships. In these models, the unit of analysis was dyadic, and included 2,695 pairwise combinations of participating students within classrooms (Neal, Neal, & Cappella, 2014). Data were clustered by both dyad and classroom, violating statistical assumptions of independence. Therefore, the authors used grouped dyadic standard errors to assess statistical significance (Fafchamps & Gubert, 2007).

Key Findings and Conclusions

Using the CSS data, Neal, Neal, and Cappella (2014) found that there was considerable agreement between self-reported relationships from the *locally aggregated structures* and peer-inferred relationships from the *consensus aggregations*. Specifically, peer inferences about relationships were consistent with self-reported relationships for about 80% of the pairwise combinations in the sample. However, there were some cases of disagreement. For example, in 6% of pairwise combinations, peers inferred a relationship that was not self-reported. Additionally, in 14% of cases, children self-reported a relationship but peers did not infer its existence.

Study results indicated that spatial and social proximities were positive predictors of both self-reported and peer-inferred relationships. However, certain proximities were more strongly predictive of peer-inferred relationships than self-reported relationships. Specifically, peers tended to overestimate the importance of sex similarity in relationship formation. Same-sex children were nine times more likely to have a self-reported relationship but peers were more than 50 times more likely to infer that same-sex children had a relationship.

This study illustrates the power of CSS for comparing and contrasting self-reported relationships with those reported by classroom peers. Unlike other measures of social networks (e.g., self-reported surveys, observations), CSS data allows researchers to assess children's perceptions of relationships and to examine accuracies between self-reported and peer-reported network data. Substantively, results suggest that peers may rely heavily on visible markers like sex similarity to infer the existence of relationships. Methodologically, results suggest that peer-reported data on relationships is not identical to self-reported data and may lead to inflated estimates of sex segregation in classrooms (Neal, Neal, & Cappella, 2014).

Example 2: What Predicts Teacher and Student Agreement on Classroom Networks?

Neal, Cappella, Wagner, and Atkins (2011) investigated associations between classroom characteristics and teacher–student agreement on children's classroom peer networks. Substantively, answering this research question contributed to a growing body of literature on teacher attunement to children's peer relationships, which has been linked to positive outcomes like school bonding (e.g., Gest, Medill, Zadzora, Miller, & Rodkin, in press; Pearl et al., 2007; Pittinsky & Carolan, 2008). Methodologically, answering this research question provides information about the classroom conditions under which teacher and student reports of classroom peer networks are likely to match.

Setting and Sample

This research project used the same CSS dataset used in Neal et al.'s (2014) study described in Example 1. Specifically, CSS data were collected in 34 second through fourth-grade classrooms located in five urban schools in low-income neighborhoods. However, teacher data were not available for one of the 34 classrooms resulting in a sample of 33 teachers, 418 participating students, and 251 students who served as secondary participants. Secondary participants are individuals who do not actively participate in the study as respondents but whose network data are provided by participating respondents. In the context of CSS, participating respondents report on all individuals in a particular setting (e.g., classroom), generally requiring secondary participation (see Klovdahl, 2005 and Neal, 2008 for ethical considerations).

The majority of teachers in the sample were female (85%) and identified as African American (46%) or White (42%). The student sample was predominantly African American (99%) and eligible for free or reduced price lunch (99%). In the student sample, 49% were female and there was a relatively even percentage of second (30%), third (37%), and fourth (33%) graders.

Data Collection and Measures

To measure teacher's perceptions of children's classroom peer networks, Neal et al. (2011) collected CSS survey data on each teacher's perception of his/her students' "hanging out" relationships in the class (see example in Figure 8.1). These CSS data were then used to determine each teacher's *slice*, or view of the whole classroom network. As noted in Example 1 above, CSS survey data were also collected from students. In this study, student CSS data were aggregated within classroom using *consensus aggregations*. Within each classroom, undirected data from respondents k were summed and divided by the total number of respondents to create a matrix where each cell R_{ij} indicated the proportion of respondents who identified a "hanging out relationship" between classmates i and j. Next, this matrix was dichotomized using a binomial test that determined how many respondents needed to indicate a "hanging out relationship" between each pair of classmates to exceed random chance (i.e., $p \leq .05$; see Neal, 2008).

Neal et al. (2011) also collected additional classroom data on grade level, class size, classroom normative behaviors, and teacher classroom organization in their sample. First, they used classroom rosters to collect data on grade level and class size. Second, they conducted peer nomination surveys with students to collect data on classroom norms of aggression perpetration (i.e., five items assessing both physical and verbal aggression) and prosocial behavior (i.e., three items). Classroom norms of aggression perpetration and prosocial behavior were calculated by tallying the number of peer nominations all children in the classroom received (excluding self-nominations) and dividing by the total number of possible peer nominations. Finally, Neal et al. (2011) used the Classroom Assessment Scoring System (CLASS) to conduct structured observations of three dimensions of teachers' classroom organization: behavior management, productivity, and instructional learning formats (Hamre, Mashburn, Pianta, Locasale-Crouch, & La Paro, 2006). In the CLASS, behavior management indicates how well teachers proactively and efficiently promote positive behavior and deal with misbehavior in their classrooms. Productivity indicates the degree to which teachers maximize learning time and establish routines in their classrooms. Finally, instructional learning formats indicate the degree to which teachers actively engage their students in learning (Pianta, La Paro, & Hamre, 2008).

Data Analysis

To determine the level of agreement between teacher's perceptions of the classroom network (i.e., teacher slices) and aggregate peer perceptions of the classroom network (i.e., peer consensus aggregations), Neal et al. (2011) calculated Jaccard similarity coefficients. Jaccard similarity coefficients range between 0 and 1, and can be interpreted as the proportion of agreement in the report of present relationships between teacher slices and peer consensus aggregations. For example, a Jaccard similarity coefficient of 0.3 would indicate that of all of the relationships reported in the teacher slice or peer consensus aggregation, 30% are reported in both. Neal et al. (2011) used regression analysis to determine whether classroom factors (i.e., grade level, size, behavioral norms, teacher classroom organization) predicted Jaccard similarity coefficients indicating teacher–student agreement on classroom networks.

Key Findings and Conclusions

Descriptively, Neal et al. (2011) found that Jaccard similarity coefficients ranged from 0.1 to 0.71, indicating a wide amount of variation in teacher–student agreement on classroom networks. Results of the regression analysis indicated higher levels of teacher–student agreement on classroom networks in higher grades. Teacher–student agreement on classroom networks was lower in larger classrooms and in classrooms with high levels of normative aggressive behavior. High levels of behavior management were associated with less teacher–student agreement on classroom networks while high levels of productivity were associated with more teacher-student agreement on classroom networks.

These findings suggest that teacher attunement to classroom networks depend on a variety of classroom contextual factors. Some of these classroom factors such as class size, behavioral norms, and features of teachers' classroom organization may be malleable points of intervention. For example, reductions in class size have many benefits for teacher–student relationships and peer relationships (see Tseng & Seidman, 2007). As an added benefit, results of this study suggest that reductions in class size may also be associated with more teacher attunement to students' classroom relationships. Findings also suggest that teacher reports of classroom peer networks can be good proxies of student reports, but only under certain circumstances (e.g., when classrooms are small, aggressive norms are low, etc.).

Example 3: How Are Children's Network Positions Associated with Behavior?

Neal (2009) examined how children's positions in their social network position (i.e., degree centrality and ego network density) were associated with relationally aggressive behaviors (i.e. social exclusion and rumor spreading). Substantively, answering this research question contributes to a growing body of literature on childhood relational aggression by exploring where relationally aggressive children are located in their grade-level social network. Previous research had primarily focused on individual characteristics of relationally aggressive children like gender and age (see Archer & Coyne, 2005). Methodologically, answering this research question illustrates how *consensus aggregation* can be used to examine how network positions are associated with childhood classroom behaviors. Furthermore, this example highlights the utility of *consensus aggregation* in samples where typical response rates preclude traditional forms of network analysis (Neal, 2008).

Setting and Sample

Neal (2009) collected CSS data from 144 third- through eighth-grade students located in one urban public elementary school. The sample was racially diverse with African American (34%), White (25.7%), and Latino (29.9%) students. Approximately half of the sample was female (48.6%), and 79.9% were eligible for free or reduced lunch. A total of 99 (68.7%) students providing parental consent and assent served as primary participants by completing CSS surveys. An additional 45 students (31.5%) did not actively participate in the CSS surveys but served as secondary participants (i.e., their names were included in the CSS surveys and their relationships were included in the resulting CSS data).

Data Collection and Measures

To assess grade-level peer social networks, Neal (2009) collected CSS survey data on classmates' "hanging out" relationships from participating children in each classroom. Of particular concern were parent consent rates within grades, which ranged from 53.3% to 86.4%, thus preventing traditional forms of network analysis. Therefore, *consensus aggregation* was crucial for collecting complete whole network data for each grade. Using the same procedures described in detail above in Example 2, *consensus aggregations* were created by aggregating reports of relationships across participating students in each grade level and dichotomizing the resulting matrix based on a binomial test. Using the grade-level consensus aggregations, Neal (2009) calculated Freeman's (1978/1979) normed degree centrality for each child. Normed degree centrality indicates each child's proportion of direct ties to other students out of all possible ties in their grade. Ego network density was calculated for each child as the proportion of present to possible hanging out ties between each of his or her individual alters (i.e., the individuals to whom s/he "hangs out" with).

Demographic characteristics (i.e., race/ethnicity, gender, grade) were determined using self-reported data of participating students, and teacher report of secondary participants. In addition, Neal (2009) collected both teacher-rated and peer-nominated measures of relational aggression. Teacher- and peer-nominated aggression consisted of the following scales: Children's Social Behavior Scale-Teacher Report (CSBS-T; Crick, 1996), Children's Social Behavior Scale-Peer Report (CSBS-P; Crick & Grotpeter, 1995) and an additional item adapted from other peer measures of relational aggression (Henington, Hughes, Cavell, & Thompson, 1998; Osterman et al., 1994).

Data Analysis

To test the effects of centrality and ego network density on teacher-rated and peer-nominated relational aggression, Neal (2009) employed two hierarchical regression models. Children with only one relationship (N = 13) were removed from the analysis as they had undefined scores on the ego network density variable. Demographics were entered into the first block of the regression, and network features were entered into the second block. The dependent variables in the two hierarchical regression models, teacher-rated and peer-nominated relational aggression, were each log transformed prior to analysis to improve normality and linearity.

Key Findings and Conclusions

Neal (2009) found that ego network density explained a significant portion of variance in teacher rated relational aggression. Specifically, children with high levels of ego network density (i.e., tight social clustering among the peers they hang out with) had higher levels of teacher-rated relational aggression. Notably, normed degree centrality was not significantly associated with teacher-rated relational aggression. In contrast, Neal (2009) did not find a significant effect of ego network density on peer-nominated aggression, but did find a significant quadratic effect of centrality, with peer-nominated relational aggression peaking at moderate levels of centrality (i.e., ties with 27% of grade-level peers). Discrepant results for teacher-rated and peer-nominated aggression may indicate that different stakeholders (teachers and peers) have different perceptions of relational aggression.

This study illustrates that where children fall in their peer networks can explain unique variance of relational aggression beyond common demographic features like gender and grade level. Substantively, these results suggest that peer social structures play an important role in the process of relational aggression. Notably, simple changes to classrooms and schools that encourage larger more inclusive networks and less dense networks could be possible points of intervention for reducing children's relationally aggressive behaviors. Methodologically, this study illustrates how consensus aggregation can be used to construct grade level peer networks, and examine associations between children's network positions and behavior.

Benefits and Drawbacks of CSS

The example studies featured in this chapter showcase the rich, multifaceted nature of CSS data and the potential of this method for understanding peer relations in childhood and adolescence. CSS methods are flexible, allowing researchers to concurrently assess children's self-reported ties (i.e., *locally aggregated structures*), children's perceptions of the entire peer network in a setting (i.e., *slices*), and networks that triangulate across multiple reports of relationships by peers (i.e., *consensus aggregations*) (Krackhardt, 1987). Indeed, as illustrated in this chapter, these different types of CSS data can be mixed and matched in studies to answer a myriad of questions about peer relations.

Perhaps one of the biggest benefits of CSS is its ability to rigorously assess individuals' perceptions of social networks. In peer relations research, this opens new doors for understanding children's and adolescents' perceptions of peer networks in settings like classrooms, schools, and neighborhood. CSS can also be used to understand how attuned important adults like teachers are to children's peer relationships. Finally, *consensus aggregations* of CSS data yield full network data even in settings where response rates are low, a common occurrence in research on youth settings in the United States (Neal, 2008). Indeed, as Neal (2008) notes, consensus aggregations not only combat the problem of missing data in peer relations research but also triangulate network data across multiple respondents, reducing the potential for measurement error (Neal, 2008).

Despite several benefits, CSS is not without drawbacks. CSS data collection is time-consuming and can be burdensome for respondents. In a CSS survey, respondents are asked to report on a total of $N \times (N - 1)$ relationships. For example, in a classroom of 30, respondents report on a total of 870 relationships between peers (30 × 29). Researchers interested in using CSS must therefore prioritize this assessment in their data collection efforts, and anticipate that CSS surveys will require a substantial time commitment (i.e., at least 30–40 minutes for a classroom of 25–30 students). In larger settings like high schools, CSS data collection may be impractical given the sizeable number of data points that each respondent would need to report (Neal, 2008). There are some potential solutions that may ease the response burden of CSS data collection. First, if relationships are theorized to be undirected (e.g., "hanging out" relationships), children may only need to report a relationship one time for each dyad. This would reduce the total number of data points respondents would have to report on from 870 to 435 in a classroom of 30 (i.e., $[N \times (N - 1)]/2$). Employing this solution might require abandoning the paper/pencil roster survey illustrated in Figure 8.1 in favor of a web or computer-based survey that asks respondents to report on each possible dyad in the setting just once. Second, some network methodologists have proposed sampling designs that would only require respondents to report on a subset of relationships within a setting (e.g., Butts, 2003). While these sampling designs would reduce response burdens and potentially allow for some types of CSS data (i.e., consensus aggregations), it is important to note that they would not allow for the collection of CSS *slice* data.

Overall, CSS offers a versatile approach to measuring children and adolescent peer networks that can be sliced and diced to understand self-reported relationships, perceptions of whole networks, and peer consensus on relationships. Although CSS has been employed in the adult literature (e.g., Krackhardt, 1987), few researchers have just started to explore its utility for research on childhood and adolescent peer relations in the last decade (e.g., see Neal, 2008). This chapter showcases the promise of this technique for understanding childhood and adolescent peer relationships in an attempt to increase the use of CSS methods among peer relations researchers in the future.

References

Archer, J., & Coyne, S.M. (2005). An integrated review of indirect, relational, and social aggression. *Personality and Social Psychology Review, 9*(3), 212–230.

Butts, C.T. (2003). Network inference, error, and informant in(accuracy): A Bayesian approach. *Social Networks, 25*, 103–140.

Cairns, R. B., & Cairns, B. D. (1994). Social networks and the functions of friendships. In R. B.Cairns & B. D. Cairns (Eds.), *Lifelines and Risks: Pathways of Youth in our Time* (pp. 91–129). New York: Cambridge University Press.

Cappella, E., & Neal, J.W. (2012). A classmate at your side: Classroom characteristics, peer victimization, and network connections in urban schools. *School Mental Health, 4*, 81–94. doi:10.1007/s12310–012–9072–2.

Cappella, E., Neal, J.W., & Sahu, N. (2012). Children's agreement on classroom social networks: Cross-level predictors in urban elementary schools. *Merrill-Palmer Quarterly, 58*(3), 285–313. doi:10.1353/mpq.2012.0017.

Cappella, E., Kim, H.Y., Neal, J.W., & Jackson, D. (2013). Classroom peer relationships and academic engagement in elementary school: The role of social network equity. *American Journal of Community Psychology, 52*(3–4), 367–379. doi:10.1007/s10464–013–9603–5.

Cillessen, A.H.N., & Bellmore, A. (1999). Accuracy of social self-perceptions and peer competence in middle childhood. *Merrill-Palmer Quarterly, 45*(4), 650–676.

Crick, N.R. (1996). The role of overt aggression, relational aggression, and prosocial behavior in the prediction of children's future social adjustment. *Child Development, 67*, 2317–2327.

Crick, N.R., & Grotpeter, J.K. (1995). Relational aggression, gender, and social-psychological adjustment. *Child Development, 66*(3), 710–722.

Ennett, S.T., & Baumann, K.E. (1993). Peer group structure and adolescent cigarette smoking: A social network analysis. *Journal of Health and Social Behavior, 34*(3), 226–236.

Fafchamps, M., & Gubert, F. (2007). The formation of risk sharing networks. *Journal of Development Economics, 83*, 326–350. doi:10.1016/j.jdeveco.2006.05.005.

Faris, R., & Felmlee, D. (2011). Status struggles: Network centrality and gender segregation in same- and cross-gender aggression. *American Sociological Review, 76*, 48–73. doi: 10.1177/0003122410396196.

Farmer, T.W., Leung, M.C., Pearl, R., Rodkin, P.C., Cadwallader, T.W., & Van Acker, R. (2002). Deviant or diverse peer groups? The peer affiliations of aggressive elementary students. *Journal of Educational Psychology, 94*(3), 611–620.

Farmer, T.W., Lines, M.A., & Hamm, J.V. (2011). Revealing the invisible hand: The role of teachers in children's peer experiences. *Journal of Applied Developmental Psychology, 32*, 247–256. doi:10.1016/j.appdev.2011.04.006.

Freeman, L.C. (1978/1979). Centrality in social networks conceptual clarification. *Social Networks, 1*, 215–239.

Gest, S. D. (2008). Social cognitive map. In W. A. Darity Jr. (Ed.), *International Encyclopedia of the Social Sciences* (2nd ed., pp. 575–576). Woodbridge, CT: Macmillan Reference USA.

Gest, S.D., Medill, R.A., Zadzora, K.M., Miller, A.M., & Rodkin, P.C. (in press). Teacher management of elementary classroom social dynamics: Associations with changes in student adjustment. *Journal of Emotional and Behavioral Disorders*. doi:10.1177/1063426613512677.

Gest, S.D., & Rodkin, P.C. (2011). Teaching practices and elementary classroom peer ecologies. *Journal of Applied Developmental Psychology, 32*, 288–296. doi:10.1016/j.appdev.2011.02.004.
Gifford-Smith, M.E., & Brownell, C.A. (2003). Childhood peer relationships: Social acceptance, friendships, and peer networks. *Journal of School Psychology, 41*, 235–284.
Hallinan, M.T., & Smith, S.S. (1989). Classroom characteristics and student friendship cliques. *Social Forces, 67*(4), 898–919.
Hamm, J.V., Farmer, T.W., Dadisman, K., Gravelle, M., & Murray, A.R. (2011). Teachers' attunement to students' peer group affiliations as a source of improved student experiences of the school social-affective context following the middle school transition. *Journal of Applied Developmental Psychology, 32*, 267–277. doi:10.1016/j.appdev.2010.06.003.
Hamre, B. K., Mashburn, A. J., Pianta, R. C., Locasale-Crouch, J., & La Paro, K. M. (2006). *Classroom Assessment Scoring System: Technical Manual.* Charlottesville, VA: University of Virginia.
Hanish, L.D., Martin, C.L., Fabes, R.A., Leonard, S., & Herzog, M. (2005). Exposure to externalizing peers in early childhood: Homophily and peer contagion processes. *Journal of Abnormal Child Psychology, 33*(3), 267–281. doi: 10.1007/s10802–005–3564–6.
Henington, C., Hughes, J.N., Cavell, T.A., & Thompson, B. (1998). The role of relational aggression in identifying aggressive boys and girls. *Journal of School Psychology, 36(4)*, 457–477.
Henry, D.H., & Kobus, K. (2007). Early adolescent social networks and substance use. *Journal of Early Adolescence, 27*(3), 346–362. doi: 10.1177/0272431607302005.
Klovdahl, A.S. (2005). Social network research and human subjects protection: Towards more effective infectious disease control. *Social Networks, 27*, 119–137. doi:10.1016/j.socnet.2005.01.006.
Kobus, K., & Henry, D.H. (2010). Interplay of network position and peer substance use in early adolescent cigarette, alcohol, and marijuana use. *Journal of Early Adolescence, 30*(2), 225–245. doi: 10.1177/0272431609333300.
Krackhardt, D. (1987). Cognitive social structures. *Social Networks, 9*, 109–134.
Lease, A.M., McFall, R.M., Treat, T.A., & Viken, R.J. (2003). Assessing children's representations of their peer groups using a multidimensional scaling technique. *Journal of Social and Personal Relationships, 20*, 707–728. doi:10.1177/0265407503206001.
Marsden, P.V. (1990). Network data and measurement. *Annual Review of Sociology, 16*, 435–463.
Moody, J. (2001). Race, school integration, and friendship segregation in America. *American Journal of Sociology, 107*(3), 679–716.
Neal, J.W. (2007). Why social networks matter: A structural approach to the study of relational aggression in middle childhood and adolescence. *Child and Youth Care Forum, 36*(5–6), 195–211. doi:10.1007/s10566–007–9042–2.
Neal, J.W. (2008). Kracking the missing data problem: Applying Krackhardt's cognitive social structures to school-based social network research. *Sociology of Education, 81*, 140–162. doi: 10.1177/003804070808100202.
Neal, J.W. (2009). Network ties and mean lies: A relational approach to relational aggression. *Journal of Community Psychology, 37*(6), 737–753. doi:10.1002/jcop.20328.
Neal, J.W. (2010a). Hanging out: Features of urban children's peer networks. *Journal of Social and Personal Relationships. 27*(7), 982–1000. doi:10.1177/0265407510378124.
Neal, J.W. (2010b). Social aggression and social position in middle childhood and early adolescence: Burning bridges or building them? *Journal of Early Adolescence, 30*(1), 122–137. doi:10.1177/0272431609350924.
Neal, J.W., & Cappella, E. (2012). An examination of network position and childhood relational aggression: Integrating resource control and social exchange theories. *Aggressive Behavior, 38*, 126–140. doi: 10.1002/ab.21414.
Neal, J.W., & Cappella, E. (2014). The bright side of positive perceptual bias: Children's estimations of network centrality and aggression. *Aggressive Behavior, 40*, 140–151. doi:10.1002/ab.21511.
Neal, J.W., & Neal, Z.P. (2013). The multiple meanings of peer groups in social cognitive mapping. *Social Development, 22*(3), 580–594. doi:10.1111/j.1467–9507.2012.00656.x.
Neal, J.W., Neal, Z.P., & Cappella, E. (2014). I know who my friends are, but do you? Predictors of self-reported and peer-inferred relationships. *Child Development, 85*(4), 1366–1372. doi: 10.1111/cdev.12194.
Neal, J.W., Cappella, E., Wagner, C., & Atkins, M.S. (2011). Seeing eye to eye: Predicting teacher–student agreement on classroom social networks. *Social Development, 20*(2), 376–393. doi:10.1111/j.1467–9507.2010.00582.x.

Osterman, K., Bjorkqvist, K., Lagerspetz, K.M.J., Kaukiainen, A., Huesmann, L.R., & Fraczek, A. (1994). Peer and self-estimated aggression and victimization in 8-year-old children from five ethnic groups. *Aggressive Behavior, 20*(6), 411–428.

Pearl, R., Leung, M., Van Acker, R., Farmer, T.W., & Rodkin, P.C. (2007). Fourth and fifth grade teachers' awareness of their classrooms' social networks. *Elementary School Journal, 108*, 25–39.

Pianta, R. C., La Paro, K. M., & Hamre, B. K. (2008). *Classroom Assessment Scoring System: K-3 Manual.* Baltimore, MD: Brookes Publishing.

Pittinsky, M., & Carolan, B.V. (2008). Behavioral versus cognitive classroom friendship networks. *Social Psychology of Education, 11*, 133–147.

Ryan, A. (2001). The peer group as a context for the development of young adolescent motivation and achievement. *Child Development, 72*(4), 1135–1150.

Schaefer, D.R., Light, J.M., Fabes, R.A., Hanish, L.D., & Martin, C.L. (2010). Fundamental principles of network formation among preschool children. *Social Networks, 32*, 61–71. doi:10.1016/j.socnet.2009.04.003.

Shrum, W., Cheek, N.H., & Hunter, S.M. (1988). Friendship in school: Gender and racial homophily. *Sociology of Education, 61*(4), 227–239.

Tseng, V. & Seidman, E. (2007). A systems framework for understanding social settings. *American Journal of Community Psychology, 39*, 217–228. doi:10.1007/s10464–007–9101–8.

Veenstra, R., Djikstra, J.K., Steglich, C., & Van Zalk, M.H.W. (2013). Network-behavior dynamics. *Journal of Research on Adolescence, 23*(3), 399–412. doi:10.1111/jora.12070.

9

CHAINS OF AFFECTION

The Structure of Adolescent Romantic and Sexual Networks

Peter S. Bearman, James Moody, and Katherine Stovel

Research Question: What do adolescent romance networks look like and how do they form?
System Science Method(s): Networks
Things to Notice:

- Using simulation to uncover network formation mechanisms
- Implications of global network structure for individual health risks

This chapter describes the structure of the adolescent romantic and sexual network in a population of over 800 adolescents residing in a midsized town in the midwestern United States. Precise images and measures of network structure are derived from reports of relationships that occurred over a period of 18 months between 1993 and 1995. The study offers a comparison of the structural characteristics of the observed network to simulated networks conditioned on the distribution of ties; the observed structure reveals networks characterized by longer contact chains and fewer cycles than expected. The chapter identifies the micromechanisms that generate networks with structural features similar to the observed network. Implications for disease transmission dynamics and social policy are explored.

This chapter describes the structure of adolescent romantic and sexual networks in an American high school, accounts for the emergence of this structure, and links the observed structure to the diffusion dynamics of disease. Our goal is to show how local preferences governing partner choice shape the macrostructures in which individuals are embedded and hence affect both the potential for disease diffusion and the determinants of individual risk.[1] Because the structure of sexual networks is critical for understanding the diffusion of sexually transmitted diseases (STDs), it is surprising that epidemiologists have only a limited idea of what such networks look like. The insight we do have is generally restricted to that provided by a set of ego-centered network surveys (Morris and Kretzschmar 1995, 1997; Laumann et al. 1994; Laumann and Youm 1999) and snowball samples of populations of highest risk to HIV acquisition, such as male homosexuals (Klovdahl 1985) and IV drug users (Rothenberg, Potterat, and Woodhouse 1996; Rothenberg et al. 1997; Friedman et al. 1997). While they may reveal much about the characteristics of the local networks in which individuals are embedded, ego-centered and snowball samples provide limited information on the global network properties that determine disease spread.

In this chapter, we describe extensive partnership patterns and network structure for one population of interacting adolescents in a midsized American town, thereby providing detailed

DOI: 10.4324/9781315748771-10

images of, and measurement for, key structural characteristics of a largely complete romantic and sexual network through which STDs may diffuse. As background, we begin by describing some models of sexual networks that are implicit in the existing literature on STDs. We then report the structure of the network generated by the romantic and sexual partnership nominations provided by most of the adolescents in the study community. We consider both cross-sectional and temporal views of this network, and we discuss the extent to which the cross-sectional view obscures the potential for disease diffusion. We then turn to how such a structure could emerge. Because it is theoretically possible that homophily in partner selection—the tendency for individuals with similar attributes, characteristics, or practices to form partnerships—could generate the network structure we observe, we explore the determinants of partnership choice and show that the observed structure is not *solely* a by-product of preferences for particular attributes. We subsequently propose a parsimonious micromodel that, given the determinants of partnership choice, accounts for the structure we observe. Implications for public policy are considered in the conclusion.

Below, we show that (1) current models of disease diffusion rest on sexual network structures that differ in fundamental ways from what we observe, (2) preferences governing partner choice combined with a simple normative proscription against cycles of length 4 (Don't date your old partner's current partner's old partner) induce the structure we observe, (3) partnership preference models that ignore the proscription against completing cycles of length 4 induce incorrect structural representations, and (4) consequently, current intervention efforts that assume the existence of cores may be poorly conceived.

Models of Disease Diffusion

The fundamental quantity in models of disease diffusion is the basic reproductive rate R_0.[2] When $R_0 > 1$, a self-sustaining epidemic occurs; when $R_0 < 1$, the disease dies out. In models of disease diffusion, the reproductive rate is a function of three parameters: the infectivity of the microbe given contact between an infected and a susceptible (β), average duration of infectiousness (D), and the structure of disease-relevant contact within a population (c). The critical sociological parameter is C, the network structure that governs contact.

The simplest epidemiological models assume *random mixing* among all members of the population. Under random mixing, the number of new infections at time t is easily calculated as the number of susceptibles times the number of infecteds times the proportion of contacts between susceptibles and infecteds that result in infection. The result of a random mixing model is the classical S-shaped diffusion curve, where one observes a slow start, followed by exponential growth, and then a decline, either from recovery or death (Sattenspiel 1990).[3]

One can think of random mixing as the statement "people choose partners independent of their characteristics." For many diseases, random mixing captures the essential aspects of the diffusion process. The sneeze of a flu-ridden person on a transatlantic plane sends viral and bacterial material through the air, potentially infecting all the passengers, though those sitting next to the sick person are at greatest risk. Although we may feel otherwise in our less gracious moments, we know that the airlines did not select *us* to sit next to a sneezer and that he or she did not sneeze on us because of our characteristics. For STDs, however, pure random mixing provides a poor approximation of the underlying contact structure.[4]

As sociologists have long noted, partner-selection processes count. Thus, models that explicitly consider bias in partner choice may more closely reflect the social and behavioral processes that give rise to disease-relevant contact structures. For example, the obvious bias relevant for diseases spread via heterosexual contact is toward partners of the opposite sex. Among two-sex models of disease diffusion, the best-known class of partner-bias models are preferred-mixing models that assume disproportionately high levels of contact between individuals who share some attribute (Koopman et al. 1989; Sattenspiel 1990; Jacquez et al. 1988; Hethcote 2000).[5] Based on the homophily principle, these models recognize that, given opposite-sex partnerships, persons often prefer contact with those who are similar to themselves with respect to race, religiosity, sexual preference, activity level, and so on.[6] In such models, leftover contacts occur between people of different groups proportional to the level of sexual activity of these groups. Depending on the values of specific mixing parameters, these preferred-mixing models predict different levels and patterns of disease spread.

Though systematic differences in connectivity patterns have been shown to have striking implications for disease transmission (Morris 1997; Newman 2002; Dezso and Barabasi 2002; Moody et al. 2003), preferred-mixing models do not consider sexual network structure in a direct way. Yet representing the models as networks is a useful way to reveal their assumptions about contact patterns.[7] Three stylized images of sexual networks can be derived from the literature on the diffusion of STDs.[8] The first, and most influential, is that of a *core*. According to standard models, a core is a group of high activity-level actors (e.g., those with multiple partners or who are frequent drug users) who interact frequently and pass infection to one another (often causing reinfection for treatable STDs), and diffuse infection out to a less densely connected population (Phillips, Potterat, and Rothenberg 1980; Hethcote and Yorke 1984a, 1984b; St. John and Curran 1978). Under the general diffusion model, cores are predicted to sustain endemic pockets of disease, since the pattern of intense interaction among members of the core pushes R_0 in the core above 1.

We represent the network structure implied by a core model in Figure 9.1A; here, circles represent individuals and lines represent (disease-relevant) relationships. High-activity actors (core members) are indicated by black circles, and the core is circled. Here we do not differentiate by sex: core membership is determined by activity level, and the core is assumed to contain both males and females. Translating a core-based preferred-mixing model into a network structure highlights specific measurable properties of the resulting graph. In a core, it is likely that an individual's past partner is tied through multiple chains to his or her current or future partner. Thus, if cores exist in a population, cyclicity will be extremely high in the network, and the length of chains connecting pairs of individuals in the population (geodesics) will be low.

While core-based models have been used to account for the diffusion of bacterial STDs like gonorrhea (Hethcote and Yorke 1984a, 1984b; Hethcote and van den Driessche 2000; Hethcote and Van Ark 1987; Aral et al. 1999; St. John and Curran 1978), core models offer a poor description of sexual contact patterns in many contexts. For example, when a key mode of transmission is male long-distance truck drivers having sex with female commercial sex workers (CSWs), members of the groups that constitute possible infection reservoirs (like CSWs) are structurally disconnected from one another and do not transmit infection *directly* to one another. Capturing such dynamics—which may be more characteristic of two-sex diffusion processes—requires more complex switching models, often called *inverse core* models (Garnett et al. 1996). In an *inverse core*, a central group of infected persons pumps disease out to others but does not pass infection directly among themselves. For instance, prostitutes might be infected by previously infected johns and then pass infection on to other johns.

We represent the network associated with an inverse core in Figure 9.1B. Here we distinguish actors according to their *role* in the diffusion process (commercial sex workers are black, sex customers are white; the inverse core is circled). The key difference between a core and an inverse core stems from the social organization of sexual relations, since johns are more likely than other potential carriers to spread infection to individuals not in the graph (specifically,

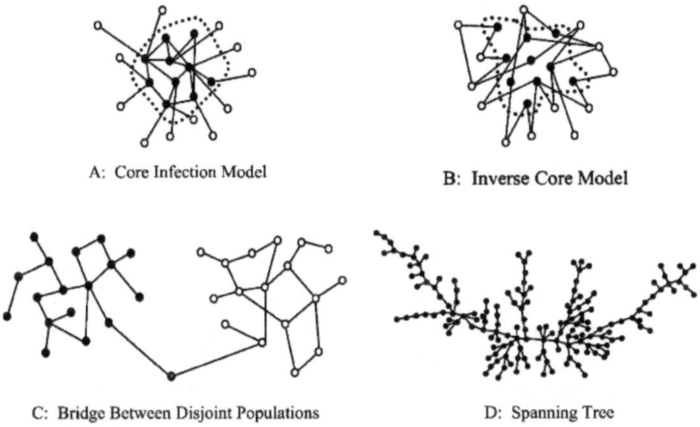

Figure 9.1 The network structure of four models of infection

their regular sex partners). Structurally, however, the two networks are quite similar; though cores may be smaller and denser, both structures are associated with high cyclicity and low path distance between individuals. Since viruses are hardly attentive to the social details that occupy us, the two structures hold similar potential for disease diffusion.

A third model in the epidemiological literature describes disease diffusion dynamics as driven by *bridging processes* (Aral 2000; Gorbach et al. 2000; Morris et al. 1996). These models posit two populations of persons engaged in different behaviors (i.e., a high-risk and a low-risk population) linked by a few individuals who bridge the boundary between each world (e.g., an IV drug user who shares needles with his drug partners and who has sex with non-IV-drug users). A network consistent with this model is shown in Figure 9.1C, where the black circles denote actors who engage in high-risk behavior and the white circles denote actors who engage in lower-risk behavior. Bridges are those who link the two worlds.[9]

Each of these graphs represents the network foundation of a preferred-mixing model. In the core and inverse core models, mixing results from in-group preference with respect to risk status (or some other attribute, like IV drug use, associated with risk status). Even in the bridging model, parameters for cross-group contacts are estimated from individual-level data, and random mixing within groups is presumed. In all cases, disease potential is contingent upon the extent to which (1) there is at least one local pocket of densely interconnected persons (connected either via direct connections as in the core or via short and redundant cycles as in the inverse core) and (2) the pocket of high density is connected to the remainder of the population through bridging chains that reach into the periphery of the network. However, these models are useful only to the extent that any empirically observed network structure matches that implied by the in- and out-group contact parameterization. Obviously, if the network one actually observes bears little relationship to the structure implied by these models, we must radically reconsider their usefulness.[10] For our data, this is the situation.

Specifically, in the context we study we observe a network structure that has the appearance of a *spanning tree* – that is, a long chain of interconnections that stretches across a population, like rural phone wires running from a long trunk line to individual houses (Hage and Harary 1996, 1983). The global structure of a chainlike spanning tree is characterized by a graph with few cycles, low redundancy, and consequently very sparse overall density.[11] The shortest distance between any two randomly selected individuals (geodesic) is significantly higher than that observed in either the core or inverse core structures.[12] A typical spanning tree structure is represented as Figure 9.1D.

Random-mixing dynamics and positive preferences for partners do not produce spanning tree structures.[13] Rather, this network structure appears when formal or informal rules *preclude* the enactment of specific relations. In the language of kinship structures, spanning trees are the product of negative proscriptions: sets of rules about whom one *cannot* be in a relationship with. Consequently, they are most frequently observed in large and complex generalized exchange systems, as in the exchange of valuables in the Kula ring (Hage and Harary 1996; Schweizer and White 1998).

As noted above, the extant models of sexually transmitted disease diffusion implicitly assume network structures that correspond to one of the first three images in Figure 9.1. Yet we have essentially no complete population data from which to conclude that any of these models are empirically appropriate. Fundamental at this point is the need to learn more about actual networks and the structural characteristics that are relevant for disease diffusion. In this article, we describe these characteristics in an observed romantic and sexual network in a population of adolescents. The network structure we find closely approximates a spanning tree. Since such structures are the result of rules restricting partnership choice, we focus on identifying a parsimonious rule that could produce the structure we observe empirically. Conditional on simple homophily preferences in partnership choice, the structural properties of networks simulated according to this rule closely correspond to what we observe.

Context and Data

Data for this chapter are drawn from the wave 1 component of the National Longitudinal Study of Adolescent Health (hereafter, Add Health), a longitudinal study of adolescents in grades 7–12. In 1994, in-school questionnaires were administered to approximately 90,000 students in 140 schools. Almost a year later, a nationally representative sample of over 20,000 of these students completed extensive interviews in their homes. In 14 saturated field settings composed of two large (N = 1,000; N = 1,800) and 12 small (N < 300) schools, Add Health attempted home interviews with *all* enrolled students. The two large schools were selected with the intent of capturing typical high school experiences in urban and less urban communities. The adolescent in-home interview was conducted using audio-CASI technology for all sensitive health status and health risk behavior questions. Adolescents listened to the questions through earphones and directly entered their responses into a computer, thereby eliminating interviewer or parental effects on their responses (Turner et al. 1998). Adolescents were asked to identify their sexual and romantic partners from a roster of other students attending their school. Consequently, in the saturated field settings, we have almost complete sexual and romantic network data.

Context: "Jefferson High"

In this chapter we report data from the 832 respondents who attended a school we identify as "Jefferson High School," one of the two large high schools where Add Health attempted in-home interviews with all students. Jefferson High is an almost all-white high school of roughly 1,000 students located in a midsized midwestern town. Jefferson is the only public high school in the town. The town, "Jefferson City," is over an hour's drive from the nearest large city. While densely settled, Jefferson City is surrounded by beautiful countryside and is home to many agricultural enterprises. At one time the town served as a resort for city dwellers, drawing an annual influx of summer visitors, though this is no longer the case, and many of the old resort properties show signs of decay. At the time of the fieldwork, students were reacting to the deaths of two girls killed in an automobile accident. Despite this, fieldwork proceeded exceptionally well. Adolescents frequently approached interviewers wearing yellow Add Health buttons and asked when they would be invited to participate in the study.[14] In all, 90% of the students on the school roster participated in the in-school survey, and over the course of the interview period, 83% of all students in the school completed in-home interviews.

Jefferson is a close-knit, insular, predominantly working-class community, which offers few activities for young people. In describing the events of the past year, many students report that there is absolutely nothing to do in Jefferson. For fun, students like to drive to the outskirts of town and get drunk. For our purposes, the relative isolation of the community is an important factor, significant for the patterns of romantic partnership and sexual partnership choices we observe. The context provides a good setting in which to look for the networks suggested by preferred-mixing models, for if redundant structures (and therefore, cores) exist, they are most likely to appear in island populations not permeated by the currents of larger, more cosmopolitan settings.

Table 9.1 describes the tenth- to twelfth-grade students at Jefferson High across a broad spectrum of characteristics.[15] It also contains comparisons with all other high schools in the sample (col. 2); all disproportionately white schools (over 75% white; col. 3); high schools of comparable size (col. 4); and finally, the small set of other disproportionately white high schools of similar size in Add Health (col. 5).

Sign tests reveal that, in general, Jefferson High is similar to other U.S. schools across most of the comparison variables.[16] However, Jefferson students earn lower grades, are suspended more often, feel less attached to school, and come from poorer families than those at comparable schools. They are more likely than students in other high schools to have trouble paying attention, and they have lower self-esteem, pray more, have fewer expectations about college, and are more likely to have a permanent tattoo. Compared to other students in large, disproportionately white schools, adolescents at Jefferson High are more likely to drink until they are drunk. In schools of comparable race and size, on average 30% of tenth- to twelfth-grade students smoke cigarettes regularly, whereas in Jefferson that figure is 36%. Drug use is moderate, comparable to national norms. More than half of all students report having had sexual intercourse, a rate comparable to the national average and only slightly higher than observed for schools similar with respect to race and size. Jefferson is not Middletown, but it looks an awful lot like it.

Romantic Partnerships and Sexual Partnerships

During the in-home interview, adolescents were asked if they were in or had been involved in a *special romantic relationship* at some point during the past 18 months. Adolescents in such relationships were asked to describe their three most recent relationships, including any current relationships.[17] In addition, adolescents were asked to identify up to three individuals with whom they had a *nonromantic sexual relationship* in the past 18 months. A nonromantic sexual relationship was defined as a relationship involving sexual intercourse that the respondent did not identify as special and in which the partners did not kiss, hold hands, or say that they liked each other. A large number of sexual, nonromantic relationships were reported. For the vast majority of reported partnerships, start and end dates for all romantic and nonromantic sexual partnerships were collected. Slightly less than one-quarter of all Jefferson students reported no romantic or nonromantic sexual relationship during the preceding 18 months.

After collecting detailed information about partnerships, respondents were asked if their partner attended their school (or the middle school that fed students into the high school). If their partner

Table 9.1 School-level comparisons: Jefferson High and other high schools in the Add Health sample

	Jefferson High	*All*	*Mainly White*	*600–1,000 Students*	*White, 600–1,000 Students*
Family SES	5.59	5.78 (1.16)	5.95 (.91)	5.73 (1.20)	5.86 (.86)
Proportion in poverty	.13	.16 (.14)	.09** (.09)	.16 (.15)	.13 (.13)
Log(family income)	3.61	3.56 (.42)	3.65 (.29)	3.55 (.47)	3.54 (.35)
GPA	2.49	2.83*** (.24)	2.93*** (.18)	2.78*** (.19)	2.82** (.10)
Expect college graduation	3.77	4.09*** (.48)	4.18 (.36)	4.09*** (.30)	4.09* (.25)
School attachment	3.27	3.67*** (.26)	3.68*** (.29)	3.59*** (.23)	3.49* (.17)
Trouble in school	1.20	1.01*** (.13)	1.02*** (.11)	1.04*** (.12)	1.05* (.13)
Drunk	1.14	.86*** (.30)	.94** (.30)	.93*** (.27)	1.02* (.11)
Delinquency	.29	.26** (.07)	.25* (.06)	.27 (.05)	.25 (.05)
Hours watching TV	10.29	13.7*** (3.52)	11.81 (2.74)	13.87** (4.27)	12.03 (2.39)
Religiosity (praying)	2.71	2.09*** (.42)	2.22*** (.49)	2.18*** (.38)	2.32 (.37)
In-degree	5.32	4.39*** (1.02)	4.85* (.77)	4.61** (.92)	5.00 (.41)
Self-esteem	2.99	3.18*** (.11)	3.17*** (.09)	3.18*** (.12)	3.14** (.08)
Sexually active	.59	.53 (.16)	.51 (.16)	.57 (.14)	.55 (.10)
Autonomy	.86	.80*** (.05)	.82*** (.04)	.82*** (.05)	.84* (.03)
Expects to get AIDS	1.52	1.53 (.15)	1.53 (.11)	1.58 (.10)	1.58 (.09)
Marry by 25	3.06	3.18*** (.25)	3.32*** (.21)	3.13** (.17)	3.21** (.13)
Attractiveness	3.45	3.59*** (.21)	3.59** (.18)	3.59 (.22)	3.57 (.22)
AH_PVT	105.32	101.05*** (8.96)	105.29 (3.06)	102.06 (6.67)	105.11 (3.62)
Two biological parents	.46	.53** (.13)	.58*** (.11)	.54 (.13)	.58* (.08)
Smokes regularly	.36	.26*** (.11)	.32* (.09)	.28*** (.08)	.32* (.03)

Table 9.1 Continued

	Jefferson High	*All*	*Mainly White*	*600–1,00 0 Students*	*White, 600–1,000 Students*
School suspension	.40	.27***	.22***	.30**	.24**
		(.14)	(.12)	(.14)	(.11)
Tattoo	.10	.06***	.05***	.05***	.06*
		(.04)	(.03)	(.03)	(.03)
Number	1	75	28	23	9

Notes: Numbers in parentheses are SDs (Students in tenth, eleventh, and twelfth grades only). $*p < .05$, whether the median of the sample distribution equals the Jefferson value, based on a sign test. $**p < .01$. $***p < .001$, whether the median of the sample distribution equals the Jefferson value, based on a sign test.

attended either school, respondents were asked to identify their partner from a roster by a unique ID. Through this process we collected data on 477 partnerships between respondents at Jefferson High and one of the two sampled schools in Jefferson.[18] We use these partnerships to generate a snapshot of the network of romantic and sexual relations among adolescents attending high school in Jefferson—the first such image that does not rely solely on egocentric reports from a small fraction of the relevant population.

Observed Romantic and Sexual Networks at Jefferson High

Figure 9.2 maps the actual network structure that links the 573 students involved in a romantic or sexual relationship with another student at Jefferson High.[19] Circles denote individual students; romantic or sexual relations are flows between nodes. Time is suppressed in this representation.

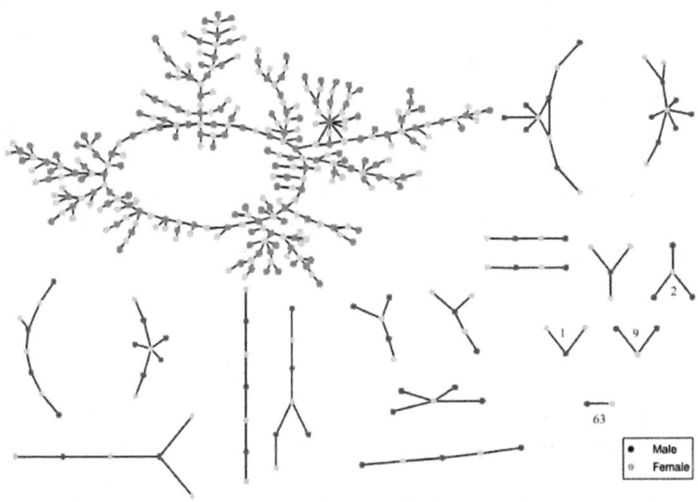

Figure 9.2 The direct relationship structure at Jefferson High

We begin by examining the distribution of components in this network. A component is a subgraph of a network in which all nodes are reachable from other nodes in the subgraph (Wasserman and Faust 1994). Components are significant for disease transmission, since individuals who are not in the same component of a sexual network cannot infect each other with sexually transmitted diseases. A few simple components occur with some frequency in Jefferson High. For example, the simple dyadic structure (two individuals whose only partnership is with each other) occurs 63 times at Jefferson. Thus 126 students are involved in isolated dyadic relations. It is important to note, however, that far more than 126 students at Jefferson report only one relationship; many of the more complex components also include students with only one partner. However, the partners of these students have other partners. This illustrates the importance of collecting data extending beyond ego-centric networks, for it is only by learning directly about the behavior of partners' partners that we can map the structure of connectivity through which disease must flow.[20]

Components involving three students are also fairly prevalent at Jefferson. Triads composed of one male and two females occur 12 times, and triads composed of one female and two males occur nine times. All told, a total of 189 students at Jefferson (35% of the romantically active students) are embedded in sexual and romantic network components containing three or fewer students. There are very few components of intermediate size (4–15 students).

The most striking feature of the network is the existence of a very large component involving 52% (N = 288) of the romantically involved students at Jefferson. While this large component involves the vast majority of individuals with multiple partners, it has numerous short branches. Further, it is very broad: the two most distant individuals are 37 steps apart. Most surprising, it is characterized by the almost complete absence of short cycles. Thus the network closely approximates a chainlike spanning tree.

The size of the large component of connected nodes identifies the worst-case scenario for potential disease diffusion within the population. While one-third of all students are embedded in small, disjoint dyads and triads, in an 18-month period more than 50% of the students at Jefferson were chained together through romantic and sexual relationships that could have involved the exchange of fluids. Recall that there are many individuals at the end of small branches in the large component who have only one partner. While these adolescents have only had one partner, their risk for contracting an STD may be significantly greater than an individual with multiple partners who is embedded in a smaller, disjoint component. Consequently, STD risk is not simply a matter of number of partners.

While it is reasonable to think that an individual might have some sense of their own partners' relationships, the structure of the larger components, and certainly the largest component, is not likely to be visible, or meaningful, to the students at Jefferson. These structures reflect relationships that may be long over, and they link individuals together in chains far too long to be the subject of even the most intense gossip and scrutiny. Nevertheless, they are real: like social facts, they are invisible yet consequential macrostructures that arise as the product of individual agency.

Temporal Unfolding

Figure 9.2 depicts the direct relationship structure linking individuals together. Disease transmission, however, rests on temporally ordered relationships, and these determine the indirect pathways that can put individuals at risk for disease. Thus, if A and B are partners at time 1, and B and C are partners at time 2, from a viral or bacterial perspective a meaningful directed path with the capacity to transmit disease exists between A and C. In contrast, given this pattern

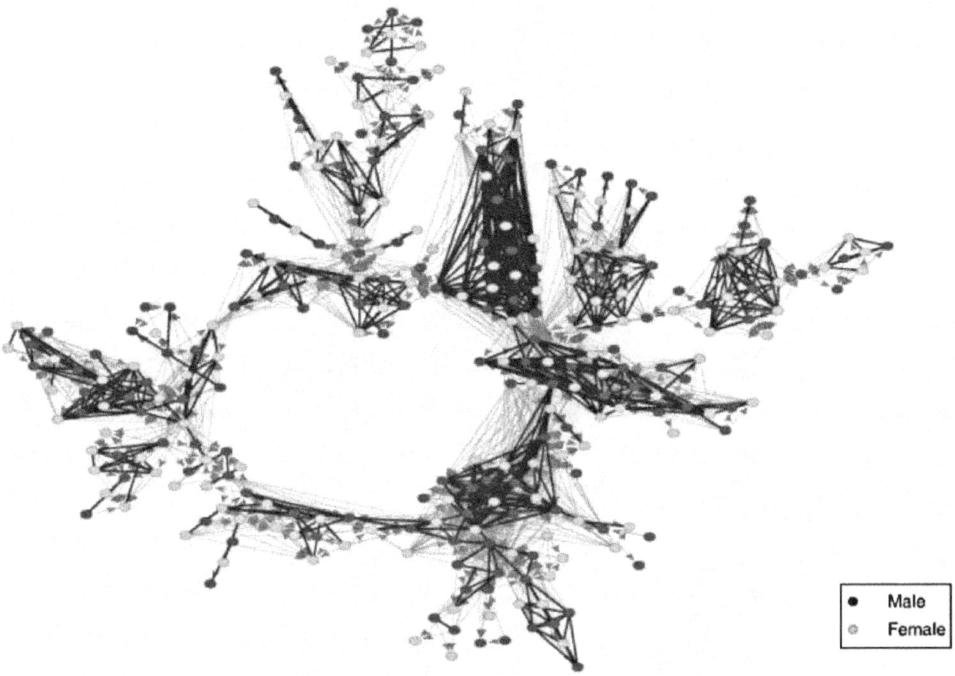

Figure 9.3 Temporally ordered ties in the Jefferson High partnership network

of relationships, disease cannot flow from C to A. Taking into account data describing the temporal ordering of relationships, Figure 9.3 reports all indirect and direct ties that could potentially transmit disease within the major component of the Jefferson network.[21] Note that compared to the direct graph in Figure 9.2, the indirect graph is quite dense and contains many regions with interacting adolescents. As a comparison, consider Figure 9.4, the graph of a simulated network containing the *fewest* possible indirect relations derived from the original component. This is the minimal arrangement from a disease perspective; the difference in density between Figure 9.3 and Figure 9.4 suggests the extent to which the actual dynamic unfolding of partnerships at Jefferson increases the potential for widespread disease diffusion (Moody 2002).

Structural Fragility

Examining the pattern of indirect ties reveals the level of connectivity and redundancy of the network through which disease could travel. While Figure 9.3 reveals the existence of clusters of romantically involved students, it does not reveal how robustly connected these clusters are to one another. In general, structures like spanning trees are considered structurally fragile because the deletion of a single tie or a single node can break a large component into disconnected subgraphs.[22] Consider again the analogy to phone lines: if phone lines are laid out as a spanning tree, a break in the major trunk line separates a single component into two disjoint components and prevents calls from traveling from one component to the other. Engineers protect against such failure by adding lines that build redundancy into the system. The essential structural fragility of spanning trees reveals how subtle changes in local network structure (deleting or adding a relationship, e.g.) can have profound effects at the macrolevel.

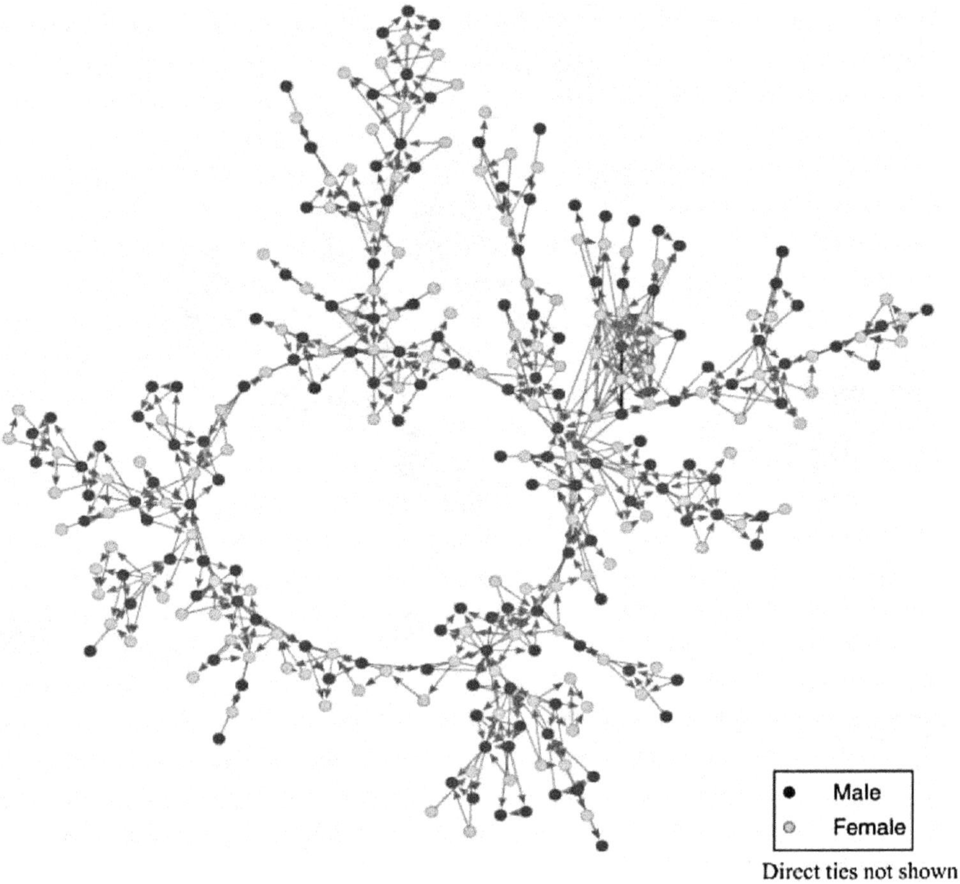

Figure 9.4 Simulated minimal temporal connectivity in the Jefferson High partnership network

Building from the temporally ordered indirect network shown in Figure 9.3, Figure 9.5 reveals how the structure of indirect ties breaks into a set of smaller, mutually reachable sets when cut-points (single pathways between nodes) are eliminated. While each of the remaining smaller components appears to be dense and corelike, simply removing ties at the cut-points fractures the structure into separate components. For sexual networks, redundant lines provide the foundation for cores, the incubators of epidemics. Thus, in a sparse treelike contact structure with many cut-points, failure to transmit disease within a partnership that happens to be a cut-point can break the larger connected components into separate, unconnected subcomponents, thereby fragmenting the potential epidemic.

Generating the Structure: Comparison to Simulated Networks

Data describing the complete structural mapping of a romantic/sexual network in an interacting population has not been previously collected, so there is no obvious baseline against which to evaluate whether what we observe is unusual. Further, the distributional properties of many of the network statistics we are interested in are not well known. Thus, while the graph of the observed network at Jefferson appears to stand in clear contrast to the structures implied by

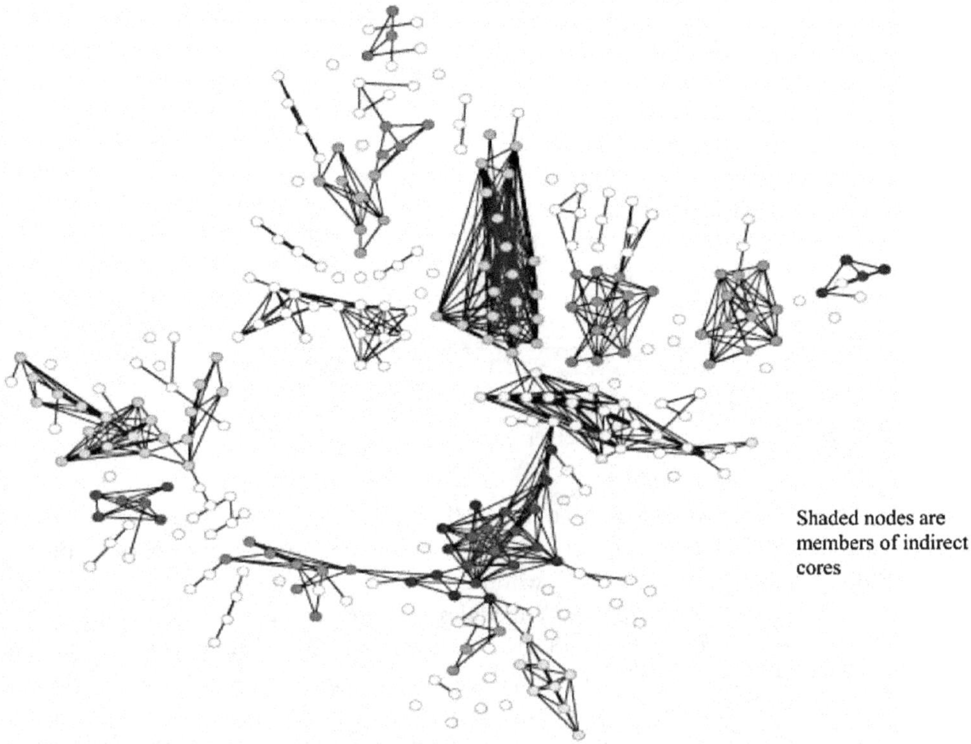

Figure 9.5 Temporally ordered partnerships: cutpoints removed

most epidemiological models, there is some possibility that it is simply a stochastic realization of one of the random or preferred-mixing models. To test against these alternatives, we simulate 1,000 networks consistent with various constraints characteristic of these other models and examine whether the relevant structural characteristics of the Jefferson network are statistically likely, given the distribution of simulated networks.

We begin with the simplest model: Jefferson students select their partners at random.[23] To test this, we simulate 1,000 random networks with the same size and degree as observed in Jefferson,[24] and then we consider where the network at Jefferson falls relative to the distribution of simulated networks.

Figure 9.6 presents box plots comparing the Jefferson High network to the simulated networks across six measures relevant for STD diffusion dynamics: density at maximum reach, centralization, mean geodesic length, maximum geodesic length, skew of reach distribution, and number of cycles. The values for each network measure are standardized (mean = 0; SD = 1). The cross within each box plot reports the median value from the simulated nets; the interquartile range is shaded. A dark circle indicates the value we observe for Jefferson High. Not surprisingly, across *all* of these basic measures of network structure, the sexual and romantic network in Jefferson is an outlier relative to the simulated networks generated by random mixing. We discuss each measure in turn.

The first measure, *density at maximum reach*, assesses the extent to which the overall network is connected. Here, we measure the density of the network that arises when ties link all pairs of ever-reachable individuals.[25] All things being equal, heightened connectivity is associated with

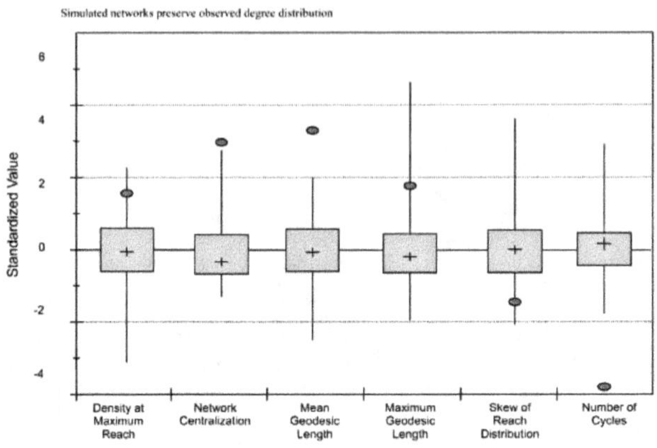

Figure 9.6 Simulated networks, preserving observed degree distribution

more efficient disease spread. Compared to the 1,000 simulated networks, the Jefferson network is highly connected. This means that students at Jefferson are more likely to have partners at school who have other partners at school who have other partners at school. The alternative is many dyads, or other small groups of linked adolescents, that are ultimately disjoint from the rest of the population: for example, a core disconnected from a set of smaller components. The highly connected structure at Jefferson, therefore, poses a greater disease risk than would exist if the partnerships were formed at random.

We next consider *network centralization*, a measure of the inequality of the centrality of persons in the network. We calculate centralization using Bonacich's centrality algorithm with a negative beta value (Bonacich 1987). This parameterization gives extra weight to individuals who connect otherwise unconnected individuals and less weight to those whose bridges are redundant to other bridges in the network. Compared to simulated networks with the same degree distribution, the Jefferson network is highly centralized, suggesting that some actors play a central role in linking disjoint clusters.

Continuing to move from left to right, the next box plot compares the *mean geodesic length* of the Jefferson network to the simulated networks. A geodesic is the shortest path between two connected persons in a network: mean geodesic length is the mean of the shortest path between every connected pair in the network. Large geodesics are characteristic of spanning trees, the sparse chainlike structures with few alternate paths directly connecting persons. In contrast, holding connectivity constant, networks containing redundant links between actors (cycles, a core, or starlike structures) will have smaller mean geodesic lengths. With respect to STD diffusion, the absence of redundancy places pressure on the values of the β (probability of transmission given contact) and D (duration of infectiousness) parameters discussed previously. If β and/or D are low, spanning trees are inefficient structures for diffusion of STDs. Compared to the simulated networks, the network at Jefferson High has very long mean geodesics. This is the result of the extremely large component and the overall absence of "short-cuts," or redundant ties, within the large component.

While the Jefferson network is highly connected, this connectivity is the result of very long chains. It follows that as the mean geodesic length is large, that the *maximum geodesic length*—a measure that captures the number of steps between the most distant pair of connected persons—will also be large. And compared to the simulated networks, the most distant pairs of connected individuals in Jefferson are quite distant from one another. In fact, they are not likely to know that they are involved in the same romantic web, which exists as a social fact beyond the reach of ordinary cognition.

By definition, every person in both the simulated and observed networks is connected to at least one other person (their romantic relationship partner). In addition to their own direct relations, however, they may be indirectly connected to others through the relations of their partners and their partners' partners. Extending this logic, we can calculate for each actor an individual-level measure of the number of "reachable" alters in the network. We then can consider the *skew of the reach distribution* how unequally the number of reachable partners is distributed across the population. If most of the population were in isolated dyads, the distribution would show a strong positive skew, and the structure would contain few efficient pathways for disease transmission. In contrast, a network that includes a very large component would show a strong negative skew. This is the case in Jefferson. Negatively skewed reach distributions are a trace of contact structures with heightened potential for disease spread.

Among the most structurally characteristic feature of the graph of the Jefferson network is the pronounced absence of *short cycles*. The absence of short cycles guarantees that we do not observe a densely interconnected core that has the capacity to function as a disease reservoir. In comparison to the simulated networks, the romantic and sexual network at Jefferson is characterized by significantly *fewer* cycles than occur when partnerships are chosen randomly. Consequently, STD models that assume a core or inverse core structure are not appropriate here. Using such models in contexts such as Jefferson could result in underestimation of the potential for disease spread—especially if β or D is moderately high, as is the case when treatment rates are low or asymptomatic cases are frequent.

Compared to randomly simulated networks of similar size and degree, the empirical sexual network we observe is quite distinctive. The Jefferson network is dominated by an extraordinarily large component that connects more than half of all the students who are romantically and sexually active in the school. Yet while this component ties individuals together into long chains of potential infectivity, it is extremely fragile. This fragility is largely due to the striking absence of cycles (redundant paths) in the large component.

Preferences for Partners

Because the spanning tree structure we observe is extremely unlikely to be the result of random mixing, some other set of processes governing partnership selection must account for it. It is obvious that, when individuals choose partners, they do not base their choice on its contribution to the global macrostructure. Put most starkly, adolescents do not account for their partner choice by saying, "By selecting this partner, I maximize the probability of inducing a spanning tree." First, they cannot see the global structure, and second, they do not care about it. They do care, however, about the more immediate local structure in which the partnership is embedded, and they care about the attributes their potential partner has.

One possibility is that there is a simple micropreference governing choice that, if followed by most individuals, would naturally produce a spanning tree. This is the solution we ultimately consider. We propose a specific rule that, if followed, induces the macrostructure we observe, given the conditions of partnership preference in Jefferson. Later in this article we provide reasons for thinking that this preference is enacted, even if adolescents do not articulate it. Getting to this point requires examining the empirical determinants of partnership choice at Jefferson, which we consider immediately below.

Attribute-Based Selection Preferences

Everyday experience, a cursory glance at personal advertisements in the classified section of any newspaper, a brief inquiry into the underlying logic of dating or matchmaking services, and a wide body of research all suggest that individuals select partners on the basis of characteristics, and that persons tend to prefer partners who are similar to them. While the number of attributes and behaviors that could provide a foundation for preferential partnership selection is enormous, in Table 9.2 we report the level of homophily across a set of attributes and behaviors that might reasonably be expected to govern partnership formation among adolescents.[26] To assess the extent of homophily on selected attributes within romantic partnerships in Jefferson, we generated 500 permutations of the attribute distance/matching matrix with the romantic relation matrix, and then we used QAP to evaluate the difference in attribute means between actual romantic pairs and the randomly simulated partnerships.[27] For continuous variables, the test statistic compares the mean of the difference in the absolute value of the attribute measure for romantic pairs with the mean of the difference between the randomly assigned pairs.[28] Thus, for example, the 0.367 value for grade means that, on average, romantic pairs are about a third of a grade closer to each other than are randomly assigned pairs. For categorical variables (i.e., smoking), the test compares matching scores between real and randomly assigned pairs ($X_{ij} = 1$ if $x_i = x_j$ where X is the matching indicator, x is the attribute, and i and j are the members of the pair). The difference formula is then the proportion of nonromantic pairs that match minus the proportion of romantic pairs that match. Thus, the value of –.11 for smoking means that the proportion of similar-smoking-status romantic pairs is .11 larger than that for randomly assigned pairs.

Table 9.2 demonstrates clear evidence of homophily in romantic partnerships.[29] Adolescents at Jefferson tend to select partners with similar socioeconomic status, grade point average, college plans, attachment to school, trouble in school, drinking behavior, IQ, and grade. With respect to categorical attributes, partners tend to be similar in terms of sexual experience, suspension from school, and smoking. Less important is religious denomination. Evidently, students who smoke prefer other students who smoke. Alternatively, students who smoke induce smoking in their partners, perhaps because only smokers can tolerate kissing smokers.

While homophily is strong, the preference for similarity does not extend to all characteristics, most obviously sex and age. Almost every single reported romantic relationship at Jefferson is a cross-sex relationship, and as is true in most high schools, girls at Jefferson tend to be involved with older boys. Ninth-grade girls tend to be in relationships with ninth- and tenth-grade boys, tenth-grade girls with boys in the tenth and eleventh grades, and so on. Among all partnerships involving Jefferson students, we observe a mean grade difference of .9, less than expected if relationships were formed independent of age (mean difference = 1.23 in the randomly assigned pairs), but evidence of a female preference for older boys (or male preference for younger girls).[30]

Homophily in Partnership Experience

Even given these revealed preferences for attributes, adolescents have a great deal of leeway in terms of selecting potential romantic partners. Among adults, we know that experienced partners prefer experienced partners (homophily on experience), a preference that can give rise to cores (Laumann et al. 1994). Visual inspection of the graph shown in Figure 9.2 suggests that many of the differences between the Jefferson sexual and romantic network and the simulated networks may be the result of the large number of isolated dyads we observed in Jefferson. Thus, we ask whether the single large component involving half of all students is a mathematical byproduct of homophily on this one partnership characteristic: the number of previous partners

Table 9.2 Homophily in student pairs

Variable	*Qap Mean Difference*[a]	
	Full Network	*Cross-Sex Only*
Family SES	.299***	.295***
Grade	.331***	.367***
GPA	.096**	.102***
Expect to graduate college	.202***	.222***
School attachment	.118***	.132***
Trouble in school	.029	.019
Gets drunk	.180***	.195***
Delinquency[b]	–.058	–.070
Hours watching TV	–.149	–.027
Religiosity (praying)	–.006	–.012
Popularity (in-degree)	–.377*	–.211
Self-esteem	.004	.008
Autonomy	.008	.002
Expect to get HIV	.003	–.007
Expect to marry by 25	.025	.020
Attractiveness	.013	.047
Vocabulary (AH_PVT)	1.508***	1.671***
Religion	–.034*	–.043*
Sexually active	–.100***	–.124***
Smoking	–.087***	–.110***
School suspension	–.028	–.066**
Tattoo	–.003	–.016

a Significance reflects exact *P*-test comparison to 500 permutations of the attribute distance/matching matrix with the romantic relation matrix.

b Delinquency is standardized by gender and age.

*$p < .05$; ** $p < .01$; *** $p < .001$.

an individual has had. If a majority of the individuals with only one partner are involved with individuals who also have only one partner, it would follow that those with multiple partners are constrained to be involved with persons who have also had previous partners. The catenation of these individuals should, all things being equal, generate large interconnected components.

To test this idea, we again simulate 1,000 networks with fixed size and degree distribution, this time removing the 63 isolated dyads (involving the 126 persons whose single partner has only a single partner) and prohibiting the creation of new isolated dyads. We then compare the large component from the Jefferson network to the structural characteristics of network simulated with the prohibition against isolated dyads. Adding this single additional constraint has a stunning impact on the structure of our simulated networks. Specifically, the mean size of the largest component in the simulated networks is now very close to the size of the large component in the Jefferson network (mean of 283 nodes vs. 288 nodes in Jefferson). As suspected, homophily in partnership selection among less experienced partners (those with only a single romantic involvement) *provides an efficient micromechanism for the generation of a large component.* Thus, homophily on experience is a key element in generating the structure we observe.

To consider network features other than component size, we again compare the simulated networks with the Jefferson network across the set of network measures salient for disease diffusion previously discussed in Figure 9.6. These results are shown in Figure 9.7.

As before, values are standardized (mean = 0; SD = 1). The cross-hatch within each box plot reports the median value, and the interquartile range is shaded. Dark circles indicate the values we observe for Jefferson. Across all six network measures, the structure of the Jefferson sexual network remains significantly different than expected, although less so than under the less constrained simulations, where we did not prohibit isolated dyads. The improved fit results from smaller variances in the simulated networks.

Although homophily on experience appears to account for the size of the largest component, the structural characteristics of the observed Jefferson network are still unusual relative to the simulated networks. Thus, this micromechanism is not sufficient to reproduce the structural properties of the observed network. The main differences between the simulated and real networks, in mean geodesic length, network centralization, reach, and skew of maximum reach, are the product of the absence of cycles. Thus, while preferential selection on partnership experience level provides an efficient foundation for generating large components in adolescent sexual networks, it fails to generate a spanning tree. Among romantically active students, random mixing produces more redundant ties than exist in Jefferson. As a consequence, the simulated networks reveal core structures rather than spanning trees.

Uncovering Governing Norms

Our analyses thus far demonstrate that the macrolevel network structure at Jefferson is neither the simple product of random mixing nor of individual preferences for partners with particular attributes. Because we find many cycles of length 4 in the simulated networks, but few in Jefferson, we believe that there must be a prohibition against partnerships that involve the creation of short cycles.

We adopt a new strategy to investigate just how unusual short cycles are at Jefferson. Earlier we showed that while spanning trees may be efficient for disease transmission, they are structurally fragile. Whereas our investigation of structural fragility was based on the consequences of *removing* relationships from the graph, we now consider the effects of *random rewiring* of the network—that is, we randomly reassign partnerships from one pair of nodes to another pair. Since the new partnerships we introduce are formed at random, they are insensitive to any existing norms or preferences that may govern partner choice at Jefferson. By analogy, consider the existence of an incest taboo that restricts available partners to those who are a culturally agreed upon distance from ego. "Rewiring" the resulting marriage graph means that some relations prohibited by this rule will be added to the network, and therefore some structural properties of the new graph may no longer match the original data structure. Structural features

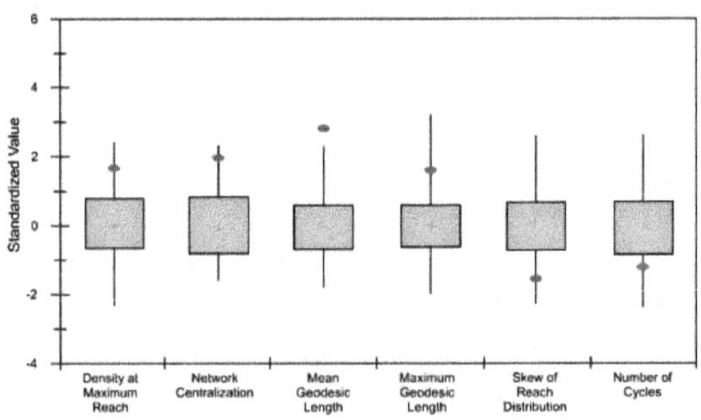

Figure 9.7 Simulated networks, preserving observed degree and isolated dyad distribution

of the rewired graph that deviate from the original graph help us identify behavioral rules that govern that specific parameter.

To rewire the empirically observed graph, we select 5% of the relationships at random and reassign them conditional only on the degree distribution of the original graph. In this way, we ensure that individuals with many partners continue to have many partners and that individuals with few partners do not suddenly gain partners. Table 9.3 reports the effect of rewiring and is based on a comparison of the observed network with 1,000 rewired graphs.

Compared with our earlier simulations, the rewired graphs are quite similar to the observed network. This is what we would expect, since we change only 5% of the ties at random, while holding the distribution constant. Consequently, all the network centralization measures are fit well, as are the reach measures. The difference between the observed number of components in Jefferson and those arising from the simulations is trivial. *The only statistic that is fit poorly is the number of cycles.* The rewired networks have almost twice as many cycles as are observed in Jefferson. Since we observe a spanning tree in the Jefferson network, it is not surprising that rewiring produces redundant ties, which appear here as cycles. Thus, rewiring isolates the single structural feature we have to account for—in this case, the absence of cycles. Thus, the only puzzle is, Why are they absent?

The Basis for a Spanning Tree Structure: Unarticulated Partnership Prohibitions

To explain why cycles are absent at Jefferson, recall that spanning trees are theoretically produced by negative proscriptions. What kinds of relationships are prohibited? The simple answer is that the prohibited relationships are those that induce short cycles. The *smallest possible heterosexual cycle* has a length of 4. Consider four individuals: Bob, Carol, Ted, and Alice. Imagine that Bob and Carol were once partners, but that Carol left Bob for Ted. Further, imagine that Ted and Alice were partners, but that Ted dumped Alice for Carol. Should Bob and Alice date? From Bob's perspective, Alice was his former partner's current partner's partner, or the former "lover" of his former girlfriend's current lover. Alice looks at Bob with the same lens. Her former boyfriend is dating the girl who left Bob.[31] These scenarios can be summarized by a graph, as in Figure 9.8, where lines indicate a relationship between nodes (here, persons), yielding a potential cycle of length 4.

Using the simulation strategy introduced earlier, we can operationalize a normative rule that persons do not date the former (or current) partner of their former (or current) partner by prohibiting all cycles of length 4. We simulate 1,000 random networks, this time conditional

Table 9.3 Robustness of the observed statistics of random rewiring

	*Observed Network**		*Random Rewire Distribution*	
Variable	*Coefficient*	*P*	*Coefficient*	*SD*
Density at maximum reach	.42	.34	.43	.06
Network centralization	.024	.23	.023	.003
Mean geodesic length	15.94	.10	13.8	1.71
Maximum geodesic length	37	.30	34.9	5.52
Reach distribution skew	−.61	.34	−.67	.21
N of cycles	5	.04	9.03	2.26

Notes: *N* of rewired nets = 1,000. Excludes isolated dyads.

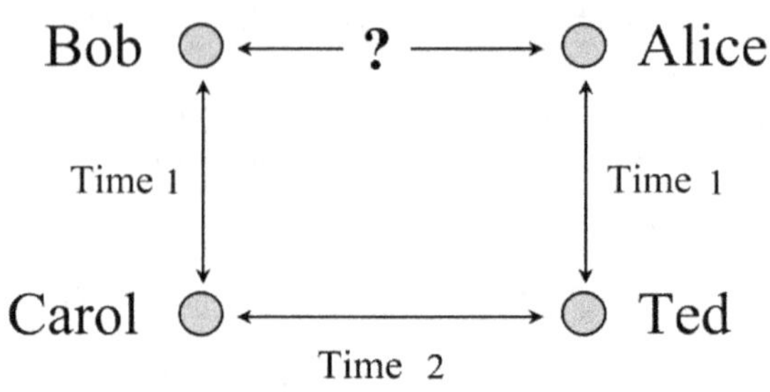

Figure 9.8 Hypothetical cycle of length 4

on the following constraints: fixed degree distributions matching those observed at Jefferson, no isolated dyads, and a single parameter that prohibits cycles of length 4. The question is whether this set of constraints generates graphs with structural features similar to those observed at Jefferson. Figure 9.9 shows that they do: on all the structural parameters we consider, the Jefferson network is quite close to the central tendency of the distributions generated by the simulated networks.

Comparison of the internal structure of these random networks and the observed Jefferson network shows that they are essentially isomorphic. This similarity is illustrated in Figure 9.10, which shows graphs of the largest components from four randomly selected networks simulated by this model. One immediately sees network structures strikingly similar to the structure observed in Jefferson. Given fixed degree and homophily in experience, the sufficient condition for generating a spanning tree is the prohibition against cycles of length 4. Such a prohibition may operate in Jefferson.

Status Dislocation and Closeness

Given the conditions of homophily described previously, Figures 9.9 and 9.10 show that a simple rule—the prohibition against dating (from a female perspective) one's old boyfriend's current girlfriend's old boyfriend—accounts for the structure of the romantic network at Jefferson. Why might this negative proscription operate in a medium-sized community of essentially homogeneous adolescents?

The explanation we offer only makes sense for short cycles. From the perspective of males or females (and independent of the pattern of "rejection"), a relationship that completes a cycle of length 4 can be thought of as a "seconds partnership," and therefore involves a public loss of status.[32] Most adolescents would probably stare blankly at the researcher who asked boys: "Is there a prohibition in your school against being in a relationship with your old girlfriend's current boyfriend's old girlfriend?" It is a mouthful, but it makes intuitive sense. Like adults, adolescents choose partners with purpose from the pool of eligible partners. But beyond preferences for some types of partners over others—for example, preferences for partners interested in athletics who do not smoke, or who will skip school to have more fun—adolescents prefer partners who will not cause them to lose status in the eyes of their peers. In the same way that high-status students avoid relationships with low-status students, by selecting partners on the basis of the characteristics that have resonance for the local determination of prestige, students avoid relationships whose *structure* would lower their status in the eyes of their peers. In a large and essentially homogeneous school like Jefferson, the pool of potential partners with the "right" mix of attributes is relatively large, so students can fairly easily avoid taking "seconds" and still

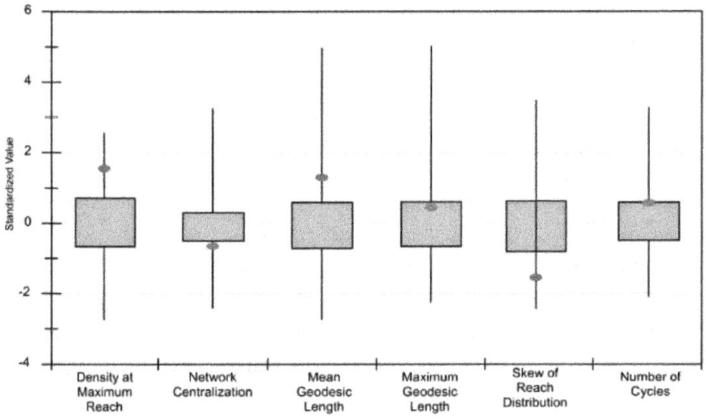

Figure 9.9 Simulated networks preserve observed degree, isolated dyad distribution, and four-cycle constraint

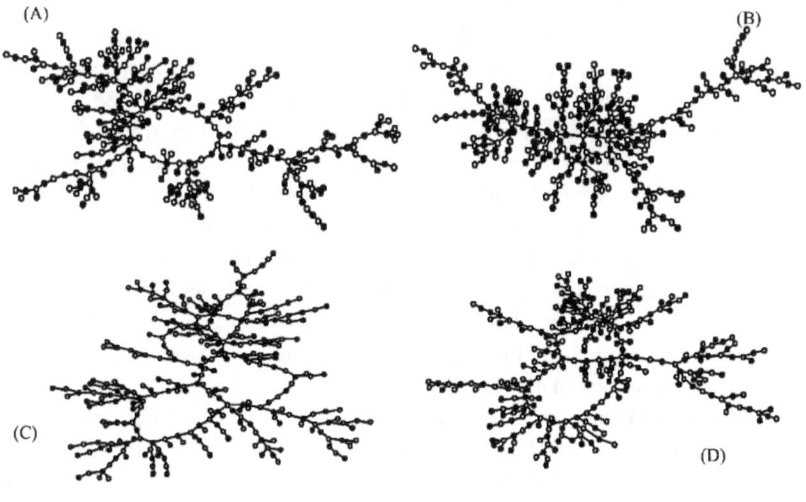

Figure 9.10 Simulated networks preserve observed degree, isolated dyad distribution, and cycle constraint

preserve their basic attribute or experience preferences. More generally, in intact communities where observation of temporally proximate partnerships is possible, we should expect to see successful avoidance of relationships that complete cycles of length 4. Such avoidance should not, however, extend to larger cycles, since larger cycles typically involve relationships in the distant romantic past that cannot be systematically observed.[33]

For adolescents, the consequence of this prohibition is of little interest; what concerns them is avoiding status loss. But from the perspective of those interested in understanding the determinants of disease diffusion, the significance of a norm against relationships that complete short cycles is profound. The structural impact of the norm is that it induces a spanning tree, as versus a structure characterized by many densely connected pockets of activity (i.e., a core structure). As a consequence, this prohibition, combined with existing homophily preferences, both shapes the potential for local disease diffusion and affects what social policy interventions will be effective at stemming disease spread.

Discussion

Disease diffusion is widespread among adolescent populations. The standard models that epidemiologists use to describe the dynamics of diffusion carry implicit ideas about the contact structure through which disease travels. These ideas are associated with distinct structural features of sexual networks. The most critical feature in STD epidemiology is the idea of a core, which is associated with cycles in networks. Moody et al. (2003) have demonstrated that very low average degree networks can give rise to densely interconnected cores, characterized by high cyclicity. In our data, we find that this key structural feature is largely absent. We have proposed a reason for its absence, specifically a norm against seconds partnerships. From this norm, combined with basic homophily preferences, we generate networks that are structurally isomorphic to the one we observe empirically. This suggests that in adolescent society—where partner choice is salient for local status—it seems reasonable to think that such a rule operates.[34]

Nonetheless, the scope conditions for this article are implied in the central finding and the mechanism we claim accounts for it. Specifically, our mechanism presumes that actors can watch each other, that they are capable of recording immediately prior partnerships, and that they are susceptible to collective assessment of their personal choices. One can only fear losing status in the eyes of others if the others are watching and if one cares about their assessments.

These conditions may be absent for adults who are embedded in worlds larger and more disjoint than adolescents. More than adolescents, adults may be capable of segregating audiences across the various settings in which they are embedded (work, leisure, play, school, etc.) and are therefore less subject to the scrutiny and sanctioning of their peers than are adolescents. While entering into a partnership that completes a cycle of length 4 may result in a loss of face for an adult, it is more likely that among adults such cycles are generated without anyone ever knowing. This is unlikely in a high school, where much social energy is devoted to understanding who is going out with whom.[35] All this suggests that we would be less likely to observe spanning trees among adult populations than among adolescents.

In theory, spanning trees are among the most efficient structures for diffusion since the absence of redundant lines maximizes reach at lowest density. Yet their efficiency is counteracted by their fragility: spanning trees are highly susceptible to breaks in transmission. Electric provision systems would be set up as spanning trees if service providers did not worry about failing to deliver power to some customers. But since they worry about small breaks in the line, they establish more densely connected power grids. It follows that for highly infectious diseases with long periods of infectivity, transmission is also quite efficient under a spanning tree. Yet if the duration of infectivity is short, or if the disease is not particularly infectious, the probability of transmission within any given partnership is low. From the perspective of disease spread, failure to transmit disease within a given partnership is effectively a structural break in the network (Watts 2003). Since the natural infectivity and duration of infectiousness varies across STDs, we believe that the most effective strategy for reducing disease diffusion rests on creating structural breaks.

We might then ask a new question: What kinds of policy intervention will be most effective at *inducing* structural breaks in the sexual networks of adolescents? Here the answer is exceedingly simple. Assume that some proportion of actors who are "reached" through an intervention decide to change their behavior. Under core and inverse core structures, it matters enormously *which* actors are reached, while under a spanning tree structure the key is not so much *which* actors are reached, just that some are. This is because given the dynamic tendency for unconnected dyads and triads to attach to the main component, the structure is equally sensitive to a break (failure to transmit disease) at any site in the graph. In this way, relatively low levels of behavior change—even by low-risk actors, who are perhaps the easiest to influence—can easily break a spanning tree network into small disconnected components, thereby fragmenting the epidemic and radically limiting its scope. Obviously, similarly low levels of change in cores will have little impact.

This example highlights how having an accurate sense of the real structure of a network matters for the effectiveness of an intervention. If cores exist, one develops interventions that target core members. But if there are no actual structural cores, interventions targeted primarily at high-risk individuals will do less to stem the overall spread of disease than will broadcast interventions directed toward all actors. Ironically, early HIV prevention strategies that utilized broadcast diffusion techniques may have been more effective at reducing overall incidence of disease than more recent interventions focused on isolating those seen as being at a higher risk for infection (though these targeted interventions may reduce risk of disease acquisition at the individual level).

Epidemiologists, unable to observe or measure directly the structure of sexual networks, have tended to latch onto a single idea: specifically, the idea that the number of partners matters for STD diffusion dynamics. If, as their models assume, the real contact structures are well approximated by core or inverse core network structures, degree distributions are meaningful, and the number of partners will be a key parameter. But this need not be the case theoretically, and, as we empirically show, it is not likely the case for adolescents. The fact that the relevant contact structure is a spanning tree explains why the rates of bacterial STDs have been so high among adolescents in the past decade, and why most social policy, which focuses on high-risk individuals within the adolescent community, has failed to stem the flood of new infections. Our data suggest that a shift in social policy toward comprehensive STD education for all adolescents, not just those at highest risk, would be significantly more effective than current intervention models.

Acknowledgments

This chapter originally appeared in *American Journal of Sociology*, 110, 44–91 (2004). Data for this article are drawn from the National Longitudinal Study of Adolescent Health (Add Health), a program project designed by J. Richard Udry and Peter Bearman, and funded by a grant from the National Institute of Child Health and Human Development (HD31921). The authors thank Douglas White, Martina Morris, Mark Handcock, J. Richard Udry, and the *AJS* reviewers for helpful comments on previous drafts of this article.

Notes

1. As background, each year in the United States over 12 million individuals discover that they carry a sexually transmitted disease (STD). The two leading STDs, herpes and human papillary virus (HSVT2 and HPV, respectively), are chronic and, although subject to palliative treatment, not curable. Adolescent STD acquisition rates outpace those of all other groups, with no change in sight. Roughly 5% of all sexually active adolescents have acquired chlamydia or gonorrhea (Aral et al. 1999). Among sexually active black adolescents, 25% are likely to be infected with herpes (CDC 2000), and probably 40–50% of all sexually active females have had a previous HPV infection, now known to account for most cases of adult cervical cancer (Holmes et al. 1999). The literature identifies three reasons for these gloomy facts. First, one-half of all adolescents over 15 years old report being sexually active, and a significant proportion of these adolescents are inconsistent in their use of condoms, therefore

heightening the risk of STD acquisition and transmission (Bearman and Brückner 1999). Furthermore, many adolescents who have not had intercourse are sexually active in a substantively meaningful (if technically ambiguous) way, and most do not use condoms during non-coital sex. Specifically, of adolescents who report that they are virgins (i.e., have not had sexual intercourse) roughly one-third have had genital contact with a partner resulting in fluid exchange in the past year. Thus, virginal status does not mean that adolescents are not engaging in behaviors that are free of risk for STD transmission. Second, the majority of adolescents with an STD have no idea that they are infected (Holmes et al. 1999); consequently, they may fail to protect their partners even if they would prefer to do so. And third, relative to adults, adolescents tend to form romantic partnerships of short duration, on average only 15 months, but with a strong skew toward relationships of extremely short duration (less than four months; Laumann et al. 1994). Most sex in adolescent relationships occurs, if it is to occur, within the first two months (Bearman, Hillmann, and Brückner 2001). This combination of short duration partnerships, inconsistent safe-sex practices, and incorrect assessment of STD status provides a partial account for the diffusion of STDs among the adolescent population. As fundamental is the role that sexual contact structures play in STD transmission dynamics.

2. R_0 is defined as the number of new infections produced by an infected individual over the duration of infectivity (Anderson and May 1991).
3. The S-shaped curve is not specific to random mixing. As noted by a reviewer, the S-shape may result from many different contact structures.
4. This is not to suggest that such models have no utility. For example, as one reviewer notes, mixing models in which groups are based on the number of sexual partners show that the probability of having sex with an infected person exceeds the prevalence of infection in the population. However, this would not be a *pure* random-mixing model.
5. Preferred-mixing models operate on persons classified by attribute rather than by structural position. The models we develop subsequently identify position as the critical element, such that attributes of persons are substitutable across positions. But, in an abstract sense, they are also preferred-mixing models.
6. Preferred-mixing models need not assume homogeneous mixing within groups. Variants include assortative-matching models that incorporate out-group preferences, e.g., age-skewed models that match older males with younger females, or role separation models. The implications of skewed age matching for HIV diffusion are explored in Morris (1993).
7. An important development has been the focus on strongly skewed positive degree distribution in scale-free networks for diffusion dynamics (Newman 2002; Dezso and Barabasi 2002; Newman et al. 2001; Barabasi and Albert 1999; Watts 2003). As noted subsequently, the structure we observe is not a scale-free network with a power law distribution of degree. Moody et al. (2003) show that a large densely connected core can emerge in populations with low degree, demonstrating that the structural conditions for large epidemics are possible even in populations without skewed partnership distributions.
8. Each of these models has been refined in the literature to incorporate both theoretical and empirical advances in our understanding of the process by which sexual partnerships are formed. Thus, e.g., there are variants of these models that emphasize role separation in male homosexual populations, temporal overlap of partnerships (con-currency), and multiple or complex preference structures in heterogeneous populations.
9. This graph expands on the triads linking the core to the periphery in the core and inverse core networks, though this image draws attention to the fact that the bridging triad is embedded in a macrostructure different from either of these other two models.
10. Recent work on the structure of large networks—e.g., those linking nodes on the World Wide Web—reveals starlike nodes with very high degree. Such a structure is not replicated in the data we observe.
11. Here and throughout we refer to chainlike structures that are not dominated by cycles or small numbers of highly central nodes as spanning trees, though technically a spanning tree is any connected noncycle graph.
12. A super star graph is technically a spanning tree, but is associated with short geodesics (Barabasi and Albert 1999).
13. One reviewer suggests that if sex "has an element of contagion in which only sexually experienced actors recruit new participants," the resulting graph would be a spanning tree. See app. C in the original version of this chapter, published in *American Journal of Sociology*, 110, 44–91 (2004), for our response to this suggestion.
14. Adolescents were given $20 in appreciation for completing the interview. Just before Mothers' Day and the prom, many adolescents were eager to be interviewed.

15. Additional information about these measures is provided in app. A in the original version of this chapter, published in *American Journal of Sociology*, 110, 44–91 (2004).
16. Reviewers suggested a nonparametric test for these comparisons. Following Conover (1980), we use a sign test to compare the median value of each of these characteristics at Jefferson to each of the four subsets of other schools. The assessment is quite sensitive, especially with respect to col. 5, where the *N* of comparison schools is small. The differences between Jefferson and other schools in the sample can be qualitatively described simply: Jefferson is an all-white school that is largely working class; most all-white schools in the country are composed of upper middle-class adolescents who reside in segregated suburbs. Consequently, social class predominantly drives differences in behavior, academic orientation, and achievement.
17. Adolescents who did not identify that they had a special relationship were asked if in any relationship over the past 18 months they had "held hands, kissed, or told someone that they liked or loved them." If an adolescent was in such a relationship, then they were asked to identify their partner and describe their relationship. Both self-identified and behavior-induced "partnerships" could, but did not necessarily, involve sexual intercourse. Of adolescents who reported being a virgin (i.e., had not had sexual intercourse) one-third had had genital contact with a partner resulting in fluid exchange in the past year (Schuster, Bell, and Kanouse 1996). Thus, not having intercourse does not mean refraining from behaviors that are risky for HIV or STD transmission (although the noncoital fluid-exchange behaviors they do engage in carry less risk for both partners than intercourse).
18. Nominations to students account for 51.2% of all romantic nominations and 39.4% of all nonromantic sexual partnership nominations. These partnerships involve roughly 75% of all students who reported having a romantic relationship. In the other large saturated field setting in the Add Health sample, which is located in an ethnically heterogeneous metropolitan area, only 11% of all partnership nominations were directed toward other students. Thus, in Jefferson, the school community provides the key focal context (Feld 1981) for adolescent social and sexual relations. Given the relative isolation of the community, this orientation is expected.
19. In Figure 9.2, and in all discussions presented here, all romantic and sexual relationship nominations linking students are included, whether or not the nomination from i to j was reciprocated with a nomination from j to i.
20. Even if each respondent had simply reported on their number of partners in the past 18 months, without their having selected them from a roster of possible partners, we would not have been able to generate this structure, for we would not have known how—or whether—these partnerships connected into a macrostructure. Attribute-based matching schemes are similarly unable to specify the interrelationship of partners' partners.
21. A "movie" of the Jefferson network unfolding through time is available at: www.columbia.edu/iserp/people/bearman/chains
22. Structural fragility is also referred to as "1-connectedness" in the technical literature on graphs.
23. The idea seems far-fetched, but, as one reviewer notes, the context is already essentially homophilous with respect to age, education, race, social class, citizenship, religious orientation, and ethnicity—the major determinants of partnership choice for adults. Given this, random choice makes sense as a baseline.
24. Specifically, we generate 1,000 networks with the same number of nodes and the same distribution of number of partners as observed in the Jefferson net, with ties assigned randomly between nodes. Details about the algorithm used to generate these conditional random graphs can be found in Moody 1998.
25. Alternatively, one could represent this measure as the mean number reachable in a network. Likewise, from the same framework one can assess whether the maximum reach of the largest component, in our case involving roughly one-half of all students in the school, is to be expected by chance. Relative to the simulated samples, the largest component is almost 2 SDs greater than expected by a random-mixing model.
26. Obviously, judgment is required here. Adolescents may select partners on the basis of unobserved characteristics (or unobserved to us) that vary across individuals in a completely unsystematic way. This is, in one sense, what the idea of romantic love suggests. Our strategy is to identify a set of characteristics that are observable, common, and have face validity as salient attributes. We consider homophily on these, and then simulate the global structure that would arise, should these elements provide the basis for choice. We do not include one of the most salient attributes for partnership choice, race, since Jefferson is all white.
27. We thank an *AJS* reviewer for suggesting a nonparametric analysis strategy (QAP) in this context.

28. Specifically, mean[abs(X_i — X_j) | ij = 0] – mean[abs(X_i — X_j) | ij = 1].
29. It is important to recall that the context is already quite homogeneous. So, among white students of roughly the same social class, ethnicity, citizenship, and so on, these are the salient determinants of partnership choice.
30. Recall that not all partners, sexual or romantic, are drawn from school. On average, out-of-school partners were 3.21 years older than the respondent at the start of their relationship, although we observe a pronounced skew in the age-difference distribution. For example, one out-of-school sexual partner of two girls was 39 years old. Aside from a few exceptions like this, most of the students involved in out-of-school relationships have partners slightly older than themselves. Together with other evidence in the survey describing where the respondents met their partners, we conclude that many of these out-of-school partners attended Jefferson prior to our survey, and most continue to live in the same neighborhood.
31. Here, preferences for partner attributes break down: if Carol is attracted to both Bob and Ted, then they must be similar with respect to attributes, and yet clearly Carol and Alice are not equivalent substitutes from Bob's perspective.
32. The status-loss hypothesis competes with other potential micromechanisms, e.g., "jealousy" or the avoidance of too much "closeness," a sentiment perhaps best described unscientifically as the "yuck factor." The status-loss hypothesis involves significant scope limitations: namely, status loss is limited to contexts where actors by virtue of their relational density can watch each other relatively closely. By contrast, the "yuck factor"—which is essentially individualized—could operate in more diffuse contexts.
33. In a sense, the prohibition against cycles of length 4 suggests shifting from the cross-sectional perspective considered in Figure 9.2 and focusing instead on the temporally ordered graph shown in Figure 9.3, since a cycle unfolds over time. The picture is made somewhat more complex by instances in which relationships are concurrent. While not uncommon at Jefferson, concurrency appears to be more common among adults than adolescents.
34. We note again that not all of the sexual partnerships in Jefferson are directed toward other students. It is possible that the spanning tree structure we observe is a by-product of missing data on the prior partnerships of the out-of-school partners. We consider this unlikely for three reasons. First, adolescents with any out-of-school partners are disproportionately older and more likely to be female than are those with only in-school partnerships. Second, those with out-of-school partners are less likely than those with in-school partners to have other in-school partnerships. And third, analysis of temporal sequencing indicate that most out-of-school partnerships are temporally subsequent to in-school partnerships, should there be any. Consequently, their impact on the structure of the observed network is necessarily modest, since if present, they do not link (from a viral/bacterial perspective) nodes in the in-school graph we consider.
35. Because out-of-school partnerships tend to be more privatized—i.e., involve seeing less of friends—the scope conditions are relevant here as well, as privatized relationships are less likely to be observed and hence could, theoretically, not be subject to the four-cycle constraint.

References

Anderson, R. M., and R. M. May. 1991. *Infectious Diseases of Humans' Dynamics and Control.* New York: Oxford University Press.

Aral, S. O. 2000. "Behavioral Aspects of Sexually Transmitted Diseases: Core Groups and Bridge Populations." *Sexually Transmitted Diseases* 27: 327–28.

Aral S. O., J. P. Hughes, B. Stoner, W. Whittington, H. H. Handsfield, R. M. Anderson, and K. K. Holmes. 1999. "Sexual Mixing Patterns in the Spread of Gonococcal and Chlamydial Infections." *American Journal of Public Health* 89 (6): 825–33.

Barabasi, Albert-Laszlo, and Reka Albert. 1999. "Emergence of Scaling in Random Networks." *Science* 286: 509–12.

Bearman, Peter S., and Hannah Brückner. 1999. *Peer Effects on Adolescent Girls' Sexual Debut and Pregnancy.* Washington, DC: National Campaign to Prevent Teen Pregnancy.

Bearman, Peter S., Henning Hillmann, and Hannah Brückner. 2001. "PAIRS: Adolescent Relationship Dynamics." Manuscript. Columbia University, Institute for Social and Economic Research and Policy.

Bonacich, Phillip. 1987. "Power and Centrality: A Family of Measures." *American Journal of Sociology* 92: 1170–82.

CDC (Centers for Disease Control and Prevention). 2000. *Tracking the Hidden Epidemics: Trends in STDs in the United States*. Atlanta, GA.

Conover, W. J. 1980. *Practical Nonparametric Statistics*, 2nd ed. New York: John Wiley.

Dezso, Zoltan, and Barabasi, Albert-Laszlo. 2002. "Halting Viruses in Scale-Free Networks." *Physical Review E* 65: 055103.

Feld, S. L. 1981. "The Focused Organization of Social Ties." *American Journal of Sociology* 86: 1015–35.

Friedman, S. R., A. Neaigus, B. Jose, R. Curtis, M. Goldstein, G. Ildefonso, R. B. Rothenberg, and D. C. Des Jarlais. 1997. "Sociometric Risk Networks and HIV Risk." *American Journal of Public Health* 87: 1289–96.

Garnett, G. P., J. P. Hughes, R. M. Anderson, B. P. Stoner, S. O. Aral, W. L. Whittington, H. H. Handsfield, and K. K. Holmes. 1996. "Sexual Mixing Patterns of Patients Attending Sexually Transmitted Disease Clinics." *Sexually Transmitted Diseases* 23 (3): 248–57.

Gorbach P. M., H. Sopheab, T. Phalla, H. B. Leng, S. Mills, A. Bennett, K. K. Holmes. 2000. "Sexual Bridging by Cambodian Men: Potential Importance for General Population Spread of STD and HIV Epidemics." *Sexually Transmitted Diseases* 27: 320–26.

Hage, P., and F. Harary. 1983. *Structural Models in Anthropology*. Cambridge: Cambridge University Press.

——. 1996. *Island Networks: Communication, Kinship, and Classification Structure in Oceania*. New York: Cambridge University Press.

Hethcote, H. 2000. "The Mathematics of Infectious Diseases." *SIAM Review* 42: 599–653.

Hethcote, H. W., and J. W. Van Ark. 1987. "Epidemiological Models for Heterogeneous Populations: Proportionate Mixing, Parameter Estimation, and Immunization Programs." *Mathematical Biosciences* 84: 85–118.

Hethcote, H. W., and P. van den Driessche. 2000. "Two SIS Epidemiologic Models with Delays." *Journal of Mathematical Biology* 40: 3–26.

Hethcote, H., and J. A. Yorke. 1984a. "Gonorrhea Modeling: A Comparison of Central Methods." *Mathematical Biosciences* 58: 93–109.

——. 1984b. *Gonorrhea Transmission Dynamics and Control*. Berlin: Springer Verlag.

Holmes, K. K., P. A. Mardh, P. F. Sparling, S. M. Lemon, W. E. Stamm, P. Piot, J. N. Wasserheit, Eds. 1999. *Sexually Transmitted Diseases*, 3rd ed. New York: McGraw Hill.

Jacquez, J. A., C. P. Simon, J. Koopman, L. Sattenspiel, and T. Perry. 1988. "Modeling and Analyzing HIV Transmission: The Effect of Contact Patterns." *Mathematical Biosciences* 92: 119–99.

Klovdahl, A. S. 1985. "Social Networks and the Spread of Infectious Diseases: The AIDS Example." *Social Science Medicine* 21: 1203–16.

Koopman, J., C. P. Simon, J. A. Jacquez, and T. S. Park. 1989. "Selective Contact within Structured Mixing with an Application to HIV Transmission Risk from Oral and Anal Sex." Pp. 316–49 in *Mathematical and Statistical Approaches to AIDS Epidemiology*, edited by C. Castillo-Chavez. Berlin: Springer-Verlag.

Laumann, Edward O., John H. Gagnon, Robert T. Michael, and Stuart Michaels. 1994. *The Social Organization of Sexuality: Sexual Practices in the United States*. Chicago: University of Chicago Press.

Laumann, Edward O., and Yoosik Youm. 1999. "Racial/Ethnic Differences in the Prevalence of Sexually Transmitted Diseases in the United States: A Network Explanation." *Sexually Transmitted Diseases* 26: 250–61.

Moody, James. 1998. "A General Method for Creating Approximate Conditionally Uniform Random Graphs." Paper presented at the Sunbelt Social Networks Conference. Sitges, Spain.

——. 2002. "The Importance of Relationship Timing for Diffusion: Indirect Connectivity and STD Infection Risk." *Social Forces* 81: 25–56.

Moody, James, Martina Morris, Jimi Adams, and Mark Handcock. 2003. "Epidemic Potential in Low Degree Networks." Manuscript. Ohio State University.

Morris, Martina. 1993. "Epidemiology and Social Networks: Modeling Structured Diffusion." *Sociological Methods and Research* 22: 99–126.

——. 1997. "Sexual Networks and HIV." *AIDS 97: Year in Review* 11: S209–S216.

Morris, Martina, and M. Kretzschmar. 1995. "Concurrent Partnerships and Transmission Dynamics in Networks." Special issue, Social Networks and Infectious Disease: HIV/AIDS. *Social Networks* 17: 299–318.

——. 1997. "Concurrent Partnerships and the Spread of HIV." *AIDS* 11: 641–48.

Morris, M., C. Podhisita, M. Wawer, and M. Handcock. 1996. "Bridge Populations in the Spread of HIV/AIDS in Thailand." *AIDS* 10: 1265–71.

Newman, M. E. J. 2002. "Spread of Epidemic Disease on Networks." *Physical Review* E 66 016128.

Newman, M. E. J., S. J. Strogatz, and D. J. Watts. 2001. Random Graphs with Arbitrary Degree Distributions and Their Applications. *Physical Review E* 84: 026118.

Phillips, L., J. Potterat, and R. Rothenberg. 1980. "Focused Interviewing in Gonorrhea Control." *American Journal of Public Health* 70: 705–8.

Rothenberg, R. B., J. J. Potterat, and D. E. Woodhouse. 1996. "Personal Risk Taking and the Spread of Disease: Beyond Core Groups." *Journal of Infectious Diseases* 174 (Suppl. 2): S144–S149.

Rothenberg, R. B., et al. 1997. "Using Social Network and Ethnographic Tools to Evaluate Syphilis Transmission." *Sexually Transmitted Diseases* 25: 154–60.

Sattenspiel, L. 1990. "Modeling the Spread of Infectious Disease in Human Populations." *Yearbook of Physical Anthropology* 32: 245–76.

Schuster M. A., R. M. Bell, and D. E. Kanouse. 1996. "The Sexual Practices of Adolescent Virgins: Genital Sexual Activities of High School Students Who Have Never Had Vaginal Intercourse." *American Journal of Public Health* 86: 1570–76.

Schweizer, T., and D. R. White. 1998. *Kinship, Networks and Exchange*. Cambridge: Cambridge University Press.

St. John, R., and J. Curran. 1978. "Epidemiology of Gonorrhea." *Sexually Transmitted Diseases* 5: 81–82.

Turner, C. F., et al. 1998. "Adolescent Sexual Behavior, Drug Use, and Violence: Increased Reporting with Computer Survey Technology." *Science* 280: 867–73.

Wasserman, S., and K. Faust. 1994. *Social Network Analysis*. Cambridge: Cambridge University Press.

Watts, D. 2003. *Six Degrees: The Science of a Connected Age*. New York: W. W. Norton.

10

THE ART OF BUILDING DYNAMIC SYSTEMS MODELS

Saskia Kunnen

Research Question: How can a theory of adolescent commitment development be captured by a dynamic systems model?
System Science Method(s): System dynamics
Things to Notice:

- Model building as a stepwise process
- Linkage between theory and model

In this chapter we aim to provide the reader with guidelines to build one's own quantitative dynamic systems model. To be able to build such a model, two types of knowledge are important. The first type concerns the knowledge of how to translate theoretical notions into a conceptual dynamic systems model, and how to translate that conceptual model into quantitative expressions. The chapter will focus on this first type of knowledge. The second type of knowledge concerns the technical part – knowledge about which software to use, the type of equations that represent different types of development, etc. In addition, knowledge is needed about how to enter quantitative mathematical expressions in a spreadsheet, how to generate simulations and how to make graphs. In our examples and explanations we use spreadsheets, because our aim is to make the art of model building available for all colleagues, not only for those highly skilled in computer programs and mathematics. Spreadsheet programs have proved to be very useful for building dynamic systems, and most people have these programs.

So, in this chapter we will describe the process from general theories to a quantitative dynamic systems model. We will do that by means of an example: the building of a model of commitment development. The chapter will describe the modeling process in different steps.

The first step in building a quantitative dynamic systems model is to develop a conceptual model. In other words, we have to explicate our ideas about which variables affect commitment development and how they affect them. In general, these ideas are based on a mixture of empirical knowledge, theories and common sense. As we will see, the common sense is especially needed when it comes to the question of how other factors affect the development of the target grower. The next step is the actual building of the model. The conceptual model is the basis for the quantitative model. In fact, the quantitative model is a translation of the conceptual descriptions in formulas and equations. As we will see, in that translation a lot of choices have to be made. Especially for the value of the parameters in the model these choices are seemingly arbitrary. As a third step it is therefore necessary to explore with different sets of parameters to find out which value sets "work." We call this the calibration. The final step is to evaluate and validate

 DOI: 10.4324/9781315748771-11

the model, which can be done in different ways. The most straightforward way is to gather data that closely resemble the simulated process of commitment development and compare the simulated trajectories with the empirically found trajectories, but this is practically and theoretically difficult. Often the first steps in validation are more indirect, caused by lack of longitudinal data. In the final section we will discuss both the possibilities and the limitations of this technique, illustrated by the model that has been developed in this chapter.

Developing a Conceptual Model of Identity Development

In this first step we have to define the core characteristics of the model, starting with defining the target variables and the time span. To do that, we start with a short description of the theory that is directly relevant for the model. Theories on identity development consider the development of commitments in different domains of life as one of the core characteristics of identity development. A commitment can be described as identity choices that give the individual clarity about who he is and what he wants to be and help to define his place in society both for himself and for others. Marcia (1980) distinguishes two core processes in identity development: commitment formation and exploration. Strong commitments give one a strong sense of knowing who one is. Exploration means the process of considering alternatives, of exploring different possible commitments. The development of a mature identity concerns the development of strong and flexible commitments following a period of exploration (Marcia, 1980). The element of actively making identity choices is central in Marcia's identity status model, which is the empirical approach to identity in adolescence and adulthood that is most often used (Marcia, Waterman, Matteson, Archer and Orlafsky, 1993). Identity statuses are "modes of dealing with the identity issue characteristic of late adolescents" (Marcia, 1980, p. 161), and they form an extension of Erikson's bipolar description of the outcome of the identity crisis in adolescence (identity versus identity diffusion). According to the identity status model, by the end of adolescence, adolescents have either actively made identity choices and in this way solved the normative identity crisis (Achievement), or have not made such choices (and end up in a state of Identity Diffusion), or are still actively exploring identity choices (Moratorium), or have never experienced an exploratory period (Foreclosure).

The change of commitment strength and level of exploration over time will be our target. In dynamic systems language, commitment strength and exploration are the main growers in our model. Depending on the conceptual model, there may be other growers. Commitment change and development is not restricted to adolescence, but most identity work does take place in late adolescence and early adulthood. With our model we aim to describe the development of commitments in a specific domain in the period between about age 16 and age 24.

A conceptual model of the development of commitments means that we make a scheme or a flow chart that shows which variables affect commitment strength, and how they do so. It should also include how commitment strength affects the other variables in the model. To develop such a model we start with an exploration of existing research and theories.

Theoretical Background

In the last decades the development of commitments has received increasing attention in research. However, actual data in the mechanisms and processes of commitment development are still scarce. Moreover, not all findings concerning the relation between commitment development and other variables are useful: we need information concerning the actual mechanisms, and the usefulness of, for example, correlational data is limited. A major limitation of the usefulness of existing empirical findings is the problem of ergodicity. Most data on identity development concern group data. These data may be useful in a later stage, in the evaluation of the model, but they are not helpful in the model building itself. We need insight into the question of what exactly happens during the development of commitments. Which variables interact with commitment development? What makes the strength of commitments stronger or weaker? As a basis for our model we will use the conceptual model discussed in a paper by Bosma and Kunnen (2001, p. 63):

> Let us resume the model described above: the development of identity could be seen as an iterative process. Each iteration concerns a transaction between person and context. In these transactions, a conflict may occur. Initially, people will try to resolve this conflict by means of assimilation, by adjusting their interpretation of the situation in such a way that it can be assimilated into their existing identity. If this fails, the conflict will remain and weaken the existing commitments; until accommodation, or change of identity occurs. Development results from such accommodational changes. People differ as to how long they stick to assimilation and how easily they change their commitments. Personal and contextual determinants determine the ratio between assimilation and accommodation, and the optimal balance. As discussed, we cannot separate the effects of different determinants. Their effects can be seen only in relation to each other.

Although rooted in theory, the model described above remains to be proven. A first step could be an intensive and longitudinal study of commitments. Regular assessments of commitments and of conflicts people encounter, will give some insight into the validity of the process described in the model.

Since the publication of that paper in 2001, at least two longitudinal studies have appeared on the development of commitments: one by Luyckx, Goossens and Soenen (2006) and our own research (Kunnen, 2009). In both studies university students were followed for a period of at least four years, with identity assessments every six months. The findings in these studies largely confirm and elaborate the model presented in 2001. We demonstrated that conflict is an important factor in the development of commitments (Kunnen, 2006). We have found some evidence for the assumption that in conflicts the strength of commitments decreases. A factor that is closely related to the development of commitments is exploration. Exploration means the active exploring of different commitments by gathering information, talking with people, behavior, etc. Kunnen, Sappa, van Geert and Bonica (2008) state that crises can be described by both an increase in exploration and a decrease in commitment strength. Luyckx, Goossens and Soenen (2006) differentiate between exploration in depth, which means that the commitment that has been chosen is explored, and exploration in breadth, which means that a whole range of potential commitments is explored. They found that:

> Commitment making and exploration in breadth were negatively interrelated, indicating that both are at odds with each other, at least concurrently. Identification with commitment and exploration in depth were positively interrelated at each measurement time.
>
> Commitment making was positively related to both identification with commitment and exploration in depth. Finally, exploration in breadth and exploration in depth were positively interrelated, as an indication of their common focus on gathering identity-relevant information.
>
> *(p. 372)*

Luyckx, Goossens, Soenen and Beyer (2006) differentiate between two cycles: commitment-formation and commitment-evaluation. The commitment-formation cycle consists of exploration in breadth, which is related negatively to commitment making and unrelated to identification with commitment. This exploration dimension seems to be associated with a period of crisis and existential doubt about important life-choices that precedes the actual formation of commitments. The commitment-evaluation cycle consists of exploration in depth, which is related positively to both commitment making and identification with commitment, emphasizing that it serves the strengthening and evaluation of commitments. During crises, a strong negative intra-individual relation between commitment strength and level of exploration was found (Bosman, 2009).

The Conceptual Model

Now, based on the theory and findings described above, we see some contours of our model. Our model will describe the sequence of interactions between person and context. The development of commitments can be assumed to be driven by events. These events consist of examples, experiences, pieces of information, and they affect the strength of a commitment in either a negative or a positive way. Basically, this interaction can have two outcomes: a fit or confirmation of the commitment, or a conflict or crisis. A conflict that concerns a commitment or a domain that is important for the person will trigger strong negative emotions and a sense of urgency: something should be done (Frijda, 1986). Thus, a conflict triggers exploration. However, we do not know yet whether exploration has a negative effect on the growth of commitment, or whether the conflict, or the negative emotions accompanying conflict, has a negative effect on commitment growth/making. Based on common sense, we assume that level of exploration and commitment strength directly mutually affect each other. It seems plausible that an active exploring of different possibilities undermines the existing commitment, and that the loss of a strong commitment makes exploration urgent.

The choices that we make here clearly demonstrate the lags in our knowledge concerning mechanisms and processes. There is – as far as we know – no empirical knowledge concerning the mechanisms in the relation between exploration and strength of commitment in the case of crisis.

The shape of commitment development is different in different people (Kunnen et al., 2008). Bosma and Kunnen (2001) suggest that stable differences exist between people in how they cope with identity conflicts: in how long they stick to assimilation and how easily they change their commitments. Personal and contextual determinants determine the ratio between assimilation and accommodation, and the optimal balance. This means that the model also needs some parameter that represents this stable inter-individually different preference. Research shows, for example, relations between the development of an achieved identity (which implies that commitments have been developed) and personality factors such as openness to experience (Marcia, 1993) and identity styles (Berzonsky, 1990).

Each iteration starts with the outcome of an interaction. This outcome can be a conflict or a fit. However, the outcome is not a simple dichotomy. Conflicts can be small or big, and positive outcomes can be very confirmative for a commitment or more or less neutral. Therefore, we will express the outcome of each interaction by a number in the range between –1 and +1. Numbers below zero express a negative outcome, thus a conflict, and numbers above zero express a positive outcome, thus a fit. The other variables in the model are the commitment strength, the exploration in breadth and the exploration in depth. We assume that commitment making and identification of commitment can both be expressed in one variable:

the strength of commitment. The making of commitment is a process that is related to the growing of the commitment. The identification of commitment refers to the process of evaluation. It is the process that keeps the commitment strong. With regard to both types of exploration this is less clear. They may have different roles, and we will start by including them as two different growers.

Based on the theory and research described above we can formulate relations that should be included in the model:

1. The commitment strength.
   (a) Affected by outcome: a positive outcome increases the strength; a negative outcome decreases the strength.
   (b) Negatively affected by exploration in breadth.
2. The exploration in breadth.
   (a) Affected by outcome: a negative outcome triggers exploration.
   (b) Affected by changes in levels of commitment strength: increase in strength reduces the exploration in breadth.
3. The exploration in depth.
   (a) Active only in the case of high levels of commitment.
   (b) Affected by a stable personal characteristic (openness to experience).
4. The outcome.
   (a) Determined by chance and normative environmental events.
   (b) Affected by stable personal characteristics such as openness to experience. This can be seen as a tendency to assimilate or accommodate: higher tendency to assimilate means lower openness, which reduces the chance of conflict.

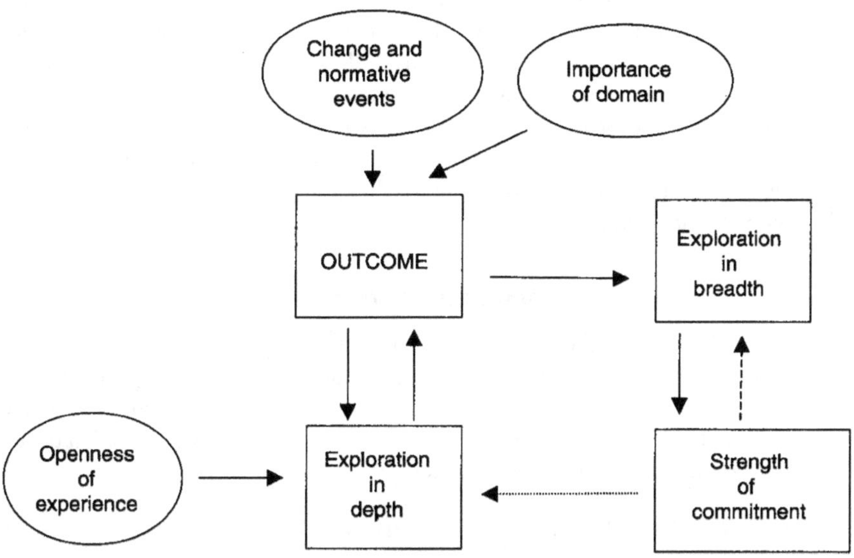

Figure 10.1 A schematic drawing of the conceptual: dotted arrows indicate a conditional relation, the broken arrow indicates an effect of the change of the grower

(c) Influenced by the importance of the domain: a relatively irrelevant domain will contain smaller conflicts and successes.
(d) Negatively affected by the commitment strength: strong commitments reduce the openness to conflicting information.

Figure 10.1 shows a schematic representation of the conceptual model.

Now that we have sketched the contours of the model, the next step is to specify the type of relations in a quantitative way.

Developing a Quantitative Model of Commitment Development

As the next step, we have to choose the type of equation that describes the change of the variables and the relations between them. Let us start with the equation for the growth of commitment strength. The simplest form is the linear one. However, as argued in the first chapter in this book, linear relations in psychology are almost non-existent. With regard to commitment growth, it is evident that it cannot be described in a linear way. There is a limit to the strength of a commitment; it cannot grow endlessly. The same holds for the decrease of commitment strength. The most plausible shape of the growth of a commitment is the S-shaped curve: a slow start, growth and then slowing down at the end. However, fluctuating change also is possible (Kunnen, 2009; Kunnen et al., 2008). Starting from the principle that we choose the most simple equation that works, we select the logistic growth equation. This equation allows for different shapes of development, and it meets the demands concerning the limitations in growth and decay. The basic form of the logistic growth equation is given below:

$$C_{t+1} = C_t + a \times C_t - a \times C_t^2 / C_{max} \tag{10.1a}$$

where C_t is the commitment strength at time t and a is the growth rate. In normal language, Equation 10.1a says that the value of the commitment strength C at some time t+1 is the result of the value of C at the previous time (C_t) plus a growth rate times the actual level of C. The part after the minus sign includes C_{max}, which represents an upper limit for the value of the commitment strength. The part of the equation that follows the minus sign has the characteristic that it increases as the level of C approaches this upper level.

The behavior over time of this equation depends on the value of the growth rate a. For values of a below around 1 the equation generates a smooth S-shaped curve. For values above 1 the curve fluctuates once the value of C approaches C_{max}. A value of a above 2.7 results in chaotic behavior (Figure 10.2).

Commitment Strength

Equation 10.1a is the basis of the equation for commitment growth in our model. The next step is to integrate the two factors assumed to affect commitment growth into the equation. The outcome of an experience is expected to have a positive effect and the level of exploration in breadth is expected to have a negative effect. These variables have been included in the equation by integrating them in the growth rate. The new equation then becomes:

$$C_{t+1} = C_t + (\mathrm{OUT}_t - \mathrm{EB}_t) \times C_t - (\mathrm{OUT}_t - \mathrm{EB}_t) \times C_t^2 / C_{max} \tag{10.1b}$$

where C_t is the commitment strength, OUT_t is the outcome and EB_t is the exploration in breadth, all at time t.

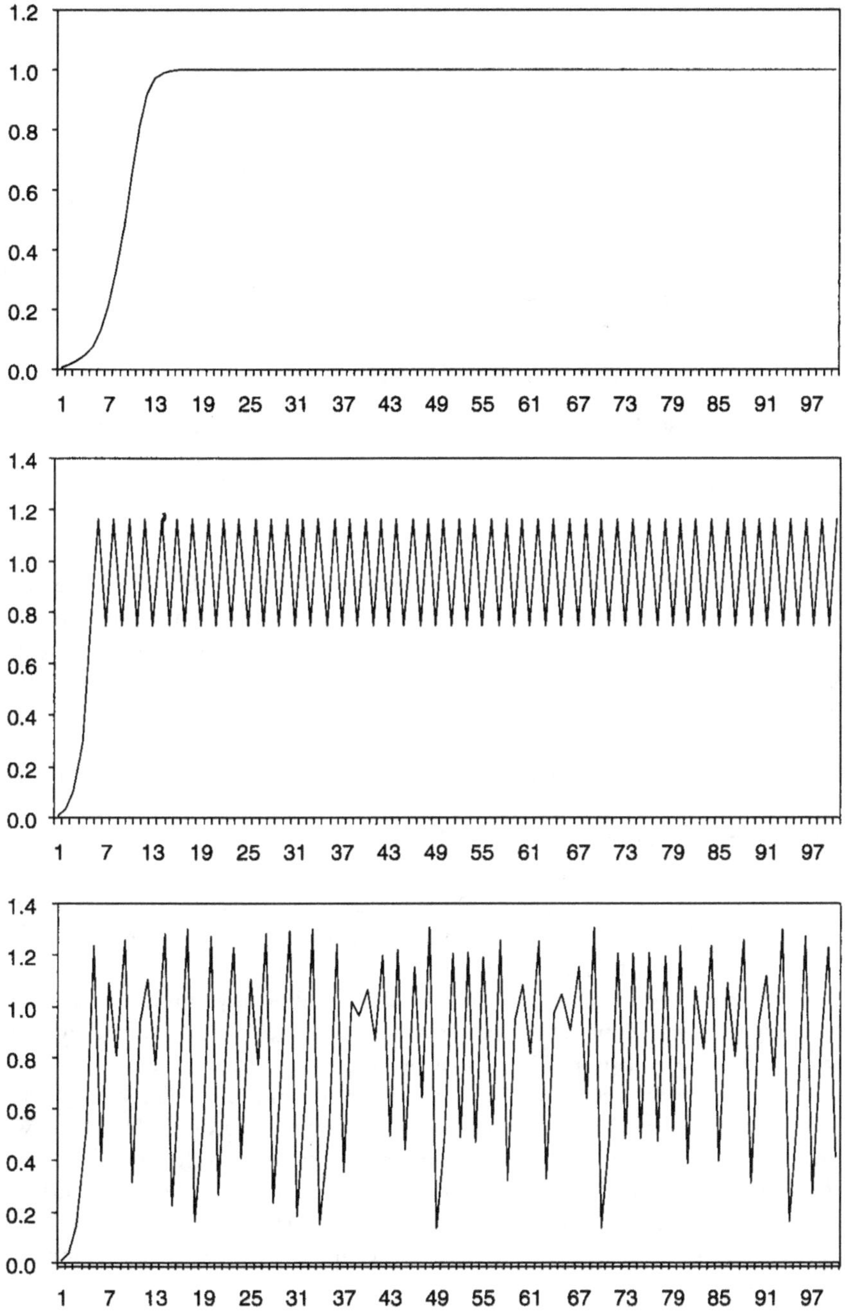

Figure 10.2 Trajectory generated by a basic logistic growth equation with growth parameter values of 0.7, 2.2 and 2.9, respectively

As a final step we add, for technical reasons, parameters that we can use to regulate the relative impact of each of the components of the equation:

$$C_{t+1} = C_t + (h \times \text{OUT}_t - k \times \text{EB}_t) \times C_t - (h \times \text{OUT}_t - k \times \text{EB}_t) \times C_t^2/C_{\max} \quad (10.1c)$$

where h regulates the impact of OUT and k regulates the impact of EB.

In the same way, we choose equations for the change of exploration in depth (ED) and exploration in breadth (EB). These growers also have an upper limit and can be expected to show S-shaped changes or fluctuations, therefore the logistic growth equation seems a good choice for these growers as well.

Exploration in Depth

The basis for the equation describing the levels of exploration in depth over time becomes:

$$\text{ED}_{t+1} = \text{ED}_t + b \times \text{ED}_t - b \times \text{ED}_t^2/\text{ED}_{\max} \quad (10.2a)$$

where ED is the exploration in depth and b is the growth rate.

We hypothesized that the growth of exploration in depth is affected by one factor: the openness to experience or tendency to accommodate. Because we have conceptualized the tendency to assimilate and the tendency to accommodate as two poles of the same construct, the growth rate should consist of the negative pole of the assimilation tendency s. We include this in the equation as f-s:

$$\text{ED}_{t+1} = \text{ED}_t + (f\text{-}s) \times \text{ED}_t - (f\text{-}s) \times \text{ED}_t^2/\text{ED}_{\max} \quad (10.2b)$$

where f is a parameter that regulates the impact of s and s is the tendency to assimilate. We assume that the growth of exploration in depth is conditional: it can grow only in the case of a high level of commitment strength. We include this condition in the equation by means of an IF statement (IF $C_t < 0.75$, THEN 0, ELSE f), which says that the growth rate is 0 as long as the level of commitment strength is below 0.75.

Compared to commitment growth, exploration is a time- and energy-consuming activity. As a consequence, high levels of exploration cannot continue forever due to exhaustion. This has to be integrated into the equation. We do that by adding a component that consists of the average value of exploration over, say, the previous 100 iterations. A high value of this component should have a negative effect on the level of exploration in breadth. However, we want an effect of exhaustion only if the exploration in breadth has been really high for a long period. We do not want a median effect in the case of a prolonged period of a median level of exploration. This can be achieved by not simply using the average of the level of exploration over the previous 100 iterations, but to raise this average to the second power. In this way, an average exploration level of, for example, 0.4 results in the value 0.026, while an average level of 0.9 results in the value 0.666.

$$\text{ED}_{t+1} = \text{ED}_t + ((f\text{-}s) - w \times \text{EDA}_t) \times \text{ED}_t - ((f\text{-}s) - w \times \text{EDA}_t) \times \text{ED}_t^2/\text{ED}_{\max} \quad (10.2c)$$

where w is a parameter that regulates the impact of ED and EDA is the average of the previous 100 values of ED.

Exploration in Breadth

The change of the third grower, exploration in breadth, is represented by the logistic growth equation based on the same arguments as the two other growers:

$$EB_{t+1} = EB_t + g \times EB_t - g \times EB_t^2/EB_{max} \tag{10.3a}$$

where EB is the exploration in breadth and g is the growth rate.

We assumed that two factors influence the change of commitment in breadth: a negative outcome triggers exploration and an increase in commitment strength reduces the exploration in breadth. These two factors are integrated in the growth rate g:

$$g = (m + (C_t - C_{t+1}) \times n) - (p \times OUT_t) \tag{10.3b}$$

where m is a constant, n regulates the impact of the change in C and p regulates the impact of OUT.

The first part of Equation 10.3b describes the effect of the change in commitment strength $(C_t - C_{t+1})$, and the second part describes the effect of outcome (OUT). The parameters p, m and n have been included in order to be able to regulate the relative influence of both factors. We find the complete equation for the growth of exploration in breadth when we combine Equations 10.3a and 10.3b:

$$EB_{t+1} = EB_t + (m + (C_t - C_{t+1}) \times n) - (p \times OUT_t) \times EB_t - (m + (C_t - C_{t+1}) \times n) - (p \times OUT_t) \times EB_t^2/EB_{max} \tag{10.3c}$$

where $C_t - C_{t+1}$ is the change in commitment strength. Also here we add an exhaustion factor:

$$EB_{t+1} = EB_t + (m + (C_t - C_{t+1}) \times n) - (p \times OUT_t - EBA_t \times z) \times EB_t - (m + (C_t - C_{t+1}) \times n) - (p \times OUT_t - EBA_t \times z) \times EB_t^2/EB_{max} \tag{10.3d}$$

where EBA_t is the fourth power of the average of the previous 100 values of EB and z regulates the impact of EBA.

Outcome

Finally, we have to define an equation for the outcome. As mentioned, the outcome of identity-relevant events is determined by chance and normative environmental events. In the model each event is generated separately by a random number in each iteration. We assume that most events do not have a strong impact: there are more events with a low impact value than with a high value. Most spreadsheet programs generate random values between 0 and 1. To generate random values that can have a positive and a negative value, and with a distribution that peaks around the middle (i.e. low values around zero), we choose the equation:

$$OUT_t = \text{random number} - \text{random number} \tag{10.4a}$$

We assume that openness to experience, or the tendency to assimilate or accommodate, affects the outcome values. A higher tendency to assimilate (i.e. lower openness) reduces the chance of conflict and thus the chance that high negative outcomes occur. Furthermore, the outcome is influenced by the importance of the domain: a relatively irrelevant domain will contain smaller conflicts and successes. Finally, the outcome will be negatively affected by the commitment strength: strong commitments reduce the openness to conflicting information. Thus, the equation becomes:

$$OUT_t = r \times (RANDOM - RANDOM) + s + t \times C_t \qquad (10.4b)$$

where OUT_t is the value of the outcome at time t; r determines the possible range of the outcome values, and thus represents the importance of the domain; RANDOM is a random number between 0 and 1; s represents the tendency to assimilate; C_t is the level of commitment strength at time t; and t regulates the relative impact of the value of the commitment strength.

All four equations described above (Equations 10.1–10.4) together form the model. The next step is to enter these equations in a spreadsheet file in such a way that we can simulate the behavior of the model – that is, we can generate sequences of values over a period of time for each of the growers in the model.

In this phase we have to decide on the time period we want to simulate and the number of iterations the model should include. In our model, each iteration represents an identity-relevant event. These events may be small. Let us assume that there will be such an event about once in every three days – say, 100 in a year. We want to simulate the development of commitment and exploration over a period of five years, ranging from age 17 until age 22. This means that our model should contain 500 iterations. The trajectory of each variable thus consists of a sequence of 500 values.

Calibration of the Model

Until now, the choice of equations was mainly theory driven. In the next step we have to choose values for the different parameters, and this step is purely technical. The values of the parameters have no theoretical meaning. We cannot say that a value of 100 for parameter t is a high value because it depends on the scale we use. Later on we will show that the differences between values of the parameters do have a theoretical meaning. The goal of this calibration is to find values for the parameters in the model for which the model behaves more or less in the expected way. This means that the simulations should not explode into infinity or get stuck at zero, but result in trajectories for the parameters that are theoretically possible. The upper graph in Figure 10.3 shows an example of an exploding model. The values of commitment strength and exploration in breadth fluctuate between extremely high and extremely low and then collapse. Exploration in depth remains low. The second graph in Figure 10.3 shows an example of a simulation that is theoretically possible. This simulation shows a period of fluctuation of commitment strength, followed by an increase in value. Exploration in breadth is high in the period when the commitment strength fluctuates, but decreases after the growth of commitment strength. At that time, exploration in depth starts to grow.

Basically, finding a plausible parameter set is a matter of trying different values and analyzing the sequences of values to get an idea of how the simulated trajectory evolves over time. For example, if a simulated trajectory explodes, one should look at the iterations preceding the explosion and analyze which grower starts – for example, to grow very fast or to fluctuate between high positive and negative values, and which influence causes this. It is best to start with low values for the parameters (close to one if they are multiplied with a grower, and zero or close to zero if they are added). For the model described here we chose the set of parameters presented

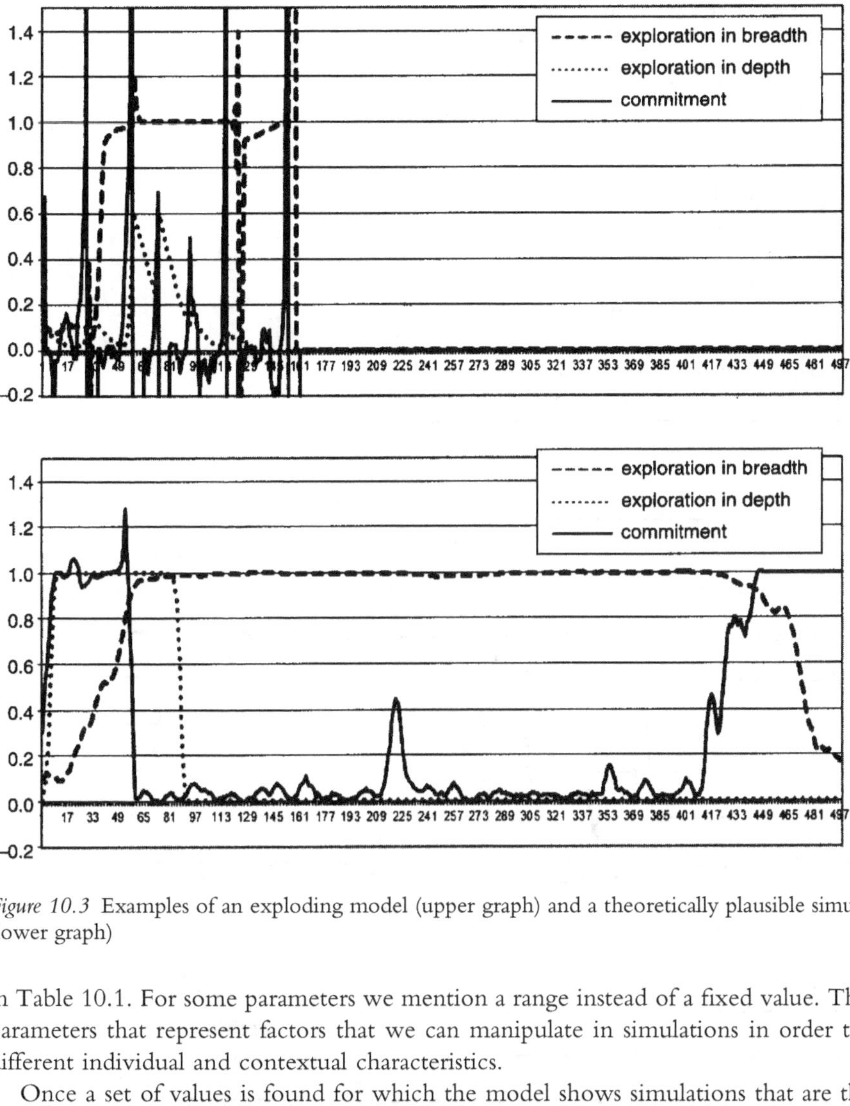

Figure 10.3 Examples of an exploding model (upper graph) and a theoretically plausible simulation (lower graph)

in Table 10.1. For some parameters we mention a range instead of a fixed value. These are the parameters that represent factors that we can manipulate in simulations in order to represent different individual and contextual characteristics.

Once a set of values is found for which the model shows simulations that are theoretically possible, we can start to fine-tune the model. In this step, the theoretical meaning of the growers and the parameters becomes important again. We have to make assumptions about the effects of slight differences in the parameters. For example, what do we expect to happen when we slightly increase the value of parameter s in Equation 10.4b? Parameter s represents the tendency to assimilate. Its theoretical meaning is that it represents stable differences between people with regard to their preference to assimilate or to accommodate. We assume that these differences in preference manifest themselves in types of identity development. A high tendency to assimilate means a high chance that the person develops foreclosed commitments: commitments that become strong and remain strong without much exploration. A low preference for assimilation and thus a high preference for accommodation means that there is a high chance that the person will develop commitments only after periods of exploration, and will show fluctuations in commitment strength. An extremely high tendency to accommodate may result in a prolonged period of exploration without the development of strong commitments.

Table 10.1 Chosen parameter values in the model

Parameter	*Description*	*Value*
m	Growth rate of EB	0.015
h	Effect of OUT on C	0.6–1
k	Effect of EB on C	0.1
n	Effect of C on ED	5
s	Assimilation tendency	0–0.06
p	Effect of OUT on EB	0.3
t	Effect of C on OUT	0.05
f	Regulates impact of *s* on ED	0.06
r	Range of OUT	1.5–2.5
z	Effect of previous levels of EB	0.3
w	Regulates impact of time lag on ED	1

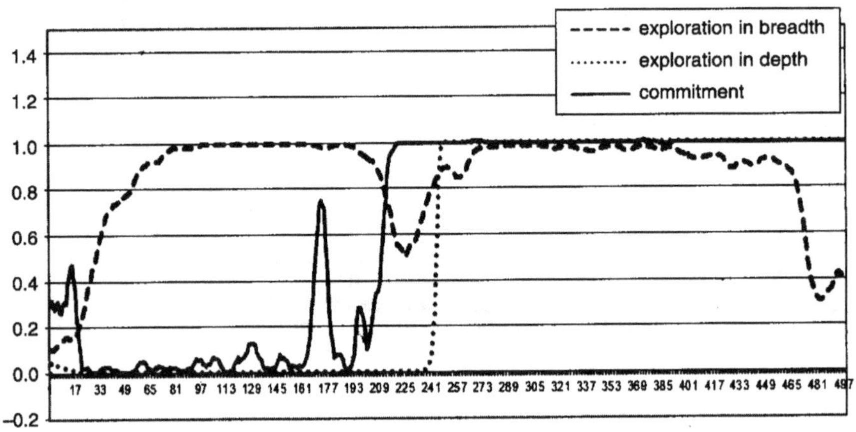

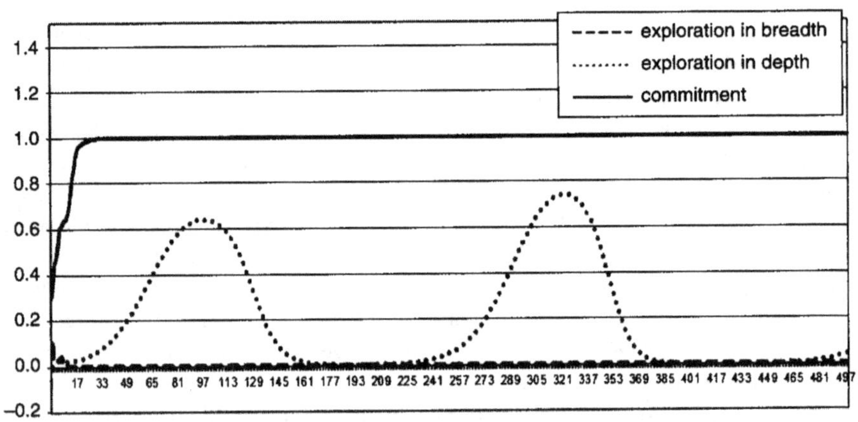

Figure 10.4 Simulations with two different values of *s*: in the upper graph $s = 0.01$; in the lower graph $s = 0.1$

The mathematical effect of parameter *s* is that the outcomes of commitment-relevant events become a little biased towards the positive side. This means that relatively more events stimulate the formation of a commitment, whereas there are less negative experiences related to the commitment. We thus expect an accelerated growth of the commitment strength. Figure 10.4 shows two simulations as typical examples. In the upper graph the value of s is less than in the lower graph. The graphs show what was theoretically expected. Due to the random values in the model, no two simulations are the same.

In the process described above, we looked for a set of parameter values that result in plausible simulated trajectories. What "plausible" is depends on general theoretical notions and expectations. For example, we expect the model to generate trajectories with rapidly growing levels of commitment strength more often if the assimilation level is high, but we also expect the same parameter set to generate different trajectories due to the random influence of the outcomes (i.e. the unpredictable influence of the context). In some simulations high levels of commitment strength should decrease again, representing a period of crisis, when existing commitments have to be reconsidered. In some simulations no commitments should develop, especially if the ratio of assimilation to accommodation is low. This phase in model development is basically the playing phase. One has to imagine the real-life meaning of differences in parameter values, and to make assumptions about what that could mean for the generated trajectories. Now that we have found such a set of parameters, we are ready for the next step.

Evaluation and Validation of the Model

Now our model behaves more or less in the way that we expect on the basis of our common sense and theoretical expectations. This is already a first and important step in the validation of the model. Some critics argue that this is a trivial step, because the equations and parameter values have been manipulated and changed until it behaves in the desired way. However, this argument is valid only in linear models, such as those in regression analysis. Our model building is different in two ways. Technically, it is not possible to manipulate the model into desired behavior by just selecting the right parameter values. Because of the non-linear iterative character, it is not possible to determine the generated trajectories in any desired way: the model is underdetermined by the parameters. Second, and still more importantly, as discussed in Chapter 7, our aim is not to fit an existing set of empirical data. Our aim is to make a model that represents theoretically plausible variables and relations, and that generates trajectories that fit with empirical data. With regard to the manipulation of the equations, our main restriction is the theoretical assumptions underlying it. All our equations are chosen in such a way that they represent theoretically postulated mechanisms of commitment development. The adjustments we made to correct undesired outcomes were done in such a way that they represent the theory. This means that every adjustment of the equations is an adjustment of the theoretical assumptions that we want to test. Our initial model was based on a theoretical model that was too simple, because we did not take into account the relevance of factors that result in reduction of the level of exploration. Thus, the components we added in Equation 10.3d are the result of a refinement of our theoretical assumptions.

As a next step, the model has to be validated in a more stringent way. But how? We may gather longitudinal data on the development of commitment strength and levels of exploration that can be compared with simulated trajectories, but that is not so easy. Practically, we have no methods to assess commitment strength and explorations in breadth and depth in a direct, unobtrusive and easy way. We cannot think of methods to gather 500 data points over a period of five years without driving the subjects crazy and disturbing the development we want to observe. Theoretically, also, this way of validation is problematic because of the role of chance and unpredictable events. When we run the model ten times with the same parameter values it generates ten different trajectories, and most probably none of these trajectories is based on a patterning of events that is the same as the life events encountered by the real subject. Thus, we need to develop more indirect methods to validate the model.

One way to do so is to use the model to simulate groups of people and compare simulated outcomes with existing data from group studies. For example, we may simulate a group of, say, 1,000 young adults at age 20. Let us assume that the value of the assimilation tendency is more or less evenly distributed in the range between 0 and 0.06. Let us also assume that the level of challenges in the adolescents' environment is normally distributed. This means that we run 1,000 simulations with different values for the parameters representing the assimilation tendency and the level of challenges. For each simulation we register the level of commitment strength and exploration at the data point that represents age 20. (Remember that our model represents a period of five years and starts at age 17, so data point 300 represents age 20.) As a next step, we may assess the commitment strength and exploration levels in 1,000 20-year-olds and compare the distribution of values. A more detailed example of this way of validating a quantitative model can be found in Kunnen and Bosma (2000).

There are still other ways to validate the model. For example, even though it is not possible to fit empirical and simulated trajectories in a one-by-one way, it is possible to compare the types of trajectory generated by the model with those found in empirical research. We can compare distributions of simulated trajectories with trajectories found in a population. We can assess differences in the ratio of assimilation to accommodation in that population, and compare the trajectories of different empirical subgroups with simulations of the same subgroups. We can simulate and thereby predict the effects of sudden life events, such as a sequence of events that challenge one's commitments. In the context of a university, such predictions could be tested by investigating study commitment trajectories in students who repeatedly fail their exams.

The Ultimate Model of Identity Development?

Given that we succeed in all validations of the model, does that mean that our model is the model of commitment development? We do not think so. Reality is too complex to achieve that, and the development of commitments concerns many different aspects. This means that other models, based on (slightly) other assumptions, are also possible and may be equally valid. A model represents only a few growers and parameters. Which growers are included in the model depends, among others, on the research questions one wants to address. For example, in 2001 we published a model that describes commitment development from a slightly different perspective: as a development of competing commitments (Kunnen, Bosma and van Geert, 2001). Of course, the basic notions concerning the effect of positive and negative events on the growth and decrease of commitment strength are the same in both models. These are the basic assumptions about the underlying mechanisms of commitment development, and of course different models that simulate the same time span of the same grower should generate comparable trajectories. The proof of all models lies in the validation of simulated trajectories with empirical data.

The Gain in Knowledge Offered by this Technique

Translating theoretical assumptions into a mathematical model allows us to generate detailed and testable hypotheses and it necessitates and allows for far more specific questions and assumptions than a theory or a conceptual model can ever do (Nowak and Vallacher, 1998). Playing with the model – exploring different plausible sets of parameters – generates trajectories that can be the basis of new hypotheses. The simulations help to specify and evaluate the plausibility of theoretical ideas and thus restrict the number of alternative hypotheses that have to be put to the empirical test. The confirmation of hypotheses by the simulations with the mathematical model will provide us with evidence for the plausibility of the model and its underlying theoretical assumptions. We should realize that verbal models alone do not allow us to arrive at predictions beyond a few steps in the developmental process. The point is that the wealth of – often mutual – relationships discerned by those models prevents us from verbally anticipating the intended effect of those relationships beyond a very short time window. We need a "mechanical" device in order to infer the results of the developmental mechanisms. Such a device takes the form of a simple computer program – a spreadsheet model – that allows us to run numerical experiments with various types of – psychologically plausible – conditions. For instance, our very simple model has shown under which circumstances (the type of sequence of events) an individual tendency to assimilate may result in a foreclosed, diffused or achieved identity trajectory.

One major advance in building mathematical models is, thus, that it forces researchers to be very explicit about their assumptions concerning the developmental mechanisms and processes. Although these assumptions underlie the theories, they often remain hidden. By exposing them, and comparing the simulations with the expected trajectories, these assumptions can be tested and challenged. As for the development of commitments, a well-accepted and explicit assumption is that differences in the tendency to assimilate are important. In our model, we represented the strength of this tendency as a bias in the interpretation of events. The stronger the bias, the more events were seen as supporting one's existing preferred commitment. In our simulations, the effects of differences in the tendency to assimilate accorded with theoretical expectations and empirical findings: higher assimilation results in more foreclosed trajectories. However, alternative representations of the assimilation tendency are possible. The simulations of models including these alternative representations could be compared with those presented here. Moreover, further research could be carried out to test different representations of the tendency to assimilate more directly.

Acknowledgment

This chapter originally appeared in *A Dynamic Systems Approach to Adolescent Development* (New York: Psychology Press, 2012).

References

Berzonsky, M. D. (1990). Self-construction over the life span: A process perspective on identity formation. In G. J. Neimeyer & R. A. Neimeyer (Eds.), *Advances in Personal Construct Psychology* (Vol. 1, pp. 155–186). Greenwich, CT: JAI Press.

Bosma, H. A., & Kunnen, E. S. (2001). Determinants and mechanisms in identity development: A review and synthesis. *Developmental Review, 21*, 39–66.

Bosman, K. R. (2009). The role of exploration and emotions in commitment development. Master's thesis, Department of Developmental Psychology, University of Groningen, The Netherlands.

Frijda, N. H. (1986). *The Emotions*. Cambridge, UK: Cambridge University Press.

Kunnen, E. S. (2006). Are conflicts the motor in identity change? *Identity, 6*, 169–186.

Kunnen, E. S. (2009). Qualitative and quantitative aspects of commitment development in psychology students. *Journal of Adolescence, 32*, 567–584

Kunnen, E. S., & Bosma, H. A. (2000). Development of meaning making: A dynamic systems conceptualization. *New Ideas in Psychology*, *18*, 57–82.

Kunnen, E. S., Bosma, H.A., & van Geert, P. L. C. (2001). A dynamic systems approach to identity formation: Theoretical background and methodological possibilities In: J.-E. Nurmi (Ed.), *Navigating Through Adolescence: European Perspectives* (pp. 247–274). New York: Garland Publishing.

Kunnen, E. S., Sappa, V., van Geert, P. L. C., & Bonica, L. (2008). The shapes of commitment development in emerging adulthood. *Journal of Adult Development*, *15*, 113–131.

Luyckx, K., Goossens, L., & Soenen, B. (2006). A developmental-contextual perspective on identity construction in emerging adulthood: Change dynamics in commitment formation and commitment evaluation. *Developmental Psychology*, *42*, 366–380.

Luyckx, K., Goossens, L., Soenen, B., & Beyer W. (2006). Unpacking commitment and exploration: Preliminary validation of an integrative model of late adolescent identity formation, *Journal of Adolescence*, *29*, 361–378.

Marcia, J. E. (1980). Identity in adolescence. In J. Adelson (Ed.), *Handbook of Adolescent Psychology* (pp. 159–187). New York: John Wiley & Sons.

Marcia, J. E. (1993). The status of the statuses: Research review. In J. E. Marcia, A. S. Waterman, D. R. Matteson, S. L. Archer, & J. L. Orlofsky (Eds.), *Ego Identity: A Handbook for Psychosocial Research* (pp. 22–41). New York: Springer Verlag.

Marcia, J. E., Waterman, A. S., Matteson, D. R., Archer, S. L., & Orlofsky, J. L. (1993). *Ego Identity: A Handbook for Psychosocial Research*. New York: Springer Verlag.

Nowak, A., & Vallacher, R. R. (1998). *Dynamical Social Psychology*. New York: Guilford Press.

PART 2

Environment and Sustainability

Introduction

The second set of chapters in this volume explores applications of system science methods to issues of the environment and environmental sustainability, ranging from the complex interactions between humans and their environments to attempts at influencing local and national environmental policy. System dynamics models have found application in environmental research for several decades, tracing initially to Jay Forrester's WORLD model, designed in 1970 to examine the consequences of exponential population growth given the planet's ostensibly finite carrying capacity. Likewise, agent-based models are widely used in this area because they are particularly well suited to capturing spatial features like terrain, land use, and distance. Network analysis is perhaps less common than other system science methods for environmental research, but its use is rapidly growing, especially in combination with other system science methods as Chapters 14 and 19 highlight.

Coupled Systems

Significant attention within environmental studies has turned to investigating coupled human and natural systems or CHANS, for which the U.S. National Science Foundation now maintains a standing program. The CHANS approach views human systems like housing mobility and labor markets as inextricably linked with natural systems like the water cycle and food chains. As a theoretical framework, it insists that both types of systems must be examined together, and thus is well matched with the logic of system science methods, which focus on understanding how individual elements are interrelated in a larger, holistic system.

Richard Dudley describes one example of a coupled human and natural system, where the coupling leads to seemingly irrational behaviors. Indonesian villagers engage in illegal logging activities, despite the fact that these activities are rapidly depleting the forests from which they derive a significant part of their livelihood. Why? He decomposes this coupling of illegal logging activity (a human system) and forest depletion (a natural system) into a series of five causal loops that interact in a single system dynamics model. For example, one causal loop specifies that as some villagers participate in these activities, they encourage others to do so, which interacts with another loop specifying that as villagers' need for money increases, they are more likely to engage in illegal logging. By exploring interactions between these different causal processes,

 DOI: 10.4324/9781315748771-12

which link human and natural systems together, he develops a tool for evaluating the potential effectiveness of strategies intended to control illegal logging.

The two chapters by An and colleagues present two additional CHANS models, focused on the interplay between human development activity and the habitats of endangered species in the Wolong Nature Reserve in China and the Chitwan National Park in Nepal. Their analysis of the Wolong site is similar to Dudley's because their model is built up from the combination of a series of subsystems – for example, building separate but interacting models of human demographic dynamics and natural landscape dynamics. However, unlike Dudley who used a system dynamics model, they use an agent-based model to achieve this, thereby illustrating how different system science methods can be used to model CHANS. In the second chapter, they compare their Wolong model to an agent-based model of the Chitwan site, highlighting the similarities and differences that emerge when the same system science method is applied in two different CHANS contexts. Here, they illustrate the ODD (overview, design, details) framework for describing agent-based models. Their comparison informs a series of recommendations for using system science models to understand CHANS, and serves as an example for developing best practices for the use of system science models in other contexts.

Environmental Policy

The first three chapters in this part focus on using system science to understand the complex dynamics of coupled systems. But, armed with such an understanding, how can public policy be used to manage these systems? System science methods are useful here too. Barton and colleagues take a close look at the European Union Water Framework Directive, a set of policies designed to improve the quality of ground and surface water. More narrowly, they focus on the case of Vanemfjord lake in South-Eastern Norway, and on how to combine multiple perspectives on the value of eutrophication abatement policies adopted under the directive. They rely on an innovative combination of system dynamics and networks to build a meta-model. The model begins with a series of separate system dynamics models each capturing different aspects of eutrophication abatement. These separate system dynamics models are then linked to each other in a unique type of network known as a Bayesian belief network, which captures how each system dynamics model depends on the others. Although this meta-modeling strategy brings us closer to capturing the full complexity of policy decisions in CHANS management, they concede that even the most sophisticated system science models will inevitably omit some key details, which simultaneously highlights the promise and challenge of using system science methods to evaluate public policy.

Also seeking to understand policies aimed at cultivating sustainable practices, Bush and colleagues develop a system dynamics model in a different context: the development of the biofuels industry. Biofuels represent one avenue toward reducing dependence on imported energy as well as reducing the emissions of greenhouse gases, but there are several potential barriers to the development of this industry. Using a system dynamics model known as the Biomass Scenario Model (BSM), developed at the National Renewable Energy Laboratory, the authors explore how different incentive-based policies might be used to facilitate biofuel industry development. While their analysis serves as another example of the use of system science methods to evaluate public policy options, their findings highlight another reason system science methods are useful: their ability to uncover synergistic effects. Traditional approaches to modeling might aim to identify the effect of each incentive on its recipient – what is the effect of an incentive to farmers, and what is the effect of an incentive to consumers – then add these effects together. But as the BSM demonstrates, this approach misses part of the story. When incentives are coordinated, their total effect is greater than the sum of their individual effects.

Working in a similar context toward sustainability, Shafiei and colleagues focus not on the fuel, but on the vehicle, seeking to understand the barriers to adoption of alternative fuel vehicles. Like Bush in Chapter 14, they also combine multiple system science methods, building a hybrid model that includes both system dynamics and agent-based components. System dynamics models focus on interactions among macro-scale variables from a top-down perspective, while agent-based models focus on interactions between micro-scale agents from a bottom-up perspective, thus each is best suited for capturing different aspects of a complex system. Like a conventional agent-based model, consumers, car manufacturers, car dealers and other key agents interact with one another. But each of these agents' own behaviors is driven by an internal system dynamics model; a system dynamics model of car manufacturers controls how car manufacturers behave as agents in the agent-based model, and likewise for the other agent types. This unique hybrid combination gets us closer to how decisions about vehicle selection, or perhaps about anything, are actually made: we each make some internal calculations (like a system dynamics model), but these are influenced by our interactions with others (like an agent-based model).

Stakeholder Engagement

Ultimately, for public policy to be effective, it requires the support or buy-in of stakeholders. The final chapters in this part explore ways to use system science methods to promote stakeholder engagement in the modeling process and to understand stakeholder influence in environmental issues. The first step in engaging stakeholders is actually locating them. Working in the Peak District National Park in the United Kingdom, Prell and colleagues use social network analysis to understand which stakeholders communicate with each other, and subsequently to identify the most influential stakeholders. Guided by this analysis, they were able to make more informed choices about who should be invited to participate in conversations about natural resource management in the park. This helps to ensure that different categories of stakeholders have a place at the table, and also that those at the table are well positioned to disseminate information and cultivate support for decisions.

While networks are useful for identifying stakeholders, D'Aquino and Bah explore the possibility of actually involving stakeholders in the development, testing, and analysis of the system science method itself. Their goal was to understand how Senegalese farmers deal with the uncertainty of rainfall in the African drylands. Who better to ask than Senegalese farmers? Through a series of meetings with farmers, they first developed an initial sense of the factors most influential on farmers' behavior and the issues the farmers most wanted to address. The authors then developed a preliminary interactive agent-based model in the form of a board game, which the farmers played. Farmers' behavior while playing the board game provided insight into how the model should be refined. This explicitly participatory approach to system science, which remains rare but is gaining attention, offers a number of potential advantages over laboratory-developed system science models. By giving the research participants a direct role in the research, they have a sense of ownership and commitment to the project, and are more likely to trust the results. Additionally, building the model in collaboration with participants offers an opportunity for the model to incorporate both indigenous and external scientific knowledge.

In many cases including those discussed by Prell and D'Aquino, stakeholders are viewed as actors that can and should be included in the policy decision-making process. But stakeholders can also represent risks that need to be managed, particularly when some stakeholders' goals conflict with others. Boutilier and colleagues explore this possibility in the context of African mining companies, whose success depends in part on diverse stakeholders' willingness to grant the companies a "social license to operate." Because such companies' operations can involve multiple risks to the environment and to the surrounding population, relations with stakeholders are often controversial and the companies require strategies for managing these relations. Similar to Prell, the authors use network analysis to identify the stakeholders with the greatest potential to be influential. They then use the network analysis findings to inform an agent-based model designed to test a series of potential stakeholder management strategies. Together, these three chapters illustrate that system science methods are useful for exploring stakeholder engagement, whether that engagement is pursued as a valuable end in itself, or as an instrumental step toward protecting a particular stakeholder's interests.

11

A SYSTEM DYNAMICS EXAMINATION OF THE WILLINGNESS OF VILLAGERS TO ENGAGE IN ILLEGAL LOGGING

Richard G. Dudley

Research Question: Despite the fact that illegal logging will result in the loss of a major source of their livelihood, why do Indonesian villagers participate in these activities?
System Science Method(s): System dynamics
Things to Notice:

- Building a complete model from multiple submodels
- Using interacting systems to understand seemingly irrational behavior

Much of the work of illegal logging in Indonesia is carried out by villagers. Several factors determine villagers' willingness to participate in such activities. Chief among these are (1) the need for income, (2) the fact that other villagers (and non-villagers) are already illegally logging, and (3) the realization of loss of community control over traditional forest areas. These factors form the basis of feedback loops, which trap villagers in illegal logging systems, which will likely result in the disappearance of a major source of livelihood. Ideas for system dynamics model structure were obtained from field reports and interviews with stakeholders. These ideas were examined using causal loop diagrams to represent different views of illegal logging. One village level view was formulated as a quantified system dynamics model using Vensim software. The model allows examination of scenarios, which might alter system behavior. The model is a tool for understanding consequences of various proposed strategies to control illegal logging. These strategies include enforcement of laws, strengthening of community rights, the prevention of outside labor in local forests, and the provision of alternate sources of income. This is part of a larger effort to describe and analyze illegal logging using system dynamics modeling.

The decline of Indonesian forests is well documented. In 1997 and 1998, during the first phases of Indonesia's economic crisis, between 3% and 50% of Indonesian timber harvest was unaccounted for in official statistics (Palmer, 2001; Scotland et al., 2000). Illegal logging was thought to account for a large portion of this shortfall. Prior to 1997 Indonesian forests were disappearing at the rate of almost 1.6 million ha per year, equivalent to an annual decline of 1.5% (World Bank, 2001). Annual rate of decline was more rapid within Sumatra (2.8%) and Kalimantan (1.9%). Several reports indicate significant increases in illegal logging since 1998 (e.g., McCarthy,

 DOI: 10.4324/9781315748771-13

2000; Obidzinski and Suramenggala, 2000; Casson, 2000), so the rate of forest loss has presumably increased considerably.

The nature of illegal logging activity has rapidly changed. Prior to 1998, much of the "illegality" was confined to large-scale timber operations owned by a well-connected business and political elite. To a significant extent these operations were technically legal because laws and regulations establishing them were created and used by the same elites. Some small-scale illegal logging took place as local people tried to gain access to traditional lands that were within the extensive forest concessions granted to powerful business and government interests. These attempts were suppressed by military and police who had ties to concession holders.

After the fall of President Soeharto in 1998, the situation changed dramatically. Large-scale concession holders, timber tycoons, and their political backers no longer had the political power to control what happened in the provinces. The effect of this failure of central government was magnified by an official, and inevitable, move toward decentralization during 2000 and 2001, which was backed by foreign donors.

A two-step change occurred. First, from 1998 to 2000, there was an explosion of illegal logging at the local level whereby entrepreneurs came in to cut trees illegally, often making deals directly with villagers and village heads. Second, by early 2000 newly empowered local governments started to assert their new authority. They created laws to permit local logging concessions. These new laws allowed local officials and entrepreneurs to create corrupt business arrangements more easily. Many local communities saw this as a challenge and an opportunity. They claimed blocks of land as traditional forest, requesting approval from local government, which then allowed them to also make deals directly with entrepreneurs. Using this approach, traditional lands can even be claimed inside established protected areas and forest concessions. Sustained yield forest management has little to do with this type of quasi-legal over-exploitation. In any case, modern forestry expertise is largely non-existent at the local level (see Obidzinski et al., 2001; Casson and Obidzinski, 2001).

When we consider the involvement of villagers in illegal logging we need to consider the above context. Villagers live in a world of uncertain laws, with the knowledge that existing laws have always been manipulated by powerful individuals for their own ends. Laws have little meaning if they are widely ignored and unenforced. Nevertheless, the emergence of local, as opposed to centralized, political power makes the use of local and traditional laws a seemingly attractive option for natural resource management. To use such an approach effectively, a better understanding of the involvement of local people in illegal logging is needed.

The purpose of the research reported here is to develop a conceptual model of village level aspects of illegal logging that explains basic causal relationships leading to villagers' willingness to engage in illegal and/or destructive activities that appear to be against their own long-term interests. A system dynamics model of this sort can be considered a theory about how a system works, and why it produces particular results. It can then be used to gain insights into the workings of the actual system. The theory can also be compared to reality and modified as better information is obtained.

This chapter represents part of a larger effort to develop system dynamics models of various aspects of illegal logging (e.g., Dudley, 2002).

Methods

Data and information used as the basis for model building were obtained from field reports (e.g., Casson, 2000; McCarthy, 2000; Obidzinski and Suramenggala, 2000; Obidzinski et al., 2001; and Wadley, 2001) and interviews with various stakeholders. Several qualitative conceptual

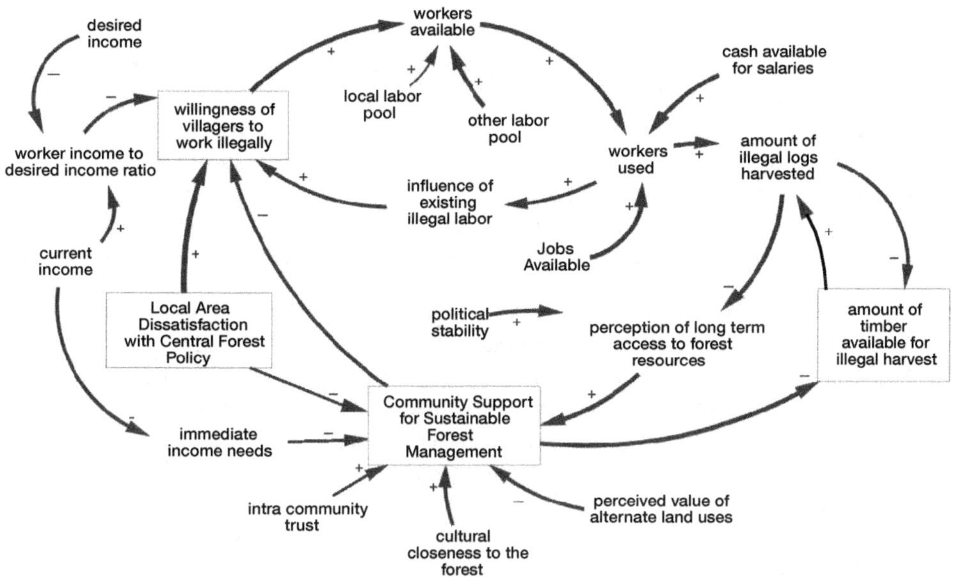

Figure 11.1 Conceptual framework for modeling was based on this diagram from Dudley (2001) which describes factors affecting villagers' perception of illegal logging

models, from different stakeholder perspectives, were earlier formulated as causal loop diagrams (see Dudley 2001 for details). Subsequently, some of these conceptual models were used as the basis for quantified system dynamics models. The model presented here is based on one of those conceptual models (Figure 11.1) that deals with one aspect of the local area view of illegal logging. Other local area aspects include the views of entrepreneurs, and the relationship between entrepreneurs and local officials. The conceptual model was used as the basis for construction of a system dynamics model. See Sterman (2000) for a discussion of the system dynamics modeling approach.

Figures presented herein are of two types: causal loop diagrams and parts of the actual model. The loop diagrams (Figures 11.1 and 11.6) show only major causal relationships and do not specifically show flows. The other figures illustrate parts of the actual model. The following conventions are used in labeling model components: stocks (sometimes called state variables or levels) are capitalized and are enclosed in boxes, flows are shown as hollow arrows with a valve and are labeled in lowercase. Auxiliary variables are also in lowercase. Constants are shown in all uppercase. Arrows are illustrated with a plus (+) or a minus (–) to indicate the general trend of the relationship between the two connected variables. A plus indicates a change in the same direction – that is, if *X* increases then *Y* also increases if this relationship is taken by itself. A minus indicates a change in the opposite direction. The actual relationship is described by the model equations. In general thicker arrows were used to illustrate more important relationships. When used in the text, names of model components are italicized.

Model Structure

The model describes a theoretical group of small communities with a total labor pool of 1,000 villagers available for logging work. The communities have 5,000 ha of well-forested traditional land holding 200 m^3 of merchantable timber per ha. Normal harvests from this forest start at about 5,000 m^3 per year (i.e., 1.0 m^3 ha^{-1}), which is calculated as a fixed fraction of available timber. The model also assumes a baseline illegal harvest of 1,400 m^3 per year for a total annual harvest of 6,400 m^3 (1.2 m^3 ha^{-1}), a reasonable sustainable harvest from these forests (see, for example, Bruenig 1996, p. 173). This creates an income of $143 per village forest laborer per year ($125 from normal logging and $1 from illegal logging). It is assumed that, initially, 1% of villagers engage in illegal logging. Villagers have other income sources so that the total annual income for members of the above labor pool averages $1,800.

The model consists of five interacting feedback loops that describe the following relationships:

1. An increase in need for money increases villagers' willingness to work illegally.
2. As more villagers work illegally they influence others to work illegally.
3. As the amount of forest still intact decreases, community support for sustainable management gradually declines. This increases willingness of villagers to work illegally.
4. As community support for sustainable management declines the amount of forest made available to entrepreneurs for exploitation increases, raising entrepreneurs' desire and ability to provide illegal salaries.
5. As the forest eventually disappears the funds from entrepreneurs decline, and jobs from logging also disappear.

Loop 1: Need for Income Forces Villagers to Work Illegally

According to the model, if villagers have adequate income then they have no need to work illegally. If their income level drops for any reason their willingness to participate in illegal logging will rise, other things being equal, and this *willingness based on income need* can be viewed as a function of the ratio of income to desired income, called the *desired income ratio*. The *desired income ratio* could also change if *desired income* changes. This might happen, for example, if villagers became aware of new desirable consumer goods, or if school fees were increased.

If the income needs of villagers increase, for example, the *current willingness of villagers to work illegally* will gradually increase as well. If this happens, some villagers will start earning illegal income. This, in turn, will raise the income of the villagers as a whole. It is assumed that the money is shared with other villagers so the average income of villagers seeking work is raised even when only some participate in illegal activities. For example, food may be purchased from neighbors. Eventually, the *desired income ratio* is raised enough so that willingness does not rise any further unless disturbed by some other factor. This is a negative feedback, or stabilizing, loop (Figure 11.2).

Loop 2: Illegal Workers Create More Illegal Workers

As illegal workers (villagers or outsiders) become more common they have a significant influence on others to participate in illegal logging. This creates a positive feedback that can spiral out of control in the absence of other controlling factors. If the *current willingness of villagers to work illegally* increases, the number of *local villagers available for work* increases. If illegal work is available and enforcement of laws is weak, the *number of illegal workers actually used* for labor will increase, and will include both villagers and outside workers if they are available. It is likely that in any village there are some *people normally working illegally* and other villagers are accustomed to this. However, at some point the *number of illegal workers actually used* rises above the normal number. As this *illegal worker ratio* increases it causes an increase in the willingness of villagers to work illegally as indicated in Figure 11.3. This is a positive, or reinforcing, feedback

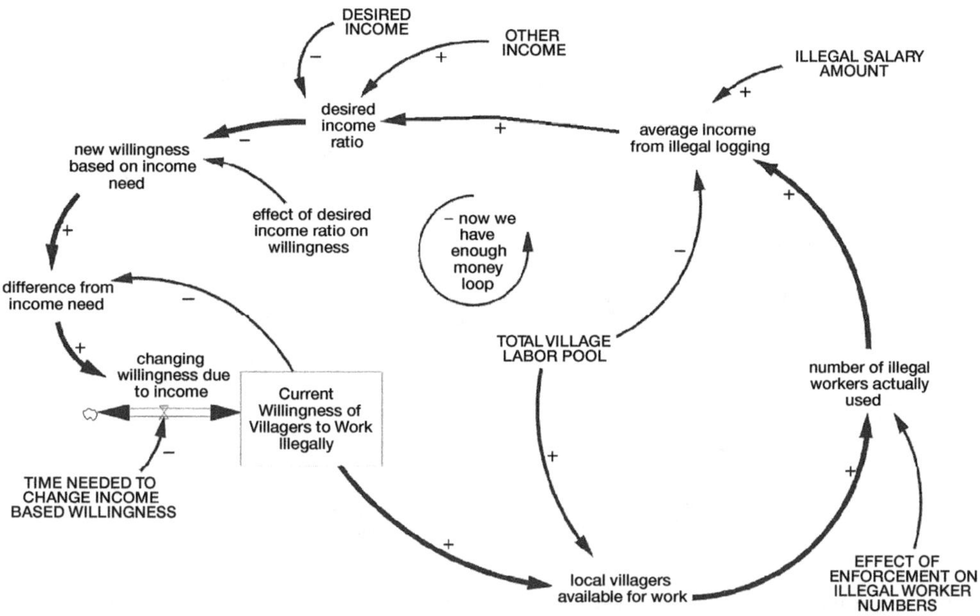

Figure 11.2 One factor affecting the willingness of villagers to work illegally is the need for income. This is a negative feedback, or stabilizing, loop, forming part of the model. As income from illegal logging rises, the willingness stabilizes. While enforcement can limit illegal workers, it will also prevent the rise of income levels so the willingness to work illegally will remain high. For clarity, some model components have been omitted in this view.

loop, which will lead to all villagers participating in illegal logging if no other factors influenced the outcome.

Loop 3: Disappearing Forest Decreases Community Support for Good Forest Management

As forests disappear in relation to what villagers see as *normal forest cover*, there will be a weakening of the community's *strength of perception of long-term access to resources*. As a community's sense of control over these resources dwindles, *community support for sustainable forest management* also decreases. If *community support for sustainable forest management* is strong, community norms, local customs and rules tend to discourage villagers from working on illegal timber operations. If *community support for sustainable forest management* weakens, then, other things being equal, the *current willingness of villagers to work illegally* will increase, causing a further increase in illegal logging and a decrease in forest cover. If no other factors come into play, this positive feedback loop will spiral out of control as forest cover disappears (Figure 11.4).

Loop 4: Decreasing Community Support Makes More Forest Available for Illegal Operations

Decreasing *community support for sustainable forest management* causes an overall weakening of traditional community control over its lands. If that happens, community leaders may become more willing to make illegal or corrupt arrangements with outsiders, or outsiders may become

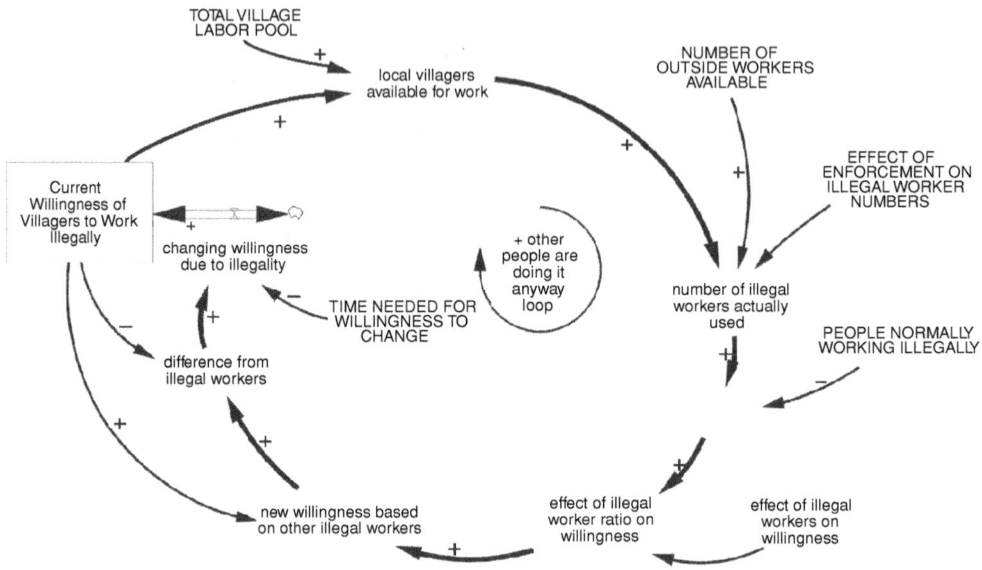

Figure 11.3 A positive feedback loop which illustrates the effect that existing illegal workers have on the willingness of villagers to work illegally. As villagers realize that others are carrying out illegal logging, they gradually lose remaining inhibitions to do so. Over time, illegal logging becomes the normal thing to do if other forces do not act on the system. For clarity, some model components have been omitted in this view.

more willing to ignore community rules and regulations that are no longer considered important by community members. As a consequence, the amount of community lands available for illegal or inappropriate exploitation will increase or decrease depending on the direction of change in *community support for sustainable forest management.* If more community lands are believed to be open for exploitation, and the risks are considered acceptable by entrepreneurs, they will provide operating capital for equipment, salaries and other needs. Workers will be hired and more illegal logging will occur, further degrading *community support for sustainable forest management.*

The *fraction of forest that can be logged illegally* also depends on the *strength of community rights* under the legal system. In some cases, community rights have been eroded as new laws give rights to other government authorities (e.g., see McCarthy, 2000). As presented here, *strength of community rights* is a constant (which can be manipulated by the model user), but other model formulations are possible. For example, we may wish to assume that as *community support for sustainable forest management* decreases, the legal support for such control would also change after some delay. That is, there is a distinction between the actual legal backing for community rights and the desire to apply those rights to manage and control logging activities on community lands.

Loop 5: Disappearing Forests Cause Disappearing Jobs

Harvest from the forest provides jobs. If forest cover becomes degraded not only will sustained yield forest productivity decline, but the temporary benefits of over-harvest will also disappear. While significant *harvestable forest cover* remains there is a large *potential illegal harvest* based on the *fraction of forest that can be logged illegally*. Eventually, *harvesting* becomes self limiting: as over-

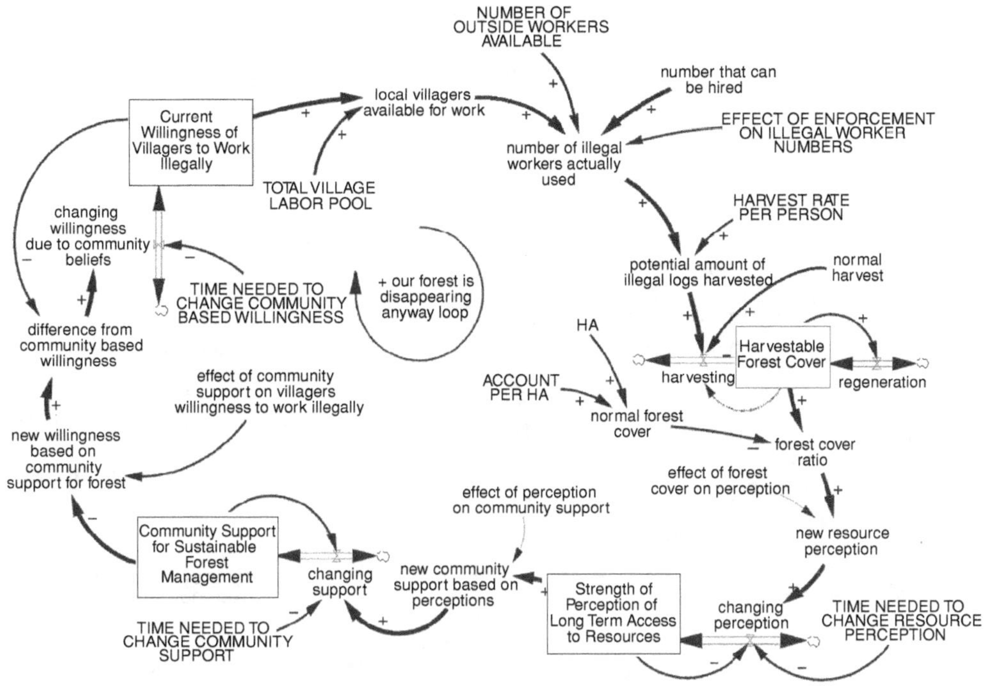

Figure 11.4 The effect of changes in *forest cover* on *community support for sustainable forest management* and, therefore, also on the willingness of villagers to work illegally. If forests disappear a community's perception of its access to resources weakens, weakening community support for long term management approaches. This, in turn, tends to weaken or remove community sanctions or restrictions on villagers working illegally. For clarity, some model components have been omitted in this view.

harvest occurs potential illegal harvest will decline if no other factors interfere. Unfortunately, this self limitation does not stop over-harvest. It merely decreases the rate at which timber is removed. Forest cover will gradually approach zero as will forest related jobs. It is important to remember that illegal harvest is not necessarily over-harvest, and legal harvest is not necessarily sustainable.

One minor difficulty arises in the selection of a simple approach for depicting the forest that is being harvested. It is not my intention here to provide a detailed forest vegetation model, but the approach used should reflect the ability of the forest to regenerate (grow and reproduce), and *regeneration* should be somewhat higher at intermediate stand densities. I have chosen a biomass approach that disregards stems per ha, size of the trees, and species composition (Figure 11.6). This is not to say that these components lack importance, but that such a level of detail is not necessary to describe the illegal logging dynamics discussed here.

The full model incorporates all the above feedback loops, which are interlinked as illustrated, in simplified form, in Figure 11.7. A complete model diagram and equations are presented at the end of the chapter.

Model Outcomes

Presented here are some model outcomes that might be expected if an increase in illegal logging initially started in response to significant drops in income levels of rural communities. A drop

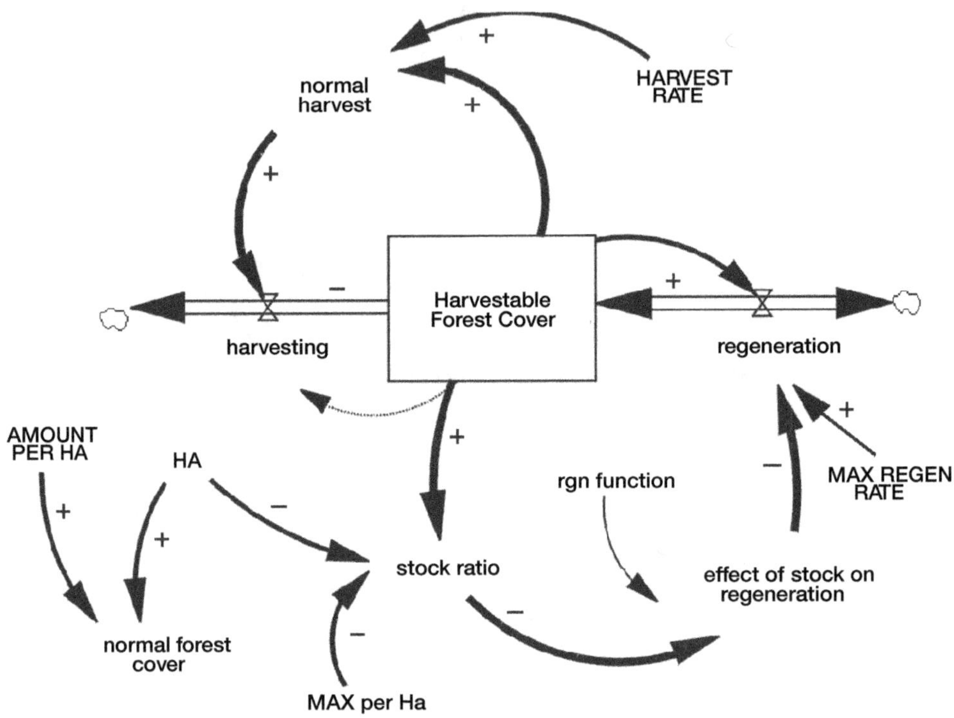

Figure 11.5 Harvestable forest is described by a simple biomass model. Net gain (*regeneration*) of harvestable forest biomass is a fraction of existing biomass. When forest cover is very low, that fraction is highest (equal to *max regen rate*). As harvestable forest cover increases, stock ratio increases, and because of the shape of the regeneration function (*rgn function*) the *effect of stock on regeneration* lowers *regeneration*. *Regeneration* reaches is zero if *harvestable forest cover* equals or exceeds *max per ha*.

of 50% in income over a two- year period (1997–1998) was used as a triggering mechanism. To accomplish this, *other income* was decreased with a ramp function at a rate of $450 per year over the two years. The model was started in approximate equilibrium with the sum of income sources equaling the annual *desired income* of $1,800. This represents the effects of the Asian monetary crisis of those years. Primary factors examined for their effect on illegal logging were *number of outside workers available*, *strength of community rights*, and *effect of enforcement on illegal worker numbers*.

General Pattern of Willingness to Work Illegally

A typical response of villagers' willingness to engage in illegal logging is illustrated in Figure 11.7. Here we see three phases of willingness while forest still exists. First, willingness rises rapidly as incomes drop and people become willing to work illegally. During this period, as more people work illegally the willingness of others to work illegally also increases. At the same time *community support for sustainable forest management* remains strong and, as income needs are partially satisfied, is able to limit a further rise of willingness.

After this period of rapid increase, willingness increases more slowly. During this period, the effects of *community support for sustainable forest management* almost balance the effects of income need and other workers' illegality. Overall, willingness rises only slightly. Even so, the effect

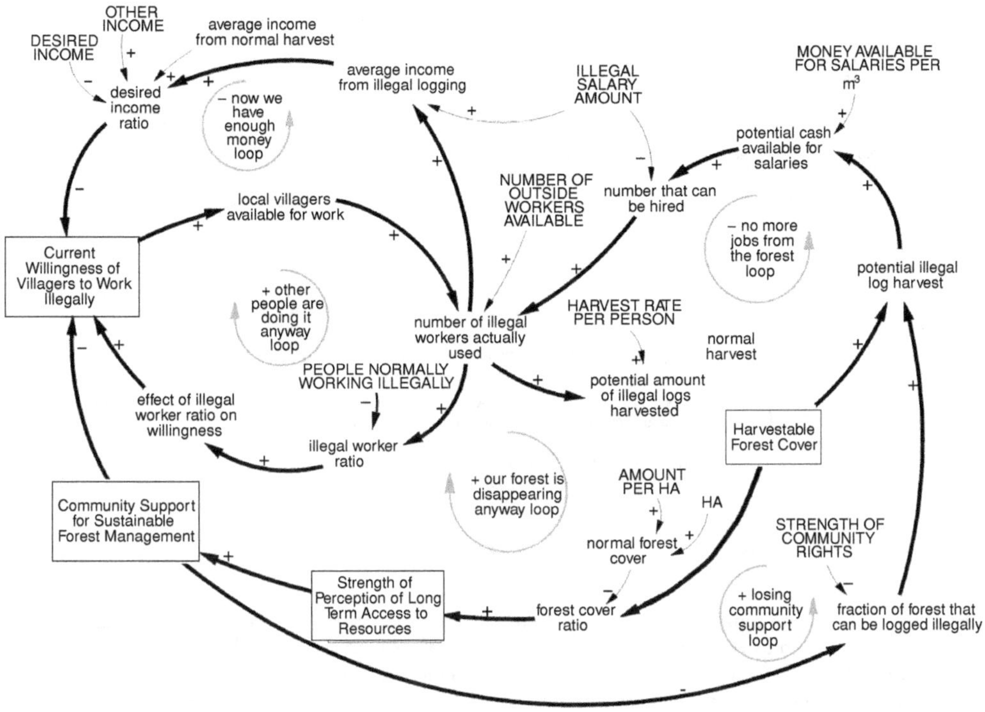

Figure 11.6 The full model is shown here as a causal loop diagram. All five major feedback loops are represented in general form. A detailed model diagram, and model equations, are presented at the end of the chapter, in Figure 11.15.

on the forest is extreme because the willingness to work illegally remains between 0.3 and 0.4; over 30% of villagers in the labor pool are willing to work illegally. These villagers, plus outside workers, are hired by entrepreneurs. By 2008 (in the scenario shown) 50% of the forest is gone, and by 2013 it has been reduced to 0% of its original amount.

Toward the end of the previous phase the strength of *community support for sustainable forest management* collapses as communities finally realize that the forest is disappearing anyway, regardless of their efforts. This causes a third phase where willingness jumps again. This jump is reinforced by the effect of illegal workers as more illegal workers take to the field. However, this period of increase is short-lived because it also increases income flowing into the village. Increased income limits further increases in willingness as income levels approach the desired income level causing willingness to remain at just above 0.5. At the end of this phase the forest is essentially gone, but further changes in willingness occur.

Following the disappearance of the forest, income levels drop precipitously causing a large jump in willingness to work illegally. However, there is no forest left to cut. Willingness jumps to over 0.9 but then drops as illegal loggers, and the idea of illegal logging, disappear along with other forest related jobs (Figures 11.7 and 11.8).

The Effect of Outside Workers

Not all illegal logging is done by villagers themselves. Some is carried out by migrants who move to forested areas specifically to find such jobs. As the number of outsiders increases, the

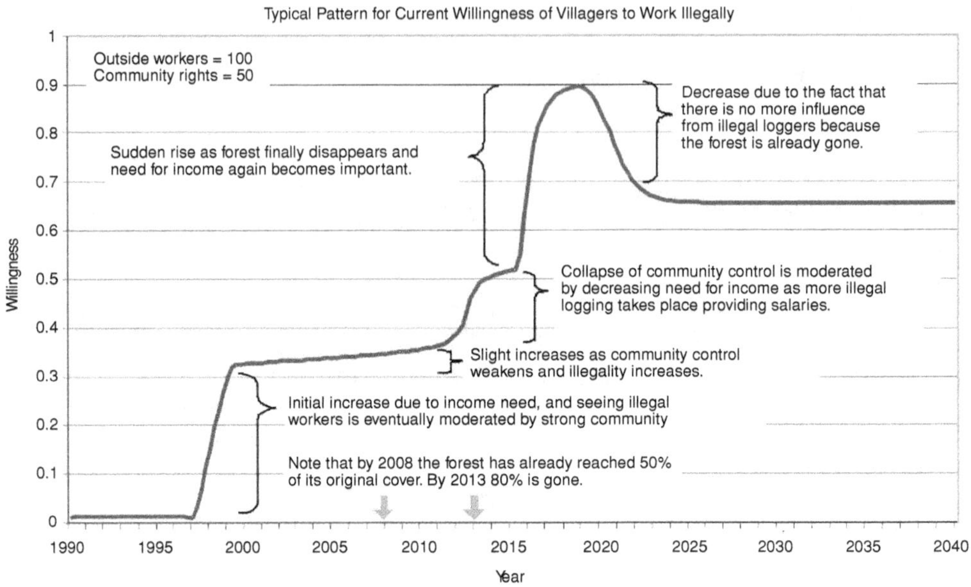

Figure 11.7 This figure represents the general pattern, over time, of villagers' willingness to engage in illegal logging. The triggering even during 1997–98 is a 50% drop in income. There is a rapid rise in willingness in response to income need, but this need is addressed somewhat as income rises. During this same period the existence of illegal workers stimulates more illegal workers. After this first growth in willingness ends in late 1999 it remains relatively stable because of residual community desire for long term management. By the end of this period forests are largely gone and remaining community control collapses. By the end of this second period incomes have jumped again and willingness starts to stabilize at a higher level. Near the end of this period income from the forest is depleted and willingness due to income need rises rapidly but no additional income is forthcoming. Eventually willingness drops as illegal logging disappears due to the disappearance of the forest.

forest disappears faster (Figure 11.9). Outsiders take illegal jobs that locals don't want, but their presence also influences locals to start working illegally (Figure 11.10). An increase in *the number of outside workers available* causes an increase in *harvesting* because there are more outside workers and also because more villagers are influenced to work illegally.

In addition, outside workers take salaries that would ultimately go to locals. Assuming that little money paid to outsiders is spent locally, the overall amount of money reaching villagers is lower if more outside workers are present. However, shortly prior to the collapse of illegal logging, villagers actually bring slightly higher amounts of money into the village when more outsiders are present. This is because more of the available village labor force is stimulated to work illegally under those conditions creating a peak in illegal incomes that precedes the collapse (Figure 11.11).

Effect of the Strength of Community Rights

Strong community rights can limit the amount of forest that illegal entrepreneurs can log. If entrepreneurs expect to have difficulty making illegal arrangements to get logs, they will be less willing to invest in illegal operations. When community rights are strong we would expect to see less illegal logging, assuming that communities want to maintain a long term management approach.

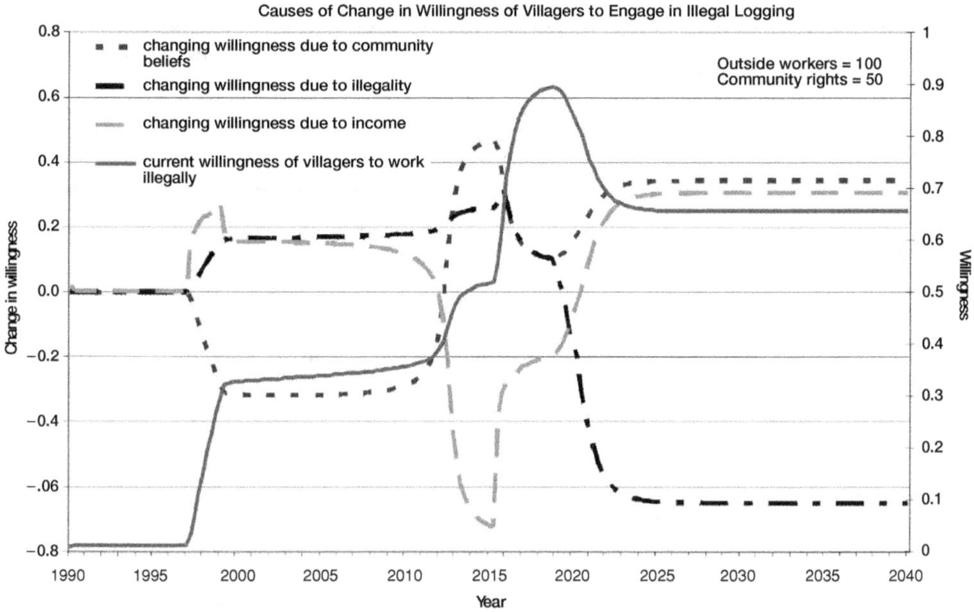

Figure 11.8 This figure details the changes in willingness illustrated in the previous figure. Here the left-hand axis gives the change in willingness. Three sources of changes are shown. When the sum of these three is zero willingness (right-hand axis) will not change. When the sum is positive willingness will rise and when negative willingness will decline. We can think of the dotted or dashed lines above zero as pulling willingness up and those that are negative as pulling willingness down. *Changing willingness due to community beliefs* refers to changes caused by strength of community support for sustainable forest management. *Changing willingness due to illegality* refers to the effect of other illegal workers, and *changing willingness due to income* refers to the effect of income need.

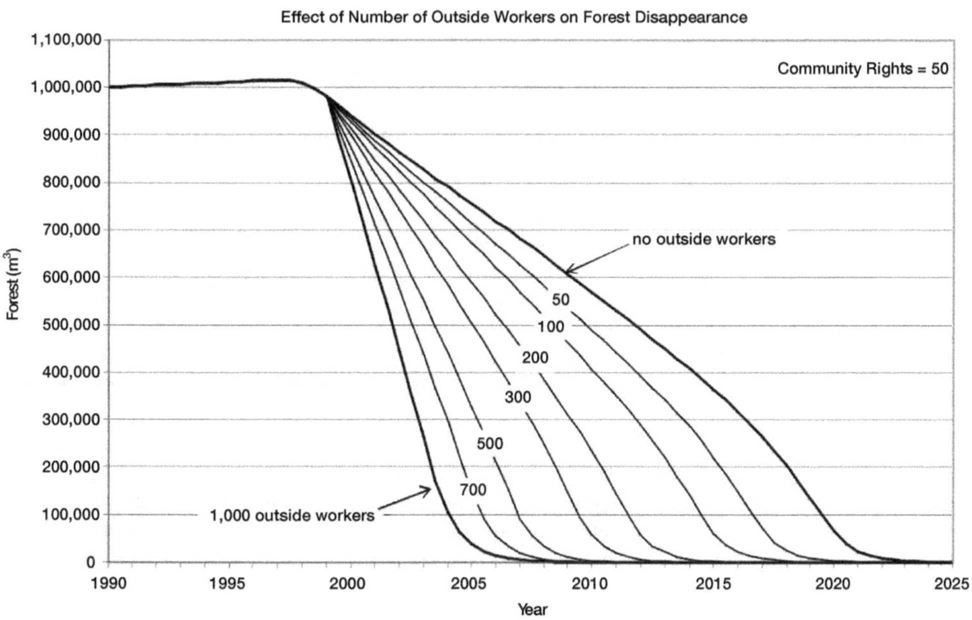

Figure 11.9 Influence of outside workers on forest disappearance. Outside workers work illegally and also influence villagers to work illegally. In addition, outsiders absorb some of the cash that would otherwise have gone to villagers.

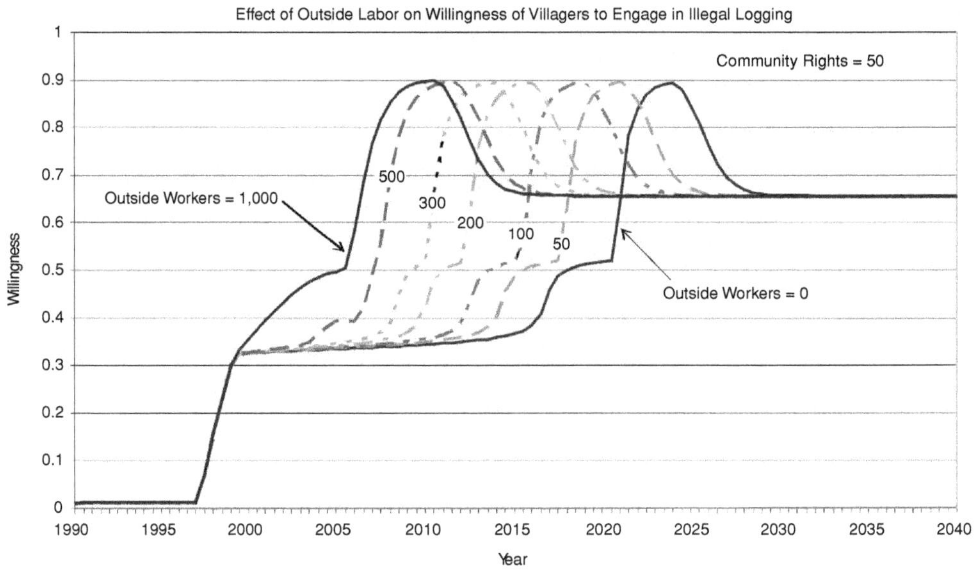

Figure 11.10 The effect of outside illegal workers on villagers' willingness to engage in illegal activities. Outside workers stimulate more villagers to participate in illegal logging earlier than they would in the absence of outsiders.

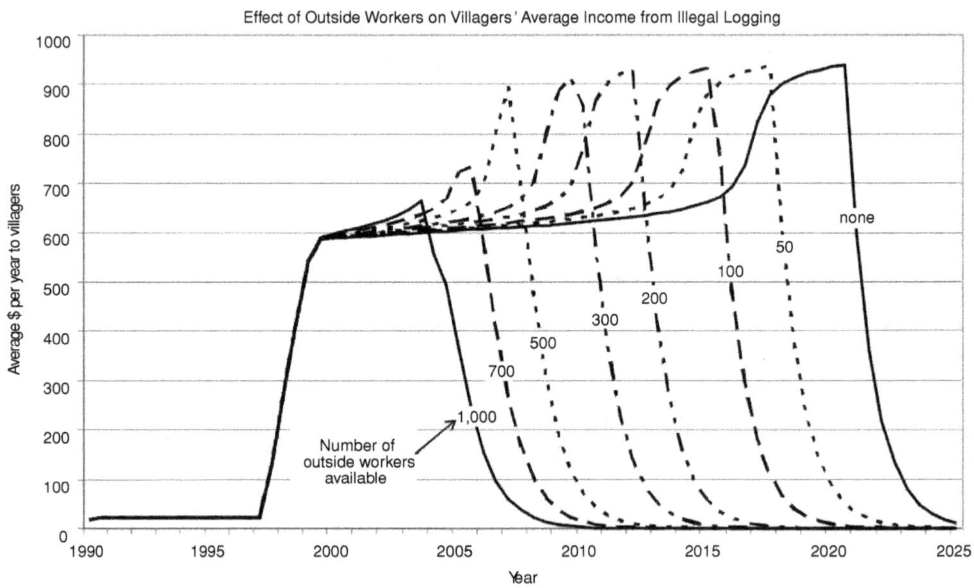

Figure 11.11 Average salary flowing to the village labor pool under situations with different numbers of outside workers. Ultimately, the more outside workers there are the less money goes to village workers. (The area under each curve can be considered the average amount earned per labor pool member.)

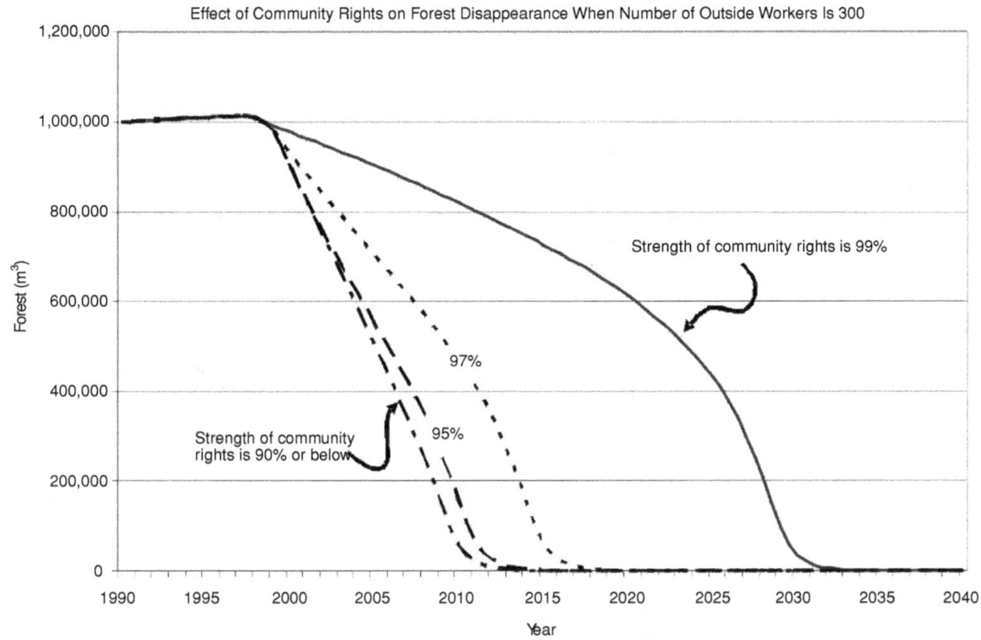

Figure 11.12 The strength of community rights can have a strong effect on forest integrity, but this effect is dependent on the proportion of outside workers available and has little effect if rights are not reasonably strong. Community rights which protect only a small portion of forest are irrelevant from the entrepreneurs' point of view since they can merely harvest other areas of the forest until community resolve weakens.

Two problems tend to counteract the role that community rights can play in preventing illegal logging. First, unless rights are very strong there is always some fraction of the forest that can be logged illegally. As that forest is logged community resolve will weaken allowing a larger fraction to be logged. Second, logging is to some extent limited by worker availability. If labor is limited, illegal entrepreneurs will be satisfied to use what labor is available on that portion of the forest not currently under community protection. In cases where labor is limited, the effects of moderate levels of community rights may not be noticeable.

Because there is a fairly large amount of forest available to be cut, community rights do not have a significant impact unless the strength of community control is high. For example, if *strength of community rights* is only 50%, entrepreneurs can work on harvesting the remaining 50% until community resolve weakens (Figure 11.12).

If community rights are weak but the number of workers is limited, then the number of workers available will limit entrepreneurs' operations. Consequently, if *number of outside workers available* is high, then *strength of community rights* plays a bigger role (compare Figures 11.12 and 11.13).

Enforcement

If enforcement is feasible, it can have a major effect on forest protection. This is especially true if *strength of community rights* is weak and the *number of outside workers available* is high – that is, under conditions where illegal logging is most likely to occur, enforcement is most likely to

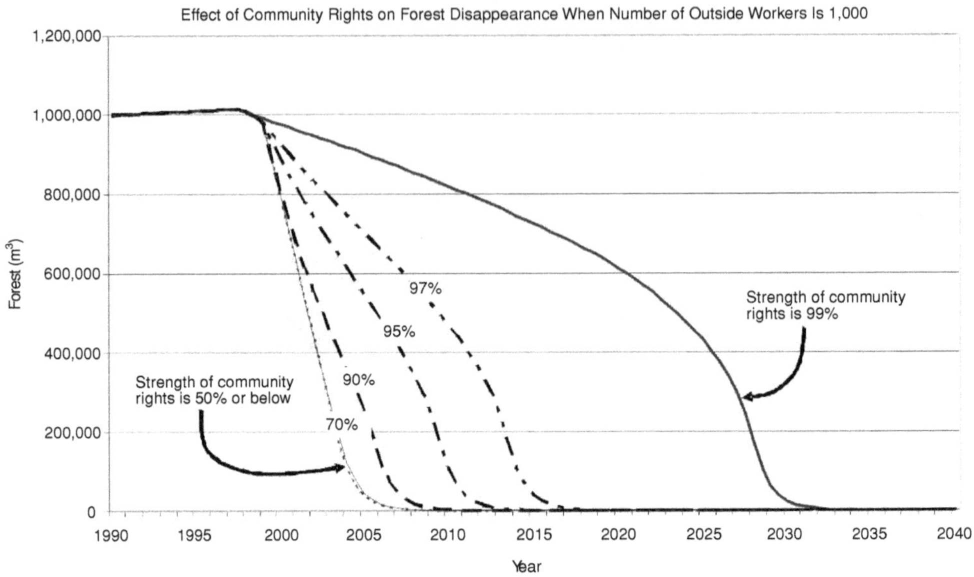

Figure 11.13 If the number of outside workers is high, the effect of community rights will be more important because labor to harvest available forest is less limiting

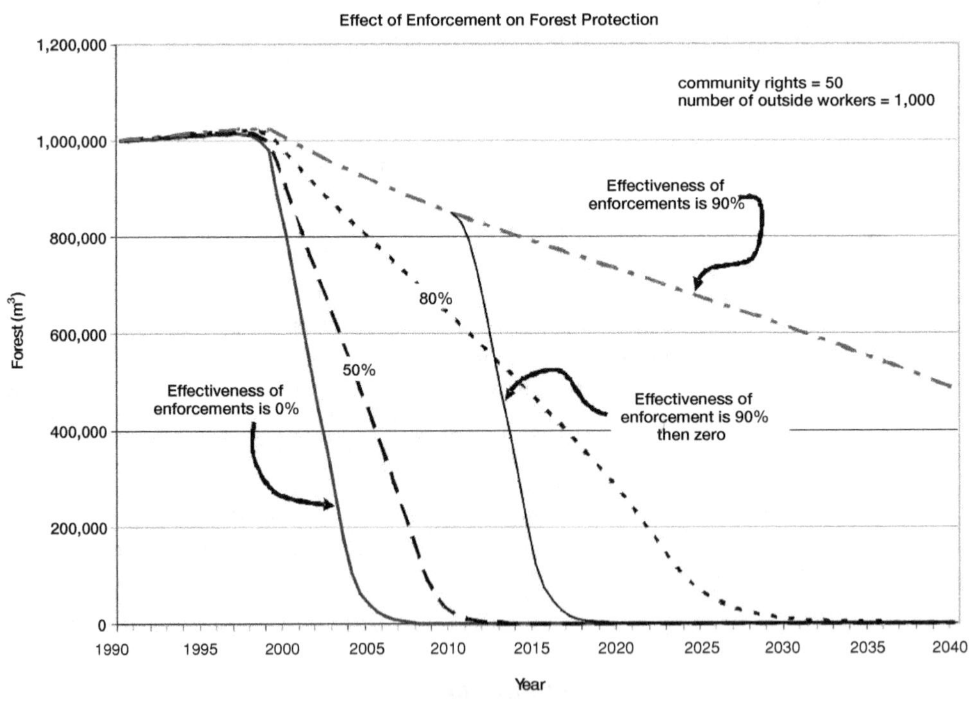

Figure 11.14 Enforcement has a major effect on protecting forest, especially if community rights are weak and the number of outside workers is high. However, with strong enforcement, villagers' willingness to engage in illegal activities due to income need remains high. Thus, if enforcement is suddenly weakened (as illustrated here in year 2010), illegal activities will rapidly increase.

have a positive effect. Figure 11.14 illustrates this situation with 1,000 outside workers and 50% community control. At low levels of enforcement the forest disappears quickly, but at higher levels the disappearance is delayed considerably. Preventing villagers from working illegally prevents their income needs from being satisfied, which raises their willingness to work illegally. On the other hand, by maintaining forest cover enforcement tends to reinforce community support for long-term management. (Factors limiting the effectiveness of enforcement present an excellent subject for another model.)

Under the conditions of the model, enforcement did not significantly lower the effect that other illegal workers in the forest had on willingness. This is be cause for most model conditions used (especially with 1,000 outsiders), regardless of enforcement effectiveness, there were always more illegal workers than the number of *people normally working illegally*. This situation tends to raise willingness. Even in a case with no outsiders, unless enforcement is very strong, enough illegal workers will be working (initially due to income need) that other workers will be influenced to want to join the ranks of illegal workers. This is, of course, dependent on how easily villagers are so influenced. Note that willingness to work illegally, by itself, does not necessarily mean the villagers work illegally. Both availability of salary money and enforcement affect whether a willing villager will actually work illegally.

Discussion

The Model and the Real World: What Is Missing?

In considering the usefulness of the modeling process we need to consider if the model is good enough for its intended purpose – that is, can this model help to explain basic causal relationships leading to villagers' willingness to engage in illegal logging? The model represents one attempt to describe and investigate these causal relationships. It simultaneously considers five feedback relationships connecting villagers' willingness to work illegally, their need for income, the availability of forest to support that income, the communities' role in good forest management, and illegal entrepreneurs' role in hiring villagers.

Within limitations the model does help to elucidate these relationships. We see more clearly that although income levels are raised by illegal logging, other factors also serve to stimulate additional willingness to work illegally even when income levels are already reasonably high. As forests disappear, willingness increases even further because community support for long-term management disappears as well. In examining the model, we are stimulated to consider other elements not included in the model's current structure. Some aspects of illegal logging at the village level are not specifically addressed by the model. Would addition of such factors improve the model, or would they merely make the model more confusing and harder to understand?

We may wonder, for example, if the amount that villagers consider as *desired income* might also rise as incomes from illegal logging rise. If this were the case, then as income expectations rose, it would be harder for *desired income* to be matched by actual income, and illegal logging levels would grow more rapidly. Also, the model includes no additional sources of income that might emerge as forests disappear. One example might be labor income from work on mono-crop plantations – typically oil palm in Indonesia. Alternate income sources would tend to lower the need for illegal employment, especially if these alternate sources grew as forests declined.

In the model, communities' views of long-term forest management remain strong until the forest is obviously disappearing. In the real world, community views may be influenced, for example, by possibilities for other uses of community land such as plantation development. Most

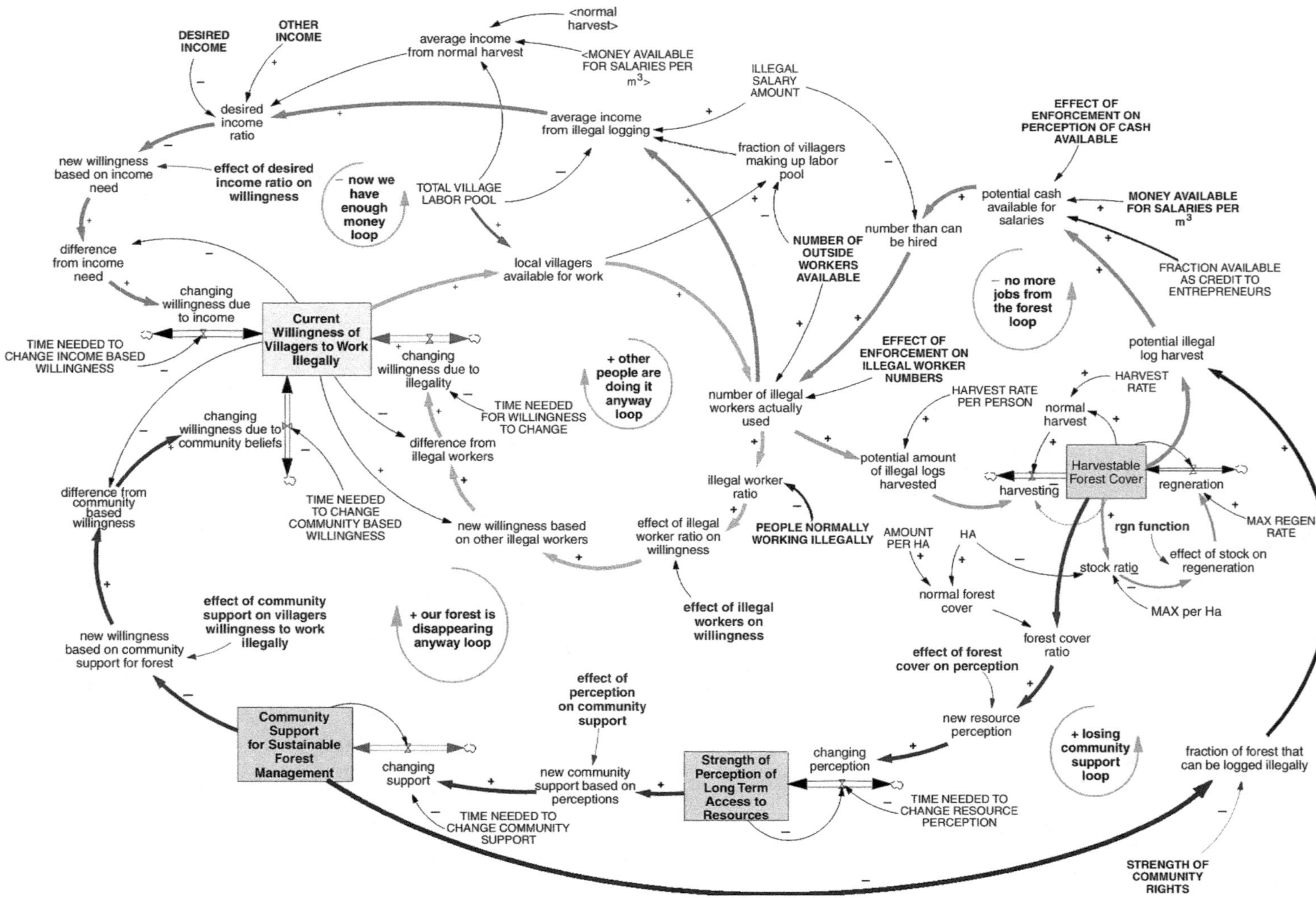

Figure 11.15 Willingness of villagers to engage in illegal logging

probably, as the value of alternate land uses rises, the communities' desire for long-term forest management would weaken more rapidly.

In the model, forest disappearance does not generate more concern for forest protection, and this seems to agree with reality, at least in practical terms – that is, there is not a resurgence of meaningful community desire to protect forest resources as they disappear. We might use the model to help us consider possible changes to this portion of the real world system. As forest cover disappears, is there a feedback mechanism whereby the desire to protect and rehabilitate forest can be strengthened? Are communities willing to merely accept conversion of forest to non-forest, or can new mechanisms stimulate the desire for long-term forest management? Can such mechanisms be self-reinforcing?

Except for competition for timber, the model does not examine specific relationships between legal and illegal logging. Where laws and regulations limit access to forest, increases in illegal logging from those same forests will remove future timber harvest originally destined for the legal operators. The model reflects this case. If, however, illegal logging is carried out in protected areas, for example, legal and illegal operations would not be in competition for timber. The model also does not reflect the situation whereby cheaper illegal timber and timber products provide competition for legally produced items. This is a subject for a related modeling effort.

Once illegal logging starts, its development in a given area seems to accelerate. This acceleration may be partly related to competition among illegal entrepreneurs and their corrupt colleagues in local government. In such an environment, those who can tend to scramble for resources creating a gold rush mentality. The urgency created by such activity would create additional pressures on communities, and community leaders may be bribed or tricked into illegal agreements. This would clearly hasten the spread of illegal logging. The model does not address such competition. Entrepreneurs in the model are motivated only by money available in standing timber, and there is no direct influence of entrepreneurial desires on weakening community resolve.

Although the model examines the role of outside workers, it is assumed that outsiders and villagers share work in proportion to their numbers in the labor pool. It is also assumed that money paid to villagers remains among villagers, and that paid to outsiders leaves the area. What proportion of outsiders become de facto villagers? Some of their money may stay within the communities and provide income for others. In the model the number of outsiders is determined externally. It is possible to allow the number to be determined by demand for labor. In this case labor would be less limiting and illegal logging would proceed more quickly.

The preceding paragraphs present a number of areas where the model may be inadequate or unfinished. However, the correction of these weaknesses could make the model less understandable and less able to provide insights. The purpose of the model is to provide a framework for thinking about illegal logging at the local level. It is not intended that this model will provide detailed management strategies. However, it can be used as a first step to examine larger, as well as more detailed, issues. Such examination can expose those secondary issues that might provide avenues for modification, not only to the model itself, but more importantly to the real illegal logging system that is our ultimate target.

Acknowledgments

This chapter originally appeared in *Journal of Sustainable Forestry*, 19, 31–53 (2004). Much of the work reported here was carried out while the author was working as a consultant with the Center for International Forestry Research in Bogor, Indonesia. Among the many people who provided advice and discussion during this work were: Joyotee Smith, Krystof Obidzinski, Carol Colfer, Patrice Levang, Herry Purnomo, Doddy Sukadri, Subarudi, Agus Purnomo, Agus Setyarso, Graham Applegate, and Eva Wolenberg.

References

Bruenig, E. F. 1996. Conservation and management of tropical rainforests: An integrated approach to sustainability. CAB International, Oxford. 339p.

Casson, A. 2000. Illegal tropical timber trade in Central Kalimantan, Indonesia. Draft report prepared for the Center for International Forestry Research, Bogor, Indonesia.

Casson, A. and K. Obidzinski. 2001. From new order to regional autonomy: Shifting dynamics of 'illegal logging' in Kalimantan, Indonesia. Resource Management in Asia-Pacific Conference on Resource Tenure, Forest Management and Conflict Resolution: Perspectives form Borneo and New Guinea, Australian National University, Canberra.

Dudley, R. G. 2001. Dynamics of illegal logging in Indonesia. In: Colfer, C. J. P., and I. A. P. Resosudarmo (Eds.). *Which Way Forward? Forests, People and Policymaking in Indonesia.* Resources for the Future. Washington, DC, pp. 35–32.

Dudley, R. G. 2002. An application of system dynamics modeling to the question of a log export ban for Indonesia with comments on illegal logging. Paper presented at the 20th International Conference of the System Dynamics Society, July 2–August 1, 2002, Palermo, Italy.

McCarthy, J. F. 2000. "Wild logging": the rise and fall of logging networks and biodiversity conservation projects on Sumatra's rainforest frontier. Center for International Forestry Research, Occasional Paper No. 31. Bogor, Indonesia.

Obidzinski, K. and I. Suramenggala. 2000. Illegal logging in Indonesia – A contextual approach to the problem. Draft paper prepared for the Center for International Forestry Research, Bogor, Indonesia.

Obidzinski, K., I. Suramenggala and P. Levang. 2001. L'exploition forestière illégale en Indonésie: Un inquiétant processus de légalization (Illegal logging in Indonesia: A disturbing legalization trend). *Bois et Forets des Tropique* 270(4):5–97.

Palmer, C. E. 2001. The extent and causes of illegal logging: An analysis of a major cause of tropical deforestation in Indonesia. CSERGE Working Paper. Centre for Social and Economic Research on the Global Environment, University College, London.

Scotland, N. (with J. Smith, H. Lisa, M. Hiller, B. Jarvis, C. Kaiser, M. Leighton, L. Paulson, E. Pollard, D. Ratnasari, R. Ravenel, S. Stanley, Erwidodo, D. Currey, A. Setyarso). 2000. Indonesia country paper on illegal logging. Report prepared for the World Bank–WWF Workshop on Control of Illegal Logging in East Asia. Jakarta, August 2, 2000. Edited by William Finlayson and Neil Scotland.

Sterman, J. D. 2000. *Business Dynamics: Systems Thinking and Modeling for a Complex World.* Irwin/McGraw-Hill, Boston, MA. 92p.

Wadley, R. L. 2001. Histories of natural resource use and control in West Kalimantan, Indonesia: Danau Sentarum National Park and its vicinity (100–2000). A Report for the CIFOR Project "Local People, Devolution, and Adaptive Co-Management." Center for International Forestry Research, Bogor, Indonesia.

World Bank. 2001. Indonesia: Environment and natural resource management in a time of transition. 129p.

12

EXPLORING COMPLEXITY IN A HUMAN–ENVIRONMENT SYSTEM

An Agent-Based Spatial Model for Multidisciplinary and Multiscale Integration

Li An, Marc Linderman, Jiaguo Qi, Ashton Shortridge, and Jianguo Liu

Research Question: How does rural population growth in China affect panda habitats, and vice versa?
System Science Method(s): Agent-based models
Things to Notice:

- Building a complete model from multiple submodels
- Simulation of a real spatial environment using empirical data

Traditional approaches to studying human–environment interactions often ignore individual-level information, do not account for complexities, or fail to integrate cross-scale or cross-discipline data and methods, thus, in many situations, resulting in a great loss in predictive or explanatory power. This chapter reports on the development, implementation, validation, and results of an agent-based spatial model that addresses such issues. Using data from Wolong Nature Reserve for giant pandas (China), the model simulates the impact of the growing rural population on the forests and panda habitat. The households in Wolong follow a traditional rural lifestyle, in which fuelwood consumption has been shown to cause panda habitat degradation. By tracking the life history of individual persons and the dynamics of households, this model equips household agents with "knowledge" about themselves, other agents, and the environment, and allows individual agents to interact with each other and the environment through their activities in accordance with a set of artificial-intelligence rules. The households and environment coevolve over time and space, resulting in macroscopic human and habitat dynamics. The results from the model may have value for understanding the roles of socioeconomic and demographic factors, for identifying particular areas of special concern, and for conservation policy making. In addition to the specific results of the study, the general approach described here may provide researchers with a useful general framework to capture complex human–environment interactions, to incorporate individual-level information, and to help integrate multidisciplinary research efforts, theories, data, and methods across varying spatial and temporal scales.

 DOI: 10.4324/9781315748771-14

Complex human–environment interactions have been increasingly attracting the attention of researchers with different backgrounds and research purposes. On one hand, characterizing the environment and the complex role that human actions play within it is challenging. This is partially due to the inherent complexity of the processes. The accumulated impact of individual decisions made by dozens, hundreds, or millions of people is the immediate cause of human-induced environmental change. On the other hand, these individual actions are shaped by the particular social, political, economic, and environmental frameworks within which they occur. These frameworks change through time as conditions change. Furthermore, the imprint of these activities varies throughout space and across different spatial scales.

The science of complexity has provided key theoretical contributions and techniques for environmental modelers wrestling with these challenges (Flake 1998). It is concerned with the manner in which fundamental processes can lead to emergent phenomena or behaviors in complex adaptive systems (CAS), focusing on many kinds of complexities such as hierarchical structures, feedback, self-organization, scaling, and time lags (Malanson 1999). Levin et al. (1997) provide a broad overview of approaches to considering complexity in ecosystem modeling. They describe several advantages:

- Incorporation of substantial local and individual characteristics.
- Recognition of the stochastic nature of complex systems.
- Explicit characterization of the impact activities at one scale have on patterns at another.

However, the challenge is also a technical or implementational one. How can researchers integrate data and models to deal with these complex processes? A variety of approaches have been adopted; a general overview of these was recently published in this journal (Parker et al. 2003) and is only briefly reviewed here. Geographers and other human–environment modelers often turn to Geographic Information Systems (GIS) to assist in data management and modeling of spatially explicit variables. GIS is a powerful tool to capture, store, manipulate, and analyze spatial data, and it has been extensively used in studying human–environment interactions. The data models employed by common GIS are inherently static, however; they do not handle time well, nor do they capture functions or dynamic processes effectively (Peuquet 1999). Environmental modelers usually resort to externally implemented methods to handle advanced modeling problems (although interesting integrated implementations exist, e.g., the PCRaster platform discussed by Wesseling et al. 1996). These methods include multivariate spatial models (e.g., Seto and Kaufmann 2003), Markov chain analysis (e.g., Brown, Pijanowski, and Duh. 2000; López et al. 2001), and cellular automata (e.g., Batty, Xie, and Sun 1994; Li and Yeh 2002; Malanson 2002). In particular, cellular automata (CA) models have been shown to be powerful in modeling many ecological processes because, as a bottom-up approach, they have a better capacity than GIS overlay or map algebra functionality to capture and represent local interactions that give rise to global complex patterns (e.g., Li and Reynolds 1997; Clarke and Gaydos 1998). CA models, however, face challenges in simulating human decision making and capturing feedback elegantly (Parker et al. 2003).

It may be worth highlighting a few of these diverse approaches employed by scientists with similar research problems and data to those of the present study. These studies, like our own, seek to model relationships between social factors and landscape change. Pan et al. (2004) report on work regressing landscape pattern metrics with data collected in farmstead surveys in Ecuador. Geographic factors, including extent and spatial scale, are shown to play an important role in model results. Walsh et al. (1999) demonstrate the effect of spatial resolution on multiple-regression models relating population per unit of cultivated land to six independent social and physical variables for a region in Thailand. Both studies integrated social data collected with surveys with land-cover and other spatial data. The same Ecuadorian study site employed by Pan et al. (2004) was also subject to a cellular-automata-based model in another study (Messina and Walsh 2001). In that model, land-cover change over time is modeled using patterns observed in data collected in five time intervals over three decades. Rules identifying the probability of cell state change based on its neighbors are developed, leading to predictive models of land-cover dynamics. While the goals and data of these different studies are similar, very different methods are employed. The first two use multivariate techniques to identity overall relationships between land cover and human processes. The third employs a bottom-up approach in which individual cells change state over time according to probabilities associated with their properties and those of neighboring cells.

Agent-based modeling (similar to individual-based modeling in many ecological studies) is another bottom-up methodology that has been specifically employed to deal with complexity, especially when coupled with GIS. The research reported in this article utilizes this approach. Agent-based modeling (ABM) predicts or explains emergent higher-level phenomena by tracking the actions of multiple low-level "agents" that constitute, or at least impact, the *system* behaviors. Agents usually have some degree of self-awareness, intelligence, autonomous behaviors, and knowledge of the environment and other agents as well; they can adjust their own actions in response to environmental changes (Lim et al. 2002; Parker et al. 2003). The concepts underlying ABM are similar to those of the object-oriented programming (OOP) paradigm in computer science, and ABM models frequently employ object-oriented programming languages like C++ and Java.

Unlike procedural programming, for which data and operations on the data are separated, object-oriented programming groups operations and data (or behavior and state) into modular units called objects and lets the user combine objects into a structured network and form a useful program (Larkin and Wilson 1999). Figure 12.1a is an illustration of an object with operations (called methods) and data bound together. The strengths of OOP lie in its modularity, software reusability, and its separation between interface and implementation. Modularity reduces programming complexity by dividing code into relatively separated "parts" or modules, each with different functional focuses. Software reusability means that one piece of code, when defined and tested, can be reused as many times as possible. The separation between interface

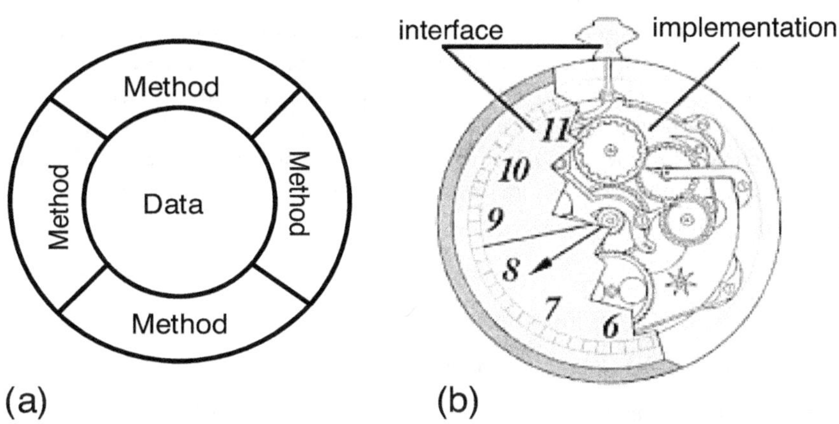

Figure 12.1 (a) An object in object oriented programming with *data* and operations (methods) combined and (b) an interface and implementation of object oriented programming (NeXT Software, Inc. 1992)

and implementation hides technical details inside the system surface, such as the parts in a clock and how these parts interact with each other (Figure 12.1b; NeXT Software, Inc. 1992). The "implementation" feature (technical details) makes the system work well. A user-friendly interface running above this provides simple data input, output, and display functions so that other objects (or users) can call or use them.

A set of agent-based modeling tools with the above OOP features is readily available for use – e.g., Swarm, RePast, NetLogo, Ascape, and StarLogo.[1] Swarm may be one of the most powerful and comprehensive toolkits (Najlis, Janssen, and Parker 2001). Swarm is a software package for multiagent simulation of complex *systems* publicly available under GNU licensing terms. Originally developed at the Santa Fe Institute and now maintained by the Swarm Development Group, Swarm decomposes emergent phenomena at a certain level into attributes and actions of collections (swarms) of concurrently interacting agents at lower levels. Swarm allows for a hierarchical structure for agent organization and management, which means a higher-level agent in the hierarchy can include and manipulate a number of lower-level agents and their actions (Minar et al. 1996).

ABM alone does not address complexity (e.g., spatial heterogeneity, structural hierarchy) well, however. A growing number of efforts to integrate ABM with traditional approaches such as equation-based models and GIS have been made in the environmental modeling arena (e.g., Bian 1997; Berger 2001; Gimblett 2002; Jiang and Gimblett 2002), but relatively few studies have been implemented to examine the complex manner in which the accumulation of individual decisions, as affected by social/political factors and economic conditions, may affect the biophysical environment across a range of spatial and temporal scales. Here we address this critical topic with the following specific objectives: (1) linking spatial patterns and temporal processes by capturing complexity (e.g., heterogeneity, nonlinearity, feedback, and time lag) via ABM in a coupled human–environment system; (2) constructing a framework to integrate data and/or methods across disciplines, spatial/temporal *scales,* and aggregation levels; and (3) providing an effective policy-analysis tool for biodiversity conservation in relation to low-level anthropogenic (e.g., household life history) and environmental (e.g., spatially varying forest volume and growth rate) characteristics and relationships.

This chapter is fundamentally about integration: the integration of social and environmental drivers, the integration of diverse modeling techniques, the integration of fundamental knowledge generation with the policy implications of that knowledge, and the integration of data about processes operating across a range of spatial and temporal scales. It addresses these issues within an explicitly geographical framework and takes advantage of spatial tools that provide a practical modeling environment for conducting this complex analysis. We recognize that the project's technical architecture might divert attention from the project's specific purpose: to represent relationships between giant panda habitat loss in the Wolong Nature Reserve (China) and local household dynamics as affected by the social, economic, and political context. For this reason, a variety of validation efforts are employed to *assess* the fidelity of model results and their sensitivity to input parameters.

Methods

Study Area

An excellent site to conduct research with the above objectives is Wolong Nature Reserve in China (Figure 12.2) for the following reasons: (1) it is recognized as a globally significant biodiversity conservation site; (2) much is known about the biology and physical environment

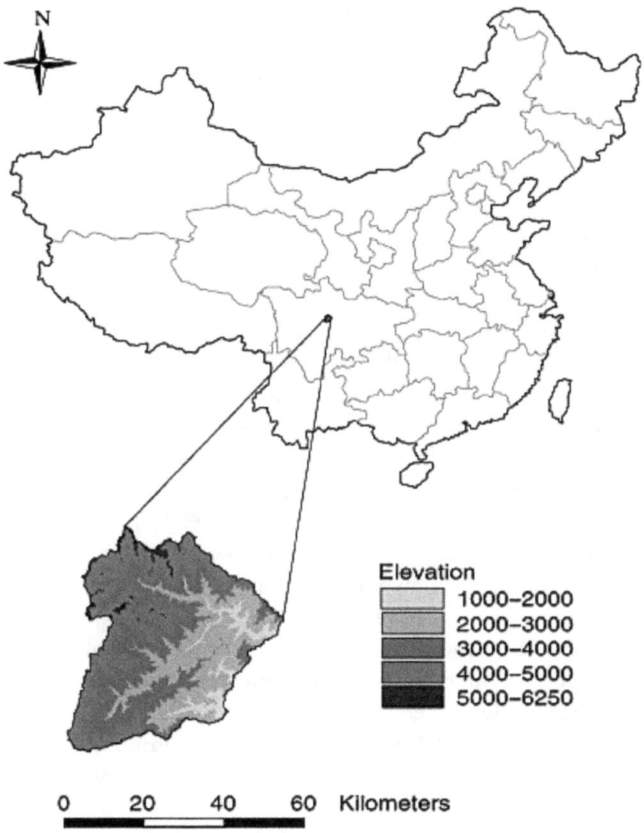

Figure 12.2 The location and elevation (m) of Wolong Nature Reserve in China

of the giant panda; (3) human impact on panda habitat is a serious problem in the reserve; and (4) the complexity of the problem necessitates an integrated approach. We discuss each of these points in turn in the following paragraphs.

First, there is broad consensus that the Wolong Nature Reserve is of global significance. Established in 1975 for conserving the endangered giant panda *(Ailuropoda melanoleuca)*, the reserve is within one of the twenty-five global biodiversity hotspots (Myers et al. 2000). Over 2,200 animal/insect species and more than 4,000 plant species (Wolong Administration 1987) cohabit with the giant panda in a diverse biophysical environment occupying approximately 2,000 km^2. Species richness, an important indicator of biodiversity, is directly linked to availability of associated habitat types based on the utter dependence of organisms on an appropriate environment (Ehrlich and Wilson 1991), especially for contexts associated with island biogeography. A famous example is the species–area relationship – i.e., $S = cA^z$, where S is the number of species that occur in a region with area A, and c and z are relevant constants (MacArthur and Wilson 1967). Thus, protecting habitat is a necessary step toward conserving any single type of organism. The implications for Wolong are clear: conserving panda habitat means that both the internationally renowned and endangered giant panda and the less widely known species that live within and comprise that habitat can be protected.

Second, we have collected extensive socioeconomic and environmental (e.g., remote-sensing) data about the Wolong area through our intensive fieldwork from 1998 to 2002.

Extensive research efforts have been invested on giant panda biology, ecology, and habitat studies, such as the relationship between the giant panda and bamboo forest, the canopy cover that serves the giant panda as shelter, and the understory bamboo that serves as a primary food source (e.g., Schaller et al. 1985; Liu, Ouyang, Tan, et al. 1999).

Third, the rural human population in the reserve threatens panda habitat. Although Wolong enjoys high domestic standing as a "flagship" reserve in China with considerable domestic and international financial and technical support, the reserve also supports a substantial human population that is growing rapidly, with an even more rapid increase in the number of households (Liu, Daily, et al. 2003). The population (approximately 4,400 local residents in 2000) comprises four ethnic groups: Han, Tibetan, Qiang, and Hui, following a traditional rural lifestyle (Liu, Ouyang, Tan, et al. 1999; An et al. 2001). In spite of the enormous time, energy, and increasing difficulty involved in collecting fuelwood (mostly due to the shrinking forest area and the extremely rugged mountainous terrain), the majority of households in Wolong cut wood from the surrounding forests to cook and heat their homes. Although electricity is available in the reserve,[2] only a small proportion of the households use electricity for cooking and heating; the primary use of electricity is for lighting and electronic appliances (An et al. 2001, 2002). Assuming that an average household consumes 15 m^3 of fuelwood per year (An et al. 2001) and that an average hectare of beech (*Fagus*), oak (*Quercus*), birch (*Betula*) and poplar (*Populus*) forest contains 80 m^3 of fuelwood (Yang and Li 1992), then a 90 × 90 m pixel of mixed forest can sustain one household's fuelwood demand for about four years.

Despite abundant economic incentives (e.g., a lower agricultural tax) and policies (e.g., prohibiting some tree species from being harvested) implemented by the reserve administration, the past two decades have still witnessed a continued increase in annual fuelwood consumption (from 4,000 m^3 to 10,000 m^3 over the past two decades), contributing to a total reduction of over 20,000 ha of panda habitat (Liu, Ouyang, Taylor, et al. 1999). Degradation of forests comprising panda habitat undoubtedly accounts for part of the documented decrease in the Wolong panda population in recent decades: from 145 individuals in 1974 (Schaller et al. 1985) to 72 in 1986 (China's Ministry of Forestry and World Wildlife Fund 1989). This degradation may arise from some combination of ineffective enforcement of existing policies, the common-property nature of the forests, and the difficulty in monitoring, given the rugged landscape of the reserve. All these factors make biodiversity of the Wolong Nature Reserve highly sensitive to human activities and policy changes and threaten the long-term viability of the ecosystem to support wild populations of giant panda and other coexistent species.

Last, the complexity underlying various ecological, socioeconomic, and demographic processes has necessitated interdisciplinary research involving a range of spatial and temporal scales. Piecemeal approaches fail to account for many important factors; nonspatial approaches cannot characterize the critical role location and relative position play for interactions between humans and the environment; evaluating habitat in single time periods ignores the dynamic nature of the processes that drive habitat destruction. Existing research efforts to characterize the relationship between panda habitat, fuelwood consumption, and human socioeconomic/demographic factors (e.g., An et al. 2001, 2002, 2003; Linderman et al. forthcoming) are inadequate because they account for too few aspects of Wolong's human–environment system.

For instance, the fuelwood model (An et al. 2001) links the household-level fuelwood demand to household demographic and socioeconomic factors, but it is basically aspatial and unable to link household fuelwood demand to its impact on the landscape. The electricity-demand model suffers from the same limitation in that it links the probability of switching to electricity to a set of socioeconomic, demographic, and geographic factors (An et al. 2002), but cannot identify the impact of this change to specific locations, or to the forests of Wolong. In addition, household demographics are exogenous in these models; no attempt to characterize the dynamic nature of the household was made. An, Mertig, and Liu (2003) develop a household formation model to address this deficit. This model attempts to characterize the movement of young adults from their parental homes to establish their own households in the Wolong region. However, it does not address the spatial location of households whose demands for fuelwood and electricity are determined as mentioned above. More importantly, these models alone still cannot effectively address the complexity in this coupled human–environment system: nonlinear interactions, cross-scale (spatial and temporal) data, feedback, and time lags between different subsystems. The model described in the subsequent paragraphs is designed to overcome these limitations and to accomplish the objectives discussed in the introduction. It is intended to be a comprehensive tool to (1) employ a valid representation of the Wolong landscape and incorporate its spatial heterogeneity; (2) incorporate the demographic dynamics of households and individuals situated within this landscape; and (3) link fuelwood/electricity demand to the changing characteristics of the landscape and households as established in (1) and (2).

Conceptual Model

With an excellent study site and a wealth of data, we have developed an Integrative Model for Simulating Household and Ecosystem Dynamics (IMSHED). As in many other studies (e.g., Deadman et al. 2001; Liu et al. 2003; Rindfuss et al. 2003; Walsh et al. 2003), households are chosen to be the fundamental unit for local people's decision making and behavior in relation to consumption and production of local resources. The conceptual framework is illustrated in Figure 12.3. The model consists of three major components: household development, fuelwood demand, and fuelwood growth and harvesting. Each of these components is addressed in turn within the remainder of this subsection.

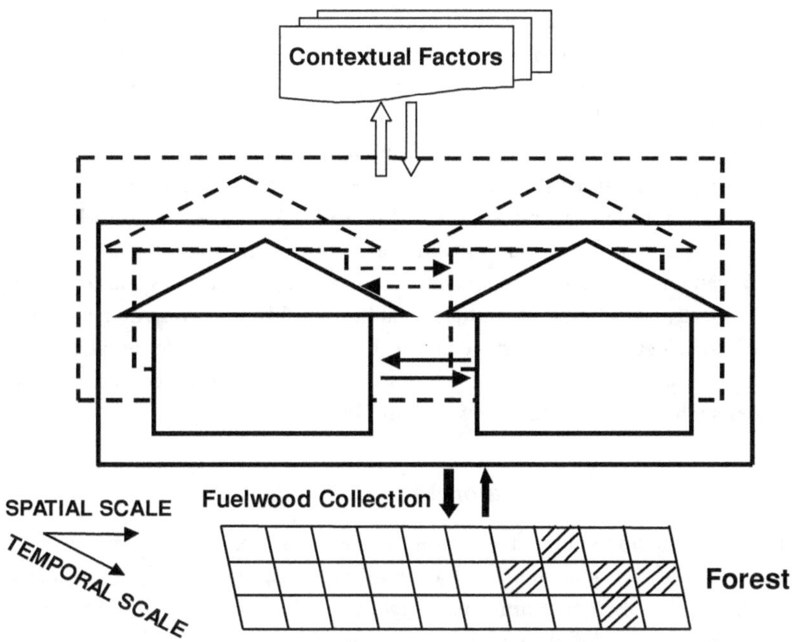

Figure 12.3 The conceptual framework of IMSHED

First, the houses within the dashed box represent households in Wolong at a particular point in time, while the solid houses represent households in the same landscape but at a later time. Household change during this interval might involve one or more of many events: households may increase or decrease in size, dissolve, or relocate; new households may be initiated as individual persons go through their life history, the details of which are illustrated in the section "Demographic Submodel." The formation of a new household, which depends upon the actions of young adults who have reached marriageable age, is explained by a set of psychosocial factors using a structural equation modeling approach based on our 220-household in-person interview data (An, Mertig, and Liu 2003). The findings from this model and our in-person interviews indicate that the intention to leave the parental home and establish a new home (the variable "leave-home intention" in Tables 12.2 and 12.3) is determined or influenced by his/her sibling status (whether he/she has siblings, and whether he/she is the youngest son or daughter (Figure 12.5b), the availability of land, and the behavior of parents and peers. This intention is the major predictor of new household formation for young adults after marriage.

Second, at each snapshot over time, the fuelwood demand of a household may be modeled as a function of household size, whether there is a senior person (60+ years old) in the household, and the area of cropland based on our extensive fieldwork and interview data (An et al. 2001). Presence of seniors is positively correlated with fuelwood demand because Wolong households with seniors use more firewood for longer heating time and higher home temperatures throughout the winter. Cropland area is also positively correlated with fuelwood demand because larger acreages are associated with hog farming. These farms grow potatoes and corn, the bulk of which are cooked in fuelwood stoves and fed to pigs. This phenomenon may arise from a traditional belief that cooked food makes pigs grow faster and healthier. The pork or bacon thus obtained is consumed and often sold to tourists and local restaurants for cash. This explains why there is a link between cropland area (for corn and potatoes) and fuelwood demand. The probability of switching from fuelwood to electricity for each household is predicted by a number of socioeconomic and demographic factors using a discrete choice logit model (An et al. 2002), in which a household's decision to switch to electricity or continue to use fuelwood is regressed against the age, gender, and education of the household head, annual household income, electricity price, outage frequency, voltage levels, and a few other factors.

Last, the forests on the landscape, given no human interference, grow and die by themselves. The interactions between humans and the environment are realized through fuelwood collection, as shown by the two lower vertical arrows in Figure 12.3. Harvesters from local households, given a certain amount of fuelwood demand derived from the model described above, travel to the most convenient set of locations (pixels in a raster grid) to cut fuelwood. Increasing distance for fuelwood collection may, in turn, reduce local households' fuelwood demand and encourage the substitution of electricity. Important physical and social factors (context factors in Figure 12.3), including distance, elevation, policy decisions, and law enforcement, exert impacts on many processes such as demand for fuelwood and electricity.

Major Agents/Objects

Major agents/objects include individual persons, households, pixels (square grid cells representing homogeneous units of the landscape), and some management agents helping us manage various objects or tasks (e.g., a list containing many agents of the same type). We only describe the major agents, starting from definitions of the corresponding classes. A collection of management agents handle mostly technical details in Swarm, and are not further elaborated on here.

Person. This class includes attribute variables such as personal ID, age, ID of the household that she/he belongs to, education level, gender, personal IDs of his/her mother and father, and his/her marital status. Also, the Person class has a few variables associated with childbirth: birth plan (how many children this person would have), birth interval (number of years between two consecutive children), marriage year (the year the person gets married), birth year (the year the person gives birth to a child), and first-child interval (the time between the marriage and the birth of her first baby). We will discuss the use of these variables in the section "Demographic Submodel." The actions (called "methods" in Java) include: give birth, die, grow, marry, move out of a household, move into a household, and cut fuelwood. Some other detailed actions (e.g., set the value for an attribute variable) specific to Java-Swarm programming are not discussed here.

Household. This class includes attribute variables such as household ID (consistent with that defined in the Person class), *x* coordinate, *y* coordinate, cropland area, household income, electricity price, outage level, voltage level, location of the household (Wolong Township for 0 and Gengda Township for 1; this is consistent with the dummy variable of location in the econometric model of An et al. 2002), distance of fuelwood transportation, and probability of switching from fuelwood to electricity. All the variables needed for predicting electricity and fuelwood demand are defined here because these demands are determined at the household level.

The actions in this class include formation of a new household, dissolution of a current household (when the number of people who belong to this household goes to zero), or an increase or decrease in household size (i.e., number of people in a household). We assume that when a new household is established, the area within 90 m (a parameter) around it has to be deforested and becomes nonhabitat for the pandas. This parameter is set in accordance with the fine spatial resolution that will be discussed in the data section. As a general rule, the value should be a multiple of the finest spatial resolution employed (90 m; another resolution is reported on in this study) unless only the pixel containing the household is to be deforested.

Pixel. The Pixel class contains all the information necessary for simulation of landscape changes. It contains attribute variables such as the *x* coordinate, y coordinate, elevation, slope, land-cover type, forest age, and forest volume (for non-forest pixels, this volume is automatically set to zero). Methods for the Pixel class include land-cover change (primarily from forest to non-forest), forest age increase, and volume growth (in forest pixels with tree species). We assume that forest volume reduction is primarily caused by fuelwood collection, because other factors such as forest fires and timber cutting are rare in the study region (M. Liu, personal communication).

Data Collection, Preparation, and Integration

The performance and application of any spatial model depend substantially on the data available for parameterization, calibration, and validation. Our data fall into the three categories of spatial environmental data, demographic data, and socioeconomic data.

Spatial environmental data. We have assembled a wealth of empirical data on both social and environmental factors and have built a database in both Microsoft Access and GIS (ArcInfo). Because remote sensing can provide views of the processes under study with adequate spatial extent, information detail, and temporal frequency (Herold, Goldstein, and Clarke 2003), we have used two time steps of high-quality, remotely sensed satellite images: Landsat TM (1997) and IKONOS (2000). We have conducted a supervised classification of the 1997 data based on 126 sample plots of 60 × 60 m, and validated them using a reserved set of 63 sample plots, resulting in an overall classification accuracy greater than 80 percent (shadow areas were classified as unclassified). We also have developed a digital elevation model of 30 × 30 m resolution, interpolated from 100-interval contour maps using the topogrid interpolation method in Grid ArcInfo. Based on a set of 313 Global Positioning System (GPS) points measured throughout the reserve, the DEM has a vertical accuracy of less than 50-m root mean square error (RMSE) and a standard deviation of approximately 37 m (in some most rugged areas the difference could be as large as 200 m).

Spatial coordinates of current household locations were obtained with Global Positioning System (GPS) measurement, and image analysis. The following steps illustrate the manner in which household locations were identified and linked with remotely sensed data and information from household surveys: (1) In the summer of 2001, we measured a total of 59 households using a Trimble GPS unit with real time differential correction (Omnistar), for which positional accuracy was estimated to be within 2–3 m. Four IKONOS satellite images (with 1 m resolution) that cover most of the area of Wolong with human settlements were also obtained. (2) We printed out a set of IKONOS-derived maps with spatial resolution chosen so that households and their interrelationships were most easily identifiable. Using these maps, we visited each household and collected the demographic information described in the agents section above and linked that information to their spatial locations. (3) Using the coordinates of the 59 households and 55 control points collected in the summers of 1998–1999 (Linderman et al. 2004) as control points, we georeferenced the four IKONOS images and recorded the coordinates of all the identifiable households. For households not identifiable in the IKONOS images, we used GPS to measure their coordinates. (4) Using the names of household heads as unique identifiers, we linked the demographic and socioeconomic data with the locational data (coordinates) on the IKONOS maps. Because nearly all Wolong houses are surrounded by their apportioned land, and land cannot be sold or traded in China (see the section "Demographic Submodel"), we did not record homestead field boundaries but assumed they were located immediately around the recorded house locations. For details of these processes, see Liu, An, et al. (2003).

Choosing an appropriate spatial resolution is a key challenge for this modeling effort. There is a trade-off between fidelity of spatial representation (an overly coarse cell resolution may mask some spatial variations) and efficiency (halving the resolution quadruples the amount of data and thus increases storage requirements and model execution times). We identified two resolutions to employ: 90 m and 360 m. For submodels requiring extensive human demographic factors (population size, number of households), the coarser (360 m) resolution was used. For submodels requiring or characterizing landscape characteristics (e.g., forest growth, distribution of panda habitats), the finer (90 m) resolution was used. Both resolutions were generated by resampling the 30 m raster DEM, slope, and Landsat TM-derived land-cover data; the methodology is reported by Linderman et al. (2004). Land-cover data were collapsed to nonforest (0), deciduous forest (1), conifer forest (2), and mixed forest (3). Processing took place in Erdas Imagine and ESRI ArcGIS, after which elevation, slope, and land-cover data were converted to ASCII text format for input to the Java-Swarm IMSHED model.

The fuelwood volume in each pixel is estimated according to the dominant tree species in that class. Class 1 consists of beech (*Fagus*), oak (*Quercus*), birch (*Betula*), and poplar (*Populus*), and the volume range is 60–100 m^3/ha, with ages ranging from 50 to 100 years old. Class 2 consists of fir (*Abies*), pine (*Pinus*), and spruce (*Picea*), and the volume is from 200 to 400 m^3/ha with ages ranging from 40 to 110 years old. Class 3 could be a mixture of any of these species and other woody ground cover; we set its volume from 125 (the average of lower bounds of Classes 1 and 2) to 250 m^3/ha (the average of upper bounds of Classes 1 and 2); and the age is set to be from 40 to 90 years (Yang and Li 1992; Ouyang et al. unpublished data; Linderman et al. 2004).

Demographic data. Socioeconomic data were obtained from a range of sources. Government data included the 2000 Population Census data of Wolong (Wolong Administration 2000), and the 1996 Wolong Agricultural census data (Wolong Administration 1996). Survey data were collected for approximately 1,000 households in Wolong, of which 220 were face-to-face interviews. Survey information included household economic status, social network (kinship relationship), and attitudes toward such issues as fertility. All these individual-based data, arranged by household, include personal ID, ID of the household that the person belongs to, gender, age, kinship relation to the household head, and other attributes of the "Person" and "Household" classes. These data cover all people in the reserve.

Each person (an object of Person class) keeps his/her father and mother IDs as attributes. In case the person's father or mother is unknown, dead, or not in the reserve, the value for the associated ID is set to zero. By doing so, we keep the kinship relations clear, which makes the simulations as realistic as possible. For instance, a brother in the single male list cannot "marry" his sister in the single female list by mistake because two people who share the same mother ID, father ID, or household ID are not allowed to "marry" each other.

These data can be employed to identify likelihoods for important household state changes: they are used to derive values for the parameters described in the following section. For example, consider the probability of in- and out-migration of young people from their parents' homes. Females emigrate from Wolong (0.28 percent); females immigrate to Wolong through marriage (0.19 percent); males emigrate from Wolong (0.043 percent); males immigrate to Wolong through marriage (0.043 percent). Since female in- and out-migration is relatively common, these parameters are employed in the demographic model. Parameters like these may not lead to changes in the amount of panda habitat over short time periods, but effects could be substantial at later times, and we examine their significance later in the section "Complexity Exploration."

Socioeconomic data. The 1996 and 2000 demographic data sets identified in the previous section also contain some useful socioeconomic information, such as the cropland area for each household, which is very important in determining household fuelwood demand (An et al. 2001). However, primary socioeconomic data were obtained from our interviews of the 220 households. The primary data include current electricity prices, outage frequencies, and voltage levels, which are used in computing the probability to switch from fuelwood to electricity (see the section "Electricity Demand"). Other questions in the same interview sessions have led to very useful information about what factors affect young adults' decisions about leaving their parental homes and establishing their own households after marriage. Those rules (to be described later in the section "Immigration and Local Movement Through Marriage"; also see Figure 12.5) about where to live after marriage are mainly based on these data.

Demographic Submodel

All individual-based data were entered into an Access database and exported as text into IMSHED. The model keeps track of the life history of individuals (objects of Person class) as follows: persons may give birth or be born, die, get married, and move into or out of a household (subsequently, into or out of the reserve in some cases) through marriage. Out-migration occurs when local residents move out of the reserve, immigration occurs when people move into the reserve from other places, and local movement occurs when local residents move within the reserve (primarily through marriage). Households are affected by these changes: they increase or decrease in size, new households form, and some existing households dissolve.

Death and out-migration. The death of each person is simulated through a random process. The likelihood of death for a person in a given year is in accordance with his/her age—0.00745 for people aged 0–5, 0.0009 for people aged 6–12, 0.00131 for people aged 13–15, 0.00196 for people aged 16–20, 0.00291 for people aged 21–60, and 0.05354 for people older than 60 (An et al. 2001). If a number drawn from a uniform distribution is less than the mortality rate associated with the person's age, he/she dies (as person on the left in Figure 12.4), and his/her spouse (if he/she has one) changes her/his marital status to "without spouse" while switching to the single male (or female) group; otherwise he/she survives the year.

Out-migration in Wolong is of two types: move-out through education and move-out through marriage;[3] the latter distinguishes between males and females. If a person survives, the model checks his/her age. If the age is between 16 and 20, and the random number generator creates a number smaller than the college attendance rate for people in this age group (0.0192 for each of the five years; see An et al. 2001), then he/she goes to college and leaves the reserve (exits from the simulation in IMSHED) permanently. Otherwise, the person remains in the household for that year. The rationale for doing so is that nearly all of Wolong young people who leave for the college find work in cities after graduation and do not return to the rural life of the reserve. For each single person above 22 (the minimal age for marriage by law), he/she will move out of the reserve (exit from simulation in IMSHED) if the random number generator creates a number smaller than a male's/female's move-out through marriage probability (0.043 percent, and 0.28 percent, respectively).

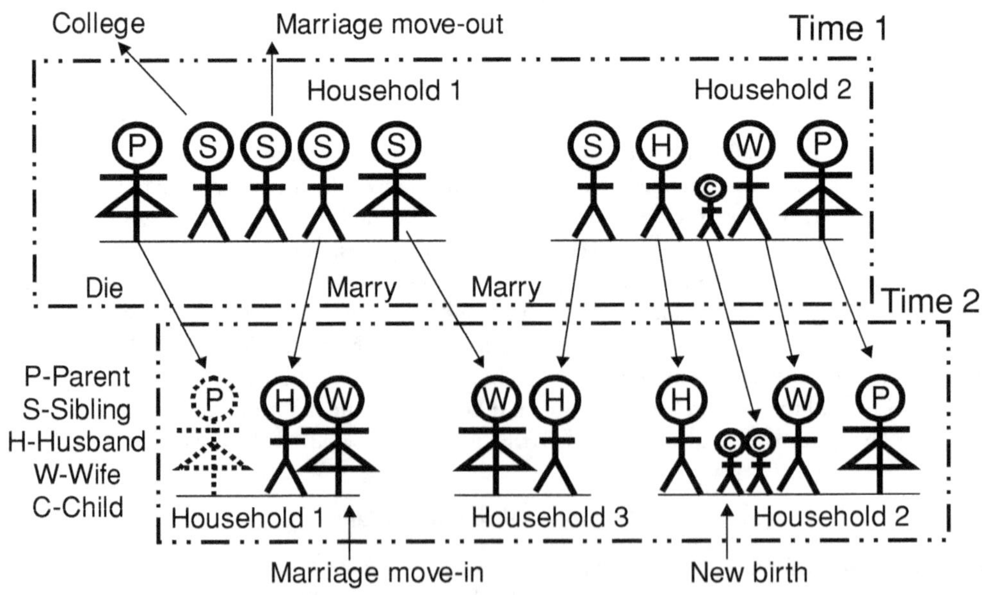

Figure 12.4 An illustration of individual-based demographic simulation

Immigration and local movement through marriage. Immigration is restricted due to Wolong's stand as a nature reserve for panda conservation. The only legal way for people outside the reserve to move in and obtain permanent residence licenses (*Hukou*) is through marriages with local people. A very important decision associated with both immigration and local movement is to determine whether a newly married couple will initiate a new household or not. This is important in IMSHED because the efficiency of fuelwood consumption differs among households as household sizes change (Liu, Daily, et al. 2003). The following situations are included in IMSHED: (1) A local male brings an outside female into Wolong through marriage, and the decision process is illustrated in Figure 12.5a. (2) A local female brings an outside male into Wolong through marriage. The decision process is similar to that in (1). Based on the findings of An, Mertig, and Liu (2003), the decision of whether to initiate a new household for these two people is: if (a) the female has no sibling, or (b) though she has siblings, all of them are females, and she is the youngest among them, then her husband and she will remain in her

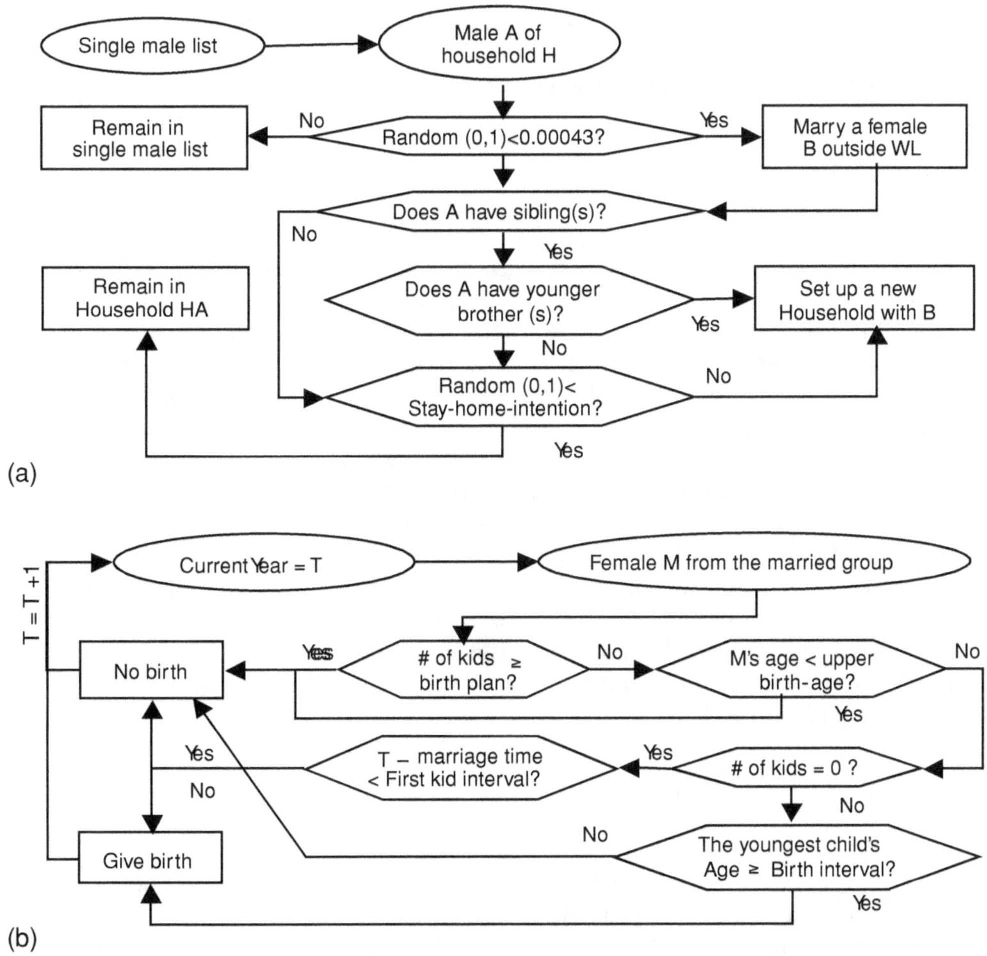

Figure 12.5 (a) Processes determining whether or not a new household is initiated when a local single male brings an outside female into Wolong through marriage. (b) Processes determining whether and when a household produces a child.

original household with a probability of 0.58, a parameter subject to change. Otherwise, they will initiate a new household. (3) A local male marries a local female: When two local singles get married, if the husband (a) has no siblings, or (b) has only female siblings (sisters), or (c) is the youngest among male siblings, then the couple lives in the husband's original household with a probability of 0.58; otherwise, check the sibling status of the wife. If she (a) has no siblings, or (b) she has only female siblings (sisters) and is the youngest among them, then the couple lives in the wife's original household with a probability of 0.58. Otherwise, the couple initiates a new household.

Based on the age and marital status of a person, a person is assigned to one of the four groups (lists in Java): (1) young group (all unmarried males and females less than 22 years old), (2) single male group (all single males over 22 years old, including the males whose spouses died), (3) single female group (all single females over 22 years old, including the females whose spouses died), and (4) married group (all females and males who have spouses with them). For example, if a male in the young group reaches 22, he will move to the single male group; at some time, if he gets married, he will move to the married group. However, if for some reason his spouse dies, he moves back to single male group again and has the potential to remarry, but the chance of doing so decreases as his age increases. The following equation is used to represent this relationship in accordance with our field observations:

$$\text{Rate of marriage at age } \chi = \frac{0.35}{(\chi - 30)^{0.4}} \; . \tag{1}$$

Childbirth. The event of childbirth only happens to females in the married group. For easier explanation, suppose that the woman under consideration is called M (indicating mother). As mentioned in the introduction of the Person class, each person has a birth plan that is used to set the number of children he/she may have. As indicated by Liu, Ouyang, Tan, et al. (1999), the number of children for each couple is 2.5. We use a binomial random variable Y to assign the number of children that M would have (Figure 12.5b). Since most families do not have more than five children (Wolong Administration 2000), we assume she would have 0, 1, 2, 3, 4, or 5 children with the probabilities of 0.03125, 0.15625, 0.3125, 0.3125, 0.15625, and 0.03125. The cumulative probabilities are 0.03125, 0.1875, 0.5, 0.8125, 0.96875, and 1, which are used to set probability intervals later. This is based on the classic problem of flipping a coin n (n = 5 here) times and observing the number of heads above Y, where the probability of success (observing heads up) is 0.5, and Y is a random variable that could take values from 0, 1, . . ., to 5. From the binomial distribution, the average of Y is $n \times p = 5 \times 0.5 = 2.5$ (number of children per mother). The probabilities are computed by the following equation, where p is the probability of "success."

$$\text{Probability } (Y = y) = \binom{N}{y} \times p^{y} \times (1 - p)^{(N-y)} \; . \tag{2}$$

As an alternative childbearing model (overloading in object-oriented programming), we simply randomly choose a number between two integer bounds with equal probability. For other parameters, we set their values based on our field observations. Birth interval (age difference between two consecutive children) is randomly chosen between 1 and 6 years because the observed average birth interval is around 3.5 years. The first-child interval (the time between marriage of a couple and birth of their first child) is set to be 1 or 2 years with even probabilities.

All these parameters are subject to change for different purposes, such as sensitivity or uncertainty analysis and policy design and test. We employed this model to simulate the childbearing for each female in the married group as illustrated in Figure 12.5b, where the above parameters (e.g., birth plan, upper birth age, and birth interval) all affect her childbirth decision.

Household dynamics. In accordance with all possible events for each individual, a household may decrease or increase in size, be initiated, or dissolve. When a new household is initiated, it is randomly assigned a site that is within a certain distance from its parental or original household, subject to two topographical restrictions of slope less than 37 degrees and elevation less than 2,610 m (He, Bearer, and Liu unpublished field data). This distance is controlled by a parameter with the default of 800 m based on our field observations. The new household is assigned a portion of the land from its parental household in proportion to its size rather than carved out from untenured land, which is based on the current Chinese land system. Farmers only have usufruct, and land can neither be traded nor developed without government permission because, based on China's constitution, the government and the collective organizations (quasi-governments) hold title to all land. The household responsibility system implemented in the late 1970s or early 1980s (the time for Wolong) assigned a certain amount of land to each rural household based on a set of criteria including household size and land quality, which has endured almost unchanged in spite of shifts in household sociodemographic factors such as household size. For this reason, new households founded by young married couples are often located in the immediate vicinity of one of the parental households. This fact explains why we did not map property boundaries for each household and why we did not employ a model to simulate land use change.

Landscape Submodel

There are several landscape-oriented components of the model; one is a forest growth model that is used to determine fuelwood volume, two are concerned with path identification and selecting forested pixels for harvest, and one is a conventional GIS/cartographic model to identify panda habitat. Each will be discussed in turn. In all cases, the basic unit is the pixel, with resolutions of 90 m and 360 m, depending on the requirements of the specific model.

Forest growth. Due to data limitations, we only consider forest growth using the simplified forest cover classification scheme presented in the section "Data Collection, Preparation, and Integration." According to Yang and Li (1992), the growth model for Class 1 is set to be 0.6, 0.8, and 1.0 m^3/ha/year if the forest is younger than 20 years, between 20 and 80 years, and older than 80 years. For Class 2, the rate is set to 2.0 m^3/ha/year regardless of the age. For Class 3 (mixed of Classes 1 and 2), the rate is set to 1.5. The maximal volumes for these three classes are set to be 350, 400, and 300; growth rate is set to zero when the volume of a pixel reaches its upper boundary.

Path finding. Finding the path to collect fuelwood is one of the primary processes in landscape simulation. Aside from the land apportioned to a household (usually adjacent to the household), the vast amount of rural land in China (including forests) is accessible to the public unless otherwise specified or regulated. Although Wolong has some habitat regulation policies, as mentioned in the section "Study Area," their implementation was ineffective, and most forest could be regarded as an open resource (later, we use a parameter "house buffer distance" to represent a fuelwood restriction policy; see the section "Model Test"). Therefore, we simply consider the effects of

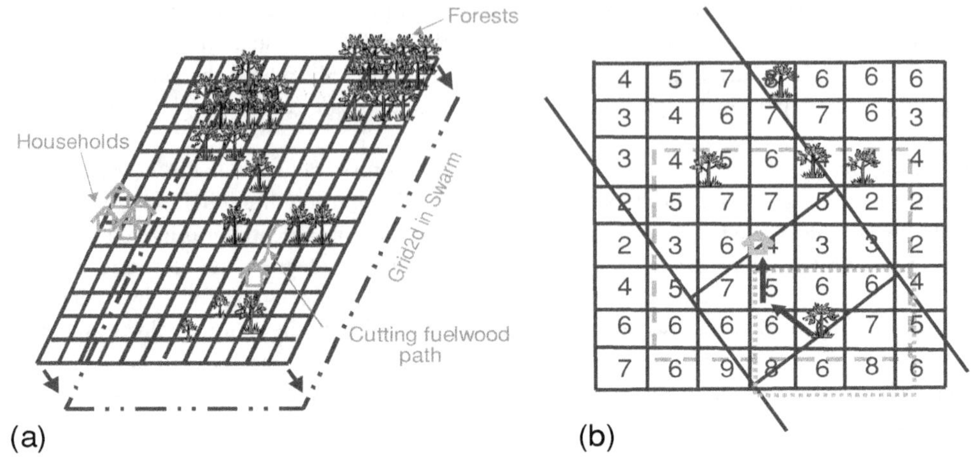

Figure 12.6 (a) An illustrative distribution of households and forests and (b) the procedure to find the least-cost path to collect fuelwood (the numbers in the pixels represent elevations)

topography and distance on the selection of routes to forested pixels, which may be visited by multiple households. In Figure 12.6(a), the household in the lower right corner needs to decide where to cut a certain amount of fuelwood, which has been determined by a number of socioeconomic and demographic factors and the probability of switching to electricity. Here we use a set of artificial intelligence rules in accordance with our in-site observations and interview data, such as the rule of limited viewing scope to be illustrated next.

To illustrate the function of the spatial fuelwood model, we consider the decision process underlying it (Figure 12.6b). (1) The fuelwood collector from the household has a limited geographical scope (the rule of limited viewing scope), so he/she only chooses among the forest pixels within the dashed window of size 5 × 5 (the window size is a parameter that may be adjusted by the user; 5 × 5 is only used for demonstration). (2) Within this window, only four pixels have forests, and the next step is to identify which pixel has the least cost to reach. Starting from the forested pixel, the fuelwood collector would deviate as little as possible from the direct path to each household pixel. We assume that his/her path-finding behavior is confined by the two southeast–northwest lines parallel to an assumed line cutting across the household and the pixel, where the distance between these two lines is a parameter. (3) Since he/she would not turn back while carrying a load of fuelwood, we assume that he/she goes northwest and does not go beyond the forest pixel; therefore, his/her path is also confined by the two lines that are perpendicular to the above two southeast–northwest lines. Within the area set by these four lines, he/she chooses the least-cost path. (4) Starting from the forest pixel, he/she chooses the pixel that has a lower elevation (or the one with the smallest elevation change if multiple pixels have lower elevations) and goes northwest as indicated by the arrow. For simplification, at some point if the household pixel is within one pixel to his/her standing pixel, he/she goes to the household directly. (5) He/she continues in this manner until the household pixel is reached.

Once the path is determined, its total length is calculated. Then it is adjusted by the slope between adjacent pixels along the route. The result indicates the cost of traversing the route between the home and a single forested cell. The same procedure is conducted on the remaining three forest pixels. Then, for that specific year, the collector chooses the pixel with the least-cost distance.

Fuelwood site selection. For each household, we identified all the forest pixels within a certain buffer distance (3,600 m as default based on unpublished field data from He, Bearer, and Liu) and put them on a list. We then calculate the cost distance between each forest pixel and the household using the method described above. We then group all these forest pixels into three categories of pixels that are 1,080 m, between 1,080 and 2,160 m, and over 2,160 m from the household, respectively. These three categories are chosen because interview data confirmed that 48.1 percent, 27.3 percent, and 24.6 percent, respectively, of the households collected fuelwood at sites corresponding to the above distances (He et al., unpublished field data). Therefore, if the random number generator creates a number smaller than 0.481, between 0.481 and 0.754, or greater than 0.754, the household will collect fuelwood in sites corresponding to the above distances.

We also apply artificial intelligence to the household under consideration: once it selects a pixel to collect fuelwood in a given year, it returns to the same pixel next year as long as the forest is still available. Doing so not only matches our field observations but also saves computer memory and time in computing and running the program, as fuel collection sites can be saved as an attribute of the household. Once this pixel is deforested, the household identifies a neighboring forested pixel.

The household "remembers" this distance (annually updated based on the location of the forest pixel), which affects its fuelwood demand by altering the household's perceived fuelwood collection proximity. Based on the travel distance employed in the previous year, proximity is set to be one of the three levels (short, moderate, and distant), corresponding to less than X m, between X m and $2.5X$ m, greater than $2.5X$ m, respectively ($X = 800$ m is the model default). Perceived proximity affects electricity demand and therefore provides a feedback effect. As fuelwood becomes harder to collect (sites become more distant), fuelwood demand drops, and households may substitute electricity for heating and cooking. The impact of the feedback effect is tested by varying the threshold X, a parameter called *perceived threshold distance* later in this chapter.

Habitat determination. To identify habitat, we use the criteria of Liu, Ouyang, Tan, et al. (1999): any pixel with an elevation between 2,250 and 3,250 m, slope less than 30 degrees, and containing canopy forest is viewed as potential habitat. We combine two classes "highly suitable habitat" and "suitable habitat" from Liu et al. (2001) into one category, "habitat," and exclude "marginally suitable habitat" (Liu et al. 2001) in our simulation because we want to provide a conservative estimate of panda habitat.

Socioeconomic Submodel

Potential fuelwood demand. Fuelwood consumption is calculated on a household basis in IMSHED. According to An et al. (2001), the fuelwood demand from a household can be modeled as a linear function of (1) household size, (2) whether a household has a senior person, and (3) area cultivated in corn and potato. (These two crops are usually grown together, and both are primarily used as fodder. This area is largely in proportion (60–80 percent) to the total land area obtained from the parental household.) The first two factors, which are characterized in the demographic submodel, are checked annually, while area under corn–potato cultivation changes only if a new household is initiated from the parental household, in which case land is partitioned proportional to the sizes of the new household and the resultant parental household.

Electricity demand. The fuelwood demand model just described does not consider the probability of switching from fuelwood to electricity and is, thus, incomplete. This switch probability is determined by the age, gender, and education of the household head, household annual income, current electricity price, outage frequency level, voltage level, perceived distance of fuelwood transportation, and location of the household under consideration (An et al. 2002). The following equation quantifies this relationship:

$$\text{Probability (switch} \mid \mathbf{x}_i, z_i, \alpha, \beta, \chi) = \frac{\exp\left[\alpha + \beta(\mathbf{x}_i^1 - \mathbf{x}_i^0) + \chi \mathbf{z}_i\right]}{\left[1 + \exp\left(\alpha + \beta(\mathbf{x}_i^1 - \mathbf{x}_i^0) + \chi \mathbf{z}_i\right)\right]} = \frac{1}{\left[1 + \exp\left(-\alpha - \beta(\mathbf{x}_i^1 - \mathbf{x}_i^0) - \chi \mathbf{z}_i\right)\right]} . \tag{3}$$

Vectors $\mathbf{x}_i^1$ and $\mathbf{x}_i^0$ represent the hypothetical and current electricity conditions (price, outage levels, and voltage levels), respectively, and β is the parameter vector associated with $\mathbf{x}_i^1$ and $\mathbf{x}_i^0$. Other nonelectricity factors, such as age and geographic locations, are described by the vector $\mathbf{z}_i$ with an associated parameter vector χ. The coefficients (χ) for the two dummy variables of low perceived distance and moderate perceived distance are – 1.24 and – 0.34 (An et al. 2002), indicating that as the fuelwood collection sites become far enough to change the household's current perceived distance (see the section "Landscape Submodel"), the switch-to-electricity probability will rise and the demand for fuelwood will accordingly decrease.

Reduced fuelwood demand. When electricity is available, the ultimate fuelwood demand is computed as the fuelwood demand derived above times the probability that the household does not switch to electricity, which is 1 minus the probability of switching from fuelwood to electricity as computed above.

Programming for Simulation

The model is programmed using Java-Swarm 2.1.1, a collection of software libraries developed by Swarm Development Group and briefly described in the Introduction. Swarm (Java version; it also supports Objective C) provides many readily useable packages for Java programmers. In addition, by resorting to a few readily made application programming interfaces (API's), IMSHED provides a user-friendly and graphical interface to set parameters and run the program.

IMSHED also provides a batch mode without graphical interfaces, where command-line arguments are allowed from the Unix/Linux shell. The modeling environment employs a command-line-based experiment manager written in Perl (Perl is a high-level programming language particularly well suited for tasks involving quick prototyping, system utilities, system management tasks, World Wide Web programming, and so on; see www.perl.com), which allows for efficient experiments by sweeping varying combinations of parameters designated in the Perl manager, leaving multiple runs progress unattended, and writing simulation results to designated directories.

Model Test

Model test, a crucial step after model calibration, is subject to many theoretical and practical challenges. Though models of any complex open system (e.g., agent-based spatial models like ours) may not be truly verified and validated (Overton 1977; Oreskes, Shrader-Frechette, and Belitz 1994), we still follow the traditional terms of model verification and validation. Both of these involve fitting the model to data or theory, but verification checks for the proper functioning of the programming, while validation investigates the correspondence between the software model and the conceptual model (structural validation) and between model outcomes and empirical data (empirical validation; see Manson 2001).

Model verification includes progressive debugging (see the paragraph below) and uncertainty testing (Table 12.1). Debugging is progressive in that model construction and calibration run in parallel with debugging/verification processes. We begin with a very simple model, and then add and test new features or algorithms progressively until we are confident in moving on. Testing involves assessing output of a series of 30 runs over a span of 20 years. This span is chosen because it is long enough for teenagers at model initiation (1996) to grow and experience nearly all the major life-history events such as marriage and household development, but short enough so that some assumptions or parameters (except the one (s) being tested) can be reasonably left unchanged, since socioeconomic and ecological uncertainties increase as we attempt to model farther into the future.

Uncertainty testing consists of extreme tests and extreme combination tests, which are employed to determine if the model becomes corrupted at some stages or returns wholly unreasonable values, which may signify potential programming bugs or design flaws (Rykiel, 1996). The former refers to setting each major parameter to minimum and maximum feasible values, conducting 30 runs, and constructing envelopes at the 95 percent confidence level over

Table 12.1 Model test methods

	Instrument	*Testing stage*	*Contents*	*Criteria*	*Data source*
Verification	Progressive building and debugging	Beginning–completion			
	Uncertainty test	Upon completion	Extreme tests	Theory, experience	Simulation results
			Extreme combination tests[1]	Theory, experience	Simulation results
Validation	Empirical validation	Upon completion	Demographic validation	*t*-test at 0.05 level	Independent government records
			Habitat validation	Change rate closeness	Independent results by other researchers
	Sensitivity analysis	Upon completion	See Table 12.3	Experience	Simulation results
	Experience/ expert opinion	Upon completion	Spatiotemporal pattern	Theory, experience and expert opinion	Simulation results

Note: 1. The selection of the variables for the combination tests depends on the results of the sensitivity analysis: the most sensitive factor in each of the four categories in Table 12.3 is selected.

20 years for the number of households, population size, and habitat area. Extreme combination tests combine sets of values of the four most sensitive parameters (see Table 12.1, Note 1) and observing model behavior (see Table 12.4). For the sake of simplicity, we only choose either the minimal or the maximal values of each parameter in each combination.

Model validation includes empirical validation, sensitivity analysis, and experience/expert opinion validation (in relation to predicted spatial pattern, Table 12.1) (Parker et al. 2003). To validate a model empirically, one may employ either spatially independent data or temporally independent data. Spatially independent data are collected at the same time as those used to calibrate the model, but from a separate region, and are not used to calibrate the initial model. Temporally independent data are collected in the same region as those used to calibrate the model, but at a different (usually later) time. Our empirical validation includes demographic validation, which is concerned with comparing predicted populations from 1997 to 2003 to observed data for that time period, and with numbers of households over 1997–2000. Our predetermined criterion is to pass the two-sample (observed and predicted data) paired *t*-test at the 0.05 α level with a null hypothesis that the differences between the model predictions and real observations are zero. We also empirically validate the habitat model by comparing predicted habitat change with results from other researchers' independent studies.

Experience/expert opinion validation is concerned with the plausibility of the model output (Manson 2001), in particular with the spatial pattern of the habitat model. We construct the probability for each cell to be deforested and become nonhabitat, map the result, and consider the map's plausibility based on our field observations and expert opinion of a few local researchers. Sensitivity analysis considers the robustness of model results to relatively small changes in input parameters. A highly sensitive model is undesirable, given the uncertainty in model input. Sensitivity may be assessed by perturbing each major parameter by a certain magnitude (here, 50 percent), and calculating the sensitivity index (Jørgensen 1986) as:

$$S_x = \frac{(dX/X)}{(dP/P)} \tag{4}$$

where P is the value of the independent variable, dP is the value for a small change of P, X is the value of the dependent variable, and dX is the corresponding change in X in response to the change in P.

Simulation Experiments

We employ two types of model experiments: scenario analysis and complexity exploration. The objective of scenario analysis is twofold. On one hand, we use it as a continuation of the model test process since unexpected outcomes may signal potential errors or bugs in the model; on the other hand, we want to provide policy makers some insights into possible outcomes under various practical conditions, as opposed to the extreme conditions investigated during validation. Here we are interested in discovering how population size, number of households, and panda habitat respond to varying conditions: (a) baseline scenario: employing status quo conditions; (b) conservation scenario: setting the sensitive factors and a few demographic factors to values that would presumably benefit panda habitat conservation; (c) development scenario: setting the factors to values that would presumably degrade panda habitat. We change each parameter in such a magnitude that would be (1) practical in the real world – e.g., fertility would be more likely to reduce to 1.5 in the conservation scenario than to 0; and (2) large enough to make a difference in model output based on our sensitivity test or field observations. For details of these scenarios, see Table 12.5. The use of these highly divergent scenarios may provide some insights into the possible trajectories of panda habitat change, and its consequent effect on the likelihood of giant panda survival.

In complexity exploration, we are mainly interested in using the model to find and test particular features of complexity, including the impact of time lags, feedback effects, and nonlinearity on key processes. We hypothesize that: (1) most of the demographic factors have substantial time-lag effects, and these effects escalate over time; (2) most of the spatial variables have complex nonlinear behavior due to human feedback and behavior adjustment. We illustrate (1) by focusing on the intention of young people to leave their parents' homesteads (a parameter called "leave-home intention" in Tables 12.2 and 12.3). This variable is important because it represents lifestyle changes, which affect the efficiency of resource utilization (Liu, Daily, et al. 2003). For (2), we use perceived threshold distance and house buffer distance (for definitions, see Table 12.2) to test how projected panda habitat changes when these two variables take either the minimal value, the value corresponding to the first quartile (1/4(max – min)), the median value, the third quartile (3/4 (max – min)), or the maximum value. The perceived threshold distance is an indicator of local households' own perceptions of the ease of fuelwood collection. A fuelwood collection site that is within this distance from a specific household is considered to have a "short" proximity (see the section "Fuelwood Site Selection"). The larger the value, the more likely a household views the current fuelwood collection distance as a "short distance." This variable may be affected by many other sociodemographic or psychological factors. For example, an increase in a specific household's annual income may lead it to value leisure time more highly, thereby decreasing the perceived ease of fuelwood harvesting and reducing fuelwood demand. The house buffer distance could be viewed as a policy control: a zero distance represents enforcement of no-cut regulations against fuelwood collection, while a very long distance (e.g., 7,200 m in Table 12.2) represents no or little restriction.

Results

We present the results in three sections. The first section shows the outcomes of the model test efforts, including both model verification and validation. The second section reports on results of the three scenarios described previously: the baseline scenario, the conservation scenario, and the development scenario. The third section illustrates the patterns of complexity detected by the model simulations.

Model Test

We verify our model in two steps: (1) extreme value tests and (2) extreme combination tests. The outcome of these tests is reported in Table 12.2. The model behaves as expected under the two extreme values of each variable. For instance, when the parameter "leave-home intention" is set to be 0 (indicating that all young adults remain in their parents' home after marriage), the final model reports total habitat of 580.78 km^2. When set to 1.0 (indicating that all young adults leave their parents' home and establish their own households after marriage), habitat area falls to 569.90 km^2. This may be caused by the great difference in the number of households: approximately 1,600 as opposed to 860 (Figure 12.8a); the human population is identical under both scenarios.

The 95 percent confidence envelopes become increasingly wide for all three variables as the model projects into the future, indicating increasing uncertainty in the prediction. For the number

Table 12.2 Extreme test design and results

	Parameters	*Definition*	*Default value*[1]	*Min, max*	$Habitat_{min}$, $Habitat_{max}$[2]
Family-planning	Max first kid interval	Maximal time between marriage and birth of the 1st child (years)	4	1, 20	576.33 (0.71) 578.40 (0.75)
	Max birth interval	Maximal time (years) between births of consecutive children	6	1, 20	575.94 (0.95) 576.07 (0.86)
	Upper birth age	The upper age that a woman gives birth to child	50	30, 60	575.16 (0.84) 574.52 (0.89)
	Marry age	The age of first marriage	22	18, 40	575.55 (0.65) 581.52 (0.74)
	Fertility	No. of children a woman may give birth to during her lifetime	2.0	0, 20	577.63 (1.05) 575.16 (0.81)
Migration	College rate	Ratio between the number of people who go to college and the total number of people between 16–22 at a year	1.92%	0.0%, 100%	575.81 (1.06) 580.48 (0.99)
	Leave-home intention	Probability that a "parental-home dweller"[3] leaves parental household and set up his/her own.	42%	0.0%, 100%	580.61 (0.96) 569.85 (0.67)
	Female marry-out rate	Ratio of the females between 22–30 who moved out of Wolong by marriage to all the females between 22–30 at a year	0.28%	0.0%, 100%	574.90 (0.93) 579.44 (0.70)
	Male bring-female-in rate	Ratio of the males between 22–30 who bring females into Wolong by marriage to all the males between 22–30 at a year	0.19%	0.0%, 100%	574.78 (0.89) 571.02 (0.92)
Electricity	Outage change	Change of outage levels (0 for low, 1 for medium, 2 for high)	0	–2, 2	582.42 (1.14) 570.11 (0.80)
	Voltage change	Change of voltage levels (0 for low, 1 for medium, 2 for high)	0	–2, 2	565.96 (0.81) 574.65 (1.42)
	Price change	Price change (Yuan)	0	–0.50, 0.50	597.72 (0.42) 558.32 (1.24)

Table 12.2 Continued

	Parameters	*Definition*	*Default value*[1]	*Min, max*	$Habitat_{min}$, $Habitat_{max}$[2]
Spatial	Perceived threshold distance	Distance (m) within which the perceived fuelwood collection distance is low	800	0, 8,000	579.31 (0.83) 565.06 (1.15)
	House buffer distance fuelwood	Maximum distance (m) within which households collect	3,600	0, 7,200	603.68 (0.45) 584.50 (1.44)
	Children-parent house distance	Maximum distance (m) between households of children and parents	800	0, 7,200	580.87 (0.78) 567.52 (1.41)

1. The default values for the associated variables based on field observations. Habitat area (km^2) under the default values at year 2016 is 575.49 (0.69). Minimal and maximal values for the associated variables used to test the model.
2. The habitat area (km') under minimal and maximal values for the associated parameter. The numbers in the parentheses are standard errors.
3. An adult child who remains in his/her parental home after marriage.

Table 12.3 Sensitivity tests for model input parameters (habitat area projected in 2016)

	Parameters	*Default Value*	*+50%*[1] *Perturbation*	*Habitat area* *(km^2)*[2]	*Different from baseline*[3]	*Sensitivity*
Family-plan	Max first kid interval	4	6	575.55 (0.63)	no	N/A
	Max birth interval	6	9	576.72 (1.35)	no	N/A
	Upper birth age	50	60[4]	574.52 (0.63)	no	N/A
	Age at marriage	22	33	580.48 (0.87)	yes	1.73%
	Fertility	2.0	3.0	575.16 (0.95)	no	N/A
Migration	College rate	1.92%	2.88%	569.8	yes	0.06%
	Leave-home intention	42%	63%	565.29	yes	–1.02%
	Female marry-out rate	0.28%	0.42%	569.43	yes	0.04%
	Male bring-female-in rate	0.19%	0.29%	568.32	no	N/A
Electricity	Outage change	0	1	573.22 (0.88)	no	N/A
	Voltage change	0	1	575.68 (0.68)	no	N/A
	Price change	0	0.05	566.09 (0.88)	yes	–3.27%
Spatial	Fuelwood-change distance	800	1,200	574.65 (1.11)	no	N/A
	House buffer distance (m)	3,600	5,400	582.29 (0.96)	yes	2.36%
	Kid-parent house distance (m)	800	1,200	574.52 (1.25)	no	N/A

1. The perturbation range of 50 percent is determined in consideration of: (1) it should be relatively small (otherwise we can use extreme tests as in Table 12.2); and (2) the response magnitude of the habitat change should be large enough for our calculation. An alternative of –50 percent perturbation is not included simply for space consideration.
2. The standard error is in the parentheses following each average value.
3. We use two-sample paired t-test at the 0.05 level to test whether the predicted habitat area is different from the baseline value at year 20.
4. Here only 20 percent perturbation because an upper birth age of 75 years old (a 50 percent increase) does not make sense in the real world.
5. All the 4 parameters under the category "Migration" have insignificant impact on panda habitat over 20 years. The numbers reported here are simulation results over 30 years, and the sensitivity index is calculated using the amount of habitat over 30 years (568.20 km^2).

Table 12.4 Extreme combination test design and results

Marry age	*Leave-home intention*	*Electricity price (Yuan)*	*Buffer Distance*[1] *(m)*	*Average habitat area (km²)*[2]			
				Year 5	*Year 10*	*Year 15*	*Year 20*
18 (s)[3]	0.0 (s)	–0.5 (s)	0 (s)	607.18 (0.00)	607.18 (0.00)	607.18 (0.00)	607.18 (0.00)
			7,200 (b)	607.18 (0.00)	606.95 (0.13)	606.92 (0.21)	606.79 (0.22)
		0.5 (b)	0 (s)	607.18 (0.05)	607.18 (0.05)	607.18 (0.05)	607.05 (0.05)
			7,200 (b)	601.47 (0.34)	588.64 (0.96)	577.11 (1.26)	571.54 (1.21)
	1.0 (b)	–0.5 (s)	0 (s)	603.03 (0.74)	601.34 (1.05)	599.79 (0.84)	598.23 (0.89)
			7,200 (b)	601.99 (0.67)	598.88 (0.90)	595.25 (0.90)	593.18 (0.85)
		0.5 (b)	0 (s)	603.42 (0.67)	599.53 (0.52)	597.84 (0.85)	596.42 (0.83)
			7,200 (b)	596.16 (0.62)	582.29 (0.57)	571.67 (0.98)	562.98 (1.07)
40 (b)	0.0 (s)	–0.5 (s)	0 (s)	607.18 (0.00)	607.18 (0.00)	607.18 (0.00)	607.18 (0.00)
			7,200 (b)	607.18 (0.00)	607.05 (0.08)	606.92 (0.17)	606.53 (0.18)
		0.5 (b)	0 (s)	607.18 (0.00)	607.18 (0.00)	607.18 (0.00)	607.18 (0.00)
			7,200 (b)	601.86 (0.32)	588.90 (0.81)	579.06 (1.11)	574.78 (1.12)
	1.0 (b)	–0.5 (s)	0 (s)	606.66 (0.40)	606.14 (0.64)	605.62 (0.66)	605.10 (0.63)
			7,200 (b)	605.75 (0.27)	604.45 (0.42)	603.29 (0.50)	600.83 (0.93)
		0.5 (b)	0 (s)	606.27 (0.36)	606.14 (0.41)	605.49 (0.60)	602.12 (0.59)
			7,200 (b)	601.47 (0.51)	589.42 (0.81)	579.44 (1.01)	571.67 (1.39)

1. House buffer distance (see Table 12.2 for its definition).
2. The standard error is in the parentheses following each average value.
3. The letters "s" and "b" stand for "small" and "big," respectively, corresponding to the minimal and maximal values of each parameter.

of households (Figure 12.8a), the envelope for the value of 1.0 (leave-home intention) is much higher than that for the baseline, which, in turn is higher than that for the value of 0.0. Because all the adults remain in their parental homes and do not establish their own houses (leave-home intention = 0.0), there is relatively little uncertainty over time as the total number of households slightly decreases. The population size dynamics do not differ among the three situations (Figure 12.8b), with all three envelopes nearly overlapping each other. This is because the leave-home intention parameter only affects the likelihood for young adults to establish their own households, rather than population size. The area of panda habitat (Figure 12.8c) decreases over time in all scenarios, but the rate varies logically between the three situations: results using a value of 1 show the most rapid decrease, as far more households are established and consume substantially more fuelwood.

Before reporting the results of the extreme combination tests, we report sensitivity test outcomes (listed as part of model validation in Table 12.1) because we use them to identify the most sensitive variables for the extreme combination tests. Table 12.3 lists the sensitivity of model parameters in four key groups. The most sensitive parameters are age at marriage in the family-planning category (1.73 percent of sensitivity, see Table 12.3), leave-home intention in the migration group (–1.02 percent over 30 years), price change in the electricity group (–3.27 percent), and house buffer distance in the spatial group (2.36 percent). These variables are examined in greater detail in subsequent extreme combination tests.

The model runs through all the extreme combination tests (i.e., a total of sixteen tests: two extreme values for each of the four variables, Table 12.4) as well and gives reasonable results. Model runs are reported for a string of four letters, such as "*bbss*"*; s* symbolizes a minimal value ("small"), while "*b*" symbolizes a large ("big") value; "*bsss*" means a combination that takes big, small, small, and small values for the first to fourth variables in the combination. The outcome habitat area ranges from 607.18 km^2 in such combinations as "*ssss*" and "*bsss*" to 562.98 in km^2 in a combination of "*sbbb*." These combination tests may also signify the relative importance or contribution of each of the variables in affecting panda habitat change. For instance, when the electricity price change is "*s*" (–0.5 Yuan, or a 0.5 Yuan decrease), the value of the house buffer distance ("*s*" or "*b*") does not make much difference; when it is "*b*" (0.5 Yuan), it makes a great difference, e.g., "*bsbs*" and "*bsbb*" give 607.18 and 574.78 km^2, respectively. This implies that, given an increase in the price of electricity, a hypothetical policy on forbidding fuelwood collection (here we do not consider its practicability because it is a model test) will substantially reduce habitat loss compared to no or little enforced collection restriction (house buffer distance = 7,200 m). This is intuitively correct and consistent with our observations.

The model validation consists of three sections: empirical validation, sensitivity testing, and validation via expert opinion/corroboration. We reported the sensitivity test results earlier in this section. For empirical validation, we use two-sample paired t-tests to decide whether the predictions are acceptable in relation to field observations. The model passes such t-tests for both demographic variables (number of households and population size), resulting in two *p*-values of 0.89 and 0.88 (Figures 12.7a and 12.7b). In addition, our predicted annual population increase rate is 0.48 percent (a total of 9.50 percent over twenty years), while the same rate from 1982 to 1996 is 1.05 percent (Liu, Ouyang, Tan, et al. 1999). This decrease in the rate of population growth could be due to the strict population control in the 1990s (Liu, Ouyang, Tan, et al. 1999). Our predicted annual rate of increase for the number of households in Wolong is 1.18 percent (a total of 23.63 percent over twenty years, Table 12.3), which is greater than the increase in population. This is consistent with the pattern from 1975 to 1999, for which the number of households increased more rapidly than did the population size as reported by Liu, Ouyang, Tan, et al. (1999).

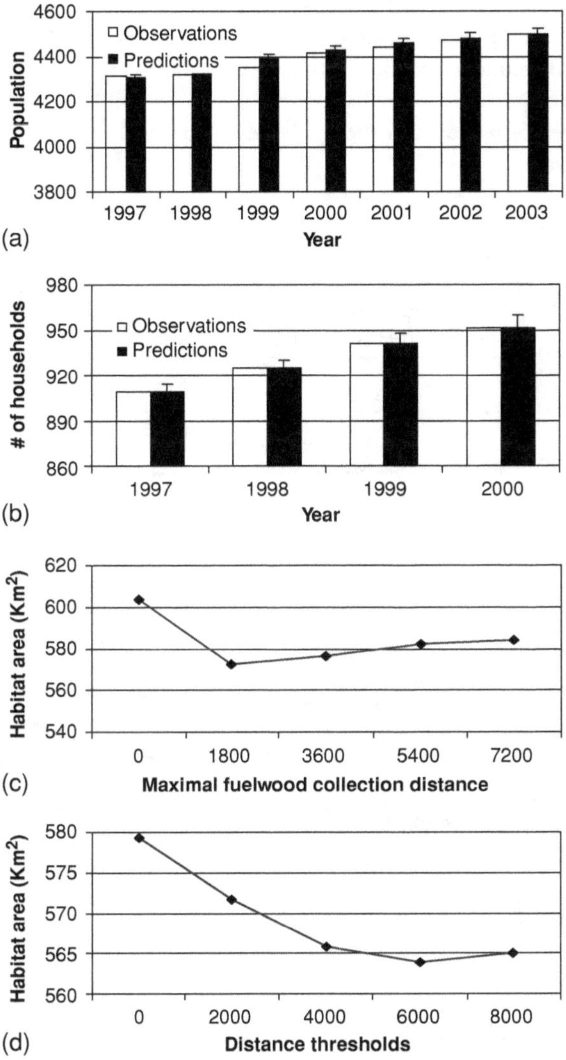

Figure 12.7 (a) Comparison between the observed and predicted population size ($t = -0.14$, $p = 0.89$). (b) Comparison between the observed and predicted number of households ($t = 0.16$, $p = 0.88$). (c) Habitat under varying maximal fuelwood collection distance (m). (d) Habitat under varying perceived distance for fuelwood collection easiness.

To corroborate our work with that of other experts, we compare our predictions about panda habitat dynamics with results from other researchers. According to Laurie and Pan (1991), the annual loss of forest area in Wolong was 2.5 km^2 prior to 1991. Because the ratio between the total area of habitat (607 km^2, see Figure 12.8c) and the total area of forest (1,249 km^2, calculated by adding all the cells of forest classes defined in Linderman et al. 2004) in 1997 was 49 percent, this forest loss of 2.5 km^2/year is largely equivalent to a 1.23 km^2/year loss of habitat if the same ratio applies. Our model shows that under the status quo, panda habitat will decline from approximately 607 km^2 to 576 km^2 from 1996 to 2016 (Figure 12.8c), which translates into an annual habitat loss of 1.55 km^2. We explain our slightly higher habitat-loss rate in this way: habitat is not evenly distributed in all types of forests; instead, it is located in forest areas within certain elevation and slope thresholds (see the section "Habitat Determination"), to which people have easy and frequent access. Cutting trees in these areas would degrade the habitat more than for an average forest plot that might be less accessible. Therefore, using the ratio between the total area of habitat and the total area of forest (49 percent) may lead to a lower value than the true value.

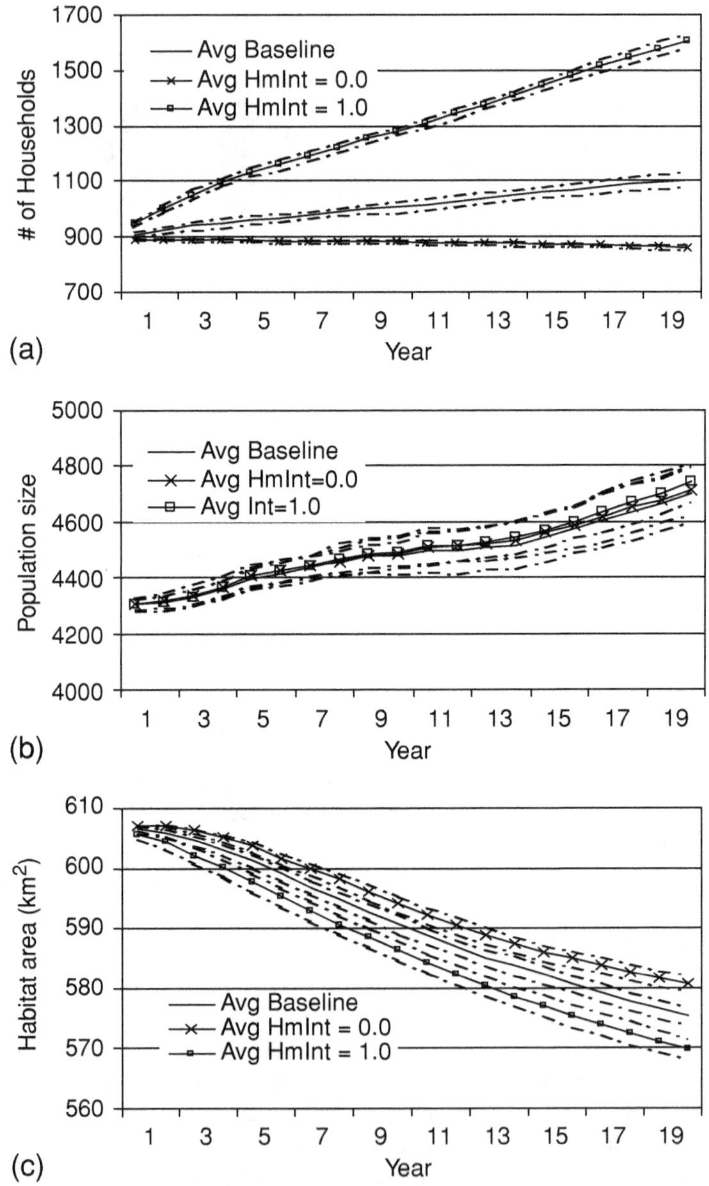

Figure 12.8 (a) The average numbers of households, (b) population size, and (c) predicted panda habitat and associated upper and lower 95 percent confidence envelopes for (1) the baseline simulation, (2) leave-home intention set to 0.0, and (3) leave-home intention set to 1.0.

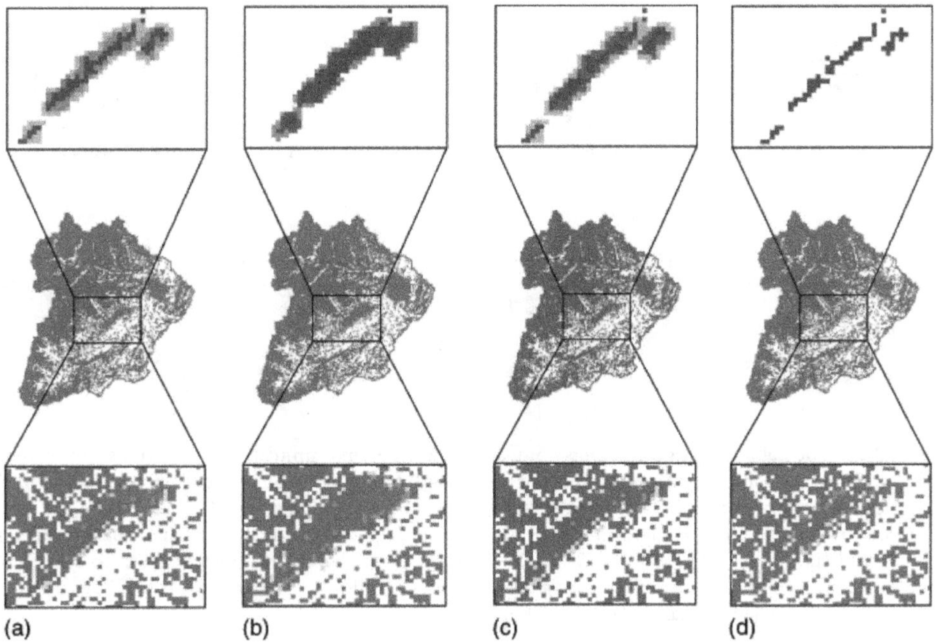

Figure 12.9 The gradients for household locations (top) and panda habitats (bottom) for model predictions employing (a) leave-home intention = 1.0 and time = 10; (b) leave-home intention = 1.0 and time = 20; (c) leave-home intention = 0.0 and time = 10; (d) leave-home intention = 0.0 and time = 20. For the household location gradients, the white, light gray, gray, and dark gray colors represent probabilities of <25%, 25%–50%, 50%–75%, and >75% for the occurrence of households. For the habitat gradients, the gray, dark gray, light gray, and white colors represent probabilities of <25%, 25%–50%, 50%–75%, and >75% for the occurrence of habitat.

Last, we report our model validation by the spatiotemporal patterns. When time is controlled, the households occupy more land when leave-home intention is 1, which is more obvious in year 20 (Figure 12.9b, top) than in year 10 (Figure 12.9a, top). This is largely due to the model assumption that all young adults establish their own households and thus use more land. Consequently, habitat loss is more severe when leave-home intention is 1.0 than when it is 0.0, in particular for year 20 (Figures 12.9b and 12.9d, bottom). From the spatial gradients, the most likely pixels for future households are close to those existing households. These are surrounded by more distant, less likely pixels, which are in turn surrounded by the least likely pixels. Farthest away are the totally unlikely ones. For habitat, we see a reversed trend: the least likely habitat cells are closest to households, then the less likely ones, and the most likely ones are the farthest. These phenomena are consistent with a diffusion model characterizing household choice in clearing land for construction and cutting fuelwood: begin with the nearest suitable sites, and then move outwards. Moreover, the model predicts the potential habitat gains or losses when leave-home intention equals 1.0 or 0.0. In the first case, habitat loss is increasingly large with respect to the baseline simulation as time progresses in the model (Figures 12.10a and 12.10b). In the second case, less habitat is lost relative to the baseline simulation (Figures 12.10c and 12.10d). These results agree with the experience of our own and local researchers (Shiqiang Zhou, personal communications[4]).

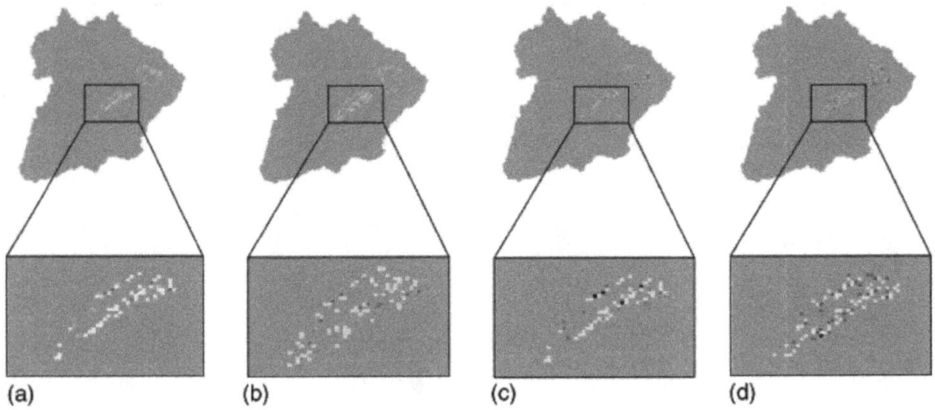

Figure 12.10 The habitat change gradients compared to the baseline simulations for various alternative scenarios: (a) leave-home intention = 1.0 and time = 10; (b) (b) leave-home intention = 1.0 and time = 20; (b) leave-home intention = 1.0 and time = 20; (c) leave-home intention = 0.0 and time = 10; (d) leave-home intention = 0.0 and time = 20. The three levels of light gray, gray, and dark gray in (a) and (b) represent the probabilities of <25%, 25%–50%, >50% for habitat loss relative to the baseline, while the three levels of light gray, gray, and dark gray in (c) and (d) represent the probabilities of <25%, 25%–50%, >50% for habitat gain relative to the baseline.

Scenario Analysis

The conservation and development scenarios lead to great differences in the three key variables. The conservation scenario predicts approximately 600 km^2 of remaining panda habitat in 2016, the final model year. The development scenario estimates just 554 km^2 of remaining panda habitat. The conservation scenario predicts approximately 873 households, with a total human population of 2,611. The development scenario predicts much larger numbers of households and people: 1,602 and 6,305, respectively (Table 12.5).

The spatial patterns corroborate the above numerical changes in the number of households and population size. Under the development scenario, the households expand rapidly outward as time progresses, destroying some previous panda habitat. Under the conservation scenario, the land occupied by local households remains nearly unchanged, resulting in little habitat loss. Figure 12.11 gives a snapshot of panda dynamics under the conservation and development scenarios, which shows that the development scenario exerts much more severe impacts on panda habitat in parallel with a big cluster of households.

Complexity Exploration

As shown in Figure 12.8a, the numbers of households differ from each other increasingly over time when the leave-home intention takes 0, 0.42, and 1.0; the same is true for panda habitat (Figure 12.8c). The differences in panda habitat among these three scenarios become more significant over time: approximately from year 15, the lower bound of the 0.0 intention envelope is substantially higher than the upper bound of the baseline envelope; from year 11, the lower bound of the baseline envelope is substantially higher than the upper bound of the 1.0 intention envelope. This supports our first hypothesis that differences in initial demographic factors have large and escalating effects on model outcome.

Table 12.5 Definition and results of conservation and development scenarios and simulation results over 20 years[a]

Category	*Variable*	*Conservation scenario*	*Development scenario*
Electricity	Price	0.05 Yuan decline	0.05 Yuan rise
	Outage levels	One level decrease	One level rise
	Voltage levels	One level increase	One level decline
Migration	Leaving parental home intention	0.42 → 0.21	0.42 → 0.95
	College rate	1.92% → 30% (16–20 youth)	1.92% → 0.0%
	Female marry-out rate	0.28% → 120%	0.28% → 20%
Family planning	Fertility	2.0 → 1.5	2.0 → 5.0
	Birth interval	3.5 → 5.5	3.5 → 1.5
	Marriage age	22 → 28	22
Fuelwood	Distance for demand change (m)	800 → 0	800 → 8000
Results	# of households	873.00 (7.48)	1,602.00 (12.12)
	Population size	2,611 (27.51)	6,305 (76.92)
	Habitat (km^2)	599.92 (0.54)	553.52 (1.13)

(a) The first numbers in the spaces are the default values in the model, and the second values are those used in the associated scenarios.

The second hypothesis concerns nonlinearity in observed spatial patterns of household processes. Here we consider the impact of house buffer distance, which is a parameter representing a possible habitat protection policy to control how far people are allowed to search for fuelwood from their homes. As the house buffer distance rises, consequent total panda habitat area falls to a certain point (approximately 573 km^2), and then rises slowly (Figure 12.7c). We then consider a different, yet related, variable: perceived fuelwood distance. As the threshold for the perceived fuelwood collection distance rises, a similar pattern occurs, except that the changing point is much larger (approximately 6,000 m; Figure 12.7d).

Discussion

Though the agent-based model presented here shows great potential for addressing practical issues about panda protection, it has been developed to address more general, theoretically important *issues* such as integrating socioeconomics, ecology, and demography, understanding complexities in some coupled society–environment *systems*, and linking ABM and GIS to study spatiotemporal dynamics of land-use and land-cover changes. The following subsections discuss several of these issues.

Model Results and Policy Implications

Model verification and validation turn out to be a theoretical and practical challenge in modeling complex systems (e.g., Manson 2001; Parker et al. 2003), and we address this challenge by designing a systematic strategy (see section "Model Test"; Table 12.1) and subjecting our model to this strategy. Though our model passes all the tests, it would be preferable to collect more detailed spatial data for forest volume and panda habitat in order to further validate the model.

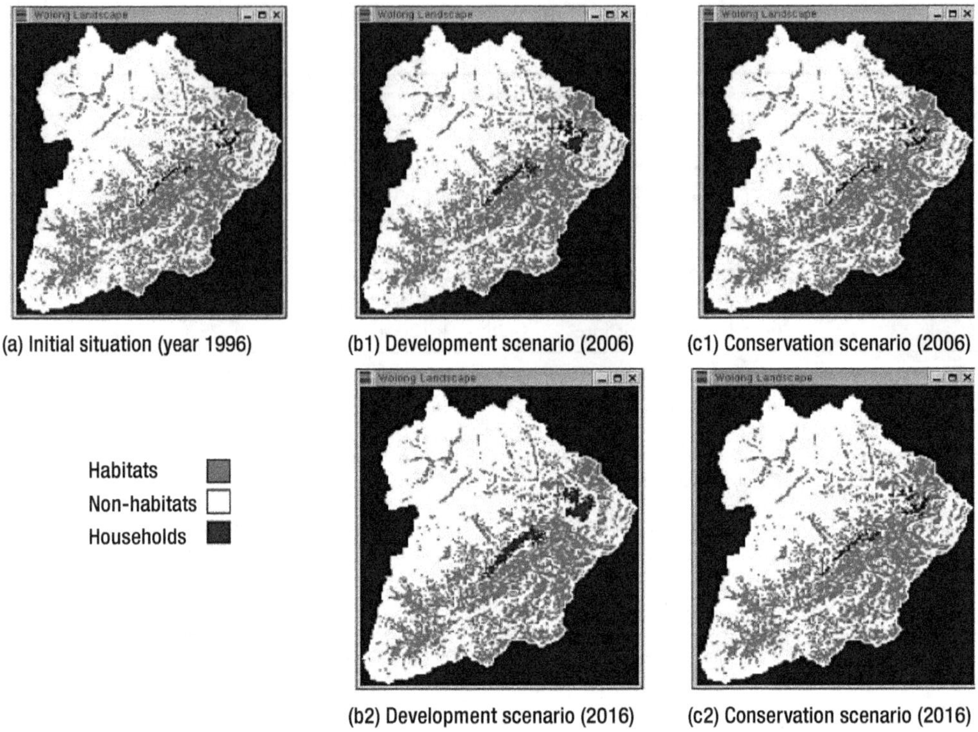

Figure 12.11 A snapshot of panda habitat fragmentation and household distribution in 1996 (a), 2006 (b1 for development and cl for conservation), and 2016 (b2 development and c2 for conservation)

Greater detail would also facilitate the application of the model in balancing the needs for panda conservation and those for human wellbeing within the reserve.

This study used total area of potential panda habitat as a primary model output to assess the impact of different scenarios on the panda. Though a decrease of 5.27 percent (about 32 km^2) under the baseline situation (the status quo scenario) over 20 years may seem moderate, the spatial distribution and fragmentation of panda habitat should be of concern. Pandas appear to prefer areas that humans also tend to visit for fuelwood collection (Schaller et al. 1985; Liu, Ouyang, Taylor, et al. 1999). Therefore, the predicted decrease in potential habitat may occur in the panda's preferred habitat. Furthermore, employing total area alone does not address the problem of habitat fragmentation, which may render potential habitat pixels unusable for pandas. According to Schaller et al. (1985), a giant panda usually occupies a home range with an area of about 2 km^2. In our model, all pixels meeting the simple spatial criteria, including fragmented areas smaller than 2 km^2, are counted as panda habitats; therefore, it is likely that our model overestimates actual viable panda habitat. However, this approximation could be viewed from another perspective: within all the potential panda habitats (green blocks in Figure 12.11), there might be pandas; within all the areas that are counted as nonhabitats, there should be no pandas. A final spatial concern for the results is that stochastic environmental shocks, such as forest fires, could lead to a substantial sudden loss of panda habitat. This model does not account for such factors.

Model results under conservation and development scenarios indicate that human socio-economic and demographic factors substantially affect panda habitat, but the impact of these factors takes time to manifest itself on the environment. Implementing policies that encourage family planning, human out-migration from the reserve, lifestyle change, or the increased use of electricity could result in subsequent preservation of panda habitat to varying degrees. For instance, the model predicts that an electricity subsidy of 0.05 Yuan could reduce total habitat loss by the year 2016 from around 32 km^2 to 18 km^2. If combined with other conservation activities, even more habitat could be spared. However, our results indicate that the environmental benefits of such policies – or the penalties should they not be implemented–are not immediately obvious.

The nonlinear and counterintuitive relationship between the amount of panda habitat and the house buffer distances as shown in Figure 12.7c may be caused by the following relationships: when this buffer distance is very small (even zero), people are allowed to harvest little or no fuelwood, and thus panda habitat is better preserved; when this buffer distance is very large, local households' fuelwood collection is scattered throughout a large buffer region, and some areas in this region may have a substantial forest volume or a rapid regeneration rate (capturing this type of uncertainty in the section "Spatial Environmental Data" is one strength of this agent-based model), so cutting wood may not cause severe habitat loss. If, however, this buffer distance is somewhere in between (around 1,800 m), local households' unrestricted fuelwood demand is satisfied through cutting all available wood in this small region (very likely going beyond its carrying capacity) and causing more habitat loss. As time goes on, the local households are likely to move outwards, as the pattern in Figures 12.10a and 12.10b shows.

Similarly, the parameter perceived threshold distance also leads to nonlinear changes in panda habitat (Figure 12.7d). This variable could be explained as the ease of adjusting household fuelwood demand based on the existing fuelwood collection distance: the longer the distance, the harder to reduce their fuelwood demand, thus more habitat loss caused by satisfying this demand. However, after a threshold (6,000 m), this effect diminishes and yields to other complexities (such as the above-mentioned habitat vs. buffer distance relationships).

Methodology: Integration, Complexity, and Coupling of ABM and GIS

The IMSHED developed for this research has utilized data and models across scales, disciplines, and time periods. Data integration included working with data at different scales (e.g., individual-level data such as age vs. population data such as mortality rate), from different disciplines (e.g., ecological data such as forest regeneration vs. sociological data such as leave-home intention), and with varying degrees of uncertainty (e.g., accurate human individual demographic data vs. forest volume data with a wider range of uncertainty). Far more so than data, however, the combination of methods and models from different disciplines reflects the breadth of the human–environment modeling challenges. For instance, IMSHED employs a fuelwood demand model (An et al. 2001) and an econometric model for electricity demand (An et al. 2002) to compute fuelwood demand on a household basis (see the section "Demographic Submodel"). By integrating methods and empirical results concerning young adults' propensity to leave their parental homes and form their own households, IMSHED is able to project household dynamics and link them to panda habitat loss. In addition, ecological uncertainties or variability (e.g., variations in forest volume and regrowth rates) are taken into consideration through assigning varying values to the attributes of the forest pixel objects, which demonstrate the utility of object-oriented programming (OOP) for capturing environmental variability.

From the above analyses, we are confident that our modeling framework effectively integrates individual-level data and transdisciplinary models for projecting spatiotemporal changes of some key response variables. This challenge has been viewed as very significant for studying environmental sustainability (Clark 2002). This bottom-up approach "starts from the 'parts' (i.e., individuals) of a system and then tries to understand how the system's properties emerge from the interactions among these parts" (Grimm 1999). Consequently, this framework can efficiently deal with many research needs that traditional approaches may find difficult or impossible to deal with, and may provide more accurate predictions. However, this increased accuracy can only be found in the aggregated results, such as human population size or number of households. Some stochastic processes are simulated at the agent level (e.g., an individual person's leaving parental home decisions), and individual simulation runs are essentially single realizations of the process. These simulation runs are useful in predicting the overall number of households and the resultant spatial pattern, but whether a particular household will be established at a specific location is not a question that our model resolves to answer.

Developing and using ABM do not discredit the traditional state variable, statistic, or analytic approaches. On the contrary, in many situations, our framework uses these approaches because it is unnecessary or sometimes impossible to account for every detail of the agents under consideration. For instance, when computing the probability to switch from fuelwood to electricity, we use a logistic regression (see the section "Socioeconomic Submodel"), and obviously this regression model is an average trend derived from a number of households. It is important to balance between using outcomes based on individual agent actions and averaged trend data to find an appropriate level of resolution and aggregation in predictions. The choice depends on research needs, the applicability of individual data, and available resources (time, budget, and other conditions like computational power).

The complexities in many coupled society–environment open systems have been barriers for effectively studying and understanding such systems. By decomposing the population-level dynamics into life histories of all the individuals and characterizing the dynamics of all households in the landscape, it becomes easier to capture any time-lag effects of demographic factors. For instance, an increase of 0.5 in fertility may be "considered" by many couples planning to have children at an appropriate time (the model "knows" the time), and this consideration may lead to an increase in population size and number of households with a cumulative effect. This effect may not cause observable habitat degradation in ten years, but may do so in thirty years. In addition, with feedback (households decrease their fuelwood demand as collection distances rise) built into our model, the environment is not simply a passive cache of resources waiting to be developed; instead, its geography imposes opportunities and limitations on the human inhabitants. This leads to a more dynamic portrayal of the human–environment system and, we believe, more representative model results.

Finally, the integration of ABM and GIS in this study has allowed for further insights into the spatial trajectories of some key ecological or socioeconomic processes, such as the gradients for household locations and panda habitats in Figure 12.9. These insights are not only important in validating the model (abnormal spatial trajectories may signify potential bugs; see Parker et al. 2003), but also may be significant for panda conservation efforts because the trajectories provide policy makers with information about where, when, and under what conditions panda habitat would be lost or conserved. However, this integration is still in its fledgling stage. Much readily usable functionality in GIS (such as finding the cost distance) has to be coded in Java Swarm by the authors, which sometimes becomes a heavy burden. As an alternative, some model outcomes have to be exported to ArcGIS for further spatial analysis (e.g., the gradient of households locations in Figure 12.9). Coupling environmental models and GIS has long been recognized as a key challenge (Goodchild, Parks, and Steyaert 1993; Wesseling et al. 1996), and our experience reflects this. A more specific issue for coupling ABM and GIS is the development of rule sets based on empirical metrics. Current metrics may not adequately reflect key aspects of the environment, and new landscape and/or spatial metrics (e.g., Herold, Goldstein, and Clarke 2003) may need to be developed to capture different spatial and temporal dynamics of landscape/habitat change.

In this study, we develop a framework to integrate geographical, ecological, socioeconomic, and demographic data into different levels or types of agents or objects (persons, households, pixels), incorporate some complex mechanisms (e.g., time lag, feedback), and project the spatial patterns of panda habitat extent over time. This framework enables us to study how changes in socioeconomic and demographic factors work in both straightforward and complex ways to affect panda habitat. This impact is characterized over time in a spatially explicit manner. Using this combined model has enabled us to develop a better understanding of the relationships between people and panda habitat in Wolong, which may, in turn, help to develop environmentally sound policies in the reserve. More broadly, we have provided a working example of a framework (including the tool) to explain or predict overall landscape patterns as a result of the actions of many agents. This framework is a powerful means for integrating data and models across varying scales and disciplines and shows promise for many human-environment studies.

Acknowledgments

This chapter originally appeared in *Annals of the Association of American Geographers*, 95, 54–79 (2005). We thank Guangming He, Jinyan Huang, Shiqiang Zhou, Zhiyun Ouyang, and Hemin Zhang for their assistance in data acquisition. We appreciate the valuable comments from Dr. Michael Goodchild and the four anonymous reviewers. We are also indebted to financial support from the National Science Foundation (NSF), National Institute of Child Health and Human Development (NICHD), American Association for Advancement of Sciences, the John D. and Catherine T. MacArthur Foundation, and Michigan State University. Special thanks go to the online help from Swarm Development Group, in particular Dr. Paul Johnson at the University of Kansas.

Notes

1. For link to these tools, go to http://wiki.swarm.org/wiki/Main_Page (last accessed on June 17, 2004).
2. Electricity is the readily available substitute for fuelwood in the reserve, subject to government price control and some quality problems (An et al. 2002). Other energy sources such as coal, charcoal, biogas, and sun/wind power are not used and no market exists for them.
3. Though we observed some people who took temporary jobs in outside areas (primarily big cities), they still had their residence registration (known as *Hukou*) license in Wolong. More importantly, they often come back to Wolong during busy agricultural seasons and Chinese spring festivals and conduct resource-related activities such as fuelwood collection. Thus, they are not treated as out-migrants.
4. Shiqiang Zhou from the Wolong Nature Reserve is an experienced researcher with extensive knowledge in local biology, ecology, and socioeconomic and demographic situations. The authors have closely worked with him to collect the data, build the models (including earlier models as published by An et al. 2001, 2002, 2003), and discuss the model outcomes during 1998–2004.

References

An, L., J. Liu, Z. Ouyang, M. A. Linderman, S. Zhou, and H. Zhang. 2001. Simulating demographic and socioeconomic processes on household level and implications on giant panda habitat. *Ecological Modeling* 140: 31–49.

An, L., F. Lupi, J. Liu, M. A. Linderman, and J. Huang. 2002. Modeling the choice to switch from fuelwood to electricity: Implications for giant panda habitat conservation. *Ecological Economics* 42 (3): 445–57.

An, L., A. G. Mertig, and J. Liu. 2003. Adolescents' leaving parental home: Psychosocial correlates and implications for biodiversity conservation. *Population and Environment: A Journal of Interdisciplinary Studies* 24 (5): 415–44.

Batty, M., Y. Xie, and Z. L. Sun. 1994. Modelling urban dynamics through GIS-based cellular automata. *Computers, Environment and Urban Systems* 23: 205–33.

Berger, T. 2002. Agent-based spatial models applied to agriculture: A simulation tool for technology diffusion, resource-use changes, and policy analysis. *Agricultural Economics* 25 (2–3): 245–60.

Bian, L. 1997. Multiscale nature of spatial data in scaling up environmental models. In *Scale in Remote Sensing and GIS*, Eds. D. A. Quattrochi and M. E. Goodchild, 13–26. New York: Lewis Publishers.

Brown, D. G., B. C. Pijanowski, and J. D. Duh. 2000. Modeling the relationships between land use and land cover on private lands in the Upper Midwest, U.S. *Journal of Environment Management* 59: 247–63.

China's Ministry of Forestry and World Wildlife Fund. 1989. *Conservation and Management Plan for Giant Pandas and their Habitat.* Beijing, China.

Clark, W. C. 2002. Research systems for a transition toward sustainability. In *Challenges of a Changing earth: Proceedings of the Global Change Open Science Conference* (Amsterdam NL, July 10–13, 2001), Eds. W. Steffen, J. Jäger, D. Carson, and C. Bradshaw. Berlin: Springer-Verlag. Available at: http://ksgnotes1.harvard.edu/BCSIA/sust.nsf/pubs/pub54 (last accessed 11 June 2004).

Clarke, K. C., and L. Gaydos. 1998. Loose coupling a cellular automaton model and GIS: Long-term growth prediction for San Francisco and Washington/Baltimore. *International Journal of Geographical Information Science* 12 (7): 699–714.

Deadman, P, K. Lim, D. Robinson, E. Moran, E. Brondízio, and S. McCracken. 2001. LUCITA: multi-agent simulations of land-use change near Altamira, Brazil (Section 3.6). In *Agent-Based Models of Land-Use and Land-Cover Change*, Eds. D. C. Parker, T. Berger, and S. M. Manson (W. J. McConnell, Man. Ed.). Available at: www.csiss.org/maslucc/ ABM-LUCC.htm (last accessed June 11, 2004).

Ehrlich, P. R., and E. O. Wilson. 1991. Biodiversity studies: Science and policy. *Science* 253: 758–62.

Flake, G. W. 1998. *The Computational Beauty of Nature: Computer Explorations of Fractals, Chaos, Complex Systems, and Adaptation.* Cambridge, MA: MIT Press.

Gimblett, H. R. 2002. Integrating geographic information systems and agent-based technologies for modeling and simulating social and ecological phenomena. In *Integrating Geographic Information Systems and Agent-Based Techniques for Simulating Social and Ecological Processes,* Ed. H. R. Gimblett, 1–20. New York: Oxford University Press.

Goodchild, M. E., B. O. Parks, and L. T. Steyaert, Eds. 1993. *Environmental Modeling with GIS.* New York: Oxford University Press.

Grimm, V. 1999. Ten years of individual-based modeling in ecology: What have we learned and what could we learn in the future? *Ecological Modeling* 115: 129–48.

He, G., S. Bearer, X. Chen, and J. Liu. Unpublished. The household fuelwood collection survey (Section II). 2003.

He, G., S. Bearer, and J. Liu. Unpublished. The household fuelwood collection survey (Section I). 2001–2002.

Herold, M., N. C. Goldstein, and K. C. Clarke. 2003. The spatio-temporal form of urban growth: measurement, analysis and modeling. *Remote Sensing of Environment.* 86 (3): 286–302.

Jiang, B., and H. R. Gimblett. 2002. An agent-based approach to environmental and urban systems within geographic information systems. In *Integrating Geographic Information Systems and Agent-Based Techniques for Simulating Social and Ecological Processes*, Ed. H. R. Gimblett, 171–89. New York: Oxford University Press.

Jørgensen, S. E. 1986. *Fundamentals of Ecological Modelling.* Amsterdam: Elsevier.

Larkin, D., and G. Wilson. 1999. *Object-Oriented Programming and the Objective-C Language.* Cupertino, CA: Apple Computer.

Laurie, A., and W. Pan. 1991. Working group report on giant pandas in the wild. In *Giant Panda and Red Panda Conservation workshop: Working Group Reports: National Zoological Park*, Eds. D. Klerman and M. Roberts, 5–12. Washington, DC: Smithsonian Institution.

Levin, S. A., B. Grenfell, A. Hastings, and A. S. Perelson. 1997. Mathematical and computational challenges in population biology and ecosystems science. *Science* 275: 334–43.

Li, H., and J. F. Reynolds. 1997. Modeling effects of spatial pattern, drought, and grazing on rates of rangelands degradation: A combined Markov and cellular automaton approach. In *Scale in Remote Sensing and GIS*, Ed. D. A. Quattrochi and M. F. Goodchild, 211–30. New York: Lewis Publishers.

Li, X., and A. G. Yeh. 2002. Neutral-network-based cellular automata for simulating multiple land use changes using GIS. *International Journal of Geographical Information Science* 16 (4): 323–43.

Lim, K., P. J. Deadman, E. Moran, E. Brondizio, and S. Mc-Cracken. 2002. Agent-based simulations of household decision-making and land use change near Altamira, Brazil. In *Integrating Geographic Information Systems and Agent-Based Techniques for Simulating Social and Ecological Processes*, Ed. H. R. Gimblett, 277–308. New York: Oxford University Press.

Linderman, M., L. An, S. Bearer, G. He, Z. Ouyang, and J. Liu. Forthcoming. Modeling the spatio-temporal dynamics and interactions of households, landscapes, and giant panda habitat. *Ecological Modelling*.

Linderman, M., J. Liu, J. Qi, Z. Ouyang, L. An, J. Yang, and Y. Tan. 2004. Using artificial neural networks to map the spatial distribution of understory bamboo from remote sensing data. *International Journal of Remote Sensing* 25 (6): 1685–1700.

Liu, J., L. An, S. S. Batie, R. E. Groop, L. Zai, M. A. Linderman, A. G. Mertig, Z. Ouyang, and J. Qi. 2003. Human impacts on land cover and panda habitat in Wolong Nature Reserve: Linking ecological, socioeconomic, demographic, and behavioral data. In *People and the Environment: Approaches for Linking Household and Community Surveys to Remote Sensing and GIS*, Eds. J. Fox, R. R. Rindfuss, S. J. Walsh, and V. Mishra, 241–63. Boston, MA: Kluwer Academic Publishers.

Liu, J., G. C. Daily, P. R. Ehrlich, and G. W. Luck. 2003. Effects of household dynamics on resource consumption and biodiversity. *Nature* 421: 530–33.

Liu, J., M. Linderman, Z. Ouyang, L. An, J. Yang, and H. Zhang. 2001. Ecological degradation in protected areas: The case of Wolong Nature Reserve for giant pandas. *Science* 292: 98–101.

Liu, J., Z. Ouyang, Y. Tan, J. Yang, and H. Zhang. 1999. Changes in human population structure: Implications for biodiversity. *Population and Environment: A Journal of Interdisciplinary Studies* 21: 46–58.

Liu, J., Z. Ouyang, W. W. Taylor, R. Groop, and H. Zhang. 1999. A framework for evaluating the effects of human factors on wildlife habitat: The case of giant pandas. *Conservation Biology* 13: 1360–70.

López, E., G. Bocco, M. M. Mendoza, and E. Duhau. 2001. Predicting land-cover and land-use change in the urban fringe: A case in Morelia city, Mexico. *Landscape and Urban Planning* 55: 271–85.

MacArthur, R. H., and E. O. Wilson. 1967. *The Theory of Island Biogeography*. Princeton, NJ: Princeton University Press.

Malanson, G. P 1999. Considering complexity. *Annals of the Association of American Geographers* 89: 746–53.

——. 2002. Extinction-debt trajectories and spatial patterns of habitat destruction. *Annals of the Association of American Geographers* 92: 177–88.

Manson, S. M. 2001. Calibration, verification, and validation (Section 2.4). In *Agent-Based Models of Land-Use and Land-Cover Change*, Eds. D. C. Parker, T. Berger, and S. M. Manson (W. J. McConnell, mng. Ed.). Available at: www.csiss.org/maslucc/ABM-LUCC.htm (last accessed June 11, 2004).

Messina, J. P., and S. J. Walsh. 2001. 2.5D morphogenesis: Modeling landuse and landcover dynamics in the Ecuadorian Amazon. *Plant Ecology* 156: 75–88.

Minar, N., R. Burkhart, C. Langton, and M. Askenazi. 1996. The Swarm simulation system: A toolkit for building multi-agent simulations (SFI Working Paper 96–06–042). Available at: www.swarm.org/archive/overview.ps (last accessed June 11, 2004).

Myers, N., R. A. Mittermeier, C. G. Mittermeier, G. A. B da Fonseca, and J. Kent. 2000. Biodiversity hotspots for conservation priorities. *Nature* 403: 853–58.

Najlis, R., M. A. Janssen, and D. C. Parker. 2001. Software tools and communication issues (Section 2.3). In *Agent-Based Models of Land-Use and Land-Cover Change*, Eds. D. C. Parker, T. Berger, and S. M. Manson (W. J. McConnell, mng. Ed.), 27–39. Available at: www.csiss.org/maslucc/ABM-LUCC.htm (last accessed June 11, 2004).

NeXT Software. 1992. *Object-Oriented Programming and the Objective-C Language*. Available at: www.gnustep.org/resources/documentation/ObjectivCBook.pdf (last accessed June 11, 2004).

Oreskes, N., K. Shrader-Frechette, and K. Belitz. 1994. Verification, validation, and confirmation of numerical models in the earth sciences. *Science* 263 (4 February): 641–46.

Ouyang, Z., J. Liu, S. Zhou, and Y. Tan. Unpublished. The vegetation survey in Wolong Nature Reserve. 1998.

Overton, S. 1977. A strategy of model construction. In *Ecosystem Modeling in Theory and Practice: An Introduction with Case Histories*, Eds. C. Hall and J. Day, 49–73. New York: John Wiley & Sons.

Pan, W. K. Y., C. M. Erlien, F. Baquero, S. J. Walsh, R. E. Bilsborrow, and B. G. Frizzelle. 2004. Farm-level models of spatial patterns of land use and land cover dynamics in the Ecuadorian Amazon. *Agriculture, Ecosystems, & Environment* 101 (2–3): 117–34.

Parker, D. C., S. M. Manson, M. A. Janssen, M. J. Hoffmann, and P. Deadman. 2003. Multi-agent systems for the simulation of land-use and land-cover change: A review. *Annals of the Association of American Geographers* 93: 314–37.

Peuquet, D. J. 1999. Time in GIS and geographical databases. In *Geographical Information Systems* (2nd ed.), Eds. P. A. Longley, M. F. Goodchild, D. J. Maguire, and D. W. Rhind, 91–103. New York: Wiley.

Rindfuss, R. R., P. Prasartkul, S. J. Walsh, B. Entwisle, Y. Sawangdee, and J. B. Vogler. 2003. Household-parcel linkages in Nang Rang, Thailand: Challenges of large samples. In *People and the Environment: Approaches for Linking Household and Community Surveys to Remote Sensing and GIS*, Eds. J. Fox, R. R. Rindfuss, S. J. Walsh, and V. Mishra, 241–3. Boston, MA: Kluwer Academic Publishers.

Rykiel, E. J., Jr. 1996. Testing ecological models: the meaning of validation. *Ecological Modeling* 90: 229–44.

Schaller, G. B., J. Hu, W. Pan, and J. Zhu. 1985. *The Giant Pandas of Wolong.* Chicago: University of Chicago Press.

Seto, K. C., and R. K. Kaufmann. 2003. Modeling the drivers of urban land use change in the Pearl River Delta, China: Integrating remote sensing with socioeconomic data. *Land Economics* 79 (1): 106–1.

Walsh, S. J., R. E. Bilsborrow, S. J. McGregor, B. G. Frizelle, J. P. Messina, W. K. T Pan, K. A. Crews-Meyer, G. M. Taff, and E Baquero. 2003. Integration of longitudinal surveys, remote sensing time series, and spatial analysis: approaches for linking people and place. In *People and the Environment: Approaches for Linking Household and Community Surveys to Remote Sensing and GIS*, Eds. J. Fox, R. R. Rindfuss, S. J. Walsh, and V. Mishra, 241–63. Boston, MA: Kluwer Academic Publishers.

Walsh, S. J., T. P. Evans, W. E. Welsh, B. Entwisle, and R. R. Rindfuss. 1999. Scale-dependent relationships between population and environment in northeastern Thailand. *Photogrammetric Engineering & Remote Sensing* 65 (1): 97–105.

Wesseling, C. G., D. J. Karssenberg, P. A. Burrough, and W. P. A. Van Deursen. 1996. Integrated dynamic environmental models in GIS: The development of a dynamic modelling language. *Transactions in GIS* 1: 40–48.

Wolong Administration. 1987. *Wolong Vegetation Cover and Plant Resources.* Chengdu: Sichuan Science and Technology Press.

—— 1996.1996 Agricultural Census, Wolong (unpublished, in Chinese).

—— 2000. Wolong 2000 nationwide population census data (unpublished, in Chinese).

Yang, Y., and C. Li, Eds. 1992. *Sichuan Forests.* Beijing, China: Yunnan, China Forestry Press.

13

AGENT-BASED MODELING IN COUPLED HUMAN AND NATURAL SYSTEMS (CHANS)

Lessons from a Comparative Analysis

Li An, Alex Zvoleff, Jianguo Liu, and William Axinn

Research Question: What are the key features that are, or should, be included in models of coupled human and natural systems?
System Science Method(s): Agent-based models
Things to Notice:
• Similarities and differences in two closely related agent-based models
• Use of the ODD (Overview, Design, Details) framework for model description

Coupled human and natural systems (CHANS) are characterized by many complex features, including feedback loops, nonlinearity and thresholds, surprises, legacy effects and time lags, and resilience. Agent-based models (ABMs) are powerful for handling such complexity in CHANS models, facilitating in-depth understanding of CHANS dynamics. ABMs have been employed mostly on a site-specific basis, however. Little of this work provides a common infrastructure with which CHANS researchers (especially nonmodeling experts) can comprehend, compare, and envision CHANS processes and dynamics. We advance the science of CHANS by developing a CHANS-oriented protocol based on the overview, design concepts, and details (ODD) framework to help CHANS modelers and other researchers build, document, and compare CHANS-oriented ABMs. Using this approach, we show how complex demographic decisions, environmental processes, and human–environment interaction in CHANS can be represented and simulated in a relatively straightforward, standard way with ABMs by focusing on a comparison of two world-renowned CHANS: the Wolong Nature Reserve in China and the Chitwan National Park in Nepal. The four key lessons we learn from this cross-site comparison in relation to CHANS models include how to represent agents and the landscape, the need for standardized modules for CHANS ABMs, the impacts of scheduling on model outcomes, and precautions in interpreting "surprises" in CHANS model outcomes. We conclude with a CHANS protocol in the hope of advancing the science of CHANS.

Many ecosystem services vital to human existence and well-being have been degraded due to population pressures and unsustainable exploitation of natural resources (Vitousek et al. 1997; Foley et al. 2005). This poses a global challenge to scientists and practitioners about how to better understand these changes and manage ecosystem services. Although some researchers have

 DOI: 10.4324/9781315748771-15

long realized the importance of, and have invested efforts in, coupling human and natural systems, until recently much research still emphasized either human systems or natural systems (J. Liu et al. 2007; An 2012), largely holding the other as exogenous or as background. This approach gives inadequate attention to the reciprocal relationships that often exist between human and natural systems over space and time. Human interference in these systems can have unexpected consequences. For instance, human activities might lead to sudden shifts in natural systems from desirable to undesirable states (Folke et al. 2004), such as from a clear lake to a lake with toxic algae blooms (Folke et al. 2002).

The division between natural and social sciences, along with the assumption that connections between natural and human systems are decomposable into a set of simple, unidirectional relationships, has hindered understanding of these systems. Empirical studies on human–nature systems (e.g., Bian, Quattrochi, and Goodchild 1997; Irwin and Geoghegan 2001; Deadman et al. 2004; An et al. 2005; Crawford et al. 2005; Grimm et al. 2005; Messina and Walsh 2005; Brown et al. 2008; Yu et al. 2009) reveal a range of features that are difficult to address using this traditional approach. These features include (1) reciprocal effects and feedback loops: humans and nature interact with each other and form complex feedback loops; (2) nonlinearity and thresholds: the relationships within or among coupled systems are often nonlinear, and there exist transition points (thresholds) between alternate states; (3) surprises: surprising outcomes are observable as a result of human–nature couplings; (4) legacy effects and time lags: prior human–nature couplings have substantial impacts on later conditions; (5) resilience: human–nature systems are capable of retaining similar structures and functioning after disturbances; and (6) heterogeneity: even within a system substantial differences exist in socioeconomic variables, human choices and behavior, and ecological conditions and should not be ignored (J. Liu et al. 2007). Corroborating evidence for these features also comes from the Amazon (Malanson, Zeng, and Walsh 2006a, 2006b), the southern Yucatán in Mexico (Manson 2005), northern Ecuador (Walsh et al. 2008), China (J. Liu 2010), North America (Lepczyk et al. 2008; Rutledge et al. 2001), and other places around the world (Rindfuss et al. 2008; An 2012).

In this context, complexity refers to the six features just mentioned along with a set of features not analytically tractable from system components and their attributes alone, such as path dependence, self-organization, difficulty of prediction, and emergence (Manson 2001; Bankes 2002; An 2012). Complexity theory, from the study of complex systems, offers great potential to address these phenomena. With partial origin from general systems theory (Von Bertalanffy 1969; Warren, Franklin, and Streeter 1998), the study of complex systems focuses on heterogeneous subsystems, autonomous entities, nonlinear relationships, and multiple interactions such as feedbacks, learning, and adaptation (Arthur 1999; Axelrod and Cohen 2000; Manson 2001; Crawford et al. 2005). Instead of a cure-all, the complex systems approach provides a unique systematic paradigm to harness complexity instead of decomposing it into (often) oversimplified unidirectional linkages. On account of this capacity, the complex systems approach helps system managers to take innovative action to steer systems of interest in beneficial directions (Axelrod and Cohen 2000).

Building on complexity theory, researchers investigating human–nature systems and ecosystem services have developed the coupled human and natural systems (CHANS) framework (J. Liu et al. 2007). This framework has evolved in parallel with many closely related concepts similar to or parts of CHANS, including coupled natural and human (CNH) systems, human–environment systems (Turner et al. 2003), social–ecological systems (SES; Ostrom 2007), and social–environmental systems (Eakin and Luers 2006). Both these other integrative approaches and the CHANS framework provide many theoretical advantages and empirical insights, including those already mentioned. Understanding CHANS is by no means a trivial task, however.

Understanding CHANS requires modeling approaches that can represent these relationships and the characteristic structures and processes of CHANS (J. Liu et al. 2007). Given this requirement, agent-based modeling is an ideal modeling tool for CHANS. Agent-based modeling builds in part on the long history in the spatial and social sciences of trying to understand and represent relationships and networks between human actors and landscape change (Latour 1996; Epstein 1999). First, agent-based models (ABMs) can integrate data across spatial, temporal, and hierarchical scales and directly capture the relationships between system components and decision-making processes of individual agents (An 2012). CHANS are defined in large part by the reciprocal relationships between their component parts, or agents. Agents might be human actors on the landscape, or agents might represent structures (e.g., governing bodies) that could constrain and shape processes on the corresponding landscape. At a basic level, CHANS ABMs represent individual agents and their environment in a computer model, where agents of the same type or agents at different levels of a scalar hierarchy (e.g., persons, households, and cities) can interact with one another. ABMs can also represent external forces (e.g., climate, policy, economic conditions) that a modeler might not choose to model directly but that could be important to the processes being modeled within a certain ABM.

Although great progress has been made in modeling CHANS using ABMs in the past decade (J. Liu and Ashton 1999; Axtell et al. 2002; Parker et al. 2003; Deadman et al. 2004; Evans and Kelley 2004, 2008; An et al. 2005; Monticino et al. 2007; Werner and McNamara 2007; Entwisle et al. 2008; An 2012; Chen et al. 2012), generalizing findings from CHANS research remains a continuing challenge. Previous synthesis of CHANS studies prioritizes effort to integrate site-specific case studies to reach broader, generalizable conclusions (Turner et al. 2003; J. Liu et al. 2007; Acevedo et al. 2008). Given the degree of site-specific detail that is often included in ABMs, and variations in the design and structure of ABMs of different sites, it can be difficult to compare ABM-related findings from case studies. To address this challenge, we present here a comparison of two CHANS ABMs: one in the Wolong Nature Reserve, China, and the other in the Chitwan Valley, Nepal (Figure 13.1). Our goal is to highlight the similarities and differences between these two models, to focus on generalizations of modeling approaches, and to raise several issues of importance for CHANS agent-based modeling. We begin by introducing the two CHANS.

The Wolong Nature Reserve, established in 1975, is one of China's flagship nature reserves for the endangered giant panda (*Ailuropoda melanoleuca*; Viña et al. 2008; Tuanmu et al. 2010). The reserve is internationally recognized as part of a global biodiversity hotspot (Myers et al. 2000; J. Liu, Linderman et al. 2001; J. Liu, Daily et al. 2003) and, until the Wenchuan earthquake in 2008, was a major tourist destination in China (He et al. 2008; W. Liu et al. 2012). Wolong is also home to more than 5,000 local villagers living within approximately 1,120 households (as of 2005; Wenchuan County 2006). These villagers live a subsistence lifestyle and collect fuelwood from forests, an activity that directly affects panda habitat (Chen et al. 2009). They might reduce fuelwood consumption if a subsidy is provided for more use of electricity (the only substitutable energy source) or if available forests become more distant from their households (An et al. 2002). Other legal activities, such as farming and husbandry, exert less direct impact on panda habitat (Viña et al. 2007). Wolong features a rugged and drastically varying terrain with elevation ranging from approximately 1,000 m to over 6,200 m, giving rise to drastic changes in vegetation and land-cover type (Liu, Ouyang, Tan, et al. 1999).

Our second case study site, the Chitwan Valley, is located in south-central Nepal along the Nepal–India border. The area, formerly densely forested, was partially deforested in the 1950s to make way for settlement and agricultural land use, and eradication of malaria in the area contributed to a rapid increase in population (Barber et al. 1997). We focus our study on the

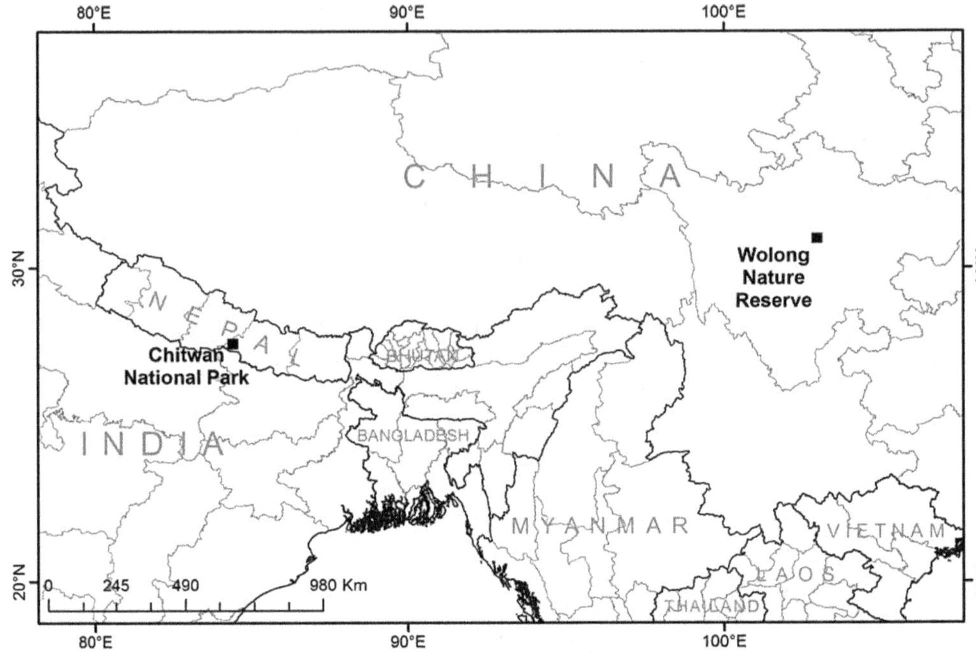

Figure 13.1 The location of our study sites: Wolong Nature Reserve in China and Chitwan Valley in Nepal

western part of the Chitwan Valley, which at the time of the last available census in 2011 supported a population of 284,939 people (Central Bureau of Statistics (CBS) 2012). The valley, part of the lowland Terai landscape at the foothills of the Himalayas, is generally flat, with a mean elevation of about 320 m. The western part of the valley we focus on here is bordered by rivers to the west and north and the Chitwan National Park and Barandabar forest (a protected buffer zone forest) to the south and east, respectively. The 932-km^2 national park (established in 1973) and its 750-km^2 buffer zone (established in 1996) provide habitat for endangered species including the Bengal tiger (*Panthera tigris tigris*) and one-horned rhinoceros (*Rhinoceros unicornis*; Carter, Riley, and Liu 2012).

Comparison Approach

The complexity of ABMs can complicate textual description. Making model code freely available is a step toward transparency, but code is often only understandable by experts (Parker et al. 2003; Grimm et al. 2005). Some CHANS researchers are modeling experts interested in duplicating model results (which is certainly important), but generalization of CHANS findings requires making model structure and results readable by a nonspecialist audience. Thankfully, there has been recent progress in development of communication methods for ecological models (Schmolke et al. 2010) and for ABMs (individual-based models, or IBMs in the ecological literature) specifically (J. Liu and Ashton 1999; Grimm et al. 2006). The overview, design concepts, and details (ODD) framework developed by Grimm et al. (2006; Grimm et al. 2010) and Grimm and Railsback (2012) is a framework for describing ABMs in a standardized format. The ODD framework has been successfully used to describe ABMs in ecology and other disciplines (Grimm et al. 2010) and has also been adopted by many models in the model library maintained by the Open ABM Consortium (Janssen et al. 2008).

Using the ODD protocol, we compare the Wolong and Chitwan ABMs and present the differences and similarities in their model structure. Low-level programming details, including software or platform-specific parameters, are skipped for brevity. The two models are described elsewhere with additional details for a technical audience (An et al. 2005; Zvoleff and An forthcoming).

As ODD is primarily designed to document ABMs rather than to discuss model reliability, we add two additional components to our description—model verification and validation, and CHANS characteristic features (under "Model Comparison"). Verification and validation are essential steps before CHANS ABMs can be put into use for purposes such as experimentation, prediction, and scenario analysis (Schmolke et al. 2010). Although debate continues regarding whether models for complex systems like CHANS can be "verified" or "validated" (Oreskes, Shrader-Frechette, and Belitz 1994), we use the term *verification and validation* in the sense that models are tested in relation to their objectives, intended use, and application domain (Overton 1981; Rykiel 1996). Our working definition for verification is to check for proper functioning of the program, and for validation is to investigate the correspondence between the software model and the conceptual model (structural validation), and that between model outcomes and empirical data (empirical validation; see Manson 2001; An et al. 2005). By adding model verification and validation and CHANS characteristic features as ODD elements, we extend ODD to ABMs of complex CHANS, enriching ODD to apply to a broader audience of CHANS researchers.

Model Comparison

Following the ODD protocol, we summarize the comparison results (Table 13.1). Guided by this framework, we compare and contrast the similarities and differences between the Wolong ABM and Chitwan ABM, and explain the rationale for alternative model design decisions. Our goal is to shed light on the lessons that we can learn for modeling CHANS using ABMs.

Purpose

The Wolong ABM aims to integrate socioeconomic, demographic, and biophysical processes operating at different spatial, temporal, and organizational scales into a systems model and to understand and envision how the habitat of the giant panda might evolve in response to changes in the preceding processes (An et al. 2005). The Chitwan ABM was constructed with similar aims of data integration across spatial and temporal scales. As prior work at the Chitwan study site has uncovered reciprocal connections between environment and human processes, the Chitwan ABM additionally focuses heavily on representing these connections.

Entities, State Variables, and Scales

Next we present a list of agents, their state variables and actions, and the similarities and differences in the two ABMs. We then compare the representations of the environment in the Wolong and Chitwan ABMs.

Structure of Agents. A real human society can include a number of hierarchical levels, such as individuals, households, and communities of different types and scales, regional institutions, and

Table 13.1 Comparison of the Wolong and Chitwan ABM models: summary

		Wolong ABM	*Chitwan ABM*
Overview	Purpose	Integrate multiscale and multidisciplinary data to envision panda habitat over time and space	Integrate multiscale and multidisciplinary data and feedback between demographic and environmental change
	Entities, state variables, and scales	Persons, households, and their major demographic attributes	Similar to Wolong, except that communities are added
	Process overview and scheduling	Initiation of agents and environment; forest growth and fuelwood collection; demographic submodels	Initiation of agents and environment; demographic submodels, land change, and fuelwood submodels
Design concepts	Basic principles	Agent's self- and environmental awareness; maximizing economic utility; minimizing energy costs	Same as Wolong
	Emergence	Habitat and population indexes vary unpredictably with demographic and socioeconomic conditions	Land use change, household size, and fuelwood consumption vary unpredictably with changing demographics
	Adaptation	Change fuelwood search radius and reduce fuelwood demand when forest becomes farther	Change marriage or first birth timing as land use changes; change consumption as household size changes
	Objectives	Minimize the cost of wood collection; maximize economic utility by switching to electricity	Heuristics used to model decision process (e.g., outmigration follows empirically derived models)
	Learning	Agents "remember" the fuelwood location and distance from their household; change fuelwood search radius when forest becomes farther	Learning not explicitly modeled in the Chitwan ABM
	Prediction	Households calculate the distance to the nearest fuelwood collection site	Prediction is not modeled in the Chitwan ABM

Table 13.1 Continued

		Wolong ABM	*Chitwan ABM*
	Sensing	Agents' awareness of demographic and socioeconomic characteristics of themselves and others	Same as Wolong
	Interaction	Marry; household formation (reducing vegetation), fuelwood collection	Marry; household formation (reducing vegetation)
	Stochasticity	Many (e.g., death submodel)	Many (e.g., death submodel)
	Collectives	Household agents are one type of imposed collective	Household and neighborhood agents are two types of imposed collective
	Observation	Personal, household, and population attributes and events; panda habitat	Personal, household, neighborhood, and population attributes and events; land use, fuelwood consumption
Details	Initiation	Create person and household agents and landscape objects using empirical data	Same; neighborhood agents
	Input data	Not used	Not used
	Submodels[a]	Demographic, socioeconomic, biophysical, and human–environment submodels	Same as Wolong

Notes: Comparison is made based on the overview, design concepts, and details protocol (Grimm et al. 2006; Grimm et al. 2010). ABM = agent-based model.
a Refer to other tables for details.

Table 13.2 Entities in the Wolong and Chitwan ABMs

	Wolong ABM		*Chitwan ABM*	
Entity	*Present (n)*	*Purpose*	*Present (n)*	*Purpose*
Person	Yes (4,314)	Represents an individual person and his or her attributes	Yes (8,245)	Same
Household	Yes (893)	Represents an individual household and its attributes	Yes (1,522)	Same
Neighborhood	No		Yes (151)	Represents the impact of neighborhood context on decision making
World	Yes (1)	The model world represents the biophysical environment and exogenous socioeconomic factors	Yes (1)	Same

Notes: Entities are arranged starting with the lowest level of the hierarchy at the top of the table. ABM = agent-based model.

national institutions. Both ABMs follow this pattern, although slight differences exist in how each model creates and manages agents: The Wolong ABM has a hierarchy of person, household, and environment agents (Table 13.2). This structure is mirrored in the Chitwan ABM with the addition of community-level agents between household and environment. Community agents in the Chitwan ABM represent a cluster of (usually10–20) households that live in close proximity and share common community context—defined as similar access to markets, employment opportunities, schools, bus stops, and health centers. The inclusion of this additional level in the Chitwan ABM is due to the abundant research in Chitwan about how community context might affect demographic and land use decisions (e.g., Axinn et al. 2007; Ghimire and Axinn 2010; Axinn and Ghimire 2011).

The agents in the Wolong and Chitwan ABMs have a set of state variables that vary depending on the type of agent. The state variables associated with each agent type also vary slightly between the two models. We discuss only the major differences between the two models here—for a complete overview of each type of agent and the associated state variables in the two models, see our online supplement (http://complexity.sdsu.edu/CHANS-ABMs).

Person agents in both models maintain a set of state variables tracking interrelationships among each other (person agents have unique person ID variables) as well as their personal life history events, preferences, and parents' characteristics. The key differences between the Wolong and Chitwan models are in (1) fertility: the Wolong ABM includes an allowed number of children parameter to reflect governmental family planning policies in China; (2) parental characteristics: the Chitwan ABM associates with each person agent information on that person's parents' employment activities; (3) ethnicity: the Chitwan ABM tracks person agent ethnicity, which explains a portion of the variability in marriage timing, fertility, fuelwood consumption, and migration activities in Chitwan. The Wolong ABM only has a parameter *not leave parental-home intention* that is derived from An, Mertig, and Liu (2003). It represents the tendency that married young people, especially those with special sibling situations (e.g., birth order among siblings), are more likely to leave their parental home and set up their own separate home. Due to lack of data, this parameter is set at an empirically derived constant without being linked to other socioeconomic, demographic, and environmental variables.

Household agents are composed of person agents (households could be considered imposed "collectives" in ODD terminology) but also possess their own unique, higher level state variables. In the Wolong ABM, households track their location (*x*, *y* coordinates) and electricity quality (price, voltage, and outage levels). In the Chitwan ABM, we track these variables at the community level. We model electricity quality variables at the household level in Wolong because they were the major concerns in local people's decision to switch from using fuelwood to using electricity (An et al. 2002). In Chitwan, household agents additionally track land ownership (used in the fertility submodels), time of last migration (used in the migration and fuelwood usage submodels), and use of nonwood fuel sources (for the fuelwood usage submodel).

In Chitwan, another type of collective agent represents community context: the community agent. Community agents in the Chitwan ABM track community location (*x*, *y* coordinates of polygon vertices in the neighborhood boundary), land use, and community context variables. Community agents contain a set of lower level household agents but also possess their own, community-level state variables. Community context is measured in the Chitwan ABM by a series of variables tracking the distance in minutes on foot to a number of key community services and organizations: markets, employers, bus stops, health centers, schools, and the major urban area in the valley. These metrics are consistent with past work in Chitwan (e.g., Ghimire and Axinn 2010; Axinn and Ghimire 2011) that has documented the influence of changing community context on individual-level decision making.

Representation of Time and Environment. We describe the definition of environment and time in our models in two domains: (1) resolution, or the smallest (spatial or temporal) unit over which the phenomenon of interest is represented, and (2) extent, or geographic scope or time span over which a process operates or is measured. We use a yearly time resolution in the Wolong ABM as most of the data collection (e.g., socioeconomic data in statistical yearbooks), as well as processes or activities (e.g., harvest of crops, collection of fuel wood), occur on an annual, occasionally multiple-year, basis in Wolong. In the Chitwan model, we choose a monthly time step to match the primary empirical data sources behind the model (see "Initialization"), and to match the time scale of demographic processes such as migration. The time span for both models is set at twenty to fifty years depending on the outcome under study. This moderate time span allows considerable change in the modeled systems (e.g., vegetation regrowth) but does not span so long as to make the assumptions underlying the models untenable.

A key difference between the two models is in their representations of the physical environment, which represents the space where the corresponding agents reside or occur, and in many instances make the associated decisions. For instance, a person agent in Chitwan might look into the environment (e.g., surrounding land use) before making marriage timing and first birth timing decisions, and a household agent in Wolong might examine the physical distance between his or her household and the nearest forest before deciding how much fuelwood to collect (sections 1 and 2, online supplement). The Wolong ABM represents the physical environment in a rectangle of 3,402 km^2 that completely covers the reserve boundaries in tangent because our survey and observations show that the impacts of human activities, primarily fuelwood collection, often do not go beyond the reserve boundaries (J. Liu, Ouyang, Taylor, et al. 1999; An et al. 2005; Bearer et al. 2008; He et al. 2009; Hull et al. 2011). This rectangle consists of a lattice of 696 rows by 602 columns if the spatial resolution (pixel size) is chosen to be 90 m or optionally a lattice of 175 rows by 151 columns if the spatial resolution is chosen to be 360 m (An and Liu 2010). Our spatial resolutions are chosen partly due to the 30-m resolution of our major remote sensing data source (Landsat, from which we resample to the preceding resolutions) and partly due to our concerns regarding simulation speed. Too fine a spatial resolution would exponentially slow down simulation, which was a great concern at the time of developing the Wolong ABM, when technological advances (e.g., faster computers, parallel computing) were more limited. At the same time, we believe that the fine resolution of 90 m should suffice in capturing the major influence of fuel-wood collection on panda habitat, and we use the coarse resolution of 360 m when higher level socioeconomic or demographic outcomes are of major interest.

The Chitwan ABM, in contrast, does not have a spatially contiguous landscape like the Wolong ABM. Instead, its physical environment is composed of 151 spatially disconnected communities that range in area from 350 to 300,000 m^2 with a mean of 75,000 m^2. These spatially disconnected communities are located within a region of 493 km^2, representing the full extent of the area north of the Chitwan National Park and west of the Barandabar buffer zone forest in Chitwan (Carter et al. 2013; Zvoleff and An forthcoming). The area of each community is calculated from a ground survey using tapes and compasses. Community-level state variables, including numerical values of the area of each land-use class (agricultural vegetation, nonagricultural vegetation, private buildings, public buildings, and other), are tracked within each community. We made this choice for the Chitwan ABM because the submodels (see "Submodels"; for details, see the online supplement) within the Chitwan ABM do not depend on the spatial distribution of land use within a community and because data limitations prevent the assignment of land parcels to particular households. Similar to the Wolong ABM, the Chitwan ABM also includes a 90-m spatial resolution grid of 319 columns and 189 rows (488 km^2) containing elevation data from a Shuttle Radar Topography-Mission (SRTM) digital elevation model. This data could support future work considering natural hazards.

Process Overview and Scheduling

Both ABMs, at the initialization step, set up the landscape and create person, household, and community (Chitwan only) agents. Each model reads in input data associated with each agent, and each agent is located on the landscape. The Wolong and Chitwan ABMs run with yearly and monthly time steps, respectively. We leave the explanation of these submodels for later but highlight the relationships between these processes (submodels) as well as differences and similarities between the two ABMs here (Figure 13.2). We can see that the Wolong ABM leans more toward forest dynamics and fuelwood collection, whereas the Chitwan ABM focuses more on land use that is primarily affected by new household formation. Also, the major demographic processes (i.e., migration, marriage, fertility, mortality) in Chitwan, unlike those in Wolong, are endogenous and thus receive impacts from the environment.

In both ABMs, a particular person agent experiences the demographic submodels within his or her lifetime, not necessarily in the order of submodel placement in the code but in accordance with a set of sociodemographic constraints such as the age and marital status of that agent. Consequently, a person agent goes through the submodels in an order that mirrors the sequence of events in his or her personal life course. The mortality and the nondemographic submodels are exceptions, which we place in an order consistent with empirical expectations (see the sections "Submodels" and "Lessons Learned" for the effects of varying the sequence of submodels). To test the potential impact of scheduling on key variables of interest such as population size, number of households, and panda habitat (Wolong) or population size, number of households, and agricultural land use (Chitwan), we reversed the order of these submodels and ran 30 simulations.

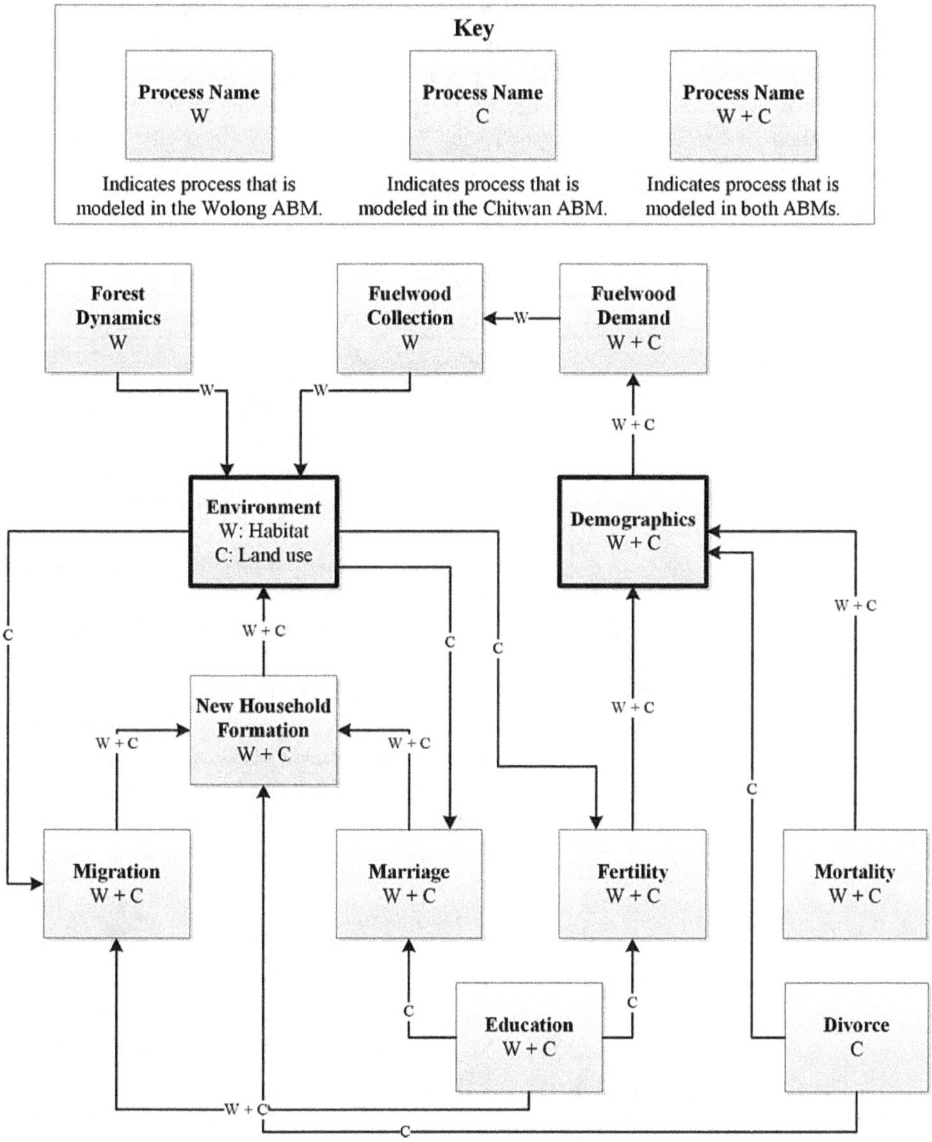

Figure 13.2 Comparison of the major processes between the Wolong ABM and the Chitwan ABM. The rectangles and arrows represent the major processes and links among them, respectively.
Note: W = Wolong; C = Chitwan; ABM = agent-based model.

Design Concepts

In this section, we discuss the similarities and differences in the design concepts guiding construction of the two ABMs, focusing on basic principles, objectives, learning, adaptation and prediction, sensing and interaction, and stochasticity.

Basic Principles. Both ABMs integrate theoretical findings from the geographic, sociodemographic, and ecological literature in a series of submodels linking human and natural systems at each study site. Both models represent agent decision making with varying degrees of stochasticity, implicitly assuming that agents make rational decisions bounded by certain knowledge or information constraints, to maximize their wellbeing. For example, we represent the energy transition in the Wolong and Chitwan models based on our understanding of household's attempts to minimize their energy costs. Our stochastic "tendency to use electricity or fuelwood" submodels (Table 13.3) implicitly represents this economic decision by assigning a probability of fuelwood usage to each household according to a number of covariates.

Table 13.3 Submodels in the Wolong and Chitwan ABMs

Category	*Submodels*	
	Name	*Brief description*
Demographic	Marriage – Marriage probability	The probability is empirically determined (Wolong) or regression-based (Chitwan).
	Marriage – Spouse choice	A qualified agent marries a local or inmigrant with consideration of kinship, age, and ethnicity (Chitwan only).
	Fertility – Birth	A married woman bears a child if several conditions are met.
	Fertility – First birth timing	This timing is empirically determined (Wolong) or regression based (Chitwan).
	Household establishment	It follows a marriage with several constraints.
	Household removal	This happens when household size becomes zero. In Chitwan, nonpermanent migrants are allowed.
	Outmigration – Decision to outmigrate	Individuals or households (Chitwan) might migrate out contingent on age, sex, and other covariates.
	Outmigration – Outmigration length	The nonpermanent migrant (Chitwan) will return when enough time (equal to outmigration length) passes.
	Inmigration	Individuals or households (Chitwan) might migrate in contingent on age, sex, and other covariates.
	Divorce	It is controlled by a fixed monthly probability (Chitwan only).
	Death	Each person may die according to age (and sex in Chitwan).
Socioeconomic	Potential fuelwood demand	This is a linear function of covariates that are different in Wolong and Chitwan.
	Tendency to use electricity or fuelwood	The tendency is modeled as a probability function of covariates (different in Wolong and Chitwan).
Biophysical	Vegetation growth	This measures how pixels of different vegetation grow until reaching age or volume upper bounds (Wolong). In Chitwan, very few collect live wood.
Human–environment interaction	Fuelwood collection	A fuelwood collector chooses a pixel with the shortest cost distance (Wolong).
	Land use change	New building construction takes land from agricultural land (preferentially) or nonagricultural vegetation (Chitwan).

Note: See online supplement for details. ABM = agent-based model.

Both models assume two-way influences between sociodemographic features (or actions) and the local environment. For instance, household-level resource demands depend on a set of socioeconomic, demographic, and geographic variables in both models. Drawing on the richer longitudinal data available in Chitwan (see "Initialization"), the Chitwan ABM allows for more built-in endogeneity between demographic decisions and environmental change than does the Wolong ABM.

Objectives. The objectives of the agents within each ABM influence the construction, validation, and application of each model (Grimm et al. 2006). In the Wolong and Chitwan models, agents are modeled to encounter a number of decision-making points rather than to maximize "success" or achieve a particular objective. At each decision point, agents make decisions in accordance with the observed data, with a set of techniques (empirically derived probability distributions, regression models, heuristic models, etc.) used to model decision-making processes. There are two exceptions: The first is the path finding model included in the Wolong model, in which agents seek to minimize the cost of wood collection (see the online supplement). The second is the decision to switch from using fuelwood to electricity, which implicitly aims to maximize household economic utility (An et al. 2002).

Learning, Adaptation, and Prediction. We group learning, adaptation, and prediction—three separate subcategories in the original ODD (Grimm et al. 2006)—into one category, as they are all related to agent decision making. Learning is a key part of the Wolong fuelwood collection model, in which agents "remember" the location and distance from their household of the last pixel on which they collected fuelwood. Adaptation is interpreted in an ecological sense – that is, as traits or rules of agents in "making decisions or changing behavior in response to changes in themselves or their environment" (Grimm et al. 2010, p. 2764), where traits or rules themselves might or might not change. Therefore, adaptation is represented in both the Wolong and Chitwan models. In the Wolong ABM, as fuelwood in nearby forest areas is depleted, agents adapt by changing their search radius to consider more distant forest patches. As available forests become farther, local people might also reduce their fuelwood demand and adapt by using more electricity. The Wolong ABM also allows a set of hypothetical conditions (related to electricity price, voltage, and outage) as policy controls, and local agents are modeled to predict the future consequences of such policies implicitly, where the policy-induced change in an unspoken household economic return is represented as inflated or lowered probabilities to use electricity in place of fuelwood.

In the Chitwan ABM, adaptation to conditions is reflected in the rules included in the stochastic submodels. For example, the marriage model uses the results of a logistic regression to predict probability of marriage in a given month for a particular person agent dependent on a set of person-level (age, gender, etc.) and community-level (including land use) state variables. As land use and community context change, marriage rates change at a magnitude determined by the coefficient of the corresponding variables in the regression model. We use similar approaches to model how fertility and migration behavior adapts to changing land use in the Chitwan Valley.

Sensing and Interaction. In both the Wolong and Chitwan ABMs, we assume that all person and household agents know their own demographic and socioeconomic characteristics as well as environmental features in their residence and surrounding areas (see "Entities, State Variables, and Scales" for details). This information informs the agent's decisions. For example, in the Wolong ABM, when the fuelwood transportation distance varies, household agents change their probability of using fuelwood.

In both models, one of the primary interactions between agents is marriage. In the marriage submodel, two eligible person agents might get married to each other and build a new house or two married agents might divorce (Chitwan ABM only) with probabilities dependent on the state variables associated with that person, household, and community. Agents also interact with their environment. Although new household formation is the primary land change process in Chitwan, agents in the Wolong model also interact with their environment through fuelwood collection.

Stochasticity. Uncertainty is prevalent in many processes in CHANS. There is uncertainty in determining if, when, and where an event will happen. To reflect this fact, CHANS models often include stochastic processes, or processes with a certain degree of randomness. There are many stochastic processes in both ABMs. One example is the mortality submodel. In both models, to decide whether a person agent may die in a given time step, the model creates a random number between zero and one and compares it with the death rate of people in the corresponding age group (the Chitwan ABM is also gender differentiated). The person dies if the random number is smaller than the rate; otherwise, he or she survives to the next time step.

Initialization

The Wolong model is initialized with 4,314 person agents and 893 household agents. The Chitwan model is initialized with 8,242 person agents, 1,522 household agents, and 151 community agents. The primary data sources for the Wolong ABM are the 1996 Wolong Agricultural census (Wolong Administration 1996), the 2000 population census (Wolong Administration 2000), and in-person surveys of 220 households (conducted in 1999; An et al. 2001). The Chitwan ABM is parameterized primarily using data sets from the Chitwan Valley Family Study (CVFS; Axinn et al. 2007), a 15-year multilevel panel study launched in 1995–1996. At the 1995–1996 baseline, a retrospective 50-year neighborhood history calendar was collected for each sample neighborhood in the study, and a matched retrospective life history calendar was constructed for each individual respondent (Axinn, Barber, and Ghimire 1997; Axinn, Pearce, and Ghimire 1999). In January 1997, the CVFS launched a demographic event registry for all households and individuals in the baseline to track demographic events (births, deaths, marriages, and migrations) in the 151 sample neighborhoods in the western Chitwan Valley. The CVFS also produced detailed maps of land use and land cover, and household agriculture and consumption measures in all study neighborhoods in 1996, 2001, and 2006. Additional fieldwork and household interviews were conducted by the authors in 2009 (80 households) and in 2011 (297 households) to further support development of the Chitwan ABM.

The two ABMs differ in the way they select agents to be included in the model. The Wolong ABM is initialized with every resident in the Wolong study area in 1996 represented as a person agent in the model. The Chitwan ABM, however, simulates only a sample of the total population of the western Chitwan study area. The sample used in the Chitwan ABM is taken from the respondents of the CVFS (Axinn et al. 2007). The CVFS sample includes 1,522 out of the 30,838 households in Chitwan as of the 1991 census (CBS 1991). This sample is distributed among 151 communities spread throughout Chitwan, each with a set of household and person agents. The 151 communities in the model act as a set of "windows" into human–environment interactions within Chitwan (Zvoleff and An forthcoming). This "sample" approach allows examination of spatial and temporal variation in demographic processes without the need to simulate all (more than 200,000) individual agents in the model (see "Lessons Learned" for additional details on initialization of CHANS ABMs).

When assigning initial values to agent state variables and submodel parameters, we make a distinction between model parameterization and model calibration. Calibration is the process of tuning the model parameters so that model output matches what is empirically observed (Oreskes, Shrader-Frechette, and Belitz 1994). Parameterization, on the other hand, refers to the process of empirically determining parameter values based on observed data from the system itself, including surveys, plot data, and household registries. To avoid the problem of tuning the model to fit our expectations, we derive the parameter values used in the Chitwan and Wolong ABMs empirically.

Submodels

The two ABMs are imbalanced in the amount of information included in the different types of submodels. Given the multiple types of interactions modeled in each of these CHANS, we break up the submodels into four key categories: demographic submodels, socioeconomic submodels, biophysical submodels, and human–environment submodels. In general, the Wolong ABM tends toward a higher degree of detail in the biophysical submodels, whereas the Chitwan ABM includes greater detail in the demographic submodels. Due to the key role of human demographic decisions in affecting CHANS dynamics and availability of the relevant data (especially in Chitwan), the majority of our submodels are about demographic processes or decisions (Table 13.3). The other types of submodels in general include less detail and are developed as part of our ABMs according to our site-specific understanding of key CHANS processes affecting our major dependent variable(s) (panda habitat in Wolong and agricultural land use in Chitwan).

The submodels in both ABMs are parameterized drawing on the census and survey data sources (see "Initialization" for details) and a set of standard statistical techniques, including empirically derived probability distributions, ordinary least squares regression, generalized linear models, multilevel modeling, and event history analysis (Zvoleff and An 2014). The brief characteristics of these submodels are summarized in Table 13.3 with details posted online at http://complexity.sdsu.edu/CHANS-ABMs. Next we present an overview and comparison of the submodels in the two ABMs.

In the Wolong ABM, the forest dynamics submodel runs first after the agents and landscape are initiated and mapped (Figure 13.3). This is a process of natural growth of the four vegetation types (derived from satellite imagery) according to empirical data (section 3 of the online supplement). Following this natural process, the fuelwood collection submodel (section 4 of the online supplement) runs based on the fuelwood demand assigned to each household during model initiation. Later all households choose the nearest (in cost-distance) forest patch and cut down trees at an amount determined by the household fuelwood demand submodel. Humans affect the environment (panda habitat in particular) through this major means (Linderman et al. 2005). Next, the Wolong ABM runs the household fuelwood demand submodel, which calculates fuelwood demand based on household socioeconomic (including electricity substitution) and demographic data (section 2 of the online supplement). Then the model runs the education (and outmigration) submodel, where all people between 16 and 20 may go to college or technical school at an empirical probability and thus migrate out of the reserve (section 1.6.1 of the online supplement). Following that, the model runs the mortality (and increment ages) submodel, in which each individual either dies if the random number generated is less than the corresponding age- and sex-based mortality rate or has his or her age incremented by one year. Then the Wolong ABM runs the marriage (and household formation, and outmigration) submodel, where each person, once eligible (or after passing several checks), could marry a local

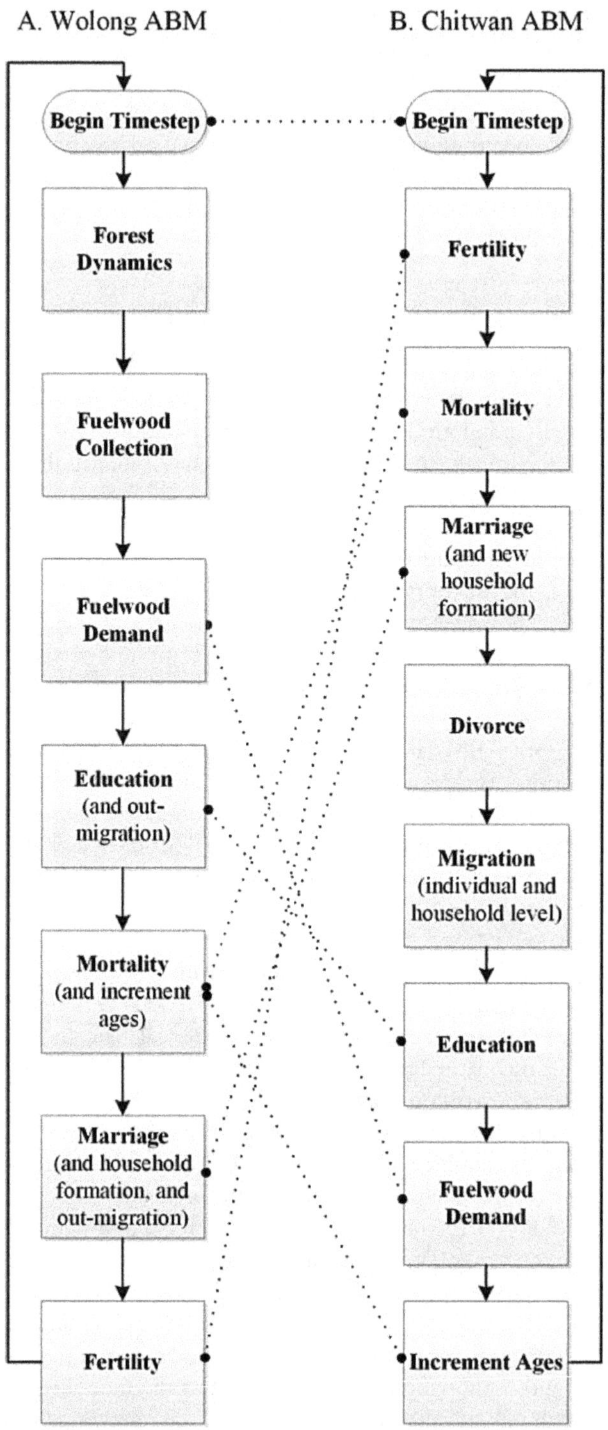

Figure 13.3 Sequence of the major submodels in the Wolong ABM and the Chitwan ABM. The arrows link submodels that are run in a sequential order (i.e., from the earliest to the latest) at each time step. The dotted lines link similar submodels in both ABMs. *Note:* ABM = agent-based model.

or outside person according to different empirically derived probabilities. This submodel, when checking all related information (age, sex, sibling order), also incorporates postmarriage establishment of new households and resource (farmland primarily) allocation. Finally, the fertility submodel is implemented, where all eligible women (married without children or married with fewer children than desired, and provided that enough time has elapsed since marriage or the last live birth) may bear children.

The Chitwan ABM runs a similar series of submodels in sequence with a scheduling order that agrees with our data and insights into the system. First, the fertility submodel runs, which handles women's first childbearing after marriage and then handles subsequent births to women who have already had their first child, are within the allowed age range to give birth (age 15 to 45), and have fewer than their desired total number of children. The mortality submodel runs next, in which each person agent is subject to a small probability of dying within the time step, dependent on the agent's age and sex. The marriage submodel follows, in which the results of an empirical model are used to calculate the probability of each eligible agent (unmarried person agents older than age fifteen) marrying within the time step (Yabiku 2006). Married couples move out of the husband's parental home with a fixed probability, the *household fission rate*, which is determined empirically. If they move out, they establish a new household on agricultural land (an empirical observation), converting the land to private infrastructure. We draw the parcel size for the new household from an empirical probability distribution. Following the marriage submodel is the divorce submodel, in which each married individual is subject to a small probability of divorce within the time step. Next, the migration submodel allows in- and outmigration at the individual and household levels. We calculate the probability of individual outmigration outside of the western Chitwan Valley study area for each eligible person agent (those older than 15 years old), again based on an empirical model including individual-, household-, and neighborhood-level covariates (Massey, Axinn, and Ghimire 2010). A separate portion of the migration submodel determines the probability a migration will be permanent or, if not, the length of time an individual will be absent from the valley. Household-level in- and outmigration is determined with simpler models (given the limitations of our empirical data). For household-level outmigration, we assign a single probability of outmigration to all households in the model. For household-level inmigration, we assign a histogram of the number of households that might inmigrate in a given time step. The number of inmigrating households is chosen from this histogram. The education submodel uses empirical data from the CVFS to model the final years of schooling each individual will achieve, based on sex, ethnicity, and community characteristics. Finally, in the fuelwood demand submodel, the probability of a household using fuelwood is modeled with an empirically derived logistic regression model, and the total fuelwood demand is determined based on a linear model taking into account household size, ethnicity, stove type, and household gender composition.

As already seen, the scheduling order in the Wolong and Chitwan ABMs differs. To test the effects of scheduling on model outcomes, we reverse the order of the submodels (Figure 13.3). For the Wolong ABM, we grouped the submodels into three categories: (1) simulating environment (SE), which includes the forest dynamics submodel; (2) simulating human–environment interaction (SHEI), which includes the fuelwood collection submodel; and (3) simulating sociodemographics (SSD), which includes, given the household-level fuelwood demand determined, education, mortality, marriage, and fertility submodels (in this order) that run at the individual level. To compare with the original order of SE → SHEI → SSD in the Wolong ABM, we reversed the order to be SSD → SHEI → SE and simulated population size, number of households, and area of panda habitat over 50 years. The *t* test (two-tailed assuming unequal variances) results show that at year 50, the population size and number of households do not change significantly (p = 0.33 and 0.98, respectively). The area of panda habitat at year 50 has experienced changes that are small in magnitude (about 0.33 percent) – that is, a decrease from M = 280.91 (SD = 1.04) in the original order (SE → SHEI → SSD) to M = 279.98 (SD = 1.31 km^2) in the reversed order (SSD → SHEI → SE). The change is statistically significant, however, with a p value of 0.0035 (two-tailed t test assuming unequal variances).

Because of the abundance in sociodemographic processes in the Wolong ABM, we specifically reverse the order of submodels in the SSD category to be fertility → marriage → education → mortality and then calculate the fuelwood demand at the household level. This reversal causes a significant change in population size (p = 0.0022, for two-tailed t test assuming unequal variances), even though the magnitude of change is still small (1.02 percent). The number of households and habitat area do not change significantly (p = 0.27 and 0.88, respectively).

To test the effect of scheduling order in the Chitwan model, we performed a similar experiment, again finding that scheduling order can lead to statistically significant differences in model outcomes. Moving the education and migration submodels to occur in sequence at the beginning of the time step, and moving the fertility model to occur after the mortality submodel, in the year 2050, we see an increase of 3.83 percent in the number of households (relative to the original scheduling order), and a decline of 0.24 percent in the total population ($p < 0.001$ and $p = 0.53$, respectively, again using two-tailed t tests assuming unequal variances). Comparing the two scheduling orders, we see the largest change in land use (expected given the change in number of households), with a decline in agricultural land of 5.35 percent when we reorder the submodels ($p < 0.001$).

Model Verification and Validation

The Wolong ABM established a protocol to verify and validate complex ABMs, including (1) progressive model building and debugging, (2) uncertainty testing (extreme tests and extreme combination tests), (3) empirical validation, (4) sensitivity analysis, and (5) experience or expert opinion (An et al. 2005). The key state variables, including panda habitat amount, human population size and composition, the number of households, and household size, pass all the above tests (An et al. 2005).

Verification and validation of the Chitwan ABM follows a protocol similar to that described earlier. To ensure the model code in the Chitwan ABM functions as expected (and that there are no software bugs), we also include in the model code simplified alternative versions of each major submodel, which we can turn on or off for testing purposes. Model outcomes from runs using the simplified instead of the more complex submodels should not be identical, but we would also not expect them to diverge radically. Special verification code is also hard-coded into the model to verify that each submodel functions as expected and that all agents have reasonable values for their state variables at all times. For example, we track the age of the oldest person in the Chitwan ABM, and of the mean age of the population, to ensure the mortality model is functioning as expected.

CHANS Characteristic Features

CHANS have been noted to have several characteristic features that can impact system structure and function: reciprocal effects and feedback loops, non-linearity and thresholds, surprises, legacy effects and time lags, resilience, and heterogeneity (J. Liu et al. 2007). These features are often observed in the outcomes of CHANS ABMs, and we summarize them in Table 13.4 (see section 5 of the online supplement for details).

Table 13.4 CHANS characteristic features

	Wolong ABM examples	*Chitwan ABM examples*
Reciprocal effects and feedback loops	Intense use of fuelwood, thus distancing nearest forest providing such fuelwood, would feed back into a decreased fuelwood demand	Women in places with more agricultural land get married earlier, bear children sooner, establish households faster, and convert more agricultural land to other land uses
Nonlinearity and thresholds	Habitat is unresponsive when the perceived fuelwood collection is beyond 5,000 m (5,000+ m habitat unresponsiveness)	Per-person fuelwood consumption is nonlinearly dependent on household size
Surprises	Number of households remain unchanged for 16 years when marriages are delayed 16 years; see above 5,000+ m habitat unresponsiveness	Fuelwood is tardy in response to population increase
Legacy effects and time lags	Population size, number of households, and habitat area respond to changes in family planning factors with increasing lags	Fuelwood usage lags population increase due to slower increase in household size, decline in fertility, and increase in marriage age as younger population ages
Resilience	Panda habitat would respond very little with increasing fertility	Land use change is resilient to moderate changes in fertility or migration rates
Heterogeneity	All the state variables	The same

One prominent issue that comes from our comparison is related to surprises. We bring attention to surprises that emerge from the unique characteristics of humans, the environment, and the ways in which humans and the environment interact. Detecting and explaining surprises of this kind, often difficult (if not impossible) by looking at the data alone, often involves some type of modeling, systems integration, or both (agent-based modeling is an excellent tool). Examples include the habitat unresponsiveness within 5,000 m of perceived fuelwood collection distance (Wolong), where less motivation to reduce fuelwood, greater environmental heterogeneity (thus allowing for more disturbance like fuelwood collection), and other reasons might help explain these surprises (see the online supplement for more details). On the other hand, there are surprises that are relatively easy to detect and understand by looking at the available data or weaving together different kinds of information. Examples of this kind include the "16-year dormancy" in Wolong (the number of households remains unchanged for 16 years when marriages are delayed 16 years) and fuelwood's delayed response to population increase in Chitwan (fuelwood demand increases slowly as population size increases, given increased efficiency of resource usage in large households). Surprises of this kind are still informative because they could stimulate in-depth thinking about the model structure, function, and interrelationships among model components and thus help in model verification (especially for novices). Finally, we bring to CHANS researchers' attention errors in data, mistakes in ABM rules, or bugs in model code that might give rise to seemingly "surprising" outcomes but are essentially not surprises that reveal CHANS characteristic features.

Lessons Learned

Our comparison of the Chitwan and Wolong ABMs leads to several lessons to take into account in the design, construction, and analysis of CHANS ABMs. We organize this section around the four lessons we learned: (1) What Should the Agents Be? (2) Can We Reuse CHANS Submodels? (3) Does the Scheduling Order of Submodels Matter? and (4) What Are the Surprises That Deserve More Attention in CHANS Research?

What Should the Agents Be?

Choosing what agents to include in a CHANS ABM, of course, hinges on many related factors, such as objectives of the study, availability of data and understanding of the CHANS of interest, and the modeler's views on the complexity of the system. Making this decision is both a science and an art, as this decision might play a fundamental role in determining the structure of the ABM and in shaping the resultant understanding of the corresponding CHANS. Given the dual nature of any CHANS (humans on the one side and the environment on the other), we recommend first drawing up a hierarchical list of potential agents. For example, this list might include persons, households, lower level communities (e.g., villages), higher level communities (e.g., districts), and environment units, up to the whole landscape. We use this type of agent-based representation in both ABMs. The Wolong ABM has a hierarchical list of persons–households–environment, and the Chitwan has its counterpart as persons–households–communities–environment. Depending on the objectives, CHANS modelers could start from any level in the list and end somewhere later, contingent on including at least one type of human (or community) agent and one type of environment agent such that the dual (human and environment) nature of CHANS is represented. Revolving around this relatively straightforward and self-evident recommendation, a few issues emerge that might deserve more attention.

First, should we choose all individuals or a subset of them in our study site as agents? From our comparison of the Chitwan and Wolong ABMs, we note the flexibility modelers might have in deciding how to set up the initial agents in a CHANS ABM. We initialize the Wolong ABM with the full population of the study site in 1996—what we call the "population" approach. The Chitwan ABM, however, only uses a subset of the people and households that are spatially scattered on the Chitwan landscape in 1996—the "sample" approach. The Chitwan ABM is, to the best of our knowledge, the first usage of this approach, which is appropriate given its goal of exploring reciprocal connections between population and environment with a heavy focus on community context. To visualize population-level outcomes (total population, etc.), we can upscale findings from our sample to the population level simply by weighting according to the sampling scheme of the original CVFS survey (Barber et al. 1997). Furthermore, given the detailed demographic and socioeconomic data that are available in Chitwan through the CVFS project, we are hesitant to create agents whose characteristics are drawn from aggregate distributions (e.g., mean, standard deviation, histogram) or relationships, as we might lose or dilute the interrelationships between the agents and agent state variables in our model. Finally, a practical concern is the huge (compared to Wolong) population size in Chitwan (284,939 people in almost 67,988 households in 2011; CBS 2012).

In parallel to this flexibility, our follow-up questions are as follows: Should we choose all of the landscape (the spatially contiguous landscape) or a subset (usually a number of spatially discontinuous locations) of our study site as the environment and how fine (or coarse) should the environmental agents be? The spatial extent of an ABM traditionally represents all of the landscape, usually in a raster format. This is reflected in the Wolong ABM. Complementary to this approach, CHANS modelers could also build their simulations on a (not necessarily spatially contiguous) subset of the landscape, as we did in Chitwan, as long as this is taken into account when interpreting, interpolating, or extrapolating the simulation results. In principle, CHANS modelers should choose a resolution and extent that are appropriate for the major processes under investigation. The spatial resolution should be fine enough to capture variability of the major processes and patterns of interest, but it is largely up to the modeler to decide what variability needs to be captured.

Similar conclusions apply to the choice of time span and temporal resolution (yearly for Wolong and monthly for Chitwan), where data availability, major processes of interest, and research goals could all play a major role. When choosing the simulation time span, we could choose a span that is long enough that all major processes can take place (e.g., a child grows up, marries, and forms his or her own household) but not so long as to allow model uncertainty to escalate, decreasing model reliability to a low level.

Can We Reuse CHANS Submodels?

In our comparison, we show that from a modeling standpoint, many of the submodels in the Chitwan and Wolong models function similarly, even given their different contexts. Given these similarities, using standardized submodels in the two ABMs, and in CHANS ABMs in general, would have several benefits. The first would be to simplify comparison of ABMs, lowering the technical barriers to CHANS ABM construction. The second would be to make the impact of model structure on model outcomes more clear—as modelers will become familiar with standardized modules. At the same time, when multiple users use and test modules, the modules are more likely to be error free and reliable.

Although many of the modules in the two ABMs are similar, some are highly dependent on site-specific context, either due to substantive differences in processes from site to site or due to the same or similar processes being measured differently between the two sites. Due to our modeling focus in this article, we focus on how different processes can be handled, leaving measurement differences to another article in preparation. Falling in the highly site-specific category are the tendency to use electricity and land use and path-finding submodels in the Wolong ABM and the marriage timing, circular outmigration, and first birth timing submodels in the Chitwan ABM. Despite such site-specific processes, many of the submodels share a fair amount of similarity while having certain site-specific features. Taking migration as an example, the Wolong ABM allows only permanent outmigration, whereas the Chitwan ABM allows both permanent and circular migration, with the migration decision based on a large number of individual-, household-, and neighborhood-level covariates. The more detailed Chitwan ABM migration model is made possible by the more extensive migration histories available at the Chitwan site (a measurement difference between the two sites).

To arrive at a module that is comparable and reusable across sites, we recommend decomposition of each process down to the lowest level at which we are able to build reusable modules. For instance, we might decompose the process of outmigration into two parts: the decision to outmigrate and the outmigration action itself. For the decision to outmigrate submodel, one model might have more factors represented in the decision-making process than another (in our case the Chitwan ABM has more factors represented than the Wolong ABM). For this example, we would suggest building a standardized migration decision submodel based on the more complicated process representation (the Chitwan ABM) that can be reused in a simplified form in other models (the Wolong ABM).

This approach follows from the fact that many of the decision-making submodels in the Wolong and Chitwan ABMs function similarly in the following aspects: (1) calculate probability of agent performing action, (2) draw random number, and (3) if random number is less than the calculated probability of the agent performing the action, then the agent will perform the action. Parts 2 and 3 do not need to be modified across sites (code tracking agent locations and ID numbers need not be site-specific); for Part 1, site-dependent regressions can be used to compute the probability needed. This modeling approach also allows incorporation of results from regression models (e.g., hazard modeling or logistic regression) that researchers from other fields might already be familiar with from their previous work.

To this end, we have composed a set of modules as pseudo-code, easily readable by nonmodeling experts, which might be adaptable for use in other CHANS ABMs. We have built a preliminary library of these modules in Netlogo and Python (see http://complexity.sdsu.edu/CHANS-ABMs), two popular programming languages, and have released them under the GNU General Public License. The goal of this library is to offer a set of modules that is transparent and reusable, and subject to improvement and modification from us and other people in the CHANS modeling community.

Does the Scheduling Order of Submodels Matter?

The potential impact of scheduling order (the order in which model processes are implemented) on model output has long been recognized in the literature (e.g., Axtell 2001; Railsback, Lytinen, and Jackson 2006) but has rarely been quantified and taken into account in existing work. Based on our simulation data, we have noticed that statistically significant differences have arisen in several key dependent variables of interest solely from changing the order of the major submodels in the Wolong and Chitwan ABMs. Although statistically significant, these differences are generally small (often around 1 percent) in magnitude for simulations over a 50-year time span. On one hand, the small magnitudes of these differences might indicate that the large-scale patterns of our key dependent variables (panda habitat in Wolong and agricultural land use in Chitwan) are controlled by the major processes and parameters in the model rather than by the order of these processes. This allows us to have confidence in the reliability and usefulness of our CHANS ABMs.

One the other hand, the scheduling-induced differences are statistically significant, suggesting that scheduling order is contributing to the model outcome in a systematic (nonrandom) manner, and in many instances, this contribution will be escalating over time. This is confirmed by our simulation results, where the percentage difference between the default scheduling order and a reordered schedule lead to differences in total population (relative to the default scheduling order) of 0.51 percent ($p < .001$) at 10 years, rising to 1.02 percent ($p < .001$) after 50 years for the Wolong model. The Chitwan model exhibits similar sensitivity to scheduling order, with a model with a reordered schedule showing declines in total agricultural land (relative to the default scheduling order) of 0.80 percent after 10 years ($p < .001$), and 5.34 percent after 50 years ($p < .001$). The accumulated effects of these scheduling-induced differences could play a key role in CHANS structure and dynamics, especially over long time scales (over 50 years in the case of the Chitwan and Wolong ABMs) or in combination with other complexity factors. For instance, the lost habitat due to a change in scheduling order, although small in amount, might be located in places that break existing corridors for pandas to move from different habitat patches. More interestingly, we posit that scheduling order could be more important for ABMs with a coarse time step, where a single time step can represent a relatively long period in the life of an agent. Our simulation data do not support this proposition, as the Chitwan ABM with a fine (monthly) time resolution has also displayed significant differences arising from scheduling order.

In summary, we recommend that investigation of the importance of scheduling order be included in the design of CHANS ABMs, such as through randomization of process order, unless we are certain that one process should come before or after others. This way we could minimize the influences of scheduling order on model outcomes, making model outcomes more likely to reveal what they are supposed to reveal. If not able or possible to specifically address the impact of scheduling order, we should at least consider it in the future as part of the model verification and validation process through, for example, showing that over the time span of simulation, the consequence of scheduling order is negligible and would not substantively affect conclusions to be made from the corresponding ABM.

What are the Surprises that Deserve More Attention in CHANS Research?

Modeling complex systems such as CHANS using an ABM approach might give rise to many surprising outcomes (J. Liu et al. 2007). Such surprises could reveal essential mechanisms underlying CHANS dynamics that might not be able to be detected using other approaches, providing insightful hints for better policy or management. We recommend paying attention to surprises that arise from unique features and interactions within (and sometimes beyond) the CHANS of interest, however, because these surprises might offer clues and opportunities to obtain unknown CHANS mechanisms and thus deserve more attention. Aside from the example of habitat unresponsiveness beyond 5,000 m in Wolong, we believe that those emerging outcomes from theoretical (e.g., the prisoners' dilemma, the El Farol Bar example; see Axelrod 1984 and Arthur 1994, respectively) and empirical (e.g., the macrolevel land use patterns arising from agent "behavior and heterogeneity in the actors and the landscape"; Brown et al. 2008, 807) ABM experiments belong to this category. The surprises that are relatively easy to detect or understand, although useful in providing understanding of the system under investigation as well as in verifying the ABM to some extent as mentioned earlier, do not provide much "hidden" insight into the CHANS. In addition to the examples presented earlier (16-year dormancy in Wolong and lagged response of fuelwood usage to population change in Chitwan), we see these types of surprises in the literature such as the fishbone (along the two sides of major road networks) style of deforestation in the Amazon (Cabrera et al. 2012).

Equally (if not more) important is to identify various surprises, which could come from errors in input data, bugs in model code, or mistakes in ABM rules. During construction of the Wolong and Chitwan ABMs, surprises of this kind occurred; for example, a dead person agent still goes to college, or a male person agent bears a child. It is relatively easy to find mistakes of this nature if the modeler pays enough attention to model verification (this is one of the reasons we propose adding model verification and validation to the ODD protocol). It is more difficult, however, to identify some "invisible" (but seemingly reasonable) mistakes. For instance, an earlier version of the Wolong ABM had a habitat jump in panda habitat in year one. This surprise was later found to arise from a bug in calculating panda habitat after landscape and agent initialization. There is no cure-all for this kind of problem, although several steps should help; for example, making the code transparent and making it subject to public screening and testing, breaking complex (often long) code or processes into individual simpler submodels and testing them separately, and running sensitivity and uncertainty tests (An et al. 2005).

Conclusion

Because of this chapter's goals, we have focused on several important issues or lessons through comparing two CHANS ABMs. Our comparison of ABMs is a necessary step toward better understanding the interrelationships between real people and real environments because of the large effect of ABM structure on the outcomes of ABM models. Without comparing ABM model structure directly, we cannot appreciate the strengths and limitations of ABM model outcomes.

This type of ABM-focused pursuit, however, by no means depreciates the importance of other non-ABM approaches, especially the so-called top-down equation-based models (Parker et al. 2008). From our work, we can actually see the essential role of different statistical models in providing parameter values or rules for both the Wolong and Chitwan ABMs and in testing model outcomes for statistical significance. Therefore, both agent-based modeling and other top-down approaches are complementary with one another (An et al. 2005; An 2012). Although not a new finding, as with any technique, agent-based modeling has its strengths and weaknesses. Although ABM can capture many characteristic features of CHANS, it is a data-intensive modeling strategy, and there is still a high barrier for novices to enter the field. Additionally, ABMs can be difficult to communicate. A natural question follows: When shall we consider using agent-based modeling in CHANS research?

This is not an easy-to-answer question. We provide some insights from our comparison work here, which does not exhaust all possible situations. First, when feedback (or interaction in a broader sense) between different components is essential to the CHANS processes under investigation, ABM has irreplaceable power and might be worthy of consideration. The Chitwan ABM focuses on many feedback loops between land use and population processes. Second (also related to the first point), when systems integration (e.g., integration of data and models from multiple disciplines and scales) and envisioning of systems dynamics under different input parameters are prevalent goals of the modeler, agent-based modeling has unique strength. The Wolong ABM is an exercise in this regard, which integrates data and models from geography, ecology, sociodemography, and other disciplines. Third, when dealing with human behavior and adaptation to social, societal, and environmental changes is of critical importance, agent-based modeling might be the best choice (An 2012). We have discussed how the Wolong and Chitwan ABMs incorporate adaptation rules based on empirically derived behavioral rules. For instance, empirical studies have shown that a decrease in agricultural land would lead to later marriages and lower fertility rates, and this can be easily programmed in the agent decision rules and implemented in the Chitwan ABM. Aside from these three major situations, we acknowledge other situations in which ABMs might also be applicable, such as a context where high heterogeneity in agent or environmental attributes should not be aggregated.

Although this chapter is based on one published model (An et al. 2005; An and Liu 2010) and one model in review (Zvoleff and An forthcoming), it is by no means simply a replication of these two models. Recent years have witnessed an increasing number of agent-based modeling applications at different sites and in different contexts (for an ABM review, see An 2012). Although these advances are important and necessary for ABM development and CHANS research, it is difficult, if not impossible, to distinguish between commonalities and site specifics of CHANS ABMs by separately considering individual case studies. This situation points to an urgent need for better synthesis of multiple ABM results to enable generalization of findings and advancement of the CHANS theory. This context has inspired us to distill commonalities in CHANS structure and processes, and to reflect such commonalities in ABM methodology. The unique contributions of this article, enumerated next, constitute a significant advance toward this aim.

First, we have shown that different CHANS share many common structures and processes of interest. Comparing the two ABMs developed with substantially different goals and contexts, we have found a large amount of modeling similarities (e.g., see Figure 13.2 and Tables 13.1–4). These similar modeling efforts could arise from similarities in CHANS processes, which might further justify our comparative approach to better understanding CHANS structure and dynamics.

Second, we have proposed and demonstrated the usefulness of a modified ODD protocol in comparing and distilling commonalities from different CHANS ABMs. The ODD protocol provides a relatively straightforward and standard way for modeling complex demographic decisions, environmental processes, and human–environment interactions in CHANS. As a common infrastructure that we aim to bring into CHANS research, the protocol helps CHANS researchers better comprehend, compare, and envision CHANS structure and process in a standard way that minimizes arbitrariness in presenting or documenting ABM components. Due to the complexity in CHANS, however, ABMs of CHANS usually have more parameters and state variables than specialist models that focus on individual components of a CHANS. Our modified CHANS ODD protocol has less detail in most areas than a full ODD description, as shown earlier and summarized in Table 13.1. The primary purpose of this modified CHANS ODD protocol, we argue, should be on description for a wider audience.

Additionally, we have enriched the standard ODD protocol by adding two essential components to enhance the accessibility of CHANS model descriptions and to ensure their applicability to a broad audience. As CHANS ABMs are often used as tools for policy recommendation or analysis, it is important that modelers and users be familiar with the process used to evaluate CHANS ABMs. For this reason, we add a "Model Verification and Validation" section to the ODD protocol. The verification and validation process ensures that models function as expected and helps to give modelers and analysts a measure of the uncertainty in model outcomes. In addition, we add another section to the ODD protocol: "CHANS Characteristic Features." We feel that it is important to provide a space in the standard protocol for CHANS modelers to outline the key complexity features of CHANS model outcomes, as these features can have great importance for policymakers. Although traditional ODD focuses on details of model implementation, evaluating model outcomes (with a focus on complexity features) is essential for policy design and implementation. With these modifications to ODD, we hope to reduce the need for readers to consult separate literatures as they make use of CHANS ABMs.

Last but not least, another contribution of this chapter is technical development that facilitates CHANS-related agent-based modeling, including the online pseudo-code and preliminary library of reusable modules in Netlogo, and our test of the importance of scheduling order in ABM modules. All of these features, not previously explored with the Chitwan and Wolong models, should be particularly useful for ABM novices. Aside from embarking on building and testing CHANS ABMs, we provide insights into interpreting agent-based modeling outcomes in the hope that more attention be directed toward surprising outcomes, as these outcomes might offer clues or opportunities to better understand CHANS structure and mechanisms.

In summary, this chapter addresses the difficulty in documenting and comparing CHANS ABMs with an aim to generalize common features from site-specific case studies in CHANS research. We have proposed a standardized approach to model documentation and comparison based on the modified and expanded ODD protocol, highlighted the commonalities of CHANS

ABMs, and pointed out the need for further work on surprises in CHANS and on several technical issues yet to be addressed by the CHANS modeling community. This chapter began to build CHANS-related pseudo-code and a preliminary library of reusable modules, a pursuit with substantial long-term potential for advancing the ABM methodology. It is our hope that future work, using a similar comparative approach, will synthesize more new and existing CHANS case studies and further development of the theory of CHANS.

Acknowledgments

This chapter originally appeared in *Annals of the Association of American Geographers*, 104, 723–745 (2014). We are indebted to financial support from the National Science Foundation through the Partnership for International Research and Education Program (OISE-0729709) and the Coupled Natural and Human Systems Program (EF-CNH 0709717 and DEB-1212183). We are grateful for the very helpful comments from the anonymous reviewers and the editor Dr. Mei-Po Kwan. We thank San Diego State University, Michigan State University, and the University of Michigan for ongoing support of our research. Thanks also go to Wolong Nature Reserve in China and the Institute for Social and Environmental Research–Nepal for their assistance in our data collection.

References

Acevedo, M. F., J. B. Callicott, M. Monticino, D. Lyons, J. Palomino, J. Rosales, L. Delgado, et al. 2008. Models of natural and human dynamics in forest landscapes: Cross-site and cross-cultural synthesis. *Geoforum* 39 (2): 846–66.

An, L. 2012. Modeling human decisions in coupled human and natural systems: Review of agent-based models. *Ecological Modelling* 229: 25–36.

An, L., and J. Liu. 2010. Long-term effects of family planning and other determinants of fertility on population and environment: Agent-based modeling evidence from Wolong Nature Reserve, China. *Population and Environment* 31 (6): 427–59.

An, L., A. G. Mertig, and J. Liu. 2003. Adolescents leaving parental home: Psychosocial correlates and implications for conservation. *Population & Environment* 24 (5): 415–44.

An, L., M. A. Linderman, J. Qi, A. Shortridge, and J. Liu. 2005. Exploring complexity in a human environment system: An agent-based spatial model for multidisciplinary and multiscale integration. *Annals of the Association of American Geographers* 95 (1): 54–79.

An, L., J. Liu, Z. Ouyang, M. A. Linderman, S. Zhou, and H. Zhang. 2001. Simulating demographic and socioeconomic processes on household level and implications for giant panda habitats. *Ecological Modelling* 140 (1–2): 31–49.

An, L., F. Lupi, J. Liu, M. A. Linderman, and J. Huang. 2002. Modeling the choice to switch from fuelwood to electricity: Implications for giant panda habitat conservation. *Ecological Economics* 42 (3): 445–57.

Arthur, W. B. 1994. Inductive reasoning and bounded rationality. *The American Economic Review* 84 (2): 406–11.

——. 1999. Complexity and the economy. *Science* 284 (5411): 107–9.

Axelrod, R. M. 1984. *The Evolution of Cooperation*. New York: Basic Books.

Axelrod, R. M., and M. D. Cohen. 2000. *Harnessing Complexity: Organizational Implications of a Scientific Frontier*. New York: Free Press.

Axinn, W. G., and D. J. Ghimire. 2011. Social organization, population, and land use. *American Journal of Sociology* 117 (1): 209–58.

Axinn, W. G., J. S. Barber, and D. J. Ghimire. 1997. The neighborhood history calendar: A data collection method designed for dynamic multilevel modeling. *Sociological Methodology* 27 (1): 355–92.

Axinn, W. G., L. D. Pearce, and D. J. Ghimire. 1999. Innovations in life history calendar applications. *Social Science Research* 28 (3): 243–64.

Axinn, W. G., A. Thornton, J. S. Barber, S. A. Murphy, D. J. Ghimire, T. Fricke, et al. 2007. *Chitwan Valley Family Study*. Ann Arbor, MI: University of Michigan, Population Studies Center and Survey Research Center.

Axtell, R. L. 2001. Effects of interaction topology and activation regime in several multi-agent systems. In *Multi-Agent-Based Simulation: Lecture Notes in Computer Science*, Eds. S. Moss and P. Davidsson, 33–48. Berlin: Springer. Available at: http://dx.doi.org/10.1007/3-540–44561–7 3 (last accessed April 26, 2014).

Axtell, R. L., J. M. Epstein, J. S. Dean, G. J. Gumerman, A. C. Swedlund, J. Harburger, S. Chakravarty, R. Hammond, J. Parker, and M. Parker. 2002. Population growth and collapse in a multiagent model of the Kayenta Anasazi in Long House Valley. *Proceedings of the National Academy of Sciences* 99 (90003): 7275–79.

Bankes, S. C. 2002. Tools and techniques for developing policies for complex and uncertain systems. *Proceedings of the National Academy of Sciences of the United States of America* 99 (Suppl. 3): 7263.

Barber, J. S., G. P. Shivakoti, W. G. Axinn, and K. Gajurel. 1997. Sampling strategies for rural settings: A detailed example from Chitwan Valley Family Study, Nepal. *Nepal Population Journal* 6 (5): 193–203.

Bearer, S. L., M. Linderman, J. Huang, L. An, G. He, and J. Liu. 2008. Effects of fuelwood collection and timber harvesting on giant panda habitat use. *Biological Conservation* 141: 385–93.

Bian, L., D. A. Quattrochi, and M. F. Goodchild.1997. Multiscale nature of spatial data in scaling up environmental models. In *Scale in Remote Sensing and GIS*, Eds. D. A. Quattrochi and M. F. Goodchild, 13–26. Boca Raton, FL: CRC.

Brown, D. G., D. T. Robinson, L. An, J. I. Nassauer, M. Zellner, W. Rand, R. Riolo, S. E. Page, B. Low, and Z. Wang. 2008. Exurbia from the bottom-up: Confronting empirical challenges to characterizing a complex system. *Geoforum* 39 (2): 805–18.

Cabrera, A. R., P. Deadman, E. Moran, E. S. Brond ızio, and L. K. Vanwey. 2012. Exploring demographic and lot effects in an ABM/LUCC of agriculture in the Brazilian Amazon. In *Agent-Based Models of Geographical Systems*, Eds. A. J. Heppenstall, A. T. Crooks, L. M. See, and M. Batty, 663–76. Amsterdam: Springer. Available at: http://link.springer.com.proxy.library.ucsb.edu: 2048/chapter/10. 1007/978–90–481–8927–4 33 (last accessed April 9, 2013).

Carter, N. H., S. Riley, and J. Liu. 2012. Utility of a psychological framework for carnivore conservation. *Oryx* 46 (4): 525–35.

Carter, N. H., A. Shortridge, A. Vina, H. Campa, III, J. B. Karki, and J. Liu. 2013. Assessing spatiotemporal changes in tiger habitat across different land management regimes. *Ecosphere* 4 (10): Article 124.

Central Bureau of Statistics (CBS). 1991. *Population Census 1991*. Kathmandu, Nepal: Nepal Central Bureau of Statistics, Government of Nepal National Planning Commission Secretariat.

——. 2012. *National Population and Housing Census 2011*. Kathmandu, Nepal: Central Bureau of Statistics, Government of Nepal, National Planning Commission Secretariat.

Chen, X., F. Lupi, L. An, R. Sheely, A. Viña, and J. Liu. 2012. Agent-based modeling of the effects of social norms on enrollment in payments for ecosystem services. *Ecological Modelling* 229: 16–24.

Chen, X., F. Lupi, G. He, and J. Liu. 2009. Linking social norms to efficient conservation investment in payments for ecosystem services. *Proceedings of the National Academy of Sciences* 106 (28): 11812–17.

Crawford, T. W., J. P. Messina, S. M. Manson, and D. O'Sullivan. 2005. Complexity science, complex systems, and land-use research. *Environment and Planning B: Planning and Design* 32 (6): 792–98.

Deadman, P., D. Robinson, E. F. Moran, and E. Brondizio. 2004. Colonist household decisionmaking and land-use change in the Amazon rainforest: An agent-based simulation. *Environment and Planning B: Planning and Design* 31: 693–709.

Eakin, H., and A. L. Luers. 2006. Assessing the vulnerability of social-environmental systems. *Annual Review of Environment and Resources* 31 (1): 365–94.

Entwisle, B., G. P. Malanson, R. R. Rindfuss, and S. J. Walsh. 2008. An agent-based model of household dynamics and land use change. *Journal of Land Use Science* 3 (1): 73.

Epstein, J. M. 1999. Agent-based computational models and generative social science. *Complexity* 4 (5): 41–60.

Evans, T. P., and H. Kelley. 2004. Multi-scale analysis of a household level agent-based model of landcover change. *Journal of Environmental Management* 72 (1–2): 57–72.

——. 2008. Assessing the transition from deforestation to forest regrowth with an agent-based model of land cover change for south-central Indiana (USA). *Geoforum* 39 (2): 819–32.

Foley, J. A., R. DeFries, G. P. Asner, C. Barford, G. Bonan, S. R. Carpenter, F. S. Chapin, et al. 2005. Global consequences of land use. *Science* 309 (5734): 570–74.

Folke, C., S. R. Carpenter, T. Elmqvist, L. Gunderson, C. S. Holling, and B. Walker. 2002. Resilience and sustainable development: Building adaptive capacity in a world of transformations. *Ambio* 31 (5): 437–40.

Folke, C., S. R. Carpenter, B. Walker, M. Scheffer, T. Elmqvist, L. Gunderson, and C. S. Holling. 2004. Regime shifts, resilience, and biodiversity in ecosystem management. *Annual Review of Ecology, Evolution, and Systematics* 35: 557–81.

Ghimire, D. J., and W. G. Axinn. 2010. Community context, land use, and first birth. *Rural Sociology* 75 (3): 478–513.

Grimm, V., and S. F. Railsback. 2012. Designing, formulating, and communicating agent-based models. In *Agent-Based Models of Geographical Systems*, Eds. A. J. Heppenstall, A. T. Crooks, L. M. See, and M. Batty, 361–77. Amsterdam: Springer. Available at: www.springerlink.com/content/h75n3g267j2l70ql/abstract/ (last accessed April 27, 2012).

Grimm, V., U. Berger, D. L. DeAngelis, J. G. Polhill, J. Giske, and S. F. Railsback. 2010. The ODD protocol: A review and first update. *Ecological Modelling* 221 (23): 2760–68.

Grimm, V., U. Berger, F. Bastiansen, S. Eliassen, V. Ginot, J. Giske, J. Goss-Custard, et al. 2006. A standard protocol for describing individual-based and agent-based models. *Ecological Modelling* 198 (1–2): 115–26.

Grimm, V., E. Revilla, U. Berger, F. Jeltsch, W. M. Mooij, S. F. Railsback, H.-H. Thulke, J. Weiner, T. Wiegand, and D. L. DeAngelis. 2005. Pattern-oriented modeling of agent-based complex systems: Lessons from ecology. *Science* 310 (5750): 987–91.

He, G., X. Chen, S. Bearer, M. Colunga, A. Mertig, L. An, S. Zhou, et al. 2009. Spatial and temporal patterns of fuelwood collection. *Landscape and Urban Planning* 92: 1–9.

He, G., X. Chen, W. Liu, S. Bearer, S. Zhou, L. Y. Cheng, H. Zhang, Z. Ouyang, and J. Liu. 2008. Distribution of economic benefits from ecotourism: A case study of Wolong Nature Reserve for giant pandas in China. *Environmental Management* 42 (6): 1017–25.

Hull, V., W. Xu, W. Liu, S. Zhou, A. Viña, J. Zhang, M. Tuanmu, et al. 2011. Evaluating the efficacy of zoning designations for protected area management. *Biological Conservation* 144: 3028–37.

Irwin, E. G., and J. Geoghegan. 2001. Theory, data, methods: Developing spatially explicit economic models of land use change. *Agriculture, Ecosystems & Environment* 85 (1–3): 7–24.

Janssen, M. A., L. N. Alessa, M. Barton, S. Bergin, and A. Lee. 2008. Towards a community framework for agent-based modelling. *Journal of Artificial Societies and Social Simulation* 11 (2): 6.

Latour, B. 1996. On actor-network theory: A few clarifications. *Soziale welt* 47 (4): 369–81.

Lepczyk, C. A., C. H. Flather, V. C. Radeloff, A. M. Pidgeon, R. B. Hammer, and J. Liu. 2008. Human impacts on regional avian diversity and abundance. *Conservation Biology* 22: 405–16.

Linderman, M. A., L. An, S. Bearer, G. He, Z. Ouyang, and J. Liu. 2005. Modeling the spatio-temporal dynamics and interactions of households, landscapes, and giant panda habitat. *Ecological Modelling* 183 (1): 47–65.

Liu, J. 2010. China's road to sustainability. *Science* 328 (5974): 50.

Liu, J., and P. S. Ashton. 1999. Simulating effects of landscape context and timber harvest on tree species diversity. *Ecological Applications* 9 (1): 186–201.

Liu, J., G. C. Daily, P. R. Ehrlich, and G. W. Luck. 2003. Effects of household dynamics on resource consumption and biodiversity. *Nature* 421: 530–33.

Liu, J., T. Dietz, S. R. Carpenter, M. Alberti, C. Folke, E. F. Moran, A. N. Pell, et al. 2007. Complexity of coupled human and natural systems. *Science* 317 (5844): 1513–16.

Liu, J., M. Linderman, Z. Ouyang, L. An, J. Yang, and H. Zhang. 2001. Ecological degradation in protected areas: The case of Wolong Nature Reserve for giant pandas. *Science* 292: 98–101.

Liu, J., Z. Ouyang, Y. Tan, J. Yang, and H. Zhang. 1999. Changes in human population structure: Implications for biodiversity conservation. *Population and Environment* 21 (1): 45–58.

Liu, J., Z. Ouyang, W. Taylor, R. Groop, Y. Tan, and H. Zhang. 1999. A framework for evaluating effects of human factors on wildlife habitat: The case of the giant pandas. *Conservation Biology* 13 (6): 1360–70.

Liu, W., C. A. Vogt, J. Luo, G. He, K. A. Frank, and J. Liu. 2012. Drivers and socioeconomic impacts of tourism participation in protected areas. *PLoS ONE* 7 (4): e35420.

Malanson, G. P., Y. Zeng, and S. J. Walsh. 2006a. Complexity at advancing ecotones and frontiers. *Environment and Planning A* 38 (4): 619.

——. 2006b. Landscape frontiers, geography frontiers: Lessons to be learned. *The Professional Geographer* 58 (4): 383–96.

Manson, S. M. 2001. Simplifying complexity: A review of complexity theory. *Geoforum* 32 (3): 405–14.

——. 2005. Agent-based modeling and genetic programming for modeling land change in the southern Yucatán Peninsular region of Mexico. *Agriculture, Ecosystems & Environment* 111 (1–4): 47–62.

Massey, D., W. G. Axinn, and D. J. Ghimire. 2010. Environmental change and out-migration: Evidence from Nepal. *Population and Environment.*

Messina, J. P., and S. J. Walsh. 2005. Dynamic spatial simulation modeling of the population-environment matrix in the Ecuadorian Amazon. *Environment and Planning B: Planning and Design* 32 (6): 835–56.

Monticino, M., M. F. Acevedo, J. B. Callicott, T. Cogdill, and C. Lindquist. 2007. Coupled human and natural systems: A multi-agent-based approach. *Environmental Modelling & Software* 22 (5): 656–63.

Myers, N., R. A. Mittermeier, C. G. Mittermeier, G. A. B. da Fonseca, and J. Kent. 2000. Biodiversity hotspots for conservation priorities. *Nature* 403 (6772): 853–58.

Oreskes, N., K. Shrader-Frechette, and K. Belitz.1994. Verification, validation, and confirmation of numerical models in the earth sciences. *Science* 263 (5147): 641–46.

Ostrom, E. 2007. A diagnostic approach for going beyond panaceas. *Proceedings of the National Academy of Sciences* 104 (39): 15181–87.

Overton, S. 1981. A strategy of model construction. In *Ecosystem Modeling in Theory and Practice: An Introduction with Case Histories*, Eds. C. A. S. Hall and J. W. J. Day, 50–73. New York: Wiley.

Parker, D. C., B. Entwisle, R. R. Rindfuss, L. K. VanWey, S. M. Manson, E. F. Moran, L. An, P. Deadman, T. P. Evans, and M. A. Linderman. 2008. Case studies, cross-site comparisons, and the challenge of generalization: Comparing agent-based models of land-use change in frontier regions. *Journal of Land Use Science* 3 (1): 41– 72.

Parker, D. C., S. M. Manson, M. A. Janssen, M. J. Hoffmann, and P. Deadman. 2003. Multi-agent systems for the simulation of land-use and land-cover change: A review. *Annals of the Association of American Geographers* 93 (2): 314–37.

Railsback, S. F., S. L. Lytinen, and S. K. Jackson. 2006. Agent-based simulation platforms: Review and development recommendations. *Simulation* 82 (9): 609–23.

Rindfuss, R. R., B. Entwisle, S. J. Walsh, L. An, N. Badenoch, D. G. Brown, P. Deadman, et al. 2008. Land use change: Complexity and comparisons. *Journal of Land Use Science* 3: 1–10.

Rutledge, D., C. Lepcyzk, J. Xie, and J. Liu. 2001. Spatial and temporal dynamics of endangered species hotspots in the United States. *Conservation Biology* 15: 475–87.

Rykiel, E. J., Jr. 1996. Testing ecological models: The meaning of validation. *Ecological Modelling* 90 (3): 229– 44.

Schmolke, A., P. Thorbek, D. L. DeAngelis, and V. Grimm. 2010. Ecological models supporting environmental decision making: A strategy for the future. *Trends in Ecology & Evolution* 25 (8): 479–86.

Tuanmu, M. N., A. Viña, S. Bearer, W. Xu, Z. Ouyang, H. Zhang, and J. Liu. 2010. Mapping understory vegetation using phonological characteristics derived from remotely sensed data. *Remote Sensing of Environment* 114: 1833–44.

Turner, B. L., P. A. Matson, J. J. McCarthy, R. W. Corell, L. Christensen, N. Eckley, G. K. Hovelsrud-Broda, et al. 2003. Illustrating the coupled human–environment system for vulnerability analysis: Three case studies. *Proceedings of the National Academy of Sciences of the United States of America* 100 (14): 8080–85.

Viña, A., S. Bearer, H. Zhang, Z. Ouyang, and J. Liu. 2008. Evaluating MODIS data for mapping wildlife habitat distribution. *Remote Sensing of Environment* 112: 2160–69.

Viña, A., S. Bearer, X. Chen, G. He, M. Linderman, L. An, H. Zhang, Z. Ouyang, and J. Liu. 2007. Temporal changes in giant panda habitat connectivity across boundaries of Wolong Nature Reserve, China. *Ecological Applications* 17 (4): 1019–30.

Vitousek, P. M., H. A. Mooney, J. Lubchenco, and J. M. Melillo. 1997. Human domination of Earth's ecosystems. *Science* 277 (5325): 494–99.

Von Bertalanffy, L. 1969. *General Systems Theory: Foundations, Development.* New York: Braziller.

Walsh, S. J., J. P. Messina, C. F. Mena, G. P. Malanson, and P. H. Page. 2008. Complexity theory, spatial simulation models, and land use dynamics in the northern Ecuadorian Amazon. *Geoforum* 39 (2): 867–78.

Warren, K., C. Franklin, and C. L. Streeter. 1998. New directions in systems theory: Chaos and complexity. *Social Work* 43 (4): 357–72.

Wenchuan County. 2006. *Statistical Yearbook of Wenchuan County.* Sichuan Province, China: Wenchuan County.

Werner, B. T., and D. E. McNamara. 2007. Dynamics of coupled human–landscape systems. *Geomorphology* 91(3–4): 393–407.

Wolong Administration. 1996. Wolong agricultural census. Wolong Administration (unpublished material).

——. 2000. Nationwide population census data. Wolong Administration (unpublished material).

Yabiku, S. T. 2006. Land use and marriage timing in Nepal. *Population & Environment* 27 (5): 445–61.

Yu, C., A. M. MacEachren, D. J. Peuquet, and B. Yarnal. 2009. Integrating scientific modeling and supporting dynamic hazard management with a GeoAgent-based representation of human–environment interactions: A drought example in central Pennsylvania, USA. *Environmental Modelling & Software* 24 (12): 1501–12.

Zvoleff, A., and L. An. 2014. Analyzing human–landscape interactions: Tools that integrate. *Environmental Management* 53: 94–111. Available at: http://dx.doi.org/10.1007/s00267–012-0009–1 (last accessed April 26, 2014).

——. Forthcoming. Modeling human–environment interactions and their ecological implications. *Ecology and Society*.

14

EUTROPIA

Integrated Valuation of Lake Eutrophication Abatement Decisions Using a Bayesian Belief Network

David N. Barton, Tom Andersen, Olvar Bergland, Alexander Engebretsen, S. Jannicke Moe, Geir I. Orderud, Koji Tominaga, Eirik Romstad, and Rolf D. Vogt

Research Question: How can the value of implementing the European Union Water Framework Directive be evaluated?
System Science Method(s): System Dynamics & Networks
Things to Notice:

- Combination of system dynamics and other models using networks
- (Challenges in) using system science methods to evaluate public policy

The term "integrated valuation" is defined and its relevance is discussed in terms of bridging the gap between cost-effectiveness analysis and economic valuation in the implementation of the European Union Water Framework Directive. We demonstrate how to integrate benefit valuation with the ecosystem services cascade framework using an Object-Oriented Bayesian Network (OOBN). The OOBN is then used to assess the benefits of nutrient abatement measures across a cascade of submodels of the driver-pressure-state-impact-response (DPSIR) chain for the Vanemfjord lake in Morsa catchment in south-eastern Norway. The lake is part of a complex lake system in a semi-urbanized catchment dominated by forest and agriculture. The catchment has highly variable seasonal climatic conditions affecting nutrient run-off and algal blooms. It has been one of the most eutrophic lakes in Norway with periodic cyanobacteria blooms, but continues to attract a large recreational user population, despite the large variations in water quality. The "DPSIR-OOBN" model is used as a case study of "integrated valuation" and evaluated for its applicability for decision support in nutrient abatement. We find that the DPSIR-OOBN model meets seven of the nine criteria we propose for "integrated valuation". The model struggles to meet the criteria that ecological, social and economic values should be defined consistently in relation to impacts on lake quality. While the DPSIR-OOBN integrates from valuation methods across an ecosystem cascade to management alternatives, it is neither a full benefit-cost analysis, nor a multi-criteria analysis. However, we demonstrate how the DPSIR-OOBN can be used to explore issues of consistency in scaling and weighting of different ecological, social and economic values in the catchment system. Bayesian belief networks offer a consistent approach to analysing how management implementation probability may determine economic valuation. We discuss the implication of our integrated valuation not being able to account for farmer responses, in particular

 DOI: 10.4324/9781315748771-16

the incentive effects of the model not being able to predict abatement effectiveness and value. The resolution of the nutrient monitoring data and modeling technologies that were at our disposal are probably better in the Morsa catchment than for any other catchment of this size in Norway. We therefore conclude that using our integrated valuation model for assessing benefits of eutrophication abatement measures as part of the EU Water Framework Directive still lies in the realm of utopia – euphemistically speaking a "eutropia".

Freshwater eutrophication is one of the major environmental challenges around the world. There is a range of known factors that are responsible for water eutrophication, though the increased flux of nutrients from the sources to the water bodies is a key factor. This loading of nutrients occurs both from point sources and non-point sources. Point sources, such as sewage water, were historically the most important sources of nutrients to surface waters. With the advancement of sewage treatment technologies the culprit nutrients in the sewage are removed effectively before discharged into the water bodies. Now scientists and policy makers in most developed countries are turning their attention to the remaining non-point sources, such as from agricultural land (Parry, 1998). The challenge lies in that the mechanisms of mobilization and transport of nutrients from agricultural land are not adequately understood (Tong et al., 2003; Yang et al., 2008). Eutrophication is still a problem in many rivers and lakes in Norway despite the introduction of best management practice and numerous abatement measures in recent years. Environmental monitoring shows that the situation has remained largely unchanged during the last ten years. Norway adhered voluntarily to the EU Water Framework Directive (WFD). It was implemented in 2009 in a number of pilot river basins, with the EU objective of achieving "good ecological status" by 2015. In defining a programme of measures for a river basin, a river basin authority should assess whether costs of achieving "good ecological status" are disproportionate to benefits of measures (EC, 2003, 2009). If costs are disproportionate to benefits, a delayed and/or lower ecological status objectives may be justified for the water body under the WFD rules. While a benefit–cost rule is a relevant approach for evaluating disproportionate costs, there are few operational examples of integrating valuation of benefits with cost-effectiveness assessment of measures (Galioto et al., 2013).

A recent review of economics and ecosystem services analysis in support of the Water Framework Directive Martin-Ortega (2012) identifies five requirements for more sophisticated approaches to dealing with uncertainty:

- to address multiple stressors acting simultaneously;
- technical improvements in the valuation of ecosystem services from water bodies;
- adequate reflection of trade-offs between environmental and social objectives;
- quantification of multiple benefits;
- co-construction of knowledge and practice with stakeholders at multiple levels.

These recommendations for improving economic analysis in the WFD have many commonalities with Gómez-Baggethun et al. (2014) proposal for "integrated valuation", which we discuss in detail below.

The aim of this study is to operationalize "integrated valuation" as a way of bridging the gap between cost-effectiveness analysis and economic valuation of benefits in the implementation of the Water Framework Directive. We illustrate how Bayesian networks (Jensen and Nielsen, 2007; Kjaerulff and Madsen, 2007) offer an operational approach to integrating benefit valuation with the ecosystem services cascade framework (Haines-Young and Potschin, 2010). More specifically we use an object-oriented Bayesian network (OOBN) to show how uncertainty can be analysed consistently across a casual chain of submodels of driver–pressure–state–impact–responses (DPSIR) in a catchment and lake system (Barton et al., 2012; Tscherning et al., 2012). We demonstrate the use of Bayesian network software by assessing eutrophication abatement decisions in the Lake Vanemfjorden within the Morsa watershed in south-eastern Norway. Lake Vanemfjorden has periodically been one of the most eutrophic lakes in Norway (Bechmann and Øgaard, 2013; Bechmann et al., 2007). Notwithstanding its periodically substandard water quality, it continues to attract a large population of recreational users (Barton et al., 2009). An evaluation of the extent to which the DPSIR-OOBN model meets criteria for "integrated valuation", defined tentatively by Gómez-Baggethun et al. (2014), is conducted. We also discuss the limitations of the integrated valuation model from a systems perspective, and how these limitations may define the role the proposed model plays as a mediator in WFD policy implementation (Morrison and Morgan, 1999). This study is an integral part of the transdisciplinary Eutropia project which aims at understanding processes and pressures governing the P-flux into the eutrophic lake Vansjø, as well as thresholds and barriers in society apposing abatement actions (Orderud and Vogt, 2013).

We begin below with a definition of integrated valuation of ecosystem services as a type of systems analysis, after which we discuss the lake and catchment system boundaries. Next we discuss the modeled system boundaries as defined by different submodel domains. We demonstrate how the submodels are linked together in a driver–pressure–state–impact–response object oriented Bayesian network (DPSIR-OOBN), and we evaluate whether the model meets the definition of integrated valuation. We discuss whether the model boundaries could be extended to capture stakeholder responses to regulation, incentives and the model findings themselves. Finally, we draw conclusions on the applicability of the DPSIR-OOBN model for "integrated valuation."

Integrated Valuation of Ecosystem Services as Systems Analysis

Gómez-Baggethun et al. (2014) propose a tentative operational definition of integrated valuation as "the process of synthesizing relevant sources of knowledge and information to elicit the various ways in which people conceptualize and appraise ecosystem service values, resulting in different valuation frames that are the basis for informed deliberation, agreement and decision". They argue that measuring multiple values, which are simply assessed independently to inform environmental decisions, without a consistent and coherent evaluation, is *hybrid* valuation. The distinction between *integrated* and *hybrid* valuation contrasts a systems approach evaluating causal relationships between components of social-ecological systems, with an approach that merely combines components that have been assessed independently.

Gómez-Baggethun et al. (2014) offer four tentative criteria to evaluate whether a method can be defined as a fully *integrated* valuation approach, versus a *hybrid* valuation approach. They start by underlining the importance of specifying the decision context of valuation, although this is not seen as part of the definition of integrated valuation. In the discussion of "ecological value" they point out that value is not merely a biophysical indicator, but needs to be a measure of subjective "importance". It is crucial that this value scaling is consistent. Furthermore, they stress the importance of explicitly addressing conflicting interests and value trade-offs in decision-making as an important feature of integrated valuation. These four tentative criteria are in this study developed into nine criteria for assessing the integrated valuation of the DPSIR-OOBN.

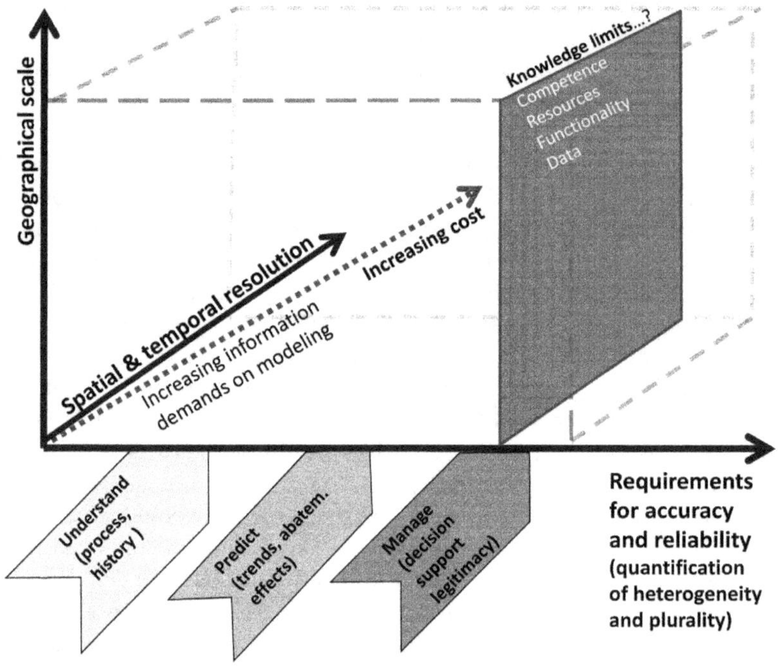

Figure 14.1 Integrated modeling can be carried out to understand or predict ecosystem function. Our definition of integrated valuation requires management decisions. Information costs of describing spatial and temporal heterogeneity of the natural system to be managed, combined with limits to knowledge may call into question the legitimacy of using integrated valuation models to inform policy.

Criterion 1. Management relevance. The ability to discern between decision alternatives and thus provide support for decisions about policy is our first criteria for integrated valuation. Integrated valuation goes beyond understanding or predicting a system, to discerning between alternative courses of management action based on the "importance" of their consequences for people. It faces information costs and knowledge limits. If the feasible accuracy and reliability of integrated valuation models is insufficient to discern between management actions, its legitimacy as a decision-support tool may be called into question (Figure 14.1).

Criterion 2. Value plurality. Integrated valuation should address ecological, social and economic value dimensions held by stakeholders. Integrated valuation thus identifies conflicts of interest across these different value dimensions.

Criterion 3. Value heterogeneity. Values vary across the time and location of decision contexts, and the location and time at which people are asked to express those values. Integrated valuation should attempt to describe systemic features of this heterogeneity by using a consistent modelling approach to describe temporal and spatial heterogeneity – or uncertainty – across submodels of the system.

Criterion 4: Interdisciplinarity. Integrated valuation should be based on contributions from several disciplines, including multiple expert domains from both social and natural sciences. Interdisciplinarity, transdisciplinarity (adding policy-makers and stakeholders), and methodological pluralism are thus key elements in integrated ecosystem services valuation.

Criterion 5: Knowledge systems. Integrated valuation of ecosystem services should be informed by different knowledge systems – i.e. the agents, practices, and institutions that organize the production, transfer and use of knowledge.

Criterion 6: Information types. Integrated ecosystem services valuation should be capable of dealing with both qualitative and quantitative information. Qualitative information includes what is generated in deliberative processes with locally defined metrics, description, public discourse and narration. A key feature is the consistent treatment of uncertainty across different information types.

Criterion 7: Levels of societal organization. Integrated valuation should cover values emerging at different scales of societal organization, from individuals, to communities, to nations. Individuals have different roles in these different contexts, mobilizing different rationalities and value systems (consumer, citizen, tax payer, voter, household representative, community resident, association member, public utility user, survey panel participant and so on).

Criterion 8. Consistent scaling of plural values. Any integrated valuation of importance for specific human interests requires individual scaling of changes in states of nature. Scaling is also an explicit step in multi-attribute utility theory used in multi-criteria decision analysis (MCDA). In terms of the ecosystem service cascade model, scaling is the equivalent of the transformation from ecosystem structure/state to ecosystem service. The identification of ecosystem services requires some form of importance scaling; they are *specific to the action context of a subject.* Value scaling therefore requires knowledge of ecosystem function connecting a decision to a service outcome. In that sense any scaling from an objective measure of a state of nature to a subjective measure of importance involves some form of (mathematical) integration across ecosystem function.

Criterion 9: Consistent comparison of plural values in decisions. Integrated valuation should inform and support decision-making processes on the basis of a consistent weighting of the relative importance of multiple types of value, explicitly addressing trade-offs between, e.g., ecological, cultural and monetary values. In MCDA terminology, this criterion requires explicit weighting of criteria and/or ranking of decision alternatives across different interests (depending on the MCDA method).

Study Area System Boundaries

This section provides a wide but brief description of the study area in terms of "study area system boundaries". The "model system boundaries" of the Bayesian belief network are presented and used to analyze eutrophication abatement measures. Our aim is to encourage the reader to consider how the model boundaries sets a limit to what can be concluded regarding the eutrophication in the study area, and what limitations this also places on the understanding of our model as integrated valuation.

Study Area Extent

The study area is the Vanemfjorden Lake – also called western Vansjø – and its local watershed with a total area of 71.5 km^2 excluding the area of water bodies (Figure 14.2). This is a part of the larger Morsa catchment – also known as Vansjø-Hobøl basin with an area of 688 km^2 – situated near Oslo in the south-eastern part of Norway (59°26′N, 10°41′E). The entire Vansjø Lake covers 36 km^2 and consists of several smaller basins separated from each other by narrow straits and shallow thresholds. The lake is divided into two main parts: one eastern part

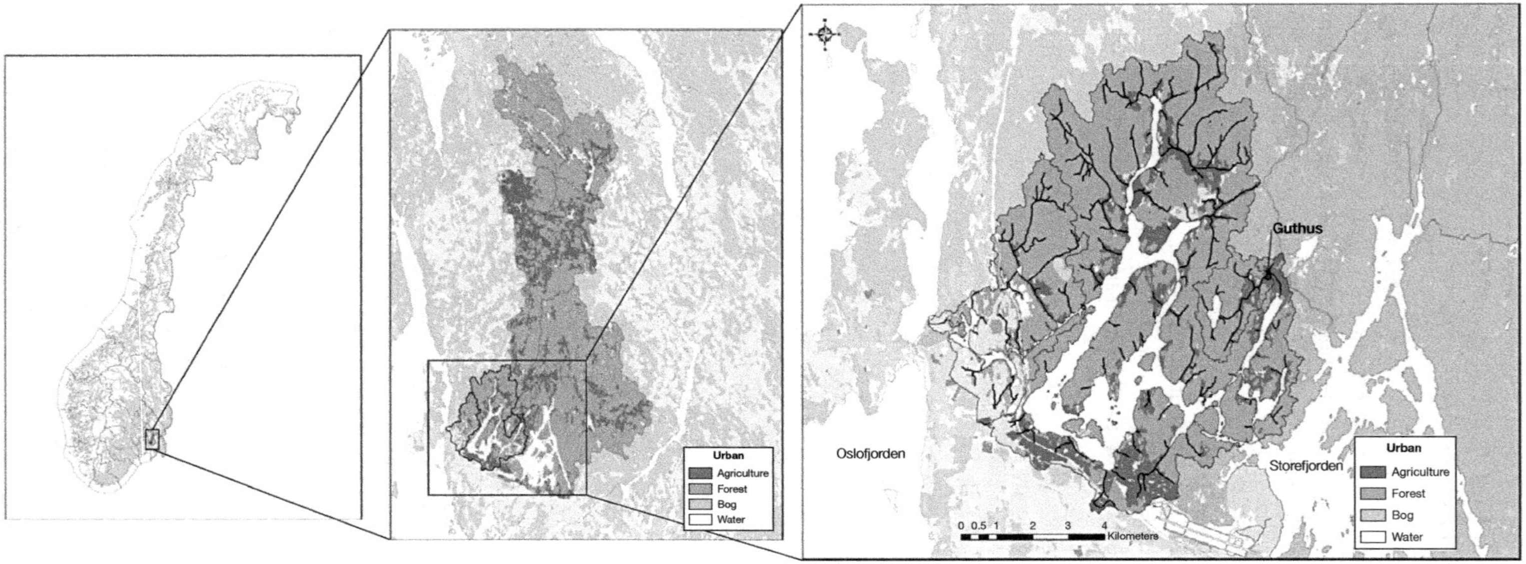

Figure 14.2 The study area of the Eutropia project lies within the Morsa catchment area, Østfold County, Norway. Integrated analysis of nutrient run-off abatement, lake eutrophication and societal response focused on the Vanemfjorden subcatchment shown in the inlay.

(Storefjorden), with an area of 24 km^2, which drains into a shallow western part (Vanemfjorden), with an area of 12 km^2. Modeling focus has been on the drainage area of Vanemfjorden, which faces the greatest eutrophication problems due to large contribution of nutrients from the local watershed.

Water Bodies

The bathymetry of Vanemfjorden is shallow (mean depth 5 m), yet it drains a large area. Consequently, water residence time is only 41 days. A negative aspect of this shallow water in terms of eutrophication is that it allows the water profile to easily mix to the bottom during the summer time, limiting the function of a nutrient sink by sedimentation. This is in stark contrast to the upstream, larger, and deeper Storefjorden lake basin, which removes most of the runoff nutrients via sedimentation, before water flows into Vanemfjorden. A positive aspect from eutrophication point of view is that water contribution from Storefjorden effectively works as a fast diluting medium for Vanemfjorden's local runoff, without which Vanemfjorden would be even more eutrophic. Long-term monitoring data for water quality in these basins exists during the ice-free seasons. Other details about the lake have been published elsewhere (Andersen and Færøvig, 2008; Saloranta, 2006; Skarbøvik et al., 2013).

Land Use

For the total Morsa catchment, 13% of the area is agricultural land, 62% is forest, 9% urban and 16% consists of lake and water (Figure 14.2). The land use distribution in the modeled Vanemfjorden subcatchment is practically the same (14% agriculture, 61% forest, 10% urban and 15% lake and water). Mostly grains and some grasslands cover 90% of the agricultural area. The northern agricultural areas in the Vanemfjorden catchment, predominantly comprising clay loam soils, are today mainly used for cereal production, whereas a narrow southern stretch consisting of sandy end-moraine has a dominance of potato and vegetable production. Animal husbandry is limited in the catchment (Bechmann and Øgaard, 2010, 2013).

There is some variation in farmers' economic characteristics within the Vanemfjorden subcatchment in terms of the type of production and the importance of farm income for households' overall well-being. These two aspects are connected in the sense that farms with more labour-intensive production, like vegetables and potatoes, also have a higher share of household income from agriculture. Off-farm employment opportunities are good in the area, implying that a large share of the farms employ less labour-intensive production that only provide a small share of household incomes. This variation has important implications for farmer responses – and presumably for attitudes – towards policies to improve the water quality, particularly because the high-income production coincides with high nutrient runoffs.

Today's farmers in the Morsa region generally have good agronomical skills, both acquired through formal education, a system of disseminating practical oriented knowledge, exchange of experiences, and ultimately learning from good farming practices (Orderud and Vogt, 2013). Most of these farmers also followed in the footsteps of their parents and are socialised into becoming a farmer and running the farm owned by the family for generations. As such, farmers might be considered conservative, but this conservatism might run counter to a political Conservatism dominated by market economics. Among the farmers there are some who are considered as models for others and running "model farms" whereas others do not perform as well. Part-time farmers are more prone to slip into the less competent category of farmers, and at the end of the day quit farming, but still live on the farm while renting out the land to other farmers.

Recreational Water Use

Vanemfjorden has experienced nuisance cyanobacteria blooms over recent decades. This has been perceived as a considerable problem by the population and municipal governments around Vansjø lakes because of significant ecosystem services provided by the lake and its surroundings (e.g., drinking-water supply, bathing, fishing, and other recreation activities, as well as an important habitat for flora and fauna) (Barton et al., 2009; Söderberg and Barton, 2013). The whole Morsa catchment draining to Vanemfjorden comprises 11 municipalities. An estimated population of approximately 40,000 households were living within the catchment in 2009. Surveys have shown that households express positive willingness to pay for nutrient abatement measures around the Storefjorden and Vanemfjorden lakes well beyond the border of municipalities intersecting the Morsa catchment (Söderberg and Barton, 2013). The largest town called Moss draws its main drinking-water supply from the Storefjorden lake. Thanks to quartiary treatment processes the plant provides some of the cleanest drinking water in the country, regardless of the quality of the raw water supplied from Vanemfjorden. About 20% of the population goes fishing and use motorized boating on about half of their trips to water bodies in the area. Almost 60% of the population goes swimming on more than half of their trips to water bodies. About 75% of the population practices some form of waterside activity on more than half their trips (walking, biking, jogging) (Barton et al., 2009).

Catchment Managers and Management Institutions

An array of management institutions define water management policies, directly or indirectly, from the local administrative level to the national level and beyond (Naustdalslid, 2014). At the local level, this can best be illustrated with those institutions represented on the board of the the Morsa Water Sub-District: the inner layer, with voting rights, is made up of mayors from the 11 municipalities; the medium layer consists of the State County Governors of the two counties Østfold and Oslo/Akershus, the two counties of Østfold and Akershus, the Norwegian Water Resources and Energy Directorate, and the Norwegian Food Safety Authority; and the outer layer (observers) comprising Oslo municipality, the Farmers' Associations in the two counties of Østfold and Akershus; the Association of Nature and Recreational Activities, the water treatment plant Movar, the Vansjø Association of Property Owners, and Moss User Rights Association. Under this structure there are two thematic groups: one on agriculture and one on sewage, both exclusively staffed by municipal officers.

The two main mitigation measures have targeted (1) diffuse runoff from farming and (2) point source sewage from dispersed settlements. Beyond general policies of establishing and enhancing public sewage system networks, farming measures were initially given the highest priority, but increasing focus has later been placed on sewage measures in dispersed settlements, with rural farmer residences facing most measures. Policies and measures are designed and implemented by departments under the Ministry of Environment, the Ministry of Agriculture and Food, and corresponding departments under the State County Governor, and lastly also at the municipal level by the agricultural offices and the water and sewage offices. Local level degrees of policy freedom are fewer than higher up in the hierarchy, but at the local level they might decide on such things as whether to connect dispersed settlements to the sewage system or demand local biological treatment. In this administrative hierarchy, the relevance and usefulness of system analyses is expected to increase the higher up one is in the policy hierarchy.

Eutrophication Management Challenges

In the Morsa catchment, and Vanemfjorden subcatchment in particular, efforts to improve lake water quality have been carried out since year 2000, but the initial mitigation efforts did not improve conditions. As a result of this, an action plan for further reducing P loads was implemented in 2007. This plan was implemented through a close collaboration between local authorities, agricultural advisors, farmers and researchers (Bechmann and Øgaard, 2013). Forty farmers in the catchment have been involved in fulfilling the action plan through contracts committing them to implement best management practices aimed at minimizing diffuse P loading from their agricultural fields. The Ministry of Food and Agriculture in Norway subsidize the farmers who participated in the project, covering expenses due to extra costs and possible loss of income, although not representing any full compensation of income losses. The farmers who signed the contract were committed to:

- Reduce P fertilizing beyond the recommended national level.
- Refrain from plowing in the autumn.
- Refrain from growing vegetable crops on fields that are at risk of being flooded.[1]
- Establish vegetative filter strips along streams.
- Plant grassed waterways in areas of high erosion risk due to surface runoff.
- Build constructed wetlands in streams draining their farmland.

The extent to which the effectiveness and benefits of such measures across the whole catchment of Vanemfjord Lake can be valued using an integrated model is a key part of the Eutropia project.

System Boundaries of the Integrated Valuation Model

We set up a system or "meta-model" spanning different temporal and spatial resolutions of submodels of the driver–pressure–state–impact–response (DPSIR) of eutrophication in the Vanemfjorden (Figure 14.3). Our systems modelling approach involved linking different models in the DPSIR chain together in an OOBN. The OOBN methodology is discussed in the next section.

A challenge for this system modelling approach lies in the confrontation of stakeholders with the uncertainty of the system described by the different scientific submodels coupled together in the OOBN. As the modelled causal chain is extended, the modelled uncertainty regarding integrated model response is expected to increase, while stakeholder comprehension of effects of abatement measures is expected to decrease (Figure 14.3).

Integrating submodels from different domains faces the challenge of different temporal and spatial resolutions. Temporal and spatial resolution defines the heterogeneity and variance that each submodel can render. For example, management measures are drivers that are implemented on an annual time scale, specific to subcatchments. Their combined effect is aggregated spatially at catchment level by the SWAT model (Arnold et al., 1998) with temporal distribution driven by daily meteorological data. As the SWAT catchment model and the MyLake model (Saloranta and Andersen, 2007) work at the same temporal resolution they can be calibrated simultaneously (see Supplementary Material S3). Based on simulation of the joint SWAT-MyLake models, daily predictions of Tot-P and Algal-P at different lake depths in June–August are summarized as probability distributions of expected lake water quality for the summer recreation season. The suitability for summer recreational use at recreational sites along the lakeshore is interpreted based on water quality provided by the Mylake model. Finally, households are asked to provide

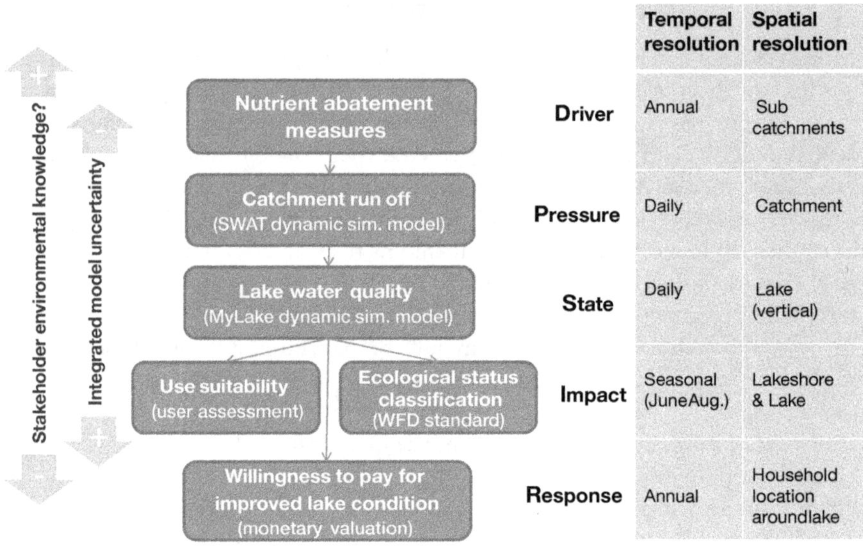

Figure 14.3 Conceptual overview of a systems modelling approach linking a series of model domains along a driver-pressure-state-impact-response causal chain

estimates of willingness-to-pay (WTP) annual sewage fees based on scenarios of summer lake condition across major lakes in the region, including Vanemfjorden. Probability distributions of WTP are generated based on household location round the lake, also extending beyond catchment boundaries. Interfacing heterogeneous temporal and spatial resolutions introduces variance due to different data aggregation procedures at model interfaces, which is additional to the variance in the data themselves. This is one possible explanation for so-called smudging or signal attenuation effects observed in serially integrated cause-effect models (Barton et al., 2008).

Presented below are the constituent submodels for nutrient abatement measures: catchment runoff; lake water quality; classification of ecological status; use suitability and willingness-to-pay for improved lake condition. From here on *italics* are used when referring to *node names* and *node states* in the object oriented Bayesian network (Figure 14.4).

Catchment Runoff Model

The Soil and Water Assessment Tool (SWAT) (Arnold et al., 1998) is a semi-distributed watershed-scale model based on physical processes that runs in continuous time, on a daily time step. Watershed heterogeneities are represented by dividing the catchment into smaller sub-basins that are spatially connected. Each sub-basin is further apportioned into hydrologic response units or HRUs. HRUs are lumped land areas comprising unique land-use, soil type and management practice combinations. Effects of management practices on hydrological processes, nutrient cycles, erosion processes and crop growth are quantified for each HRU and aggregated for each sub-basin. Water movement through the watershed (infiltration, redistribution, surface runoff, lateral subsurface flow and base flow) is modeled after deducting evapotranspiration (ET), potential ET and canopy storage (water intercepted by vegetative surfaces). Nutrient abatement measures included: (1) reduced fertilization; (2) changed plowing practices; (3) constructed wetlands; (4) vegetation buffers; and (5) reduction of point sources from dispersed settlements (septic and farm sludge tanks). Expert judgment was used to organize these measures into alternative scenarios or programs of measures which described the historical, current, and hypothetical future extreme management regimes.

Lake Water Quality Model

Among many lake ecosystem models (Mooij et al., 2010), the MyLake model (Saloranta and Andersen, 2007) was chosen due to researchers' familiarity with the model, and its documented applications to phenomena describing central mechanisms of eutrophication in Vanemfjorden. MyLake is a one-dimensional model code for the simulation of the daily vertical distribution of lake water temperature and thus density stratification, the evolution of seasonal lake ice and snow cover, and most importantly the sediment–water interactions, and phosphorus–phytoplankton dynamics. The basic idea behind MyLake has been to include only the significant physical, chemical and biological processes in a well-balanced and robust way.

Because of the unidirectional flow of water from the catchment to the lake, catchment loading simulated by SWAT is used as an important part of the input to MyLake. Thus, the downstream model MyLake depends on non-uniquely identified input before calibration is conducted. This problem was circumvented by calibrating *SWAT* and *MyLake* sequentially using monitoring data from streams and lake basins. Conditional probability tables (CPTs) for the key model interface variables were produced by running MyLake repeatedly with different parameter and input factor values from SWAT in a Monte Carlo Markov chain calibration. The resulting subnetwork and further documentation on the joint calibration of SWAT-MyLake can be found in Supplementary Material S3.

Ecological Status Classification Model

The European Union's Water Framework Directive (WFD) mandates "good ecological status" as an environmental objective for all water bodies in Member States. Norway adhered to this standard voluntarily. In the present study a classification model, developed by Moe et al. (2014), was adapted to also describe classification uncertainty. Lake ecological status is classified as "poor", "moderate", "good" or "very good" according to the classification standards for Tot-P and Algal-P defined for lakes in the marine zone. A regulatory definition of (un)acceptable chemical and biological indicator levels from a societal point of view can be interpreted as an importance scoring of lake status, i.e. as an "ecological valuation" as defined by Gómez-Baggethun et al. (2014). In MCDA terminology, this is valuation interpreted as scaling a single criterion, but without weighting relative to other criteria. The resulting subnetwork for ecological status classification is explained further in Supplementary Material S4.

Use Suitability Model

A household web-based survey conducted in 2008 presented questions about lake recreational habits and perceptions (Barton et al., 2009). The survey asked for water users' perceptions of the lake's suitability for different water uses at different eutrophication levels, using a series of illustrations of the lakeshore. A challenge in the integrated valuation was determining the water quality parameters of Algal-P and Tot-P that corresponded to the illustration of the water quality scenarios presented in the survey. This was solved by consulting three independent limnologists familiar with the Vansjø lakes, and asking each of them to evaluate the four ecological status illustrations in terms of ranges of Algal-P and Tot-P with 95% confidence. A combination of the three judgments was used for the Bayesian network, giving each expert equal weight. The lake condition illustrations and subnetwork generated by experts are explained further in Supplementary Material S5.

Willingness to Pay Model

The web-based household survey also mapped the respondents' willingness-to-pay increased water and sewage fees to finance nutrient abatement measures in the catchment. The survey conducted monetary valuation using the contingent valuation and choice experiment methods (Barton et al., 2009; Söderberg and Barton, 2013). The results from the choice experiment valuation were used in the DPSIR-OOBN model. The choice experiment asked households to compare and choose between pairwise scenarios of ecological status of Vanemfjorden and other major lakes in the region. The ecological status of the lakes was varied using an experimental design combining different lake conditions with different annual sewage fees. An econometric model was used to estimate the willingness-to-pay in Norwegian kroner (NOK) per household per year for incremental improvements in the condition of each lake. The quality of neighboring lakes was also included in order to control for respondents who prefer two or more adjacent lakes, so-called substitution effects. The choice experiment scenario maps of lake water quality and the subnetwork for household willingness-to-pay is explained further in Supplementary Material S6.

Integrated Valuation Using an Object Oriented Bayesian Network (OOBN)

Causal networks, conditional probability distributions and Bayesian statistics constitute a consistent framework for evaluating spatial and temporal variance across linked submodels (Barton et al., 2008). The Bayesian Belief Network (BBN) methodology is used to study how integrated model uncertainty increases in this causal chain due to heterogeneity in the respective sub-models. There is a substantial literature on the use of BBNs to integrate knowledge domains in environmental and resource management (Barton et al., 2012; Cain, 2001; Darwiche, 2009; Henriksen et al., 2011; Jensen and Nielsen, 2007; Kuikka et al., 1999; Marcot et al., 2006; McCann et al., 2006; Nyberg et al., 2006; Uusitalo, 2007; Varis and Kuikka, 1999). BBNs are models that graphically and probabilistically represent relationships among variables. They can be used diagnostically to study the probability of outcomes given specific causes, reasoning "top-down" through the causal chain of drivers-pressures-states-impacts- responses (DPSIR). BBNs also facilitate using Bayes' theorem for inductive or "bottom-up" reasoning in the causal chain, to determine the likelihood of different valuation outcomes given knowledge about the states of the lake, the management and context variables (Barton et al., 2012). Figure 14.4 describes the modelled system as it is represented in the Bayesian belief network software Hugin Expert.

In Figure 14.4 the network "nodes" (ovals) represent conditional probability tables, with conditional relationships represented by edges (arrows). Individual domain models are represented as subnetworks (white rectangles). Subnetworks contain a number of model variables which are nested in an OOBN in order to reduce the complexity of the visual representation of the model chain. The lake eutrophication system is represented by a cascade of driver–pressure–state–impact–response models in the object-oriented Bayesian network, hence the abbreviation "DPSIR-OOBN".

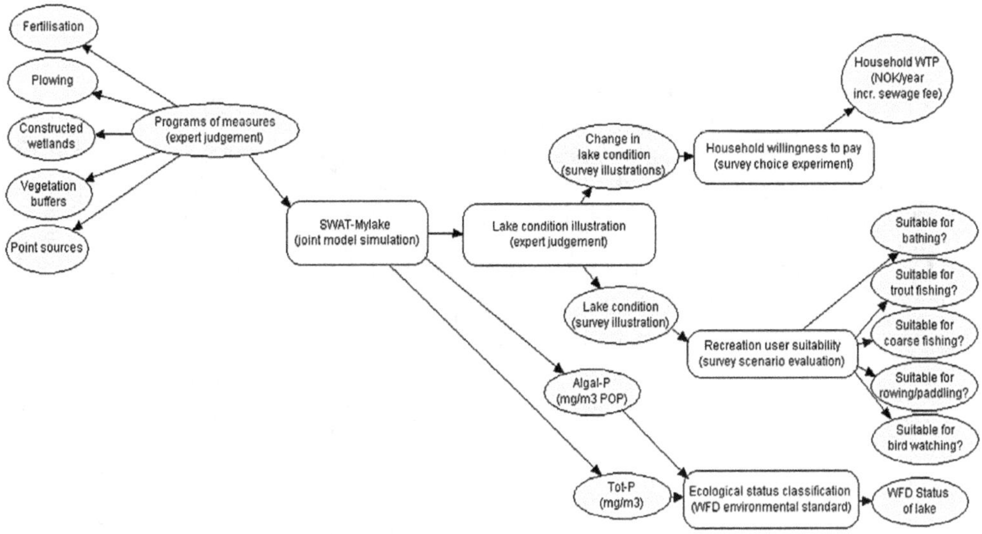

Figure 14.4 DPSIR model chain of nutrient abatement measures and their impacts modelled in an object oriented Bayesian network (OOBN) as represented in Hugin Expert software's graphical user interface

Five types of farm and point source nutrient abatement measures make up alternative "*programs of measures*" in the DPSIR-OOBN. In Figure 14.4 the representation of the SWAT and MyLake calibrated models is condensed into a single node, visualising only two key variables from the simulation (Algal-P and Tot-P). The expert judgment of the link between lake parameters and visual representation of lake condition is complex and it has been condensed here, showing only the outcome of expert judgment on the change in lake condition. Three different valuation methods are identified in the network: (1) Change in lake condition relative to the current condition determines *household willingness-to-pay*, which is a multivariate model that has been visually condensed into a single outcome variable. (2) *Recreational user suitability* is conditional on visual representation of lake condition. Suitability is disaggregated for different types of water users. (3) *Ecological status classification* is directly conditional on Algal-P and Tot-P concentration predicted by SWAT-MyLake.

The OOBN uses a "utility node" in the willingness-to-pay subnetwork to enable evaluation of any node state in the DPSIR model chain in terms of its marginal monetary importance for households. This is explained further in Supplementary Material S6. Because large non-linear integrals cannot be solved analytically, Monte Carlo simulation and Bayes' rule are used across multiple conditional probability tables in the network to assess how values "downstream" in the DPSIR causal chain *scale* to different biophysical states "upstream".

Results and Discussion

Figure 14.5 shows how the Hugin Expert software is used to calculate expected utilities of all nodes and states in the network. Using utility nodes it is technically possible to integrate any chosen "importance score" or value dimension across the whole network. However, in this section the main focus will be on whether the DPSIR-OOBN for Vanemfjorden catchments passes our proposed set of criteria for defining integrated valuation.

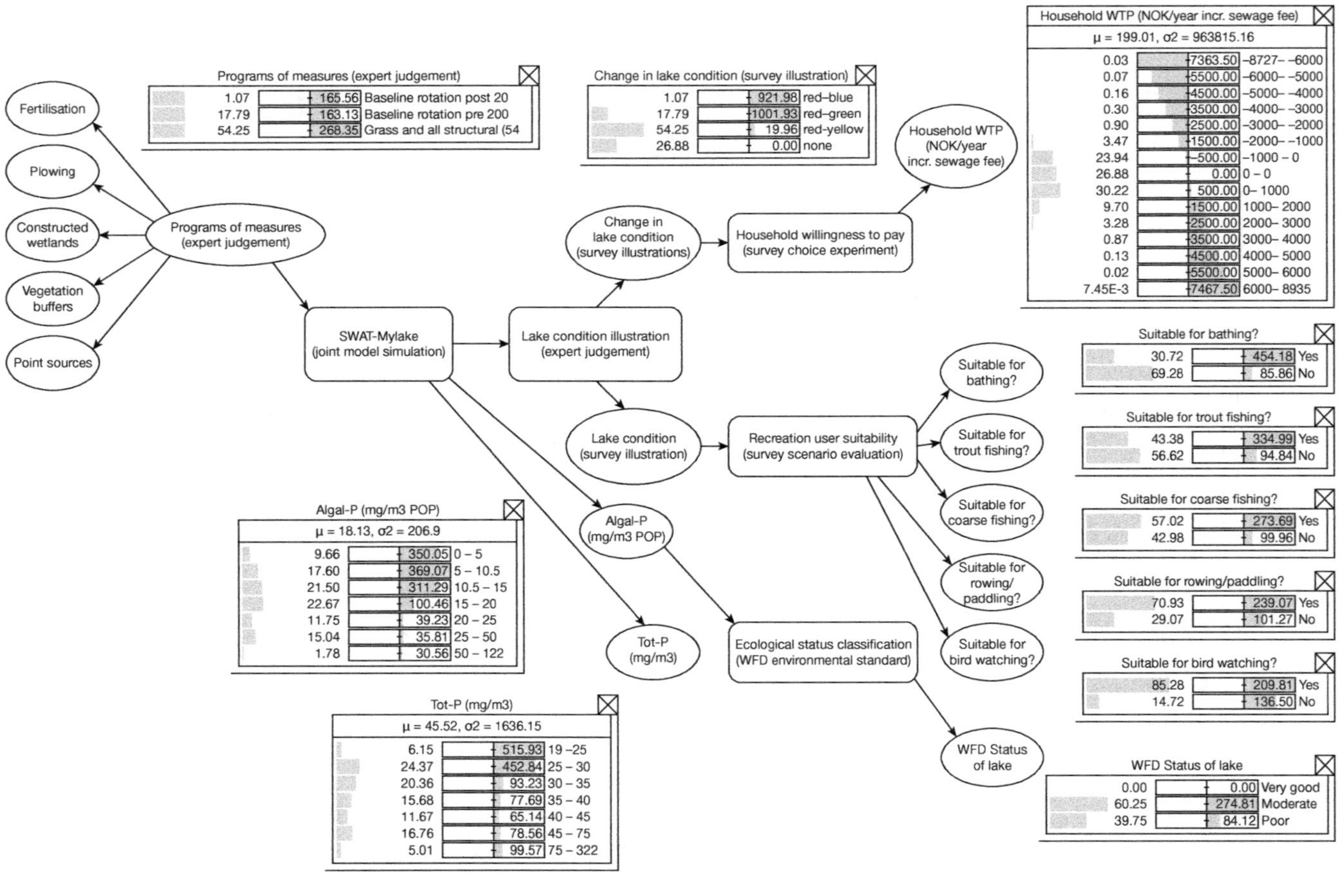

Figure 14.5 Bayesian belief network showing the probability distribution of selected variables (gray columns and number of left-hand side of the node monitors). Node resolution is shown on the right-hand side of the monitors. The gray bars in the middle of the node monitors show the expected WTP of each node state, based on integration (scaling) across the causal chain from the household WTP node (upper left hand of network). For example, the expected WTP of "Grass and all structural" abatement measures in the "programs of measures" (upper left hand of the network) is computed as 268.35 NOK/year per household.

Management Relevance

The network does not qualify fully as integrated valuation by Criterion 1, listed on page 300, because it does not specify (1) the monetary costs of measures, nor (2) the geographical and social distribution of costs and benefits across different upstream and downstream stakeholder interests. While the DPSIR-OOBN is not a full benefit-cost analysis tool with distributional impacts it can still be used to discuss management alternatives and whether model predictions are sufficiently accurate and constitute a reliable basis for action. The DPSIR-OOBN can be used for reasoning deductively about the expected benefits in and around the lake of the alternative programs of measures in the catchment. For example, converting the whole catchment to grass and implementing all nutrient abatement structures computes to an expected willingness-to-pay of 268 NOK/year per household, while baseline crop rotations post or pre 2007 only have an expected willingness-to-pay of 163 or 166 NOK/year, respectively. Apparently, only the radical land use conversion measure has a discernable effect in terms of household willingness-to-pay. Even without knowing monetary management costs, this integrated valuation also gives the impression that the predicted benefits across the other value dimensions – ecological status and recreational suitability – are relatively small compared to the physical magnitude of the abatement measures. The DPSIR-OOBN can also be used to reason inductively, or in diagnostic mode. For example, one may ask "if our management target is lake water that is 100% suitable for bathing (the most quality demanding recreational use), what is the likelihood that any of the programs of measures on the table can attain this?" The integrated valuation shows that with conversion to grass and all structural abatement measures, it is only about 6.5% more likely to achieve this than the baseline crop rotations post-2007. See Supplementary Material S7 for further explanation.

Value Plurality

The DPSIR-OOBN is an integrated valuation method with regard to Criterion 2 because it involves several different measures of the importance of lake condition. The DPSIR-OOBN explicitly handles ecological values in the node "ecological status classification", defining a regulatory threshold for good versus moderate status. Monetary values are defined in terms of household WTP for improved ecological status. Finally, it could be argued that social values are also represented in the network through the evaluation of suitability for different recreational uses.

Value Heterogeneity

The DPSIR-OOBN adheres to Criterion 3 by addressing uncertainty consistently across submodels, using conditional probability tables and Bayes' Theorem to reason across the network of beliefs. Temporal and spatial heterogeneity of runoff and lake phenomena are captured at the resolution and scale considered most appropriate to represent the different ecological, economic, social phenomena. While the difference in temporal and spatial resolutions and scales is not technically consistent, the use of conditional probability tables for defining the interfaces between submodels is consistent. The resolution of the interface nodes (i.e. how finely continuous variables are disaggregated into intervals) was not determined by a consistent statistical rule, instead it depended on modelers' judgments. Independent of the chosen resolution method, some information reflecting heterogeneity is lost at each interface, reducing the sensitivity of ecological, economic and social value responses to different programs of measures in the model.

Interdisciplinarity

The DPSIR-OOBN qualifies as an integrated valuation method based on Criterion 4 because it involved interdisciplinary research. Expert domains spanned agronomy, chemistry, hydrology, limnology, human geography, environmental economics and systems modeling. Interdisciplinarity was required in (1) the joint specification of the driver–pressure–state–impact–responses causal network; (2) specification of interfaces between submodels in the network in terms of the ranges and resolution of variables along the DPSIR chain (specification of conditional probability tables in the network). Finally (3), simultaneous calibration was carried out with two dynamic models, where the lake eutrophication model MyLake was calibrated across all predicted states of the runoff model SWAT.

Knowledge Systems

The DPSIR-OOBN is also an integrated valuation in terms of combining different knowledge systems adhering to Criterion 5. In the terminology of Bayesian belief networks, all the different knowledge systems are called beliefs, whether coded models or expert opinion. Beliefs are specified as a causal network structure and consistently described using conditional probability distributions. Scientific knowledge was used in specifying (1) the alternative eutrophication management decisions, (2) defining the system boundaries in terms of eutrophication and (3) specifying the causal network as a driver–pressure–state–impact–response system. At the level of system drivers both scientific and lay knowledge held by local practitioners was used to point out the relevant management measures to be analysed. The DPSIR-OOBN uses scientific knowledge in terms of the rules/code in the dynamic simulation models SWAT and MyLake. Expert scientific opinion of visual lake condition in terms of Tot-P and Algal-P indicators represents another form of scientific knowledge. An online web-based survey is used to collect lay knowledge held by local household representatives in their role as recreational users. The survey asks for an interpretation by recreational uses of the subject specific suitability of different visual lake conditions. A choice experiment in the survey also consults local household representatives in their role as individual consumers of lake recreational amenities.

The DPSIR-OOBN does not include traditional ecological knowledge in the sense held by local indigenous or peasant communities. The dynamic catchment run-off and lake models are in place of any traditional knowledge about algal blooms that might be found in local communities around Vanemfjorden. Notably, local farmer knowledge of the effectiveness of agricultural measures in controlling nutrient run-off is replaced by experimental scientific knowledge as part of the SWAT model.

Information Types

The DPSIR-OOBN is an example of integrated valuation according to Criterion 6 because it deals with both qualitative and quantitative information. Management measures, catchment and lake, hydrology, limnology and biochemistry processes are described as quantitative dynamic simulation models. In the BBN model results are implemented as interval conditional probability tables. Visualization of lake condition (red, yellow, green, blue scenarios) and WFD classification (very good, good, moderate, poor) are implemented as discrete categorical conditional probability tables, while the suitability for use (suitable, not suitable) is binary. Common to all the information types in the network is their specification in terms of conditional probabilities of each state (whether interval, numerical, categorical, binary). While many cultural ecosystem services may not be quantifiable numerically, they can be evaluated in a causal structure if they can be described in terms of states, and if beliefs about the conditional probability of each state can be obtained from someone. In that sense BBNs treat both quantitative and qualitative information as subjective beliefs. The BBN is not designed to deal with textual narration nor discourse, unless arguments can be simplified to a network of causal relationships, described as discrete states with conditional probabilities.

Information is more than just qualitative or quantitative. Using a Bayesian belief network emphasizes that integrated valuation requires explicit treatment of the quality of information. BBNs require specifying information types (numerical, interval, categorical), the resolution (the number of states) and states' conditional probabilities, for each variable. Resolution and probability are needed to describe temporal and spatial heterogeneity of ecosystem services and are key aspects of valuation. Information types, resolution and probabilities also have costs, in terms of collection and processing. Information resolution is known to condition valuation responses. A case in point is the amount of research hours that go into finding the right balance between information resolution and cost in the number attributes and their levels in choice experiment design and multi-criteria analysis.

Finally, Gomez-Baggethun et al. (2014) suggest that integrated valuation should also account for the articulation of social and cultural values in decision-making, generally involving some sort of deliberative process, locally defined metrics, and valuation methods based on qualitative description, public discourse and narration. On this interpretation of integrated valuation, the DPSIR-OOBN does not perform so well. The management problem, the causal structure and choice of valuation metrics (environmental standard, use suitability, willingness-to-pay) were all largely defined by researchers. While focus groups, surveys and a project reference group were consulted at different stages in the research, the network was not developed as a deliberative process with stakeholders. The development of the DPSIR-OOBN spanned several different research projects (Eutropia, Aquamoney, Refesh, Openness) over almost a decade, making a consistent deliberative process with the same stakeholder representatives very difficult in practice.

Levels of Societal Organization

The DPSIR-OOBN is an example of integrated valuation with regards to Criterion 7 in that it considers multiple types of societal organization as sources of values. The willingness-to-pay values are derived from respondents consulted first and foremost in their role as household representatives, consumers of recreational amenities, public sewage utility users, and survey panel participants. Recreational use suitability is derived by respondents consulted both as individual and household representative recreational users.

The DPSIR-OOBN also qualifies as integrated valuation because it uses values from different levels of societal organization. Values at individual and household level are expressed in terms of willingness-to-pay and user suitability. The Water Framework Directive classification of ecological status of Lake Vanemfjorden is based on an environmental standards approved by the European Commission. "Good ecological status" for pilot water bodies by 2015 was adopted by the Norwegian Parliament as a policy objective, and implemented by river basin authorities and local governments.

However, the DPSIR-OOBN makes no claim of completeness regarding accounting for different societal scales, their roles and value systems. The system boundaries were defined with the aim of conducting a benefit–cost analysis of alternative management measures within the catchment boundaries. As stated above, the research design did not involve identification and participation of all affected stakeholders in the catchment in a deliberative process – i.e. a formal consultation process was not conducted as would be required by regulatory environmental impact assessment (EIA).

Consistent Scaling of Plural Values

Are we incurring double counting of values in a DPSIR cascade model? To answer this question one must first evaluate whether values were scaled independently. In the first step, illustrations of lake condition were evaluated by experts in terms of Tot-P and algal-P parameters. Recreational use suitability as interpreted by survey respondents was based on these different lake condition illustrations. The second step was to allow households to choose in an experiment between different visual representations of lake conditions in the region, versus alternative annual sewage fees. Their choice of alternative lake ecological status outcomes is a trade-off against changes in income due to different sewage fees. The choice experiment responses are then used to calculate willingness-to-pay for marginal changes in lake condition.[2] While both "willingness-to-pay" and "use suitability" are based on the same lake condition visualisations, valuation (i.e. the scaling of their importance) is independent. In the third step the WFD thresholds for ecological status (ecological value) are used to determine bad, moderate or good status based directly on water quality parameters predicted by the SWAT-MyLake model. To conclude, the network structure shows that values are scaled independently – i.e. they are not conditionally dependent on one another in the causal links of the network.

Independent scaling of economic, social and ecological values ensures value plurality, but poses problems for the consistency of values once they are all associated to specific management alternatives – i.e. to the decision context. As discussed in the methodology section, the DPSIR-OBBN is a systems approach to valuation where the expected WTP per household can be identified for all predicted states of abatement measures (driver), nutrient loading (pressure), nutrient concentration, use suitability and ecological status (state).

Below we illustrate how this value integration capability of the DPSIR-OOBN can be used to study consistency of the three different measures of the value of eutrophication abatement. We base the discussion on Figure 14.5. Turning first to the WFD standard for good ecological status as a non-monetary measure of value; the expected utility of achieving "moderate " in WFD status of lake is 275 NOK/year per household. Achieving the current "poor" ecological status calculates to an expected utility of 84 NOK/year per household. This is the result of inconsistency between (1) the experts' judgment of how well visual lake scenarios used to find WTP in the survey translate to biophysical water quality parameters, and (2) the WFD definition of ecological status relative to those same parameters.

Turning next to recreational suitability as a non-monetary measure of value: *Suitable for bathing* shows an expected WTP of 454 NOK/year per household for "yes", but also 86 NOK/year per household for "not" suitable. In fact, all nodes for suitability show a positive expected WTP for "no". This is due to a combined mismatch between (1) experts' judgment of the water quality parameters relative to the lake condition illustration in the survey, and (2) users judgement of suitability based on the same illustrations, relative to the WFD definition of ecological status.[3] In summary, inconsistent scaling is a technical way of saying that there are differences in the subjective importance of different decision criteria. The use of DPSIR-OOBN shows that consistent scaling, adhering to Criterion 8, cannot be expected, even when different criteria are made commensurate, as we have done using expected utility calculations in a Bayesian belief network.

Consistent Trade-offs Between Plural Values in Decisions

The DPSIR-OOBN specifies different types of management measures that can be implemented in the catchment in order to control nutrient input to Vanemfjorden and it addresses different decision alternatives. The network can thus be used to assess the expected willingness-to-pay per household in the catchment, suitability for different users or WFD compliance of different combinations of individual measures. The relative utility of different programs of measures can be compared as input to a decision.

Valuation methods used in the network are expected to be internally consistent, complying partly with Criterion 9. However, the DPSIR-OOBN makes no claim of consistency across value systems. As is noted above, the willingness-to-pay is not completely consistent with the WFD classification in that the combinations of beliefs in the network assign positive willingness to pay to the status quo "poor ecological status". Furthermore, respondents to the survey were encouraged to use individual consumer rationality, but the framing of the survey setting is likely to have triggered several roles at once in the respondent. It is unclear how each role affects a particular valuation metric. While respondents can be encouraged to be aware of and explain different roles using careful survey work, it seems difficult to isolate the effect of particular roles and their value systems.

In this section it is shown that it is possible to compute expected utilities of any state of the network. We have also demonstrated how to technically conduct an integrated valuation. However, the DPSIR-OOBN makes no claim to consistently integrating the values generated by different value systems, for example by explicitly weighting their relative importance. In contrast, in a multi-criteria analysis based on multi-attribute utility theory, the final step of calculating a unique utility measure across different decision alternatives would require explicit weighting of all criteria relative to one another.

Extending Systems Boundaries

Valuation, even when integrated, must have model system boundaries. Would extension of these systems boundaries for the DPSIR-OOBN for Vanemfjorden catchment better address management issues?

Farmer Response

The DPSIR-OOBN does not account for stakeholder motivations to implement land use changes and structural abatement measures. Measures are assumed to be implemented 100% within the SWAT catchment model, without any delay, lacking effectiveness or transaction costs. Farmer response could be modelled in terms of farms being profit maximising units and responding imperfectly to increased constraints on inputs.[4] However, farmer motivations and reactions extend beyond reaction to incentives or coercion. The stakeholder narratives in the following are not easily coded in a Bayesian network. Interviews with farmers revealed the basic motivations guiding the actions of today's generations of farmers (Orderud and Vogt, 2013):

1. Farmers are socialized into farming without prospects of earning a lot of money. They cultivate a family dimension of fostering an attachment to the farm, and a stewardship mentality of handing over a good farm to the next generation.
2. They have a production mentality aiming for high output and high quality by combining agronomical competence and good knowledge about their fields (soil structure, drainage,

etc.). They work with the agricultural extension service, disseminating knowledge that is trusted by farmers. However, understanding of dynamic natural processes of the catchment is relatively weak.

3. All other things equal, farmers will go for options providing higher income. Moreover, when taking actions, whether mandatory or not, they will prefer to use their own labour rather than pay money out of their wallet.
4. The internal interaction and status hierarchy among farmers is based on recognition of being good at farming. What the respected farmers do influences what other farmers are doing regarding production techniques, use of fertilizers, tillage, etc.
5. The interaction between farmers and local communities, as well as the wider society, shapes common norms. For example, farmers living around the Lake Vansjø and Moss consider it morally wrong to be the cause of inferior water quality.
6. The Norwegian governance structure, with policy-makers, public agencies, and farmers' organizations create a wide range of public regulations surrounding farming, making public policies a frame for what is considered acceptable farming.

On the basis of the above, we can draw the following conclusions regarding farmers' motives and the probability of actions as part of the Bayesian network analysis:

1. The farmers appear to be economic satisfiers within a bounded rationality approach displaying satisificing behaviour (Simon, 1982). They take into account a wide range of issues and concerns. For instance, they might pursue a high output without trying to maximize economic return.
2. The farmers show a high degree of compliance with agricultural policies because they have become accustomed to being part of public policies and receiving part of their income from the government. Potentially, this makes farmers internalize a societal responsibility of complying with environmental regulations, also when this might incur additional costs not automatically (fully) compensated for.
3. Most farmers possess good agronomical competence, being critical to farming techniques that run counter to what they consider "best practice", but making changes based on evidence. However, abatement measures running counter to farmers' own experiences are difficult to implement in a successful manner. Moreover, seeing that actions have an impact is part of motivation.
4. The farmers want to be an active partner in the decision-making process of designing and implementing policies at the local level due to their good agronomical competence and a generally high educational level. When farmers feel left out, they are more likely to be discontent. In turn that may have adverse impacts on willingness to take environmental-related actions beyond the required minimum.

Integrated Model Communication

How can scientific knowledge gained from the DPSIR-OOBN systems model strengthen or weaken existing stakeholder opinion regarding conflictive nutrient abatement measures? The impact on stakeholder knowledge depends on how that knowledge is communicated. An integrated monetary valuation of predicted abatement measures effectively changes the mode of communication with stakeholders. From the above narratives by farmers, we can draw the conclusion that reasonable and well-grounded policies and measure have a high likelihood of being accepted and implemented by farmers. For the farmers, trials and testing locally are crucial, but before that stage, system analyses very often will have played a role in identifying potential improvements. On the other hand, measures running counter to what is considered best practice of farming will meet resistance, often regardless of being accompanied by economic incentives or not. They will have low likelihood of being implemented and accepted. In Morsa, accepting mineral fertilizers with a lower phosphorus content is an example of a change that was accepted because "model" farmers tried it and it was proven to work on-farm. On the other hand, changing of tilling practice in areas not prone to erosion has been met with farmer resistance, causing discontent and reduced willingness to participate. The lack of significant improvements in observed water quality has also caused discontent among farmers, reducing the likelihood of accepting policies and measures (Orderud and Vogt, 2013). This is not to say that farmers deny that any improvements in the lake have taken place following the serious flooding in year 2000 and subsequent cyanobacteria blooms. However, farmers did not see that the water quality improvements had met expectations. Farmers have also understood that lake improvements cannot unequivocally be linked to the measures taken on-farm.

Opposing narratives about the effectiveness of measures live side by side. While on-farm monitoring shows that land use change, fertilizer reduction and structural measures are effective "at field's edge", the runoff and lake dynamic models have individually shown limited responses "at lake's edge" to the combined effect of farm measures. Limitations to the understanding of nutrient run-off from the large forested areas in the catchment are still substantial (Desta, 2013; Lukawska-Matuszewska et al., 2013) (see Supplementary Material S1). Lacking effectiveness of abatement measures in the DPSIR-OOBN is a combination of the heterogeneity in catchment system exceeding the signal from human intervention (variance), information loss at submodel boundaries (error), and still unexplained ecosystem function (uncertainty). See Supplementary Material S7 for further details on model power. The integrated valuation model may compound the impression farmers have already gained from previous interaction with researchers working on subcomponents of the system. The DPSIR-OOBN "story" is one of continued uncertainty about the effectiveness of measures in a complex system. This may have a negative effect on farmer motivations to implement further measures. If that is the case, the DPSIR-OOBN becomes not only a description of system dynamics, but also a mediator of those system dynamics.

Conclusions

This chapter has presented a list of criteria for defining a systems model as an "integrated valuation of ecosystem services". This refers to management relevance, value plurality, value heterogeneity, interdisciplinarity, knowledge systems, information types, levels of societal organization, consistency in scaling of plural values, and consistency in comparison of plural values in decisions.

We presented an object oriented Bayesian network (OOBN) of the valuation of eutrophication abatement measures in a catchment in South-Eastern Norway. The OOBN was used to link biophysical, social and economic models together following the framework of a driver–pressure–state–impact–response (DPSIR) model. We have shown how Bayesian networks offer a consistent meta-modeling approach to integrated model uncertainty – both in terms of spatial and temporal heterogeneity and value plurality. We argued that the Bayesian interpretation of all causal relationships as beliefs, whether lay, expert, scientific, or model-encoded knowledge is a useful framing for value plurality in environmental management. No single type of subjective scaling of the biophysical impacts takes primacy when all knowledge systems are framed as belief.

We think that the Bayesian belief network methodology can shift focus to the consistency of valuation methods within decision contexts. It helps to evaluate the causal chain linking human actions, via socio-ecological system dynamics, to perceptions. It should therefore be a practical approach to operationalizing the ecosystem services cascade framework in the economic analysis of programs of measures under the WFD.

The DPSIR-OOBN model discussed in this chapter meets most of the nine suggested criteria for "integrated valuation". The model fails to meet the criteria that ecological, social and economic values are to be defined consistently in relation to impacts on lake quality. The DPSIR-OOBN does not meet the criteria of consistent trade-off analysis as it is does not weight different impacts of eutrophication against one another. It is neither a full benefit-cost analysis, nor a full multi-criteria analysis. However, we have shown how the DPSIR-OOBN can be used to explore issues of consistency in scaling and weighting of different values. We think that Bayesian belief networks make it possible to take a consistent approach to how risk – defined as probability multiplied by impact – conditions valuation.

We also discussed how our integrated valuation is limited by model system boundaries, which exclude potentially significant explanations of lacking abatement effectiveness. We discuss farmer narratives where incentives for implementing nutrient abatement measures depend on many things, among them the power of scientific models to predict abatement effectiveness. The integrated valuation model is not merely a model, but also a mediator. The variability that continues to characterize eutrophication and human responses, at least in the complex Morsa catchment, also suggests that our integrated valuation model does not meet the accuracy and reliability requirements of a decision-support model under the WFD. For the moment it is still "eutropia".

Acknowledgments

We are grateful for financial support from several sources over the years it has taken to develop the various models reported in this chapter, in particular from the Research Council of Norway for the EUTROPIA project (grant no. 190028/S30), and support from the European Union for the FP6 AQUAMONEY (contract no. SSPI-022723), FP7 REFRESH (grant no. 244121) and FP7 OpenNESS project (grant no. 308428). We would like to thank Helga Gunnarsdottir, former manager of the Morsa Project, for her continued support for multi-disciplinary studies in the Morsa catchment. We thank Dag Berge, Sigrid Haande and Marit Mjelde at the Norwegian Institute for Water Research (NIVA) for participating as experts in the limnological evaluation of visual water quality scenarios.

Notes

1. Not modeled in this study.
2. The choice experiment simultaneously weights the relative importance of different lakes and their quality, and scales the marginal changes against income. Bridging choice experiment (CE) and multi-criteria (MCD) terminology, MCA weights are equivalent to the CE beta-coefficients on the non-price choice attributes, while MCA scaling is equivalent to the CE alfa also called "scale coefficient".
3. Contributing to these inconsistencies between societal valuation systems for eutrophication, are survey responses that at first glance seem inconsistent. First, we observed from *Household WTP* that there is a probability that households have a negative WTP (there is a 24% probability that $-1000 < E(WTP) < 0$). This is due to an unexpected result in the choice experiment where respondents on average expressed negative WTP for small improvement (red-yellow lake condition), while positive WTP for larger improvements (red-green, red-blue condition). Barton et al. (2009) conjecture that respondents

expect any management measures implemented to have large effects, and react negatively to scenarios that show only small improvements (despite them being positive relative to the status quo).

4. In fact, a part of the research project addressed farm returns to different cropping patterns and fertilizer use. However, opportunity costs to farmers were not included in the DPSIR-OOBN to explore the concept of "integrated valuation", rather than benefit–cost analysis.

References

Andersen, T., Færøvig, P.J., 2008. *Fire års fosforjakt i Vansjø–hva er resultatet?*, Vansjøkonferansen, September 23, Parkteatret, Moss.

Arnold, J.G., Srinivasan, R., Muttiah, R.S., Williams, J.R., 1998. Large area hydrologic modeling and assessment part 1: model development. *Journal of the American Water Resources Association*, 34(1): 73–89.

Barton, D. N., Kuikka, S., Varis, O., Uusitalo, L., Henriksen, H. J., Borsuk, M., de la Hera, A., Farmani, R., Johnson, S., Linnell, J. D. C. (2012). Bayesian networks in environmental and resource management. *Integrated Environmental Assessment Management*, 8(3): 418–29.

Barton, D.N., Navrud, S., Lande, N., Bugge Mills, A., 2009. Assessing Economic Benefits of Good Ecological Status in Lakes under the EU Water Framework Directive, Case study report, Norwegian Institute for Water Research, NIVA report 5732–2009.

Barton, D.N., Saloranta, T., Moe, S.J., Eggestad, H.O., Kuikka, S., 2008. Bayesian belief networks as a meta-modelling tool in integrated river basin management — Pros and cons in evaluating nutrient abatement decisions under uncertainty in a Norwegian river basin. *Ecological Economics*, 66: 91–104.

Bechmann, M., Øgaard, A., 2010. Critical Source Areas of Nutrient Losses from Agriculture in Norway, International Society for Horticultural Science, Leuven, Belgium, 852: 10.

Bechmann, M., Øgaard, A.F., 2013. Water quality changes following intensive focus on mitigation methods to reduce phosphorus losses in the catchment of Lake Vansjø, Norway.

Bechmann, M., Pengerud, A., Øygarden, L., Øgaard, A., Syversen, N., 2007. *Tiltak mot fosfortap fra jordbruksarealer rundt vestre Vansjø Bioforsk Jord og Miljø – Ås.*

Cain, J., 2001. Planning improvements in natural resources management: guidelines for using Bayesian networks to support the planning and management of development programmes in the water sector and beyond. Centre for Ecology & Hydrology, Crowmarsh Gifford, Wallingford, Oxon, UK, 124 pp.

Darwiche, A., 2009. *Modeling and Reasoning with Bayesian Networks.* Cambridge University Press, New York, 560 pp.

Desta, Y.H., 2013. Characterization of forest soils in the Morsa watershed in order to assess the background leaching of phosphorus to the eutrophic western Vansjø Lake. 60 stp. Thesis, University of Oslo, DUO, 121 pp.

EC, 2003. Economics and the environment – The implementation challenge of the Water Framework Directive. Guidance document no. 1. Common Implementation Strategy for the Water Framework Directive (2000/60/EC). Working Group 2.6 – WATECO. European Commission., European Commission, Luxembourg.

EC, 2009. Common implementation strategy for the Water Framework Directive. In Guidance Document on Exemptions to the Environmental Objectives, Guidance Document No. 20. European Commission, Luxembourg.

Galioto, F., Marconi, V., Raggi, M., Viaggi, D., 2013. An assessment of disproportionate costs in WFD: the experience of Emilia-Romagna. Water 5: 1967–1995.

Gómez-Baggethun, E. et al., 2014. State-of-the-art report on integrated valuation of ecosystem services. EU FP7 OpenNESS Project Deliverable 4.1, European Commission FP7, 2014.

Haines-Young, R., Potschin, M., 2010. The links between biodiversity, ecosystem services and human well-being. In: Raffaelli, D., Frid, C. (Eds.), *Ecosystem Ecology: A New Synthesis.* Cambridge University Press, Cambridge, pp. 110–139.

Henriksen, H.J., de la Hera, A., P., Z., M., B., 2011. Use of Bayesian networks for communicating divergent views and understandings in adaptive groundwater management. *Integrated Environmental Assessment and Management* (this issue).

Jensen, F.V., Nielsen, T.D., 2007. *Bayesian Networks and Decision Graphs.* 2nd ed. Berlin: Springer, 448 pp.

Kjaerulff, U.B., Madsen, A.L., 2007. *Bayesian Networks and Influence Diagrams: A Guide to Construction and Analysis.* Berlin: Springer, 318 pp.

Kuikka, S., Hildén, M., Gislason, H., Hansson, S., Sparholt, H., and Varis, O. (1999). Modeling environmentally driven uncertainties in Baltic cod (*Gadus morhua*) management by Bayesian influence diagrams. *Canadian Journal of Fisheries and Aquatic Sciences*, 56: 629–641.

Lukawska-Matuszewska, K., Vogt, R.D., Xie, R., 2013. Phosphorus pools and internal loading in a eutrophic lake with gradients in sediment geochemistry created by land use in the watershed. *Hydrobiologia*, 713(1): 183–197.

Marcot, B.G., Steventon, J.D., Sutherland, G.D., McCann, R.K., 2006. Guidelines for developing and updating Bayesian belief networks applied to ecological modeling and conservation. *Canadian Journal of Forest Research-Revue Canadienne de Recherche Forestiere*, 36(12): 3063–3074.

Martin-Ortega, J., 2012. Economic prescriptions and policy applications in the implementation of the European Water Framework Directive. *Environmental Science & Policy*, 24: 83–91.

McCann, R., Marcot, B.G., Ellis, R., 2006. Bayesian belief networks: applications in natural resource management. *Canadian Journal of Forest Research*, 36: 3053–3062.

Moe, S.J. et al., 2014. Implications of climate change for ecological reference conditions, thresholds and classifiation systems for European lakes. REFRESH Deliverable 3.15–16. 108.

Mooij, W.M. et al., 2010. Challenges and opportunities for integrating lake ecosystem modelling approaches. *Aquatic Ecology*, 44(3): 633–667.

Morrison, M., Morgan, M.S., 1999. Models as mediating instruments (Chapter 2). In: Morgan, M.S., Morrison, M. (Eds.), *Models as Mediators: Perspectives on Natural and Social Science*. Cambridge University Press, Cambridge, pp. 10–37.

Naustdalslid, J., 2014. Multi-level water governance: the case of the Morsa River Basin in Norway. *Journal of Environmental Planning and Management,* 58(5): 913–931.

Nyberg, J.B., Marcot, B.G., Sulyma, R., 2006. Using Bayesian belief networks in adaptive management. *Canadian Journal of Forest Research*, 36: 3104–3116.

Orderud, G.I., Vogt, R.D., 2013. Trans-disciplinarity required in understanding, predicting and dealing with water eutrophication. *TSDW International Journal of Sustainable Development & World Ecology*, 20(5): 404–415.

Parry, R., 1998. Agricultural phosphorus and water quality: A US Environmental Protection Agency perspective. *Journal of Environmental Quality*, 27(2): 258–261.

Saloranta, T.M., 2006. Highlighting the model code selection and application process in policy-relevant water quality modelling. *Ecological Modelling*, 19(1–3): 316–327.

Saloranta, T.M., Andersen, T., 2007. MyLake: a multi-year lake simulation model code suitable for uncertainty and sensitivity analysis simulations. *Ecological Modelling*, 207: 45-60.

Simon, H.A. (1982) *Models of Bounded Rationality, Vol. 3: Empirically Grounded Economic Reason*. Cambridge, MA: MIP Press.

Skarbøvik, E., S., H., Bechmann, M., 2013. Overvåking Vansjø/Morsa 2011–2012. *Resultater fra overvåking av innsjøer, elver og bekker i perioden oktober 2011 – oktober 2012, BioForsk.*

Söderberg, M., Barton, D.N., 2013. Marginal WTP and Distance Decay: The Role of 'Protest' and 'True Zero' Responses in the Economic Valuation of Recreational Water Quality. *Environmental and Resource Economics*, 59(3): 389–405.

Tong, C., Yang, X., Pu, P., 2003. Degradation of aquatic ecosystem in the catchment of Mu-Ge Lake and its Remediation countermeasures. *Journal of Soil and Water Conservation*, 17(1): 72–88.

Tscherning, K., Helming, K., Krippner, B., Sieber, S., Paloma, S.G.Y., 2012. Does research applying the DPSIR framework support decision making? *Land Use Policy*, 29(1): 102–110.

Uusitalo, L., 2007. Advantages and challenges of Bayesian networks in environmental modelling. *Ecological Modelling*, 203(3–4): 312–318.

Varis, O., Kuikka, S., 1999. Learning Bayesian decision analysis by doing: lessons from environmental and natural resources management. *Ecological Modelling*, 119(2–3): 177–195.

Yang, X.-e., Wu, X., Hao, H.-l., He, Z.-l., 2008. Mechanisms and assessment of water eutrophication. *Journal of Zhejiang University Science B*, 9(3): 197–209.

15

USING SYSTEM DYNAMICS TO MODEL INDUSTRY'S DEVELOPMENTAL RESPONSE TO ENERGY POLICY

Brian Bush, Daniel Inman, Emily Newes, Corey Peck, Steve Peterson, Dana Stright, and Laura Vimmerstedt

Research Question: What are the barriers and facilitators to the development of the biofuels industry?
System Science Method(s): System dynamics
Things to Notice:

- Evaluation of public policy-based incentives
- Ability of system models to uncover synergistic effects

In this chapter we explore the potential development of the biofuels industry using the Biomass Scenario Model (BSM), a system dynamics model developed at the National Renewable Energy Laboratory through the support of the U.S. Department of Energy. The BSM is designed to analyze the implications of policy on the development of the supply chain for biofuels in the United States. It explicitly represents the behavior of decision makers such as farmers, investors, fueling station owners, and consumers. We analyze several illustrative case studies that explore a range of policies and discuss how incentives interact with individual parts of the supply chain as well as the industry as a whole. The BSM represents specific incentives that are intended to approximate policy in the form of selected laws and regulations. Through characterizing the decision-making behaviors of economic actors within the supply chain that critically influence the adoption rate of new biofuels production technologies and demonstrating synergies among policies, we find that incentives with coordinated impacts on each major element of the supply chain catalyze net effects of decision-maker behavior such that the combined incentives are greater than the summed effects of individual incentives in isolation.

The emerging biofuels supply chain is composed of multiple interconnected subindustries. Development of the system is the result of dynamic interaction of multiple agents, including agricultural producers, researchers, and investors in conversion pathways, as well as distributors, retailers, and consumers. In the United States, another major decision-making agent, the U.S. government, has promoted biofuels through a variety of legislation in concert with strategies to lessen dependence on imported energy as well as to reduce the emissions of greenhouse gases

 DOI: 10.4324/9781315748771-17

(Office of the Biomass Program and Energy Efficiency and Renewable Energy, 2013); a cornerstone of this policy, the Energy Independence and Security Act of 2007 (EISA) mandates 36 billion gallons of renewable liquid transportation fuel in the U.S. marketplace by the year 2022 (U.S. Congress, 2007). Producing sufficient fuels to meet such large volumetric targets involves development of biomass resources such as biomass-based (a.k.a. lingocellulosic) and/or algal feedstocks. It also entails development of new technologies that are emerging from research and development. In order to spur the growth of the biofuels industry, the U.S. government has enacted a variety of incentive programs (including subsidies, fixed capital investment grants, loan guarantees, vehicle choice credits, and aggressive corporate average fuel economy standards) that encourage other economic agents to participate in the development of the industry. At any stage in the supply chain, the decisions made and actions taken by political or economic agents accumulate over time, and can influence the decisions and actions of others in the system. Briefly, the biomass-to-biofuels supply chain consists of (1) production of biomass feedstock in the form of perennial energy crops like switchgrass or willow or the collection of residue from annual crops, (2) collection and transport of that feedstock from the energy crops or from residue from other agricultural production, forestry, and the like, (3) conversion of the biomass to biofuels at biorefineries, and (4) distribution of the biofuels to points of distribution or end use. The specifics of biomass production depend on geographic location and economics: these determine whether crops dedicated to energy production are planted or whether biomass residue from other crops (corn, wheat, etc.) or feedstock sources (forest residue, urban waste) are collected (Downing, et al., 2011). Numerous potential biomass-to-biofuel conversion pathways exist, but these can be roughly categorized as involving either biochemical or thermochemical conversion processes (Bioenergy Technology Office, U.S. Department of Energy, 2014).

Systems Dynamics Approach

Many of the physical processes, decision processes, feedbacks and constraints found in the biomass-to-biofuels supply chain are represented in the BSM (Peterson et al., 2013). The BSM is a system dynamics model developed under the auspices of the U.S. Department of Energy and the culmination of a multi-year project at the National Renewable Energy Laboratory. It is a tool designed to better understand biofuels policy as it impacts the development of the supply chain for biofuels in the United States and the economic agents influencing that development through their decisions. The model is intended to generate and explore plausible scenarios for the evolution of a biofuel transportation fuel industry in the United States, representing multiple pathways leading to the production of fuel ethanol as well as advanced biofuels such as biomass-based hydrocarbons like biomass-based gasoline, diesel, jet fuel, and butanol. The BSM, which is implemented using the STELLA (isee systems, 2010) system dynamics simulation platform, integrates representations of resource availability, physical/technological/economic constraints, behavior, and incentives so as to model dynamic interactions across the supply chain. It simulates the deployment of biofuels given technological development and the reaction of the investment community to those technologies in the context of land availability, the competing oil market, consumer demand for biofuels, and government policies over time. It has a strong emphasis on the behavior and decision making of various agents.

In developing the BSM, we have used a system dynamics approach. System dynamics has been used in a wide range of policy contexts to represent and simulate complex dynamic systems driven by multiple interacting physical and social components. The model encodes a system of coupled ordinary differential equations that are integrated forward in time, thus establishing interdependence among rates of change of key parameters and feedbacks between variables representing physical, technical, economic, and behavioral aspects of the biofuels supply chain (Lin, Bush, Newes, & Peterson, 2013). Essential to system dynamics model are stocks, flows, and feedback (for more details, see Sterman, 2006):

- *Stocks and flows*: Accumulations, and the activities that cause accumulations (flows) to rise and fall over time, are fundamental to the generation of dynamics. System dynamics models are built up from stock and flow primitives. In the BSM, we use stocks to represent concepts such as prices, inventories, conversion facilities, and station owners who are contemplating investment in E85 tankage and dispensing equipment. Corresponding flows would include price changes; production, consumption, and shrinkage of inventories; investment or obsolescence of facilities; and deciding not to invest in tankage and equipment.
- *Feedback:* Dynamic social systems can contain rich webs of feedback processes. Positive feedbacks tend to drive reinforcing growth in key quantities, while negative feedbacks support self-correcting behavior. In the BSM, we have sought to capture key feedbacks within and across each stage of the supply chain.

Randers (1980) suggests a single recipe to ensure a successful modeling effort. In our case, we have adopted a more eclectic approach. We have drawn on other disciplines in order to build confidence in model structure and insights, to define and execute experiments, and to engage stakeholders. These approaches include the following: resiliency thinking (Walker & Salt, 2006) in the design of the modular architecture of the model; statistically rigorous sensitivity analysis; quantification of uncertainty; development of parameter values and relationships using "side studies" that sometimes involve statistical analysis techniques such as regression; and group classification.

In order to gain a view into the evolution of the supply chain for biofuels, the BSM focuses on the interplay between marketplace structures, various input scenarios, and government policy sets, as shown in Figure 15.1. The BSM is particularly adept at addressing the following types of inquiries: which sources of feedstock might plausibly contribute substantially to production

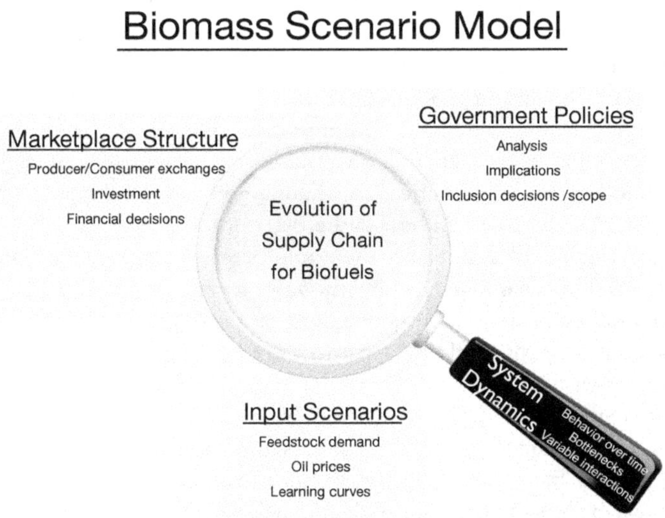

Figure 15.1 Biomass scenario model strategy and approach

in different regions of the United States; under what combination of policies does the biofuels industry experience gradual, sustained growth; and what are the implications of different gasoline price scenarios?

As these questions suggest, the BSM was developed to provide insights regarding the possible evolution of the biomass-based fuels system, not as a precise forecasting tool or predictive model. It is important to recognize that insights are subject to the limitations of data inputs, model structure, scenario design, and assumed values. Accordingly, analysis with BSM often consists of insight development, followed by careful identification of conditions within the model that would affect insight robustness.

The BSM has been designed in a top-down, modular fashion which allows material (feedstocks) to flow down the supply chain and be converted into various types of biofuels, with feedback mechanisms among and between the various modules. Systems of equations (both algebraic and integro-differential), within the sectors and spanning sectors and modules, specify the relationships between variables such as prices, costs, facilities, resources, and material. In some cases the equations represent physical or economic constraints or relationships, whereas others embody behavioral models such as investor decision-making and consumer choices (Peterson et al., 2013). In general, the BSM endogenizes the determination of prices, production, investment, and demand related to biomass and biofuel and relies on exogenously specified scenarios for boundary conditions such as petroleum prices and international trade.

In developing the model, we have taken care to create a structure that is transparent, modular, and extensible, enabling standalone analysis of individual modules as well as testing of different module combinations. As shown in Figure 15.2, the model is framed as a set of interconnected sectors and modules. Each supply-chain element has been modeled as a standalone module, but is linked to the others to receive and provide feedback. References Newes, Inman, & Bush (2011) and Peterson et al. (2013) provide details regarding the structure and function of individual modules. The feedstock production module simulates the production of biomass crops as well as other commodity crops. In the feedstock module, net-present value (NPV) drives the allocation of land to grow crops. Crops are allowed to mature at different rates depending on whether they are an annual or perennial. Once crops are harvested, they contribute to the overall inventory, which influences the price and thus NPV. The feedstock logistics module models the harvesting, collection, storage, preprocessing, and transportation of biomass feedstocks from the field (or forest) to the biorefinery. The conversion module represents more than a dozen biofuel conversion technologies at four production scales (pilot, demonstration, pioneer, and full-scale commercial). In the simulation, the biofuel produced in the conversion stage is distributed to dispensing locations and end users. The model is solved numerically at a submonthly level and typically reports output for the timeframe of 2010 to 2050. Although the description herein implies a linear flow of information between the modules, in reality the modules receive and react to information in a complex, non-linear fashion that depends on, among other things, industrial learning, project economics, installed infrastructure, consumer choices, and investment dynamics. The model is geographically stratified using the ten U.S. Department of Agriculture (USDA) farm production regions (U.S. Department of Agriculture, Economic Research Service, 2014) as a basis, which facilitates analysis of regional differences in key variables.

The following section highlights high-level insights that have been gleaned through system dynamics analysis focused on specific decision-making actors in the biofuel supply chain. Specifically, we explore the behavior of the biofuels industry in response to the interaction among policies and incentives that impact feedstock producers, investors, and consumers. In general, our analytic methodology involves (1) precisely formulating the issue being studied or question under consideration as hypotheses amenable to a systems science approach, (2)

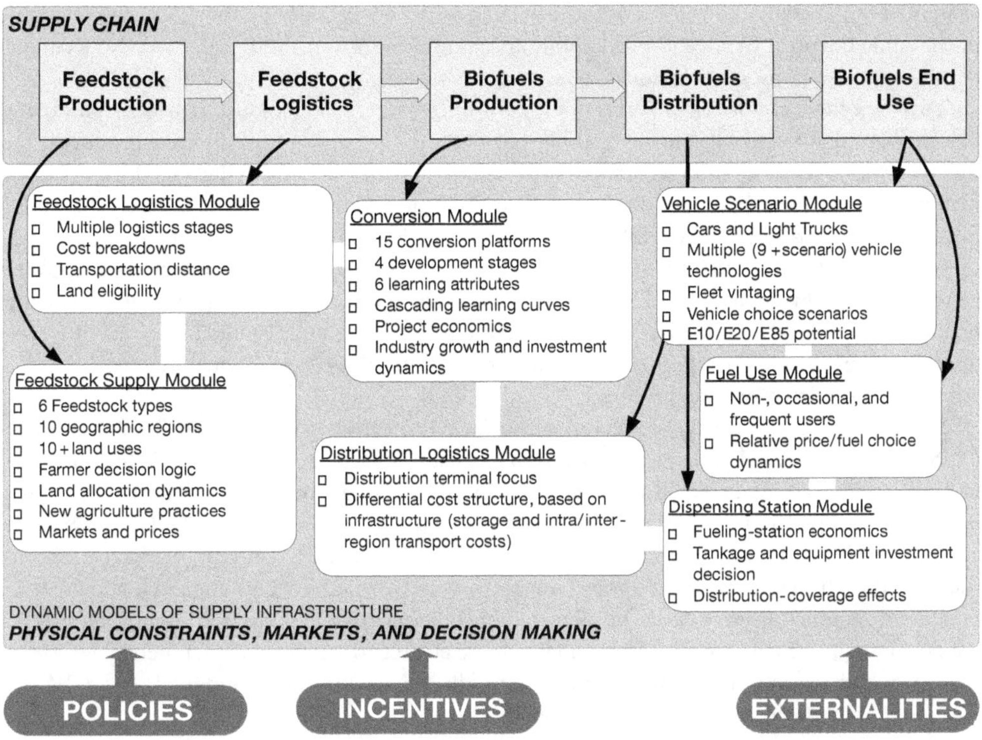

Figure 15.2 The modules in the BSM represent elements of the biomass-to-biofuels supply chain

modifying the BSM's structure or input data to support the investigation, (3) performing tens or thousands of simulations under different scenario assumptions related to the hypotheses, (4) garnering system-level insights that are robust with respect to the details of the particular input scenario assumptions and are broadly applicable to understanding system behavior in a wide variety of circumstances, and (5) communicating the conclusions in clear yet concise language while recognizing the various conditions and caveats under which those conclusions are valid.

Case Studies

In this section we present case studies that focus on specific areas of the biomass-to-biofuels supply chain: feedstock producers, biorefinery investors, retail outlets, fuel consumers, and energy-market agents. Each case presented involves interactions and feedback from the whole supply chain, but the insights and implications from these cases are presented in the context of the four areas mentioned above.

Feedstock Producers

Having a reliable feedstock supply system is essential to the growth of the biofuels industry in the United States. In an effort to incentivize biofuel feedstock production, the U.S. Congress included the Biomass Crop Assistance Program (BCAP) in the 2008 Farm Bill (U.S. Congress, 2008). We have attempted to model the BCAP as it is detailed in the 2008 Farm Bill, and have examined the overall implications of incentivizing feedstock production on the biofuel industry in the United States. In addition to exploring the implications of the BCAP, we also examine the industry's response to feedstock prices.

There are three major components of the final BCAP rule: matching payments; establishment payments; and annual payments. Matching payments are provided to individual producers for the collection, harvest, storage, and transportation (CHST) of qualifying cellulosic biomass materials. The CHST payment does not support urban residue, algal biomass, or residues used for higher value products. The CHST payment matches 50% of the costs incurred by the grower up to a maximum of $45/ton for a contract term of two years. Establishment payments are provided to producers to offset 75% of the costs associated with converting land and changing practices to produce dedicated energy crops. Annual payments are provided to qualifying producers and are based on regional agricultural cash rental rates. The annual payments have a contract term of five or fifteen years.

As modeled in the BSM, we have set the BCAP incentives to begin in 2009 with only CHST payments. We then have the transition to establishment payments in 2013 (for one year) and finally annual payments; all payments end in 2026. As part of the 2008 Farm Bill, BCAP is administered by the Farm Service Agency (FSA) of the USDA. Although the FSA may place an upper limit on the total funds that will be given to BCAP in a certain year, we have modeled the BCAP such that there is no upper limit to the funds available. This approach allows us to estimate how much money could be allocated to the program if it were not limited. However, we do constrain the land area that is considered to be eligible to 1 percent of all potential feedstock supply, which is roughly commensurate with the FSA's proposed budget for the BCAP (National Sustainable Agriculture Coalition, 2014).

- Feedstock production incentive programs such as the BCAP spur a transition in feedstock production from forest residues to dedicated herbaceous crops.

Model simulations from the BSM suggest that, in the presence of other supportive policies along the cellulosic biofuels supply chain (e.g., incentives for fuel producers and fuel consumers), the three-pronged approach of the BCAP results in increased annual feedstock production (Figure 15.3). In addition to incentivizing greater volumes of feedstock production, the BCAP initiates a shift in the type of feedstock that is produced, from forest residues to dedicated herbaceous energy crops. In the absence of feedstock production incentives, dedicated energy crops appear too risky for widespread adoption. Therefore, low-risk feedstocks such as forest residues make up the bulk of the feedstock supply. The BCAP, as modeled in the BSM, provides enough support to make transitioning to the production of dedicated herbaceous energy crops attractive to producers.

- Feedstock production incentives such as the BCAP cause regionally variable increases in feedstock production.

Implementation of the BCAP in the BSM results in a modest increase in feedstock production nationally (approximately 7%). However, much of this production occurs in the Northern Plains region, where dedicated energy crops displace conventional crops, namely wheat (Figure 15.3). The feedstock production incentives provided by the BCAP make the production of dedicated energy crops more economically attractive than the production of wheat, causing more farmers to change their practices. In addition to increasing the overall feedstock production, the BCAP has a residual effect that is observed after the termination of all BCAP-related payments; once

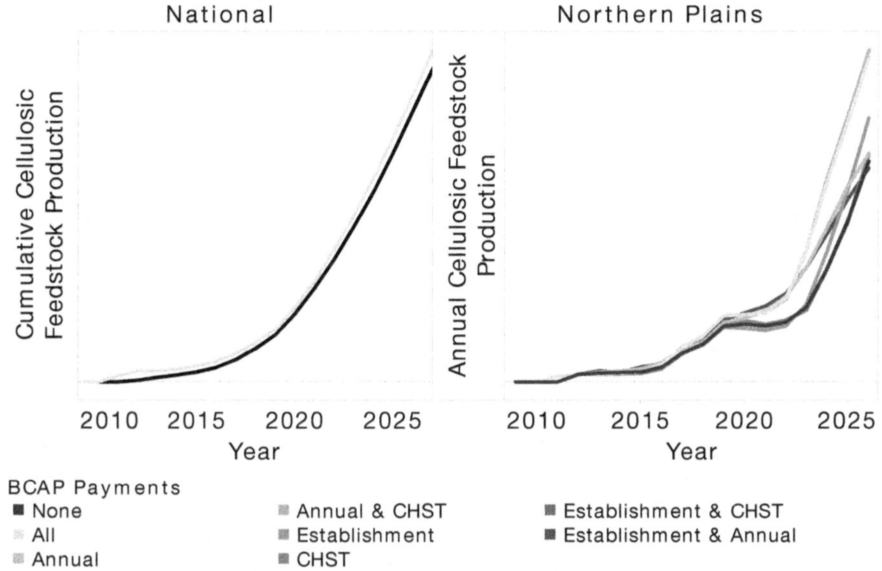

Figure 15.3 Cellulosic feedstock production under different BCAP scenarios during the period of BCAP implementation nationally (left) and for the Northern Plains (right). Note that the two panes are not displayed at the same scale or units. The different lines each represent a different scenario where only particular types of BCAP payment are included.

the BCAP payments end in 2026, feedstock production continues to increase for the duration of the simulation. Model simulations suggest that feedstock incentives such as the BCAP could help the industry overcome the initial risk associated with producing feedstock for the nascent biofuels industry. Once the biofuels industry gains inertia, the incentives may no longer be needed to ensure the supply of feedstock.

- Sustained moderate feedstock prices are sufficient to support high levels of cellulosic biofuel production. Regional feedstock markets result in prices that vary with time and geography in the BSM and are responsive to competition between energy crops and conventional crops and to the demand for biofuel feedstocks. The model uses initial feedstock prices, paid at the delivery point to the biorefinery (sometimes called the "plant gate"). During the later years of a simulation (i.e., 2030–2040), prices fluctuate regionally and can be twice as much as the initial conditions. To better understand the effect of feedstock prices on farm and conversion plant decision making, we undertook a sensitivity study in which feedstock prices are artificially held at a constant throughout the simulation: under such conditions, a feedstock plant-gate price that is approximately 50% greater than the initial conditions maximally stimulates total biofuel production. Lower feedstock prices do not provide sufficient financial incentive for typical growers to allocate land to produce cellulosic feedstock and the industry does not take off. However, results with higher prices are also ineffective at spurring biofuel production due to lower rates of conversion plant investment coupled with lower utilization rates for cellulosic plants. The higher cellulosic feedstock prices create an advantage for biofuel production technologies that utilize other types of feedstocks, such as starch-based technologies, thereby inhibiting the attractiveness of energy crops for potential feedstock producers.

- Maintaining a moderate floor on feedstock prices for growers over a long period tends to have adverse effects either on cellulosic ethanol production or on cellulosic feedstock prices.

We also investigated the effect of adding an artificial price floor (a minimum allowable price, but with potentially higher prices depending on the balance of supply and demand) sustained throughout the simulation. Overall, maintaining a price floor for feedstock growers and paid for in full by purchasers hampers biofuel production if it is set too high. This result is consistent with our previously noted observations in the fixed-feedstock-price case. For floor prices of less than the initial prices, production levels approximate those in the case where prices are set by the market.

- Regional feedstock market prices tend to do a better job of supporting the growth of the cellulosic ethanol industry than either constant-price or price-floor regimes.

We compared the results of either artificially holding feedstock prices constant or setting a floor for them against the standard BSM case where feedstock prices respond to production, demand, and inventory levels in regional markets; our simulations suggest (Figure 15.4) that the market mechanism does a better job, in the simulation, of supporting overall industry growth over long periods of simulated time, even though the feedstock prices in the BSM exhibit the cyclicality and instability typical for agricultural commodities. However, there is solid reasoning behind why the market mechanism is more effective than artificial price scenarios in achieving sustained industry growth: given the mix of downstream incentives, the BSM models a system that creates a potential demand for cellulosic feedstocks over time. The price/inventory feedbacks provide a mechanism for agricultural production to adjust to meet this demand. By artificially controlling feedstock prices in the foregoing artificial scenarios, the price floor short-circuits the adaptation process, resulting in lower amounts of cellulosic ethanol production.

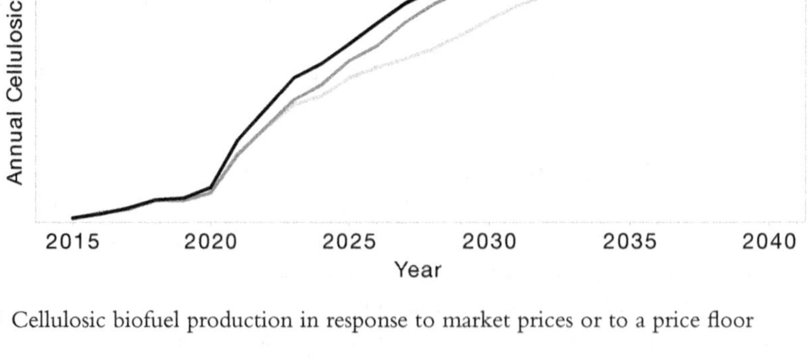

Figure 15.4 Cellulosic biofuel production in response to market prices or to a price floor

Biorefinery Investors

Providing incentives to the biofuels industry has a profound impact on investor behavior and, ultimately, the fate of the biofuels industry. Because the industry is in its early stages, biofuel conversion pathways are not always immediately attractive to potential investors. Therefore, incentives may be used to lower risk, thereby making investment more attractive and accelerating the industry's development. Complicating the biofuels industrial landscape is the fact that there are a number of potentially viable biofuel conversion pathways that are competing for market share and investment capital. Each biofuel conversion pathway has unique techno-economic parameters that interact with available incentives and investment decisions and ultimately impact the market competitiveness and longevity of any given biofuel conversion pathway. Below we explore three generic incentive scenarios in the BSM that are designed to highlight the dynamics of an industry with multiple competing technologies that have product parity yet disparate techno-economics. In the following analysis and discussion we focus only on a select group of biofuel pathways that produce a finished fuel through multiple biomass-to-hydrocarbon conversion processes. The three incentive scenarios are presented in Figure 15.5 and further discussed in Inman, Vimmerstedt, Bush, & Peterson (2014).

Output-Focused Incentives Scenario. The overall goal of the output-focused incentives scenario is to maximize biofuel production without exceeding a spending limit that is based on historic biofuel incentive expenditures. The annual spending limit is loosely based on the spending observed for the Volumetric Ethanol Excise Tax Credit (VEETC) program (Congressional Budget Office, 2010). This scenario focuses all incentives on one biomass-to-hydrocarbon fuel production technology in an effort to maximize its biofuel production. Based on the techno-economic data currently available, we have selected the most technologically attractive biofuel pathway. We have focused most of the incentives to support the construction of pioneer- and

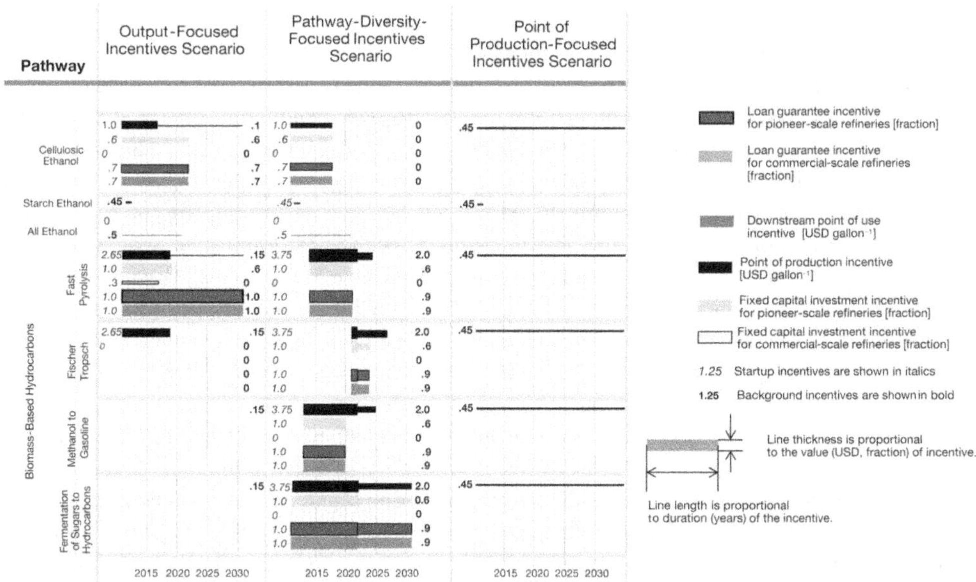

Figure 15.5 Scenario, incentive type, level, and duration

commercial-scale biorefineries by providing high levels of fixed capital investment (FCI) and loan guarantee support. The spending associated with the latter only accrues if a given biorefinery defaults on its loan.

Pathway Diversity-Focused Incentives Scenario. The pathway-diversity-focused incentives scenario utilizes a series of incentives that lead to multiple biomass-to-hydrocarbon fuel pathways gaining market share in addition to the ethanol industry retaining some market share without exceeding a spending limit that is loosely based on historic biofuels incentive spending, as detailed in the *Output-Focused Incentives Scenario* described above. The incentives' start times and durations are designed to promote multiple pathways. For example, conversion pathways that are immature receive early and substantial support for the construction of biorefineries (FCI and loan guarantee). Once these pathways have enough industrial momentum (i.e., accumulated learning), the incentives are reduced or terminated. In contrast, more mature pathways are incentivized at either lower levels or for shorter periods of time, or both. In this scenario, four pathways can be incentivized to achieve at least modest market share. Avoiding technology dominance (in which one pathway gains an insurmountable comparative advantage, thus "squeezing out" other options) is difficult and requires incentives to have staggered start times and variable durations. Some pathways have a combination of techno-economic attributes such that they require only minimal incentivizing while others need continued support to remain viable in the context of multiple competing pathways.

Point-of-Production-Focused Incentives Scenario. The point-of-production-focused incentives scenario is a case in which only a low-level production-focused incentive is available to biofuel producers. This scenario applies a minimal (i.e., less than $0.50 per gallon) point-of-production incentive to all biofuel conversion pathways, including cellulosic ethanol and starch ethanol, albeit the latter is incentivized for only one simulation year. The application and level of incentives in this scenario are somewhat similar to historical Renewable Identification Number (RIN) values and could be considered a very rough approximation of RIN market effects.

Incentives that promote the most economically attractive biofuel pathway maximize biofuel production levels. However, multiple conversion technologies can succeed and produce high levels of biofuels if incentives are deliberately designed to overcome initial maturity disparities.

The *Output-Focused Incentives Scenario* results in the greatest volume of biofuel being produced, as compared to the other two scenarios evaluated (Figure 15.6). As expected, the production associated with this scenario was predominately from one biofuel conversion pathway; the focused incentive program modeled under this scenario results in technology dominance, which prevents other biofuel conversion pathways from gaining any appreciable market share. Because the most techno-economically advantageous pathway was chosen for focused incentives, once the initial risk barriers are overcome, the pathway takes off rapidly with very little support. The early incentives reduce risk and encourage heavy investment in the pathway; early and substantial private investment is critical to its long-term market performance.

BSM simulations using the *Output-Focused Incentives Scenario* result in the highest volume of biofuel production, as compared to the other two scenarios, it is a potentially risky strategy where the development of a biofuels industry hinges on one biofuel conversion pathway. Issues such as feedstock supply disruptions, technological setbacks, and oil price fluctuations, to name a few, could have devastating implications to the industry as a whole. The *Pathway Diversity-Focused Incentives Scenario* results in multiple conversion pathways gaining market share that, when combined, result in roughly 75% of the production volume observed for the *Output-Focused*

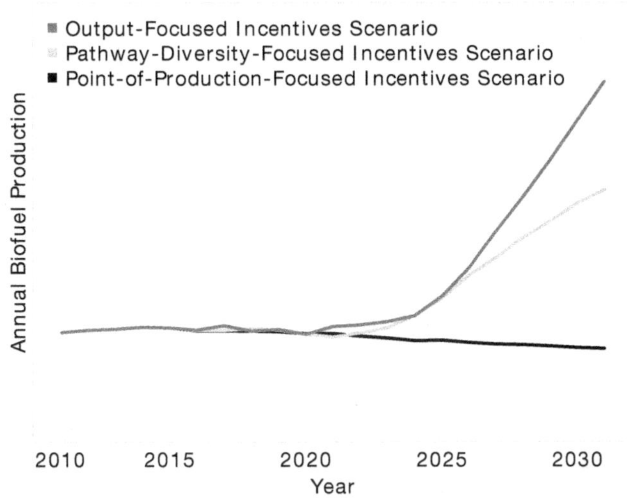

Figure 15.6 Relative biofuel production volumes by incentive scenario

Incentives Scenario. Both scenarios stay within the annual spending limit that was set. The *Pathway Diversity-Focused Incentives Scenario* takes advantage of the different maturity levels for the conversion pathways and uses strategic selection of incentives to allow conversion pathways to attract and receive investment without any one conversion pathway dominating the market. This approach is robust against disruptions to feedstock supply, and technological setbacks and failures.

In contrast to the two scenarios discussed above, model runs using the *Point-of-Production-Focused Incentives Scenario* results in stagnation of the biofuels industry. Under this scenario, industrial output declines slightly over the duration of the simulation. The decline is because of the increase in expected vehicle fleet fuel efficiency in response to Corporate Average Fuel Efficiency (CAFE) standards (U.S. Environmental Protection Agency, 2014) over the years of the simulation. New conversion technologies do not become attractive enough to garner investment and therefore never develop and gain market share; the biofuel production is created by more established conversion processes. Also, in this scenario, the biofuel that is produced is predominately used as a fuel additive and is not used as a standalone fuel.

Market Agents

In an effort to track biofuel production across fuel types and producers, the RIN (renewable inventory number) market was created as part of the regulations implementing the Energy Independence and Security Act of 2007 (U.S. Congress, 2007). In addition to the RIN market, there are also tax credits available to biofuel producers. In this exploration, we consider scenarios where there are two monetary incentives for biofuel producers: (1) tax credits and (2) the RIN market. The nuances of these policies cannot be precisely represented in the BSM; rather, we model the intent of such policies in scenarios designed for the BSM. Table 15.1 defines the simplified and idealized representation of production tax credits and RINs in this case study. This examination ignores, however, several complexities and assumptions: the tax credits do not always accrue to the producers, but may be captured by blenders and other economic entities.

Table 15.1 Idealized production tax credits and RIN values for the biomass-to-biofuel conversion pathways in the BSM

Pathway group	*Tax credit ($/gal)*	*Equivalence value (RIN/gal)*
Algae to hydrocarbons	1.00	1.6
Cellulose to butanol	0.41	1.3
Cellulose to ethanol	0.46	1.0
Biomass to hydrocarbons	1.01	1.6
Oil crop to hydrocarbons	1.00	1.6
Starch to ethanol	0.00	1.0

For this study, we assume that the full value of the tax credit is captured by the biofuel producer. (The same or other tax credits may be available to blenders and other economic entities.) The tax credits and RIN equivalence values vary somewhat within the groups of conversion pathways represented in the BSM. For this case study, we assume uniformity within those groups. The benefits of the RIN market accrue only partially to biofuel producers. We sidestep this issue by studying the evolution of the biofuels industry as a function of the effect on biorefinery revenue of the RIN market rather than as a function of RIN prices.

We created 3,456 time series for RIN prices between the years 2015 and 2030 using price-effect levels that ranged from 0 to $1.50/RIN. We constructed all possible time series that have one of the RIN prices (0 to $1.50) in 2015, changed the price linearly to another price until the year 2020 or 2025, and then changed the price linearly again to a third price in 2030; thus, the time series are piecewise linear in two segments between the years 2015 and 2030. In addition to these, we also created randomized time series where each year's price was taken randomly. We call these two types of time series "trending" and "random," respectively. Aside from varying the effect of RIN prices, we considered three pairs of options for incentive conditions:

- Starch ethanol is or is not eligible to produce RINs.
- There is a lower ("baseline deployment") or higher ("additional deployment") level of integrated biorefinery (IBR) demonstration and deployment investment. See Vimmerstedt & Bush, 2013 for a detailed definition of these scenarios, on which the current set of scenarios is loosely modeled.
- Production tax credits either are or are not available. When they are not available, the tax credits listed in Table 15.1 are equal to zero.
- RIN price trajectories can dramatically affect the long-term evolution of the biofuels industry and, in BSM simulations, have substantial synergies with other policies such as tax credits for biofuel production and investments in biorefineries.

Figure 15.7 shows the expected trend where higher average RIN prices spur greater production of biofuel and that decreasing RIN effect dampens the growth of the biofuel industry. In addition, excluding starch ethanol production from obtaining RINs somewhat discourages overall biofuel industry growth because in these simulations starch production tends to fall without profitability added by RINs. Additional investment in IBRs over the "baseline deployment" level has the most dramatic positive synergy with RIN effects on producer revenue, often doubling the growth of the cellulosic and algal biofuels industry. Loss of production tax credits significantly hampers industry growth unless RIN prices increase to a high value.

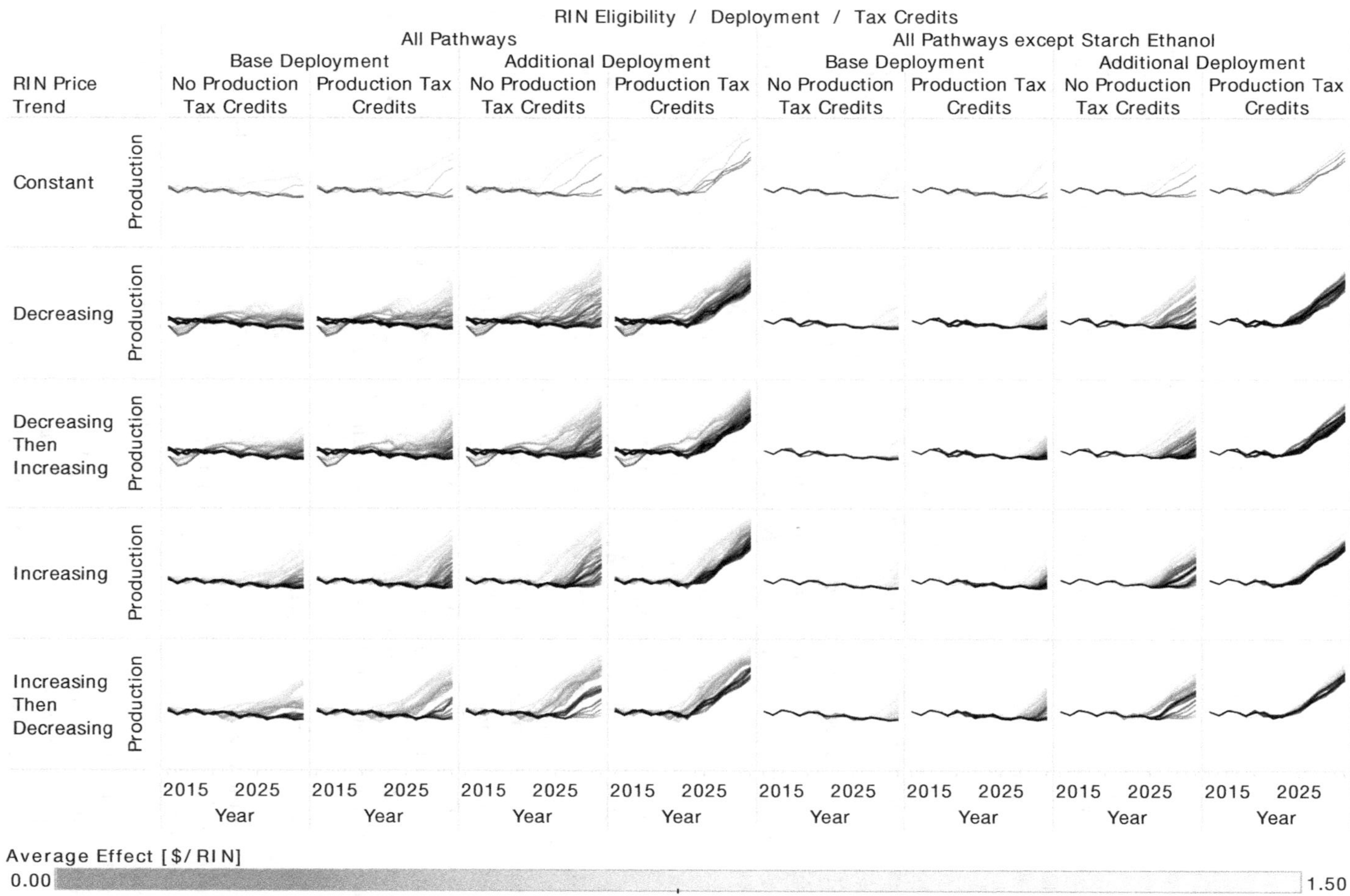

Figure 15.7 Total biofuel production as a function of RIN price scenario

Tax credits and additional deployment are sufficient—even with decreasing effects of RINs—for the eventual attainment of national biofuel volumetric goals, though not by 2022 as in EISA. RIN price trends that eventually increase seem to have a stronger influence than those that decrease, indicating that high RIN prices in early years might not be as impactful as those in later years. Highly variable RIN prices tend to destabilize the growth of the industry, and there is a greater impact when all biofuels obtain RINs than when starch ethanol is excluded from RINs.

In general, RINs alone are not sufficient to incentivize immature or techno-economically disadvantaged pathways; substantial early growth is only seen in cellulosic ethanol production. Biomass-to-hydrocarbon fuel pathways dominate later years, and cellulosic ethanol production growth stagnates unless starch-ethanol is not allowed RINs. Moreover, a substantial portion of the RIN scenarios where starch-ethanol is not allowed, RINs result in a shift away from starch-ethanol production towards cellulosic ethanol production.

Retail Outlets

Accurate modeling of the biofuel industry relies upon an understanding of retail prices and pricing strategies for biofuels, particularly ethanol. Publicly available data on ethanol rack and retail prices in the United States are neither historically nor geographically diverse: ethanol pricing data do not exist, in many cases, prior to 2007. Ethanol rack prices (i.e., the sales price at the refinery) are not available for states outside of the Midwest, Lake States, and Northern Plains regions of the BSM, which hinders broader geographic analysis. The majority of the available research literature finds gasoline and high-blend ethanol (E85) to be close substitutes. In particular, Anderson (2009) strongly supports the hypothesis that ethanol and gasoline prices are inherently linked. We performed a regression analysis on the available data, and the results suggest that there is a significant relationship between retail gasoline prices and E85 prices: for every \$0.10/gal increase in retail gasoline price, E85 price increases by \$0.077/gal.

In an effort to study the implications of various pricing strategies at refueling stations, we examined two potential retail pricing formulas for high-blend ethanol fuel. The first strategy is based on the statistical relationship observed between gasoline and E85 retail prices: E85 is priced at a set fraction of retail gas prices. The second strategy is simply a constant "markup" that is applied to the rack price of E85. Expressing the final retail price for E85 as a weighted average of these two pricing strategies allows the relevant importance of these two approaches to be investigated. The equation below illustrates this formulation.

$$P_{E85} = (b \cdot P_{gas})f + (c \cdot P_{EtOH})(1 - f) + \varepsilon ,$$

where P_{E85} is the retail price of E85, P_{gas} is the retail price of gasoline, P_{EtOH} is the rack price of ethanol, and f is the weight (between zero and one) given to the "markdown from gasoline" pricing strategy, with $(1 - f)$ representing the weight given to the "markup from ethanol" pricing approach. The quantity b is the discount on the gasoline retail price and c is the markup on the ethanol rack price; ε is the error term in the regression. Based on our data review and statistical analysis, we find f to be approximately 0.7. In other words, station owners set the retail price of E85 by applying a discount in relation to the retail price of gasoline and weighting that result by 70%, with the remaining 30% of the weighting coming from a constant markup from the ethanol rack price. E85 has approximately 75% of the energy per unit volume of gasoline; the markup on ethanol rack price varies from 35% to 39%.

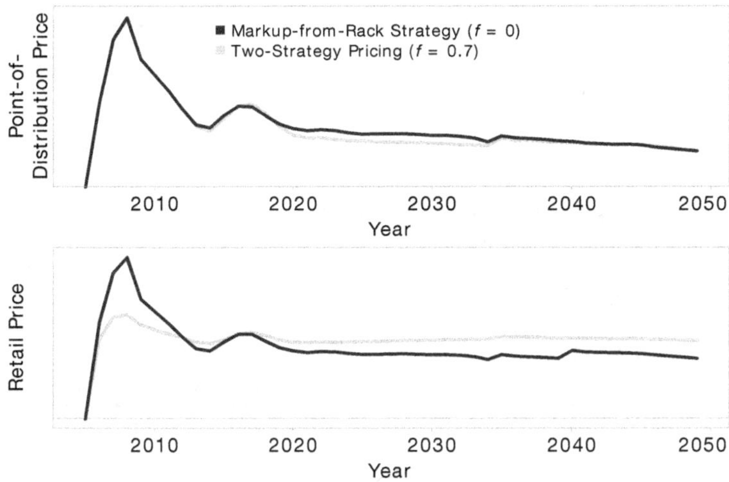

Figure 15.8 Comparison of pricing strategies and their effects on prices and deployment

- Weighted pricing with an emphasis on price coupling between gasoline to E85 results in higher prices at the pump and less ethanol production, compared to pricing E85 relative to the ethanol rack price.

We compared (Figure 15.8) the two-strategy pricing scheme ($f = 0.7$) in the BSM against the single-strategy pricing scheme ($f = 0$), which compares a pricing strategy that emphasizes price coupling (70% price coupling) and to a lesser extent (30%), a rack price markup, to a strategy that is based solely on a rack price markup. The link suggests tradeoffs. A pricing strategy that emphasizes price coupling may result in higher profits attributable to E85, which could incentivize investment in E85 tankage and dispensing equipment. On the other hand, a pricing strategy that emphasizes price coupling implies higher prices for high-blend ethanol and may portend lower market penetration than in a pricing strategy this is based on a rack price markup.

- Price coupling results in lower ethanol production than pricing based solely on ethanol rack prices.

With ethanol-gasoline price coupling ($f = 0.7$), the point of distribution price tends to be lower in most years, while the point of use price is elevated when compared to results using the uncoupled price formulation ($f = 0.7$). The impact of certain incentives and which agents receive them has an impact on the overall dynamics of the system. The blenders' tax credit (Renewable Fuels Association, 2014) with the uncoupled pricing logic resulted in the blender receiving a rebate of about \$0.50 off the excise tax on gasoline while any cost advantage was passed on to the consumer. With price coupling, the producer retains much of the benefit from the subsidy. The retailer has increased incentive for adding E85 tankage and dispensers, but consumers do not have the same incentive to purchase E85 due to the price increase.

Consumers

Understanding how various demand scenarios affect the biofuel industry is important to consider given the potential future growth in high-efficiency vehicles, flexible fuel vehicles (FFVs), and

Table 15.2 The six LDV ethanol-demand scenarios explored in this case study

Title	*Description*
E15 Transition	The proportion of E10 and E15 in the national fuel mix gradually increases over time.
RSP Transition	New RSP (renewable super premium) drivetrains that solely use E30 are gradually introduced into the vehicle fleet.
Non-liquid-fuel Transition	The proportion of new liquid-fuel LDVs introduced into the vehicle fleet is gradually reduced over time.
E85 Usage	The price-sensitivity of FFV (flex-fuel vehicle) owners to fueling with low-ethanol-blend fuel versus E85 is varied dramatically.
FFV Fleet	The proportion of new FFVs introduced into the vehicle fleet is gradually increased over time.
FFV Coverage	The propensity of FFV owners to make longer drives to fuel with E85 is varied.

higher blends of ethanol in gasoline. Using the BSM, we simulated the influence of light-duty vehicle (LDV) demand for ethanol on the overall evolution of the biofuels industry in order to develop system-level insights into the dynamic response of the biofuels supply chain at different levels of fuel demand. Note, however, that the BSM does not represent in detail all of the challenges of rapidly deploying ethanol distribution and dispensing infrastructure. We explored six sets of demand scenarios (Table 15.2) that vary the average ethanol content of fuel for gasoline-blend-fueled vehicles, the drive-train/engine-type for the LDV vehicle fleet, or the fueling behavior of FFV owners. These demand scenarios are explored in the context of the incentive scenarios displayed in Figure 15.5.

- A substantial transition to RSP (renewable super premium) drive trains or heavy use of E85 is more influential on overall biofuel production volumes than a transition to E15, although a transition to E15 incentivizes greater ethanol production in earlier years.

In cases where biomass-based hydrocarbon fuels attain significant market share, the transition to E15, RSP, or heavy FFV use of E85 increases competition for feedstock and biorefinery-construction capacity between cellulosic ethanol and infrastructure-compatible biofuels and results in moderately reduced production volumes of infrastructure-compatible biofuels and increased volumes of cellulosic ethanol, with total biofuels volumes changing only slightly (see Figure 15.9). In cases with aggressive incentives, the raising of the ethanol "blend wall" – i.e., the point where the E10 fuel market is saturated with ethanol (see U.S. Environmental Protection Agency, 2010), due to a transition to E15, RSP, or heavy FFV use of E85 alleviates the competition between starch-based ethanol and cellulosic ethanol. Starch ethanol production capacity is no longer idled as cellulosic ethanol production grows; such reduced capacity utilization is still present in the less aggressive minimal and business-as-usual policy cases. Scenarios with substantial increases in ethanol demand but no other external support do not achieve biofuel production levels that meet national goals. Transition away from liquid-fuel LDVs does severely impact both the starch ethanol and cellulosic ethanol industries; the impact is proportional to the potential low-blend ethanol consumption by the vehicle fleet. Impacts on the ethanol industry from changes in vehicle technologies are protracted and take decades for their full effects to be realized. In cases where ethanol demand increases rapidly, ethanol retail prices show large increases (i.e., doubling) for the first several years of rapid demand growth; after this, prices return to pre-transition levels because the industrial capacity expands sufficiently to meet demand.

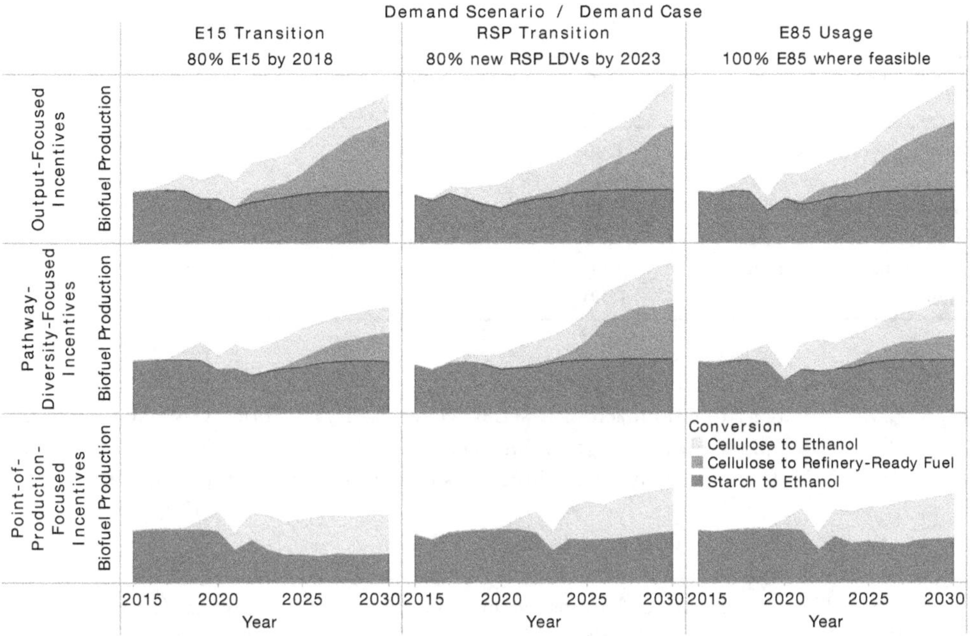

Figure 15.9 Illustration confirming that transitions to heavy RSP or E85 usage are more influential than fleetwide use of E15

- Because the price discrepancies between gasoline and higher ethanol blends can mute the impact of FFV growth on overall ethanol demand, a transition towards a fleet dominated by FFVs does not, in itself, strongly affect the biofuels industry

The greater willingness of FFV owners to aggressively seek out E85 (e.g., drive farther to find a fueling station) when it has price parity with E10 moderately accelerates the growth of the cellulosic ethanol industry. Although the details of production volumes and biorefinery construction in the various scenarios studied here are sensitive to the techno-economic assessments of conversion pathways, the insights in this case study appear to be robust with respect to those assessments.

Scenarios that increase the ethanol blend wall compel the development of the cellulosic ethanol industry and protect the starch-ethanol industry from eventual underutilization. Conversely, a transition away from liquid-fuel vehicles that use ethanol blends poses severe challenges for both the cellulosic- and starch-ethanol industries. Transitions to E15 for low-blend fuels, or a requirement that FFVs use E85 wherever possible, tend to support more rapid growth of ethanol demand than the introduction RSP vehicles requiring blends like E30; however, RSP scenarios promise larger overall ethanol demands in the longer term. Any rapid uptake in ethanol demand generates synergistic effects such as a substantial half-decade rise in ethanol rack prices, competition with infrastructure-compatible fuels for feedstock and biorefinery construction capacity, and the development of feedstock resources.

We created 3,456 time series for RIN prices between the years 2015 and 2030 using price-effect levels that ranged from 0 to $1.50/RIN. We constructed all possible time series that have one of the RIN prices (0 to $1.50) in 2015, changed the price linearly to another price until the year 2020 or 2025, and then changed the price linearly again to a third price in 2030; thus,

Table 15.3 Idealized production tax credits and RIN values for the biomass-to-biofuel conversion pathways in the BSM

Pathway group	*Tax credit (\$/gal)*	*Equivalence value (RIN/gal)*
Algae to hydrocarbons	1.00	1.6
Cellulose to butanol	0.41	1.3
Cellulose to ethanol	0.46	1.0
Biomass to hydrocarbons	1.01	1.6
Oil crop to hydrocarbons	1.00	1.6
Starch to ethanol	0.00	1.0

the time series are piecewise linear in two segments between the years 2015 and 2030. In addition to these, we also created randomized time series where each year's price was taken randomly. We call these two types of time series "trending" and "random," respectively. Aside from varying the effect of RIN prices, we considered three pairs of options for policy conditions:

- Starch ethanol is or is not eligible to produce RINs.
- There is a lower ("baseline deployment") or higher ("additional deployment") level of integrated biorefinery (IBR) demonstration and deployment investment. See Vimmerstedt & Bush, 2013 for a detailed definition of these scenarios, on which the current set of scenarios is loosely modeled.
- Production tax credits either are or are not available. When they are not available, the tax credits listed in Table 15.3 are equal to zero.

Reflections and Conclusions

Exploring the evolution of the biofuels industry using the BSM yields insights into the dynamics of the entire biofuels system as well as individual components of this system. Our methodology seeks to be pragmatic, adaptive, and parsimonious. We employ an approach that combines elements of hypothesis generation and testing, scenario development, statistical and sensitivity analysis, and exploratory data visualization. This eclectic orientation adapts to an ever-changing and developing industrial system to develop conclusions. Aspects of input assumptions may be highly uncertain, study questions imprecisely formulated, and external conditions may be variable. We address these challenges by melding traditional system dynamics methodology, sensitivity analysis and uncertainty quantification, and hypothesis testing in a combined approach that leverages the strengths of those disciplines. Most of the case studies presented above rely on traditional system-dynamics analysis and on hypothesis testing; the feedstock producers and market agents studies also involved sensitivity analysis; the retail outlet study relied on statistical analysis; and the biorefinery investor study emphasized scenario development.

Focusing on specific actors across the supply chain, as done in this chapter, reveals several insights about potential trajectories of the biofuels industry in the U.S. as well as the relative leverage of these actors on the evolution of the biofuels industry. Overall, results from our model underscore the importance of industrial learning and how policies designed to accelerate the industrial learning process move the biofuels industry forward in a lasting way. Many agents interact with the industrial learning process. In many instances, incentives may be applied in ways that do not directly interact with the industrial learning process but can have strong synergistic effects with industrial learning. The case studies presented in this chapter also illustrate the potential for incentives to be strategically timed to help lower risk at key points in the development of the industry. For example, the feedstock incentives explored in this chapter show that such incentives are important to overcome the risk of producing new feedstocks, which for the farmer may require investing in new cultivation practices and machinery. Lowering the risk of adopting new farming practices also lowers the risk to biorefinery investors which, in turn, accelerates the rate of industrial learning. Once the industry gains enough momentum, incentives may be ended without negatively impacting the overall industry.

Examining the impact of incentives on biorefinery investors, the influence of applying incentives early and in a way that lowers the overall risk of building refineries is salient. Incentives that are focused on FCI and loan guarantees greatly reduce the risk of investing in the construction of refineries. When such incentives are applied to techno-economically advantageous conversion pathways, technological dominance may occur. The existence of multiple commercial conversion pathways does not guarantee that multiple conversion pathways will gain market share. A biofuel market that is dominated by a single conversion technology is potentially undesirable, and could be avoided with significant intervention, using carefully timed incentive activation and termination. Retail outlets play an important role in the development of the biofuels industry, especially the cellulosic ethanol industry. The impact of how retail stations structure the price of high-blend ethanol fuel provides system feedback that has far-reaching implications for the biofuels industry. Pricing strategies that are based, at least in part, on coupling high-blend ethanol price with gasoline price result in higher profits (per unit volume) to the retail outlets and potentially more investment in ethanol distribution infrastructure. However, this also increases prices for the consumers and potentially lowers sales. In turn, lower sales reduce attractiveness of the cellulosic ethanol industry as an investment, slowing its development.

In general, BSM analyses have identified four keys to biofuels industry development: (1) profitability at point of production, (2) high rates of industry learning, (3) an aggressive start in building pilot, demonstration, and pioneer-scale plants, and (4) for ethanol, a high level of infrastructure investment to sustain low enough point-of-use prices (Newes, Bush, Peck, & Peterson, 2014). Even with these supportive conditions the "take off" of the industry is likely to be rough, with unstable, higher than anticipated, feedstock prices; a tendency towards a boom/bust development of production capacity; and the potential for biofuel price instability. Nevertheless, significant production volumes are feasible and could reach the levels envisioned in the Energy Independence and Security Act of 2007 (U.S. Congress, 2007).

Acknowledgments

This work was supported by the U.S. Department of Energy under Contract No. DE-AC36–08-GO28308 with the National Renewable Energy Laboratory. We appreciate Mark Paich's assistance with the analysis of ethanol pricing.

References

Anderson, S. T. (2009). *The Demand for Ethanol as a Gasoline Substitute.* East Lansing, MI: Michigan State University.

Bioenergy Technology Office, U.S. Department of Energy. (2014, June 17). *Processing and Conversion.* Retrieved from Research & Development: www.energy.gov/eere/bioenergy/processing-and-conversion

Congressional Budget Office. (2010). *Using Biofuel Tax Credits to Achieve Energy and Environmental Policy Goals.* Washington, DC.

Downing, M., Eaton, L. M., Graham, R. L., Langholtz, M. H., Perlack, R. D., Turhollow Jr, A. F., . . . Brandt, C. C. (2011). *U.S. Billion-Ton Update: Biomass Supply for a Bioenergy and Bioproducts Industry*. Oak Ridge, TN: Oak Ridge National Laboratory.

Inman, D., Vimmerstedt, L., Bush, B., & Peterson, S. (2014). *Biomass Scenario Model ScenarioLlibrary: Definitions, Construction, and Descriptions*. Golden, CO: National Renewable Energy Laboratory.

isee systems. (2010). STELLA: systems thinking for education and research software.

Lin, Y., Bush, B., Newes, E., & Peterson, S. (2013). *Biomass Scenario Model v2.0 Documentation: Data and References*. Golden, CO: National Renewable Energy Laboratory. Retrieved from: www.osti.gov/scitech/biblio/1082565

National Sustainable Agriculture Coalition. (2014, January 13). *Biomass Crop Assistance Program*. Retrieved from: http://sustainableagriculture.net/publications/grassrootsguide/renewable-energy/biomass-crop-assistance-program/

Newes, E., Inman, D., & Bush, B. (2011). Understanding the developing cellulosic biofuels industry through dynamic modeling. In M. dos Santos Bernardes, *Economic Effects of Biofuel Production* (pp. 373–404). Rijeka, Croatia: InTech.

Newes, E. K., Bush, B. W., Peck, C. T., & Peterson, S. O. (2014). Insights into the effect of policy on the potential development of the cellulosic ethanol industry. Submitted to *Environmental Science & Technology*.

Office of the Biomass Program and Energy Efficiency and Renewable Energy (2013). *Bioenergy Technologies Office Multi-Year Program Plan*. Washington, DC.

Peterson, S., Bush, B., Newes, E., Inman, D., Hsu, D., Vimmerstedt, L., & Stright, D. (2013). An overview of the biomass scenario model. 31st International Conference of the System Dynamics Society. Boston, MA.

Randers, J. (1980). *Elements of the System Dynamics Method*. Boston, MA: MIT Press.

Renewable Fuels Association. (2014, January 15). *Tax Incentives*. Retrieved from: www.ethanolrfa.org/pages/tax-incentives

Sterman, J. (2006). Learning from evidence in a complex world. *American Journal of Public Health*, 96(3): 505–14.

U.S. Congress (2007). Energy Independence and Security Act of 2007 (EISA).

U.S. Congress (2008). Food, Conservation, and Energy Act of 2008.

U.S. Department of Agriculture, Economic Research Service. (2014, January 13). *U.S. Farm Resource Regions*. Retrieved from Agricultural Resource Management Survey (ARMS): Resource Regions: http://webarchives.cdlib.org/wayback.public/UERS_ag_1/20111128195215/www.ers.usda.gov/Briefing/ARMS/resourceregions/resourceregions.htm#new

U.S. Environmental Protection Agency. (2010, July 8). *The Ethanol Blend Wall*. Retrieved from This Week In Petroleum: www.eia.gov/oog/info/twip/twiparch/100708/twipprint.html

U.S. Environmental Protection Agency. (2014, January 13). *Regulations & Standards: Light-Duty*. Retrieved from Transportation and Climate: www.epa.gov/otaq/climate/regs-light-duty.htm

U.S. White House (2011). *Blueprint for a Secure Energy Future*. Washington, DC: The White House.

Vimmerstedt, L., & Bush, B. (2013). *Effects of Deployment Investment on the Growth of the Biofuels Industry*. Golden, CO: National Renewable Energy Laboratory.

Walker, B., & Salt, D. (2006). *Resilience Thinking: Sustaining Ecosystems and People in a Changing World*. Washington, DC: Island Press.

16

INTEGRATED AGENT-BASED AND SYSTEM DYNAMICS MODELLING FOR SIMULATION OF SUSTAINABLE MOBILITY

Ehsan Shafiei, Hlynur Stefansson, Eyjolfur Ingi Asgeirsson, Brynhildur Davidsdottir, and Marco Raberto

Research Question: What are the barriers and facilitators to the adoption of alternative fuel vehicles?
System Science Method(s): System dynamics & Agent-based models
Things to Notice:

- Combination of top-down system dynamics and bottom-up agent-based models
- Application to a specific national context using empirically-informed assumptions

In this chapter, a conceptual framework for a comprehensive evaluation of the diffusion process of alternative fuel vehicles is introduced. The framework takes into account the most influencing stakeholders, including car manufacturers, car dealers, consumers, energy supply system, fuel stations and government. The underlying mathematical models of different stakeholders are then integrated in one model of the whole energy and transport system. The hybrid modeling framework links the two powerful dynamic simulation approaches of system dynamics (SD) and agent-based (AB) modelling. Integrated modelling structure gives the potential of building more accurate and computationally efficient models for simulating the transition to sustainable mobility. We specify the integration process and the most important linking variables between various energy and transport components. Then the application of the integrated model is explained through a test case and, finally, the applicability of the hybrid AB and SD approach and its potential contribution to the models of transition to sustainable mobility will be concluded.

Increasing the security of energy supply for transportation and reducing environmental emissions from vehicles are recognized as the great challenges of the twenty-first century (Wokaun & Wilhelm, 2011). It implies that the transition to a sustainable transportation system would be an essential plan during the next decades. This process is not just a matter of technological change and broader changes are needed in economy, culture, consumers' behaviours, institutions and new infrastructures (Nunen, Huijbrechts, & Rietveld, 2011). Transition towards a more sustainable mobility system entails the successful development, diffusion and adoption of both improved

 DOI: 10.4324/9781315748771-18

conventional internal combustion engines (ICEs) and alternative fuel vehicles (AFVs). In this regard, the main questions that must be answered are: Which types of vehicle technologies can support the transition to sustainable mobility? How can these technologies be introduced in the marketplace? What are the main barriers to the successful diffusion process? How can AFVs be adopted by the consumers? What are the most effective policies for the market penetration of AFVs?

Such questions are difficult to answer and there is need to provide analytical tools for modelling both energy and transportation systems to evaluate the effectiveness of decisions under different conditions. On the energy system side, possibilities to switch from conventional to alternative fuels should be evaluated. On the vehicle side, the possibility to reduce conventional fuel demand by a shift to more efficient drivetrains or AFVs can be studied (Vliet, Broek, Turkenburg, & Faaij, 2011).

To do such analysis, different quantitative models focusing on diverse aspects with different approaches have been employed. System dynamics (SD) and agent-based (AB) modelling are two well-known dynamic simulation approaches which have been widely used for studying the diffusion of AFVs. These modelling approaches take fundamentally different perspectives when simulating a system. SD uses a "top-down" modelling approach with a high level of aggregation and models a system by breaking it into its major components and interactions (Macal, 2010). AB is recognized as a "bottom-up" approach as it studies a system by modelling the individual entities and their interactions to discover their collective behaviour patterns (Macal, 2010).

The real-world problems in the transportation system, however, do not match up well with a single modelling approach. This system involves many stakeholders with different nature, thus calling for a multi-paradigm approach that includes the entire system. In this study, we propose a conceptual framework for a comprehensive analysis of the energy and transportation system. This framework suggests developing a dynamic simulation model based on the approach combining SD structures with AB techniques that has not been fully explored before. Identification of linking variables and the process of integrating energy and transport stakeholders together will therefore be the main focus of this article.

In the second section, a review of existing studies on top-down, bottom-up and hybrid simulation models is presented. In order to fully explore all aspects of transitions to sustainable AFVs, a conceptual framework is proposed in the third section. Theoretical foundations and mathematical models for the behavior of different stakeholders are presented in the fourth section. The fifth section is devoted to the discussion of an illustrative example, showing an integrated AB and SD model. Finally, conclusions and prospects for future research are provided in the final section.

Simulation of Transportation System: Review of Existing Methods and Studies

Top-Down Approaches

System dynamics is a top-down approach that looks at the process of market development as a whole and facilitates understanding the interactions of many stakeholders in complex systems. Several SD models have been developed to describe the transition to AFVs. Struben (2006a, 2006b) and Struben and Sterman (2008) developed an SD model describing a complex dynamics between consumers, vehicle manufacturers and infrastructures. Struben and Sterman (2008) considered one technology platform as the representative of different AFV technologies.

They modelled consumers' purchase probabilities by using the multinomial logit (MNL) model, which includes the utility of consumers and the effects of social exposures. In the car manufacturer section, the impact of technology improvement on the production cost of vehicles is discussed. The entrance or exit behaviour of fuel stations is studied in the infrastructure section. The main variables considered in their study are the probability of contact between consumers (as a function of consumers' utility and the effects of social exposures), costs of production for vehicles and availability of infrastructure and fuel stations.

The National Renewable Energy Laboratory in the USA has developed a spatial and dynamic model called HyDIVE (Hydrogen Dynamic Infrastructure and Vehicle Evolution). HyDIVE was founded on Struben's works and illustrated the relations between fuel cell vehicle (FCV) sales and hydrogen infrastructure build-up (Welch 2006, 2007a, 2007b). Following Struben and Sterman, a similar SD model was developed by Stephan and Feller (2009) to study the diffusion of electric vehicles (EVs) in Germany. Shepherd, Bonsall, and Harrison (2012) extended Struben's work to analyse the development of EVs in UK.

Janssen (2005), Janssen, Stephan, Lienin, Gassmann, and Wokaun (2004) and Janssen, Lienin, Gassmann, and Wokaun (2006) used the SD framework to study the market introduction of natural gas vehicles in Switzerland. In their studies, the stakeholder sectors are divided into three categories: policy-making stakeholders (gas industry and government), the international car industry sector and interdependent stakeholders. The interdependent stakeholders section, which is the heart of the model, includes customers, refuelling stations and car retail and services. Bosshardt, Ulli-Beer, Gassmann, and Wokaun (2007, 2008) and Bosshardt (2009) extended Janssen's work by improving the model structure and considering the interactions between different drivetrain technologies. Bouza, Ulli-Beer, Dietrich, and Wokaun (2009) extended the works of Bosshardt by analysing the innovation and competition dynamics in the automobile industry.

Schneider, Shade, and Grupp (2004) employed the SD approach to study the diffusion of FCVs. The purpose of the model is to depict the innovation process from a company's perspective. It reflects the three strategic approaches of pioneers, early followers and late followers, and examines the future prospects of the car manufacturers' strategies in different scenarios. Each producer can influence its competitiveness by doing R&D, marketing and pricing. In this framework, each FCV producer competes against other producers of FCVs and conventional ICE vehicles.

Stefano, Baldoni, Falsini, and Taibi (2010) proposed an SD model representing the private road transport sector to forecast the fuel demand in EU countries. The fuel consumption is estimated by means of average fuel efficiency of vehicles and services demand in terms of passenger-kilometres.

Santa-Eulalia, Neumann, and Klasen (2011) presented an SD model, which combines the Bass diffusion and the discrete choice models, to forecast the adoption rate and timing of EVs in Germany. They identified recharging technology and infrastructures as the crucial factors for successful diffusion of EVs. Park, Kim, and Lee (2011) developed an SD simulation model for penetration forecasting of hydrogen FCVs in Korea based on the generalized Bass diffusion model. Keles, Wietschel, Mosta, and Rentza (2008) developed an SD model that takes into account the actions of consumers, automotive manufacturers, fuel stations and policymakers to study the transition to FCVs. The model consists of four modules: FCV demand and supply, attractiveness, fuel stations and balance of payments. Kwon (2012) focused on strategic niche management in an SD framework to investigate the barriers to market share development of AFVs. Meyer and Winebrake (2009) used the SD approach to explore the diffusion of hydrogen FCVs and refuelling infrastructure as the complementary goods. Walther, Wansart, Kieckhafer,

Schnieder, and Spengler (2010) proposed an SD framework to assess the strategies of the automobile industry in reaction to environmental regulations. Their model includes four interactive modules: environmental regulations, vehicle stock and infrastructure, automotive industry and customers.

Several SD-based macro-economic models have been developed to assess the decisions in transport policy. Leaver and Gillingham (2008, 2010) and Leaver, Gillingham, and Leaver (2009) employed an SD model of New Zealand's energy economy to assess the impacts of AFVs. ASTRA (Assessment of Transport Strategies) is the other example of the SD-based macro-economic model, which has been developed for strategic assessment of EU transport policies (Köhler, Wietschel, Whitmarsh, Keles, & Schade, 2010; Krail & Schade, 2010). ESCOT (model for economic assessment of sustainability policies of transport) is another model to describe the development path towards a sustainable transport system in Germany (Schade & Schade, 2005).

Bottom-Up Approaches

Agent-based modelling is a bottom-up approach, which allows modeling of heterogeneous agents that interact with each other in a repetitive process (Garcia, 2005). During the past decade, the methodology of AB has been increasingly employed to simulate individuals' behaviours and the interdependencies among the key agents of the vehicle market such as manufacturers, consumers, fuel stations and governmental agencies.

Most models introduced in the literature for consumer choice behaviour are based on utility theory, which assumes that consumers purchase the vehicles in the choice set with the highest utility (Bhata, Sen, & Eluru, 2009; Mohammadian & Miller, 2003a). Some studies assume that consumer agents decide according to a simple weighted utility of individual preference and social influence (Delre, Jager, Bijmolt, & Janssen, 2007, 2010; Delre, Jager, & Janssen, 2007; Janssen & Jager, 2003). Other studies have used MNL models to determine the probability of vehicle adoption (Bhata et al., 2009; Garcia, 2007; Mohammadian & Miller, 2003a, 2003b; Mueller & Haan, 2009).

Mueller and Haan (2009) employed a two-stage micro-simulation model to analyse the individual choices of new passenger cars. Individual choice sets are constructed in the first stage and, then, household agents, who are distinguished by socio-demographic characteristics and car ownership, evaluate alternatives in their individual choice set using an MNL model. Eppstein, Grover, Marshall, and Rizzo (2011) developed an AB model for consumer vehicle choice to explore the market penetration of plug-in hybrid electric vehicles (PHEVs). The model accounts for spatial and social effects as well as media influences. Tran (2012) developed an AB model for investigating the role of individual behaviour and network influence in diffusion of AFVs. Shafiei et al. (2012) developed an AB model to study the market share evolution of EVs in Iceland. ICE and EVs compete for market penetration through a vehicle choice algorithm that accounts for social influences and consumers' attractiveness for vehicle attributes.

Some studies combine the manufacturer's perspective with the purchasing behaviour of consumers in order to simulate the market penetration of AFVs. Zhang, Gensler, and Garcia (2011) developed a multi-agent model to study the interactions among car manufacturers, consumers and policymakers. They considered three mechanisms for analysing the adoption rate of AFVs: technology push, market pull and regulatory policies.

Schwoon (2006) developed an AB model, which captures the main interdependencies among consumers, car manufacturers and fuel stations to simulate the possible diffusion paths for FCVs. Consumers, which are heterogeneous with respect to their preferred car characteristics, maximize their utility and are influenced by their neighbours' decisions and fuel availability. Fuel-station owners react towards the number of FCVs sold in the market. The producers of ICE and FCVs optimize a weighted average of the expected revenue and market share in each period. The model evaluates the impact of different tax policies on diffusion of FCVs.

Huétink, van der Vooren, and Alkemade (2010) used an AB framework to study the process of development of hydrogen vehicles from niche to market. In this model, market penetration of hydrogen vehicles is influenced by the interactions of consumers, refuelling stations and technological development. Vliet, Vries, Jager, and Turkenburg (2008) and Vliet, Vries, Faaij, Turkenburg, and Jager (2010) developed a multi-agent model to simulate the transition from petroleum product vehicles to AFVs. The model considers the entire well-to-wheel fuel system of transportation from resources to final consumption. They used techno-economic data to parametrize various fuel supply pathways and survey data to parametrize heterogeneous consumer agents. The consumer agents purchase the various fuels on the basis of their individual preferences and social influences.

Complex adaptive system (CAS) is an AB simulation model to understand how the transition to a hydrogen infrastructure might occur (Mahalik et al., 2009; Tolley & Jones, 2006). There are two types of agents in the model: potential buyers of hydrogen vehicles and potential investors in hydrogen fuel stations. The model focuses on the chicken-and-egg problem between supply of hydrogen fuel and demand for FCVs.

Stephan and Sullivan (2004a, 2004b) proposed a model that addresses refuelling concerns along actual trips together with dynamic feedbacks between hydrogen infrastructure and FCV driving. Consumers and hydrogen fuel stations are two types of agents in the model which interact on a grid representing a central metropolitan area, suburbs and a rural area. Consumers tend to purchase the FCV if they are frequently exposed to hydrogen fuel stations during their usual trips. Conversely, the number of fuel stations is increased if sufficient FCVs are observed in the market. They introduced a "worry factor" to consider the effects of long distances between hydrogen stations and consumers' usual trips.

Günther, Stummer, Wakolbinger, and Wildpaner (2011) proposed an AB model to simulate the diffusion of a second-generation biomass fuel in Austria. In their model, heterogeneous consumers are embedded in a social network and bio-fuels are characterized by the price, quality and environmental friendliness.

Hybrid Modelling

SD is a suitable approach for modeling homogeneous systems, which is described by interactions among variables and their feedback structure at an aggregate level. AB models can capture heterogeneity of individuals in an interconnected network, but with increased computational costs that may limit sensitivity analysis and model scope (Rahmandad & Sterman, 2008). No single approach is ideal for all problems, and the costs and benefits of the disaggregation determine the type of modeling approach for policy analysis (Rahmandad & Sterman, 2008). Additionally, due to conceptual and computational limitations of each approach, many problems cannot be modelled in a satisfying way by a single approach. Hence, several researchers recommend using the combined form of AB and SD approaches to handle complex systems (Grossler, Stotz, & Schieritz, 2003; Lattila, Hilletofth, & Lin, 2010; Martinez-Moyano, Sallach, Bragen, & Thimmapuram, 2007; Schieritz, 2002; Schieritz & Grossler, 2003; Scholl, 2001; Scholl & Phelan, 2004; Vincent, Giannino, Rietkerk, Moriya, & Mazzoleni, 2011).

Keenan and Paich (2004) addressed the limitations of SD for the complex models of AFV diffusion and discussed how AB models might complement it. The hybrid models for studying sustainable transportation can, however, hardly be found in the literature. To the best of our knowledge, three studies explicitly address this issue. Kieckhäfer, Walther, Axmann, and Spengler (2009) proposed a conceptual framework for integration of SD and AB to study the product strategies in the automotive industry. Kieckhäfer, Wachter, Axmann, and Spengler (2012) showed how AB and SD could be combined to support strategic product portfolio planning in the automotive industry. They addressed the potentials and limitations of hybrid modelling for practical use. Kieckhäfer's framework, however, has paid less attention to the role of the energy supply system and fuel stations. Kohler et al. (2009) attempted to use a hybrid AB–SD model for assessing transition to sustainable mobility. They defined two types of agents including a small number of complex agents, which have an internal SD structure, and a larger number of simple agents. This work did not consider the effects of the energy supply system and fuel infrastructure.

In studying the diffusion process of AFVs, modelers have to study systems composed of several components of different nature. Therefore, hybrid modeling is useful as the combination of heterogeneous and homogeneous components are observed in the transportation system. SD is able to describe the influences of factors on the processes of AFV diffusion at aggregated levels. Combined use of SD and AB approaches can provide a spatial dimension and enable the transportation system agents to interact in space (Vincent et al., 2011). Moreover, compared with the single approaches, it could lead to a satisfying computation time and accuracy.

Conceptual Framework for Hybrid Simulation of Sustainable Mobility

According to the literature surveyed above, it can be concluded that the main agents or components influencing the diffusion of AFVs can be classified as: consumers, car manufacturers, car dealers/importers, energy supply system, fuel/charging stations, government and environment. Figure 16.1 presents the proposed conceptual framework for analysing the diffusion of AFVs, showing the interconnected agents and boundaries.

Consumer agents are identified as the households, which are the private users of fuel and vehicles. Public transportation and governmental institutions can be considered as the final consumers of energy and transportation services as well. In general, the individual goal of any consumer is utility maximization. Consumers decide on vehicle and fuel purchase based on personal preferences, vehicle attributes, social influences and government incentives. Environment attributes such as availability of car services and characteristics of the built environment can influence the purchase decision. The main attributes of consumers are socio-demographic characteristics and vehicle ownership behaviours.

Car manufacturer agents design and supply the products according to consumers' purchasing behaviour. The individual goal of manufacturers is profit maximization. They decide on the type of vehicle production and vehicle attributes. Manufacturers can improve the attributes of existing vehicles through the learning process or offering new alternatives. The main attributes of manufacturer agents are type of platform, knowledge and experience level, capital stock, production capacity, market share and the role in the market (e.g. follower or leader).

Car dealer agents offer the produced vehicles according to consumers' buying behaviour. The individual goal of dealer agents is profit maximization. They decide on offering the type of vehicle and selling price in the market based on information about availability of different vehicles, consumers' preferences and the market status of fuel/charging stations. The main attributes of

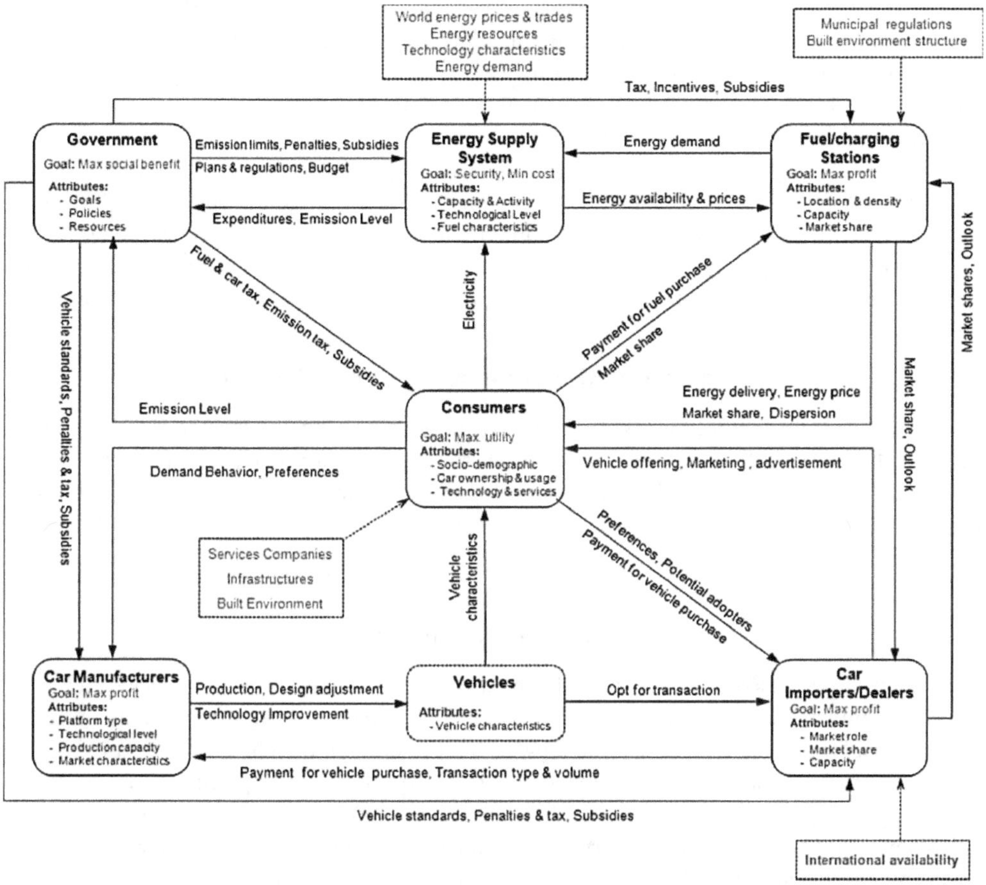

Figure 16.1 The conceptual framework for the integrated model of sustainable mobility

dealer agents are type of vehicle for transaction, offering capacity, market share and the role in the market (e.g. follower or leader).

Vehicle agents are the linkage between car industries and consumers. Car manufacturers can change the type, number and attributes of vehicle agents. Dealer agents decide on the type of vehicle to be transacted and consumers adopt vehicles based on their attributes and characteristics.

Energy supply system provides different energy carriers for the transportation sector. Demand satisfaction and insuring the security of energy supply are the main goals of this agent. Energy suppliers decide on technology type, production level and attributes of energy carriers. The main attributes are capacity and activity of technologies, knowledge level, fuel production cost and chain emissions.

Fuel station agents purchase the energy carriers from energy suppliers and sell the final energy carriers to consumers. The individual goal of any station agent is profit maximization. They decide on the type and volume of fuels by monitoring the volume and type of vehicle sales in the market. The main attributes of station agents are capacity, market share, location, density and dispersion.

Government agent acts as policymaker and legislator. Government set the market rules, fuel and vehicle standards as well as environmental emissions tax or constraints. This agent has a public goal to maximize social benefits (Zhang et al., 2011). It interacts with other agents by regulatory push/pull (Zhang et al., 2011). The main attributes of the government agent are goals, type of policies and regulations and available resources.

Environment component is defined as the world outside of the agents' boundaries. Such external factors can be identified as natural environment, macro-economic conditions, demographic indicators, national plans, geographical characteristics, the built-environment, services and maintenance companies, public transportation infrastructure and international conditions.

Mathematical Modelling of Agents' Behaviour

In this section, the theoretical foundations and the underlying mathematical models, which describe the behavioof the agents introduced in the conceptual model, are briefly explained. Our aim here is not to present the detailed and complete mathematical formulations, as it would take us out of the scope of this article. Instead, we are focusing on the basic mathematical formulations that reflect the general rationale behind the agents' behaviors and interactions.

Car Manufacturers

Car manufacturers make decisions on vehicle attributes and quantity of the vehicles they offer so as to maximize their profit (Michalek, Papalambros, & Skerlos, 2004; Schwoon, 2006; Zhang et al., 2011). Due to uncertainties of the long-term development of vehicle and energy markets, car manufacturers are assumed to maximize only their expected current profit (Schwoon, 2006). According to Equation (1), profit is calculated by total revenue minus total cost of production. According to Equation (2), vehicle attributes, production volume and manufacturer's knowledge and experience determine the unit production cost. Car manufacturers, as in Equation (3), determine the price for car dealers as a function of production cost and demand (or expected sales) for vehicles. Equation (4) implies that the rate of vehicle production is a function of manufacturer's inventory stock and dealer's demand.

$$\max \Pi^m = \sum_k Q^m_{k,t}\left(P^m_{k,t} - C^m_{k,t}\right), \tag{1}$$

$$C^m_{k,t} = f_1\left(X^m_{a,k,t}, Q^m_{k,t}, K^m_{k,t}\right), \tag{2}$$

$$P^m_{k,t} = f_2\left(C^m_{k,t}, B^d_{k,t}\right), \tag{3}$$

$$Q^m_{k,t} = f_3\left(\mathrm{ST}^m_{k,t}, B^d_{k,t}\right), \tag{4}$$

where t is the time, k the vehicle type, Π^m the car manufacturer profit, $Q^m_{k,t}$ the production quantity, $P^m_{k,t}$ the wholesale price, $C^m_{k,t}$ the unit production cost, $X^m_{a,k,t}$ the value of attribute a, $K^m_{k,t}$ the level of knowledge for production, $\mathrm{ST}^m_{k,t}$ the manufacturer's inventory stock $B^d_{k,t}$ the number of vehicles purchased by dealers.

The behavior of car manufacturers can, on one hand, be modeled as a number of flow rates (e.g. production, sale and investments) and interacting stocks (e.g. inventory level, cumulative production and knowledge level). Feedback loops from demand side influence the production type and volume. On the other hand, the adjustment of vehicle production, which can take place in the long run, influences the price and demand (Kieckhäfer et al., 2009). Hence, the SD approach is useful for building such an aggregated production and inventory system without unnecessary details. The SD approach looks at any car manufacturer from a macro-level and allows the evaluation of different production and sale strategies in response to consumers' collective behavior, energy prices, macro-economic conditions and government policies.

Car Dealers

Car dealers offer various vehicle types so as to maximize their profit according to Equation (5). They take the wholesale prices from car manufacturers (i.e. $P^{m}_{k,t}$) and set the consumer prices for each vehicle (i.e. $P^{d}_{k,t}$). Equation (6) shows that the consumer price for each vehicle, in general, is a function of the wholesale price, dealer costs and the quantity of vehicle demand by consumers. Equation (7) simply addresses the purchase behaviour of dealers (or vehicle shipment) from manufacturers as a function of dealer inventory stock and expected sales to consumers.

$$\max \Pi^{d} = \sum_{k} S^{d}_{k,t}\left(P^{d}_{k,t} - P^{m}_{k,t} - C^{d}_{k,t}\right), \tag{5}$$

$$P^{d}_{k,t} = f_4\left(P^{m}_{k,t}, S^{d}_{k,t}, C^{d}_{k,t}\right), \tag{6}$$

$$B^{d}_{k,t} = f_5\left(\mathrm{ST}^{d}_{k,t}, \mathrm{ES}^{d}_{k,t}\right), \tag{7}$$

where Π^{d} is the dealer profit, $S^{d}_{k,t}$ the sale volume in the market, $P^{d}_{k,t}$ the consumer price of vehicle k, $C^{d}_{k,t}$ the dealer costs, $B^{d}_{k,t}$ the number of vehicle k purchased by the dealer from the manufacturer, $\mathrm{ST}^{d}_{k,t}$ the dealer inventory stock and $\mathrm{ES}^{d}_{k,t}$ the expected dealer sales to consumers.

To determine the type and volume of vehicles to be transacted, the dealers mostly rely on the pattern of macro-level variables such as consumers' total demand, available vehicles and prices. Purchase and sale flow rates along with vehicle inventory build the main structure of the dealer component. These characteristics imply that the SD approach would provide a more appropriate view of dealers' behaviour. According to the conceptual framework proposed in Figure 16.1, SD representation of the car dealer makes it possible to have one integrated SD structure for the both the car manufacturer and the dealer.

Energy Supply System

Behaviour of energy suppliers can be described on the basis of producer theory (Shafiei, Saboohi, & Ghofrani, 2005). It is implicitly assumed that the security of energy supply and meeting transportation energy demand in the long run must be managed and ensured by a central planner. Equations (9)–(11) depict the rationality of a central energy supplier, who attempts to minimize the total present value of costs (i.e. Z). Equation (9) indicates that the production should meet the final demand ($D^{e}_{f,t}$). Equation (10) explains the reduction in the cost of production factors through knowledge and experience accumulation. Equation (11) reflects the availability and the bounds on the supply of production factors (e.g. primary energy and investment).

$$\min Z = \sum_{t=0}^{T}\sum_{i=0}^{N}\frac{C^e_{i,t}F^e_{i,t}}{(1+r)^t}, \tag{8}$$

$$h_1\left(F^e_{1,t}, F^e_{2,t}, \ldots, F^e_{N,t}\right) \geq D^e_{f,t}, \tag{9}$$

$$C^e_{1,t} = h_2\left(F^e_{i,t}, F^e_{i,t}\right), \tag{10}$$

$$\sum_{t=0}^{T} F^e_{i,0} \leq R^e_{i,0} + \sum_{t=0}^{T} d^e_{i,t}, \tag{11}$$

where $C^e_{i,t}$ is the price of production factor i, $F^e_{i,t}$ the amount of using production factor i, $D^e_{i,t}$ the final energy demand by fuel stations for the energy carrier f, $K^e_{i,t}$ the level of knowledge and experience for using the production factor i, $R^e_{i,0}$ the initial availability of the production factor i, $d^e_{i,t}$ the increase in availability of the production factor i, Z the total present value of energy supply system costs, r the discount rate, T the number of time periods and N the number of energy technologies.

The results of the model indicate the configuration of different technologies and energy carriers based on the least (or lower) cost of the supply system. Moreover, the model can provide the wholesale energy prices ($P^e_{f,t}$) as a function of the total cost of the supply system (Z^e_t) and energy demand $D^e_{f,t}$:

$$P^e_{f,t} = h_3\left(Z^e_t, D^e_{f,t}\right) \tag{12}$$

The future behaviour of energy suppliers and the transition to next states of development strongly depend on its current state. Moreover, the process of accumulation of energy and materials is usually observed in various stages of the energy supply system. Physical laws describe the production and consumption of energy, materials and emissions by a set of flow rates and levels. Levels of variables such as resources, energy storage, technology capacities and technological knowledge are the sources of dynamics in the system. These characteristics call for an SD approach to study the development pattern of the energy supplier. SD formulation of a complex energy supply system has another advantage in that the model can be solved faster as it can be expressed by simple sets of algebraic and ordinary differential equations.

Fuel Stations

Fuel/charging station agents distribute final energy carriers to consumers. They make decisions based on economic attractiveness and profitability of establishing and running fuel stations (Bosshardt, 2009; Janssen et al., 2004; Struben, 2006b). They maximize their profit according to Equation (13). They take the wholesale price of energy carriers from energy suppliers (i.e. $P^e_{f,t}$) and determine the consumer fuel prices (i.e. $P^s_{f,t}$). According to Equation (14), consumer energy price is a function of wholesale energy price ($P^e_{f,t}$), station costs ($C^s_{f,t}$) and consumers' energy demand ($D^s_{f,t}$):

$$\max \Pi^s = \sum_f D^s_{f,t}\left(P^s_{f,t} - P^e_{f,t} - C^s_{f,t}\right), \tag{13}$$

$$P^s_{f,t} = h_4\left(D^s_{f,t}, P^e_{f,t}, C^s_{f,t}\right). \tag{14}$$

According to the conceptual framework presented in Figure 16.1, capacity construction for fuel stations may be also influenced, in various ways, by the car manufacturer or dealer strategies.

Fuel stations are at the downstream end of any energy system, which various energy flows connect them to the energy supplier component. In the case of electric energy, charging stations are fully integrated into the electricity network. Purchasing fuel from energy suppliers increases the station stock and selling fuel to consumers reduces it. Therefore, any fuel station can be addressed by a continuous stock-and-flow representation, which is the core of the SD methodology. By using the SD approach, one integrated structure for both energy suppliers and fuel stations would be resulted. The SD model is not, however, capable of simulating spatial heterogeneous stations. Spatial distribution and availability of fuel stations may be more important than the total number of stations in a country or region. In addition, consumer demand for fuel at stations strongly depends on the location of stations and, thus, stations' capacity may vary by region. The AB approach makes it possible to consider the spatial dimension of fuel stations as well as the heterogeneity of attributes. In conclusion, application of each of these two approaches for modelling fuel stations has advantages and disadvantages, depending on the situation. No one approach works well in every situation and there is a need to choose the one that best fits the goals of the study and the system characteristics.

Consumers

Consumers are identified as heterogeneous agents with discrete decisions that can interact with each other or with the environment. The aggregate behaviour of consumers emerges from such complex interactions (Garcia, 2005; Schieritz, 2004). Consumers can also be characterized by their location and spatial distribution. Therefore, the AB approach allows a more realistic representation of consumers' behavior compared with SD models that assume homogeneous agents with average values of attributes. MNL models, introduced by McFadden (1976), are widely used to determine individual preferences probability and a decision rule for selecting the vehicle from the choice set (Train, 2009). In this framework, different vehicles are distinguished by various attributes and the consumers make choices among them so as to maximize their individual utility. Consumers' purchase probability of individual preferences can be formulated as:

$$P^c_{i,k,t} = \frac{\exp U_i(P^s_{f,t}, P^d_{k,t}, \mathrm{FA}^s_{i,k,t}, X^m_{a,k,t}, X^c_{b,t})}{\sum_{k=1}^{V} \exp U_i(P^s_{f,t}, P^d_{k,t}, \mathrm{FA}^s_{i,k,t}, X^m_{a,k,t}, X^c_{b,t})}, \tag{15}$$

where $\mathrm{PR}^c_{i,k,t}$ is the probability of individual preferences for purchasing vehicle k by consumer i, U_i the utility of consumer i, $\mathrm{FA}^s_{i,k,t}$ the fuel availability, $X^m_{a,k,t}$ the vehicle attribute a, $X^c_{b,t}$ the consumer attribute b and V the number of vehicle types.

The decision to adopt new vehicles depends also on the process of word-of-mouth and social influences. Therefore, the number of consumers who want to purchase vehicle k ($\mathrm{NW}^c_{k,t}$) can be expressed as a function of the number of potential consumers ($\mathrm{NP}^c_{k,t}$), the probability of purchasing vehicle k ($\mathrm{PR}^c_{i,k,t}$) and the strength of social network influences on agent i to buy vehicle k ($\mathrm{SN}^c_{i,k,t}$):

$$\mathrm{NW}^c_{k,t} = g_1\left(\mathrm{NP}^c_t, \mathrm{PR}^c_{i,k,t}, \mathrm{SN}^c_{i,k,t}\right),\ i = 1, 2, \ldots, \mathrm{NP}_t \tag{16}$$

The number of consumers who really purchase the vehicle type k at time t ($\mathrm{NB}^c_{i,k,t}$) depends on both the number of consumers who want to purchase ($\mathrm{NW}^c_{k,t}$) and the availability of vehicle k in the supply market (i.e. $\mathrm{ST}^d_{k,t}$):

$$\mathrm{NB}^c_{k,t} = g_2\left(\mathrm{NP}^c_{k,t}, \mathrm{ST}^d_{k,t}\right). \tag{17}$$

By determining the total number of consumers who have adopted vehicle k, total annual fuel consumption is estimated as the following equation:

$$D^c_{f,t} = \sum_k \sum_j \left(\mathrm{NP}^c_{j,k,t} \times F^m_{f,k,t}\right),\quad F^m_{f,k,t} \in X^m_{a,k,t}, \tag{18}$$

where $\mathrm{DP}^c_{j,k,t}$ is the annual kilometres driven by consumer j who uses vehicle k and $F^m_{f,k,t}$ the consumption of fuel f per kilometre by vehicle k. Equation (18) must be summed over the vehicles that use fuel type f.

Government

The government agent is assumed to be exogenous to the system and its policies can influence the variables of other components. Government influences can be considered by exogenous parameters of policy measures such as subsidies, taxes, infrastructure development and area access permits (Bosshardt et al., 2008; Zhang et al., 2011).

Linking the Components

To provide a comprehensive analysis of transition to AFVs, the mathematical models presented for each component should be integrated in a model of the whole energy and transport system. For such purpose, the linking variables and the type of information exchange among the components should be specified. Figure 16.2 depicts the structure of the interlinked components. The solid arrows represent the type of information flow, taking place in the system between various components. Dotted arrows specify the variables that may be influenced by government policies.

Energy supplier and fuel station components are linked together by means of wholesale energy price and final energy demand. Energy suppliers should satisfy the estimated energy demand ($D^e_{f,t}$) by fuel stations. On the other hand, the wholesale price for various energy carriers ($P^e_{f,t}$) can be extracted from the results of the SD model of the energy supplier and then transferred to the station component. By doing so, energy supplier and fuel station components can be modelled as two interconnected SD structures.

Linking the car manufacturer with the dealer agent is achieved through the dealer's demand for new vehicles ($B^d_{k,t}$) and the wholesale price of vehicles ($P^m_{k,t}$). This linkage provides a supply chain framework composed of manufacturer and dealer components, which can be described by one integrated SD structure. The dealer agents act as vehicle price setters for consumers (i.e. ($P^d_{k,t}$). The other vehicle attributes ($X^m_{a,k,t}$) such as vehicle fuel consumption ($F^m_{j,k,t}$), which affects

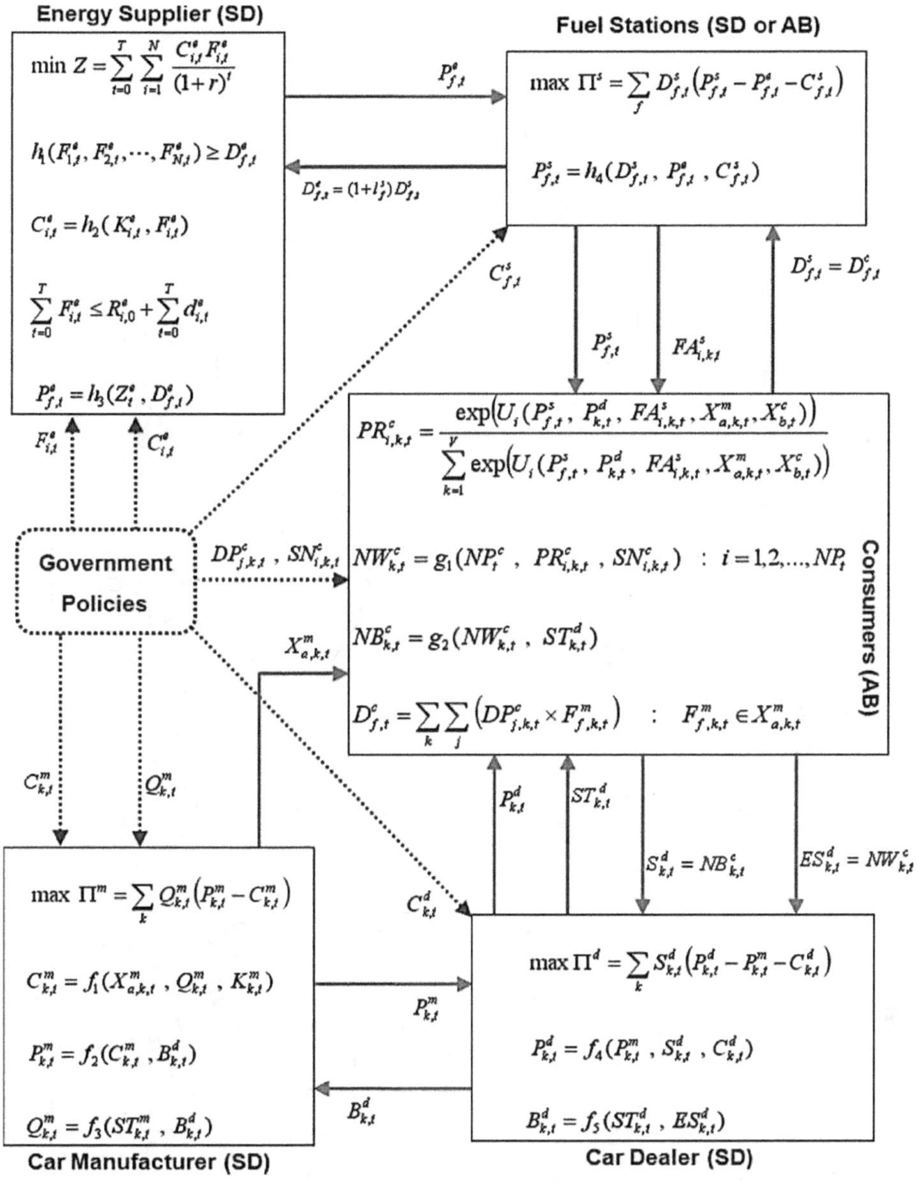

Figure 16.2 Interlinked modules for studying the diffusion of AFVs

both consumers' preferences and their energy demands, are provided by the car manufacturer component. The number of consumers who want to purchase new vehicles ($NW^{c}_{k,t}$) and the number of vehicles purchased by consumers at each time ($NB^{c}_{k,t}$) are estimated by the AB model of consumer behaviour and the results are then fed into the dealer component to specify the expected dealer's sales ($ES^{d}_{k,t}$) and the amount of sold vehicles ($S^{d}_{k,t}$) to the consumers, respectively. Dealers' inventory stock ($ST^{d}_{k,t}$) represents the availability of vehicles in the supply market. The amount of this variable at each time step affects the purchase decision of heterogeneous consumers.

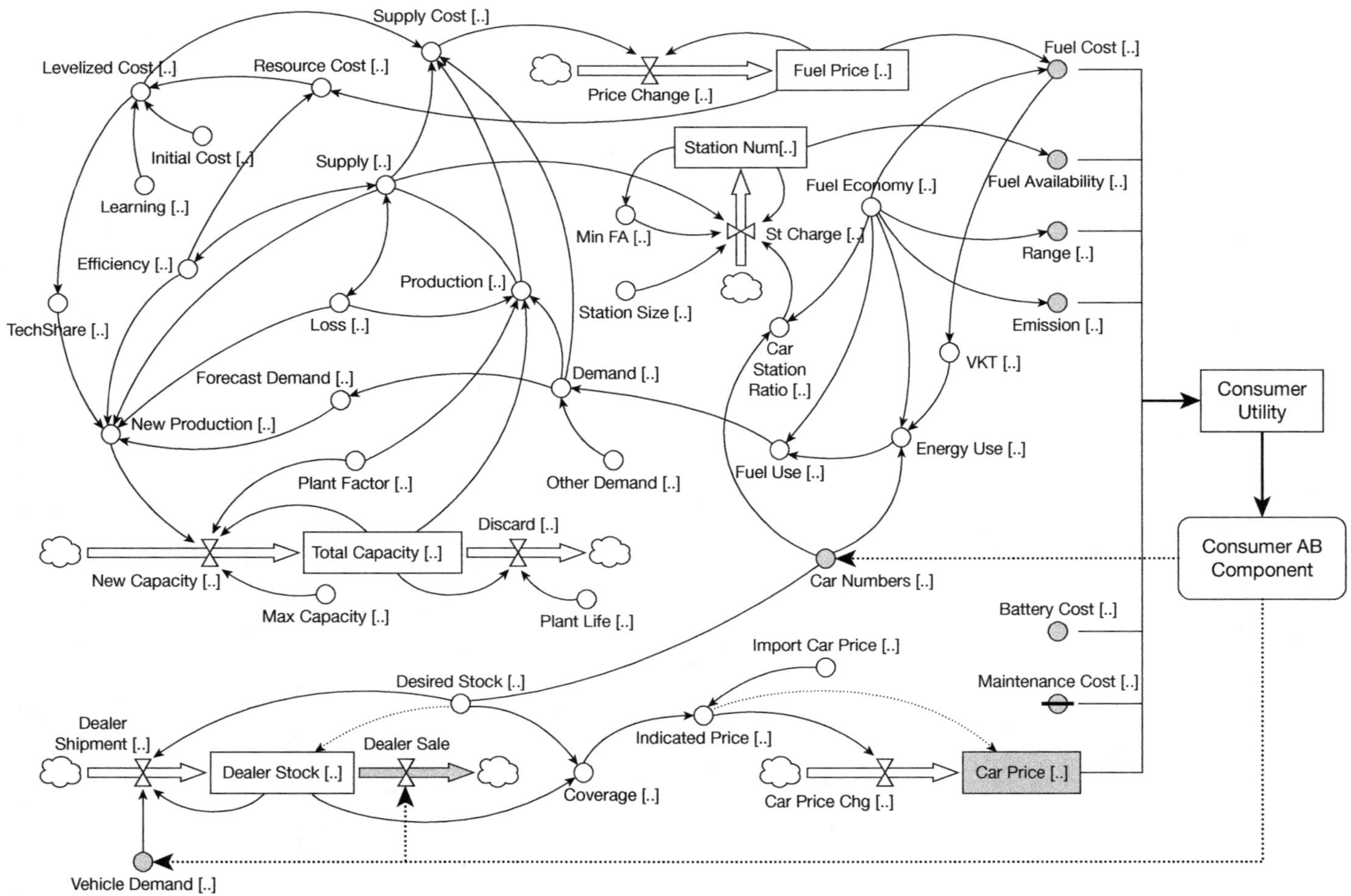

Figure 16.3 Integrated SD and AB modelling approach for studying the diffusion of AFVs

Total fuel consumption of vehicles ($D^c_{f,t}$), which reflects the consumer energy demand, is calculated by the consumer AB model and is then transferred to the fuel station component. Fuel stations, on the other hand, determine both fuel availability ($FA^s_{i,k,t}$) and price of various fuels to be sold to the consumers ($P^s_{f,t}$).

Government can establish policies such as advertisements and stimulation of word-of-mouth communications in order to increase the strength of social networks ($SN^c_{i,k,t}$), thereby changing the purchasing behaviour of consumers. In addition, government incentives or taxes may influence the driving pattern of consumers ($DP^c_{j,k,t}$). The associated costs with the activities of both fuel stations and car dealers (i.e. $C^s_{f,t}$ and $C^d_{k,t}$) can be changed by government tax or subsidy policies. Government can affect the production rate of the manufacturer agent $Q^m_{k,t}$ and its associated costs $C^m_{k,t}$ by setting the standards on emissions and fuel economy of vehicles. Such policies can also be directed at the energy supply system to change the costs and the development pattern of the energy supplier.

An Illustrative Example

In this section, a simple example of the Iceland energy and transportation system is provided to show the main aspects of the models' behavior. This example shows how to build a hybrid AB–SD model and tries to highlight the points of interaction of AB and SD components. Figure 16.3 illustrates the integrated modeling framework for studying the diffusion of AFVs in Iceland. We used Any-Logic software, developed by XJ Technologies Company, for developing the test case.

Integrated SD Structure of Energy Supply and Fuel Stations

According to Figure 16.3, the total capacity of energy technologies is increased by new installations and depreciated at a constant rate. Total capacity, plant factor and total demand (including losses) determine the amount of energy production. Fuel supply for the transport sector is estimated based on production and energy losses. Final energy demand equals the sum of transportation demand (FuelUse) and the exogenous values of non-transport energy demand. It is assumed that the gap between energy production and forecasted energy demand can be reduced by additional energy production and, thus, through new capacity installation. Production share of the technologies with the same output is expressed as a function of average production cost (AES Corporation, 1993). The levelized cost of energy production for each technology is calculated based on initial cost, resource cost and learning effects. Resource cost depends on input energy prices and technology efficiency. Energy price is controlled by price change, which is equal to the difference between current price and average supply cost. Supply cost, which is calculated based on levelized costs, is influenced by maximum supply, production and demand. Fuel price along with vehicles' fuel economy determines the fuel cost attribute of vehicles. Number of fuel stations is controlled by station change flow, which is a function of vehicle–station ratio, average station size and fuel supply. Fuel availability is configured as the number of each fuel station relative to the number of gasoline stations. Minimum availability (minFA) index is defined to facilitate the study of the effects of this attribute in different scenarios. Due to the different nature of electric charging stations, exogenous scenarios can be defined for their availability. Fuel cost and fuel availability are passed on to the AB model to calculate consumers' utility. Vehicle-kilometre traveled (VKT) is changed with fuel cost and an exogenous trend of GDP. Number of each vehicle is extracted from the consumer AB module and is used to calculate total energy requirement (EnergyUse) and the car–station ratio. Total energy use is calculated by VKT, fuel economy and car numbers and, then, is converted to fuel demand.

SD Structure of Vehicle Supply

Figure 16.3 shows a simple SD construct for vehicle supply, starting with the shipment flow to the dealer inventory. Since there is no car manufacturer in Iceland, the type and attributes of vehicles are exogenously determined in the model. Thus, instead of a complex SD structure for the manufacture sector, we have used time-varying parameters and connected variables to show the future development of vehicle attributes. Actual and desired dealer stock and the forecast of consumers' vehicle demand determine the dealer shipment for each vehicle. The AB component determines the rate of vehicle sales to the consumers, which reduces the dealer stock. The imported price of vehicles is a time-varying parameter, which is reduced by exogenous rates. The consumer vehicle price is a function of inventory coverage. If the actual dealer inventory exceeds the desired inventory, the indicated price will rise. Consumer vehicle price is controlled by price change, which is equal to the difference between indicated and current prices. Range and emission attributes are changed in proportion to fuel economy improvement over time. Vehicle price, range, emission and the exogenous trends of maintenance cost and battery replacement cost are passed on to the AB model to calculate consumers' utility.

AB Framework for Consumer Choice Behavior

We have used a modified form of the vehicle choice algorithm introduced in our previous work (Shafiei et al., 2012). Figure 16.4 shows a simple state-chart diagram of consumer behaviours. Whether or not a consumer replaces a vehicle is determined by comparing the life of the current vehicle with a stochastic vehicle lifetime. If the current age of the vehicle is greater than its lifetime, then the consumer decides to purchase a new vehicle. For simplicity and due to lack of suitable data, it is assumed that the old vehicle will vanish from the system. When a vehicle is discarded after a random lifetime period, the consumer needs to buy a replacement and enters the potential user state. Any new agent, who has not already adopted a car, is considered as a potential user as well.

Consumers purchase the vehicles based on both their own preferences and the behaviour of other consumers. We have defined different social groups based on consumers' age and income in Iceland. Each social group reveals the connected agents. Willingness to consider (WtC) each vehicle type by an agent is simply defined as the share of the vehicle in its connected agents. If a consumer adopted a vehicle, then s/he would always consider it in the choice set. If the perceived vehicle share in a social group is greater than the agent's WtC threshold, the agent will consider it. If an agent considers a vehicle, then purchase probability is defined as a function of individual preferences and social influences. Thus, we assume that the purchase probability of a vehicle in the choice set has two components: probability of individual preferences and probability of social influences.

Probability of individual preferences shows the agent's attraction towards the vehicle's attributes in the choice set (Equation (15)). We identify each type of vehicle by seven attributes: (1) vehicle price to income ratio, (2) fuel cost, (3) maintenance cost, (4) vehicle range, (5) battery replacement cost, (6) fuel availability as the number of each fuel station relative to the number of gasoline stations and (7) GHG emissions.

It is assumed that the probability of social influences equals the share of the connected agents who have already adopted a particular vehicle. Total probability of purchasing vehicle j by agent i in time t ($R_{i,j,t}$) is a weighted average of individual preferences and social influences probabilities:

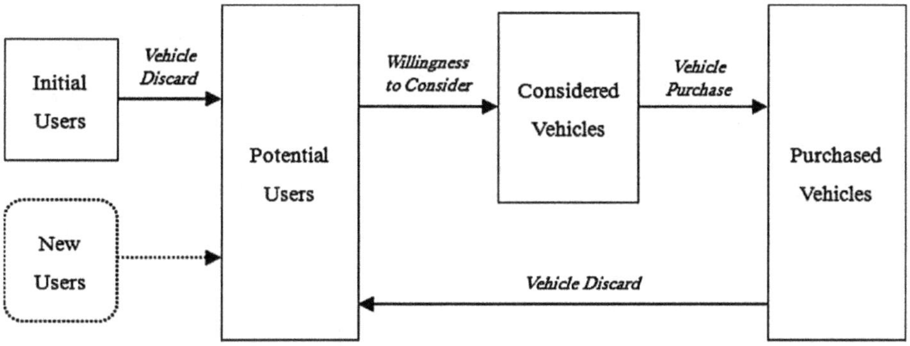

Figure 16.4 State-chart diagram for the AB model of consumer behavior

$$R_{i,j,t} = \lambda_{i,j,t} \times \mathrm{SN}_{i,j,t} + (1 - \lambda_{i,j,t}) \times \mathrm{PR}_{i,j,t}, \quad 0 \le \lambda_{i,j,t} \le 1 \tag{19}$$

where SN and PR are probabilities for social influences and individual preferences, respectively. The weighting coefficient λ, which randomly varies over the consumers, reflects the importance and strength of social influences for each individual consumer. Total purchase probabilities for various vehicles are then aggregated into a cumulative distribution function of vehicle purchases. The function is the sum of probabilities for buying all the vehicles up to and including a particular vehicle j:

$$Q_{i,j,t} = \sum_{h=1}^{j} R_{i,h,t} \; . \tag{20}$$

Next, a random uniform variable is generated and it is compared with the cumulative probability function. The input of Equation (20) that yields the closest but higher outcome than the value of the uniform random variable equals the type of vehicle that is purchased.

When the consumers prefer a certain vehicle type, they check the dealer stock. When the stock contains at least one desirable vehicle, the transition is taken and consumers buy the preferred vehicle. As a result, one unit of that vehicle type is deducted from the dealer stock. It shows a kind of interaction between consumer AB and dealer SD models. If a consumer could not find a particular vehicle in the dealer stock, it would switch to the next preferred vehicle.

Main Assumptions

The integrated model is applied for simulation of the Iceland LDV fleet. The time horizon of the study begins in 2013 and continues until 2050. The vehicle types considered in the AB module are: gasoline internal combustion engine (ICE_P), diesel internal combustion engine (ICE_D), gasoline hybrid electric vehicle (HEV_P), diesel hybrid electric vehicle (HEV_D), plug-in hybrid electric vehicle with gasoline (PHEV_P), plug-in hybrid electric vehicle with diesel (PHEV_D), battery electric vehicle (BEV), bio-ethanol ICE (E85), bio-diesel ICE (B20), biogas ICE (BGV1), dual fuel ICE with biogas and gasoline (BGV2) and fuel cell electric vehicle (FCV).

The main parameters and assumptions are presented in Table 16.1. The exogenous parameters in this table are assumed to gradually reach the specified final values. Consumers are

Table 16.1 Overview of the main parameters and assumptions

Parameters	*Assumptions and values*			
Consumer choice behaviour:				
Initial fleet composition	ICE_P = 82%, ICE_D = 18%			
Current vehicle age (years)	Truncated normal distribution: mean = 10, SD = 5, min = 0, max = 20			
Vehicle life (years)	Truncated normal distribution: mean = 12, SD = 2, min = 0, max = 25			
WtC threshold	Normal distribution: mean = 0, SD = 0.2			
Consumer social susceptibility (weight of social influences)	Truncated normal distribution: mean = 0.6, SD = 0.2, min = 0, max = 1			
Initial VKT	Average value of 12 000 km/year in the base year			
Choice model coefficients	—0.00274 ⋆ Vehicle price/income ($ /annual k$)			
	—0.14 ⋆ Fuel cost ($/km)			
	—0.00113 ⋆ Maintenance cost ($/year)			
	—52.4/vehicle range (km)			
	—0.00011 ⋆ Battery replacement cost ($)			
	—0.000667 ⋆ GHG emission (gCO2eq/km)			
	—4.98 ⋆ exp(—23.03 ⋆ Fuel availability)			
Energy supply system:				
Growth rate of fossil fuel prices	3% p.a. for imported diesel and gasoline (up to 100% increment)			
Initial cost of production ($/kW year)	Hydro = 300	Bio-ethanol = 1079	Biogas = 878	Gasoline = 764
	Geothermal = 280	Bio-diesel = 978	Hydrogen = 524	Diesel = 676
Cost of learning technologies in 2050	Hydrogen: 40% of initial cost			
	Bio-ethanol production: 90% of initial cost			
Fuel stations:				
Minimum fuel availability	10% in 2020 for hydrogen, bio-ethanol, bio-diesel and recharging stations			
Initial number of stations	Gasoline = 117	Diesel = 20	Biogas = 2	Others = 1
Initial fuel price ($ /kW year)	Gasoline = 1958	Diesel = 1742	Biogas = 1105	Hydrogen = 1894
	Bio-diesel = 1225	Bio-ethanol = 1346		Electricity = 844

Table 16.1 Continued

Parameters	*Assumptions and values*			
Vehicle supply:				
Desired dealer stock	8% of total existing cars on the road			
Initial vehicle fuel consumption (normalized IVISI/km against ICE_P)	ICE_P = 1	B20 = 0.82	PHEV P = 0.52	BGV2 = 1.0
	ICE_D = 0.82	HEV P = 0.72	PHEV D = 0.48	BEV = 0.32
	E85 = 1.0	HEV D = 0.63	BGV1 = 1.0	PCV = 0.48
Vehicle fuel consumption (MJ/km) in 2050 (% of the base year values)	ICE_P = 85%	B20 = 95%	PHEV_P = 75%	BGV2 = 85%
	ICE_D = 95%	HEV_P = 85%	PHEV_D = 80%	BEV = 55%
	E85 = 85%	HEV_D = 95%	BGV1 = 85%	FCV = 65%
Initial vehicle price (normalized price against ICE_P)	ICE_P = 1	B20 = 1.1	PHEV_P = 1.6	BGV2 = 1.1
	ICE_D = 1.1	HEV_P = 1.3	PHEV_D = 1.7	BEV = 2.0
	E85 = 1.0	HEV_D = 1.4	BGV1 = 1.1	FCV = 3.0
Vehicle price in 2050 (%, of the base year values)	ICE_P = 96%	B20 = 96%	PHEV_P = 71%	BGV2 = 96%
	ICE_D = 96%	HEV_P = 86%	PHEV_D = 71%	BEV = 62%
	E85 = 96%	HEV_D = 96%	BGV1 = 96%	FCV = 40%
Initial vehicle range (normalized against ICE_P)	ICE_P = 1	B20 = 1.2	PHEVP = 0.8	BGV2 = 1.2
	ICE_D = 1.2	HEV_P = 1.2	PHEV_D = 0.9	BEV = 0.2
	E85 = 0.7	HEVD = 1.4	BGV1 = 0.4	FCV = 0.6
Initial vehicle GHG emissions (normalized against ICE_P)	ICE_P = 1	B20 = 0.7	PHEV_P = 0.4	BGV2 = 0.9
	ICE_D = 0.8	HEVP = 0.7	PHEVD = 03	BEV = 0.0
	E85 = 0.5	HEV_D = 0.6	BGV1 = 0.8	FCV = 0.0
Battery replacement cost (1000$)	BEV = 18 (reduction rate: 3%/year)		PHEV = 9 (reduction rate: 3%/year)	

heterogeneous with respect to their income, age, WtC threshold and social susceptibility. Iceland income and age distributions are used to randomly assign to each agent. We assume that each vehicle represents an agent with independent decision behavior. Total vehicle population is scaled down, in which each agent in the model represents 100 actual agents. Assuming such a scaling factor, the population size of agents would be 2,000 in the base year. This value is increased corresponding to the growth rates of the population (0.7% p.a.) and vehicles per capita in Iceland. Vehicle per capita index in the base year is 0.65, and it is assumed to gradually approach the saturation value of 0.8.

The technologies represented on the energy supply system side are: electricity from hydropower, electricity from geothermal power plant, biogas from municipal wastes, biodiesel from waste oils, bioethanol from lignocellulose biomass and hydrogen from electrolysis. Hydrogen and bioethanol production technologies have exogenous learning for the costs, while constant costs along the horizon are assumed for the other technologies.

Simulation Results

The most important results from application of the proposed approach for the case study of Iceland are market share evolution of AFVs and consumers' fuel demands during the period 2013–50. Figure 16.5 shows the market share evolution of various vehicles in Iceland. Since the consumers' utility and social influences are heterogeneous in the simulated populations, varieties of purchase probabilities for different consumers are resulted. Nevertheless, to help explain the results for each vehicle type over time, the utility of a randomly selected consumer is presented in Figure 16.6. This figure along with Equations (15) and (19) can explain the diffusion pattern of AFVs.

During the initial periods, the number of ICE_P vehicles is increased. However, the attractiveness of ICE_D and the successful market introduction of HEV_P, HEV_P and BGV2 vehicles stop the growth of conventional ICE_P in 2020. Afterwards, the share of ICE_P is gradually reduced and mainly replaced by hybrid and biogas vehicles. The share of AFVs starts to rise after 2021 on account of both increasing consumers' utility and growing social influences (see Figure 16.6 for the utility of a sample consumer). Until 2040, the majority of AFVs in the market are hybrid and biogas vehicles. The results indicate that both dedicated and dual-fuel modes of biogas vehicles (BGV1 and BGV2) would be attractive alternatives in the long run. However, their market growth is slowed down due to the limited resource potential in Iceland. After the period 2040, the share of BEV and FCV rises because of their competitive purchase price, fuel cost and fuel availability.

According to Table 16.1, a minimum fuel availability of 10% in 2020 has been assumed for hydrogen, bioethanol, biodiesel and recharging stations. Sensitivity analysis showed that the fuel availability of these fuels in 2020 should be at least 4% to guarantee the market introduction of FCV, E85, B20 and EVs.

Figure 16.7 presents the trends of projected fuel demand for LDVs in Iceland. Gasoline, diesel and biogas make up the majority of fuel demand. While the total number of vehicles, according to Figure 16.5, is increased by 65%, total fuel demand is only increased by 15%. Improvement of the fuel economy of vehicles and market penetration of AFVs with efficient powertrains are the main reasons for the slight growth of fuel demand.

Finally, different simulations with different random seed indicated that the results are statistically robust, as we observed negligible variations across different runs.

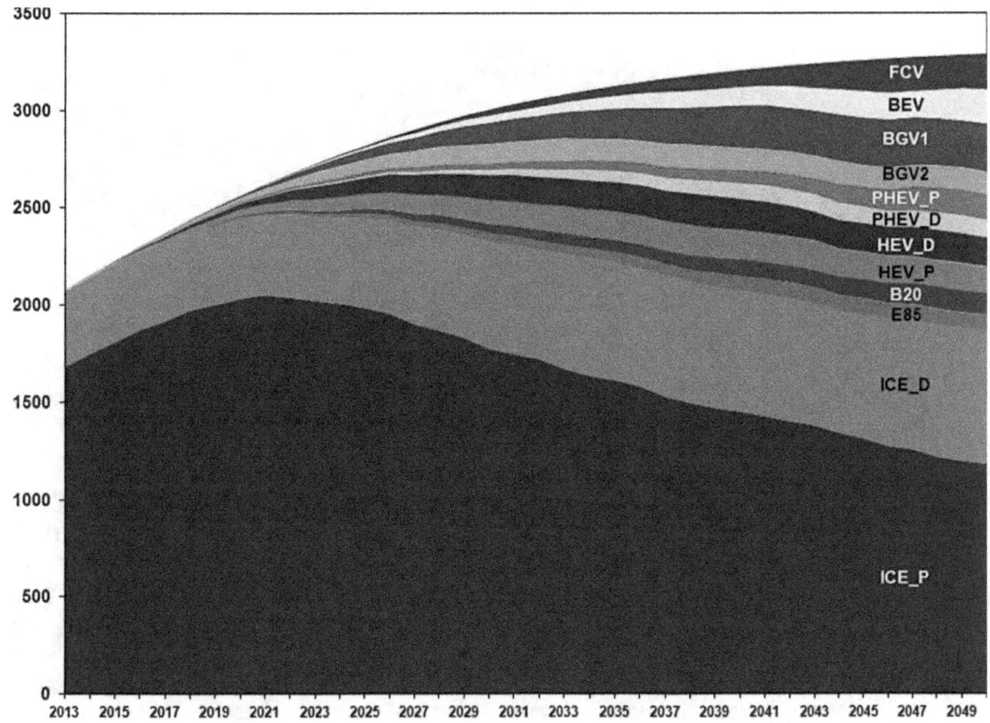

Figure 16.5 Market share evolution of different vehicles

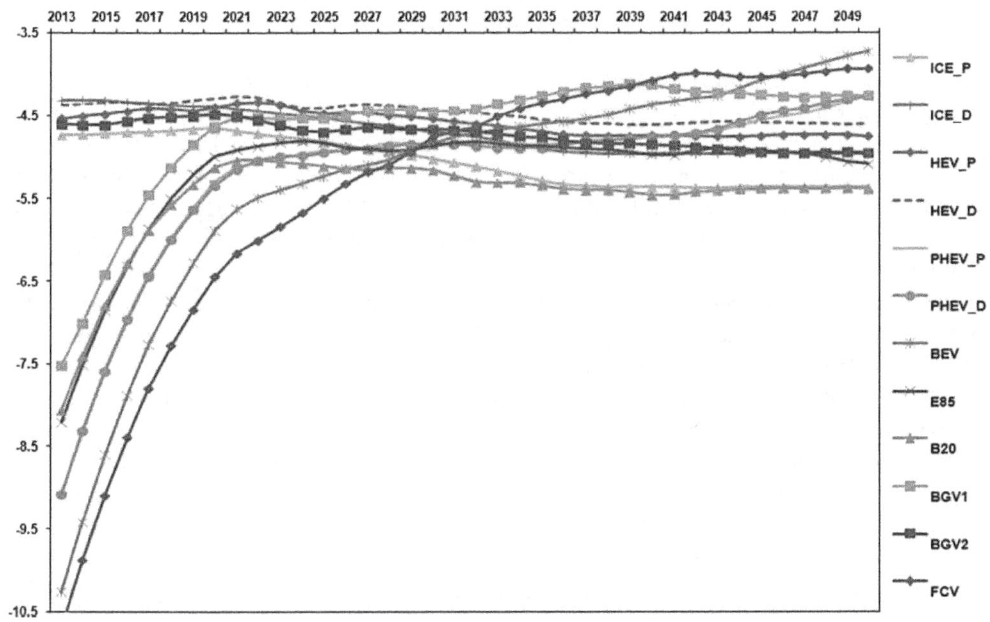

Figure 16.6 Utility of a randomly selected consumer for different vehicles

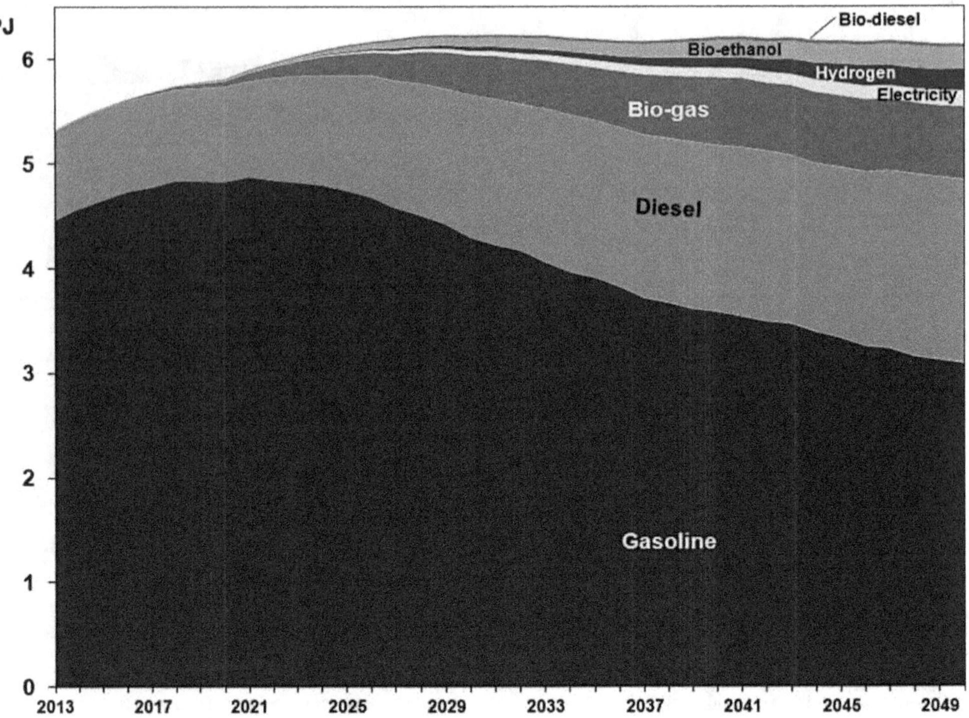

Figure 16.7 Forecasted fuel demand for LDVs

Conclusions

We introduced a conceptual framework for evaluating the transitions towards the sustainable AFV system, which takes into account supply and demand of both vehicles and fuels in the market. To provide a comprehensive analysis of sustainable AFVs, the mathematical models of different components were integrated into one model. We concluded that the combination of SD, which uses a top-down modelling approach with a high level of aggregation, and AB modeling, which uses a bottom-up approach to capture heterogeneities, enables such comprehensive analysis.

AB and SD frameworks for different transportation agents can act in a complementary way inside one integrated model. SD components include a set of algebraic and differential equations, which are calculated continuously. The AB components are usually founded on state charts, simple rules and events that occur when they are scheduled. Integration of components needs an endogenous specification of AB–SD interactions. Therefore, both AB and SD components should be solved simultaneously. The decision logic of agents and the transition from/to different states can be affected by SD variables. On the other hand, the agents' statistics or decisions can modify or update the values of SD variables.

The integrated AB–SD model can show how the decisions made by car industries or energy companies may influence consumers' preferences and vehicle adoption. The integrated model could provide an appropriate analysis of the "chicken-and-egg" barrier to diffusion and adoption of AFVs. On the one hand, impact of the collective behaviour of consumers on strategies of car industries and development of energy infrastructure can be evaluated. On the other hand, consumers' reaction to vehicle and fuel attributes in the market can be analysed. In fact, the linkages between AB and SD components can determine the effects of different scenarios to overcome the "chicken-and-egg" barrier.

The applications of the hybrid AB–SD modeling approach have potential to provide important policy insights as the model enables policy making at different levels and spatial scales. It can simulate any likely impact of different policy instruments such as taxes and subsidies on fuels and vehicles for the both supply and demand sides. Combining SD and AB methodologies could be a promising approach for studying transition to a more sustainable mobility system, as it gives the potential for building a more accurate and computationally efficient model. In general, it can be more accurate than a pure SD structure as it can capture the interactions of heterogeneous agents. On the other hand, it can be solved faster than a pure AB framework as some components can be expressed continuously by simple algebraic and ordinary differential equations.

Acknowledgments

This chapter originally appeared in *Transport Reviews*, 33, 44–70 (2013). The preparation of this chapter has been partially supported by Reykjavik University and the Norden Top-level Research Initiative sub-programme "Effect Studies and Adaptation to Climate Change" through the Nordic Centre of Excellence for Strategic Adaptation Research (NORD-STAR).

References

AES Corporation. (1993). An overview of the IDEAS model: A dynamic long-term policy simulation model of U.S. energy supply and demand. Prepared for the U.S. Department of Energy, Office of Policy, Planning and Evaluation. Retrieved February, 2012, from www.onlocationinc.com/ideas-so.pdf

Bhata, C.R., Sen, S., & Eluru, N. (2009). The impact of demographics, built environment attributes, vehicle characteristics, and gasoline prices on household vehicle holdings and use. *Transportation Research: Part 13–Methodological*, *43*(1), 1–18.

Bosshardt, M. (2009). Fleet dynamics: Identifying the main micro processes of technological change within the European passenger car feet (Ph.D. thesis, ETH Zurich).

Bosshardt, M., Ulli-Beer, S., Gassmann, F, & Wokaun, A. (2007). Developing a Diffusion Model of Competing Alternative Drive-Train Technologies. Proceedings of the 25th International Conference of the System Dynamics Society, July 29–August 2, 2007, Boston, MA.

Bosshardt, M., Ulli-Beer, S., Gassmann, F., & Wokaun, A. (2008). The Effect of Multi-Incentive Policies on the Competition of Drivetrain Technologies. International Conference of the System Dynamics Society, July 20–24, Athens, Greece.

Bouza, M., Ulli-Beer, S., Dietrich, P., & Wokaun, A. (2009). Endogenous Transformation in the Automobile Industry. Proceedings of the 27th International Conference of the System Dynamics Society, July 26–30, Albuquerque, NM, USA.

Delre, S.A., Jager, W., & Janssen, M.A. (2007). Diffusion dynamics in small-world networks with heterogeneous consumers. *Computational and Mathematical Organization Theory*, *13*(2), 185–202.

Delre, S.A., Jager, W., Bijmolt, T.H.A., & Janssen, M.A. (2007). Targeting and timing promotional activities: An agent-based model for the takeoff of new products. *Journal of Business Research*, *60*(8), 826–835.

Delre, S.A., Jager, W., Bijmolt, T.H.A., & Janssen, M.A. (2010). Will it spread or not? The effects of social influences and network topology on innovation diffusion. *Journal of Product Innovation Management*, *27*(2), 267–282.

Eppstein, M.J., Grover, D.K., Marshall, J.S., & Rizzo, D.M. (2011). An agent-based model to study market penetration of plug-in hybrid electric vehicles. *Energy Policy*, *39*(6), 3789–3802.

Garcia, R. (2005). Uses of agent-based modeling in innovation/new product development research. *Journal of Product Innovation Management*, *22*(5), 380–398.

Garcia, R. (2007). *Modeling vehicle choice behavior using agent-based approach*. Ann Arbor, MI: University of Michigan. Retrieved February, 2012, from www.essa.eu.org/Spedal%20Interest%20Groups/Second WorkshopOnMarketDynamics2007/Garcia.pdf

Grossler, A., Stotz, M., & Schieritz, N. (2003). A Software Interface Between System Dynamics and Agent-based Simulations: Linking Vensim and Repast. 21st International Conference of the System Dynamics Society, July 20–24, New York.

Günther, M., Stummer, C., Wakolbinger, L.M., & Wildpaner, M. (2011). An agent-based simulation approach for the new product diffusion of a novel biomass fuel. *Journal of the Operational Research Society*, *62*, 12–20.

Huétink, F.J., van der Vooren, A., & Alkemade, F. (2010). Initial infrastructure development strategies for the transition to sustainable mobility. *Technological Forecasting & Social Change*, 77(8), 1270–1281.

Janssen, A. (2005). Modeling the market penetration of passenger cars with new drive-train technologies (Ph.D. thesis, ETH Zurich, No. 15855).

Janssen, A., & Jager, W. (2003). Simulating market dynamics: The interactions of consumer psychology and structure of social networks. *Artificial Life*, *9*, 343–356 (4).

Janssen, A., Lienin, S., Gassmann, F., & Wokaun, A. (2006). Model aided policy development for the market penetration of natural gas vehicles in Switzerland. *Transportation Research Part A*, *40*(4), 316–333.

Janssen, A., Stephan, F., Lienin, S., Gassmann, F., & Wokaun, A. (2004). Assessing Strategies for the Market Introduction of Natural Gas Vehicles in Switzerland. International System Dynamics Conference, July 25–29, 2004, Oxford.

Keenan, P., & Paich, M. (2004). Modeling General Motors and the North American Automobile Market. The 22nd International Conference of the System Dynamics Society, July 25–29, 2004, Oxford.

Keles, D., Wietschel, M., Mosta, D., & Rentza, O. (2008). Market penetration of fuel cell vehicles—analysis based on agent behavior. *International Journal of Hydrogen Energy*, *33*(16), 4444–4455.

Kieckhäfer, K., Walther, G., Axmann, J., & Spengler, T. (2009). Integrating Agent-based Simulation and System Dynamics to Support Product Strategy Decisions in the Automotive Industry. Proceedings of the 2009 Winter Simulation Conference, December 13–16, 2009, Austin, TX.

Kieckhäfer, K., Wachter, K., Axmann, J., & Spengler, T.S. (2012). Model-based Decision Support for Future OEM Power-train Portfolios: Academic Solutions for Practical Requirements. Proceedings of the 30th International Conference of the System Dynamics Society, July 22–26, 2012, St. Gallen, Switzerland.

Köhler, J., Whitmarsh, L., Nykvist, B., Schilperoord, M., Bergman, N., & Haxeltine, A. (2009). A transitions model for sustainable mobility. *Ecological Economics*, *68*(12), 2985–2995.

Köhler, J., Wietschel, M., Whitmarsh, L., Keles, D., & Schade, W. (2010). Infrastructure investment for a transition to hydrogen automobiles. *Technological Forecasting and Social Change*, 77(8), 1237–1248.

Krail, M., & Schade, W. (2010). The Potential of Alternative Fuel Cars for Achieving CO2 Reduction Targets in EU27. The 12th World Conference on Transportation Research, July 11–15, 2010, Lisbon, Portugal.

Kwon, T. (2012). Strategic niche management of alternative fuel vehicles: A system dynamics model of the policy effect. *Technological Forecasting & Social Change*, *79*(9), 1672–1680.

Lattila, L., Hilletofth, P., & Lin, B. (2010). Hybrid simulation models—when, why, how? *Expert Systems with Applications*, *37*(12), 7969–7975.

Leaver, J.D., & Gillingham, K.T. (2008). Systems dynamics modeling of pathways to a hydrogen economy in New Zealand: final report, Unitec New Zealand. Retrieved February, 2012, from www.unitec.ac.nz/fms/research/1_10%20Pathways%20Report.pdf

Leaver, J.D., & Gillingham, K.T. (2010). Economic impact of the integration of alternative vehicle technologies into the New Zealand vehicle fleet. *Journal of Cleaner Production*, *18*(9), 908–916.

Leaver, J.D., Gillingham, K.T., & Leaver, L.H.T. (2009). Assessment of primary impacts of a hydrogen economy in New Zealand using UniSyD. *International Journal of Hydrogen Energy*, *34*(7), 2855–2865.

Macal, C.M. (2010). To Agent-based Simulation from System Synamics. Proceedings of the 2010 Winter Simulation Conference, December 5–8, 2010, Baltimore, MD.

Mahalik, M., Stephan, C.H., Conzelmann, G., Mintz, M., Tolley, G., & Jones, D. (2009). Modeling Investment Strategies in the Transition to a Hydrogen Transportation Economy. National Hydrogen Association Conference and Expo, March 30–April 3, 2009, Columbia, SC.

Martinez-Moyano, I., Sallach, D., Bragen, M., & Thimmapuram, P. (2007). Design for a Multi-layer Model of Financial Stability: Exploring the Integration of System Dynamics and Agent-based Models. International Conference of the System Dynamics Society, July 29–August 2, 2007, Boston, MA.

McFadden, D. (1976). Quantal choice analysis: A survey. *Annals of Economic and Social Measurement*, *5*(4), 363–390.

Meyer, P.E., & Winebrake, J.J. (2009). Modeling technology diffusion of complementary goods: The case of hydrogen vehicles and refueling infrastructure. *Technovation, 29*, 77–91.

Michalek, J.J., Papalambros, P.Y., & Skerlos, S.J. (2004). A study of fuel efficiency and emission policy impact on optimal vehicle design decisions. *Transactions of the ASME, 126*, 1062–1070.

Mohammadian, A., & Miller, E.J. (2003a). Empirical investigation of household vehicle type choice decisions. *Transportation Research Record, 1854*, 99–106.

Mohammadian, A., & Miller, E.J. (2003b). Dynamic modeling of household automobile transactions. *Transportation Research Record, 1831*, 98–105.

Mueller, G., & Haan, P. (2009). How much do incentives affect car purchase? Agent-based micro simulation of consumer choice of new cars, Part I: Model structure, simulation of bounded rationality, and model validation. *Energy Policy, 37*(3), 1072–1082.

Nunen, A.E.E., Huijbrechts, P., & Rietveld, P. (2011). Transitions towards sustainable mobility, new solutions and approaches for sustainable transport systems. Heidelberg: Springer.

Park, S.Y., Kim, J.W., & Lee, D.H. (2011). Development of a market penetration forecasting model for hydrogen fuel cell vehicles considering infrastructure and cost reduction effects. *Energy Policy, 39*(6), 3307–3315.

Rahmandad, H., & Sterman, J. (2008). Heterogeneity and network structure in the dynamics of diffusion: Comparing agent-based and differential equation models. *Management Science, 54*(5), 998–1014.

Santa-Eulalia, L.A., Neumann, D., & Klasen, J. (2011). A Simulation-based Innovation Forecasting Approach Combining the Bass Diffusion Model, the Discrete Choice Model and System Dynamics an Application in the German Market for electric Cars. The Third International Conference on Advances in System Simulation, October 23–29, 2011, Barcelona.

Schade, B., & Schade, W. (2005). Evaluating economic feasibility and technical progress of environmentally sustainable transport scenarios by a backcasting approach with ESCOT. *Transport Reviews, 25*(6), 647–668.

Schieritz, N. (2002). Integrating System Dynamics and Agent-based Modeling. Proceedings of the 20th International Conference of the System Dynamics Society, Palermo.

Schieritz, N. (2004). Exploring the Agent Vocabulary: Emergence and Evolution in System Dynamics. 22nd International Conference of the System Dynamics Society, July 25–29, 2004, Oxford.

Schieritz, N., & Grossler, A. (2003). Emergent Structures in Supply Chains—A Study Integrating Agent-based and System Dynamics Modeling. The 36th Annual Hawaii International Conference on System Sciences, Washington, DC.

Schneider, M., Shade, B., & Grupp, H. (2004). Innovation process fuel cell vehicle: What strategy promises to be the most successful? *Technology Analysis and Strategic Management, 16*(2), 147–172.

Scholl, H. (2001). Agent Based and System Dynamics Modeling: A Call for Cross Study and Joint Research. Proceedings of the 34th Annual Hawaii International Conference on System Sciences (Vol. 3). IEEE Computer Society, Washington, DC.

Scholl, H.J., & Phelan, S.E. (2004). Using Integrated Top-down and Bottom-up Dynamic Modeling for Triangulation and Interdisciplinary Theory Integration: The Case of Long-term Firm Performance and Survival. The 22nd International Conference of the System Dynamics Society, July 25–29, 2004, Oxford.

Schwoon, M. (2006). Simulating the adoption of fuel cell vehicles. *Journal of Evolutionary Economics, 16*(4), 435–472.

Shafiei, E., Saboohi, Y., & Ghofrani, M.B. (2005). Model for R&D Planning in Energy Sector. IEEE International Engineering Management Conference, September, 2005, St. John's, Canada.

Shafiei, E., Thorkelsson, H., Asgeirsson, E.I., Raberto, M., Davidsdottir, B., & Stefansson, H. (2012). An agent-based modeling approach to predict the evolution of market share of electric vehicles: A case study from Iceland. *Journal of Technological Forecasting & Social Change, 79, 1638–1653.*

Shepherd, S., Bonsall, P., & Harrison, G. (2012). Factors affecting future demand for electric vehicles: A model based study. *Transport Policy, 20*, 62–74.

Stefano, A., Baldoni, F., Falsini, D., & Taibi, E. (2010). A System Dynamics Energy Model for a Sustainable Transportation System. The 28th International Conference of the System Dynamics Society, July 25–29, 2010, Seoul, Korea.

Stephan, C., & Sullivan, J. (2004a). An Agent-based Hydrogen Vehicle/Infrastructure Model. Congress of Evolutionary Computation (CEC) 2004, Portland, OR.

Stephan, C., & Sullivan, J. (2004b). Growth of a Hydrogen Transportation Infrastructure. Presented at Agent 2004: Social dynamics: Interaction, Reflexivity and Emergence, Chicago.

Stephan, M., & Feller, A. (2009). Migrating from oil to electricity-powered vehicles: Modeling Germany's transition to the EV until 2040 in system dynamics (Bachelor thesis, WHU–Otto Beisheim School of Management, Germany).

Struben, J. (2006a). Identifying challenges for sustained adoption of alternative fuel vehicles and infrastructure (MIT Sloan Research Paper No. 4625–06). Retrieved November, 2012, from http://ssrn.com/abstract=927012

Struben, J. (2006b). Essays on transition challenges for alternative propulsion vehicles and transportation systems (Ph.D. thesis, Sloan School of Management, Massachusetts Institute of Technology).

Struben, J., & Sterman, J. (2008). Transition challenges for alternative fuel vehicle and transportation systems. *Environment and Planning B, 35, 1070–1097.*

Tolley, G., & Jones, D. (2006). Analysis of the hydrogen production and delivery infrastructure as a complex adaptive system (Annual Progress Report for the DOE Hydrogen Program, Department of Energy, 1120–3). Retrieved February, 2012, from www.hydrogen.energy.gov/pdfs/progress06/viii_11_jones.pdf

Train, K.E. (2009). *Discrete Choice Models with Simulation.* Cambridge: Cambridge University Press.

Tran, M. (2012). Agent-behaviour and network influence on energy innovation diffusion. *Communications in Nonlinear Science and Numerical Simulation, 17*(9), 3682–3695.

Vincenot, C., Giannino, F., Rietkerk, M., Moriya, K., & Mazzoleni, S. (2011). Theoretical considerations on the combined use of system dynamics and individual-based modeling in ecology. *Ecological Modelling, 222*(1), 210–218.

Vliet, O., Broek, M.V., Turkenburg, W., & Faaij, A. (2011). Combining hybrid cars and synthetic fuels with electricity generation and carbon capture and storage. *Energy Policy, 39*(1), 248–268.

Vliet, O., Vries, B., Faaij, A., Turkenburg, W., & Jager, W. (2010). Multi-agent simulation of adoption of alternative fuels. *Transportation Research Part D: Transport and Environment, 15*(6), 326–342.

Vliet, O., Vries, B., Jager, W., & Turkenburg, W. (2008). Preliminary results from simulating adoption of alternative fuels by heterogeneous motorists. DIME International Conference Innovation, Sustainability and Policy, September 11–13, 2008, GREThA, University Montesquieu Bordeaux IV, France.

Walther, G., Wansart, J., Kieckhafer, K., Schnieder, E., & Spengler, T.S. (2010). Impact assessment in the automotive industry: Mandatory market introduction of alternative powertrain technologies. *System Dynamics Review, 26*(3), 239–261.

Welch, C. (2006). Lessons learned from alternative transportation fuels: Modeling transition dynamics (Technical Report, NREL/TP-540–39446, National Renewable Energy Laboratory).

Welch, C. (2007a). System dynamics: Hydive (hydrogen dynamic infrastructure and vehicle evolution) model, Presentation file. Retrieved November, 2012, from www.nrel.gov/docs/fy07osti/41552.pdf

Welch, C. (2007b). Quantifying consumer sensitivity to hydrogen refueling station coverage (Annual Progress Report for the DOE Hydrogen Program, Department of Energy, 1126–9).

Wokaun, A., & Wilhelm, E. (2011). *Transition to Hydrogen: Pathways Toward Clean Transportation.* Cambridge: Cambridge University Press.

Zhang, T., Gensler, S., & Garcia, R. (2011). A study of the diffusion of alternative fuel vehicles: An agent-based modeling approach. *Journal of Product Innovation Management, 28*(2), 152–168.

17

STAKEHOLDER ANALYSIS AND SOCIAL NETWORK ANALYSIS IN NATURAL RESOURCE MANAGEMENT

Christina Prell, Klaus Hubacek, and Mark Reed

Research Question: Who should have a seat at the table for making decisions about natural resource management?
System Science Method(s): Networks
Things to Notice:
- Using networks to identify key actors in a setting
- Each network concept or metric can have a unique impact, which may be positive or negative

The increasing use of stakeholder analysis in natural resource management reflects a growing recognition that stakeholders can and should influence environmental decision making. Stakeholder analysis can be used to avoid inflaming conflicts, ensure that the marginalization of certain groups is not reinforced, and fairly represent diverse interests. We present a case study from the Peak District National Park in the United Kingdom, where we used social network analysis to inform stakeholder analysis. This information helped us identify which individuals and categories of stakeholder played more central roles in the network and which were more peripheral. This information guided our next steps for stakeholder selection. The chapter ends with a discussion on the strengths and limitations of combining social network analysis with stakeholder analysis.

Many conservation initiatives fail because they pay inadequate attention to the interests and characteristics of stakeholders[1] (Grimble and Wellard 1997). As a consequence, stakeholder analysis has gained increasing attention and is now integral to many participatory natural resource management initiatives (Mushove and Vogel 2005). However, there are a number of important limitations to current methods for stakeholder analysis. For example, stakeholders are usually identified and categorized through a subjective assessment of their relative power, influence, and legitimacy (Mitchell et al. 1997; Frooman 1999). Although widely varied categorization schemes have emerged from the literature, such as primary and secondary (Clarkson 1995), actors and those acted upon (Mitchell et al. 1997), strategic and moral (Goodpaster 1991), and generic and specific (Carroll 1989), methods have often overlooked the role communication networks can play in categorizing and understanding stakeholder relationships. Social network analysis (SNA) offers one solution to these limitations.

Environmental applications of SNA are just beginning to emerge, and so far have focused on understanding characteristics of social networks that increase the likelihood of collective action

 DOI: 10.4324/9781315748771-19

and successful natural resource management (Schneider et al. 2003; Tomkins and Adger 2004; Newman and Dale 2004; Bodin et al. 2006; Crona and Bodin 2006). In this chapter, we harness and expand upon this knowledge to inform stakeholder analysis for participatory natural resource management. By participatory natural resource management we mean a process that engages stakeholders on multiple levels of decision making and facilitates the formation and strengthening of relationships among stakeholders for mutual learning (Grimble and Wellard 1997; Dougill et al. 2006; Stringer et al. 2006). To enhance stakeholder analysis, we use SNA to identify the role and influence of different stakeholders and categories of stakeholder according to their positions within the network. We do this using case study material from the Peak District National Park, United Kingdom.

Stakeholder Analysis

Selecting relevant stakeholders for participatory processes is challenging. For example, certain categories of stakeholder may be historically marginalized from management decisions, and may therefore be difficult to identify or involve; pre-existing conflicts between different groups may preclude a willingness to join a deliberative process; and participatory processes tend to focus on small groups for in-depth deliberation and mutual learning which can lead to a lack of representativeness (Daniels and Walker 2001; Grimble and Wellard 1997; Stringer et al. 2006).

The growing popularity of stakeholder analysis in natural resource management partly reflects an increasing recognition of the extent to which stakeholders can and/or should influence environmental decision-making processes (Burroughs 1999; Varvasovszky and Brugha 2000; Duram and Brown 1999; Selin et al. 2000). Stakeholder analysis can be used to understand environmental systems by defining the aspects of the system under study; identifying who has a stake in those aspects of the system; and prioritizing stakeholders for involvement in decisions about those aspects of the system (Grimble and Wellard 1997; Mushove and Vogel 2005).

In order to identify stakeholders, it is first necessary to define the aspect(s) of the system, problem(s), or issue(s) under study. This is an important initial step, but one that is rarely considered explicitly in stakeholder analyses. This may partly be due to the difficult dialectic between issue definition and stakeholder identification. Without knowing the issues, it is difficult to know which stakeholders should be involved in identifying relevant issues (Dougill et al. 2006). As a consequence, issues are typically identified in a top-down manner by the team leading the stakeholder analysis and may therefore reflect their interests and biases (Clarkson 1995; Varvasovszky and Brugha 2000).

As relevant issues start to emerge, one can then start identifying, characterizing, and prioritizing stakeholders for future involvement in the project. One of the most common approaches is to assess the urgency, legitimacy, and power of potential stakeholders in relation to the issues under question (Mitchell et al. 1997). This may involve evaluating and ranking the type, source and level of power that different stakeholders possess. Such a process has been criticized for prioritizing top-ranked (often more powerful) stakeholders, leading to underrepresentation of lower ranked groups (Grimble and Chan 1995; Calton and Kurland 1996; MacArthur 1997). An alternative approach is to explicitly include those who are remote, weak, disinterested, or considered "nonlegitimate."

Social Network Analysis (SNA)

Social networks comprise actors who are tied to one another through socially meaningful relations. These relations can then be analyzed for structural patterns that emerge among these actors.

Thus, an analyst of social networks looks beyond attributes of individuals to also examine the relations among actors, how actors are positioned within a network, and how relations are structured into overall network patterns (Scott 2000; Wasserman and Faust 1994; Wellman and Gulia 1999).

Both the social network and resource management literature discuss ways in which networks influence individual actors and groups. For example, research on the strength of ties between actors shows how strong versus weak ties relate to different kinds of outcomes. As Granovetter (1973) notes: "The strength of a tie is a (probably linear) combination of the amount of time, the emotional intensity, and intimacy (mutual confiding), and the reciprocal services which characterize the tie" (p. 1361). Thus, the higher a tie scores on each of these attributes, the stronger the tie is. Actors sharing a strong tie tend to: (1) influence one another more than those sharing a weak tie; (2) share similar views; (3) offer one another emotional support and help in times of emergency; (4) communicate effectively regarding complex information and tasks; and (5) be more likely to trust one another (e.g., Coleman 1990; Crona and Bodin 2006; Cross and Parker 2004; Friedkin 1998; Kadushin 1966; Newman and Dale 2004; Wellman and Frank 2001). The advantages of strong ties for resource management are obvious: Stakeholders with strong ties are more likely to influence one another, and thus, creating strong ties among diverse stakeholders can enhance mutual learning and the sharing of resources and advice (Crona and Bodin 2006; Newman and Dale 2004, 2007). Benefits of strong ties may be countered, however, by the redundancy of information that typically runs through such ties, as stakeholders who have shared a strong tie for a long period of time tend to have the same information and knowledge regarding resource management. In contrast, diverse information and new ideas have been shown to travel best through weak ties. A weak tie is often characterized by less frequent communication. Research has shown that weak ties tend to exist between dissimilar others and, as such, they offer individuals and the network as a whole access to diverse pools of information and resources. They do so primarily through performing bridging roles between otherwise disconnected segments of a network (Burt 2001; Granovetter 1973). Within the context of resource management, weak ties can make a network more resilient and adaptive to environmental change. A potential drawback to weak ties, however, is that they may be easy to break. In addition, actors sharing weak ties may lack the trust and understanding needed for in-depth dialogue over environmental issues; (Burt 1992, 1997, 2000; Newman and Dale 2004; Volker and Flap 1999).[2]

Closely related to this discussion regarding strong and weak ties are the ways stakeholders' attributes can influence which ties get established within a network. Homophily, a situation where similar actors are attracted to one another and thus choose to interact with each other, is a well-documented occurrence in social networks (Friedkin 1998; Skvoretz et al. 2004). Stakeholders who are similar to one another are better able to communicate tacit, complex information, as there tends to be higher mutual understanding between such actors. Conversely, such homogeneity can be problematic, as successful natural resource management projects require different views and opinions to be recognized and brought into the discussion (Crona and Bodin 2006; Newman and Dale 2007). In such situations, it may be beneficial to increase the diversity of stakeholders engaged in the project.

Centralization is another network concept discussed in the resource management literature. A highly centralized network is one characterized by one or a few individuals holding the majority of ties with others in the network. Centralized networks are helpful for the initial phase of forming groups and building support for collective action (Crona and Bodin 2006; Olsson et al. 2004). However, research suggests that such centralized networks are disadvantageous for long-term planning and problem solution. These more long-term goals require a more decentralized structure: one holding more ties, both weak and strong, among more actors and stakeholder categories (Crona and Bodin 2006).

Just as strength of ties and network centralization can affect resource management practices, so the position of individuals within a network can affect how information and resources circulate and get exchanged in the network. The concept of centrality has recently received attention in the resource management literature (Bodin et al. 2006: Crona and Bodin 2006), but the distinction between the different kinds of centrality and their potential impacts on resource management has been largely ignored. We distinguish between two types of centrality: degree centrality and betweenness centrality. Degree centrality refers to how many others a stakeholder is directly connected to; stakeholders with a high degree centrality can be seen as important players for mobilizing the network and bringing other stakeholders together. However, because such stakeholders must exert a lot of energy to maintain a large number of ties, these ties are often weak. Thus, highly (degree) central stakeholders can be trusted to use their links to diffuse information and potentially mobilize the group to action, but there is no guarantee that they can significantly influence those to whom they are tied. On the other hand, betweenness centrality refers to how many times an actor rests between two others who are themselves disconnected (Freeman 1979; Wasserman and Faust 1994). Stakeholders holding high betweeness centrality are important for long-term resource management planning; as such, actors perform a broker role of bringing together disconnected segments of the network, thus bringing diversity and new ideas to the network (Bodin et al. 2006; Brass 1992; Prell 2003). However, it should be noted that such "brokers" may feel torn between the different elements of the network and feel forced to take sides (Krackhardt 1992), particularly in situations of resource or land use conflicts.

Thus, the resource management community is beginning to realize that social networks matter, and that such networks can be studied with a great deal of analytical precision, given the tools and concepts afforded by social network analysis (Ramírez 1999; Dougill et al. 2006; Lockie 2006). Table 17.1 summarizes these social network concepts in relation to resource management.

As Table 17.1 shows, there are trade-offs between the different network properties we have discussed. However, by understanding these properties in any given network, it is possible for those working and engaging with stakeholder networks to make better informed decisions about how to engage with and involve stakeholders in meaningful deliberation. The next section of this article shows how we have applied SNA as part of a stakeholder analysis in the Peak District National Park.

The Peak District National Park (PDNP): A Case Study

We are involved in ongoing research that aims to combine knowledge from local stakeholders, policymakers, and social and natural scientists to anticipate, monitor, and sustainably manage rural change in UK uplands (Dougill et al. 2006; Prell et al. 2007). This region is typical of the UK uplands and many marginal mountain areas of Europe that are facing pressures resulting from demographic change, policy reform, and environmental problems, such as soil erosion, biodiversity loss, and climate change.

To enhance the sustainability of upland management in this region, we have been engaging with groups of stakeholders to identify sustainability goals, strategies that could be used to reach these goals, and indicators to measure progress toward these goals. In addition, we are developing tools to evaluate the management options that emerge from this process in a multistakeholder, participatory framework. In this chapter, we report on the early stages of this process. In particular, we discuss:

- How to identify stakeholders: representing diverse stakeholder communities, accounting for divergent stakeholder opinions about who should be considered, and addressing the dialectic between stakeholder and issue identification.
- How social network analysis (SNA) can supplement qualitative information with more in-depth and quantitative data about stakeholder relationships.

Table 17.1 Network concepts relevant for natural resource management

Network concept	*Effect on resource management*
Strong ties	+ Good for communicating about and working with complex information + Hold and maintain trust between actors + Actors more likely to influence one another's thoughts, views, and behaviors + Encourage creation and maintenance of norms of trust and reciprocity – Encourage the likelihood that actors sharing strong tie hold redundant information – Actors less likely to be exposed to new ideas and thus may be less innovative – Can constrain actors
Weak ties	+ Tend to bridge across diverse actors and groups + Connect otherwise disconnected segments of the network together + Good for communicating about and working with simple tasks + New information tends to flow through these ties – Not ideal for complex tasks/information – Actors sharing weak ties are less likely to trust one another – Can break more easily
Homophily	+ Shared attributes among social actors reduces conflict, and provide the basis for the transference of tacit, complex information – Can also result in redundant information, i.e., actors have similar backgrounds and therefore similar sources of knowledge
Centrality	*Degree centrality:* + Actors with contacts to many others can be targeted for motivating the network and diffusing information fast through the network, i.e., these are the focal actors in a centralized network – These actors do not necessarily bring together diverse segments of the network – Because of their many ties to others, these ties are often weak ones, thus decreasing influence over others *Betweenness centrality:* + Actors that link across disconnected segments of the network have the most holistic view of the problem + As with degree centrality, they can mobilize and diffuse information to the larger network – They can feel constrained or torn between two (or more) positions
Centralization	+ As only a few actors hold the majority of ties linking the network together, only need reach these well-connected few to reach entire network – Reliance on only a few is not the optimal structure for purposes of resilience and long-term problem-solving

Context

The Dark Peak area of the Peak District National Park (PDNP) was chosen for its diversity of land use activities (for conservation, farming, tourism, water supply, and game/fishing), and the range of social, economic, political, and environmental pressures it faces. It is situated at the southern end of the Pennine upland range, between three large cities (Figure 17.1). With an estimated 22 million recreational visitor days annually, the PDNP is Britain's most visited National Park (Peak District National Park 2004). The PDNP contains a number of villages and towns, but only 17.2% of its 38,000 population live in the Dark Peak area (Office for National Statistics 2003). PDNP residents are more reliant on agriculture, game birds, and tourism than the national average (Office for National Statistics 2003). Most moorland is privately owned and managed for a combination of grouse and sheep production.

The Dark Peak contains a number of internationally important habitats (UK Biodiversity Steering Group 1995; English Nature 2003) that add to the list of competing demands of conservation, water supply, recreation and tourism, agriculture, and game management that have led to a conflict of interests between many upland stakeholders. English Nature (2003) attributed

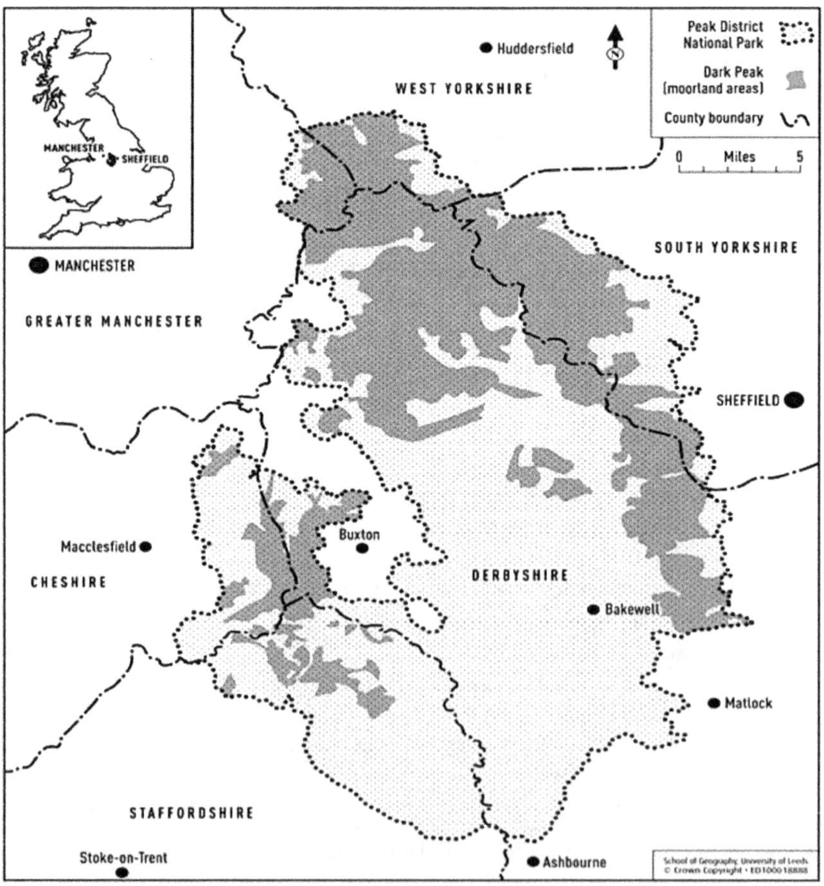

Figure 17.1 Map of Peak District National Park showing the Dark Peak area (shaded) (from Dougill et al. 2006)

the high proportion of Peak District Sites of Special Scientific Interest in unfavorable condition predominantly to a combination of overgrazing and "inappropriate" burning. Such burning is carried out to create a mosaic of heather stands to maximize grouse populations (Holden et al. 2007). These factors (compounded by historic atmospheric pollutant deposition) have also been blamed for increased erosion, with consequent effects on water quality (Tucker 2003). Grazing levels have declined significantly (mainly as a result of Environmentally Sensitive Area agreements), but rotational burning continues to be practiced widely (Dougill et al. 2006).

Methods

Identifying Stakeholders and Issues

We started by conducting a focus group with members of Moors for the Future (MFF), which is a partnership of organizations in the PDNP. Representatives from the Moors for the Future partnership were chosen to take part in this initial focus group because it had already brought together many of the key stakeholder organizations as part of their partnership, including the Peak District National Park Authority, two water companies, Natural England, National Trust, Sheffield City Council, Moorland Association, Derbyshire County Council, and the Environment Agency. In addition, two individuals who MFF had identified as relevant stakeholders were also invited to the focus group. To avoid bias arising from initial group composition, focus-group data were triangulated through semistructured interviews with eight stakeholders identified during the focus group to represent different land management perspectives. The aim of the focus group and subsequent interviews was to evaluate and adapt the proposed aims of the project in order to ensure it was focusing on relevant issues and identify and categorize stakeholders.

The focus group and interviews identified over 200 relevant organizations and groups of individuals. These organizations and individuals were initially categorized during the focus group into stakeholder categories based on the perceived role of these individuals and organizations in the PDNP. In addition, information was elicited about the most effective way to gain the support and involvement of these individuals and organizations. Successive interviews resulted in the addition and subdivision of stakeholder categories. The final categorization was then checked with participants from the initial focus group and those who had been interviewed at the beginning of the interview process.

These categories were then used to guide our snowball sample: One to two individuals from each stakeholder category were interviewed, and these interviews led to further nominations and interviews until both names and land management issues began to repeat. In total, 22 interviews representing all categories were thus conducted. These interviews were used to deepen our knowledge of the current needs and aspirations of those who work, live, and play in the park.

Social Network Analysis

After we had identified and categorized relevant stakeholders, we conducted structured telephone interviews with these individuals to gather network data that would indicate how stakeholders were socially related to one another (88% response rate). To start identifying this social network, we asked the following "name generator" question:[3] "Do you communicate with anyone from [stakeholder category named here] on upland management issues in the Peak District National Park? Please list up to five names."

We asked our respondents this question for each of the eight main stakeholder categories. Respondents nominated individuals in each stakeholder category, resulting in a total number of 147 nominations. A follow-up question was then asked, based on this name-generator question, to elicit tie strength among these stakeholders:[4] "How often do you communicate with this person?" (daily, weekly, monthly, 1–2 times/year).

This question was repeated for each name generated in each of the eight stakeholder categories. The resulting data were then analyzed in UCINET.[5] The analyses chosen were the following:

Density: This is the proportion of possible ties in a network that are actually present, and a network's density is commonly used to measure the extent to which all actors in a network are tied to one another (Wasserman and Faust 1994). A density score of 1 indicates that all actors in the network are directly tied to one another, and a density score of 0 indicates the network is fully disconnected.

Centralization: A centralization score of 1 indicates that the maximum number of ties concentrated around one actor is present, and a score of 0 indicates a fully connected network, where all actors are directly connected to each other.

Degree centrality: Refers to how many others an actor is directly connected to.

Betweenness centrality: Refers to how many times an actor rests on a short path connecting two others who are themselves disconnected.

Results

The eight stakeholder categories that emerged from our stakeholder analysis were water companies; recreational groups; agriculture; conservationists; grouse moor interests (consisting of owners/managers and game keepers); tourism-related enterprises; foresters; and statutory bodies. In the 22 interviews, the issue of heather burning continued to emerge as the most pressing issue pertaining land management issue due to the government's ongoing and highly contentious review of the Heather and Grass Burning Code.

The social network analysis, which followed from the stakeholder analysis, resulted in a network composed of 147 stakeholders from eight different stakeholder categories, linked together through differing strengths of ties. This network is shown below in graph A in Figure 17.2.

In graph A, the thicknesses of the lines depict the varying frequencies of communication, where thicker lines represent more communication between two stakeholders. The varying shapes of nodes represent the various stakeholder categories. The size of the nodes represents the relative betweenness centrality of the actors, a point we discuss in detail shortly. From these data, we were able to collapse ties occurring on a monthly or more frequent basis to provide a network composed of "strong ties." This strong ties network appears in graph B.

As noted in Table 17.1, uncovering the strength of tie can suggest which stakeholders are more likely to influence one another, which ones are more likely to hold similar views, which ones are marginalized, and which play a brokering role. An initial comparison between graphs A and B shows that once we concentrate on the stronger ties (graph B), the network breaks apart into several components, as well as a large number of isolated individuals who appear on the left side of the graph. This suggests that this stakeholder network is dependent on its weak ties for remaining fully connected, a situation that reflects the "strength of weak ties" argument (Granovetter 1973). Although the weak ties are performing the bridging roles one would expect

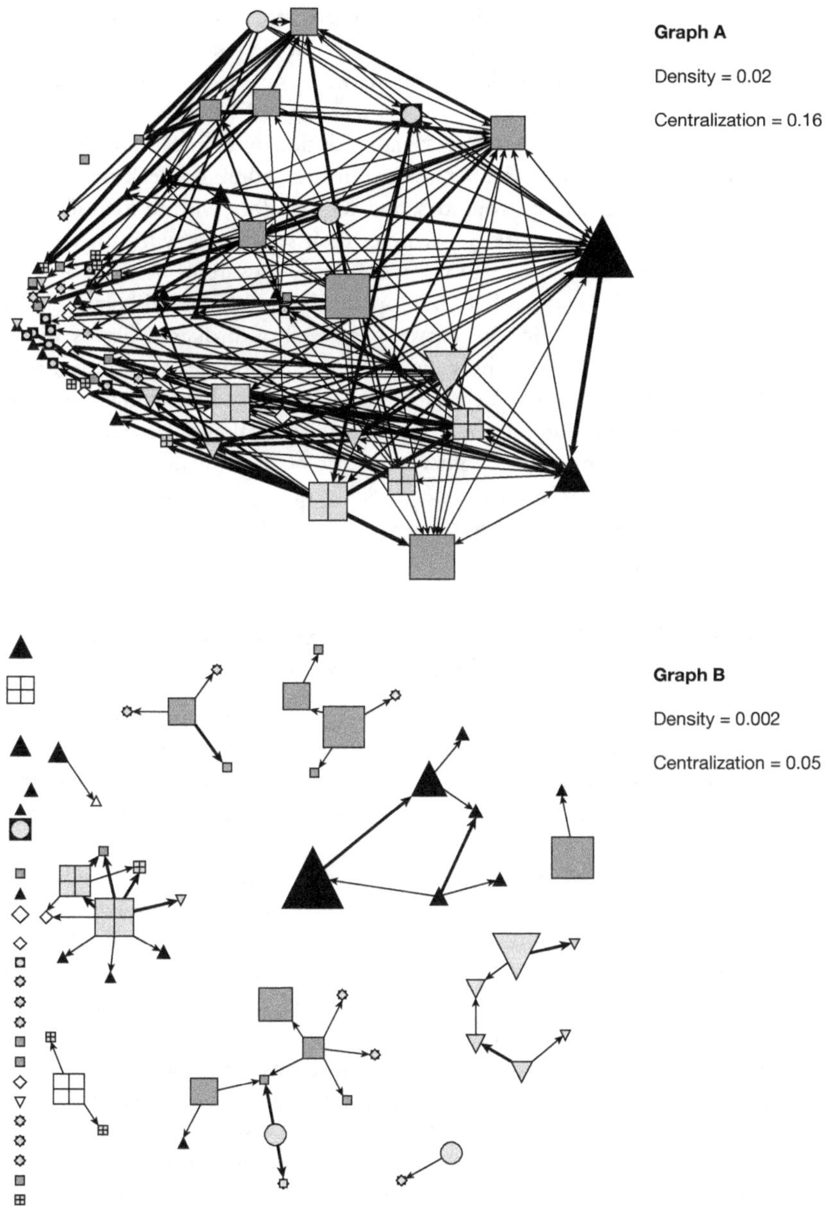

Figure 17.2 Two graphs showing social network of stakeholders

in holding otherwise disconnected segments of the network together, this reliance on weak ties also suggests potentially vulnerable areas in this network (please refer to Table 17.1).

The importance of weak ties in the network is further illustrated by comparing the overall structure of graph A with graph B. Two network analyses were chosen for this purpose: Density and centralization. The results, which appear to the right of each graph, show that graph A has a higher density score and higher centralization score than graph B. This again shows that weak ties are leading toward more connectivity in this network. In addition, the higher centralization score indicates that certain actors, through their numerous weak ties to others, are emerging as key figures in holding this network together. This is a point we take up again in our discussion on centrality. Thus, we know that weak ties are important for this network, and as such, they are also important for our natural resource management project. However, there are a plethora of stakeholders who are linked to the network via weak ties, and we cannot possibly invite all of these stakeholders into our future deliberations. Further, although weak ties are important, we also recognize the importance of strong ties, and we therefore wish to identify stakeholders who are also prominent in the network by virtue of their strong ties with others.

In addition to the preceding considerations, we were also aware of the role of centrality in identifying stakeholders, and also the role of homophily. Centrality would help us locate which stakeholders generated more ties in the network as well as brokered across disconnected segments of the network. In addition, the stakeholder categories unearthed in our stakeholder analysis would help us identify stakeholders according to issues of homophily. Thus, analyzing the centrality of stakeholders according to strength of tie and stakeholder category would help us narrow down our selection of stakeholders to a list of individuals that played important communication and brokering roles in the network. In doing so, these individuals would be more likely to bring holistic views to the discussions and diffuse information outward to the wider social network.

Locating Central Actors: Degree, Betweenness, Tie Strength, and Stakeholder Category

As summarized in Table 17.1, two forms of centrality can play important roles in resource management. Degree centrality refers to how many others an actor is directly connected to, and betweenness centrality refers to how many times an actor rests between, two others who are themselves disconnected. Table 17.2 below shows those stakeholders holding the top 10 degree and betweenness values.

These scores were calculated based on the network found in graph A, where we then dichotomized the data, and in the case of betweenness centrality, also converted all directional ties to nondirectional ties.[6] In addition, Table 17.2 shows the category each stakeholder belongs to, and a breakdown of that stakeholder's immediate neighbors in the network according to stakeholder category. This breakdown also took into consideration whether a stakeholder was strongly or weakly tied to their neighbor.

Linear regression shows that stakeholders with high betweenness scores (Table 17.2) tend to also have high degree scores ($p < .01$; $r^2 = .57$). So those stakeholders who are investing time in a great many ties also tend to form these ties across disconnected others. However, a closer look at these stakeholders' neighbors reveals a slightly different story. Recall from Table 17.1 the role of strong ties and homophily: Stakeholders sharing a strong tie are more likely to influence one another, yet they are also more likely to share many similarities. Table 17.2 shows that, by and large, these stakeholders' strong ties are with ones largely from their own stakeholder category. As such, these highly central individuals tend to be embedded within and more influenced by members of their own category, which reflects much of the literature discussed earlier on homophily and strong ties (Friedkin 1998; Newman and Dale 2004, 2007; Skvoretz et al. 2004). For example, actor 8 has a high betweenness centrality, yet the immediate neighbors to whom he is strongly tied consists mostly of actors within his own stakeholder category (i.e., Grouse Moor Managers). Thus, while actor 8 does connect many different areas of the network together and is reinforced by the diversity of his weakly tied neighbors, his immediate strong connections tend to be with people similar to himself. In contrast, actor 21,

Table 17.2 Centrality scores for stakeholders and stakeholders' network breakdown

ID	*Group*	*Degree*	*Between*	*Strong Ties*	*Weak Ties*
2	Grouse	14	752	Ag (2), Gro (1)	Gro (2), SB (2), Rec (2), Ag (1), H_2O (^{1}), For (1), Con (1)
14	H_2O	14	1180	H_2O (3), SB (1), Con (1)	Con (3), H_2O (1), Rec (1), For (1), SB (1), Gro (1), Ag (1)
1	Ag	15	702	Ag (2), SB (1)	Con (5), Gro (3), Rec (2), H_2O (1), SB (1)
8	Grouse	15	760	Gro (5), Con (1), Ag (1)	Con (3), H_2O (2), Ag (2), SB (1)
5	Grouse	16	921	Gro (2), Ag (2), Con (1), For (1)	Con (4), SB (1), Rec (1), For (1), H_2O (1), Ag (1)
12	H_2O	17	749	Con (2), SB (1)	Con (4), H_2O (3), Gro (2), Rec (2), For (2), Ag (1)
7	Grouse	19	1190	Gro (4), Con (1)	Gro (3), Con (5), Ag (4), SB (3), Rec (1)
17	Rec	20	1718	Rec (3)	Con (5), Gro (4), H_2O (3), For (2), Rec (1), SB (1), Ag (1)
6	Grouse	20	1494	Gro (4), Ag (3)	Con (6), H_2O (3), Rec (2), For (1), SB (1)
21	H_2O	21	1390	Con (5), H_2O (2), For (1), Gro (1), SB (1), Ag (3), Rec (3)	Gro (2), For (2), Con (1)
13	H_2O	21	1004	H_2O (2), Con (3), For (2), Gro (1)	SB (3), H_2O (2), Con (2), Gro (2), For (1), Ag (2), Rec (2)
9	Grouse	21	1658	Con (1), Gro (1)	H_20 (5), Gro (2), Rec (4), Ag (3), Con (3), For (1), SB (1)
11	Con	24	1070	Con (4), H_2O (1), For (1), SB (1)	Con (3), Gro (4), Ag (3), For (3), Rec (2) H_2O (2), SB (1)
18	Con	27	2299	Con (4), Rec (1), For (1), H_2O (^{1})	Gro (6), Ag (4), Rec (^{3}), For (3), Con (^{1}), H_2O (2), SB (1)

Notes: Gro = Grouse; Con = Conservation; Ag = Agriculture; H_2O = Water; Rec = Recreation; For= Forestry; SB = Statutory Body. Analyses for this table were based on Graph A, where all ties where then dichotomized and (for betweenness) they were also made undirectional.

from the water stakeholder category, does not have as high a betweenness score as actor 8, yet his immediate strongly linked neighbors comprise a more diverse mix.

Taken together, Table 17.2 helps us understand which kinds of actors might be important to involve in our resource management dialogues. Actors with high centrality scores are important for the bridging roles that they play (e.g., Granovetter 1973; Burt 2001). In addition, however, we suggest also considering the issues of (1) strength of tie and (2) homophily, thus also looking at the stakeholder categories from which actors and their immediate neighbors come from, and how strongly tied central stakeholders are to these others. This additional information can help one distinguish whether an actor is linking across similar or dissimilar others, an important distinction to be made for natural resource management purposes.

Table 17.2 also shows individuals and stakeholder categories that are not playing central roles – for example, no foresters or statutory body representatives appear as "highly central," suggesting that these categories of stakeholders could be brought more actively into dialogue about resource management.

Refining and Verifying Our Selection

We conducted a number of analyses on our social network data, and each analysis has informed decisions about whom to involve in future resource management dialogues:

1. Separating weak from strong ties: This gave us an initial view as to which stakeholders were more heavily involved in the network than others. Such stakeholders are important to identify and involve, as these stakeholders have a more durable presence in the network and their ties with others in the network hold more trust. Thus, these stakeholders' presence will most likely be felt for quite some time in the future, and the influence they have on others will be more than for those who are connected solely through weak ties. Stakeholders who "disappeared" from the network once we concentrated on strong ties are only linked through weak ties to the network. They are stakeholders who, because they do not have such an active communicative role in the network, most likely hold diverse opinions and potentially different values from those stakeholders linked together through strong ties. Thus, they are important to involve in future discussions, and our next question became, which actors from the stable and peripheral sections of the network ought to be invited to further deliberations?
2. Locating central actors: Because a simple analysis of strong and weak ties still left us with a large amount of "peripheral" stakeholders from which to select, we calculated degree and betweeness scores to highlight which particular stakeholders were playing a more active, communicative role in the network. This analysis was based on our network composed of both strong and weak ties, where all ties were then converted to 1s and 0s, and then used in calculating centrality scores for individual actors. A careful comparison of degree and betweenness centrality scores, alongside the composition of actors' neighbors' categories, revealed some central stakeholders who held strong, immediate ties with diverse others as well as performed broker roles in the network. Involving such brokers in our dialogue will potentially result in involving stakeholders who have a more "holistic" view of resource issues and who can better diffuse information and new ideas outward into the network. In addition, such actors could potentially help mediate between different conflicting groups of interest.
3. Centrality analyses resulted in a smaller number of stakeholders on which to focus our attention, and in doing so, it became very obvious that few to no stakeholders from certain

categories were playing central roles (e.g., stakeholders from forestry and statutory bodies). Such noncentral stakeholder categories represent areas of the network where more tie formation can be encouraged through inclusive dialogue.

These findings were shared with some of the stakeholders at a conference (Reed et al. 2004) where feedback from the audience reinforced our interpretations of how certain groups and individuals were isolated while others played a more central role and were well connected. The overall reaction from those present at this conference suggested to us that our findings largely coincided with stakeholders' own perceptions.

Conclusions

Environmental applications of social networks are just beginning to emerge, and so far have focused on understanding the characteristics of social networks that increase the likelihood of collective action and successful natural resource management (Bodin et al. 2006; Crona and Bodin 2006; Newman and Dale 2004; Schneider et al. 2003; Tompkins and Adger 2004). These discussions focus on linking well-known social network analysis concepts to issues and theories found in the resource management literature. In this chapter, we move beyond these discussions to demonstrate how knowledge gained from analyzing the social networks of stakeholders can be used to select stakeholders for participation in natural resource management initiatives.

This chapter has proposed methods for improving stakeholder representation in participatory processes. Such information can be critical for natural resource management initiatives that require small group sizes for deliberative processes. For example, in the United Kingdom, many landscapes outside National Parks are protected as Areas of Outstanding Natural Beauty. They are managed by committees composed of rural communities and local authorities that need to balance fair representation of diverse interests with a group size that can effectively take management decisions. Our proposed combination of stakeholder analysis and SNA can help identify stakeholder categories, ensure key groups are not marginalized, and specify representatives that are well connected with and respected by the groups they need to represent. This is done by identifying which individuals and categories of stakeholder play more central roles and which ones are more peripheral, and by gaining a sense for how the overall network is shaped. Such information is also crucial for natural resource management initiatives that aim to influence the behavior of stakeholder categories through key players. For example, agricultural extension services around the world employ agricultural demonstrators from among farming communities to showcase innovative crops and methods to their neighbors. The diffusion of innovative practices in this way requires that those selected are sufficiently well connected and respected in the local community (Rogers 1995).

We conclude by discussing some of the limitations in using social network analysis and stakeholder analysis in the way proposed here. By locating central actors according to our initial measure of frequency of communication, we potentially overlook how some stakeholders might derive their influence from sources other than their communication roles in the network. For example, statutory bodies do not appear as very central in our network, but they have a lot of influence over the ways policies are written and enacted, and thus influence the day-to-day lives of stakeholders. Taken together, we advise that social network analysis is a tool to be used in conjunction with other methods and approaches. In our case, we had qualitative data against which to compare our SNA findings. Rather than using the numbers from the SNA at face value, we used these findings as heuristics and as an input in discussions with stakeholders on how to interpret and use the data (Prell et al. 2008). This approach to stakeholder analysis is more time-consuming and costly than focus groups, but can provide more in-depth information than traditional approaches. Thus, SNA is a sophisticated technique that brings precision and a deeper understanding of social relations among stakeholders, but used in isolation from other data the results may lead to simplistic decisions about stakeholder involvement in natural resource management.

Acknowledgments

This chapter originally appeared in *Society and Natural Resources*, 22, 501–518 (2009). This research was funded through the Rural Economy and Land Use (RELU) programme, co-sponsored by Defra and SEERAD (project RES-224-25-0088).

Notes

1. By "stakeholders" we refer to individuals who affect or are affected by certain decisions and actions (Freeman 1984). These individuals can be clustered into stakeholder categories according to their similarity in views, position(s) on an issue, and/or how they affect or are affected by the issue under discussion.
2. This discussion of weak and strong ties relates to and (partially) draws from an extensive discussion on social capital. For example, the following authors discuss social capital by linking strong and weak ties to the accruement of certain benefits and/or disadvantages for individuals, groups, and society: Ron Burt (e.g., Burt 1997, 2000, 2001), Volker and Flap (1999), and Putnam (1993, 2001). These authors refer to strong and weak ties within a discussion of "closure," "cohesion," or "bonding" for strong ties and "brokerage," "structural holes," or "bridging" for weak ties (for an in-depth discussion see, e.g., Prell 2006). This chapter is not a social capital chapter, but rather draws on the ideas (taken from this literature and from other literature cited in the chapter) that tie strength coincides with the ability of actors/groups to accrue certain resources for themselves. A chapter that wishes to truly tackle the issue of how social capital relates to resource management would, among other things, consider the norms of trust and (generalized) reciprocity among this group of actors. This would make for an interesting study, and in fact, the authors are currently gathering data on social capital of stakeholders from a network perspective.
3. Name generators are a common method for gathering SNA data (Wasserman 1994; Knoke and Kuklinski 1982).
4. We are aware that additional measures for tie strength could have been developed to measure the other dimensions of tie strength as outlined by Granovetter (1973; and quoted earlier in this chapter). Unfortunately, three main constraints barred us from pursuing this: (1) We were not the only researchers approaching this group of respondents, i.e., time was of the essence; (2) SNA data are gathered through asking the same question multiple times regarding others—a more traditional questionnaire is not faced with this constraint and can thus hold multiple items for one concept; and (3) focusing on communication ties was deemed most pragmatic, as questions about emotional intensity etc. were deemed too private/sensitive for this research context.
5. This is a software package for analyzing social network data.
6. Please see Wasserman and Faust (1994) for explanation on calculating centrality on valued, directional data. Also, one needs to keep in mind that there is a certain level of measurement error in our identification of central actors: Our stakeholder analysis was iterative, and we triangulated our data to provide as complete a view as possible of the network boundary before approaching respondents with our name generator question. However, determining network boundaries is a common problem in SNA, similar to other social research where sampling frames are unavailable or the population number unknown. Although some of the individuals from our name generator turned out to be central nodes, several of them did not, suggesting that although one can never fully rule out error and bias, our central nodes were not simply an artifact of the sampling technique, but rather our stakeholder analysis had adequately outlined our network boundary.

References

Bodin, Ö., B. Crona, and H. Ernstson. 2006. Social networks in natural resource management: What is there to learn from a structural perspective? *Ecology and Society* 11: r2. Available at: www.ecology andsociety.org/vo111/iss2/resp2/ (accessed October 3, 2008).

Brass, D. J. 1992. Power in organizations: A social network perspective. In *Research in Politics and Society,* Eds. G. Moore and J. A. Whitt, 295–323. Greenwich, CT: JAI Press.

Burroughs, R. 1999. When stakeholders choose: Process, knowledge, and motivation in water quality decisions. *Society & Natural Resources* 12: 797–809.

Burt, R. 2001. Structure holes versus network closure as social capital. In *Social Capital: Theory and Research,* Eds. K. C. N. Lin and R. Burt, 31–56. New York: Aldine de Gruyter.

Burt, R. S. 1992. *Structural Holes: The Social Structure of Competition.* Cambridge, MA: Harvard University Press.

Burt, R. S. 1997. A note on social capital and network content. *Social Networks* 19: 355–373.

Burt, R. S. 2000. The network structure of social capital. In *Research in Organizational Behavior,* Eds. B. M. Staw and R. I. Sutton, 31–56. Greenwich, CT: JAI Press.

Calton, J. M. and N. B. Kurland. 1996. A theory of stakeholder enabling: Giving voice to an emerging postmodern praxis of organizational discourse. In *Postmodern Management and Organization,* Eds. D. M. Boje, R. Gephart, Jr., and T. J. Thatchenkery, 154–177. Thousand Oaks, CA: Sage.

Carroll, A. B. 1989. *Business & Society: Ethics and Stakeholder Management.* Cincinnati, OH: South-Western.

Clarkson, M. B. E. 1995. A stakeholder framework for analyzing and evaluating corporate social performance. *Academy of Management Review* 20: 92–117.

Coleman, J. S. 1990. *Foundations of Social Theory.* Cambridge, MA: Belknap Press of Harvard University Press.

Crona, B. and O. Bodin. 2006. What you know is who you know? Communication patterns among resource users as a prerequisite for comanagement. *Ecology and Society* 11. Available at: www.ecologyandsociety.org/voll1/iss2/art7 (accessed October 3, 2008).

Cross, R. L. and A. Parker. 2004. *The Hidden Power of Social Networks: Understanding How Work Really Gets Done in Organizations.* Boston, MA: Harvard Business School Press.

Daniels, S. E. and G. B. Walker. 2001. *Working Through Environmental Conflict: The Collaborative Learning Approach.* Westport, CT: Praeger.

Dougill, A. J., E. D. G. Fraser, J. Holden, K. Hubacek, C. Prell, M. S. Reed, S. T. Stagl, and L. C. Stringer. 2006. Learning from doing participatory rural research: Lessons from the Peak District National Park. *Journal of Agricultural Economics* 57: 259–275.

Duram, L. A. and K. G. Brown. 1999. Assessing public participation in U.S. watershed planning initiatives. *Society and Natural Resources* 12: 455–467.

English Nature. 2003. English Nature, England's best wildlife and geological sites: The condition of SSSIs in England in 2003. Peterborough, UK: English Nature.

Freeman, L. C. 1979. Centrality in social networks. *Social Networks* 1: 215–239.

Freeman, R. E. 1984. *Strategic Management: A Stakeholder Approach.* New York: Basic Books.

Friedkin, N. E. 1998. *A Structural Theory of Social Influence.* Cambridge, MA: Cambridge University Press.

Frooman, J. 1999. Stakeholder influence strategies. *Academy of Management Review* 24: 191–205.

Goodpaster, K. E. 1991. Business ethics and stakeholder. *Business Ethics Q.* 1: 53–72.

Granovetter, M. 1973. The strength of weak ties. *American Journal of Sociology* 78: 1360–1380.

Grimble, R. and M. K. Chan. 1995. Stakeholder analysis for natural resource management in developing countries: Some practical guidelines for making management more participatory and effective. *Natural Resources Forum* 19: 113–124.

Grimble, R. and K. Wellard. 1997. Stakeholder methodologies in natural resource management: A review of concepts, contexts, experiences and opportunities. *Agricultural Systems* 55: 173–193.

Holden, J., L. Shotbolt, A. Bonn, T. P. Burt, P. J. Chapman, A. J. Dougill, et al. 2007. Changing environmental conditions in UK moorlands: A review. *Earth Science Review* 82: 75–100.

Kadushin, C. 1966. The friends and supporters of psychotherapy: On social circles and urban life. *American Sociology Review* 31: 786–802.

Knoke, D. and J. Kuklinski. 1982. *Network Analysis.* London: Sage.

Krackhardt, D. 1992. The strength of strong ties: The importance of philos in organisations. In *Networks and Organizations: Structure, Form, and Action,* Eds. N. Nohtia and R. G. Eccles, 216–239. Boston, MA: Harvard Business School Press.

Lockie, S. 2006. Networks of agri-environmental action: Temporality, spatiality and identity within agricultural environments. *Rural Sociolology* 46(1): 22–39.

MacArthur, J. 1997. Stakeholder analysis in project planning: Origins, applications and refinements of the method. *Project Appraisal* 12(4): 251–265.

Mitchell, R. K., B. R. Agle, and D. J. Wood. 1997. Toward a theory of stakeholder identification and salience: Defining the principle of who and what really counts. *Academy of Management Review* 22: 853–886.

Mushove, P. and C. Vogel. 2005. Heads or tails? Stakeholder analysis as a tool for conservation area management. *Global Environmental Change* 15: 184–198.

Newman, L. and A. Dale. 2004. Network structure, diversity, and proactive resilience building: A response to Tompkins and Adger. *Ecology and Society* 10. Available at: www.ecologyandsociety.org/ vol10/iss1/resp2 (accessed October 3, 2008).

Newman, L. and A. Dale. 2007. Homophily and agency: Creating effective sustainable development networks. *Environment, Development and Sustainability* 9: 79–90.

Office for National Statistics. 2003. Census 2001: CD supplement to the national report for England and Wales and key statistics for local authorities in England and Wales. London: Office for National Statistics.

Olsson, P., C. Folke, and T. Hahn. 2004. Social-ecological transformation for ecosystem management: The development of adaptive co-management of a wetland landscape in southern Sweden. *Ecology and Society* 9. Available at: www.ecologyandsociety.org/vol9/iss4/art2/ (accessed October 3, 2008).

Peak District National Park. 2004. State of the park report (update). Available at: www.peakdistrict.org (accessed November 5, 2006).

Prell, C. 2003. Community networking and social capital: Early investigations. *Journal ofComputer-Mediated Communication* 8. Available at: http://jcmc.indiana.edu/vo18/issue3/prell.htrn1 (accessed October 3, 2008).

Prell, C. 2006. Social capital as network capital: Looking at the role of social networks among not-for-profits. *Sociological Research Online 11*. Available at: www.socresonline.org.uk/11/4/prell.html (accessed October 3, 2008).

Prell, C., K. Hubacek, C. Quinn, and M. Reed. 2008. Who's in the social network? When stakeholders influence data analysis. *Systemic Practice and Action Research* 21: 443–458.

Prell, C., K. Hubacek, M. S. Reed, T. P. Burt, J. Holden, N. Jin, et al. 2007. If you have a hammer everything looks like a nail: 'Traditional' versus participatory model building. *Interdisciplinary Science Review* 32: 1–20.

Putnam, R. D. 1993. *Making Democracy Work: Civic Traditions in Modem Italy*. Princeton, NJ: Princeton University Press.

Ramirez, R. 1999. Stakeholder analysis and conflict management. In *Cultivating Peace: Conflict and Collaboration in Natural Resource Management*, Ed. D. Buckles. Ottawa, Canada: International Development Research Centre. Available at: www.idrc.ca/en/ev-27971–201–1–DO TOPIC.html (accessed October 3, 2008).

Reed, M. S., C. Prell, and K. Hubacek. 2004. Managing uncertainty in dynamic socio-environmental systems: An application to upland management. Moors for the Future Conference, October, Castleton, UK.

Rogers, E. M. 1995. *Diffusion of Innovations*. New York: Free Press.

Schneider, M., J. Scholz, M. Lubell, D. Mindruta, and M. en. 2003. Building consensual institutions: Networks and the National Estuary Program. *American Journal of Political Science* 47: 143–158.

Scott, J. 2000. *Social Network Analysis: A Handbook*. Newbury Park, CA: Sage.

Selin, S. W., M. A. Schuett, and D. Carr. 2000. Modeling stakeholder perceptions of collaborative initiative effectiveness. *Society and Natural Resources* 13: 735–745.

Skvoretz, J., T. J. Fararo, and F. Agneessens. 2004. Advances in biased net theory: Definitions, derivations, and estimations. *Social Networks* 26: 113–139.

Stringer, L. C., C. Prell, M. S. Reed, K. Hubacek, E. D. G. Fraser, and A. J. Dougill. 2006. Unpacking 'participation' in the adaptive management of socio-ecological systems: A critical review. *Ecology and Society* 11. Available at: www.ecologyandsociety.org/vo111/iss2/art39/(accessed October 3, 2008).

Tompkins, E. L. and W. N. Adger. 2004. Does adaptive management of natural resources enhance resilience to climate change? *Ecology and Society* 9(10). Available at: www.ecologyandsociety.org/vol9/iss2/art10 (accessed October 3, 2008).

Tucker, G. 2003. Review of the impacts of heather and grassland burning in the uplands on soils, hydrology and biodiversity. Research report no. 550. Peterborough, UK: English Nature.

UK Biodiversity Steering Group. 1995. The UK Steering Group report—Volume II: Action plans. London: HMSO.

Varvasovszky, Z. and R. Brugha. 2000. How to do (or not to do) . . . A stakeholder analysis. *Health Policy Plan* 15: 338–345.

Volker, B. and H. Flap. 1999. Getting ahead in the GDR: Human capital and social capital in the status attainment process under Communism. *Acta Sociologica* 42: 1, 17–34.

Wasserman, S. and F. Faust. 1994. *Social Network Analysis: Methods and Applications.* Cambridge, MA: Cambridge University Press.

Wellman, B. and M. Gulia. 1999. Virtual communities as communities: Net surfers don't ride alone. In *Communities in Cyberspace*, Eds. M. A. Smith and P. Kollock, 167–194. New York: Routledge.

Wellman, B. and K. Frank. 2001. Network capital in a multi-level world: Getting support in personal communities. In *Social Capital: Theory and Research*, Eds. N. Lin, K. Cook, and R. Burt, 233–237. New York: Aldine de Gruyter.

18

THE MULTILEVEL PARTICIPATORY MODELLING OF LAND USE POLICIES IN AFRICAN DRYLANDS

A Method to Embed Adaptability Skills of Drylands Societies in a Policy Framework

Patrick d'Aquino and Alassane Bah

Research Question: How can farmers in the African drylands cope with environmental uncertainty?
System Science Method(s): Agent-based models
Things to Notice:

- Involving stakeholders in model development, testing, and analysis
- Incorporating indigenous and external scientific knowledge

Embedded in deeply risk-prone environments, farmers progressively developed particular ways of thinking about how to "surf on uncertainty" rather than contending with it. These drylands societies' way of thinking may be useful in the search for new forms of adaptability. A participatory modelling approach using agent based modelling, called "self-design", has been experimented in Senegal with the aim of letting farmers design their own conceptual model of their environmental management issues. The method focuses on how to enable stakeholders to incorporate their own perception of environmental uncertainty and the way to deal with it in a policy simulation. This approach uses role-playing games and agent-based modelling to ensure that a range of different points of view are preserved in the shared modelling of resources management, with outcomes in terms of mutual learning and management innovations. A twofold multiscale scope is supplied within the "game". The whole Sahelian logic of multiscale, or even open-scale, can then be expressed and on another hand stakeholders are able to deal with region-wide changes of implied by a policy decision. Participants create a qualitative model that nevertheless reflects the complexity of drylands environmental uncertainty, then use themselves the model to shape unusual uncertainty management principles for policy design, which is currently enriching the debate about the value of local worldviews for environmental modelling. Furthermore, the specific inclusionary method which is implemented involves and interlinks stakeholders from local to national levels, which is led to a novel multilevel participatory policy design.

DOI: 10.4324/9781315748771-20

Contrary to common misconceptions, rural people in southern countries have efficiently dealt with ecological and socioeconomic scarcity for several centuries. Embedded in deeply uncertain contexts, local communities progressively developed particular ways of thinking about how to organize access to nature in a way that "surfs on uncertainty" rather than contending with it. This is particularly true in risk-prone environments such as drylands. Sahelian societies are a prime example of such "surfing", as for centuries. Drylands farmers may use different practices in the same field, for example, spreading manure on only one part, and hoeing another, i.e. increasing their range of practices to be sure to have included the practice that is best adapted to the uncertain climate that year. In the same way, in response to the spatial uncertainty of rainfall and to avoid uncontrolled access, herders shaped collective rules for open access to land based on subtle social agreements that may vary from family to family and from village to village. These rules enable many different uses for each piece of land, as they allocate specific rights of access for each possible use of each natural resource, such as cultivating annual crops, planting an orchard, creating a pasture, hunting, gathering wild fruits, collecting firewood or fodder, or harvesting wood for crafts. In addition, rights of access may change with the season, and with the duration and the economic value of the activity, an annual or perennial crop, fruit trees, or gum trees, among others. Rather than being interpreted as land ownership, these complex access rights should be seen as a "bundle of rights" that control the appropriation, exploitation and use of natural resources in a given space. On the whole, traditional practices and livelihood strategies are based on diversity, whether of seeds, livestock breeds, technical practices, or land uses, and on the resulting flexibility, with the aim of improving their adaptability.

Drylands societies' rules and practices may be less suited to contemporary demographics and climate changes, but their way of thinking about adaptability may still be useful in the search for new forms of adaptability. However, designing new policies using this adaptability only makes sense if new policy paradigms are created in which flexibility is a key value. This could be achieved by more efficiently embedding the specific worldviews of drylands societies in the current policy framework paradigm. The participatory modelling approach presented here takes up this challenge.

Participatory modelling experiments have expanded considerably in the last decade, especially for the management of socioecological systems. The purpose is to reach mutual understanding between the modellers and stakeholders on their knowledge and points of view (Gaddis et al., 2010; Purnomo et al., 2009; Voivnov and Bousquet, 2010). Many methodological options have been explored, the diversity of which can be classified in different ways (e.g. Lynam et al., 2007; Mendoza and Prabhu, 2006; Parker et al., 2002; Renger et al., 2008; Rouwette et al., 2002). Participatory modelling approaches may have a variety of goals, which can range from involving stakeholders in the choice of goals and modelling agenda, methods that incorporate empirical knowledge in the modeller's knowledge structure (e.g. Argent and Grayson, 2003; Martinez-Santos et al., 2008), others that let the stakeholders explore and test the modeller's knowledge (Barreteau et al., 2003a), and yet others that let stakeholders try to model their own worldviews (d'Aquino et al., 2003 and d'Aquino and Bah, 2013; Simon and Etienne, 2010). The scientific value of including local subjective knowledge is often questioned: first, because some stakeholders' intuitive assertions may appear to be unsound and to threaten the quality of the model; second, because managing stakeholders' conflicting views in the model may be difficult and thus call into question the efficiency of the whole process; finally, many modellers simply wonder what is the point of this unusual methodological option. Here, we test and then discuss the value of incorporating stakeholders' worldviews in modelling, based on the outcomes of ten years of experience in participatory modelling in Africa. We describe a modelling method designed to highlight stakeholders' worldviews, then discuss the value of the modelling outputs produced by the stakeholders.

A participatory modelling approach using agent-based modelling, called "self-design", has been experimented in Senegal with the aim of letting farmers design their own model of the local natural resources management issues. The success of the experiment and its outputs led to a new participatory modelling based on the central principle of letting stakeholders design and use their own conceptual model of environmental management. In a scientific perspective, the relevance of this endogenous model is currently enriching the debate about the value of local worldviews for environmental modelling. In a policy perspective, the participatory use of the modelling approach led to an inclusionary multilevel policy approach. The modelling approach is used into an inclusionary method which involves and interlinks stakeholders at local to national levels. Participants create a qualitative model that nevertheless reflects the complexity of drylands environmental uncertainty. Then, participants, gathered into a multilevel participatory modelling approach, use the model to shape unusual uncertainty management principles for policy design.

The hypothesis of our approach is that local communities around the world have tailored worthwhile flexible ways to manage natural resources, especially in order to deal with their deep uncertainty about the environment, which lacks consideration and verification of official rules and policies. The first goal of our scientific input is to meet these communities' growing demands to provide a way to express their uncertain behaviour, rules, and institutional practices in a form that is able to fit into official rules and policy frameworks. The second objective of this elicitation method is to add these communities' experience and knowledge some uncommon uncertainty management principles which may enrich the scientific framework about how to make more flexible the management of their environmental uncertainty.

The method focuses on how to enable stakeholders to incorporate their own perception of environmental uncertainty and the way to deal with it in a policy simulation. This approach uses role-playing games and agent-based modelling to ensure that a range of different points of view are preserved in the shared modelling of resources management, with outcomes in terms of mutual learning and management innovations (Barreteau et al., 2003b; Etienne, 2011). From the outputs of this step, the structural components of a meta-conceptual model are built progressively (d'Aquino and Bah, 2012). The outputs of this workshop are used to prepare a more operational role-playing game and a computerized version of the game using an agent-based model (ABM), which has exactly the same features as the game (d'Aquino et al., 2002). Following the self-design method, the multiscale representation of the issue is the second methodological thrust which helps participants to elicit their background worldview on how to manage uncertainty. A multiscale representation of the Sahelian environment is supplied within the "game" with the aim of encouraging participants to reflect on management rules which are not only appropriate for the participants' very local location but also for other places and at other scales – in other words, a policy scope. For this reason, the game map provides a multiscale representation of the drylands (Figure 18.2) as a simulation support which enables participants to handle both the logic of uses and environmental management options at different scales (especially mobility). Given participants can adapt the content of the maps to their context and scales, this multi-scale representation is able to model different ranges of scales, depending on the decision levels chosen by participants: the basic spatial unit may represent a farm parcel, several parcels the land base of a farm, while the geographical level two (in bold lines) will be community territory, a "map" of the regional area, and several "maps" of national territory. But in other simulations, the basic unit may be considered right as a community territory (and the players will thus play communities), the level 2 will represent a region, a "map" the country, and several maps a multinational region. In other usable interpretations, a "map" is a village territory, the sub-level (level 2) the lineages land bases, while a group of maps will be a small inter-communities area. Players can design regional maps and add them to the game (see Figure 18.2) if they wish, in an "open-scale" way to assess their environmental policy options. Therefore, the multiscale scope is twofold: on one hand the whole Sahelian logic of multiscale, or even open-scale, can then be expressed and on the other hand stakeholders are able to deal with regionwide changes of implied by a policy decision.

Method

"Self-design" Modelling

It has been increasingly recognised that modelling and participatory approaches can be mutually reinforcing when applied to complex environmental issues (e.g. Reed et al., 2008; Voinov and Bousquet, 2010; Dougill et al., 2010). A wide range of participatory modelling approaches exists, from those which incorporate empirical knowledge in a scholar's prior knowledge structure to those which let the stakeholders test the scholar's knowledge, and yet others which focus on eliciting local knowledge (Stringler et al. 2006; Reed et al., 2008). The method described here belongs to the last category. It focuses on how to enable stakeholders to incorporate their own perception of environmental uncertainty and how to deal with it in a simulation. This approach uses role-playing games and agent-based modelling to ensure that a range of different points of view are preserved in the shared modelling of resources management, with outcomes in terms of mutual learning and management innovations (Barreteau et al., 2003b; Etienne, 2011).

Since 1999, we have been working on a particular kind of participatory modelling we call "self-design". "Self-design" means letting participants design their own conceptual framework of issues and goals with no inputs from facilitators, modellers, or scholars' perceptions (d'Aquino et al., 2003; d'Aquino and Bah, 2013). The process has two main phases which specifically focus on letting participants decide on all the crucial elements (Figure 18.1).

A first "suggesting" meeting. This first meeting is held in many different locations to reach out to a wide panel of potential local partners. During the meeting, the participatory simulating approach is presented in detail including a detailed explanation of its objective – i.e. to support people in designing their own land policy views, and of the method – i.e. the self-design of a role-playing game and a computerized model. Participants are then asked to contact the team if they are interested in implementing this approach on their own.

A "self-eliciting" workshop. An endogenous self-appraisal (B) is held with the local partners who recontacted the team. During this workshop, the participants themselves identify the aims of the process – i.e. (1) the policy stakes they wish to target; (2) the stakeholders they think they will need to take into account in their self-policy design, (3) the information they think they will need to tackle the policy issues on their own and (4) the constraints they think could be critical for these issues. Participants are made aware of the level of description they will be asked to provide – i.e., detailed enough to capture their local needs but sufficiently summarized to enable analysis at the national scale.

A second participatory workshop. A second workshop (C) is then held during which participants "self-design" their own conceptual model. For this purpose, the outputs of the previous "self-eliciting" workshop are structured by the research team into a first simple role-playing game, as a way to let the participants design a conceptual model of their issues.

The settings of this first game are basic but nevertheless very subtle. The challenge is to summarise the major stakeholders' needs and constraints and the main policy stakes they identified in the previous workshop in a qualitative support. First, a spatial grid is provided to highlight the simplest environmental typology that can be used without concealing the structural components of the issue. Coloured pawns are provided to represent the different potential uses of each type of landscapes; the different colours represent the range of possible activities. Tokens are provided as a way of qualitatively assessing indicators of the major policy stakes. The tokens are removed from the landscape as the players consume the natural resources of the landscape parcels and keep them as a cash stock. The tokens thus enable both the qualitative assessment of the natural resources available on each landscape parcel and the capacity for self-sufficiency of the different stakeholders. Lastly, "events" cards represent crucial factors and trends (for example, climate events, demographic pressure, arrival of agribusiness) in the game.

The background structure of the role-playing game must not embody the modeller's perception of how to improve the environmental situation. The goal is not to have participants progress towards the modeller's knowledge system but to let them design their own conceptual model based their own worldview (Figure 18.1). This is why the structure of the game is intentionally kept simple. The board game combines different maps in order to highlight the possibility of the seasonal movements – pastoralism, for example. The geographic structure of each map is open to modification by the players: four maps are provided for the first exploratory game, and the participants are asked to modify the maps on their own – i.e. the typology of landscapes and the layout of the parcels. The time component of the game is a year, which comprises three seasons (rainy, dry cold, and dry hot seasons). In each season, the players choose

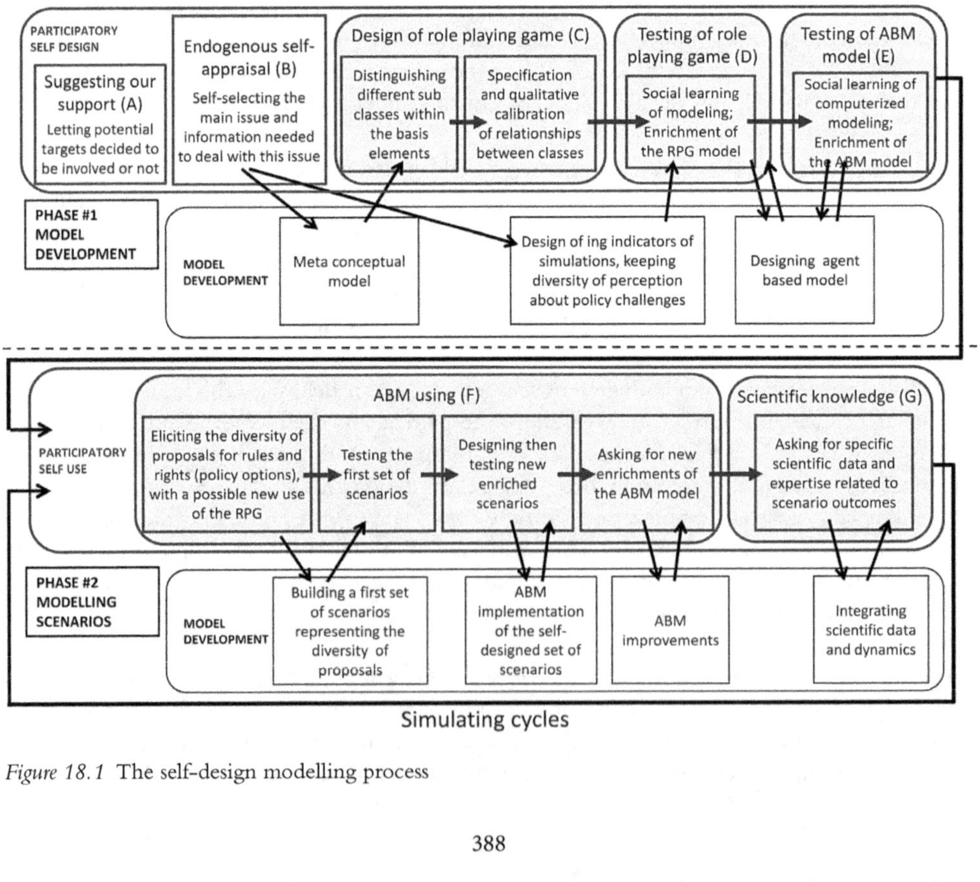

Figure 18.1 The self-design modelling process

the activities they want to carry out, place their activity pawns on the appropriate landscape parcel, and remove the appropriate amount of resource tokens from the parcel, if the climate in the year concerned combined with the type and state of the landscape parcel they have chosen allows agricultural production. The game rules concerning the different activities must be simple but nevertheless incorporate elements that would be affected by land policy – i.e. a participatory calibration about how many resources tokens players can remove, depending on the activity they implement, the landscape they used, and the annual rainfall on the parcel concerned. This means the rules that apply to the farmers' production activities and environmental impacts are very qualitative, with the sole aim of allowing qualitative comparisons between different land uses scenarios. The rules that apply to land access are also kept very simple, to leave the frame sufficiently open to the players' conception of collective rules. This means that players can place a pawn representing a particular activity anywhere they choose among the several board maps provided, but cannot place two pawns representing agricultural activities on the same parcel at the same time, because contrary to pastoralism and gathering activities, it is physically impossible to have two fields on the same place. So no collective rules are pre-established, because that would mean imposing the designer's conception of the relevant collective rules. For example, the risk of crop injury cause by livestock is incorporated as follows: when a pawn representing a livestock activity is placed close to a pawn representing an agricultural activity, the risk of damage occurring (quantified by throwing a dice) is incurred by the agricultural activity and the pastoralist is not affected as long as the players themselves set up a collective rule about sanctions in the case of damage. In this way, participants have the opportunity to thoroughly check – and if necessary improve – the initial structural elements of the game (see stages D and E) – for example, by enriching the spatial legend or extending the list of potential uses or assessment indicators.

A participatory simulation. At this point, thanks to the previous learning-by-designing process, participants are able to handle the participatory simulation (F) support satisfactorily. Consequently, in the third workshop, participants use the final simulation support to think among themselves about how to improve land policies and test new environmental management options: collective rules, new forms of land rights, new infrastructures, new practices, etc. The outputs of the previous self-design workshop are used to provide a more complete support for this simulation – i.e. a more complex role-playing game but also a computerized version of the game, using an agent-based model (ABM) which has exactly the same features as the board game (d'Aquino et al., 2003). As the computerized model (Figure 18.2) is based on the game they themselves designed, the participants can use it on their own. They are thus able to continue testing some of the scenarios they started testing in the game but this time testing on the computer. When playing the board game, the participants play the role of local users who consume natural resources and who also draw up the rules of access which apply to the players-users. In the computerized version of the game, computerized agents act as users who have the same incentives as the players in the role-playing game, and the participants no longer play but only define the collective rules which apply to the agents. These two forms of the same conceptual model are complementary. The board game helps stakeholders to tailor their own representation of the issue and its challenges, while the computerized version allows them to test more detailed and operational scenarios. While social complexity is clearly more efficiently comprehended by playing the game (because players can try out new behaviours and practices), biophysical and long-term dynamics are more efficiently comprehended using the computer. In other words, the role-playing game supports the self-design of the participants' conceptual model of environmental uncertainty, and the computerized version supports a more accurate but simulated use of this conceptual model.

The two supports (i.e. the game board and the computer simulation, Figures 18.3 and 18.4) are intentionally left sufficiently open so they can be enriched and contextualized in a continuous and iterative "companion" process (Etienne, 2011). Participants can incorporate new rules and items including risk events (climate, bush fires, prices of goods, etc.), new forms of land use (intensive farming, hunting, tourism, etc.), social behaviours (users' or managers' strategies, forms of negotiation for access to land, etc.), or collective rules and organisation (decentralization, common pool resources, etc.).

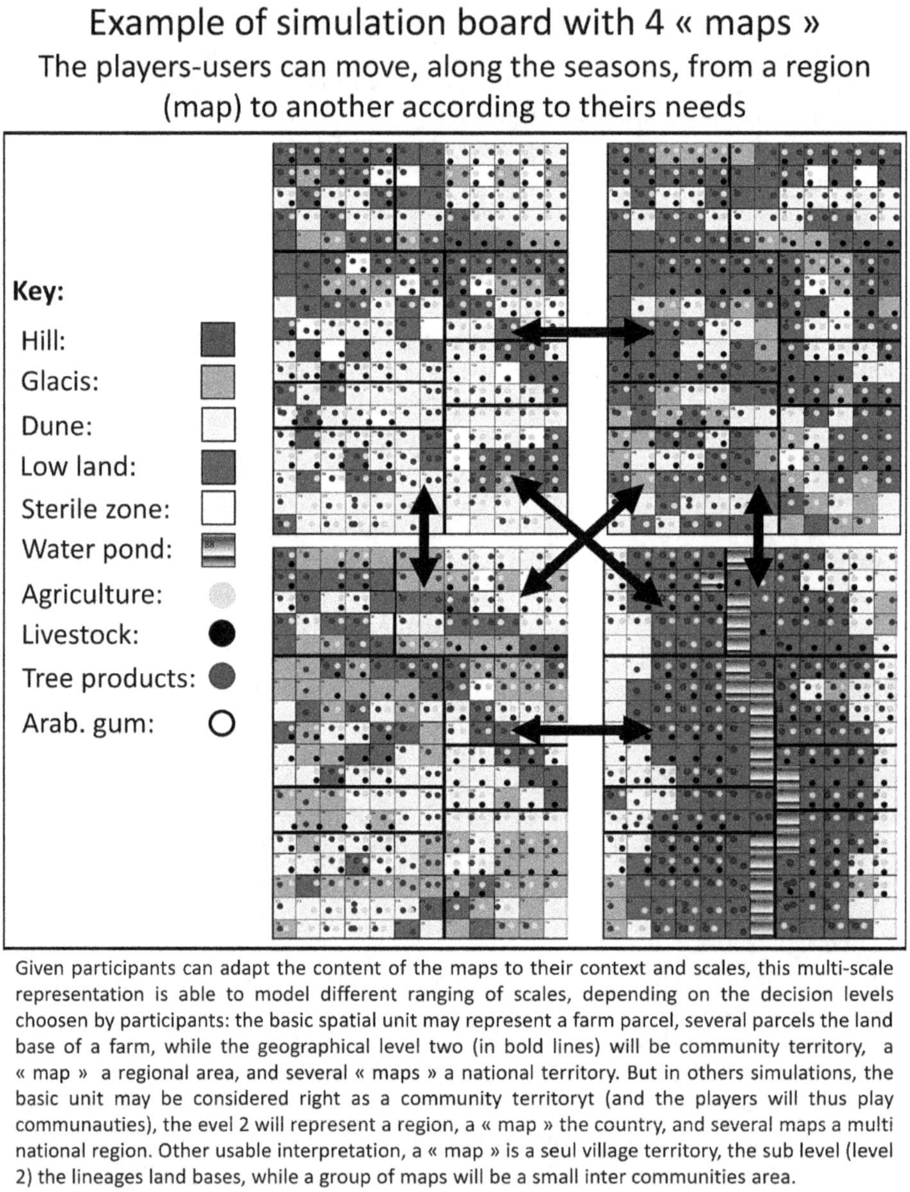

Figure 18.2 Structure of model maps (resulting from the co-design process with stakeholders)

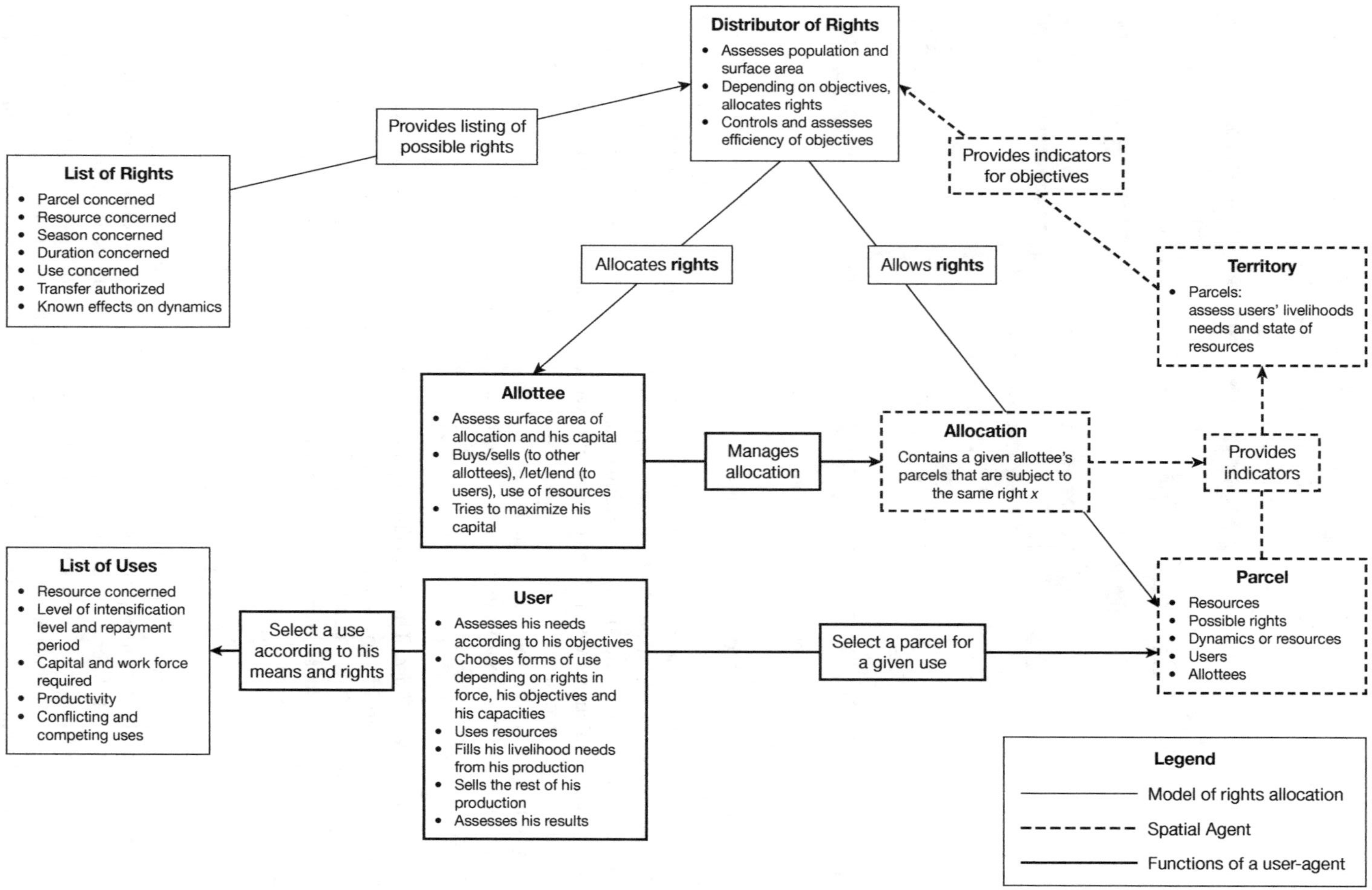

Figure 18.3 The conceptual model emerging from the stakeholders' self-design of the game

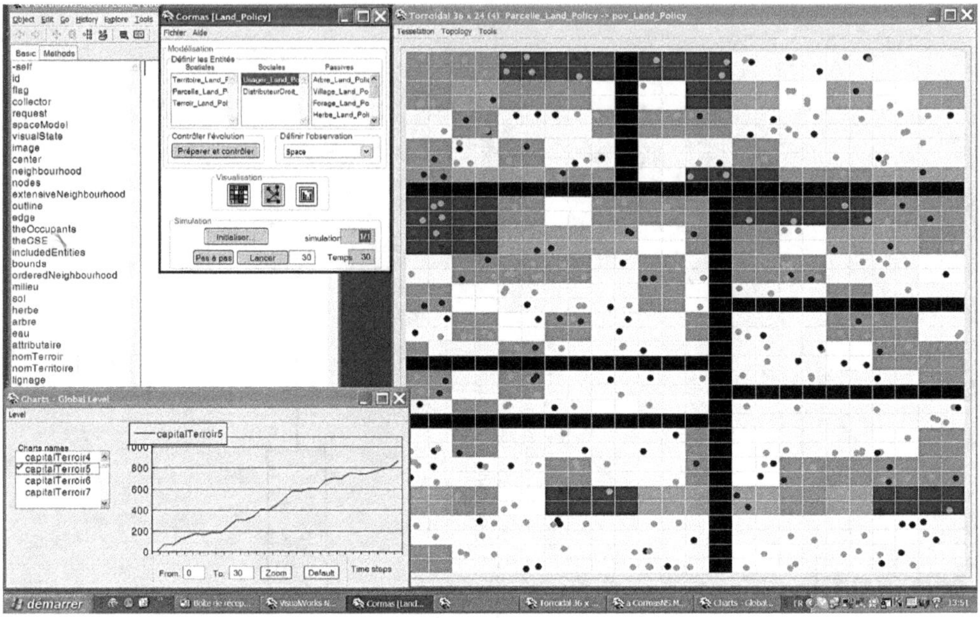

Figure 18.4 Key variables and features of the computerized model

The use of the participatory simulation support enables participants to design the simulation scenarios themselves (see stage F) by combining (1) a *climate scenario*, i.e. a sequence of "high", "moderate", and "low" annual rainfall years in the model; (2) a *socioeconomic scenario*, including user densities for each social scale, the user's workforce and starting capital; (3) a *regulatory scenario* – i.e. the different rights and rules concerning access to land and to the natural resources in each spatial cell, which are defined by combining two cell attributes: "right of use", which defines the uses allowed in the cell in each season, and a "right of access to land", which defines who has right of access to the cell. Participants are then left to iteratively explore and modify not only the scenarios but also the model while they are actually using it, and in this way, to increase the complexity of their representation of the issue.

Introducing scientific knowledge. The modelling settings are at this point sufficiently "self-framed" to allow *external* scientific knowledge and points of view to be incorporated without masking "local uniqueness" (G). Given that facilitators also have their own unconscious perceptions which affect the way they conduct a participatory appraisal, one of the aims of the self-design process is to mitigate this influence.

A Multi-scale Focus

The game board map provides a multi-scale representation of the drylands (Figure 18.2), as a simulation support which enables participants to handle both the logic of uses and environmental management options at different scales (especially mobility, as players are able to move from one board map to another to place the pawns representing their different activities on any of the maps). This multiscale representation of the Sahelian environment also aims to encourage participants to reflect on management rules which are not only appropriate for their own particular location but also for other places and at other scales: in other words a policy scope. This multiscale

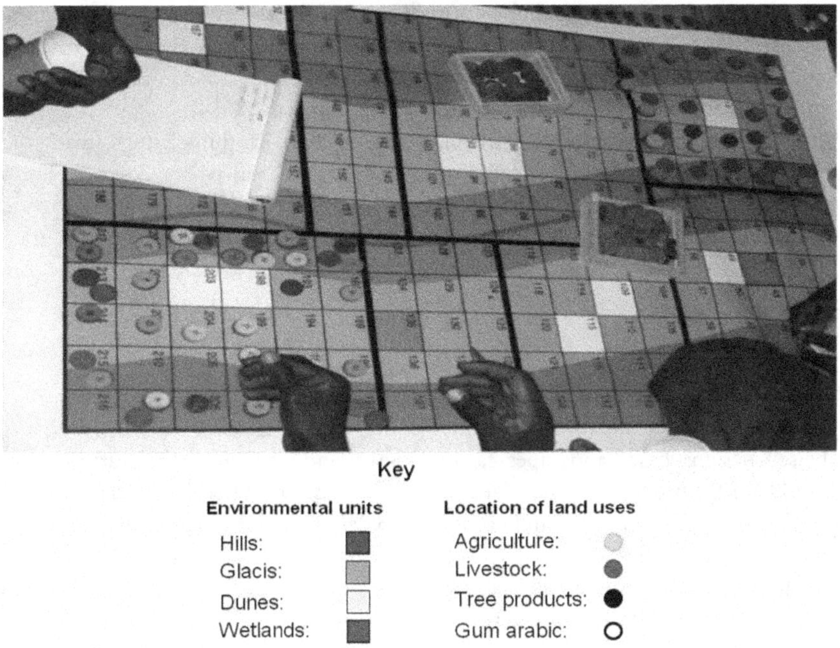

Figure 18.5 The role-playing game board

feature supplements the set of indicators we provide to encourage the participants to take everybody else's needs and interests into account – in other words, the "common interest", while engaging the participants in a multilevel assessment of their own scenarios (see, for example, Figure 18.5). The aim of the multiscale focus is twofold: on one hand to enable the whole multiscale Sahelian logic to be expressed and on the other hand to encourage the players to deal with possible regionwide changes implied by their policy scenarios.

We have also implemented a "multilevel inclusionary strategy" (d'Aquino and Bah, 2014). The objective is to be very explicit with the stakeholders about an inclusionary challenge: letting the participants depict uncertainty issues from their own point of view before incorporating the policy makers' points of view, so that they are then able to design and test some policy options together, in order to comply with both their own uncertainty management principles and current policy needs. As our strategic frame focus to be embedded into local collective actions and needs (d'Aquino and Papazian, 2014), our methodological support inserted into a Senegalese civil society group action to develop and negotiate land policy proposals with the Senegalese government. Thus, our methodological support has been proposed to the civil society,[1] which is committed to the approach and used it, with our methodological advice, to implement participatory workshops in different parts of the country (Defalt, 2014).

Then, in order to balance power relationships between local and national stakeholders, each step of the self-design process (see above) is first organized at the local level and afterwards at the national level. As a result, a same single "self-simulating" process brings together two target groups, local and national stakeholders, but the process involves two separate but parallel "modeling arenas". The design and use of the model is nevertheless shared thanks to structural links between the workshops and the tools: all the components incorporated in a simulation session – for example, a tool which has been enriched, or a new simulation scenario, are also incorporated in the equivalent parallel session. This means creating a sort of single exchange arena but with two different interfaces, one for the local workshops and one for the national workshops. This shared design has proved to be capable of mixing appropriate policy goals both from the national and local points of view (d'Aquino and Bah, 2014).

To recap, the self-design approach includes several "self-handling" steps, from the initial "self-commitment" to launch the process to the final step of "self-policy design". This process results in a simple qualitative representation but which is nevertheless fine enough to accurately capture the complexity of drylands uncertainty, as shown by the results.

Results

Key Variables and Features of the Final Computer Model

Different features must be distinguished. First, some "basic" features are fixed by participants during the self-design phase (landscapes, farmers' needs and strategies, rainfall impact on natural resources). These features are considered fixed but can be modified by participants if needed be. Second, some variables are used and combined to design "environment" modelling scenarios: demographic densities (different from one landscape to another), average level of annual rainfall (which is afterwards differently applied according to northern – drier-region – and southern – wetter-region – maps), and the possibility to change some parts of landscape units. Third, land uses rules can be modified for each cell, with this simulating grid:

- One can specify the resources of the cell which are allowed to be used (soil, water, grass, or tree), the kind of use which is allowed (agriculture, pastoralism, gathering), and the seasons concerned. Every combination of these three elements is possible.
- One can also specify which social categories of users are allowed to use this cell: only locals, only people from a specific lineage (synthetizing a customary land tenure), or for a nuclear family.

The desired combination of these two variables is implemented and conserved as a "land rules option map" and can be mobilized for simulations. Consequently, a simulation scenario gathers a specific combination of environmental features and a specific combination of land rules access and tenure. Then, a simulation is monitored by the follow-up of few variables, along the simulation and for the different landscapes (see one of these graphs at the bottom left of Figure 18.4):

- The evolution of the quantity of the four natural resources.
- The production of each use.
- The economic success of every user.

Stakeholders' Ownership of the Method

Participants revealed their incentives and demonstrated their ability to incorporate an eclectic list of relevant policy challenges (see Figure 18.1) in order to ensure everyone's point of view was included. Both local stakeholders and policy makers (quite surprisingly on the part of the latter) showed great interest in incorporating the other stakeholders' indicators. Indeed, they were interested in designing innovative policies by mixing local and policy frameworks. Last but not least, their use of self-modelling led to some unusual principles of uncertainty management.

The Emergence of Specific Uncertainty Features from the Self-Design Process

The first fundamental element participants introduced in the game settings was a qualitative calibration of the relationships between the amount of rainfall in one year and production. They were asked to qualify the difference in productivity between years with "high", "moderate" and "low" rainfall. As the computerized version of the game enables variations in the different qualitative calibrations of the game, the effects of selecting the stakeholders' ratio can be simulated. Yet, while running the model, the only way to reveal the difference in productivity between "high", "moderate" and "low" rainfall years (defined by the stakeholders) was not simply to reduce the direct impact of rainfall on the production of resources but also to increase the scarcity of resources (Figure 18.6). The graph in the figure shows the results of a series of computer simulations in which natural resources in the landscape vary (see x axis: rate of available resources). The y axis shows the rate of production depending on the yearly rainfall: a higher curve means a greater effect of rainier years on productivity. Consequently, the peak of the greatest impact of rainier years is in the middle of the graph when resources are least scarce (following better environmental conditions when a rainy year is less useful, and before resources become so scarce that even rainy year cannot have really beneficial impact). Actually, the qualitative model calibrated by local stakeholders in the game sessions matches the period when the impact of rainfall on productivity is highest, and the rainiest years provide the greatest benefit: participants instinctively shape a model which summarizes the specific conditions of land uses in the Sahel, and thus produce a user-friendly model that can be used to help design policies to fit these particular conditions.

Moreover, Figure 18.6 shows that the conditions described by the stakeholders resulted in a simulation support in which a slight difference in resources availability caused a considerable difference in productivity. Thus, behind its apparent qualitative simplicity, the model simulation support provides an interesting representation of Sahelian uncertainty conditions based on a fine balance between the scarcity of resources and rainfall (see Bourgoin, d'Aquino, and Bah, 2014). Because of the way the stakeholders calibrated this fragility, the most advantageous use can vary even in the case of limited environmental variability (Figure 18.7). The x and y axes in the figure are the same as in the previous graph (x axis: the varying parameters of natural resources in the computer model; y axis: users' production according to the availability of resources). The box highlights the same area as in the previous graph: the qualitative conditions of available resources designed by stakeholders during the self-designed game. While the previous graph focuses on the greater impact of climate in the stakeholders' model, this graph highlights another feature that also only comes to light in the environmental conditions designed by the stakeholders (i.e. in the box): the shift between agriculture and livestock as the most productive use; in other words, with the environmental conditions self-depicted by stakeholders, a slight difference in annual climate causes a shift in the most productive use between agriculture and livestock.

This variability results in highly variable spatial productivity, which is the product of many different sources of diversity and uncertainty (d'Aquino and Bah, 2012): climatic uncertainty, the range of different landscapes, the varying location of key resources, the variety of resource uses, and, depending on the season, the users' mobility, and finally, changes in user density. Thus, the self-designed simulation support describes a situation in which a particular land use may have certain advantages depending on the prevailing environmental conditions, and a particular landscape may be advantageous for a particular use, but only with a combination of particular environmental conditions, such as rainfall in the year concerned, user density, or the location of certain key resources. Certain ecological units may be a key resource for a particular type of production but only if used in a particular season. The overall long-term productivity balance relies on certain key resources, such as the use of wetlands for agriculture in wetter years, wetlands being unevenly distributed in space, but also the use of wetlands for gathering natural products in drier years, dry years being unevenly distributed in time.

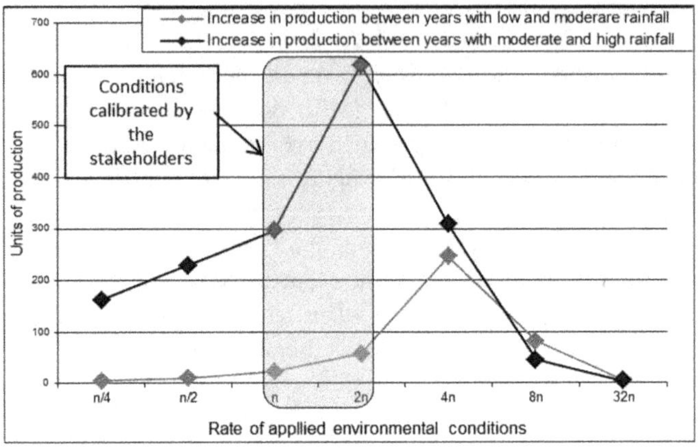

Figure 18.6 Impact of variation in rainfall on different types of production according to the scarcity of natural resources in the computerized version of the game

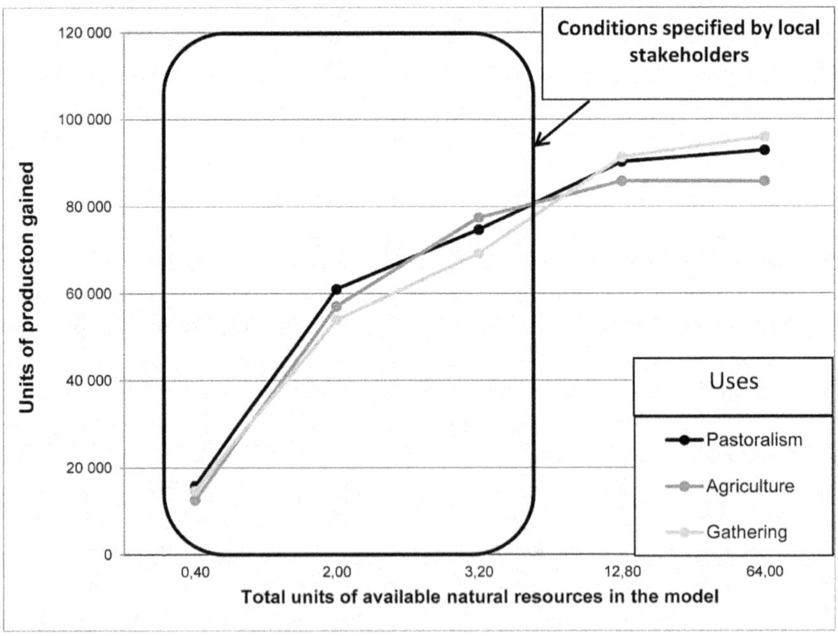

Figure 18.7 Productivity of the three major uses according to the level of resources scarcity (including the impact of rainfall)

Qualitative and Easy to Use, but Nevertheless Appropriate for Complex Management of Uncertainty

Despite its simple qualitative structure, the first outputs showed that the model manages to conserve the fundamental subtle features of complex Sahelian uncertainty. First, the specific combination of environmental conditions (semi-aridity and climatic uncertainty, scarcity of natural resources, specific spatial high variability, etc.) crafted by participants, led to a modelling situation in which natural resources are so rare that in the worst cases, the intensive use of natural resources may be economically less efficient than extensive use. This claim has been made by Sahelian researchers for many years but was not often accepted by decision makers until they themselves designed the game. In fact, decision makers and stakeholders designed a model in which they did not "come up with" these dynamics, but in which the dynamics "emerged" as logical outputs of their world view: an emergent feature that matched scientific reality.

Last, the most interesting feature that emerged from the modelling outputs is the fact that overall yield is correlated with a very rare and localized condition: the agriculture use of lowland parcels in years with the highest rainfall. A particular aspect that land tenure rights and management need to take into account by incorporating a multi-annual perspective: how can the use of these vital resources during these particular years be preserved for all the users and not only for some owners? This is the kind of issue our modelling support can help decision makers and stakeholders resolve together.

Indeed, a more detailed run of the computerized version of the game demonstrated that the self-designed model is a true model of non-equilibrium ecological dynamics, shaped by a complex combination of specific features characteristic of drylands uncertainty:

- There is no general economic advantage of any particular use. Model outputs underlined the fact that the most profitable use differs not only depending on annual rainfall, resource scarcity, and the density of users, but also on the spatial structure of the ecological landscape, confirming the value of multi-purpose use in this kind of uncertain environment.
- No specific type of landscape has an overriding economic advantage. The spatial combination of variability and uncertainty results in complex variability of landscape potentialities that depend on the type of use and exploitation rate, as well as on resource scarcity and annual rainfall. This means that the potentially best parcels of land and landscapes change from one year to the next depending on subtle environmental conditions and uses.
- The environmental potentialities are unforeseeable because the above features are combined in such complex and varying ways.
- Last but not least, the future sustainability of the whole agrarian system depends on specific access to some restricted space-time resource niches. A very remarkable Sahelian environmental uncertainty feature was revealed by this self-designed model: long term yield depends entirely on a few resource niches that are spatially and temporally rare, such as specific wetlands that can be used for agricultural purposes in the wet season in years with high rainfall and for livestock in years with moderate rainfall, or sandy land for grazing livestock in the dry season in years with low rainfall. This specific environmental context has long been reported by drylands researchers where key high value resources are found alongside low value extensive resources (Scoones, 1994; Mehta et al., 1999; Dougill et al., 2010). Thus the stakeholders' intuitive modelling proved its relevance, even based on such fine features. Moreover, it now provides a simulation support that summarizes this environmental specificity in a user-friendly frame and hence makes it more appropriate for inclusion in policy frameworks . . . and resilience modelling.

The Emergence of First Indigenous Principles for Collective Rules from the Self-Simulation Process

As the modelling platform is shaped by the stakeholders themselves, it is easy for them to use and is consequently a powerful support to help stakeholders to reflect among themselves on the best environmental policies to enhance drylands sustainability. At the same time, it tests drylands people's ability to design innovative principles of environmental management, drawing on the historical ability of their society to surf on uncertainty. Thus, after self-designing a multi-scale "model" of environmental uncertainty, stakeholders use the model they have crafted to test (in the form of a game) scenarios of environmental policies they would like to implement to improve their current situation.

The outputs of these first experiments are rules the participants tailored and tested. Participants naturally introduced unconventional environmental management principles: first, they intentionally kept multi-use and multi-user access to land because of the spatial uncertainty and variability of their environment. They then agreed on a "priority principle" for collective regulation: each area had a priority use or user but with a "soft" restriction, meaning all users can access the area but are responsible and answerable for not disturbing the priority use or user. As one participant remarked, this means "freeing up the zoning".

Another highly innovative proposal shaped by the participants that emerged was distinguishing a soft flexible "common law", similar to the "priority zoning" mentioned above, to be applied in standard areas and years, and on the other hand, common ownership with strict collective rules for rare and vital space-time resource niches. Participants listed the following vital space-time niches:

- exceptional rainfall in drought years;
- certain local wetlands in years with high rainfall;
- other types of wetlands in drought years;
- particular regional spots of pasture biodiversity that play an essential role in pastoral productivity;
- bush resources for gathering in the dry season.

These critical space-time resources belong to a common pool and are controlled by strict allocation rules in such a way that everybody profits from partial access. The distinction between the two regulation systems applies at all management scales, from local districts to natural regions and beyond, to the international Sahel. The details of these regulations and their overlapping regimes are not yet finalized and require further investigations in a new set of simulation workshops, but they already describe general natural resources management which distinguishes between two regulation systems: the first applies in normal situations, i.e. reasonable environmental conditions, when rules of access can be softened and controlled using the original "priority principle", and the second to be applied in a critical situation that can occur at a seasonal, annual or regional scale.

This unusual indigenous proposal leads us to a peculiar multi-level perspective. While some of these micro spots of resources may be too small to be integrated efficiently at regional or even at local management scales, by combining particular rules for micro specificities within a generally flexible regulation, these Sahelian stakeholders propose an interesting multi-level form of natural resources management, and they appear to have the necessary experience to put it into practice.

Obviously, these initial results (the emerging policy principles) require deeper collective adjustments to become operational. Scholars' expertise and policy-makers' points of view will also need to be incorporated (see perspectives below). Yet the indigenous principles of environmental management drawn up by the stakeholders are already sufficiently innovative to fuel the current Senegalese debate about land tenure reforms.

Discussion

The kind of role-playing game we used has been successfully tested all over the world in the last decade (see commod.org). When participatory approaches succeed, the main question that arises is whether they are reproducible, as success is always embedded in the local features of a society. Of course, like any other participatory approach, the success of the self-design process depends on the facilitator's awareness of the social background, and this usually takes a few months to acquire. However, mobilizing the participants can only succeed if the issue proposed for discussion is a "real" issue for them (Kok et al., 2007): what we are describing here is not an awareness approach but a support for a collective discussion about the participants' own issues. Indeed, the only true obstacles to this kind of approach are first, that the participants must already have the same aims and feel the need to be involved. In the case of policy design, in some societies, very local users may be not interested in being involved in policy design, even though the policy will have an impact on their livelihoods. The second obstacle is the difficulty for facilitators to avoid incorporating their own points of view when framing the process, either intentionally – for example, with an environmental aim, or subconsciously.

The self-design approach presented above takes up this methodological challenge by supporting people in designing their own conceptual settings, and then using these endogenous settings to define their own regulation options. We believe that a major requirement of this methodological approach is limiting the influence of external scientific point of view on the stakeholders' eliciting process. The first milestone is then making sure that very limited external scientific data and knowledge are incorporated during the diagnostic process, so that the participants' framework is not "spoiled" by exogenous points of view. In the self-design approach, adding scientific knowledge and data is only appropriate when the local framework and model are sufficiently solid to withstand the influence of prevailing scientific opinion. The second milestone is still trickier: limiting the influence of the facilitator, like that of the scientist, on the stakeholders' eliciting process. In fact, whatever the approach, simply by establishing a dialogue, the facilitator already influences the participants' reactions. This is a fact that scientists simply must understand. Consequently, the only scientific way to tackle this influence is to acknowledge it, and then carefully and rigorously limit and control it. We need to rigorously check the very limited questions we toss out in front of participants.

However, some tricky epistemology issues arise from this kind of maieutic process (d'Aquino and Bah, 2012): faced with such a deep iterative analysis, researchers find it very difficult to adhere to a sufficiently rigorous process. For this reason, from our ten years' experience with this kind iterative modeling, we have extracted some sound principles to ensure a rigorous procedure (Etienne, 2011), applying a monitoring framework that makes every researcher's choice of social setting explicit and expressible in a refutable form: why and how should each form of knowledge be used and at what stage, why and how should the different stakeholders' points of view and goals be incorporated in the development of the appraisal process, and so on. Even though, as stated above, this self-modelling process enables the expression of indigenous frameworks, we still need to know how to incorporate scientific knowledge at a later stage, in our particular case, knowledge of the dynamics of natural resources depletion under increasing pressure. Indeed, in our experience, the stakeholders themselves often want scientific knowledge when they reach a stage in their self-appraisal process where this type of knowledge serves a purpose; for instance, when in their simulations, increasing the production of fodder becomes indispensable for sustainability (Corniaux et al., 2003). However, in some cases, local people may not acknowledge the authenticity of certain environmental facts (Dray et al., 2007). In such cases, the first part of the solution may be helping people first to assimilate a multiscale view of their problem (see our multiscale settings), as this will reveal aspects that are not visible at their usual scale of perception. Another part of the solution may be that scholars reorganize their approach in a more comprehensive framework (d'Aquino and Bah, 2012), by starting with a true co-definition of the priority issue that really takes local priorities and points of view into account, not only the scholars' economic and ecological viewpoints (see post-normal attitude: Funtowicz and Ravetz, 1994).

Back to the Policy Challenge

The last hurdle in this kind of policy design is embedding local proposals in the policy-making process. Indeed, it is difficult to change policy-makers' ways of thinking about regulations – i.e. privatization and closure of landscapes, zoning land for separate uses, enforcing static carrying capacities, corporate management linked to territorially delimited pastures, formalized nested regulatory structures, all measures that restrict flexibility and adaptability. Despite the successes of the first self-designing process experimented at the local level in 2000, which subsequently publicized a new form of local land use management and zoning to other Sahelian countries (d'Aquino and Papazian, 2012), the basic structure of the policy, such as the legal access rights, has still not changed. Thus, the success of the self-designed approach in developing local management tools has highlighted the need to change the environmental management paradigm at a more general level. This is why a bottom-up self-design, like the one described here, is called for. Indeed, if policy makers are involved in the local stakeholders' design, as in the process we have described here, the chances of succeeding in embedding indigenous skills about uncertainty in the policy debate will be greater.

Conclusions

The multilevel self-design process tested in the Sahel succeeded in eliciting the background principles of adaptability. The results, which confirm the relevance of the method and of the simulation support produced, mean that this simulation support can be used to enable stakeholders to design their own operational ideas of policies and then to analyse the outputs of the policies using scientific adaptability frameworks.

Another option is the use of this kind of "paradigm exchange" between indigenous knowledge and scientific knowledge. If indigenous thinking about adaptability can improve our management of adaptability, it should be included in the resilience thinking framework. Pursuing this goal may lead to the use of the self-designed process not only to elicit indigenous points of view, but also to facilitate constructive exchanges with other bodies of knowledge on environmental management. Indeed, the self-design process translates a part of indigenous knowledge into a qualitative language, and could do the same with other forms of knowledge with the aim of achieving better mutual understanding and exchange. On one hand, theories of environmental management can be formalized in the form of a rules scenario which can then be used in the self-designed game, and subsequently easily debated with the players. On the other hand, empirical scenarios formalized by indigenous players can be assessed in an economic and juridical framework, and can fuel scientific debate about adaptability and resilience. Thus, the next step in our work in Senegal is to analyse to what extent these empirical principles of adaptability management can be transformed into practical rules and institutions for resilience and co-adaptative management policies.

In conclusion, more work remains to be done than the work accomplished up to now. Nonetheless, the very first results reported in this chapter confirm the value of this approach: first, some innovative participatory methods enable stakeholders to use their indigenous way of thinking to design a "modern" model of environmental management; second, their model may provide new insights into how to design flexible rules to manage uncertainty. In point of fact, the entire methodological framework described in this chapter is an attempt to find a better way for future hybridization of scientific knowledge and indigenous capacity for adaptability, by taking the first steps towards creating an amenable arena for a more comprehensible exchange between different sources of knowledge, towards the co-building of new "post-normal" knowledge.

Note

1. National civil society platform about land tenure and policies (CRAFS).

References

Argent, R. M. and Grayson, R. B. (2003). A modelling shell for participatory assessment and management of natural resources. *Environmental Modelling & Software*, 18(6), 541–551.

Barreteau O., Antona, M., P. d'Aquino, S. Aubert, S. Boissau, F. Bousquet, W. Daré, M. Etienne, C. Le Page, R. Mathevet, G. Trébuil, and J. Weber. 2003a. Our companion modelling approach. *Journal of Artificial Societies and Social Simulation*, 6(2) (online). Available at: http://jasss.soc.surrey.ac.uk/6/2/5.html

Barreteau, O., Le Page, C. and D'Aquino, P. 2003b. Role-playing games, models and negotiation processes. *Journal of Artificial Societies and Social Simulation* 6(2). Available at: http://jasss.soc.surrey.ac.uk/6/2/3.html

Bourgoin, J., d'Aquino, P. and Bah, A. Modeling outputs from a stakeholders' self-modeling to catch environmental uncertainty : A case study in the Sahel. Available at: http://hal.archives-ouvertes.fr/hal-01068080

Corniaux C., P. D'Aquino, P. and G. Molenat. 2003. Towards an adaptation of rangelands management and scientists interventions in a new agro-pastoral context: case study in rice-based farming systems in the Delta of Senegal River. VIIème International Rangeland Congress, Durban, South Africa, 26 July -–1 August.

d'Aquino, P. and Bah, A. 2012.A bottom-up participatory modelling process for a P multi-level agreement on environmental uncertainty management in West Africa. *Journal of Environmental Planning and Management*, 56(2): 271–285. Available at: http://dx.doi.org/10.1080/09640568.2012.665361

d'Aquino, P. and Bah, A. 2013. Land policies for climate change adaptation in West Africa: a multi-level Companion Modeling approach. *Simulation and Gaming*, 20 (10), 1–18.

d'Aquino, P. and Bah, A. 2014. Multi-level participatory design of land use policies in African drylands: a method to embed adaptability skills of drylands societies in a policy framework. *Journal of Environmental Management*, 132, 207–219.

d'Aquino, P. and Papazian, H., 2012. A ten-years after impact analysis of a companion modelling approach. Ifad, Cirad, Rome.

d'Aquino, P. and Papazian, H., 2014. An inclusionary strategy reaching empowering outcomes ten years after a two-year participatory land uses management. *Environmental Management and Sustainable Development*, 3, 2. Available at: http://dx.doi.org/10.5296/emsd.v3i2.6595

d'Aquino, P., Le Page, C., Bousquet, F. and Bah, A., 2003. Using self-designed role-playing games and a multi-agent system to empower a local decision-making process for land use management: the SelfCormas experiment in Senegal. *Journal of Artificial Societies and Social Simulation*, 6 (3), 5.

Defalt, Q., 2014. From communities to policies. *Movie 26mn*. Available at: www.dailymotion.com/video/x28uyd9_from-communities-to-policies_creation

Dougill, A.J., Fraser, E.D.G. and Reed, M.S., 2010. Anticipating vulnerability to climate change in dryland pastoral systems: using dynamic systems models for the Kalahari. Working Paper 32, Centre for Climate Change Economics and Policy, University of Leeds, Leeds.

Dray, A., Perez, P., Le Page, C., d'Aquino, P. and White, I.M. 2007. Who wants to terminate the game? The role of vested interests and meta players in the ATOLLGAME experience. *Simulation and Gaming*, 38(4): 494–511.

Étienne, M. (Ed.), 2011. *Companion Modelling: A Participatory Approach to Support Sustainable Development.* QUAE editions, Collection Update Sciences & Technologies.

Funtowicz, S.O. and Ravetz, J.R., 1994. The worth of a songbird: ecological economics as a post-normal science. *Ecological Economics*, 10, 197–207.

Gaddis, E.J.B., Harp Falk, H., Ginger, C. and Voinov, A., 2010. Effectiveness of a participatory modelling effort to identify and advance community water resource goals in St.Albans, Vermont. *Environmental Modelling & Software.* 25, 1428–1438.

Kok, K., Biggs, R.and Zurek, M., 2007. Methods for developing multi-scale participatory scenarios: insights from Southern Africa and Europe. *Ecology and Society*, 13(1): 8 (online). Available at: www.ecologyand society.org/vol13/iss1/art8/

Lynam, T., De Jong, W., Sheil, D., Kusumanto, T. and Evans, K., 2007. A review of tools for incorporating community knowledge, preferences and values into decision-making in natural resources management, *Ecology and Society* (12)1: 5. Available at: www.ecologyandsociety.org/vol12/iss1/art5/

Martínez-Santos, P., Llamas, M. R. and Martínez-Alfaro, P. E. (2008). Vulnerability assessment of groundwater resources: a modelling-based approach to the Mancha Occidental aquifer, Spain. *Environmental Modelling and Software*, 23(9), 1145–1162.

Mehta, L., Leach, M., Newell, P., Scoones, I., Sivaramakrishnan, K. and Way, S.A. 1999. Exploring understandings of institutions and uncertainty: new directions in natural resource management. Institute of Development Studies, Brighton.

Mendoza, G. and Prahbu, R., 2006. Participatory modelling and analysis for sustainable forest management: overview of soft system dynamics models and application. *Forest Policy and Economics*, 9, 179–196.

Parker, P., Letcher, R., Jakeman, A., Beck, M. B., Harris, G., Argent, R. M., et al., 2002. Progress in integrated assessment and modelling. *Environmental Modelling and Software*, 17, 209–217.

Purnomo, H., Guizol, P.and Muhtaman, D. R., 2009. Governing the teak furniture business: a global value chain system dynamic modelling approach. *Environmental Modelling & Software*, 24, 1391–1401.

Reed, M.S., Dougil, A.J., and Baker, T., 2008. Participatory indicator development: what can ecologists and local communities learn from each other? *Ecological Applications* 18, 1253–1269.

Renger, M., Kolshoten, G. and De Vreede, G., 2008. Challenges in collaborative modelling: a literature review and research agenda. *International Journal of Simulation and Process Modelling*, 4, 248–263.

Rouwette, E.A.J.A., Vennix, J.A.M. and van Mullekom, T., 2002. Group model building effectiveness: a review of assessment studies. *System Dynamics Review,* 18(1): 5–45.

Scoones, I., 1994. *Living with Uncertainty: New Directions in Pastoral Development in Africa.* Intermediate Technology Publications, London.

Simon, C. and Etienne, M., 2010. Modelling for stakeholders: a companion modelling approach applied to forest management planning with the Société Civile des Terres du Larzac. *Environmental Modelling & Software*, 25, 1371–1384.

Stringer, L.C., Prell, C., Reed, M.S., Hubacek, K., Fraser, E.D.G. and Dougill, A.J., 2006. Unpacking 'participation' in the adaptative management of socio-ecological systems: a critical review. *Ecology and Society*, 11(2): 39 (online). Available at: www.ecologyandsociety.org/vol11/iss2/art39/

Voinov, A. and Bousquet, F., 2010. Modelling with stakeholders. *Environmental Modelling and Software*, 25, 1268–1281.

Von Korff, Y., P. D'Aquino, K. A. Daniell and R. Bijlsma. 2010. Designing participation processes for water management and beyond. *Ecology and Society*, 15(3): 1 (online). Available at: www.ecologyand society.org/vol15/iss3/art1/

19

A SYSTEMS APPROACH TO STAKEHOLDER MANAGEMENT

Robert Boutilier, Witold Henisz, and Bennet Zelner

Research Question: How can African mining companies manage stakeholders with diverse interests?
System Science Method(s): Agent-based models & Networks
Things to Notice:

- Combination of networks and agent-based models
- Evaluation of potential management strategies to guide strategic decision making

We introduce a data-driven "systems approach" for managing the risks and opportunities arising from a focal organization's stakeholder relations: relations with the external organizations and groups that are either affected by, or can affect, the organization's activities. Our approach revolves around the notion of the "social license to operate" (SLO), defined as the extent to which stakeholders accept or approve of a firm and its operations. Stakeholders that withhold or rescind a firm's SLO may depress profitability and shareholder value by obstructing access to vital resources such as financial capital, markets, legal permissions, skilled labor, and land. We demonstrate our approach using data from interviews with over 600 stakeholder organizations at six mines in five African countries that feature some of the riskiest and most controversial stakeholder environments in the world. We used measures of the SLO that incorporated each stakeholder groups' priorities and concerns through their effects on the perceived quality of the relationship between the group and the company. We integrated multiple measures to develop measures of the salience of mine-related issues to the stakeholder and the stakeholder's power to influence other members of the stakeholder network. We then designed several alternative stakeholder engagement initiatives for each site that, by addressing the issues deemed most important to influential stakeholders, appeared most likely to increase the mining company's SLO. To compare the predicted performance of each initiative, we submitted the data to an agent-based model of stakeholder interaction incorporating information on social preferences and structure. Though we cannot yet compare actual to predicted outcomes, we report multiple instances in which our approach yielded insights or predictions that were surprising or counterintuitive to us and to the mining company's management. We conclude with observations on the approach's strengths and weaknesses as well as notes on the general strategies that appear most promising for continued development of the field of stakeholder management.

We introduce a data-driven "systems approach" to stakeholder management—management of the risks and opportunities arising from a focal organization's relations with the external organizations and groups that are either affected by or can affect the organization's activities (Freeman, 1984). The systems approach described here combines social network analysis

 DOI: 10.4324/9781315748771-21

(Borgatti, Mehra, Brass & Labianca, 2009; Wasserman & Faust, 1994) and agent-based modeling of societal phenomena (Axelrod, 1997; Axelrod & Cohen, 2002; Epstein & Axtell, 1996) and business phenomena (Bone, Dragicevic, & White, 2011; North & Macal, 2007) with recent advances in the conceptualization and measurement of the "social license to operate" (Black, 2013; Boutilier, Black & Thomson, 2012; Thomson & Boutilier, 2011).

Following Thomson and Boutilier (2011), we define an organization's "social license to operate" (SLO) as the extent to which stakeholders accept or approve of a firm and its operations. The social license to operate ties stakeholder relations to financial consequences by recognizing stakeholders' ability to withhold or obstruct access to vital resources, including financial capital, legal permissions, skilled labor, land and other natural resources, and markets. In theoretical terms, Thomson and Boutilier view the SLO as a stakeholder network-based interpretation of the resource dependence theory of why organizations thrive or fail (Barney, Wright, & Ketchen, 1991; Ireland, Hitt & Vaidyanath, 2002; Pfeffer & Salancik, 1978). When networks of stakeholders diminish access to resources, costs rise, share prices drop, and profits suffer. Stakeholder management can therefore also be viewed as risk management (Pojasek, 2008) focused on sociopolitical risks and opportunities.

In contrast to philosophical, theoretical, and normative treatments of topic of stakeholder engagement (de Colle, 2005; Elkington, 2001; Greenwood, 2007) or the development of stakeholder typologies (Jawahar & McLaughlin, 2001; Mitchell, Agle, & Wood, 1997), a systems approach to the practice of stakeholder management relies on rigorous data analysis to produce strategic recommendations for specific engagement initiatives. Interviews with stakeholders produce data about their concerns, their networks and the influence relations in those networks. These data are then used to model the decision processes of stakeholders choosing their level of support for a project. The result is a simulation of the sociopolitical dynamics surrounding the project. The simulation produces predictions about how different initiatives for engaging with stakeholders are likely to affect the overall level of social license accorded the project. Thus, largely by way of suggesting priorities for engaging with stakeholders on their concerns, it provides a guide to strategies for reducing the project's risks and costs arising from sociopolitical problems.

After explaining recent advances in the conceptualization of the social license, we discuss how viewing stakeholders through a network lens avoids futile debates and instead emphasizes the kind of information managers need in order to *manage*. We then describe the assumptions behind the computer simulation we have developed to predict how the social license would change given one course of action versus another. We describe how we applied these techniques at five mines in Africa and report some of the surprising findings and practical strategies that they produced. We conclude by discussing the strengths and weaknesses of the approach and some directions for future development.

Social License to Operate

The social license to operate (SLO) began as a metaphor for the power of communities to shut down mining operations. Thomson and Boutilier (2011) elaborated the concept to include four levels of social license and four factors that contribute to movement up or down through the levels (see Figure 19.1). The vertical shading on Figure 19.1 is meant as a reminder that the levels are not discrete even though they are qualitatively distinct from one another. The lowest level is the loss of the social license. On Figure 19.1 it is labeled the "withheld or withdrawn" level and is represented by the darker region at the bottom. It falls below the boundary criterion labeled "legitimacy." When stakeholders withdraw the social license they frequently

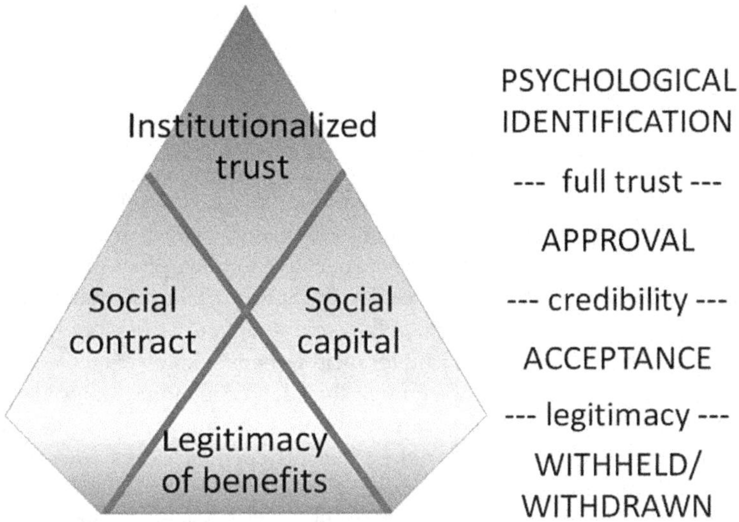

Figure 19.1 Levels and factors of the social license to operate

use the word "illegitimate" to describe the project or the company's actions. Above that is the lowest social license granted, which is called the "acceptance" level. It appears wider on Figure 19.1 because it is the most common level encountered. At this level a project has legitimacy but has not yet gained credibility. If credibility is gained, the project's social license rises to the next highest level, which is called "approval." Stakeholders granting this level see the project as having legitimacy and credibility. They therefore actively support the project. If the credibility evolves into full trust, the social license can rise to the level of "psychological identification," indicated by the darker region at the pinnacle of Figure 19.1. It is the level at which the company and the stakeholders see their futures as intertwined and maintain an institutionalized relationship that survives changes in personnel.

A factor analysis of agree/disagree statements measuring the social license at mining projects in Bolivia, Mexico, and Australia produced four factors apparently driving changes in the level of social license (Black, 2013). They roughly corresponded to the boundary criteria between levels of the social license and are represented by the regions delineated by the diagonals on Figure 19.1. The elaboration of the three original boundary criteria into the four factors adds diagnostic power to the SLO model. The project's profile of factors with each stakeholder cluster suggests what kind of initiatives are necessary to further raise the level of social license.

The first factor is called "legitimacy of benefits." It reflects the stakeholder's perceived net personal benefit or harm from the project. When no legitimacy of benefits is perceived, the stakeholder withholds or withdraws the social license. At the other extreme, the "institutionalized trust" factor, reflecting the stakeholder's perception that the company would take account of the community's interests in all its decisions, corresponds to the highest level of the social license, "psychological identification."

The transition between the middle two levels of the social license depends on stakeholders' perceptions of credibility. Credibility is captured in the two factors labeled "social capital" and "social contract." The social capital factor deals more with the quality of communications and interpersonal relationships, including qualities such as reciprocity, goodwill, respect, and dependability. The social contract factor deals more with issues of fairness and how well the project fits into the socioeconomic ecology of the stakeholder network. It involves questions like shared use of land and water, as well as the ratio of costs to benefits for the whole community.

Network View of Who Grants the Social License

The question of who grants the social license often becomes a debating point inside companies. If it is granted by the community, and the community is divided, then which side should be deemed the grantor? We propose that it is more productive to think of the grantors as a network. A social network approach (Scott, 2000) includes all clusters, coalitions, and cliques, as well as all isolated groups and organizations, located in the community or in a distant capital city. Instead of using stakeholder characteristics as the basis for inclusion or exclusion from the granting group, a network approach relies on the size of a group's "stake." To avoid including all groups in the world as stakeholders, we operationalize this attribute as the impact that a stakeholder might experience from the project or impose on the project. Experience has shown that for mining projects, 35–65 organizations typically encompass the core group of most influential and affected stakeholders. The number can be lower if the project is very remote, or higher if there are more distinct jurisdictions (e.g., villages) nearby or many ethnic divisions in the region. The average social license granted by the most influential and affected stakeholder can be taken as the project's average social license. For example, if the spokespersons for 40 stakeholder organizations were interviewed (e.g., village mayors, religious leaders, state government departments, youth group leaders, women's group leaders, environmental groups, prominent small businesses, farmers' groups, etc.), each organization would be assigned an SLO score based on the average of its spokesperson's ratings on agree/disagree statements about the organization's relationship with the company operating the project (Black, 2013). The resulting 40 scores could then be averaged to produce an overall SLO score for the project.

By itself, a number indicating an overall average social license score is not very useful to a manager aiming to maintain or improve access to resources. What managers really need to know is which groups represent a credible threat to resource access and which strategies would reduce the threat. The network approach to viewing the distribution of social license scores provides the level of detail needed to devise strategies for raising the level of social license.

Developing Strategic Scenarios

By combining information on the social license granted by each stakeholder with information on their concerns and priorities, and their collaborative relationships in the network, it is possible to develop scenarios for stakeholder engagement that improve the project's overall social license. The information on concerns and priorities indicates which kinds of community initiatives would appeal to which groups. This knowledge can be used strategically, for example, to design initiatives that would bring influential high social license granters into contact with ambivalent groups with the aim of moving the ambivalent groups towards approving of the project. The information on the social license factors helps in the design of projects so that, for example, they focus either on improving the complementarity of the project with the existing local socioeconomic system (i.e., social contract) or, alternatively, on improving communications and interpersonal trust between stakeholders and the company. Information about which stakeholder organizations spend more of their resources dealing with issues related to the project helps to select partners for collaborative initiatives and helps predict which stakeholders will spontaneously get involved.

Simulating the Political Dynamics

Once several scenarios for strategic stakeholder engagement have been developed, an agent-based simulation can be used to see what impact that scenario would have when placed in among all the strategies being pursued by all the stakeholders as well. This step acknowledges that companies are not the only ones in the network trying to produce their desired political outcomes.[1]

At the core of the simulation is Bruce Bueno de Mesquita's dynamic expected utility model, which

> is concerned with explaining how policy positions of competing players evolve over time and shape policy outcomes. Therefore, it leads to predictions about policy outcomes and identifies strategic opportunities for altering them. As such it can be used to explain and predict political decisions at any level of analysis, including, of course, foreign policy and international relations. It can also be used by policy makers to anticipate outcomes and to reshape them to be more in line with their own interests.
>
> *(De Mesquita, 1992, p. 51)*

The "BDM model"—which developed over two decades in dozens of academic publications—provides a rigorous, game-theoretic assessment of the political strategies of utility-maximizing stakeholders characterized by four attributes: (1) level of power, (2) location of "ideal point," (3) degree of issue salience, (4) beliefs about the issue's salience to their peers. The model was originally developed to forecast interstate or intrastate negotiations that could lead to conflict, but its application has expanded to a wider array of policy issues as well as other collective decisions and the formation of group opinions.

As do other agent-based models used in the field of political economy (Acemoglu & Robinson, 2006; Grossman & Helpman, 2001; Persson & Tabellini, 2002), the BDM model assumes rational behavior by a set of utility-maximizing stakeholders who possess perfect information about their interactions with other players or, alternatively, imperfect information of a type that can be formally specified or parameterized. The model also makes two key assumptions for analytical tractability: (1) one-dimensional issue preferences that can be represented on a line segment, and (2) monotonic utility functions, meaning that actors' preferences over feasible outcomes decline steadily with the Euclidean distance between an outcome and the actor's ideal point (De Mesquita, 1992, p. 51). Stakeholders in the BDM model choose how much effort to expend on trying to influence others to support their preferred policies and also how to respond to such entreaties from others, under the assumption that all stakeholders are behaving similarly. The game continues through multiple rounds until the expected change in outcome from one round to the next falls below a pre-specified threshold value.

The BDM model's key strengths include the variation its permits in stakeholder attributes and the scope of strategic stakeholder behaviors it accommodates. Weaknesses include the model's assumption that stakeholders' utility functions do not reflect different outcomes' expected impact on others; and its conceptualization of power as a stakeholder characteristic, versus a function of the (time-varying) structure of social relations. To address the latter issue, we follow a growing body of social science research (Freeman, White, & Romney, 1989; Knoke and Yang, 2008; Wasserman & Faust, 1994) by augmenting the classic BDM model to allow for such relations to influence stakeholders' beliefs and consequent actions (Knoke & Yang, 2008). Specifically, we incorporate well-established positional metrics of different stakeholders' level of structural equivalence (Wasserman & Faust, 1994, ch. 9) and beta centrality (Bonacich, 1987), as has occurred in other domains where network analysis has long been applied (e.g., Owen-Smith & Powell 2004; Powell et al., 2005).

In contrast to most studies using social network analysis—which typically examine the diffusion of a practice—we use the methodology to analyze how a given stakeholder can best alter the position of network peers with the aim of altering an outcome. Each stakeholder is influenced by peers but also seeks to influence the same peers to maximize own utility. In this regard, the model shares many features with the dynamic network models developed by Tom Snijders and his collaborators (Burk, Steglich, & Snijders, 2007; Mercken, Snijders, Steglich, Vartiainen, & de Vries, 2010; Snijders, van de Bunt, & Steglisch, 2010; Steglich, Snijders, & Pearson, 2007). As in these systems, we explicitly allow for various forms of endogenous network formation and restructuring. Dynamic network processes that either allow for stakeholders with similar behavior to form ties or for stakeholders with preexisting ties to adopt similar behavior (or both) have been powerful tools for the analysis of outcomes, such as the transmission of sexually transmitted diseases or greater interdisciplinary collaboration among academics. The critical role played by certain highly connected agents in patterns of diffusion of disease or ideas carries over into the context of stakeholder influence strategy.

Formally, we define a stakeholder's utility as a linear combination of the distance from their ideal point and the distance of peers from their ideal points (i.e., stakeholders have social preferences). In this manner we capture the intuition that stakeholders' own utility varies by the extent to which their closely tied peer stakeholders are able to realize their ideal points. We construct measures that allow for the weight assigned to each peer stakeholder to be positively correlated with:

- the strength and polarity of ties between the focal and peer stakeholder (i.e., stakeholders place a greater weight on the impact of their actions on the utility of their 'close friends');
- peer status (i.e., stakeholders place a greater weight on the impact of their actions on the utility of peer stakeholders who possess more ties to alters with more ties);[2] and
- similarity in the structural position within the network with the focal stakeholder (i.e., given uncertainty over the appropriate action, stakeholders place a greater weight on the impact of their actions on the utility of peer stakeholders who have similar structural positions in the stakeholder network, as they are inferred to share similar interests) (Burt, 1987).

In this manner, two political parties or NGOs or companies that agree on one policy but disagree on many others, and also have a long history of enmity with limited communications and few common relationships, are less likely to collaborate effectively than an alternative pair with identical preferences on the policy in question but a wider scope of agreement, long-standing collaboration with rich communication networks that share many common partners.

We also allow network structure to alter each stakeholder's power. In contrast to the communication or diffusion processes of social networks, ties to (powerful) stakeholders that have fewer outside options (i.e., possess fewer ties) in this case convey greater power. This insight regarding the importance of brokerage positions providing privileged connections across structural holes (Burt, 1987, 1992, 1997) has been validated in studies of entrepreneurship, innovation, organizational learning, and performance (Burt, 2004). Total power is, thus, a weighted function of the stakeholder's own power and the power it derives from connections to (powerful) peer stakeholders with few other ties.[3] For example, a ruling government party may derive power in domestic political negotiations through its connection to the World

Bank or International Monetary Fund. These multilateral agencies possess substantial resources and typically interact primarily with the sovereign government of a nation. By invoking the threat of sanction from the World Bank or International Monetary Fund, a Prime Minister or Finance Minister may be able to successfully create a threat point that drives other stakeholders to compromise rather than maintain their opposition to change in law or regulation. If the World Bank or International Monetary Fund had multiple points of contact into the domestic political system, the opposition might approach them and try to offer a compromise more amenable to their position. The lack of alternative ties creates a beneficial ability for the ruling part to craft a take it or leave it offer that is favorable to their interests.

We combine advances in conceptualizing and measuring the social license with what we call the Geopolitical Influence and Strategy Tool (GIST) agent-based simulation, which itself combines decision influences from the embeddedness of the stakeholder in its social network and the stakeholder's dynamic expected utility. The level of social license is taken as the outcome variable for comparing the strategic scenarios.

Methods

Figure 19.2 summarizes the methodology that integrates these various components. It begins with data collection from surveys, (social) media, situation reports, emails, financial records and human intelligence. From this unstructured text, analysts supplemented by information extraction software including natural language parsing and data mining algorithms extract the identity of key stakeholders, issues and relationships. This process generates structured quantitative data that can be modeled using the decision-process system which then generates predictions about how various actions (or inactions) will shift the positions of key policymakers and, potentially, the revenue and costs of the firm.

In a recent engagement, we relied primarily on survey data collected from the spokespersons for stakeholder groups and organizations of five mines in four countries in sub-Saharan Africa. The interviews were conducted by local teams with research experience and fluent in the local languages. The numbers of interviews per mine ranged from a low of 87 of a high of 227, depending on the remoteness of the location and the complexity of the interests in the operation.

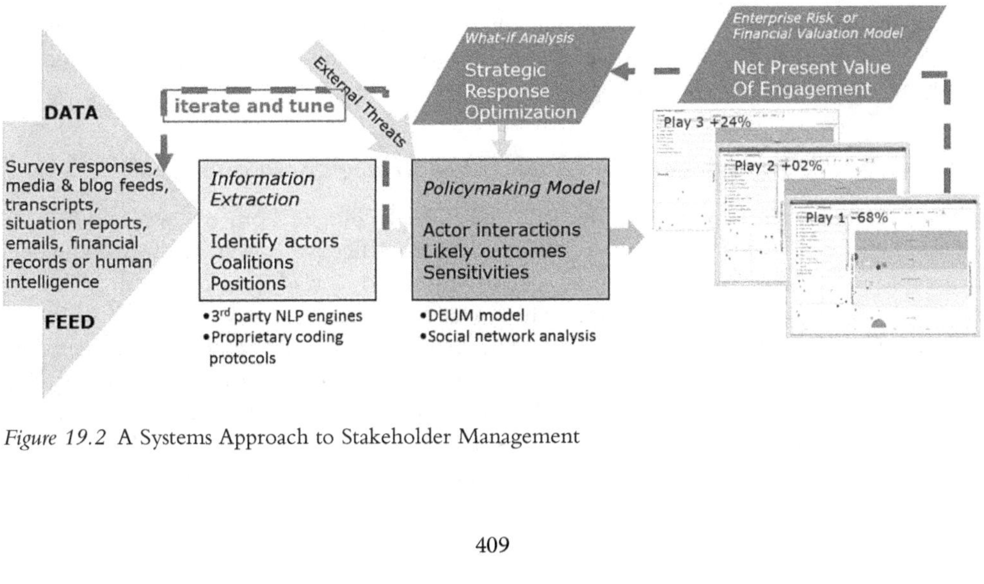

Figure 19.2 A Systems Approach to Stakeholder Management

Interviews

The 50-minute interviews began with open-ended questions about the stakeholders' concerns, the impacts they perceived to mine to have on their organizations, their priorities, and the contributions they thought they could make towards realizing those priorities. The next section of the questionnaire presented 15 statements to be rated on an agree/disagree scale ranging from 1 to 5. That was followed by a question about the reputation of the project within the stakeholder's organization or group. Then interviewees estimated how much time they spend dealing with issues related to the mine. Next they were asked which other organizations they thought might be stakeholders. This snowball sampling section was followed by probing for the names of organizations or groups with whom the interviewee's organization had a relationship and who were also stakeholders of the mine. The names were followed by rating scale questions on the satisfaction with the relationship, the degree to which the two organizations shared common goals, the degree to which the partner organization consulted the interviewee's organization before taking action on mine related issues, and the degree to which the interviewee's organization followed the lead of the partner organization on mine-related issues. The interview ended with an open-ended request for any final comments.

Interviewee Selection

Although there was a series of interviews, there was no sample in this study. Therefore, there was no sample size. There was, however, a census of a population. The target population was defined as stakeholders with enough involvement in the network to have been mentioned by at least one other stakeholder as a partner. The names of such organizations were sought from the community relations departments of the mine, from Internet searches of organizations and issues about the mine, from previous independent studies commissioned by the company, and from interviews with stakeholders themselves. The latter "snowball" sampling technique saw organizations added to the list of target interviewees if they were mentioned at least three times by other stakeholders.

A census is considered successful if at least 90 percent of the members of the target population are interviewed. Combined calculations from all five mines showed that 125 percent of the target population was interviewed. This over-inclusion allowed for the inclusion of some stakeholders outside the population boundary. Only by having stakeholders on both sides of the population boundary is it possible to identify where the boundary is located. Therefore, the over-inclusion provided evidence that the census was successful in capturing the opinions of nearly all the members of the target population.

Analyses and Development of Strategic Scenarios

The open-ended responses were classified, coded, and counted both by raw numbers of mentions per capital and by mentions weighted by the effective power of the stakeholders making the comments. Several different systems were used to classify stakeholders, including the level of social license they granted. This yielded a description of which kinds of stakeholders cared about which kinds of issues. The network data were converted into network graphs which showed things like whether the granters of low social licenses had strong relationships among themselves, which stakeholders had the most influence over others, and what cliques or clusters existed in each mine's stakeholder networks. These kinds of analyses were integrated in order to devise two or three strategic scenarios for each mine. The scenarios typically included (a) maintain the status quo, (b) address only the complaints that presented the highest sociopolitical risk, and (c) build a broad coalition around action on issues of concern to potential coalitions that have the greatest potential to raise the level of social license for the project.

Strategic Scenario Evaluation

The scenarios were then used to adjust the social license, involvement, and power scores according the effect that proposed initiative would have on each stakeholder. These parameters were then used as input for the GIST model, which produced predictions of how the scenario would affect the social licenses granted by the most powerful stakeholders individually and by stakeholders overall. Figure 19.3 traces out the key elements of this model which takes in the quantitative data from the unstructured text and uses it to generate key parameter values in the following manner:

- the stakeholder's social license score is used as a proxy for their ideal point in the policy space (i.e., the policy or proposal outcome that makes them happiest);

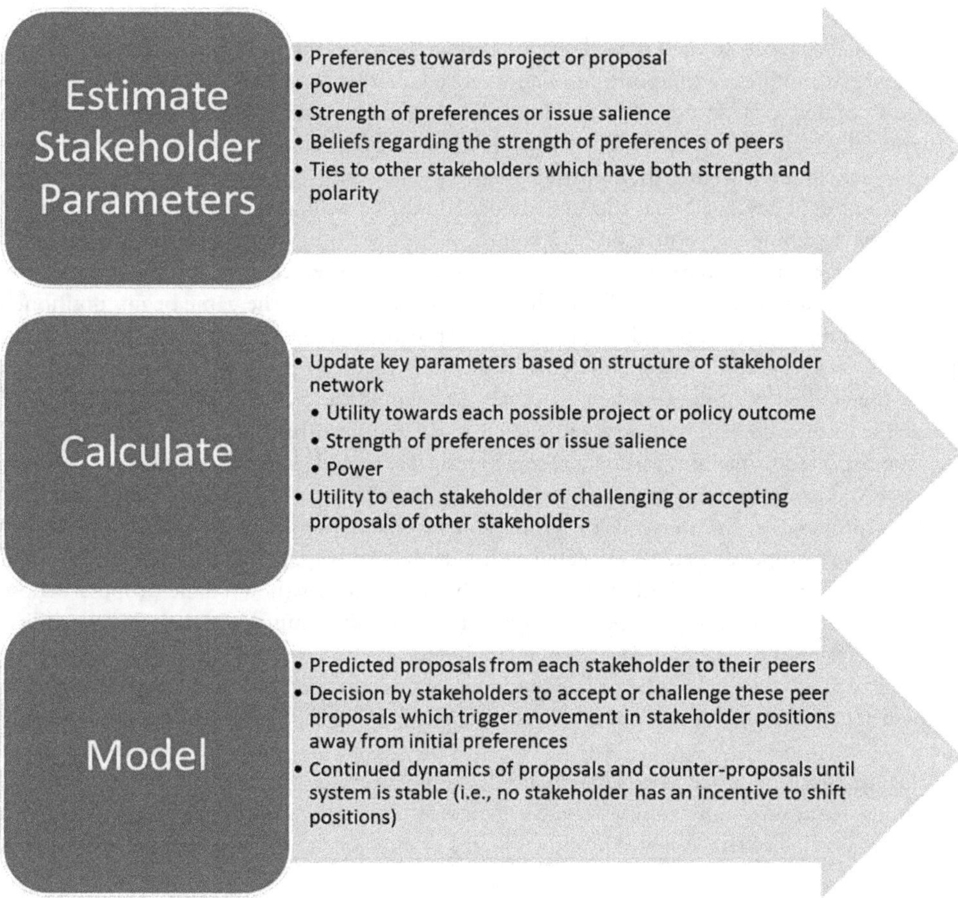

Figure 19.3 The GIST Policymaking Model

- data on the frequency with which a stakeholder is referenced in the surveys and the relative ratio with which they are perceived to lead vs. follow other stakeholders is used to generate estimates of each stakeholder's power (i.e., the ability of a stakeholder to influence the position of its peers and, as a result, the policy or proposal outcome);
- interview responses on the time allocated to this issue over recent weeks and months were used as a proxy for preference strength (i.e., the degree to which a stakeholder is willing to deploy their available resources on this issue relative to ALL other issues);
- interview responses on the number of peers a stakeholder would consult or follow were used to generate estimates of their preference dependence (i.e., the degree to which a given stakeholder's utility is influenced by the utility of the stakeholders with whom they are tied);
- we assumed that all stakeholders correctly estimated the peer stakeholder's preference strength;
- interview responses on the extent to which stakeholders consulted with, led or followed a given peer (on a Likert scale) were used to identify the network of ties between stakeholders and the strength of those ties;
- interview responses on the extent to which stakeholders agree on goals with or cooperate with vs. are in conflict with (on Likert scales) were used to identify the degree of cooperation or conflict between a pair of stakeholders.

It then assumes that, in each period, each stakeholder makes the utility-maximizing decision as to whether to influence the position of another stakeholder, adapt their position in response to a peers' influence or do nothing. The joint decisions by all stakeholders lead to the offering of a series of proposals, counter-proposals, and, in some cases, concessions. These concessions involve a stakeholder moving their current position – not their preferences or ideal point but the point in the policy space that they are advocating for or willing to accept in a given round of play. The stakeholders' choices on how to proceed result from a comparison of the expected utilities of their various strategies taking into account the expected winning proposal of any competition between offers. Formally, each of the n stakeholders in the game begins positioned at their ideal points. In each round, each stakeholder broadcasts their current position (i.e., makes a proposal) to every other stakeholder. Based on these received proposals, each stakeholder i assesses, the credibility of the received proposals (i.e., would the received proposal be expected to beat the status quo policy in a vote); their expected utility from challenging the proposal (i.e., (the expected probability that the received proposal would defeat *i*'s proposal × *i*'s utility if the received proposal wins) + (1 – the expected probability that the received proposal would defeat *i*'s proposal × *i*'s utility if *i*'s proposal wins); and their expected utility from not challenging the proposal (i.e., *i*'s expected utility at the status quo – *i*'s expected utility if all other stakeholders play their best strategy). Stakeholder *i* chooses the credible proposal whose net expected utility for $i < 0 <$ net expected utility for j that minimizes their loss in utility. Play continues until no stakeholder has an incentive to seek to influence any other stakeholder nor concede to another stakeholder. The outcome, at this point, is equal to the effective-power weighted mean position of the stakeholders. See Henisz (2013) for the full equation set and process description.

While this decision process obviously differs from the more complex pattern of conversations, accusations, stand-offs, compromises and conflicts that characterize corporate interactions with their external stakeholders, it nevertheless captures key elements of that interaction. Specifically,

- firms and their external stakeholders compete to form coalitions of support (or opposition) to a project or proposal;

- the choice by a stakeholder on whom to support is a function of both how they perceive the project and proposal and how they perceive those who support or oppose it;
- few stakeholders fight pointless battles opposing projects or proposals that are clear winners but some, who care sufficiently, will continues to do so.

As in any formal model or decision process tool, the question a reader or user needs to ask is whether the inclusion of these key features is sufficient for the outcome of the model or decision process to provide insight into, and mirror, actual outcomes. Repeated use of the core of this model by the intelligence community, the World Bank and numerous private sector consulting clients of multiple providers, including McKinsey, Decision Insights and SENTIA, highlight the utility of this modeling approach. Furthermore, the Central Intelligence Agency has subjected the core model to backwards testing, finding that it generated valuable predictions on coalition dynamics and outcomes in over 90% of the cases analyzed, as compared to alternate models or processes (Feder, 1995).

Results and Discussion

One of the strategic scenarios always tested was the status quo. For most of the other mines, the status quo scenario predicted that nothing much would change if nothing were done. However, in two cases these analyses produced predictions that surprised mine management. In one case, the status quo (i.e., keep doing what you are doing) produced nearly as much improvement as additional initiatives. This encouraged management to focus on more improvements in internal coordination and processes for the initiatives they already had in place. What they had in place responded to the concerns of a wide range of stakeholders, possibly because there was more consensus in this region about priorities. It was a more stable population with a traditional governance structure that was perceived as legitimate and that represented stakeholders' concerns.

In another case, the status quo scenario predicted a decline in the level of social license to the point of complete withdrawal. Moreover, the scenarios designed to raise the social license had only temporary effects. Their impacts eventually gave way to a negative momentum, leading again to a withdrawal of the social license. This confirmed the fears of a minority of executives that the best course of action would be a voluntary halt to operations either until the political climate improved or until a buyer could be found. This was a less stable region where the company faced competition from artisanal mining and where many people lamented the paucity of tangible benefits from the revenues the government received from formal sector mining companies.

At all the mines there were over a dozen major issues that various stakeholders promoted. The results made it possible to choose from these the ones that would do the most to raise the level of social license. At different mines the most strategic issues varied among health and sanitation infrastructure, potable water supply, road improvements, drainage needs, support for the local hospital, education and training needs, land compensation, and the resolution of problems with illegal and artisanal miners. For each issue at each mine the research findings suggested the names of specific organizations that would be to best to include in a collaborative forum that could raise the social license granted by stakeholders. It also suggested communications themes that would attract the best clusters within the network in order to successfully address each issue.

At one mine, the findings showed that a recently launched health initiative had created a partnership with a health-oriented stakeholder that was neither the most influential nor the neediest. However, the findings also showed that the company had a reputation for failing to fulfill commitments. Therefore, it was recommended that the commitment be fulfilled but that the more needy and influential health sector stakeholders be engaged with an additional program.

Although each mine had unique issues, there was one issue that was ranked among the top three at all mines. Local communities complained that they were not seeing economic benefits commensurate with the benefits extracted by the company. Communities wanted jobs at the mine, or failing that, local economic stimulus through more intensive local purchasing programs. They wanted the education and training that would allow youth to get jobs at the mine or to start successful businesses. They also wanted direct assistance with their existing agricultural economy, often in the form of water and transportation projects. Addressing this ubiquitous issue was complicated by the desire of the company to avoid accumulating government responsibilities, especially since most of the mines had already paid billions to the national government in taxes and royalties. The whereabouts of these billions remained unknown to local stakeholders, despite calls by elected local officials in every country for transparency regarding mining agreements. Stakeholders in one country had become so cynical that some explicitly stated that a legal license from the national government demonstrated that the company did not deserve a social license. We interpreted these findings as reflecting a risk facing the whole mining industry in countries that lack transparency in government finance. It was evidence that the social contract factor of the social license to operate (see left side of Figure 19.1) provides a valuable conceptual link to the whole industry's social license.

The approach described here has strengths and weaknesses. The strengths include the capacity to develop stakeholder engagement strategies from quantifiable data submitted to analyses with explicitly stated assumption. Progress can be tracked. The strategy recommendations are specific, to the point of naming stakeholders for partnership, but are also embedded in an analysis that takes account of the impacts each stakeholder relationship has on the whole network of relationships. The procedures can be applied across time and cultures to yield comparisons among an entire portfolio of operations at different phases of the commodity cycle and the mine life cycle.

These studies also revealed some weaknesses that need attention. The entire analysis depends on stakeholders giving frank answers in their interviews. In one interview, for example, the interviewer noted loud noises from an illegal mining operation on the stakeholder's land, yet the stakeholder mentioned nothing about the issue of illegal mining in his interview. In other cases, government officials said that answering the questions placed their personal security at risk. Some apparently suspected that the interviewers were undercover informants for other contingents within the government. The findings from interviews conducted in such circumstances cannot be taken literally but can still be very useful when interpreted in context. They can contribute to a more politically realistic overall strategy. Nonetheless, we have yet to develop a formal quantitative way of assessing this type of information.

The data-driven systems approach to stakeholder strategy development is currently being expanded in two main directions. First, work involving the IFC's "FV Tool" permits conversion of the level of social license into dollar values of cost or benefit for a company, facilitating assessment of different initiatives as well as comparisons with competing activities. In the current example, we began by valuing the impact of the various scenarios by calculating an effective power-weighted issue list (i.e., we considered the priority issues for each stakeholder and then tabulated what percentage of the total effective power in the stakeholder system was interested in each issue). Next, we subjectively assessed whether risks on the mine's risk register were related to each of these issues. For example, the risk of disruptions to operations was linked to the issue of perceptions of water contamination. Finally, we linked the initiatives modeled in GIST to these issues. In this case, a water-treatment program had the potential, if effectively implemented and communicated, to alter perceptions of water contamination and thus to reduce the probability of disruptions to operations. The magnitude of this potential impact was determined not by the stakeholder engagement team but the manager accountable for the risk. In this case, the Chief Operating Officer and the Security team assessed the likely impact on the risk of disruptions to operations of a water treatment program. The short-term costs of implementing and communicating the program could then be set against both the short-term benefits and the medium- to long-term impact on the reduction in the incidence of risks and the magnitude of their impact on cash flows. In this example, the short-term benefits might include a reduction in lost workdays due to water-borne illness for workers and their families whereas the medium- to long-term benefits would also include a reduction in the incidence of operational disruptions and a reduction in the average length of disruptions that do occur. Such analysis enables a comparison and prioritization across various sustainability initiatives as well as a comparison between a select group of sustainability initiatives and more traditional investments in physical or social capital. Critically, it shifts the tenor of the discussion on sustainability initiatives from their cost to a more balanced assessment of their costs and benefits. Through this process cross-functional conversations on the best means to organize operations, human resources and other functions so as to maximize overall value of the operation multiplied. Counterintuitively, by overcoming their inherent resistance to translating the impact of sustainability on net present value, champions of sustainability found that they were able to generate more support for, new ideas on and progress in implementing sustainability.

Second, work is under way to extract estimates of the variables obtained from the interviews from non-interview texts (e.g., Dorobantu, Henisz, & Nartey, 2013; Henisz, Dorobantu, & Nartey, 2013; Nartey, Dorobantu, & Henisz, 2013a, 2013b). This would allow analyses to be performed on archival data, and could also permit the inclusion of non-interviewable stakeholders into the analyses (e.g., criminal or rebel group spokespersons, hostile stakeholders, inaccessible government leaders, etc.). Information extraction software at the intersection of computer science and linguistics is on the frontier of being able to read unstructured text (e.g., news feeds, press releases, speech transcripts, and blogs), consistently identify subject-verb-object triples, and code them with respect to affect. For example, if the policy in question is the munificence of the policy environment for a gold mine or its "social license to operate," a search for all relevant news articles on the mine and on mining in the focal country could be executed on line. The resulting tens of thousands (or more) of news articles could be parsed so as to identify thousands of subject–verb–object triples. The subjects and objects are normalized so that alternative names or members are recognized. The information extraction tool would return a series of the number of press mentions of each stakeholder (i.e., a stakeholder-specific measure of power), the ratio of the frequency of discussion of this mine by each stakeholder relative to other mines or other policy issues (i.e., a stakeholder-specific measure of issue salience), the number of press mentions between each stakeholder dyad (i.e., tie strength), the average affect embedded in the intervening verbs linking each stakeholder dyad (i.e., tie affect), and the average affect embedded in each verb when the object is the mine in question (i.e., stakeholder-specific positions).

Conclusion

We evaluated the practical utility of basing stakeholder engagement strategies on a data-driven "systems" approach that combines (a) a risk management conceptualization and measurement of the social license to operate, (b) a social network approach to identifying stakeholders, and (c) an agent-based modeling approach to predicting mutual influences among stakeholders.

The approach produced insights and predictions that were surprising both to us and to mine management.

The strengths of the approach include its replicability, its universal applicability, its foundation in empirical evidence, and its capacity to evaluate alternative courses of action. Although the approach was applied to help the management of companies, there is no logical reason why it could not be used with equal effectiveness by managers in government or non-governmental organizations. All organizations benefit from having sound stakeholder engagement strategies. In terms of the weaknesses of the approach, we found that it does not work as well in situations heavily laden with covert illegal stakeholder activity or corruption, although overt illegal activity that stakeholders perceive as deserving of legitimization presents no problem.

Future direction for the development of the data-driven systems approach include formulae for converting social license scores into financial impacts on the company and procedures extracting relevant data from text sources such as media reports, government archives, and websites.

Notes

1. This section is a modified excerpt from Henisz (2013).
2. We adopt the Bonacich (1987) measure of centrality with β initially set to + 0.5. To ensure β is less than the reciprocal of the largest eigenvalue of R, we reduce the value of β in 0.01 increments until this constraint is met.
3. We adopt the Bonacich (1987) measure of centrality with β initially set to − 0.5. To ensure β is less than the reciprocal of the largest eigenvalue of R, we reduce the value of β in 0.01 increments until this constraint is met.

References

Acemoglu, D. & Robinson, J. A. (2006). *Economic Origins of Dictatorship and Democracy*. New York: Cambridge University Press.

Axelrod, R. M. (1997). *The Complexity of Cooperation: Agent-based Models of Competition and Collaboration*. Princeton, NJ: Princeton University Press.

Axelrod, R. M. & Cohen, M. D. (2002). *Harnessing Complexity: Organizational Implications of a Scientific Frontier*. New York: The Free Press.

Barney, J. B., Wright, M., & Ketchen, J. (2001). The resource-based view of the firm: Ten years after 1991. *Journal of Management, 27*(6), 625–641.

Black, L. (2013). *The Social License to Operate: Your Management Framework for Complex Times*. London: Do Sustainability.

Bonacich, P. (1987). Power and centrality: A family of measures. *American Journal of Sociology, 92*(5), 1170–1182.

Bone, C., Dragicevic, S., & White, R. (2011). Modeling-in-the-middle: Bridging the gap between agent-based modeling and multi-objective decision-making for land use change. *International Journal of Geographical Information Science, 25*(5), 717–737.

Borgatti, S. P., Mehra, A., Brass, D. J., & Labianca, G. (2009). Network analysis in the social sciences. *Science 323*(5916), 892–895.

Boutilier, R. G., Black, L., & Thomson, I. (2012). From metaphor to management tool: How the social license to operate can stabilise the socio-political environment for business. *International Mine Management 2012 Proceedings* (pp. 227–237). Melbourne: Australian Institute of Mining and Minerals.

Burk, W. J., Steglich, C. E. G., & Snijders, A. B. (2007). Beyond dyadic interdependence: Actor-oriented models for co-evolving social networks and individual behaviors. *International Journal of Behavioral Development 31*(4): 397–404.

Burt, R. S. (1987). Social contagion and innovation: Cohesion versus structural equivalence. *American Journal of Sociology 92*: 1287–1335.

Burt, R. S. (1992). *Structural Holes: The Social Structure of Competition*. Cambridge, MA: Harvard University Press.

Burt, R. S. (1997). The contingent value of social capital. *Administrative Science Quarterly, 42*, 339–365.
Burt, R. S. (2004). Structural holes and good ideas. *American Journal of Sociology 110*(2): 349–399.
de Colle, S. (2005). A stakeholder management model for ethical decision making. *International Journal of Management & Decision Making, 6*(3/4), 1.
de Mesquita, B. B. (1992). *Predicting Politics*. Columbus, OH: Ohio State University Press.
Dorobantu, S., Henisz, W. J., & Nartey, L. (2013). Stakeholder capital and performance in tough times. Working paper.
Elkington, J. (2001). *The Chrysalis Economy: How Citizen CEOs and Corporations Can Fuse Values and Value Creation*. Oxford: Capstone.
Epstein, J. M. & Axtell, R. (1996). *Growing Artificial Societies: Social Science from the Bottom up*. Washington, DC: Brookings Institution.
Feder, S. A. (2002). Forecasting for policy making in the post-cold war period. *Annual Review of Political Science 5*, 111–125.
Freeman, R. E. (1984). *Strategic Management: A Stakeholder Approach*. Boston, MA: Pitman.
Freeman, L. C., White, D. R., & Romney. A. K. (1989). *Research Methods in Social Network Analysis*. Fairfax, VA: George Mason University Press.
Greenwood, M. (2007). Stakeholder engagement: Beyond the myth of corporate responsibility. *Journal of Business Ethics, 74*(4), 315–327.
Grossman, G. M. & Helpman, E. (2001). *Specia Interest Politics*. Cambridge, MA: MIT Press.
Henisz W. J. (2013). Preferences, structure and influence: The engineering of consent. *Global Strategy Journal 3*(4), 338–359.
Henisz, W. J., Dorobantu, S., & Nartey, L. (2013). Spinning gold: The financial and operational returns to external stakeholder engagement. *Strategic Management Journal*. doi: 10.1002/smj.2180
Ireland, R. D., Hitt, M. A., & Vaidyanath, D. (2002). Alliance management as a source of competitive advantage. *Journal of Management, 28*(3), 413–446.
Jawahar, I. M. & McLaughlin, G. L. (2001). Toward a descriptive stakeholder theory: An organizational life cycle approach. *Academy of Management Review, 26*(3), 397–414.
Knoke, D. & Yang, S. (2008). *Social Network Analysis* (2nd ed.). Thousand Oaks, CA: SAGE Publications.
Mercken, L., Snijders, T. A., Steglich, C., Vartiainen, E., & de Vries, H. (2010). Dynamics of adolescent friendship networks and smoking behavior. *Social Networks 32*(1): 72–81.
Mitchell, R. K., Agle, B. R., & Wood, D. J. (1997). Toward a theory of stakeholder identification and salience: Defining the principle of who and what really counts. *Academy of Management Review, 22*(4), 853–886.
Nartey, L., Dorobantu, S., & Henisz, W. J. (2013a). A participatory approach to stakeholder engagement: Defining a hierarchy of strategic action. Working paper.
Nartey, L., Dorobantu, S., & Henisz, W. J. (2013b). Pipes and prisms of political influence. Working paper.
North, M. J. & Macal, C. M. (2007). *Managing Business Complexity: Discovering Strategic Solutions with Agent-based Modeling and Simulation*. New York: Oxford University Press.
Owen-Smith, J. & Powell, W. W. (2004). Knowledge networks as channels and conduits: The effects of spillovers in the Boston biotechnology community. *Organization Science 15*(1): 5–21.
Persson, T. & Tabellini, G.E. (2002). *Political Economics: Explaining Economic Policy*. Cambridge, MA: MIT Press.
Pfeffer, J. & Salancik, G.R. (1978). *The External Control of Organizations*. New York: Harper & Row.
Pojasek, R. B. (2008). Risk management 101. *Environmental Quality Management, 17*(3), 95–101.
Powell, W. W., White, D. R., Koput, K. W., & Owen-Smith, J. (2005). Network dynamics and field evolution: The growth of interorganizational collaboration in the life sciences. *American Journal of Sociology 110*(4): 1132–1205.
Scott, J. (2000). *Social Network Analysis: A Handbook*, (2nd ed.). London: Sage.
Snijders, T. A. B., van de Bunt, G. G., Steglich, C. E. G. (2010). Introduction to stochastic actor-based models for network dynamics. *Social Networks 32*(1): 44–60.
Steglich, C. E. G., Snijders, T. A. B., & Pearson, M. (2010). Dynamic networks and behavior: Separating selection from influence. *Sociological Methodology 40*(1): 329–393.
Thomson, I. & Boutilier, R. G. (2011). Social license to operate. In P. Darling (Ed.), *SME Mining Engineering Handbook* (pp. 1779–1796). Littleton, CO: Society for Mining, Metallurgy and Exploration.
Wasserman, S. & Faust, K. (1994). *Social Network Analysis: Methods and Applications*. Cambridge, UK: Cambridge University Press.

PART 3

Communities and Social Change

Introduction

The final set of chapters in this volume explores applications of system science methods to understanding communities and social change, ranging from urban dynamics like segregation and sprawl to the intentional and unintentional changes communities experience over time. Given its focus on understanding relationships between members of a population, network analysis has a long history of being used to understand communities, starting with (among others) Jacob Moreno's 1934 study of the relationships among teenage girls at a Husdon, New York reform school. Likewise, given its ability to explicitly capture interactions in space, agent-based models also have long been used to examine community and urban dynamics, tracing initially to Thomas Schelling's deceptively simple but powerful model of segregation. It is only more recently that scholars have recognized the potential for system dynamics models in community-based research, with Chapters 27 and 28 highlighting some recent examples.

Movement and Friction in Urban Communities

Nearly every aspect of cities and neighborhoods, including their very existence, emerges from individuals' choices about where to live. But how do people make such decisions? Starting with observed patterns of household mobility in Danville, Illinois, Sara Metcalf uses an agent-based model to identify the resident behaviors that could explain the observed outcomes. Known as pattern-oriented modeling, this approach begins with empirical data and uses simulation to make inferences about the micro-level mechanisms that were responsible for producing the patterns observed in the data. While the model's focus is on agents' (i.e. residents') movement through the space, it also simulates the formation of social networks among the residents, which influence their movement and choice of neighborhood. Indeed, Metcalf finds that social network factors including race-based and income-based homophily, as well as the probability of forming local relationships, are critical to explaining the moves residents actually made in Danville. This analysis thus offers a clear example of how combining system science methods can supplement more traditional demographic data and analysis by uncovering the underlying behaviors.

One much discussed consequence of urban residential mobility is urban sprawl, but there has been little agreement on the precise causes of sprawl or on effective management techniques.

 DOI: 10.4324/9781315748771-22

To move this discussion forward, Paul Torrens develops an agent-based model that also simulates residential mobility, but at a larger geographic scale than Metcalf's model. As a model of mobility, the key is the simulation of residents' movements, which Torrens captures in five distinct forms: immediate movements outward from a starting point, nearby but longer-distance movements, irregular movements around barriers like mountains or rivers, leap-frog movements to entirely new locations, and road-like movements outward along a sequentially growing path. The model combines these five movement types to explore patterns of sprawl that might develop under three different scenarios: an area with a dominant central city like Indianapolis, an area with multiple centers like Los Angeles, and a megalopolitan region with multiple major cities like the greater Chicagoland area. Understanding the unique causes and consequences of sprawl in each scenario provides planners with more fine-tuned strategies to manage sprawl.

A second widely studied consequence or urban residential mobility is segregation. Indeed, residential mobility and segregation were the focus of one of the earliest agent-based models: Schelling's model of segregation.However, with the rise in mixed-race households, segregation in the United States is more complicated today than it was in 1969. To explore how mixed-race households impact segregation patterns, Ellis and colleagues adapt Schelling's model to include these new kinds of families. Building on past findings is a hallmark of science, but it is particularly common in the case of agent-based models; researchers routinely take existing and well-understood models as a starting point and incrementally add new kinds of agents or different types of behaviors. Their model, which is paired with a more conventional analysis of census data, reveals a trend that could easily have been missed by other analytic approaches. The overall level of segregation in a neighborhood may appear to decline as mixed-race households move in, but this can mask the persistence of segregation among single-race households.

As Ellis and colleagues find, neighborhoods can remain segregated even as mixed-race households become more common, which means that understanding the potential consequences of neighborhood segregation remains important. Many empirical studies have suggested that one consequence of segregation is greater social cohesion. But why do segregated neighborhoods tend to be more cohesive, and why do integrated neighborhoods tend to be more fragmented? And, perhaps more importantly, is there anything we can do about the latter case? These are the questions Neal and Neal explore in their agent-based model. Like Metcalf, they adopt a pattern-oriented approach to answer this question: we know that segregation tends to yield cohesion, but what micro-level mechanisms might be responsible for producing this pattern? And, like Ellis, they adopt Schelling's model as a starting point. They find that the apparent incompatibility of integration and cohesion is driven by how local residents form social network ties with one another, mainly via mechanisms of homophily (befriend similar others) and proximity (befriend nearby others). Moreover, after examining different social network-formation scenarios, it seems that this pattern may be here to stay.

At first glance, the finding that integrated neighborhoods will tend to be less cohesive may sound unfortunate. But is social cohesion necessarily positive? In their study of Los Angeles street gangs, Radil and colleagues highlight a context where network cohesion may be detrimental. In many applications of network analysis, the relationships under consideration – friendship, communication, or social support – are positive or beneficial. However, these authors examine a network of negative relations: the network of street gang rivalries. These network data are also unique because they are explicitly spatialized; relationships of rivalry are linked to each gang's territory, which in turn is associated with the locations of gang-related violence. Uniting two key features of street gangs – territory and rivalry – using system science offers a promising tool for understanding how gangs work, and potentially for reducing instances gang-related violence.

(Un)intentional Social Change

Research on the range of problems and frictions that emerge in communities often leads to calls for intentional social change, but the consequences of efforts toward intentional social change can be difficult to predict. At the same time, most social changes are unintentional, the outcome of naturally unfolding interactions among people in their communities. These too can be difficult to predict and explain. The final chapters in this section highlight how system science methods can be used to explain and anticipate the consequences of both intentional and unintentional social changes.

Changes in how we speak and write can be quite subtle, and can occur slowly over decades or centuries, which makes them difficult to detect and even harder to explain. Wagner and Ravindranath review how network analysis has been used by sociolinguists to understand why, for example, those over 50 say "groovy," while those under 50 do not. Focusing on their role in diffusion, networks have been used to explain how language changes spread from one place to another over short (across neighborhoods) and long (across states and regions) distances. Like Prell's use of networks to identify influential stakeholders in Chapter 17, they have also been used to identify the individuals – whether a teenager in California, or Queen Elizabeth I – who influence others to alter their speaking and writing patterns. Although networks can lead to changes in language, dense and cohesive networks like those Neal and Neal find in segregated neighborhoods in Chapter 23 can also allow subpopulations to maintain their distinctive speech patterns. Moreover, while network analysis has proven useful for understanding language, Wagner and Ravindranath note that linguists have recently begun exploring the use of other system science methods including agent-based models.

While changes in language transmitted through networks are almost always unintentional, in other cases networks can be used to bring about intentional behavioral changes. Warren and Doogan describe the intentional use of networks to change behaviors among ex-offenders participating in therapeutic communities. As members in these communities affirm others' positive behaviors and correct others' negative behaviors, a network of affirmations and corrections unfolds over time. They find that members tend to reciprocate affirmations directly, but also participate in generalized reciprocity: a member whose behavior has been affirmed is likely to affirm the behavior of others in the future – that is, members help one another by "paying it forward" with affirmations of positive behavior, which advances the goals of the therapeutic community. On the other hand, they also find evidence of cohort homophily – affirmations tend to come from others of a similar age – which means senior members tend not to mentor junior members and which may be a barrier to the community's goals. Using system science methods to shift the discussion of ex-offenders' behavior from the individuals to the relationships between them, Warren and Doogan are able to identify both successes and challenges for the therapeutic community model that might otherwise have been overlooked.

Both Wagner and Doogan focus on cases where individual behavior is changing, but in many cases the goal is not individual change but large-scale system change. When the goal is to change an entire social system, although the focus is on intentional and purposive changes, the risk for unintentional and possibly detrimental changes is amplified. Because system dynamics models are designed to understand how parts of a system affect one another, they are ideal for determining how best to fix a broken system, as well as to anticipate the potential unintended consequences of these "fixes." Connor and Levine focus on the case of community-based food systems, exploring how a range of health, economic, and social problems might be alleviated by intervening on one of five causal loops associated with the provision of healthy food. By unpacking each of these causal loops, they arrive at a series of concrete recommendations, but also observe that, if implemented, these recommendations could also spur undesirable gentrification in low-income urban neighborhoods. Similarly, Hirsch and colleagues focus on the case of encouraging the adoption of innovative curriculum in K-12 schools. Preliminary findings suggested, not surprisingly, that simply implementing a new curriculum in schools was unsuccessful; there was nothing to support teachers' adoption of the innovation. This might have led the researchers to recommend simultaneously implementing additional structural supports to facilitate adoption, but by using a system dynamics simulation, they recognized this approach would have its own unintended consequence: diverting attention from the curriculum innovation itself. As these chapters illustrate in the context of food systems and education, system dynamics models can be helpful in considering the positive and negative outcomes of an attempt at intentional social change in advance.

20

MODELING SOCIAL TIES AND HOUSEHOLD MOBILITY

Sara S. Metcalf

Research Question: How do urban problems like urban sprawl and socioeconomic segregation arise from individuals' choices about where to live?
System Science Method(s): Agent-based models & Networks
Things to Notice:

- Uncovering micro-level mechanisms that produce a macro-level phenomenon
- Simulation of a real spatial environment using empirical data

Underlying the aggregate phenomena of persistent problems such as urban sprawl and spatial socioeconomic disparity is the individual choice of where to live. This study develops an agent-based model to simulate social and economic influences on neighborhood choice. With Danville, Illinois, as an empirical context, a pattern-oriented approach is employed to examine the role of social ties in shaping intraurban household mobility. In the model, household agents decide whether and where to relocate within the community based on factors such as neighborhood attractiveness, affordability, and the density of a household's social network in the prospective block group. Social network and neighborhood choices are encoded with logit utility functions. The relative influence of factors affecting the formation of social ties in the simulated social network, such as geographic proximity, similarity of income, race, and presence of children, are adjusted using parameter variation to create alternative model settings. Simulated migration patterns resulting from different network and neighborhood choice coefficients are compared with observed migration patterns over a two-year period. Based on 1,000 simulation experiments, a regression of homeowner migration error (the difference between simulated and observed migration) relative to the parameter settings revealed components of social network choice such as income, race, and probability of local ties to be significant in matching observed migration patterns. A nonlinear effect of simulated social networks on household mobility and thus migration error was exhibited in this study.

The concept of spatial disparity is so embedded in urban America that residents might not recognize the problems implicit in referencing the "wrong side of town" in casual conversation. The problem of socioeconomic disparity has been particularly acute in industrial cities of the U.S. Rust Belt, where population declined and poverty increased as middle-class jobs were outsourced in the late twentieth century. Even so, many of these communities continue to expand in area while shrinking or stagnating in population. Such a dynamic intensifies the dichotomy between new and old, or rich and poor, parts of town.

 DOI: 10.4324/9781315748771-23

Fundamental economic forces ultimately drive the production of inequality, but individual choices about where to live are proximate influences on the spatial manifestation of disparity in local urban geographies. This research explores the relationship between social network structure and these household-level choices, as expressed in intraurban migration patterns. Because the relationship between social ties and location is complex and therefore difficult to untangle, an agent-based model is employed to simulate local community ties as they change over time. With this model, hypothesized effects of social networks on neighborhood choice are simulated dynamically. The model is constructed to allow systematic exploration of potential relationships between households' choices of where to live and the extent of their social networks.

A social network is an analytical abstraction of relationships between people that includes trusted channels for communication such as family, friends, and advisors. Homophily, or self-sorting according to similarity, has been shown to be a significant factor in shaping social network structure along dimensions of race, age, religion, education, occupation, and gender, with geographic proximity and family ties creating opportunities for such self-sorting connections to form (McPherson, Smith-Lovin, and Cook 2001). Because humans are mobile and idiosyncratic, in this globalized digital age we are capable of forming and maintaining diverse social obligations and attachments that span significant distances (Rainie and Wellman 2012).

Similarities have been drawn between dynamics of social and residential mobility, such as the propensity of a vacancy chain mechanism to operate. A home vacancy in a neighborhood, like a job vacancy in an organization, creates an opportunity for another to move up, into a more affluent position (White 1970). In the residential setting, the vacancy chain involves the out-migration of affluent households from a given neighborhood, which creates opportunities for lower income households to purchase homes in a neighborhood previously unaffordable (Hoyt 1939). Although the vacancy chain is a logical mechanism, it does not operate in isolation. Ethnographic geographies of social networks reveal complex dynamics such as the dialectical interplay of social influence and life opportunities for the case of women who are further marginalized in terms of class and race (Rowe and Wolch 1990; Peake 1995).

Computational advances have rendered it feasible to include social networks in geospatial analysis and therefore to accommodate the increasing availability of networked data about human interactions. Although distance continues to be relevant to human interaction, Kwan (2007) emphasized that the nature of its relevance has changed with the widespread adoption of information and communication technologies. New methods of geographic analysis explicitly account for the possibilities and constraints that are created by the intersection of geographic space with social network structure. For example, structural analyses of empirical, georeferenced social network data have informed the study of gang behavior in Los Angeles (Radil, Flint, and Tita 2010) and diarrheal disease in Bangladesh (Emch et al. 2012). Similarly, simulation studies of sociospatial network dynamics have been employed to examine influenza diffusion (Bian et al. 2012) and hurricane evacuation (Widener, Horner, and Metcalf 2013). These analyses help to untangle social and spatial influences on the dynamics of human behavior and provide ways to better apprehend the transient dimensions of mobility and communication that are consistent with the hypertext model proposed by Kwan (2007).

Modeling methods that recognize autonomy, interactivity, and contextual constraints help reconcile macro and micro interpretations of persistent social dilemmas (Schelling 1971, 1978). Schelling's segregation model, played as a game or simulated using cellular automata, demonstrates how even slight preferences for similar neighbors can induce aggregate patterns of residential segregation. An insight from such an individual-based analysis is that preferences for diversity, rather than simply a tolerance of difference, must be cultivated to overcome systemic tendencies toward segregation.

Agent-based models are increasingly used to simulate human–environment interactions, test effects of heterogeneity, and simulate mobile objects in dynamic landscapes (Westervelt and Hopkins 1999; Batty 2005). In agent-based models, individual choices shape aggregate outcomes. Information generated by changes inaggregate characteristics, such as social norms and neighborhood affluence, can then feed back to influence other agents' choices as they reevaluate their environment over time.

Bottom-up simulation of local interactions among diverse agents in a dynamic environment offers a way to explore complex social systems (Epstein and Axtell 1996). To simplify model interpretation and ease implementation, abstract representations of space are utilized in many agent-based models (Resnick 1994; Epstein and Axtell 1996). Simulation research has demonstrated ways to integrate agent-based models with empirical geographic information systems (GIS) data (Gimblett 2002; Parker et al. 2003). For example, recent work by Yin (2009) and Crooks (2010) demonstrates the continued utility of the Schelling (1971) model in formulating agent-based models of residential preferences in a dynamic urban landscape parameterized using GIS analysis.

Despite recent improvements in computational ease, the calibration and interpretation of agent- or individual-based models remains a challenge (Osgood 2009). To address this difficulty, a pattern-oriented approach was adopted for this research as a means of ground-truthing the simulation model (Grimm and Railsback 2005). The use of observed patterns to guide model development and analyze model output enables decoding of essential system information (Wiegand et al. 2003). Simulation studies enable indirect estimation of parameters for abstract models from patterns of spatial socioeconomic data. For example, in analyzing unemployment, Conley and Topa (2003) used census tract–based spatial proximity to structure a social network of local interactions, with the majority of interactions occurring within and adjacent to residents of a given tract. They demonstrated that this spatial algorithm anticipates shifts in unemployment better than an aggregate, nonspatial approach would. Similarly, this study employs the technique of indirect estimation by simulating the influence of hypothetical social networks on migration patterns, using pattern-oriented modeling to compare the simulated results with observed patterns of household mobility.

A dynamic simulation of social networks could involve connections that change over time (as in Bian et al. 2012) or processes of information diffusion within network structures (as in Widener, Horner, and Metcalf 2013). This study emphasizes the former, rendering social networks dynamic by changing connections among simulated household agents according to homophily-based reassessments of friendship utility that depend in part on geographic proximity. Model parameters for moving behavior, individual attributes, and network structure are estimated from different empirical sources. Household agents' choice algorithms are specified using utility functions to evaluate social connections and prospective residences. Simulated results are compared with observed homeowner migration patterns to explore the relative role of social influence on neighborhood choice. The next section describes the empirical context of Danville, Illinois, that is used to ground-truth this simulation study.

Empirical Context: Danville, Illinois

Empirical observations for the case of Danville, Illinois, provide a basis for specifying the household characteristics that shape social network and neighborhood choice in the agent-based model. Danville was selected as a reference case because its longitudinal patterns of population growth, decline, and stagnation, and its spatial pattern of areal expansion echo the trajectories of other Rust Belt cities.

Figure 20.1 Danville population trend

Located 215 km (134 miles) south of Chicago, Danville is a small city with an industrial legacy, with a population that declined from its peak of 42,570 people in 1970 to 33,027 people in 2010. Figure 20.1 illustrates the population trend over time for Danville on the basis of estimates from the decennial U.S. Census. In the decades prior to World War II, Danville's land area was less than 8 square miles. Since then, Danville has grown to nearly 18 square miles, with a steadily declining population density associated with post-World War II growth in the automotive era. Although Danville's land area has more than doubled since 1940, its population is now less than it was in 1920. Notably, Danville has continued to expand its geographic footprint despite population decline by annexing adjacent communities. As one Danville official explained, "We're a town of 30,000 with an infrastructure for 60,000."

The exodus of industry opportunities in the late twentieth century created a tension between attachment to hometown roots and the struggle to find work. Many middle-class workers left to find jobs elsewhere. Nearly 30 percent of Danville's population presently lives in poverty, although the city maintains a powerful class of affluent professionals and doctors who serve the community's hospitals and retirement homes.

This Rust Belt urban industrial dynamic was an important reason for selecting Danville for this study, but another reason was its manageable size for simulation. As a smaller city, Danville's size enabled computation of each household as a separate agent, so that population sampling was not necessary. An additional, practical reason for selecting Danville was its geographic proximity to the author when fieldwork was conducted in 2005. Because the simulation model was developed alongside data collection, heuristics for modeling individual choices drew on observations made during field interviews with residents of a Danville neighborhood.

Neighborhood Associations

With limited fiscal resources stretched further by an expanding areal footprint, the city of Danville actively encourages residents to form neighborhood associations. The geographic boundaries of these associations are based on the judgment of the residents who organize them; there is no predetermined size or scope. The city provides guidance and transitional support for newly formed neighborhood associations. There are seventeen active and five inactive neighborhood associations in Danville. Most active associations hold monthly meetings during the spring, summer, and fall seasons.

This study is informed by observations of participants in the Kentucky–Tennessee–Delaware (KTD) neighborhood association, named for the three north–south streets spanning its width, and extending three blocks south of Main Street in the southeast part of Danville (highlighted in Figure 20.2). The original neighborhood organizers restricted the boundaries of the association to correspond with the street pattern and keep the size manageable. Participant observation and in-depth interviews with residents were undertaken in 2005 to develop an understanding of social factors affecting household mobility. All the interviewees had participated in the neighborhood association at least once and conveyed generally positive views about the role of the association in building trust and strengthening neighborhood connections.

Although the interviewees were invariably homeowners, many of their neighbors and prior experiences included renting. Characterizations of renters as transient (and therefore disinclined to maintain their homes and neighborhood) were revealed by some homeowners at

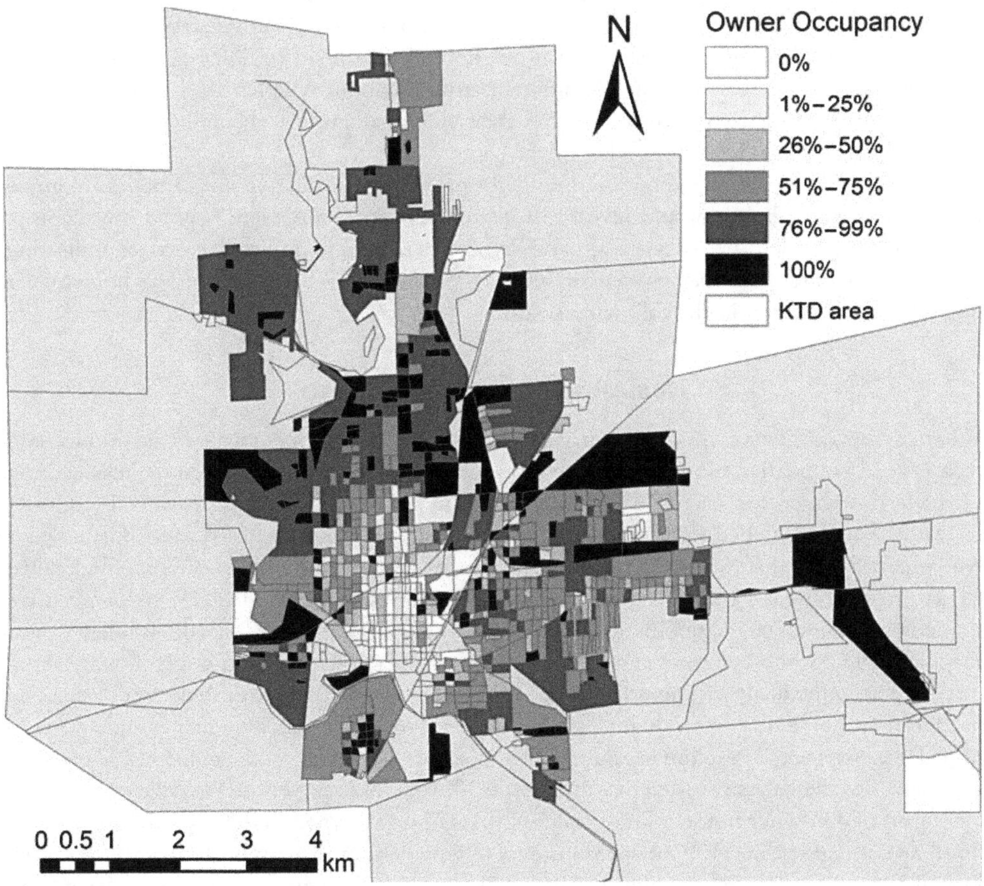

Figure 20.2 Owner occupancy in Danville census blocks (color figure available online)

neighborhood association meetings and in interviews, although others encouraged neighborhood renters to attend association meetings and neighborhood events. Danville city officials underscored the potential for the renter–owner ratio in a given neighborhood to exceed a tipping point of instability, inducing neighborhood outmigration. The map in Figure 20.2 illustrates the prevalence of owner-occupied households within Danville census blocks. The darker shades indicate places where owner-occupied households are more prevalent than renter-occupied households. The percentage shown does not include vacant households, reflecting the fraction of homeowners out of all occupied households.

The KTD neighborhood is racially diverse. As one African American resident observed, the neighborhood is "mixed. It's like black, white, black, white." Household experiences were elicited from homeowners who are women and men, African Americans and whites, who span the age range and family stages that make up the neighborhood. These interviews highlight the relevance of renter or owner status, and racial diversity among households in the neighborhood, as well as resident unease about changes in neighborhood composition and differences in family status.

The importance of children in neighborhood social networks was emphasized in the interviews. As shared by one homeowner with grown children, "My daughter knows more people around here [than me]. Like [my back-yard neighbor]—she met him because she would go out [to the back alley] and take out the garbage. They would be out there in the garage, because my grandchildren would ride their tricycles around the garage." Although the friendships children form are significant social ties for a neighborhood, some youth engage in activities—such as playing in the street—that concern neighbors accustomed to different norms of child safety and discipline. Residents shared different attitudes about whether they would confront their neighbors about a concern or whether they would sooner call the authorities to resolve a disturbance, such as noise from a late-night party.

These interviews with neighborhood residents provided a qualitative source of data to guide heuristics for agent behavior and reveal reasons for relocating that were beyond the scope of the simulation model, such as marriage and divorce dynamics. Another data source indicating homeowner records at the parcel scale for 2001, 2003, and 2005 was used to analyze homeowner migration, as described in the following section.

Homeowner Migration Patterns

Whereas the qualitative insights from interviews helped to conceptualize the model and recognize its boundaries, homeowner migration data were used explicitly to test the model using the pattern-oriented approach. To generate migration patterns, parcel data provided by the city of Danville were matched to homeowner data for 2001 and 2003 and used to identify owners who moved during that two-year time period. Changes in the inclusion or exclusion of individual family members (e.g., Jane and John Smith in 2001; John Smith in 2003) in consecutive homeowner records suggested the frequency of marriages, divorces, and deaths as likely events shaping neighborhood tenure and homeowner mobility patterns. Although it was labor-intensive to individually distinguish among new owners, departures, and those who remained in their own home during the two-year interval, the process of matching names and addresses provided a novel way to capture intraurban migration data at the scale of the individual household.

Analysis of these data revealed that during the 2001–2003 time period, Danville experienced a net gain of 43 owner households overall, with 1,472 new owners, 1,429 departures from the homeowner register, and 139 intraurban moves (8.6 percent of all arrivals). From 2003 to 2005, Danville lost 22 owner households, corresponding with 1,041 new owners, 1,063 departures, and 119 intraurban moves (10.2 percent of all arrivals). Figure 20.3A shows the intraurban migration patterns from 2001 to 2003 aggregated to the scale of the census block group. Blue shades signify areas with net gains in homeowners, such as the lake area in the relatively affluent northwest part of Danville, whereas red shades represent areas of outflow. As shown in Figure 20.3B, though, if all the homeownership changes—new owners as well as owners who left Danville—are included, the pattern nearly inverts. An increase in the overall number of homeowners includes new arrivals to Danville as well as renters who become owners. So the overall increase in ownership in lower income areas (e.g., the southeast section of Danville) likely reflects a significant number of transitions from renting to owning a home.

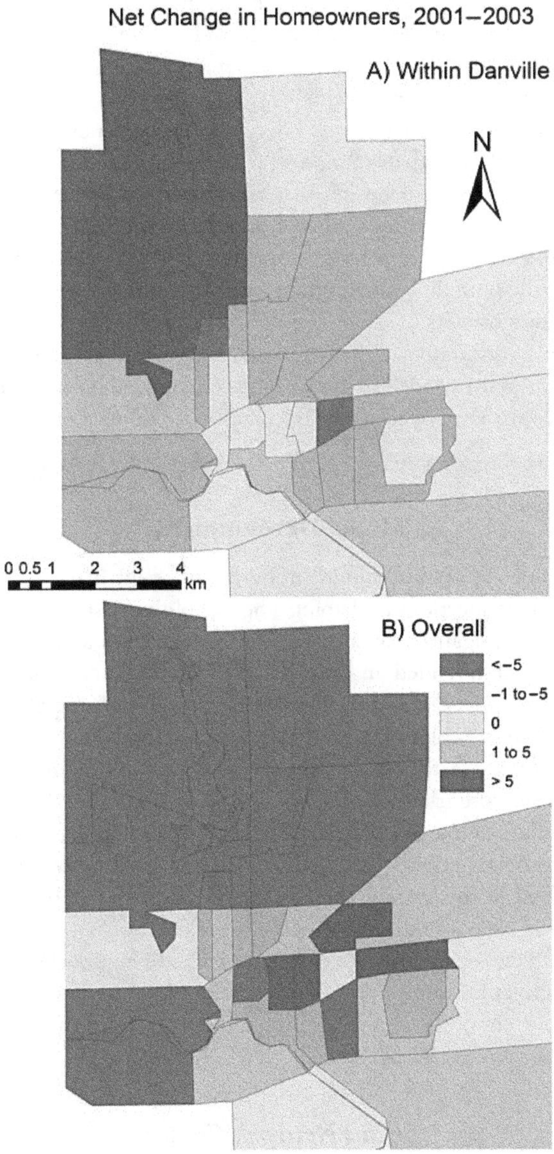

Figure 20.3 Home ownership changes (A) within Danville and (B) overall from 2001 to 2003 (color figure available online)

Tracking these owner-occupied household data over a two-year interval revealed a northwest migration pattern within Danville, as congruent with informal appraisals of Danville neighborhoods offered by city officials who were consulted during this study. This internal migration pattern contrasted with the overall flows of outmigration and new ownership. Figure 20.3 demonstrates that despite its appeal to residents of Danville, the attractive northwest lake area experienced net losses in ownership overall. Outmigration from Danville was evident in the more affluent neighborhoods, ostensibly where property values could be recouped as mobile professionals relocated for better job opportunities. Balancing this outflow of relatively affluent owners was a net increase in homeownership in the southeast part of town, where homes are more affordable. An influx of ownership in lower income areas (e.g., the KTD neighborhood) revealed the significance of renter to owner transitions. Because the property values had reached bottom in lower income areas, some landlords were willing to sell houses to tenants on contract after they made enough money from renting.

Via a musical chair kind of mechanism, Hoyt's (1939) vacancy chain hypothesis helps to explain the intraurban and interurban migration patterns observed for Danville in Figure 20.3. Hoyt's vacancy chain mechanism proposes that affluent homeowners who leave the lake area make vacancies available to upwardly mobile households within Danville who aspire to that neighborhood. These transitions create additional vacancies, as more modest homes become available to lower income households, who then vacate deteriorating homes that might be leased to renters or sold to new owners.

For this study, both qualitative and quantitative data provided patterns that shaped the formulation and analysis of the model. In particular, the empirical homeowner migration patterns previously described were employed as a direct basis of comparison for simulated migration outcomes.

Model Development

To simulate social influence on household mobility, an agent-based model is constructed to explore the role of social connections in shaping a household's decision to move to a new location in Danville. As household agents relocate over time, social ties adjust, shaping the nature of neighborhood networks. Embedded and replicated in the main class of the model, household agents make two dynamic decisions: social network choice and neighborhood choice. These decisions are described in the following sections. The full model code is available online.[1]

The process of identifying factors of household choice was informed by the empirical context of the Danville study. Because friendships among children were revealed through interviews to be a significant factor in forming neighborhood networks, the family status factor was included in the formulation for social network utility. According to the homophily heuristic, the agent-based model constructed in this study enables social connections to be formed and broken on the basis of proximity as well as similarity of demographic attributes. As specified in the sections that follow, income affects social network choice as well as neighborhood choice, through the affordability and attractiveness terms. The distance effect in choosing a social network and the social network effect in choosing a neighborhood together constitute a reflexive relationship between neighborhoods and social networks.

Social Network Choice

A discrete choice methodology is widely applied to urban and transportation problems (Ben-Akiva and Lerman 1985; McFadden 1991). Extensions demonstrate the use of logit models for discrete choice selection of social connections (Van de Bunt, Duijin, and Snijders 1999) and for choices made in the urban context (Waddell et al. 2003; Paez and Scott 2007). The social network choice in Equation 1 employs a binary logit expression, such that the probability of a household i connecting with household j is based on the exponential of the utility of that connection, divided by one plus the same exponential term (Ben-Akiva and Lerman 1985; Train 2003).

$$P_{ij} = \frac{\exp\,(U_{ij})}{1 + \exp\,(U_{ij})} \tag{1}$$

where P_{ij} is the probability of household i connecting to household j and U_{ij} is the utility of connecting household j to household i.

The logit expression assumes that unobserved factors are independent over time in a repeated choice situation and represents systematic variation of preferences (Train 2003). As is the case in this implementation, a normalization constant is often applied to the denominator of the utility term to account for the variance of the unobserved portion of utility and scale the elasticity of choice (Train 2003, 44).

Household agents evaluate the utility of their social network at a stochastic and asynchronous frequency. At the moment of network reevaluation, each household picks another household and tests whether the probability from the binary logit expression in Equation 1 passes a stochastic satisfaction threshold. Each simulated household determines the utility of a prospective social tie using the expression in Equation 2:

$$U_{ij} = C + \alpha_D \cdot D_{ij} + \alpha_I \cdot \frac{|I_i - I_j|}{I_{\text{avg}}} + \alpha_R \cdot R_{ij} + \alpha_F \cdot F_{ij} \tag{2}$$

where C is the constant average utility; D_{ij} is the distance between centroids of census blocks containing households i and j; I_i, I_j is the income of household i and household j, respectively; I_{avg} is the average income across all households in the city; R_{ij} is the binary term for similarity of racial category; F_{ij} is the binary term for similarity of family status; α_D is the utility weight for the block distance effect [0, –1]; α_I is the utility weight for the similarity of income effect [0, –1]; α_r is the utility weight for the effect of racial similarity [0, 1]; and α_F is the utility weight for the effect of family status similarity [0, 1].

Equation 2 describes multiple effects on social network choice. The utility of a social connection is expressed as a set of alpha α weights applied to various homophily effects. For the effects of block distance and income, a greater difference decreases the likelihood of connection, and thus the α_D and α_D weights range between 0 and –1. Household income is employed as a proxy for socioeconomic status. Additional effects in Equation 2 include binary categories for racial similarity (set to one if both households identify as white or if both households identify as nonwhite) and similarity of family status (set to one if children are present in both households). If the prospective household identifies with the same racial category, the value of the term R_{ij} is one. Likewise, if both households have children, the binary term F_{ij} is encoded as one. If either household does not have children, F_{ij} is zero.

Probability of Local Ties

The probability of connection as generated by the logit formula in Equation 1 and the utility expression in Equation 2 is part of an agent-level algorithm that also invokes a probability of local ties (localProb), where "local" is defined as within the same census block group. The localProb parameter establishes the odds of selecting a connection from within the same block group. When the connection algorithm is executed, a random number between zero and one is compared to the localProb odds. For a prospective social tie, if the probability of local ties exceeds the random number and both households reside in the same block group, or if localProb is less than the random number and they reside in different block groups, then the utility of the connection is evaluated in terms of the utility (Equation 2) and thus probability (Equation 1) of forming a particular connection. If the probability of connecting according to Equation 1 exceeds another random number between zero and one, then the connection is formed between the two households. Otherwise, additional prospective households are considered from within (or outside) the home block group until this probability threshold is exceeded.

Via this connection algorithm, the probability of local ties broadly determines what fraction of ties will be made within the same block group. The probability of local (intra-block-group) ties is one of two geographic factors shaping whether a social connection forms between two households. Although block-level centroids are used to establish a coarse measure of distance between households in Equation 2, the probability of local ties imposes a spatially variable, block-group-level "local" container on the formation of social ties. The reason for including this probability in the connection algorithm is to impose a different geographic component on social network choice from the block centroid-based distance between households.

Neighborhood Choice

In considering where to live, the first step each household agent takes is to assess, via a heterogeneously assigned "happiness" threshold, whether it is satisfied in its current location. If the household agent is dissatisfied, it selects a set of vacant parcels at random from the larger set of all available (vacant) parcels in Danville. If the utility of these alternative parcels exceeds the current location utility, the parcels become part of a consideration set. The limited size of the consideration set (up to fifteen parcels under base case assumptions) is a proxy for a scan of the real-estate section of the newspaper. Once the consideration set is defined, the location with the highest neighborhood utility is chosen.

For neighborhood choice, the utility of a parcel p to household h invokes a constant C term and a set of weighted effects, as expressed in Equation 3.

$$U_{p,i} = C + \alpha_{SN}\left(\frac{\mathrm{SN}_{i,bg(p)}}{\mathrm{SN}_i}\right) + \alpha_{\mathrm{Attract}}\left(\frac{I_{bg(p)} - I_{\mathrm{avg}}}{I_{\mathrm{avg}}}\right) + \alpha_{\mathrm{Afford}} \cdot A_{p,i} \tag{3}$$

where $U_{p,i}$ is the utility of parcel p to household i; α_{SN} is the utility weight for the social network effect [0, 1]; SN_i is the size of the social network for household i as number of connections; $\mathrm{SN}_{i,bg(p)}$ is the size of the social network for household i within the block group containing parcel p; $\alpha_{\mathrm{Attract}}$ is the utility weight for the effect of neighborhood attractiveness [0, 1]; $I_{b(p)}$ is the average income of the block containing the prospective parcel p; α_{Afford} is the utility weight for the affordability effect [0,1]; and $A_{p,i}$ is the neighborhood affordability for parcel p to household i.

After the constant C term on the right side of Equation 3, the first effect on the utility of a prospective parcel is that of the social network, expressed as the proportion of a household's social ties that reside in the destination block group. The composition of the social network is determined by the utility (Equation 2) and consequent probability (Equation 1) of forming a social tie with another household. The social network effect in Equation 3 completes a reinforcing, or positive, feedback mechanism in which the longer a household resides in a neighborhood, the more likely it is to have social ties in the neighborhood, and the utility of staying in the neighborhood therefore increases.

The second effect on neighborhood choice in Equation 3 assesses neighborhood attractiveness as the ratio of the average income of the parcel's block relative to the average income of the entire community (I_{avg}). As with the social network effect, the attractiveness effect also creates a positive feedback mechanism: As more affluent households move to a neighborhood, the neighborhood becomes more affluent and therefore more attractive to other households. This attractiveness is counterbalanced by the third term in Equation 3, which represents the affordability constraint, A_{pj}, as defined in Equation 4.

$$A_{p,i} = \max\left(\frac{I_i - I_{b(p)}}{I_{b(p)}}, 0\right) \tag{4}$$

The affordability term is negative if the household's income (I_i) is less than the average income of the block containing the prospective parcel. If the household's income is sufficient, affordability is not a constraint. The affordability constraint is also excluded when evaluating the utility of staying in the current location. Therefore, the relative affordability of staying in place reflects the inconvenience or "cost" of relocating. This induces a balancing, or negative, feedback mechanism on the propensity of households to move to an affluent neighborhood.

Although income is part of both attractiveness and affordability terms, the former is via aggregate assessment of block income relative to the city, whereas the latter compares household income to prospective block income. An alternative indicator of neighborhood affluence would be a measure such as property value. The use of household income enabled a computationally efficient model design, however, because census-derived income was encoded as a factor in both social network formation (Equation 2) and neighborhood choice (Equations 3 and 4). Associating household income with mobile agents facilitates the completion of both balancing and reinforcing feedback mechanisms across the individual and aggregate spatial scales of the model.

Decision Dynamics

The simulated rate at which households decide whether to move and connect with each other is probabilistic, sampling from an exponential probability distribution to facilitate asynchronous computation of household agent algorithms. The exponential form of probability distribution corresponds to a Poisson process in which events recur at a constant average rate. When aggregated, these individual events reflect a first-order time delay. Although the exact timing of household decisions is probabilistic, the frequency of considering a move is normalized to once per decade for owners and annually for renters. Therefore, the frequency with which households revisit the decision of whether and where to move depends directly on their current status as renters or owners.

The logic for simulating owner migration at a slower rate than that of renters is a heuristic drawn from interviews. For example, one homeowner noted that "a lot of the homes in this area are rentals, and rentals are going to attract more transient type of people."

In contrast to the frequency of the household's relocation decision, social network behavior is not directly affected by renter or owner status, as the model simulates the frequency with which all households evaluate their social networks on an annual basis or when relocating.

Indirectly, however, the initial distribution of renters and owners influences network connections through the proximity effect and the probability of local, intra-block-group ties.

Data Integration

The first phase of model development involved the construction of an abstract but spatially explicit prototype, non-GIS, two-neighborhood model of household network and neighborhood choice (Metcalf and Paich 2005). As consistent with the mixed-methods research design, GIS data processing was undertaken alongside model development and ethnographic fieldwork so that the final form of the agent-based model was empirically grounded. The process of data integration is illustrated in Figure 20.4, in which model objects were created from Danville data for model initialization. This process began at left with unstructured, or raw, data files. Raw data are considered unstructured because they lack semantic meaning. Geographic data from the city of Danville and the U.S. Census were processed to create four unstructured text files with rows that correspond to the number of objects.

As displayed at the left of Figure 20.4, the unstructured data were organized by census block group, census block, parcels, and households. Household income, presence of children, and migration patterns were contained by 28 block groups. Race and housing status (rent, own, or vacant) were contained by 764 blocks; location coordinates were specified for the centroids of 13,166 parcels. Initial homeowner assignments were contained by the subset of 7,576 owner-occupied parcels.

The unstructured data were imported into the Eclipse Java development platform for processing. In this stage, shown in the center panel of Figure 20.4, the first task was to properly link the data so that block groups contain blocks, which in turn contain parcels. Parcels were linked to initial owners from the 2001 owner-occupied household list. Once the structured data were properly linked, the remaining objects and attributes were identified.

For the 764 selected blocks within Danville, attributes from the 2000 U.S. Census data were extracted for tenure (number of housing units owned, rented, and vacant) and for racial identity by tenure status (owners and renters). After owner-occupied parcels were specified from city

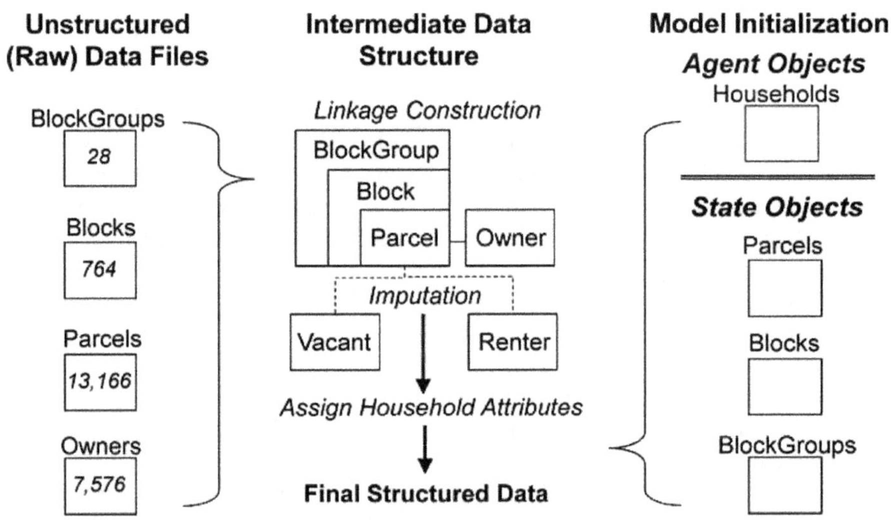

Figure 20.4 Process for data integration

data, the number of renter-occupied parcels was imputed from the ratio of rental to vacant housing units in each census block. A racial status for each household was imputed on the basis of tenure at the census block level. To do so, the probability of a household being classified as white was set to the fraction of white households within the owner-occupied and renter-occupied households. To represent race in binary form, simulated household racial categories were reduced to "white" or "nonwhite." The census racial category "Black or African American" constitutes the majority of nonwhite households in Danville. Because cross-tabulations were available at the census block group level by race, and race was first assigned at the finer grain block level, household racial status was also used to impute whether children were present and to assign household income. Specifically, census data at the block group level were used to determine the prevalence of children among white and nonwhite households, as well as the distribution of income among white and nonwhite households.

The right panel of Figure 20.4 distinguishes between agent objects and state objects. In this nomenclature, state objects reflect attributes alone, whereas agents include a capacity for choice. This distinction streamlines the dynamic computational requirements of the model. Spatially fixed state objects such as parcel, block, block group, and household were serialized and structured in AnyLogic (2006) software. During model initialization, each dynamically active household agent was created from attribute information contained in the corresponding household state object.

Analysis of Simulation Results

Pattern-oriented modeling involves the use of empirical patterns to guide the process of both building and testing individual-based models (Grimm and Railsback 2005). The pattern-oriented strategy for model calibration used in this study is illustrated in Figure 20.5, as adapted from Grimm and Railsback (2005). Beginning at the top left in Figure 20.5, model parameters are set where possible using available data for attributes such as income, race, and parcel location (as described earlier). Then broad ranges are initially chosen for the remaining uncertain parameters, such as whether they are negative or positive. In this study, the distance and income parameter weights shaping social network choice (α_D and α_I in Equation 2) were bounded between 0 and –1, so that a larger difference between households would reduce the perceived utility (and thus likelihood) of forming a social connection. All other parameter weights and linear constants in the logit functions specified in Equations 2 and 3 were bounded between 0 and 1. To provide a range of elasticity for the choice mechanism, normalizing constants applied to the denominator of the utility equations were bounded between 0.1 and 2. In addition, the probability of evaluating social connections within the home block group, localProb, was varied between a 20 and 70 percent chance of selecting a household from within an agent's block group.

Another step in the pattern-oriented process, as indicated in the top right of Figure 20.5, is to specify alternative model structures. In this study, such structures are specified using different parameter settings for the utility weights of social network and neighborhood choice. After specifying initial conditions and alternative model structures, observed patterns are selected to serve as filters for reasonable model settings. The intraurban homeowner migration data described earlier and illustrated in Figure 20.3A are used to define an objective function that minimizes the difference between simulated and observed migration (see the next section).

The model testing phase involved simulating outcomes from different model settings to evaluate which parameter combinations best matched the observed migration patterns.

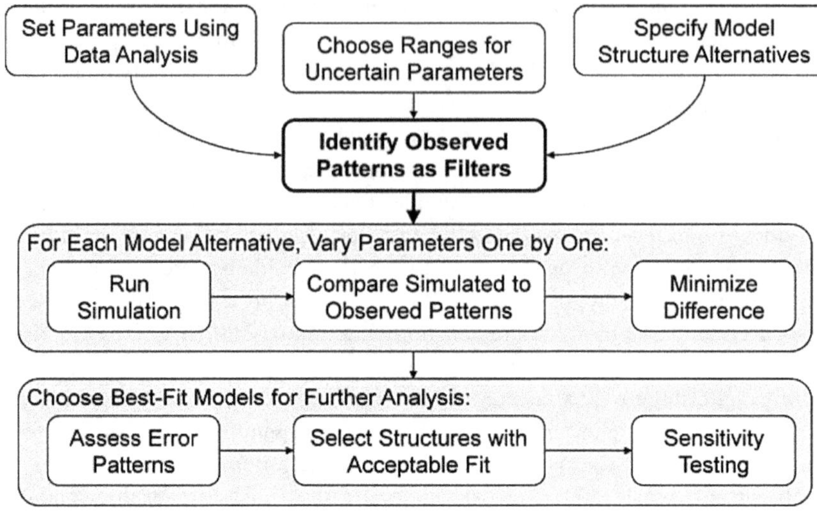

Figure 20.5 Pattern-oriented calibration strategy

To do so, a Monte Carlo method of optimization was undertaken in which agent choice parameters were varied systematically using repeated random sampling within the specified bounds. This process was performed using AnyLogic's OptQuest algorithm for 1,000 consecutive, unique parameter sets to facilitate simulation experiments under uncertainty. Each experiment included a batch of 25 stochastic simulation runs from a single initial seed and set of parameter values. Therefore, for the 1,000 experiments conducted, a total of 25,000 individual simulations were run. Average statistics for each batch were computed to account for stochastic variations between individual simulation runs. Sources of stochasticity include the timing of household choice and thresholds for forming connections.

Measuring Migration Error

Each simulation was run for a two-year period and evaluated relative to the observed homeowner migration patterns from 2001 to 2003. Directional migration patterns were assessed using a matrix to represent moves from and to each of Danville's 28 block groups. A directional move error was computed as the sum of the move error for each cell in the 28 × 28 block group matrix. In addition, aggregate move error was computed as the difference between overall simulated and observed homeowner migration patterns in Danville.

$$\mathrm{dirErr} = \sum_{i=1}^{n}\sum_{j=1}^{n} |\,\mathrm{simM}_{ij} - \mathrm{obsM}_{ij}| \tag{5}$$

where dirErr is the directional homeowner migration error; simM_{ij} is the simulated migration from block group *i* to block group *j;* and obsM_{ij} is the observed migration from block group *i* to block group *j.*

The aggregate migration error provides a check on the overall accuracy of simulated intraurban migration volume. As expressed in Equation 6, the cumulative simulated moves are compared with cumulative observed moves to determine a total, aggregate measure of move error. For the period from 2001 to 2003, there were a total of 139 homeowners who moved within the city. Therefore, the right-most expression in Equation 6 sums to 139 moves.

$$\text{totErr} = \left| \sum_{i=1}^{n} \sum_{j=1}^{n} \text{simM}_{ij} - \sum_{i=1}^{n} \sum_{j=1}^{n} \text{obsM}_{ij} \right| \tag{6}$$

where totErr is the total, aggregate homeowner migration error.

The average move error was calculated as a summary statistic for each batch of 25 simulation outcomes by combining the directional and overall moves. The combined, average migration error provides the objective function expressed in Equation 7.

$$\min \text{avgErr} = \frac{\sum_{s=1}^{r} \text{dirErr}_s + \sum_{s=1}^{r} \text{totErr}_s}{r} \tag{7}$$

where avgErr is the average homeowner migration error; r is the batch size (25), number of replications for a single parameter combination; and s is the index for an individual simulation run in an experiment.

This objective function was designed to minimize the average homeowner migration error by adjusting the parameter settings to create 1,000 unique simulation experiments. The optimization algorithm used in this study specifically tests the boundaries of different parameters in combination, "learning" to avoid parameter spaces that result in the most mismatches relative to the observed migration data (April et al. 2004).

Regression Analysis

As an indicator of how well the simulated migration patterns matched the Danville homeowner moves, the average homeowner migration error (avgErr in Equation 7) was used as a dependent variable in a regression analysis of the simulation results. In this analysis, the independent variables are household choice factors (α terms in Equations 2 and 3) that were adjusted to create the 1,000 experiments simulated. Because the agent-based model is inherently dynamic, the regression analysis is exploratory in nature. Therefore, both nonlinear and linear regressions were estimated. With the simulation experiment as the unit of analysis, Table 20.1 presents the regression results, reporting the coefficients and t statistics for the two specifications. The nonlinear regression included squared and linear terms for the independent variables. Although the nonlinear form of regression fit the data better ($R^2 = 0.701$ vs. 0.478), it also carried the explanatory burden of having twice as many independent variables.

Lower average migration error suggests that the agent-based model is better accounting for the Danville migration decisions. So, for the linear regression, the choice factors that best account for the observed migration are those that have a statistically significant negative effect. For the nonlinear regression, though, the interpretation is more subtle. Factors with a significant positive coefficient for the linear term and a significant negative coefficient for the nonlinear term could either be irrelevant or highly important because midrange values (across the experiments) produce the highest migration error. In contrast, logit choice utility weights with a significant negative coefficient for the linear term and a significant positive coefficient for the nonlinear term are factors that clearly matter in explaining the Danville migration choices, but their impact is bounded because setting the value too high might result in an increase in migration error.

Table 20.1 Coefficients from regression of neighborhood and network effects on migration error

	Nonlinear regression				*Linear regression coefficient*	
	Coefficient for linear term		*Coefficient for squared term*			
Utility weights in neighborhood choice (Equation 3)						
Affordability (locAfford; α_{Afford})	**−2.007**	(−4.94)	**0.015**	(4.44)	−0.129	(−0.90)
Attractiveness (locAttract; $\alpha_{Attract}$)	**−1.955**	(−5.40)	**0.039**	(10.65)	**1.526**	(9.81)
Social network density (locSN; α_{SN})	**1.934**	(3.96)	**−0.014**	(−3.00)	0.237	(1.27)
Utility weights in social network choice (Equation 2)						
Distance (nwDist; α_D)	−0.265	(−0.55)	0.483	(0.54)	−0.196	(−0.85)
Income (nwInc; α_I)	**2.828**	(5.82)	**−0.034**	(−7.63)	**−0.766**	(−4.85)
Race (nwRace; α_R)	**1.149**	(2.92)	**−0.014**	(−4.30)	**−0.629**	(−4.05)
Family status (nwChild; α_F)	−0.339	(−0.62)	−0.003	(−0.63)	−0.305	(−1.69)
Probability of local ties (localProb)	**−5.297**	(−3.73)	**0.045**	(3.01)	**−0.762**	(−2.83)

Note: For each of the regression coefficients listed, the t statistic is indicated in parentheses. Coefficients with statistical significance ($|t| > 2$; p value < 0.05) are shown in bold type. To avoid squaring fractions, all independent variables were multiplied by 100 before conducting the regression. The dependent variable was the average migration error, avgErr, as defined in Equation 7.

Considering the geographic factors first, Table 20.1 shows that the probability of local (intra-block-group) ties reduced migration error, with large negative, statistically significant coefficients in both the linear and nonlinear regressions (even though the coefficient of the squared term is positive, the net effect is negative over most of the parameter's range from 20 to 70 percent across the experiments). The influence of distance on network formation (α_D) also reduced migration error, but its coefficients are weak substantively and statistically. Rather than centroid-based block distance, therefore, the most significant geographic influence was the probability of forming social ties within the same block group.

As with the geographic effects, utility weights for other components of social network choice also had negative coefficients, indicating that they reduce simulated migration error. Among these, utility weights for racial similarity (α_R) and income difference (α_I) reduced migration error with statistically significant negative coefficients in both the linear and nonlinear specifications. Utility weights for family status (α_F) and neighborhood affordability (α_{Afford}) also help to account for homeowner migration patterns. The effect for homophily of family status, inferred by the presence of children, consistently reduced migration error in both regressions (with an inflection point outside the parameter range in the nonlinear specification), even though its t statistic does not reach the standard level of statistical significance. Similarly, the neighborhood affordability constraint has a negative but insignificant coefficient in the linear regression while lowering migration error over most of its range in the nonlinear regression (only increasing error for values above 0.67).

The utility weight for the social network effect (α_{SN}) on neighborhood choice resulted in a positive coefficient in the linear regression, indicating that increases in the social network influence on neighborhood choice would decrease the ability of the simulation to match observed homeowner migration. Although it is not significant in the linear regression, the influence of the social network on location choice is highly significant in the nonlinear regression. The coefficients signify that the nonlinear social network effect mitigates migration error when α_{SN}

exceeds 0.69. This nonlinearity indicates that some parameter combinations reduce simulated error under a strong social network effect, whereas other settings reduce error with a weak network effect.

Finally, Table 20.1 indicates that neighborhood attractiveness ($\alpha_{Attract}$) has a statistically significant positive coefficient in the linear regression, and the positive coefficient for its squared term in the nonlinear regression is large enough that its net effect is positive over most of the parameter range across experiments. More specifically, above the value of 0.25, increases in the utility weight on neighborhood attractiveness exacerbate simulated migration error. For this statistically significant influence, smaller weights produce better matches with observed migration patterns.

Analysis of Filtered Monte Carlo Experiments

The relationship between aggregate and directional migration error is shown in Figure 20.6 for the set of 1,000 Monte Carlo simulation experiments. This relationship reveals that if directional error were minimized completely, the trivial solution of no migration would be preferred over more meaningful scenarios. As shown in the left-most results of Figure 20.6, if no intraurban migration occurs, the comparison to each of the 139 observed homeowner moves for the time period from 2001 to 2003 produces equivalent errors in both the aggregate and directional dimensions, resulting in a combined error of 278 moves (200 percent of observed migration). If aggregate moves were exactly matched to the observed moves (zero aggregate error) but all of the moves were from and to the wrong block groups, directional error would be 278 moves.

As illustrated in Figure 20.6, aggregate migration error (totErr in Equation 6) approaches zero when directional error (dirErr in Equation 5) reaches approximately 220 moves. A threshold of 230 moves for the combined (directional and aggregate) migration error as averaged for the

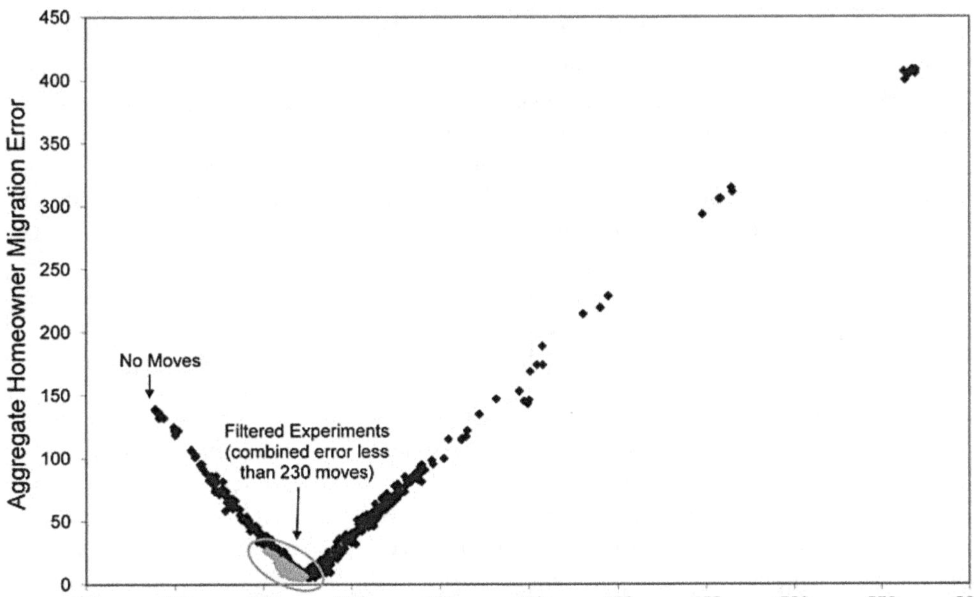

Figure 20.6 Aggregate and directional homeowner migration error (color figure available online)

25 realizations of each experiment (avgErr in Equation 7) was used to filter 486 of the 1,000 experiments for further analysis. Figure 20.6 highlights the aggregate and directional migration error of these filtered experiments.

A cluster analysis of parameter weights within this set of filtered experiments reveals the relative performance of simulated combinations of effects. This analysis was performed in SPSS (2006) as a two-step cluster classification using the Bayesian Information Criterion with a log-likelihood measure of distance in parameter space. Table 20.2 shows the average error and parameter settings for each cluster, along with the best and worst cases produced in the set of 1,000 Monte Carlo experiments.

Three clusters of similar size (N = 148, 163, 175) emerged from analysis of the 486 filtered solutions. Cluster 3 resulted in a slightly lower average migration error than the other two clusters. Cluster 3 is similar to Cluster 2 in terms of parameters such as the affordability (α_{Afford}) and social network (α_{SN}) effects on location choice. Examination of these clusters shows that the probability of selecting social ties within one's block group (localProb) approximated or exceeded 50 percent for all three clusters. In contrast, the distance effect on social network choice varied by cluster, which is consistent with the weak effect of distance in the regression analysis described earlier.

As indicated by the best case scenario in Table 20.2, the smallest average migration error achieved in a simulation experiment was 219 moves over two years. Although this simulated migration error is 158 percent of the 139 observed moves, it produces a much better match than the worst case scenario of 978 moves. The large error of the worst case is induced by excessive intraurban homeowner migration, creating 408 extraneous moves during the two-year time span. As shown in Table 20.2, the worst case effectively turns off the affordability constraint by setting α_{Afford} to zero while maximizing the attractiveness and social network effects on neighborhood choice. The dynamic behavioral consequences of these differences in parameter settings are explored in the following section.

Simulation of Spatial Inequality

To reveal how simulated household behavior impacts spatial inequality, the Gini coefficient was applied to incomes averaged at the block group level. The Gini coefficient is evaluated as a fraction between zero and one, with higher values representing greater inequality (Gini 1921). The spatial dimension of this inequality was provided by comparing the 28 block groups containing households. Equation 8 articulates the formula used to assess the Gini coefficient across the 28 block groups in the Danville model.

$$G = \frac{1}{n} \cdot \left(n + 1 - 2 \cdot \frac{\sum_{k=1}^{n} (n + 1 - k) \cdot I_k}{\sum_{k=1}^{n} I_k} \right) \tag{8}$$

where G is the Gini coefficient, n is the number of block groups (28), k is the index of block group in nondecreasing rank order, and I_k is the average income of the kth block group.

Figure 20.7 illustrates the dynamics of this spatially defined Gini coefficient over a period of 20 years under the average parameter settings for Clusters 1, 2, and 3 (see Table 20.2). These three scenarios exhibit a qualitative similarity in that the inequality between block groups increases from a starting Gini coefficient of approximately 0.12 in the first few years of the simulation and then declines to a level similar to the starting point. Because it reflects the differences in average income by census block group, the block group Gini coefficient is lower (more equal) than it would be if household incomes were directly compared to each other. For example,

Table 20.2 Average error and parameter settings for clusters and extreme cases

	Cluster 1[a]	*Cluster 2*[b]	*Cluster 3*[c]	*Best case*	*Worst case*
Components of migration error (Equations 5–7)					
Average, combined migration error (avgErr)	226.63	226.90	225.43	219.20	977.76
Directional migration error (dirErr)	217.53	216.56	215.97	210.80	569.64
Total, aggregate migration error (totErr)	9.10	10.34	9.46	8.40	408.12
Utility weights in neighborhood choice (Equation 3)					
Affordability (locAfford; α_{Afford})	0.9713	0.6922	0.9677	0.9754	0
Attractiveness (locAttract; α_{Attract})	0.4621	0.3138	0.1957	0.2847	1
Social network density (locSN; α_{SN})	0.6831	0.4167	0.6826	0.7373	1
Utility weights in social network choice (Equation 2)					
Distance (nwDist; α_D)	−0.6721	−0.3468	−0.2116	−0.3566	−1
Income (nwInc; α_I)	−0.7150	−0.4797	−0.4089	−0.4648	0
Race (nwRace; α_R)	0.9324	0.5918	0.3262	0.5318	1
Family status (nwChild; α_F)	0.9238	0.6114	0.8720	0.9035	0
Probability of local ties (localProb)	0.5343	0.4659	0.6305	0.5912	0.2

a N = 148. b N = 163. c N = 175

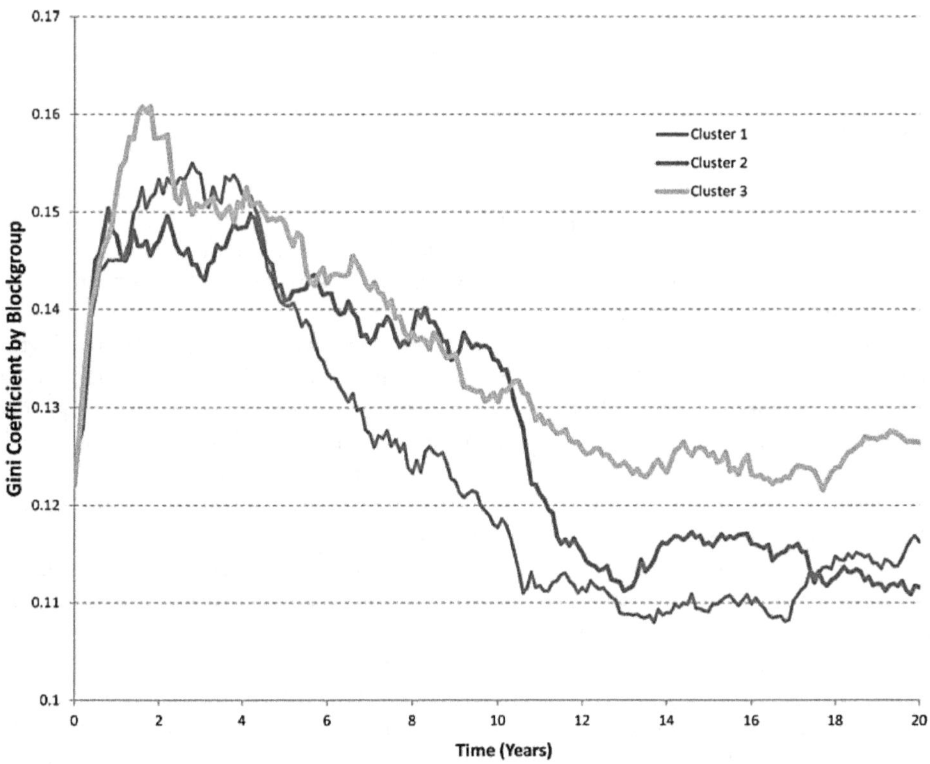

Figure 20.7 Inequality dynamics under average parameter settings for three clusters (color figure available online)

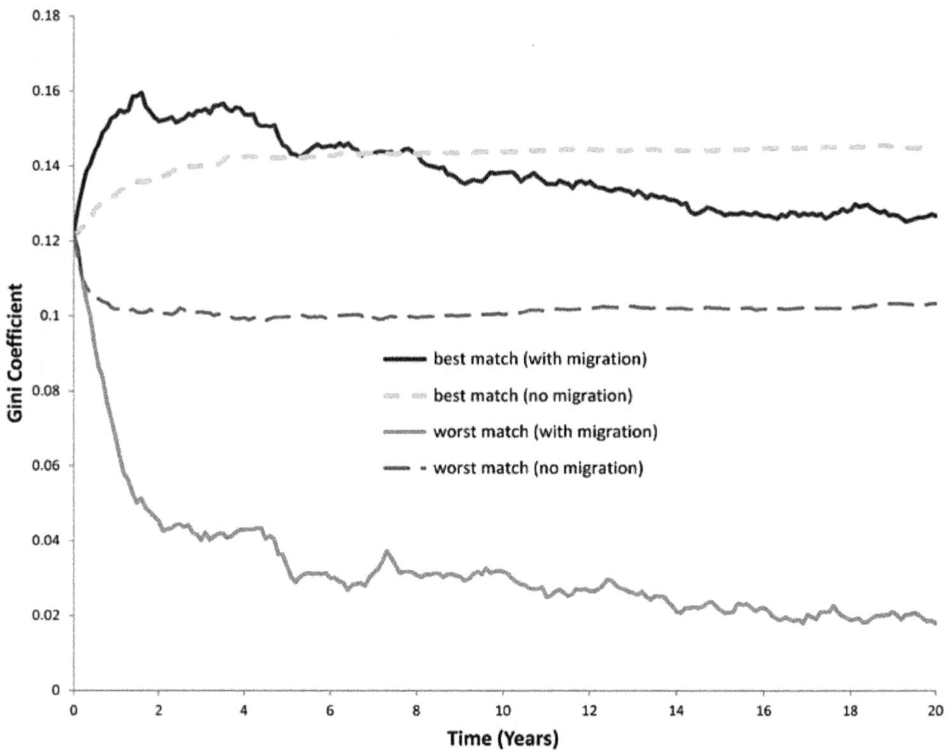

Figure 20.8 Block group inequality for best and worst cases with interurban migration effect (color figure available online)

the 2010 American Community Survey of the U.S. Census indicates that the Gini coefficient for income variation between households in Danville is 0.43, slightly less than the Gini coefficient for households nationwide and statewide, which is over 0.46 and has been trending toward increased inequality in recent decades.

The model's simulated return to a dynamic equilibrium near the starting level of block group inequality in Figure 20.7 reflects the impact of interurban migration. Under the base case assumption of 500 annual household arrivals and departures, random (and thus equalizing) effects were induced from the stochastic parcel assignments for immigrants and stochastic selection of emigrants. In simulations for which there were no arrivals to or departures from Danville, block group inequality for the filtered solutions increased to a new steady state more rapidly and with less noise than for the base case of interurban migration.

Inequality dynamics for the three clusters shown in Figure 20.7 are qualitatively similar, revealing an initial transient to a state of heightened inequality that eventually returns to a dynamic equilibrium closer to the initial condition. The transitional period illustrated in Figure 20.7 for the dynamic realization of inequality under Cluster 2 settings lasts approximately ten years, whereas the other two cluster settings produce a steadier trend of reduced inequality. In contrast to the cluster comparison, more dramatic differences in dynamic behavior are evident in Figure 20.8, which shows trajectories of inequality resulting from the parameter settings that generated the best and worst matches to observed data in the Monte Carlo optimization experiments (Table 20.2 lists the corresponding parameter settings).

The dynamic behavior of the best matched scenario in Figure 20.8 continues the qualitative pattern of the cluster dynamics in Figure 20.8, in which block group inequality increases at first but eventually returns to a level similar to the starting point. In contrast, the worst-matched scenario in Figure 20.8 shows a sharp drop in block group inequality during the first two years, followed by a more gradual decline over the remainder of the time horizon. In addition to contrasting best and worst cases, Figure 20.8 illustrates the dynamics associated with a scenario of no migration into or out of Danville. In contrast to the base case assumption of 500 households per year arriving to and departing from Danville, the absence of interurban migration creates a steadier transition to a higher equilibrium value for the Gini coefficient.

Simulated inequality dynamics reveal that increases in rates of both intraurban migration (as in the worst matched scenario) and interurban migration (to and from Danville) have the effect of lowering the inequality between block groups. As households leaving Danville break social ties, arriving households become integrated into the simulated social network. Transient model dynamics thus reflect the movement of households and adjustment of their social connections.

Limitations and Extensions

In developing the model, simplifying assumptions were made so that computation would be efficient during the iterative Monte Carlo optimization process. For example, the use of 764 census block centroids as a basis of distance between households enabled a much smaller distance matrix to be referenced during computation than the full set of 13,166 parcels. This simplification improved computational ease and eliminated the need for distance to be computed dynamically, but the resolution lost from using block centroids limited the significance of the distance effect on simulated migration error. Although the probability of local ties was a significant influence on migration outcomes, the use of block groups to bound local ties introduced limitations from variability of the geographic area occupied by different block groups.

Home ownership was the most accurately initialized attribute in the model, as it was linked to the parcel data used to deduce household migration patterns. Initialization of other household agent attributes (race, income, presence of children) from census data produced limitations of disaggregating from frequencies reported at the block or block group level. Although homeowner migration data were available at the parcel level, migration patterns were aggregated to the block group level in part to compensate for potential errors induced by model initialization of the other household attributes.

The difficulty of matching directional migration patterns is largely due to the sparseness of the observed homeowner migration matrix aggregated to the level of the block group. Just 11 percent (86 of 784) of the cells in the matrix were nonzero, containing the 139 total intraurban homeowner moves from 2001 to 2003. The sparseness of this matrix reflects the relatively few homeowner moves in the two years used to assess homeowner migration patterns. Inclusion of renter moves would provide a richer pattern to match, but such data were not available for this study.

Although the Monte Carlo parameter variation undertaken herein is in keeping with the pattern-oriented approach for exploring an underdetermined system, further experiments could be designed to vary, test, and assess the transition between renting and owning, move evaluation frequency for renters and owners, rates of Danville in- and out-migration (assumed to be in equilibrium for this study), size of the vacant parcel consideration set, and network size (a minimum number of connections was encoded at three times for this study, although the algorithms produce a range of household network sizes). Because the simulated inequality dynamics exhibit initial transients, extensions of this work would benefit from allowing a transition prior to the calibration period.

For the sake of parsimony, the model simulates social networks using a logit choice formulation that enables indirect parameter estimation but does not reflect the ways that social ties between households actually form and influence migration patterns. As with dynamics of marriage and divorce, the model excludes social institutions, yet affiliations with schools, churches, and places of work clearly facilitate connections between households. The model also omits institutional constraints such as discriminatory lending in the housing market that would exacerbate the inequalities simulated. Although the U.S. foreclosure crisis was beyond the scope of the model as designed, aspects of the crisis could be proxied by adjusting migration rates, household income, and odds of owner–renter transitions.

Although institutional constraints are not explicitly represented in the model, household agent choices are constrained significantly by the particular parameter settings governing the simulation. These reflect tendencies toward homophily in terms of the attributes simulated (proximity, race, income, presence of children). The nature and intensity of a homophily effect was tested by varying logit choice utility weights to create alternative model structures reflecting distinct social norms. Thus, although agents are assigned heterogeneous attributes and are free from institutional constraints, they are subject to universal social norms in the form of their choice functions. Because agent choices are also probabilistic, an important element of chance is embedded in the simulated household decisions.

Conclusion

This research develops an agent-based model for examining the dynamics of spatial and social influences on neighborhood and network choice. This study demonstrates the use of pattern-oriented modeling to explore simulated relative to observed migration patterns. Results show how a simulation model can be used to assess the importance of different socioeconomic factors in shaping migration patterns. Inclusion of choice at the scale of the household in a spatial dynamic simulation model is an important step in representing multidimensional complex systems (Agarwal et al. 2002). Drawing on patterns from qualitative and quantitative data sources, the model functions as a computational laboratory to test effects of alternative assumptions for household heuristics and social structures.

In the pattern-oriented calibration strategy, a Monte Carlo optimization algorithm was designed to test unique combinations of utility weights for each experiment, or batch of 25 simulation runs. The objective function minimized homeowner migration error as averaged for each experimental setting. With an average migration error of 219 moves in the best matched scenario (Table 20.2), an error threshold of 230 moves was used to filter 486 reasonable solutions from the 1,000 simulation experiments, indicating that the Monte Carlo algorithm produced many alternative parameter settings with error in the range of 220 to 230 moves. Effectively halving the parameter space precludes confidence about a singular extrapolation into the future, but the filtering process enabled a finer analysis of utility weights and inequality dynamics among relatively fit solutions.

Regression of simulated results relative to factors shaping household choice provides insight on hypothesized socioeconomic dynamics. Although increasing the influence of social networks on neighborhood choice exacerbated simulated migration error, the effect was not statistically significant in a linear regression model of the simulation results. The social network effect was significant in a nonlinear form of the regression, however. Both regression and cluster analyses reveal the role of local (intra-block-group) ties in matching observed migration patterns.

Social network factors of racial similarity, income difference, and probability of local ties had significant mitigating effects on the homeowner migration error in both forms of regression.

The finer block-level resolution of the household race assignment from the census data, along with the use of racial category in imputing other household attributes such as tenure (rent or own) status, presence of children, and income at the block group scale, might have contributed to the significance of race in the regression of simulation results against observed migration patterns. These results, along with the significance of income-based homophily, suggest that such social sorting can contribute to intraurban inequalities.

The model's emphasis on individual choice, exclusion of institutional constraints, and application of preference settings to all agents are consistent with microlevel, bottom-up explorations such as the classic Schelling (1971) segregation model and the suite of models developed by Epstein and Axtell (1996). These models demonstrate the potential for simple rules to explain complex patterns of aggregate behavior. Specifically, Schelling (1971) considered a situation in which location preferences are shaped by an adjustable parameter for tolerance of diversity, finding that a preference for homophily with just a minority of one's neighbors nevertheless resulted in segregation. Continuing this line of inquiry, the model described herein examines linkages between social network structure and household migration using an experimental design that adjusts parameters for the strength of network homophily and the relative role of social networks, neighborhood attractiveness, and affordability in neighborhood choice.

By explicitly linking social network structure with neighborhood choice and integrating GIS data into a dynamic model, this study contributes to a growing body of research that simulates social networks to illuminate a wide range of issues in geographic context, from health (e.g., Bian et al. 2012) to hurricane evacuation (e.g., Widener, Horner, and Metcalf 2012). Using both quantitative and qualitative patterns to guide model design, this research demonstrates how simulation models can be empirically grounded while used to explore hypothetical scenarios that deal with persistent social problems. The pattern-oriented modeling approach and agent-based framework employed for this work can be adapted to other urban areas facing issues associated with economic decline, such as community fragmentation, polarization, and disparity.

Acknowledgments

This chapter originally appeared in *Annals of the Association of American Geographers*, 104, 40–59 (2014). As a reflection of original research conducted for my dissertation (Metcalf 2007), this study benefited from the guidance of my advisor, Bruce Hannon, and committee members Mark Paich, Sara McLafferty, Mu Lan, and Jim Westervelt. Lyle Wallis assisted with technical aspects of model implementation. Peter Rogerson, Teri Metcalf, Harvey Palmer, and anonymous reviewers provided valuable input into the presentation and interpretation of the results in this article. This work was supported in part by National Science Foundation award HSD-0433165 (model development and analysis activities) and National Institutes of Health (NIDCR/OBSSR) awards 1R21DE021187–01 and 1R01DE023072–01 (synthesis and documentation activities).

Note

1. Code for the model developed in this study is provided at the following link: www.acsu.buffalo.edu/~smetcalf/resources/ModelCode.htm

References

Agarwal, C., G. M. Green, J. M. Grove, T. P. Evans, and C. M. Schweik. 2002. A review and assessment of land-use change models: Dynamics of space, time, and human choice. General Technical Report NE-297. Newton Square, PA: U.S. Department of Agriculture, Forest Service, Northeastern Research Station.

AnyLogic, Version 5.5.2006. St. Petersburg, Russia: XJ Technologies.

April, J., F. Glover, J. P. Kelly, and M. Laguna. 2004. The exploding domain of simulation optimization. *Newsletter of the INFORMS Computing Society* 24 (2): 1–14.

Batty, M. 2005. *Cities and Complexity: Understanding Cities with Cellular Automata, Agent-based Models, and Fractals.* Cambridge, MA: MIT Press.

Ben-Akiva, M., and S. R. Lerman. 1985. *Discrete Choice Analysis.* Cambridge, MA: MIT Press.

Bian, L., Y. Huang, L. Mao, E. Lim, G. Lee, Y. Yang, M. Cohen, and D. Wilson. 2012. Modeling individual vulnerability to communicable diseases: A framework and design. *Annals of the Association of American Geographers* 102 (5): 1016–25.

Conley, T. G., and G. Topa. 2003. Identification of local interaction models with imperfect location data. *Journal of Applied Econometrics* 18 (5): 605–18.

Crooks, A. 2010. Constructing and implementing an agent-based model of residential segregation through vector GIS. *International Journal of Geographical Information Science* 24 (5): 661–75.

Emch, M., E. D. Root, S. Giebultowicz, M. Ali, C. Perez-Heydrich, and M. Yunus. 2012. Integration of spatial and social network analysis in disease transmission studies. *Annals of the Association of American Geographers* 102 (5): 1004–15.

Epstein, J. M., and R. Axtell. 1996. *Growing Artificial Societies: Social Science from the Bottom Up.* Washington, DC: Brookings Institution Press.

Gimblett, H. R., Ed. 2002. Integrating geographic information systems and agent-based modeling techniques for simulating social and ecological processes. New York: Oxford University Press.

Gini, C. 1921. Measurement of inequality and incomes. *The Economic Journal* 31:124–26.

Grimm, V., and S. F. Railsback. 2005. *Individual-based Modeling: Lessons from Ecology.* Princeton, NJ: Princeton University Press.

Hoyt, H. 1939. The structure and growth of residential neighborhoods in American cities. Washington, DC: Federal Housing Administration.

Kwan, M.-P. 2007. Mobile communications, social networks, and urban travel: Hypertext as a new metaphor for conceptualizing spatial interaction. *The Professional Geographer* 59 (4): 434–46.

McFadden, D. 1991. Advances in computation, statistical methods, and testing of discrete choice models. *Marketing Letters* 2 (3): 215–29.

McPherson, M., L. Smith-Lovin, and J. M. Cook. 2001. Birds of a feather: Homophily in social networks. *Annual Review of Sociology* 27:415–44.

Metcalf, S. 2007. Simulating the social dynamics of spatial disparity through neighborhood network evolution. Ph.D. dissertation, University of Illinois at Urbana-Champaign, Urbana, IL.

Metcalf, S., and M. Paich. 2005. Spatial dynamics of social network evolution. In Proceedings of the 23rd International Conference of the System Dynamics Society, Eds. J. D. Sterman, N. P. Repenning, R. S. Langer, J. I. Rowe, and J. M. Yanni. Boston, MA: System Dynamics Society. Available at: www.systemdynamics.org/conferences/2005/proceed/papers/METCA487.pdf

Osgood, N. 2009. Lightening the performance burden of individual-based models through dimensional analysis and scale modeling. *System Dynamics Review* 25 (2): 101–34.

Paez, A., and D. M. Scott. 2007. Social influence on travel behavior: A simulation example of the decision to telecommute. *Environment and Planning A* 39 (3): 647–65.

Parker, D. C., S. M. Manson, M. A. Janssen, M. J. Hoffman, and P. Deadman. 2003. Multi-agent systems for the simulation of land-use and land-cover change: A review. *Annals of the Association of American Geographers* 93 (2): 314–37.

Peake, L. J. 1995. Toward a social geography of the city: Race and dimensions of urban poverty in women's lives. *Journal of Urban Affairs* 19 (3): 335–61.

Radil, S. M., C. Flint, and G. E. Tita. 2010. Spatializing social networks: Using social network analysis to investigate geographies of gang rivalry, territoriality, and violence in Los Angeles. *Annals of the Association of American Geographers* 100 (2): 307–26.

Rainie, L., and B. Wellman. 2012. *Networked: The New Social Operating System.* Cambridge, MA: MIT Press.

Resnick, M. 1994. *Turtles, Termites, and Traffic Jams: Explorations in Massively Parallel Microworlds.* Cambridge, MA: MIT Press.

Rowe, S., and J. Wolch. 1990. Social networks in time and space: Homeless women in Skid Row, Los Angeles. *Annals of the Association of American Geographers* 80 (2): 184–204.

Schelling, T. C.1971. Dynamic models of segregation. *Journal of Mathematical Sociology* 1 (2): 143–86.

——. 1978. *Micromotives and Macrobehavior.* New York: Norton. SPSS, Version 15.0 2006. Chicago: SPSS.

Train, K. 2003. *Discrete Choice Methods with Simulation*. New York: Cambridge University Press.

Van de Bunt, G., M. V. Duijin, and T. Snijders. 1999. Friendship networks through time: An actor-oriented dynamic statistical network model. *Computational and Mathematical Organization Theory* 5 (2): 167–92.

Waddell, P., A. Borning, M. Noth, N. Freier, M. Becke, and G. Ulfarsson. 2003. Microsimulation of urban development and location choices: Design and implementation of UrbanSim. *Networks and Spatial Economics* 3 (1): 43–67.

Westervelt, J. D., and L. D. Hopkins. 1999. Modeling mobile individuals in dynamic landscapes. *International Journal of Geographical Information Science* 13 (3): 191– 208.

White, H. C. 1970. Matching, vacancies, and mobility. *The Journal of Political Economy* 78 (1): 97–105.

Widener, M., M. Horner, and S. Metcalf. 2013. Simulating the effects of social networks on a population's hurricane evacuation behavior. *Journal of Geographical Systems* 15 (2): 193–209.

Wiegand, T., F. Jeltsch, I. Hanski, and V. Grimm. 2003. Using pattern-oriented modeling for revealing hidden information: A key for reconciling ecological theory and application. *Oikos* 100 (2): 209–415.

Yin, L. 2009. The dynamics of residential segregation in Buffalo: An agent-based simulation. *Urban Studies* 46 (13): 2749–70.

21

SIMULATING SPRAWL

Paul M. Torrens

Research Question: Where does urban sprawl come from, and how can we manage it?
System Science Method(s): Agent-based models
Things to Notice:

- Simulation of a real spatial environment using empirical data
- Incorporation of space and multiple movement patterns

Suburban sprawl, a relatively recent phenomenon, is among the most important urban policy issues facing contemporary cities. To date, a well-accepted rationale has not been settled on for explaining and managing the causes of sprawl. Our contention is that consideration of geography is essential—that geographical explanations offer much potential in informing the debate about sprawl. Similarly, spatial simulation could support sprawl-related research, offering what-if experimentation environments for exploring issues relating to the phenomenon. Sprawling cities may be considered as complex adaptive systems, and this warrants use of methodology that can accommodate the space-time dynamics of many interacting entities. Automata tools are well suited to representation of such systems, but could be better formulated to capture the uniquely geographical traits of phenomena such as sprawl. By means of illustrating this point, the development of a model for simulating the geographic dynamics of suburban sprawl is discussed. The model is formulated using geographic automata and is used to develop three sprawl simulations. The implications of those applications are discussed in the context of exploring geographic explanations of sprawl formation and the potential for managing sprawl by geographic means.

Urban systems are evolving and emerging in surprising ways. This is particularly true in the United States; its urban geography has essentially been redrawn over the past fifty years. The phenomenon of suburban sprawl is the poster-child for these kinds of transformations. Sprawl is a relatively new form of urbanization, falling somewhere between Ebenezer Howard's ideas for Garden Cities and Le Corbusier's notions of a ubiquitous urban form, yet it is altogether different—a "geography of nowhere," as authors have referred to it (Kunstler 1993).

Sprawl is among the most important topics in urban studies. Its implications are well understood, but the factors behind the phenomenon are less so. Several drivers are suggested in the literature but are not easily experimented with on the ground. Growth management policies are present in several U.S. cities, but their efficacy has yet to be determined. This chapter discusses construction of a computer model of suburban sprawl drivers used to test ideas about the geographical factors underlying sprawl formation. Automata-based tools are used, with an extension

DOI: 10.4324/9781315748771-24

to the conventional automata scheme, intended to represent geographic dynamics of agents of change that are responsible for sprawl. The resulting simulations are used as an artificial laboratory for exploring scenarios for urban growth.

There are several motivations underpinning the work that we present here. First, simulation is crucial to understanding sprawl and exploring alternative growth scenarios. Second, an automata approach has advantages in this regard, particularly in representing the agents of change that may be responsible for sprawl. Third, geography is essential to the phenomenon and must be incorporated directly into the methodology and model design.

The chapter is organized as follows. The topic of sprawl is introduced through discussion of its consequences and causes, as well as the need for a geographical perspective. Existing work relating to sprawl is described, with a focus on our contention that automata models are the most appropriate tools for simulating sprawl and should be the foundation for further research in this area. We present a conceptual model of sprawl based on sprawl drivers and agents of change, followed by description of the design of a computer model of sprawl. Several simulation experiments were run using the model, each exploring different aspects of sprawl formation from a geographical perspective. The empirical measurement of sprawl in those simulations is described, and implications of the experiments for understanding sprawl are discussed.

Sprawl Characteristics and Consequences

Sprawl is a new form of urbanization with characteristics that are distinct when compared to the urbanization that came before it or the urbanization that is developed under alternative (smart growth) regimes. A number of attributes are important in defining sprawl.

First, sprawl is a process of urbanization—urban growth by suburbanization. This process is quite rapid and is characteristic of the expansion of some of the fastest-growing cities in the United States (Table 21.1). The dynamics of sprawl often leave the phenomenon open to interpretation in the literature.

> The sprawl of the 1950s is frequently the greatly admired compact urban area of the early 1960s . . . The concept of time span is important in the identification and measurement of sprawl. The application of static measures to dynamic areas can easily result in the misidentification of an area as sprawl when it is really a viable, expanding, compacting portion of the city.
>
> *(R. O. Harvey and Clark 1965, 6)*

Second, sprawl manifests on the periphery of cities, often in previously nonurban areas on the metropolitan fringe. (In Europe, similar phenomena are referred to in the context of peri-urbanization.) Third, sprawl is commonly characterized as low-density in development (Peiser 1989; Ewing 1997; Gordon and Richardson 1997). Specifically, sprawl is considered to be lower in density than smart growth, urbanization in older cities, or development in central cities. Fourth, sprawl is a piecemeal form of development. The urban morphology of sprawl is scattered and fragmented in pattern—areas of sprawling suburbs in active use are often interspersed among tracts of land out of active use, or with little functional use (Lessinger 1962; Benfield, Raimi, and Chen 1999). Fifth, sprawl may be characterized by homogeneity of land use. Single-family uses lead the activity patterns of its residential landscape; commercial uses are more likely to be arranged as ribbon-sprawl (R. O. Harvey and Clark 1965) or retailscape (Gordon and Richardson 1997)—swaths of activity buffering highways and highway entry/exit ramps, with relatively little provision for nonautomobile access.

Table 21.1 The top ten fastest-growing cities in the United States

Rank	Metropolitan area name	Census population		Change 1990–2000	
		April 1990	April 2000	Number	%
1.	Las Vegas, NV–AZ	852,737	1,563,282	710,545	83.3%
2.	Naples, FL	152,099	251,377	99,278	65.3%
3.	Yuma, AZ	106,895	160,026	53,131	49.7%
4.	McAllen–Edinburg–Mission, TX	383,545	569,463	185,918	48.5%
5.	Austin–San Marcos, TX	846,227	1,249,763	403,536	47.7%
6.	Fayetteville–Springdale–Rogers, AR	210,908	311,121	100,213	47.5%
7.	Boise City, ID	295,851	432,345	136,494	46.1%
8.	Phoenix–Mesa, AZ	2,238,480	3,251,876	1,013,396	45.3%
9.	Laredo, TX	133,239	193,117	59,878	44.9%
10.	Provo–Orem, UT	263,590	368,536	104,946	39.8%

Source: Original data taken from the U.S. Census Bureau, *Census 2000 Redistricting Data* (PL. 94–171) Summary File and 1990 Census.

Sixth, sprawl has well-argued aesthetic characteristics. The urban form associated with suburban sprawl often garners criticism for the blandness of its design (Duany, Plater-Zyberk, and Speck 2000; Calthorpe, Fulton, and Fishman 2001; Duany, Speck and Plater-Zyberk 2001). Lessinger's (1962, 169) commentary in this regard is particularly illustrative of this: "Urban sprawl, roller-painted across the countryside, is often without form, grace, or a sense of community. Planning philosophies aimed to strike down this amorphous creature should only gladden our hearts." Seventh, sprawl exists under a relatively loose planning regime compared with that which operates in central urban areas or suburbs under growth management policy (Pendall 1999; Carruthers 2003).

Sprawl is understood to be problematic for several reasons. These include the direct costs of providing infrastructure and services over low-density areas on the urban periphery that often hold a minority of the city's total population. A series of indirect externalities are associated with sprawl: poor water and air quality, increased travel and accessibility costs, and unwelcome social justice costs (Real Estate Research Corporation 1974; Frank 1989; James Duncan & Associates et al. 1989; Environmental Protection Agency 1993, 2000; Downs 1994; Ewing 1994; American Farmland Trust 1995; Burchell et al. 1998; Benfield, Raimi, and Chen 1999; Johnson 2001). At the same time, sprawl satisfies residential demand (National Association of Home Builders 1999), and in some cases researchers have argued in favor of sprawl on the grounds that it provides relatively affordable housing (OTA 1995). Also, in areas like Los Angeles, the scattered and low-density nature of sprawl is useful in dispersing air pollutants (Bae and Richardson 1994).

Geography is essential to understand the factors that drive sprawl. Sprawl operates within the space-time dynamics of the city and behavior of its inhabitants. It is prevalent in some cities, but not others. Sprawl is present in distinct locations within a metropolitan area or systems of cities. It is also unforgiving in its consumption of space and may be characterized with distinctive spatial patterns and structure. Moreover, plans and policies to manage sprawl are overwhelmingly geographical in nature. European green belts introduce an absolute spatial constraint on the outward expansion of suburban growth, whereas much of the growth management policy in the United States attempts a similar goal by geographical means, dictating where development may take place and what uses land may be put to in specific locations, and introducing activity-place incentives and disincentives to influence urbanization.

Research regarding sprawl generally falls under the remit of urban planning, design, and public policy and reflects those perspectives. Much of the work relates to issues such as tabulation of the economic, social, and environmental costs of sprawl (Benfield, Raimi, and Chen 1999; Johnson 2001); case studies regarding the role of the planning regime in fostering sprawl or alternative growth regimes (Pendall 1999; Carruthers 2003); the urban design of sprawling suburbs and New Urbanist alternatives (Duany, Plater-Zyberk, and Speck 2000; Calthorpe, Fulton, and Fishman 2001; Duany, Speck, and Plater-Zyberk 2001); and identification of the most sprawling cities (Ewing, Pendall, and Chen 2002). Geographers have contributed to the debate (Gottmann and Harper 1967; Yeh and Li 1999; Herold and Clarke 2002; Hasse and Lathrop 2003a, 2003b; Herold, Liu, and Clarke 2003; Wilson et al. 2003; Hasse 2004). However, explanatory work examining geographical determinants is relatively less well developed when compared to research into other sprawl drivers.

Modeling Approaches to Sprawl

We regard simulation as essential to the study of sprawl. Our assertion is based on several motivations. Modeling and simulation may serve as generative science (Epstein 1999). We can gain understanding of the phenomenon of sprawl, and the factors that combine to produce it, by piecing elements of sprawling systems together in simulation, and studying the ways in which they interact to form system dynamics. Moreover, sprawl is not easily experimented with on the ground. It is infeasible to think that sections of the city could be reduced in density or set upon alternative growth regimes en masse without popular upheaval. Realistic but synthetic computer simulations can be built, however, as a laboratory for exploring ideas and plans that we would not otherwise be able to effect on the ground. Modeling can be used as a planning support system (PSS), to pose what-if questions and evaluate likely or alternative outcomes.

Simulation may also be used to examine future, unforeseen consequences of actions. The implications of urban policies and plans may take decades to manifest. However, in simulation, time can be accelerated or decelerated, into the past or the future, at will. Models may also be used as tools to think with. They can help to convey key properties of a problem or phenomenon to affected parties, stakeholders, policymakers, students, and other researchers. Moreover, this can be done in an interactive and visual context.

Models of urban growth abound, but exploration of sprawl is not generally the primary motivation for construction of those models. There are some exceptions, however, and a variety of models have been developed that touch on various characteristics of sprawl individually. Urban modeling in PSSs is quite relevant. Three such systems stand out in particular: the California Urban Futures models (Landis 1994, 1995, 2001; Landis and Zhang 1998a, 1998b), the What If? system (Klosterman 1999, 2001), and UrbanSim (Waddell 2000, 2001, 2002; Waddell et al. 2003). None of these PSSs are designed to simulate sprawl, although they might be employed for that task and UrbanSim comes particularly close in this regard.

Relatively recently, a series of automata models—either cellular automata (CA) or agent automata (agent-based models, agent models, multi-agent systems) in form—have been developed and applied in contexts of relevance to consideration of sprawl. These include cellular and CA models built around a development and/or land-use perspective, focusing on the conversion of land from nonurban to urban use. Early models were developed by Chapin and Weiss (1962, 1965, 1968), Tobler (1970, 1979), and Nakajima (1977). More recent models have been developed in a similar tradition and include the Dynamic Urban Evolution Model (Batty and Xie 1994, 1997; Xie 1996; Batty, Xie, and Sun 1999); the Research Institute for Knowledge Systems models (White and Engelen 1994, 1997, 2000; Engelen et al. 1995; White,

Engelen, and Uljee 1997; Power, Simms, and White 2000; Engelen, White, and Uljee 2002; Straatman, White, and Engelen 2004) and models built on the same scheme (Arai and Akiyama 2004); models developed by Yeh and Li (Li and Yeh 2000, 2002; Yeh and Li 2000, 2001, 2002), by Wu and Webster (Wu 1996, 1998b, 1999; Webster and Wu 1998; Webster, Wu, and Zhou 1998; Wu and Webster 1998, 2000), and by Semboloni (1997, 2000); the Queensland models by Ward, Murray, and Phinn (2000); and the SLEUTH model developed by Clarke and colleagues (Clarke, Hoppen, and Gaydos 1997; Clarke and Gaydos 1998; Silva and Clarke 2002; Goldstein, Candau, and Clarke 2004).

A handful of automata models deal with urbanization as a polycentric process. These models treat sprawl in terms of the formation of subcenters outside dominant urban cores. This includes work by Krugman and Fujita (for an overview, see Fujita, Krugman, and Venables 2001; Krugman 1996) and by Wu (1998a). Yeh and Li (2002) also developed models of polycentricity to explore compact growth. Other automata models deal with peripheral urbanization. Examples include models for South Australia (Bell, Dean, and Blake 1999) and Guangzhou, China (Wu 2002). Models have also been developed that consider the ecological effects of fringe sprawl in abstract cities (Brown et al. 2002; Rand et al. 2002). There are also several fringe urbanization models relating to land cover change as a result of urbanization (see Parker et al. 2003 for a review).

The existing foundation of modeling work, and the literature relating to sprawl causes, characteristics, and consequences suggest a likely conceptual model of sprawl formation on which we may build our work.

A Conceptual Model for Sprawl

Several suggestions have been offered to explain the causes of sprawl. These are multifaceted for the most part, and causes are generally understood to be tightly bound to characteristics and consequences of the phenomenon. Several of these causes are geographical in nature or have strong geographical implications. We can consider these causes generally; we might also consider sprawl from the perspective of the agents of change that are responsible for building and populating sprawling cities.

General Causes of Sprawl

At a broad level, sprawl can be considered as a mature stage in the evolution of a city toward a compact urban structure. Hall (1983), for example, discusses sprawl in the context of a city passing from a condition of primary industrialization to absolute centralization, relative centralization, relative decentralization, and absolute decentralization. Sprawl, he argues, is characteristic of the latter two stages.

Population growth is one of the most important engines of change in any urban system and this is also true of sprawl. The expansion of a city beyond its periphery requires, at a minimum, population growth and/or spatial redistribution of that growth. There are at least three ways in which population growth has contributed to sprawl: absolute growth, increasing urbanization, and restructuring in the dynamics of household demography. First, cities in North America—with only a handful of exceptions—are growing in terms of absolute population. Even the infamous Detroit metropolitan area, long observed as the preeminent example of the withering American Rust Belt, has been gaining population on aggregate. Second, at the same time, the percentage of the population living in what can be classified as urban areas is also growing. Of that urban population, the numbers residing in small cities is swelling at a striking rate. Third, and in parallel, there has been an associated decrease in household sizes and a related increase in the number of housing units.

If urban populations swell, the city must expand upward or outward, and sometimes beyond its previous boundaries, stretching into agricultural or resource land. This is not news. However, at the same time that urban populations have been growing in absolute terms, the distribution of that growth has been allocated in a spatially distinct manner, largely on the urban fringe as sprawl.

The downtown's pull on location has also been weakened by the growth of the highway system in the United States. No longer indebted to central cities as interchange points for raw material and finished goods, industry has diffused rapidly through the city to the suburbs, following its labor forces and pursuing cheap land and easy access to an expanding network of interstate highways. Suburban highways have become the new centers of gravity around which urbanization has begun to orbit. Coupled with these developments, there has been a dramatic growth in the use of the automobile and the dominance of its position in American society. Prolific use of automobiles facilitates dispersion of activities, making lower densities possible. This has been reinforced by a long-term trend of decline in gas prices in the United States (although recently the trend has been on an upward trajectory). The inflation-adjusted price of gasoline in the United States in 1996 was lower than that in 1974 (Gordon and Richardson 1997). This has allowed households to substitute housing for transportation costs by moving to the suburbs and living at lower sprawl-type densities. Rather than having a dampening effect on trip-making, suburban dwellers are shopping and recreating in record numbers.

Internet and communications technologies may well reinforce these trends. Although calls for the death of distance likely overestimate the degree to which this is the case (Cairncross 1995), there is general agreement that technological advances have greatly extended the effective radius of the city (Gordon and Richardson 1997). Disparate parts of the city may be separated spatially but linked functionally.

Agents of Change

Households. Why have urban populations been steadily redistributed toward the periphery? A simple and obvious explanation is that people want to live in these areas, whether or not planners and academics consider it to be sustainable. "[L]ike it or not, the great majority of mankind is praying for [sprawl] to come, to develop and satisfy them" (Gottmann 1967, 5). For all the criticism leveled against suburban living, it is still the preferred living arrangement for many; at least 80 percent of some survey groups prefer sprawl over other types of setting (Morrill 1991).

There are a number of likely motivating factors underlying these preferences. Some authors have accused outwardly mobile city dwellers of being racially and socially motivated in their decisions to move to the periphery. It has also been argued that suburban preferences are rooted in the long-standing tradition of ideals based on the exclusion of lower-income groups (Audirac, Shermyen, and Smith 1990), and it has been suggested that white households are moving even further out on the urban fringe and into exurbs (Galster 1991), although older studies had suggested otherwise (Farley et al. 1978). Public perception is another likely motivation, particularly regarding the inner city. Public sentiment is of worsening conditions in large cities in some cases, and there is evidence to suggest that opinion matches reality. Data from the U.S. Congress's Office of Technology Assessment (OTA 1995) show, for example, that crime rates have risen in the Baltimore area since 1985, but at faster rates in the inner city (+32.6 percent) than in the suburbs (+13.4 percent).

Employers. The movement and redistribution of population toward suburban locations may have a positive feedback influence on economic activity. Jobs are understood to follow population and in this sense population redistribution has a pull factor on urban economic activity. There are further draws to the suburbs for employers, including lower land and development costs compared to more central locations, and transport networks that facilitate lower costs of movement in outer suburban and exurban locations. This has been supported by shifts in the U.S. economy toward service industry, which is more mobile than other industries.

Developers. Developers have been blamed for encouraging scattered development in expanding suburban areas of North America. For the most part, in growing cities developers act independently in their development decisions (R. O. Harvey and Clark 1965), which promotes a discontinuity in the spatial pattern of their developments. It encourages speculation, the withholding of land for development, which means that large areas of land in the suburbs may become priced out of any market save urban use (Clawson 1962). Pendall (1999) has argued that fragmentation in the ownership of agricultural land exacerbates this problem.

Planners and Policymakers. Planners and policymakers might be considered as agents of change in sprawling systems. For the most part, planning and public policy act to control sprawl through zoning constraints, development caps, historic preservation orders, or growth management legislation such as green belts, transit-oriented development, and developer impact fees (Downs 1994; National Association of Home Builders 1999; Duany, Plater-Zyberk, and Speck 2000; Calthorpe, Fulton, and Fishman 2001; Duany, Speck, and Plater-Zyberk 2001). However, there is a general concern that planning and policy can also act to encourage sprawl, directly and indirectly.

Audirac, Shermyen, and Smith (1990) argue that the agency of planning practice in the United States can be connected to sprawl. Barnett (1995) makes a similar argument, that outdated planning regulations are responsible in large part for sprawl. Commercial strips, a design from the 1920s, were painted over the landscape with vigor in the 1950s; "apparently no one stopped to contemplate the effect of mapping commercial land exclusively in narrow strips along highways where the only means of access was the automobile" (Barnett 1995, 47). Similarly, lot-by-lot zoning and subdivision was not intended to become the only development control over large sections of the city (Barnett 1995). The geography of land-use controls exercised by planners may also be to blame. When applied spatially with varying degrees of enforcement, land-use controls can create an imbalance in the attractiveness of competing areas. If there is a discrepancy between controls inside and outside a city's boundary, for example, land-use planning may make the less-controlled area—the urban fringe—more attractive (R. O. Harvey and Clark 1965; Pendall 1999).

Bahl (1968) makes the claim that tax policy that fosters speculation in the sale of land is a factor in promoting sprawl. Others point to tax policies that essentially subsidize the costs of home-ownership over renting, with a bias toward new homes and single-family housing: "It is generally agreed that in the past the public sector encouraged low-density suburbanization through tax deductions, mortgage guarantees, and depreciation formulas favoring new construction over the upgrading and repair of existing structures" (OTA 1995, 200). Peterson has made a similar argument: "The new, low-density construction favored by tax laws is obviously most suitable for location outside the central metropolitan core" (Peterson 1980, 48–49). The federal tax code also emphasizes the creation of subdivisions in small and discontinuous increments. Land is commonly sold to developers in installments so as to minimize capital gains on income tax returns. In addition, subdividers and developers may limit their projects for any taxable year so as not to slip into higher tax brackets that might incur increased taxation of their profits (R. O. Harvey and Clark 1965).

There are more direct examples of public policy as an agent of change in the fostering of sprawl. State incentives—free land grants, subsidized training, tax breaks, tax-exempt industrial development bonds, low-interest loans—may be biased against central cities. There are well-documented examples where public policy has intervened in land markets with the intent of suburbanizing large employers; the relocation of Sears to Hoffman Estates in Illinois garnered $100 million in subsidies for the company (OTA 1995).

The Geography of Sprawl

The factors that drive sprawl are relatively difficult to isolate, simply because so much contributes to the phenomenon. Nevertheless, taken together, the characteristics, causes, and consequences of sprawl that have been discussed in the literature suggest a conceptual model of the phenomenon that we can use as a foundation for model-building. Not all of these factors can be simulated tractably; several do not lend themselves to empirical measurement or representation and likely lend themselves to other forms of analysis (D. Harvey 1969). Nonetheless, we can make use of several others in simulation.

First, it is important that sprawl be represented in space and time as a dynamic phenomenon. American sprawl is voracious in its appetite for land. Moreover, sprawled areas of the city may develop into relatively sustainable urban areas with time, as larger single-lot land parcels become subdivided and developed at higher densities, and previously fragmented areas are subject to in-fill.

Second, geography is essential to considering the phenomenon and describing its space-time dynamics. Geographical inertia is important; the future development of a city is a function of its history. The spatial pattern of sprawl as peripheral, low density, scattered, transport-adjacent development and settlement is also crucial to understanding its impacts. Similarly, the mechanisms of sprawl are geographical in nature: fringe urbanization, decentralization, and leap-frog development and settlement.

Third, growth is important. Sprawling cities exist under regimes of absolute population growth, by in-migration or through endogenous dynamics, or of relative growth and redistribution of population to the urban fringe.

Fourth, we may identify several geographical agents of change that might be considered as driving sprawl, and as mechanisms by which growth is allocated and distributed, spatially and temporally, over an urban area. These include developers, responsible for manufacturing the urban physical environment, and relocating households that populate that environment and drive its geography through demand. Employers are also important, as are planners and policymakers.

A Computer Model for Simulating Sprawl

Our sprawl model is based on the conceptual model offered in the preceding section. The model includes exogenously and endogenously considered growth, which is distributed over a simulated landscape using mechanisms designed to represent geographic drivers of sprawl: geographical inertia, diffusion, and mobile agents of change. The methodology is based around an automata core, extended as geographic automata (GA). The modeling scheme is illustrated in Figure 21.1; details of the model are discussed in the following subsections.

Geographic Automata

A basic automaton (A) (a Turing machine, finite state machine, central processing unit) is generally characterized with state variables (S) that describe its condition at a finite moment in time (t), and state transition rules (R_s) that govern how those state variables change in time, based on current state information $S(t)$ and current information input $(I(t))$ from an external source or from other automata:

$$A \sim (S, R_S, I);\ R_S : (I(t), S(t)) \rightarrow S(t+1) \qquad (1)$$

CA are a class of basic automata, defined within the discrete confines of a cellular boundary. When many CA are considered together, they may be understood to form a lattice-like configuration, with each discrete automaton representing an individual unit in the lattice. State information exchange between automata is considered within the context of neighborhoods

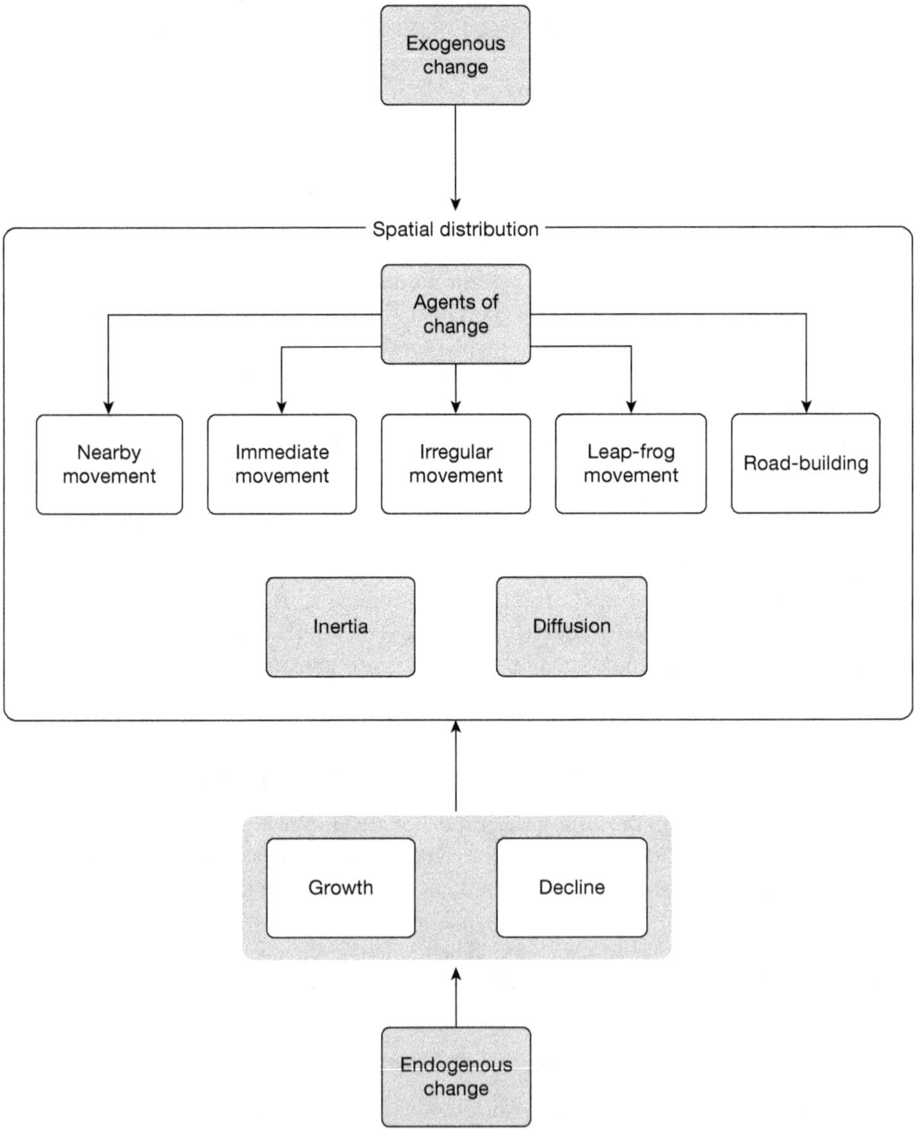

Figure 21.1 Schematic overview of the model engine

(N)—localized areas of the lattice, composed of several CA neighboring a target automaton. The only source of external information for a single automaton A in the CA is the set of its neighbors; that is, $I = I_N$, and for each automaton A in the CA:

$$A \sim (S, R_S, I_N);\ R_S : (I_N(t), S(t)) \rightarrow S(t+1) \tag{2}$$

Agent automata constitute another class of automaton, with origins in Artificial Intelligence and many varying specifications (*see* Ferber 1999; Russell and Norvig 1995; Benenson and Torrens 2004a). States are generally interpreted with respect to characteristics of agency: proactivity, goals, intent, and so forth. In addition, agents are often endowed with the ability to roam freely, as automata, within a lattice of other automata or another environment; this is characteristic of agents used in animat research (Meyer and Guillot 1994) and animation (Reynolds 1987).

The GA methodology that we propose begins with a basic automaton skeleton and adds components from CA (neighborhoods) and animat agent automata (movement). Additional spatial functionality is added: GA are located by means of geo-referencing conventions L and are endowed with the ability to move through the spaces in which they reside, by locomotion or other movement regimes (R_L). Geo-referencing conventions (L) allow GA to be registered, spatially, to environments in which they reside in time—that is, $L = L(t)$. This may be performed by direct means. GA may be located based on their actual position in the environments, allowing them to be registered to a Cartesian space, a network space, and so forth, at a finite moment in time. Indirect conventions may also be employed, relating GA's locations relative to other objects in the space, or tracking their progression through space and time. The introduction of a typology or ontology (K) of GA entities mediates the nature of L and R_L. At a basic level, K is defined with respect to fixture in space. Information input streams from the general automaton approach are considered geographically as neighborhoods of interaction and influence (we deal with neighborhoods N instead of input I). Additional neighborhood functionality is added; neighborhoods of different automata may differ in extent and shape and may change in time so that $N = N(t)$. A set of neighborhood rules is also introduced (R_N), and these rules determine how neighborhoods should change over time. Neighborhoods in the CA scheme are fixed and static. The introduction of neighborhood rules allows for a more flexible treatment of relationships between automata; neighborhood relationships, expressed as geometric areas, network links, far-from-local pointers, and so forth, can be introduced and allowed to vary over space and time.

Many GA may be combined in a *systems* context, formulated as follows:

$$\begin{aligned} &\mathrm{GA} \sim (K, S, R_S, L, R_L, N, R_N) \\ &R_s : S(t) \rightarrow S(t+1) \\ &R_L : L(t) \rightarrow L(t+1) \\ &R_N : N(t) \rightarrow N(t+1), \end{aligned} \tag{3}$$

where GA refers to a collective of individual geographic automata G. The rules are applied to each G from the GA collective. The methodology, and its connections with GIS and GIScience, are explained more fully in Torrens and Benenson (2005).

Geographic automata offer several advantages over existing automata tools commonly used in geographic simulation. The methodology is based on use for exploring geographical phenomena; theories about such phenomena dictate the components of the methodology, rather than having the tool constrain the theory that it may support. The methodology is actually capable of supporting all CA and multiagent urban models that we are aware of (Torrens and Benenson 2005). GA are also consistent with complex adaptive *systems*; they support the emergence of novel spatial ensembles. They may be designed with a relatively seamless interface to raster, and more importantly, to vector-GIS. Also, it is possible to form a symbiotic relationship between GA and object-oriented programming paradigms, object-oriented database management systems, and entity-relationship models (Benenson and Torrens 2004b; Torrens and Benenson 2005). GA offer much potential in modeling sprawl, particularly in capturing its geographical components.

External Change

Sprawl is a dynamic reaction to urban growth. This growth may come in many forms: growth that migrates to the city system from outside, as well as growth endogenous to the system. The relationship between external change and urban growth may be handled in modeling through the use of some form of allocation or spatial assignment mechanism. Commonly, these are formed as Markov, raster, or CA models. Markov models allocate change (e.g., land-use transition) over space in a city, based on existing land use in a previous time-step. Raster models determine allocation based on a vector of multivariate influences. CA models add consideration of neighboring states, on a proximity basis, to these general schemes (Tobler 1970).

The source of external change may be derived from a variety of sources. Change can be a parameter of the model, to be defined by the user (Xie 1996). It may also be extrapolated from historic land-use maps or remotely sensed data (Clarke and Gaydos 1998; Herold and Clarke 2002). In other cases, change is derived from loose- or tight-coupling of an allocation model to exogenous demographic or cohort-survival models (White and Engelen 1997).

External change is accommodated in our models at a macrolevel. It enters the model as population growth or decline (which may be expressed as a rate or absolute volume), which is distributed spatially thereafter at a more microscale. Change is a parameter to be defined by the user in our abstract simulations; it is derived from historical Census Bureau data in our real-world applications. Rates are normalized in our model, such that the rate of change of a major or central city is proportional to change in smaller cities in the simulation. These rates may be positive or negative, with the possibility of decline.

External change may be introduced to a simulated city-system as a volume D(t) of growth or decline at a given time-step t (positive or negative vales of $D(t)$ respectively), such that the population P_i of a land unit i at a subsequent time-step $t + 1$ is derived using the following equation:

$$P_i(t + 1) = P_i(t) + D_i(t), \tag{4}$$

where $D_i(t)$ is a part of $D(t)$ assigned to land unit i, $\sum_i^n D_i(t) = D(t)$.

Rates of growth or decline (above unit if growth, below unit if decline) may also be used to introduce change in a land unit's population:

$$P_i(t + 1) = P_i(t) \cdot \lambda_i(t). \tag{5}$$

External change may also be introduced on a *per-city* basis, rather than system-wide. This allows specification of differential growth and decline within the city-system. Once again, this may be specified as a volume of change or as a rate of change per city. These per-city growth/decline rates may be scaled relative to each other such that a balance is maintained between the fastest- and slowest-growing cities in the city-system.

Once the volume of externally derived change has been determined in absolute or rate terms, that growth or decline is distributed, in space-time, over the simulated city-system using dedicated GA designed to function as agents of change responsible for sprawl. Details of these automata are discussed in a later subsection, "Mobilizing Agents of Change."

Geographical Inertia

The model also includes historical, autoregressive functionality to represent geographical inertia in urban dynamics. Simulated urbanization proceeds based on development established in previous time-steps of a simulation run. Agents of change in the model thus observe a reality as established by the previous iteration (generation). If a land unit enjoys consistent development, urbanization has an opportunity to take hold over time and establish a spatial presence.

Endogenous Change

The model also incorporates functionality for representing growth or decline that originates within the system. We make a distinction between urban population and urbanizing population. The former are dormant agents of change that are counted toward the population density of a land unit; the latter are active agents of change, mobilized as GA or diffusing population.

Individual land units are endowed with the ability to decrement and generate population endogenously. We follow Sanders and colleagues (Sanders et al. 1997) in this way, affording a level of agency to the state values associated with automata units in the model. Land units in the model are automata at that level of geography, with a population state descriptor; but we also consider the automaton to contain a limited microworld within the geography of the land unit, which is composed of newly birthed population and newly declining population. This allows us to establish a spatial ecology within the land unit, as the basis for endogenous change to be communicated between automata units at a higher level of geography. A similar mechanism is employed in the urban growth models developed by Batty and Xie (1994). The mechanism provides means for incorporating birth and death dynamics in simulation, synonymous with demographic dynamics at an intra-urban scale within a city.

User-defined endogenous birth and death rates are set at the level of land units i, across land units belonging to a given city, or across the entire system. Formulation of these rates mimics that of Equation (5), but we consider endogenously derived change $\delta_2(t)$ in terms of immigration (im) of population to the land unit, emigration (em), the birth rate (b), and the death rate (d):

$$P_i(t + 1) = P_i(t)\delta_i(t), \text{ where } \delta_i(t) = im(t) - em(t) + b(t) - d(t). \qquad (6)$$

Alternatively, this growth or decline may be mobilized beyond the land unit, into the surrounding neighborhood N_i of a land unit i by diffusion or by GA that are beyond the neighborhood (action-at-a-distance). This may be determined by a user-controlled set of threshold capacities. If the population exceeds a land unit's maximum population capacity, then the excess is mobilized into the neighborhood (or farther if action-at-a-distance is employed).

Diffusion

A diffusion mechanism is used to represent very local neighborhood change—either the diffusion of urbanization or urban decline. The inclusion of diffusion also serves to introduce a decentralization mechanism in the model, which is important in representing sprawl.

Let us consider a land unit i within a neighborhood N_i. From the perspective of a neighboring land unit $j \in N_i$, the population change at j is a balance between the inflow of population by diffusion from land unit i and the diffusion outflow from j itself, and is formulated as

$$P_j(t+1) = P_j(t) + \frac{P_i(t)}{A_i} - \frac{P_j(t)}{A_j}, \qquad (7)$$

where A_i denotes the number of i's neighbors (e.g., $A_i = 5$ for a von Neumann neighborhood).

Mobilizing Agents of Change

GA in the model operate as urbanizing agents of change under several movement regimes designed to mimic development and settlement patterns known to be important to consideration of sprawl formation on the ground. A volume of GA are activated under the movement rule and released in the immediate neighborhood of a land unit. These automata proceed through their surroundings with a heading and defined length of movement, carrying population growth or decline with them and thereby distributing that growth or decay spatially over the simulated landscape by means of action-at-a-distance. The movement rule contains several parameters, which can be adjusted to make the resulting patterns of development and settlement more or less compact in form, or may be used to mimic sprawl-like patterns. Moreover, the resulting nodes of development or decline can be endowed with a greater or lower propensity for survival in subsequent simulation steps—new edge cities or urban blight can take root or not.

Generally, the movement rule R_L: $L(t-1) \rightarrow L(t)$, as applied to the geographic automaton G located at $L(t-1)$ at time t–1, can be formulated as follows:

$$R_L : L(t-1) = L_0^{t-1} \rightarrow L_1^{t-1} \rightarrow L_2^{t-1} \rightarrow L_3^{t-1} \rightarrow \ldots \rightarrow L_m^{t-1} = L(t), \qquad (8)$$

where L^{t-1} = < $L_0^{t-1}, L_1^{t-1}, L_2^{t-1}, L_3^{t-1}\ L_0^{t-1}, \ldots, L_m^{t-1}L(t)$, is the series of locations that G passes through during movement, from initial to final position.

In the case of endogenous change, the geographic automaton is mobilized with a volume of endogenously derived growth or decline $\delta_i(t)$. If the GA are distributing externally derived change, they are mobilized with a volume of externally derived change, either growth or decline $D_i(t)$. Positive values of D yield a likelihood that the resulting pattern of development and settlement will expand through subsequent time-steps; negative values of D have a shrinking effect.

The trajectory L^t of G's movement (i.e., the rule R_L that determines the series of locations L_i^t) could be designed to take place within the neighborhood N_i of a land unit i. Varying the nature of L_i during the interval $(t, t + 1)$ allows the introduction of different levels of action-at-a-distance beyond the neighborhood, and, further, accommodates different levels of action-at-a-distance when L_t takes G beyond the neighborhood. In doing so, well-known sprawl patterns can be derived in simulation, with related volumes of growth or decline.

Movement takes place during $(t, t + 1)$, and the number of locations that are visited by the geographic automaton can vary. For example, geographic automaton G may move, preserving the trajectory with length l accumulated during a given time unit. We can think of movement over space-time, from the start of the process until t, as being composed of t subtrajectories during the entire period $(0, t)$ with varying rules. An overall volume of growth or decline (D or δ) is distributed spatially over the locations traversed by G, with proportions of that growth or decline being deposited in land unit automata as packets of change within that simulation step. As G moves, it distributes growth or decline over a larger or smaller area, depending on the distance G covers during its movement.

Immediate movement. The immediate movement rule mimics initial development processes, whereby a site is settled very locally. GA move in a very confined range of an origin cell under this rule. If we consider GA defined on a two-dimensional grid, then the neighborhood of the GA located at *ij* is formulated as

$$N_{ij}(t) = (i+1, j), (i-1, j), (i, j+1), (i, j-1), (i+1, j-1), (i+1, j+1), (i-1, j-1), (i-1, j+1) \quad (9)$$

(see Figure 21.2). Generally, the movement rule confined to such a neighborhood results in very compact pattern of growth or decline in a confined radius around a target site.

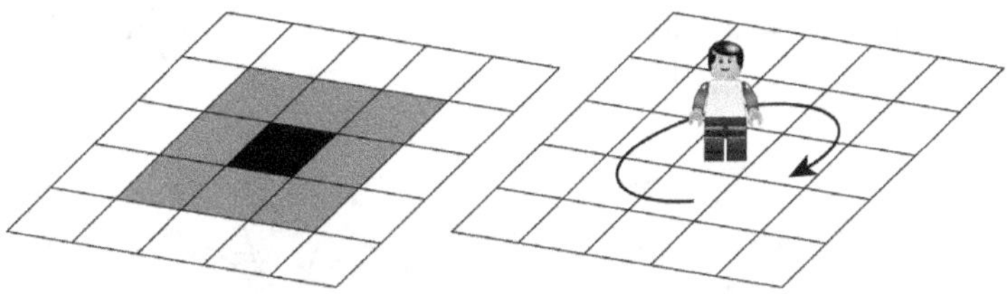

Figure 21.2 Immediate movement: black cells are origin automata; gray cells are affected units in the movement space around that origin; arrow denotes the path an automaton takes

Nearby movement. The nearby movement rule is similar in specification, except the neighborhood window for movement is much larger in size (25-automata) and the movement of G takes place within $N(t) = ij, i \pm 2, j \pm 2$ over every *t*. Generally, the rule yields dusters of growth equivalent to New Urbanist (Calthorpe, Fulton, and Fishman 2001) or transit-oriented village (Cervero 1998) types of patterns, or patches of decline synonymous with urban blight (Knox 1989; Figure 21.3).

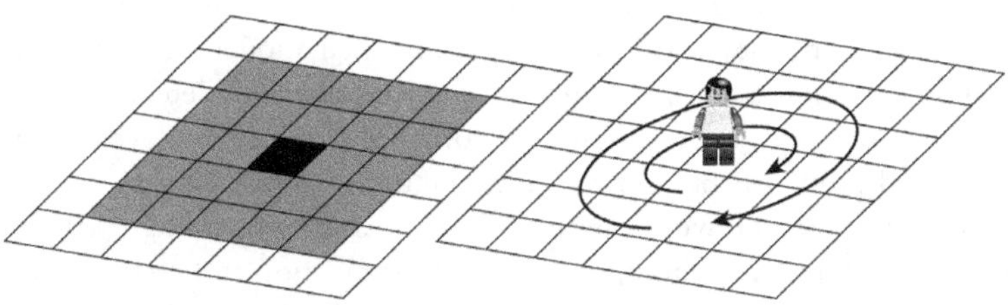

Figure 21.3 Nearby movement: black cells are origin automata; gray cells are affected units in the movement space around that origin; arrows denote the path an automaton takes

Irregular movement. The irregular movement rule is used to mimic the irregular patterns of urbanization associated with natural barriers such as mountains, rivers, wetlands, etc., or administrative confines. Under the irregular movement regime, the locations of geographic automaton *G* at $t + 1$ is randomly assigned within a user-defined range around $L(t)$. This results in varying sinuosity of *G*'s trajectory, the rule of choice of consecutive positions $L(t)$ of each move, and the range of perturbation for those values (Figure 21.4).

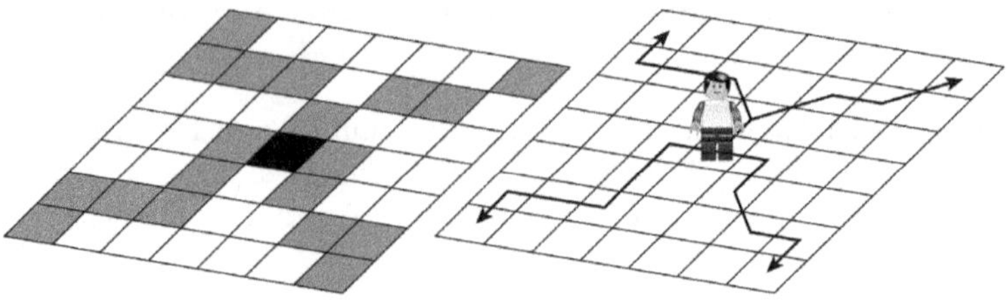

Figure 21.4 Irregular movement: black cells are origin automata; gray cells are affected units in the movement space around that origin; arrows denote the path an automaton takes

Leap-frog movement. GA may also move by leapfrogging. Under this rule, G moves in hops and L^t_{i+1} not necessarily confined to the nearest neighborhood of L^t_i. Growth or decline is deposited in land units at the termination of each hop. This mimics the land speculation leap-frog development patterns associated with sprawl that are the subject of much discussion in the literature (Lessinger 1962; Figure 21.5).

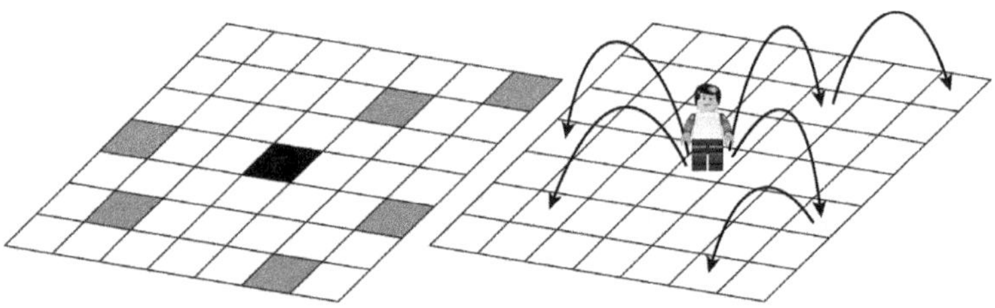

Figure 21.5 Road-like movement: black cells are origin automata; gray cells are affected units in the movement space around that origin; arrows denote the path an automaton takes

Road-like movement. The road-like movement rule is used to mimic road-building. Previous automata-based models of urban growth have introduced road development as an accretive process—roads grow, sequentially, by diffusion-limited aggregation (Xie 1996). There is some debate about growing roads in urban models (see Ward, Murray, and Phinn 2000 for a discussion of the problems they had growing roads in their models); an alternative approach might be to construct roads as links, but only open them once completed. In this model, roads are developed first as nodes, then those nodes are connected by strips of development, indicative of transport-oriented growth flanking road infrastructure. GA move by means of the leapfrog or irregular movement rules. However, instead of depositing population growth or decline, GA lay down nodes as they progress during $(t, t + 1)$. At simulation time step $t + 1$, those nodes are connected with a strip of population that is determined based on the growth at origin i of the geographic automaton's journey and may be perturbed by a parameter to be defined by the user. This results in ribbon patterns of growth radiating from land-units (Figure 21.6). Upon being laid down, those ribbons may continue to urbanize.

In addition, rules may be combined—a geographic automaton can exercise rules in isolation or can execute a sequence of rules within t before terminating its movement. For example, after moving by leapfrog, a geographic automaton might initiate either an immediate or nearby movement. Depending on which rule followed the leap-frog, the resulting pattern would be a sprinkling of isolated settlements or more polycentric forms consisting of adjacent clusters that may fill in through diffusion.

Figure 21.6 Leap-frog movement: black cells are origin automata; gray cells are affected units in the movement space around that origin; arrows denote the path an automaton takes

Constraints

A variety of constraints are introduced to the model to confine simulation runs within specified bounds and this facilitates the introduction of what-if scenarios in simulation.

Land units within the simulation may be coded as either "developable" or "non-developable," allowing for certain areas of the simulation to be withheld from transition. This follows the introduction of fixed and functional cells in the CA models developed by Engelen, White, and Uljee. Moreover, the specification of gateway automata introduces a spatial constraint, binding state transition to certain seed sites in the simulation.

A hierarchy of land-use transition is also imposed, ensuring realistic transition of land units between uses. This follows similar hierarchies in urban growth CA (Semboloni 1997; White and Engelen 1997). Developable areas may become urbanized, with a population. They may only return to nonurbanized form if the population count decrements to zero.

Simulation Experiments

The purpose of this work is to explore the geography of sprawl through simulation. Using the model, simulations were built based on two scenarios for sprawling urbanization within an abstract city-system. Both simulations evolve a city-system in a realistic fashion, with emphasis on the processes driving space-time dynamics, the patterns generated by the simulation, and the rate of simulated urbanization. A third simulation is also described, as applied to a real city-system (the Midwestern megalopolis of the United States). In these simulations, it is assumed that the rate of growth is known a priori. (In the Midwestern simulation, growth rates are based on population data from the United States Census.)

General Growth Scenario

In the first example, the model is used to build a simulation in which a dominant central city evolves in the context of a larger city-system with two additional, competing, urban centers (Figure 21.7).

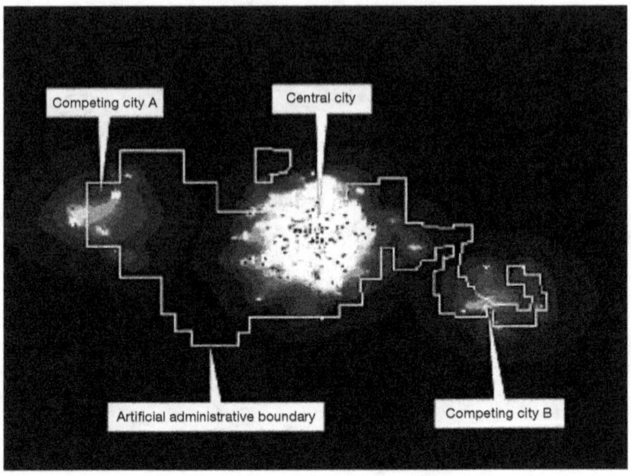

Figure 21.7 The central city and its two competing neighbors (t = 213): light gray/white areas denote densities that are higher than dark gray/black areas

The simulation is programmed with initial seed conditions that introduce gateway sites in five locations: the center of the lattice, two sites in almost immediate proximity, and two other gateways on the right and left areas of the lattice space (Figure 21.7). The ability for the growing cities to compete for space as they sprawl is specified in two ways. First, the central city is afforded an advantage from the start of the simulation by virtue of the introduction of two adjacent gateway sites; as hinterlands of the adjacent cities merge with that of the central core, they add population to the central urban mass. Second, the growth rates of the cities are treated differently, thereby influencing the temporal evolution of the urban system as well as its spatial development. The supply of growth to competing cities (denoted as A and B in Figure 21.7) is cut off roughly 75 percent of the way through a simulation run, mimicking conditions whereby the critical mass of a dominant central city begins to draw incoming migration and activity away from cities with comparatively less attraction. In the simulation, this occurs when the hinterlands (suburbs) of the competing cities meet those of the central city. At this point, exogenously derived growth (in-migration) ceases in the peripheral cities and only endogenous growth continues in those areas.

The patterns of growth generated in the simulation are synonymous with those that would be expected in a real city-system, both visually and empirically (empirical measurements of this fit are described in the following section). In addition, the timing of evolution of the system is sensible (Figures 21.8 and 21.9).

The three cities begin their early evolution as compact cities: dense monocentric masses with a surrounding lower-density suburban hinterland. As the density of settlement in the centers grows, the expanse of the suburban hinterland extends further in the simulated space (Figure 21.10B), and at an increasingly rapid rate (the diffusion rule actively disperses a greater volume of settlement as the mass of settlement in the system grows).

At t = 186 (roughly 50 percent of the way through the simulation run), the effect of the leapfrog rule becomes more pronounced; the urban mass has grown, spawning a greater number of subcenters on the periphery of the cities (Figure 21.10B).

By t = 200, the peripheral cities have become largely dispersed, with the remnants of formerly-dominant central seed areas barely visible (Figure 21.9). At t = 222, the hinterlands of the central city and competing city B have sprawled to such an extent that the two urban *masses* begin to merge (Figure 21.10C). At this point, the supply of growth to competing cities A and B is

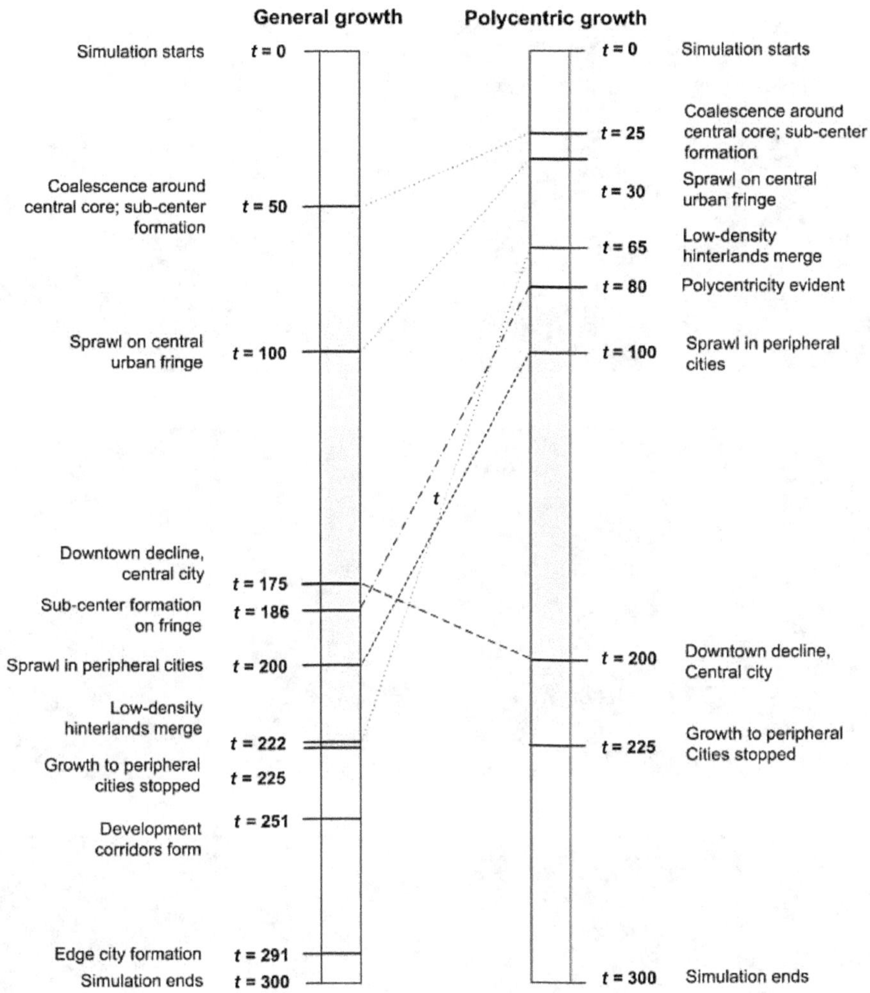

Figure 21.8 Timeline for the simulations

stopped. The downtown areas of the competing cities rapidly begin to decline in density, as growth is distributed through the system without a replenishing supply to the gateways of competing cities A and B.

Around this time, the road rule also begins to generate visible patterns—"fingers" of dense development begin to appear (Figure 21.10D), manifesting as corridors of growth extending from the main urban mass (Figure 21.10E). Competing cities A and B actually begin to develop a linear-like development pattern, succumbing to path-dependence because of initial road-like development. By t = 291, some suburban subcenters have begun to evolve as growth centers in their own right, and the overall structure of the central city becomes largely irregular, with pockets of lower-density settlement that have been by-passed by the urbanization process evident within the evolving city mass (Figure 21.10F).

Overall, the city-system sprawls dramatically, while maintaining a realistic pattern of regional-scale urbanization. It is particularly noteworthy that the spatial extent of the entire city-system

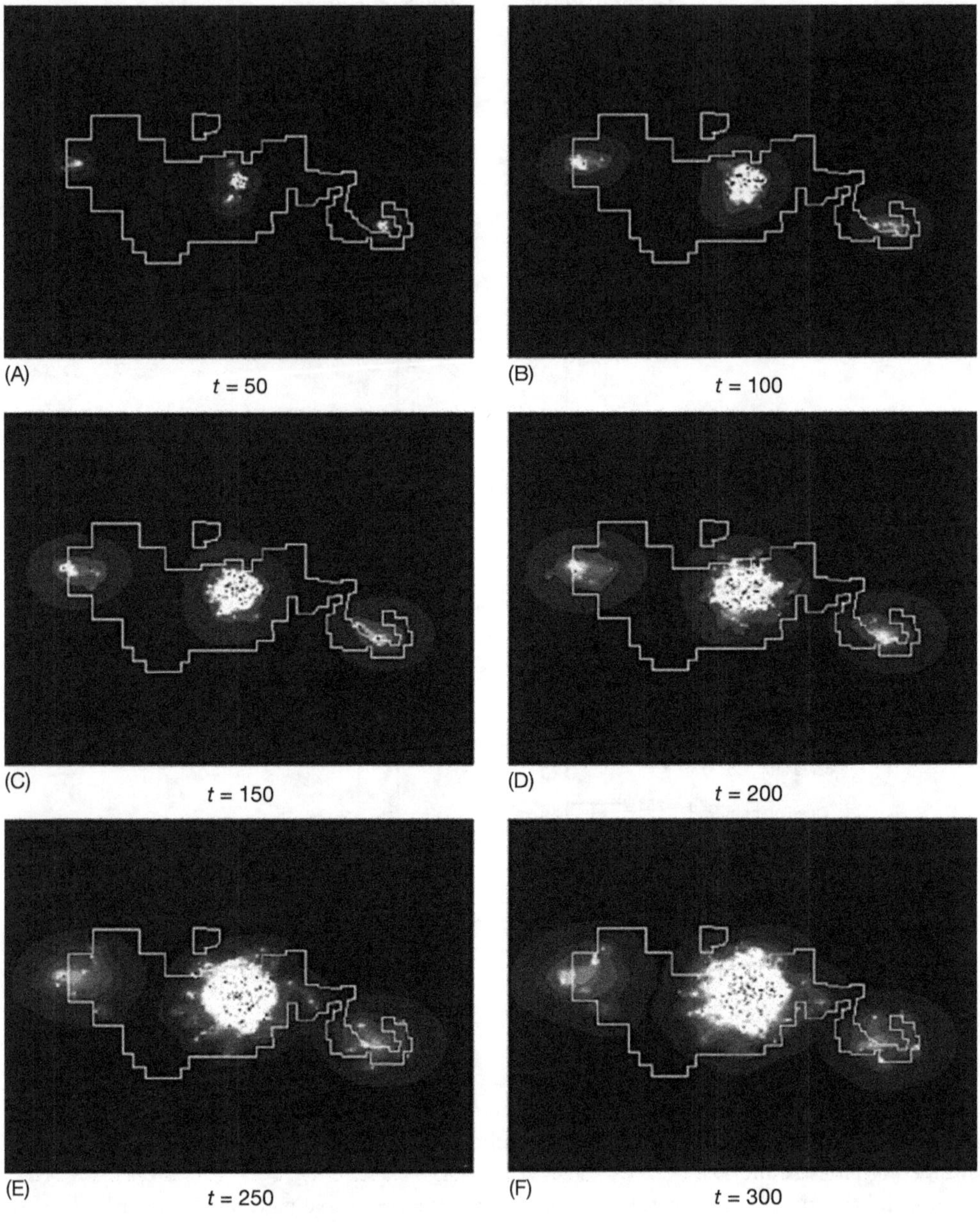

Figure 21.9 The evolution of the general growth simulation: light gray/white areas denote densities that are higher than dark gray/black areas

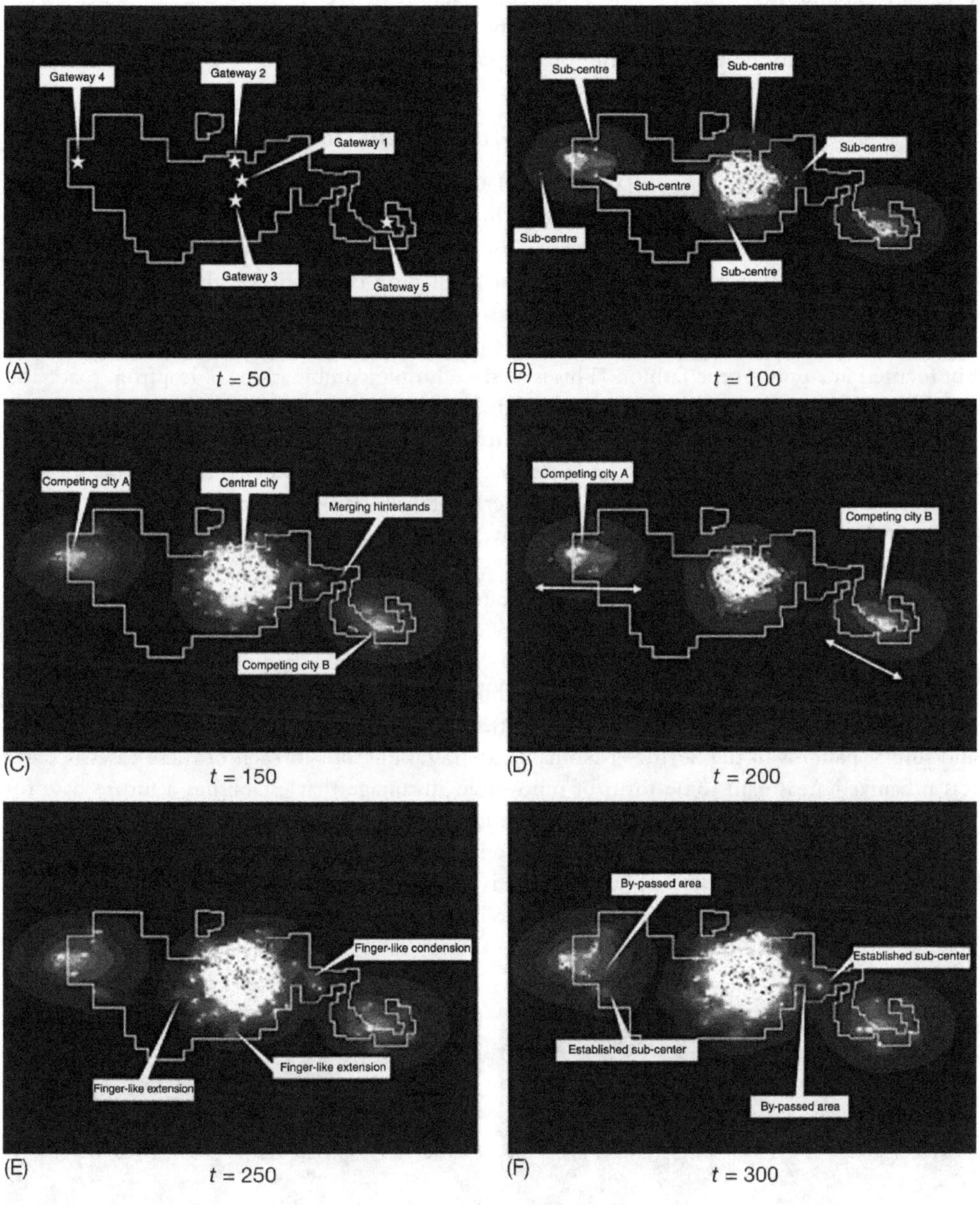

Figure 21.10 Noticeable features in the general growth simulation: (A) gateway sites (t = 0); (B) the formation of subcenters (t = 186); (C) merging hinterlands (t = 222); (D) linear development (t = 251); (B) corridors of settlement (t = 260); (F) well-established subcenters, with by-passed interstitial areas (t = 291). Light gray/white areas denote densities that are higher than dark gray/black areas

evolves to a condition whereby the low-density suburbs cover roughly the same area as the denser central cores. Of course, the low-density of that sprawled area means that those sections of the simulated city house a minority of the population.

Polycentric Growth Scenario

In the second simulation, a growth scenario is devised in much the same way as the last example, with identical growth rates and seed conditions, and the termination of growth at a point in the evolution of the simulation. However, in this scenario, the model is parameterized to encourage more polycentric development. The simulation is specified with greater propensity for the formation of peripheral clusters.

This simulation essentially operates under a smart growth regime. Growth is accommodated, but focused in a polycentric fashion. This is achieved using combinations of leapfrog, road, and irregular movement rules as part of a combined sequence that terminates in a nearby movement rule. The propensity for these clusters to generate internal and diffusing population is also greater. This combined regime is used alongside normal execution of the other rules in isolation. This approach establishes a large number of dense peripheral clusters—edge cities—as the simulation proceeds (Figures 21.11 and 21.12). Essentially, this is sprawl in characteristic form, but with emphasis on polycentricity.

The city-system evolves at a much faster rate, due to internal growth. In fact, the central city and competing city B begin to merge very early in the simulation, at $t = 65$ (Figures 21.9 and 21.11). A significant number of successful clusters are established early and these incubate a volume of internal growth that diffuses within the system. This is roughly equivalent, in a sense, to similar phenomena in real-world contexts, as in Silicon Valley in Northern California, and similar patterns in the Seattle–Tacoma area of Washington. In each of these cases, former less urbanized areas gain some form of innovative advantage that establishes a future base for impressive growth—Palo Alto and Santa Clara in the California example and Redmond in the Washington example. Growth in the simulation is still cut off 75 percent of the way through the simulation run, but at that stage there is more than enough internal momentum in competing cities A and B, and the cut-off has relatively little impact, compared to its use in the general growth simulation.

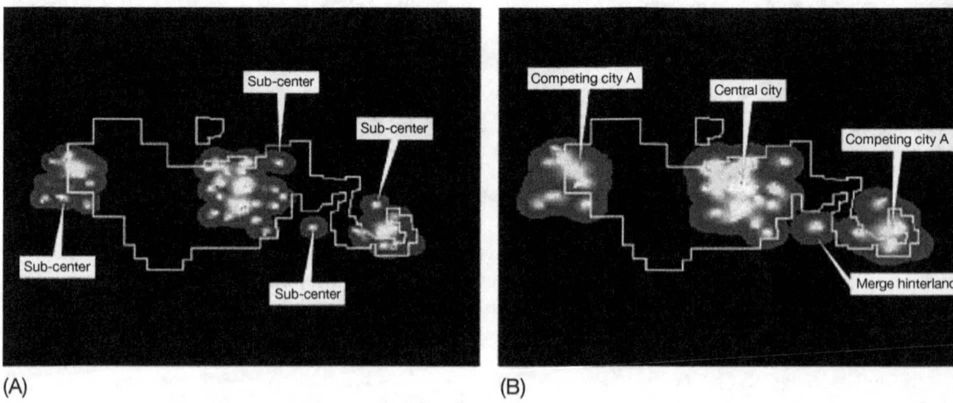

Figure 21.11 Noticeable features in the polycentric growth simulation: (A) subcenter formation ($t = 525$); (B) merging cities ($t = 565$). Light gray/white areas denote densities that are higher than dark gray/black areas

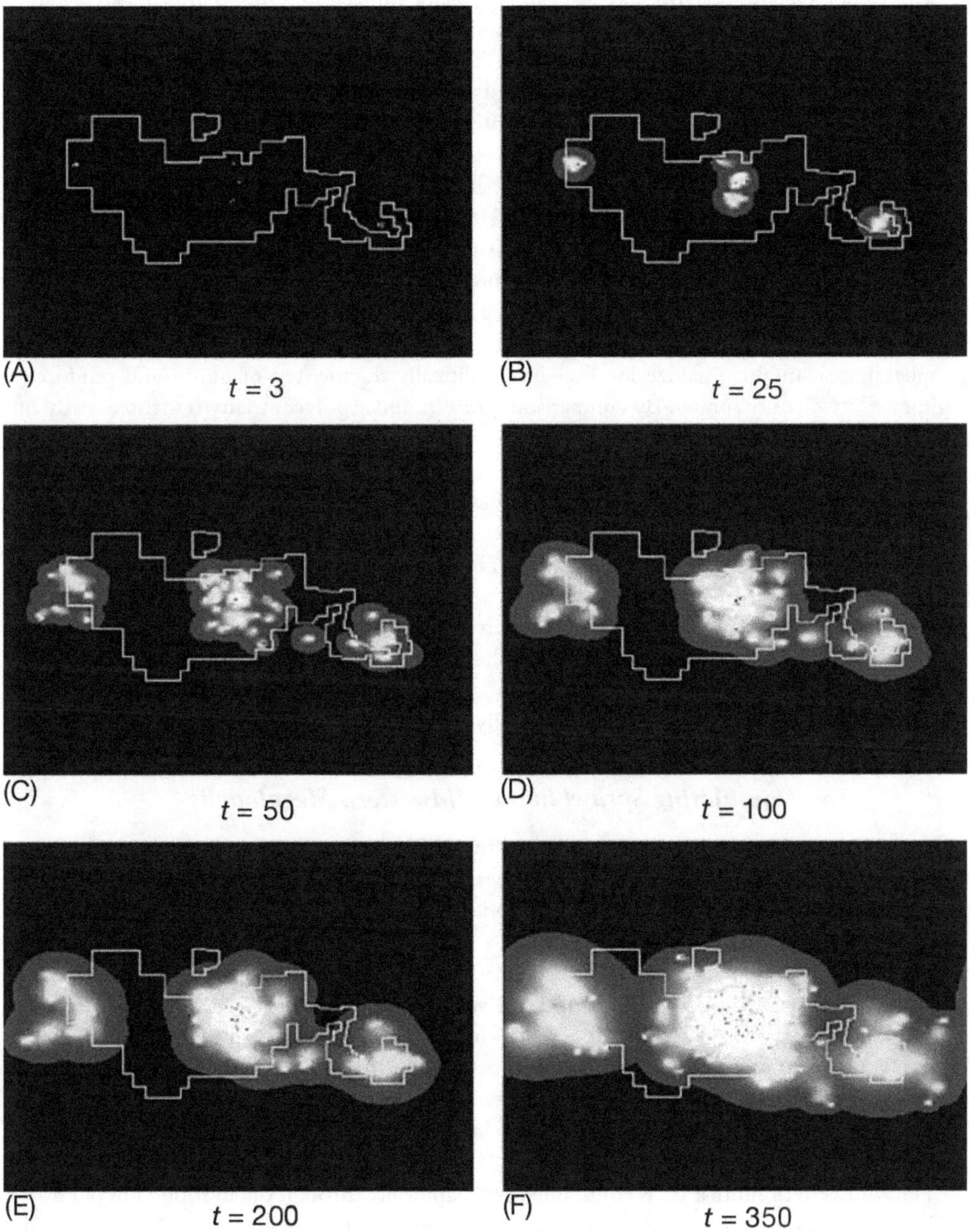

Figure 21.12 The evolution of the polycentric growth simulation: light gray/white areas denote densities that are higher than dark gray/black areas

This is much like events that take place in many sprawling cities. Once peripheral areas garner enough of a foothold they often incorporate as independent townships, with independent control over local land-use and zoning. (Gilbert and Chandler in Arizona are examples. These former suburbs of Phoenix are among the top five fastest-growing cities in the United States.) Invariably, the status quo—low-density sprawl—is protected, rather than more compact forms of development.

The polycentric approach generates suburbanization as in the general growth simulation, but the generated urban structure is much different. The city-system is surrounded by a buffer of low-density sprawl, as before, but the main urban mass exhibits a much more polycentric structure with many well-established cores (Figure 21.12). This generates a different urban future to that observed under general growth. In the general growth scenario, low-density peripheral sprawl dominated, and it was mentioned that this was synonymous with situations whereby peripheral areas might organize locally—in a politically fragmented manner—and reinforce a regime of low-density sprawl. By comparison, growth under polycentricity is focused, early on, in peripheral cores. While sprawl is present, the overall spatial structure is much more cohesive due to polycentricity in dense core distribution.

The implication for sprawl costs would, likely, be significant. The urban pattern in the polycentric case could be associated with greater system-wide accessibility and potentially lower vehicle miles traveled and vehicle emissions. The general growth scenario generated a city in which the population living in dense urban settings was roughly equal to that housed in low-density sprawl. If we assume that sprawl dwellers may follow a particular socioeconomic profile commensurate with "white flight" scenarios, the social justice implications are significant. The general growth example is indicative of large-scale system-wide socio-spatial segregation; the polycentric scenario accommodates a potentially more balanced distribution.

Simulating Sprawl in the Midwestern Megalopolis

In the next example, the model is applied to the Midwestern megalopolis region (Gottmann 1967) around Lake Michigan in the United States. The area provided some unique characteristics for applying the model, in particular the boundary formed by Lake Michigan.

The simulated landscape was derived from a Landsat TM image (Figure 21.13). Each pixel in the image was coded as an individual automaton in a regular lattice structure. The simulated region occupies a 52,125 km^2 area in the real world. The automaton lattice comprises a grid 520 automaton units wide and 630 long—327,600 units in total, with a real-world resolution of 180,093 m^2 per automaton. The Midwestern simulation is specified in much the same way as the abstract simulations described earlier. The simulation is based on the same model engine. The simulation is distinct from the general growth and polycentric simulations in its constraint parameters, however.

The Midwestern simulation is constrained geographically through the introduction of known seed sites for development. The seed sites are specified with respect to those locations in the area that came to dominate as urban centers in the region—namely, the city centers with the largest current population. Seven such sites were identified and introduced: Madison, WI; Milwaukee, WI; Chicago, IL; Gary, IN; South Bend, IN; Lansing, MI; and Grand Rapids, MI (Figure 21.13). Each of these sites serves as a gateway for the introduction of exogenous change to the simulation, thereby ensuring that the simulation retains some basic regional (and geographic) similarities with conditions in the real world.

The simulation is constrained in one additional way, and this relates to both geography and rates of change in the model. The volume of growth introduced at each time step is designed

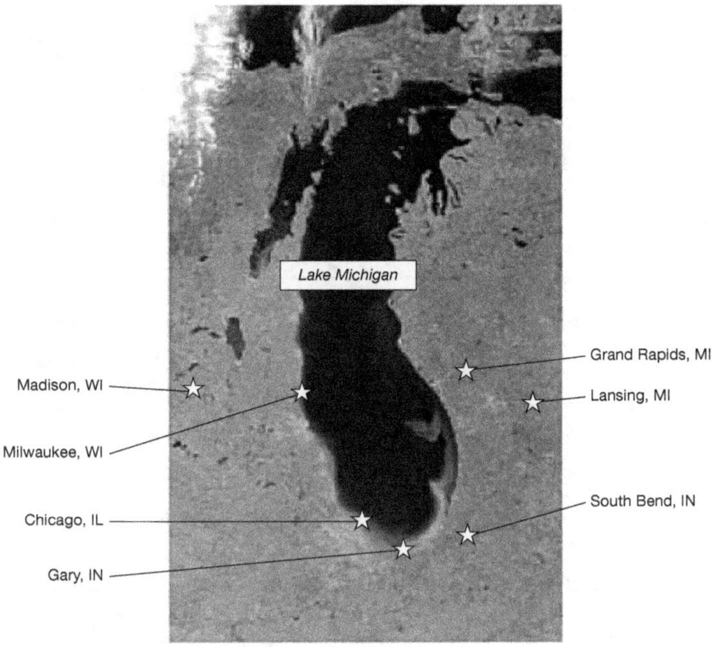

Figure 21.13 Seed sites in the Midwestern Megalopolis model

to roughly match known growth values for the particular cities (Figure 21.14). This allows us to track the development of individual cities (Figure 21.15). The growth rates were varied for different simulation runs to examine the patterns generated, but in the run illustrated in Figure 21.16 growth rates were scaled relative to known growth. Agents of change originating from these gateways are geo-referenced to the sites through which they are introduced. A greater volume of growth was introduced through Chicago, relative to the other cities; Milwaukee had more growth than Madison, and so forth. This ensures that the rate of evolution in the simulation is plausible and allows the simulation exercise to focus on the relative impact of the general state transition rules and movement rules in the model.

Using these specifications, the simulation was run with varying parameters. The example illustrated in Figure 21.16 was run with equal weighting of transition rules, for 200 iterations, from a state of only minor settlement in the seed sites (roughly synonymous with conditions in the area at the turn of the nineteenth century). These specifications generated a plausible pattern of urbanization (plausibility is discussed in the next section). The simulated city-system began developing as a loose constellation of urban clusters, scattered in the immediate vicinity of the seed sites identified in Figure 21.13. By t = 50, the relative dominance of Chicago and urbanized lower Wisconsin is evident in the system (Figure 21.16). By t = 100, the city-system has begun to coalesce, with road-influenced fingers of growth connecting spatially separated spheres of development. By t = 200, the system has begun to sprawl, with fragmented and lower-density settlement on the urban periphery, while expanding into previously undeveloped areas.

In another simulation, the model was run far ahead into the future as a speculative exercise to examine what the pattern of urbanization might look like if growth continued unchecked (Figure 21.17A), The end result was decentralization without end, reminiscent of forecasts written about in the 1980s (Hall 1983).

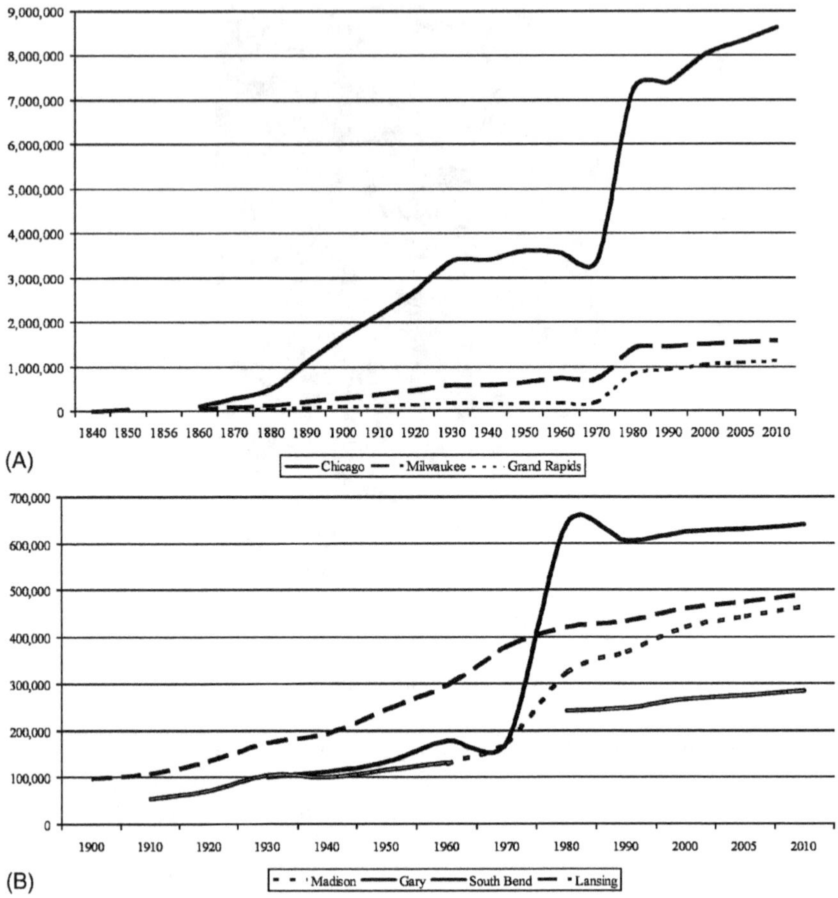

Figure 21.14 Population growth in America's Midwestern Megalopolis. Dates beyond 2000 are projected. Data before 1970 refer to town and city boundaries; data after 1970 refer to Metropolitan Statistical Area boundaries, which creates an artificial ramp in the illustration.

Note: Gaps in data for Chicago and South Bend are indicated by breaks in the line graphs.

Source: U.S. Bureau of the Census.

The relative impact of movement rules—as proxies for the behavior of agents of change—was also tested. By emphasizing one or more movement rules over others, it is possible to explore potential growth scenarios under alternative development regimes. Setting the road-like and irregular rules as the prevailing force in a simulation generates a pattern dominated by linear strips of urbanization (Figure 21.17B). Density within those strips is relatively high, but the overall pattern of adjacent growth is very scattered, with infill only occurring in areas where there is a dense network of strips in physical proximity to each other. Emphasizing the leap-frog rule relative to other rules generates an altogether different pattern of urbanization, dominated by small isolated clusters of dense settlement, with little to bind them within the urban system (Figure 21.17C). Combinations of clustering rules—the immediate, nearby, and leap-frog rules—lead to a very polycentric urban structure, characterized by a tight jigsaw of urban clusters, loosely merged by their respective bands of peripheral low-density hinterland (Figure 21.17D).

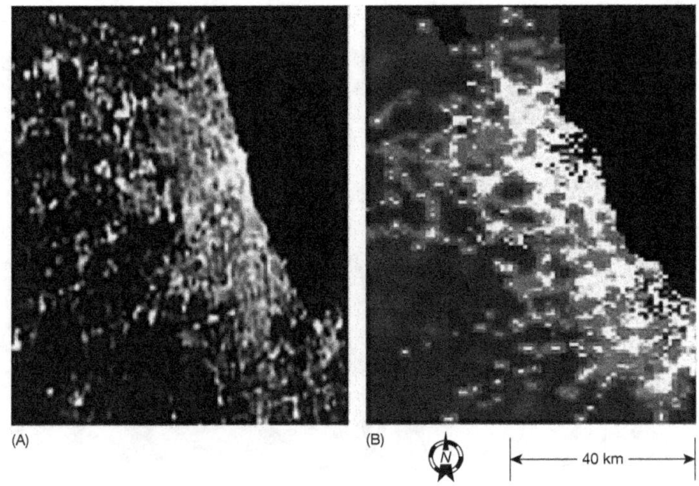

Figure 21.15 Observed and simulated conditions in Chicago. (A) The pattern of urbanization as revealed by night lights (*source:* NASA; http://science.nasa.gov/headlines/images/lights/chicago_lights.jpg); (B) a section of the simulated world corresponding to the Chicagoland area. Light gray/white areas in (B) denote densities that are higher than dark gray/black areas.

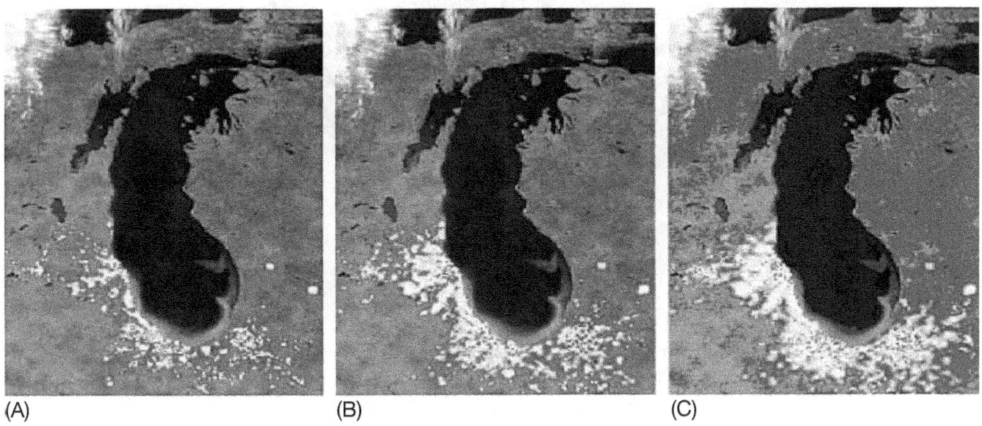

Figure 21.16 Simulated Midwestern growth at various stages: light gray/white areas denote densities that are higher than dark gray/black areas

Measuring Sprawl

The nature of sprawl generated in simulation was analyzed based on its composition and configuration, using landscape metrics and fractal dimensionality (Turner 1989; Turner and Gardner 1991; White and Engelen 1993; Batty and Langley 1994). *Composition* refers to the presence and amount of different patch types (urban, nonurban) within a landscape, without explicit reference to their spatial features. *Configuration* refers to the spatial distribution of patches within a landscape. *Patches* are distinct spatial agglomerations—blobs of urbanization in this *case.*

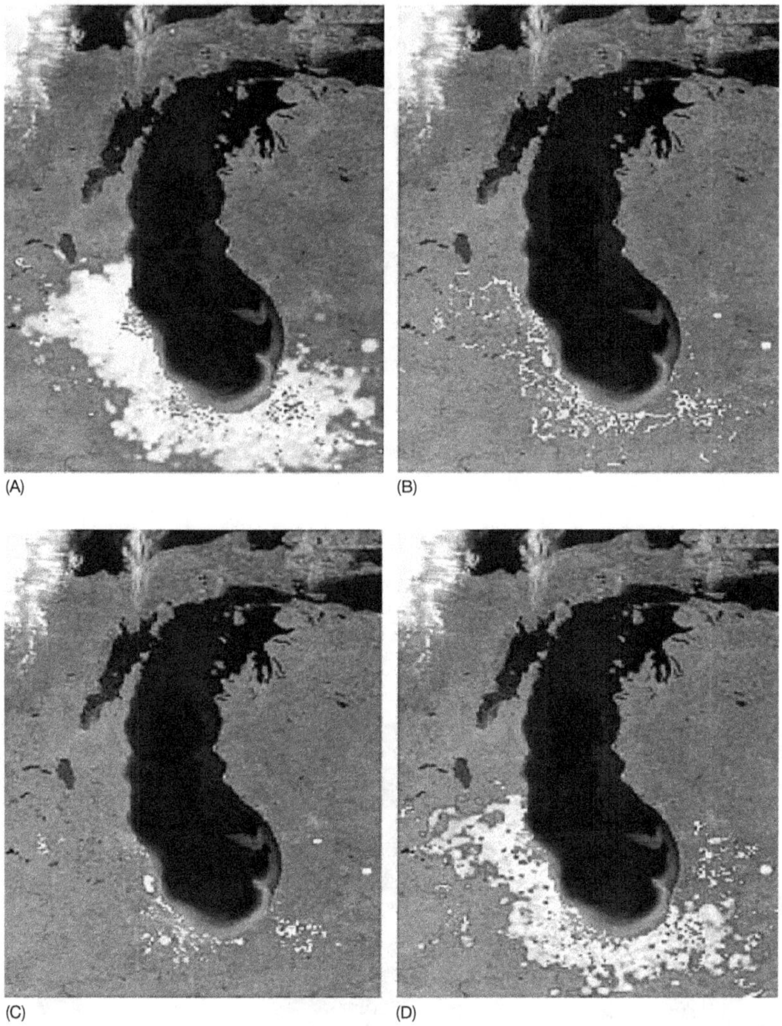

Figure 21.17 Simulated Midwestern urbanization under different scenarios: light gray/white areas denote densities that are higher than dark gray/black areas

Configuration metrics have advantages as a measure of sprawl, providing an index of the amount of space-filling and fragmentation in a city's urban pattern. Three configuration measures are used here to *assess* the degree of sprawl in simulated scenes, each at a landscape scale.

Perimeter-area fractal dimension (PAFRAC) measures the extent to which patches fill a landscape. Differences in PAFRAC value can suggest differences in the underlying pattern-generating process (Krummel et al. 1981). PAFRAC ranges in value from one to two, and is calculated using the slope of a regression line obtained by regressing the log of patch area against the log of patch perimeter. It is calculated as a double-log fractal dimension. A PAFRAC value greater than one for a two-dimensional landscape denotes a departure from Euclidean geometry and an increase in patch shape complexity. High values of PAFRAC denote situations in which patches fill-up a space; low values are synonymous with cases in which patches fill space to a lesser extent (i.e., sprawl).

$$\mathrm{PAFRAC} = \frac{2}{\dfrac{\left[N\sum_{i=1}^{m}\sum_{j=1}^{n}\left(\ln p_{ij} - \ln a_{ij}\right)\right] - \left(\sum_{i=1}^{m}\sum_{j=1}^{n}\ln p_{ij}\right)\left(\sum_{i=1}^{m}\sum_{j=1}^{n}\ln a_{ij}\right)}{\left(N\sum_{i=1}^{m}\sum_{j=1}^{n}\ln p_{ij}^{2}\right) - \left(\sum_{i=1}^{m}\sum_{j=1}^{n}\ln p_{ij}\right)^{2}}} \tag{10}$$

In the formula above, a_{ij} is the area of patch j of type i, p_{ij} is the perimeter of patch j of type i (urban/nonurban), m is the number of patch types, n is the number of patches of type i, and N is the total number of patches in the landscape.

Contagion is the probability that two randomly chosen adjacent cells belong to the same class (state). It is calculated on a cell-by-cell basis rather than a patch-by-patch basis. Contagion is the product of two probabilities: the probability that a randomly chosen cell belongs to category type i, and the conditional probability that, given a cell belongs to category i, one of its neighboring cells belongs to category j (McGarigal and Marks 1995). Where contagion is low, a landscape can be said to be composed of many small and dispersed clusters of cells—that is, it is fragmented. High contagion values are indicative of more compact landscapes.

$$C = \left[1 + \frac{\sum_{i=1}^{m}\sum_{j=1}^{m}\left[\left(P_i \dfrac{g_{ij}}{\sum_{j=1}^{m} g_{ij}}\right) \cdot \left(\ln P_i \dfrac{g_{ij}}{\sum_{j=1}^{m} g_{ij}}\right)\right]}{2 \ln (m)}\right] \cdot 100\,, \tag{11}$$

where C is the percentage of contagion, P_i is the proportional abundance of category type i, g_{ij} is the number of adjacencies between cells of category type i and all other category types, and m is the total number of category types.

The *Interspersion and Juxtaposition Index* (IJI) measures adjacency on a patch-by-patch basis. Higher values are synonymous with landscapes in which patch types are well interspersed (equally adjacent to each other). Lower values occur when landscapes contain patches that are poorly interspersed (there is a disproportionate distribution of patch type adjacencies). When IJI is zero in value, there is an uneven distribution of adjacencies between patch types. A value of 100 is indicative of a situation in which all patch types are equally adjacent to each other (McGarigal and Marks 1995). High values of IJI thus represent a relatively greater degree of homogeneity in a landscape. IJI is expressed as a percentage, and can be calculated using the following formula,

$$\mathrm{IJI} = \frac{-\sum_{i=1}^{m}\sum_{j=i+1}^{m}\left[\left(\dfrac{e_{ij}}{E}\right) - \ln\left(\dfrac{e_{ij}}{E}\right)\right]}{\ln\left(\dfrac{1}{2\left[m(m-1)\right]}\right)} \cdot 100 \tag{12}$$

Table 21.2 Fractal and landscape metrics for the simulation scenarios

Metric	*General growth*	*Polycentricity*	*Midwestern example*
No. of patches	14,375	3,066	3,782
PAFRAC	1.5305	1.5321	1.5479
Contagion	48%	65%	45%
IJI	54%	37%	20.15%

Notes: PAFRAC = perimeter-area fractal dimension; IJI = Interspersion and Juxtaposition Index.

where IJI is the value of the *Interspersion and Juxtaposition Index;* e_{ij} is the total length of edge in the landscape between patch types *i* and *j*, including landscape boundary segments representing true edge only involving patch type I; E is the total length of edge in the landscape; and m is the number of patch types in the landscape. The results of measuring simulated sprawl are shown in Table 21.2. The two abstract simulations demonstrate very different sprawl-like characteristics. The general growth simulation generated more patches than the polycentric simulation. The patch total was 14,375 for the general growth scenario and 3,066 for the polycentric scenario. This indicates that the landscape generated by the polycentric simulation was relatively less fragmented than its counterpart. The values for PAFRAC support this contention. The general growth example had a fractal dimension of 1.5305; the value for the polycentric scenario was higher at 1.5321. Both of these values are commensurate with the fractal dimension of cities in real-world contexts (Table 21.3). The higher value for the polycentric example also suggests that the simulated city in that experiment did a better job of filling the space it occupied, although the values are not dramatically different. The values for contagion further support the hypothesis that the two simulations generated cities with different spatial structures and patterns of sprawl. There is a dramatic difference in the percentage of contagion recorded for the two simulations. The general growth simulation yielded a contagion value of 48 percent; the figure for the polycentric example was much higher at 65 percent. Higher contagion is indicative of a greater degree of compaction of cells in a landscape. The city generated in the polycentric scenario can thus be considered less sprawling than its general growth counterpart. The results for interspersion and juxtaposition produced similar results: general growth demonstrated a relatively high IJI (54 percent), indicative of a landscape in which patches are well interspersed. Polycentricity produced a much lower IJI (37 percent), suggesting poorer interspersion between patches. The higher value of IJI for the case of general growth suggests that landscape is more homogeneous than that generated under a polycentric scenario.

Overall, then, the city generated by the polycentric simulation can be regarded as more compact and less sprawled than that generated under a more general growth scenario.

The number of patches generated by the Midwestern simulation was consistent with the abstract polycentric example—the Midwestern scenario produced 3,782 patches by the end of the simulation run. The fractal dimension was also consistent with the abstract simulations, and with real-world cities, at a value of 1.5479 at the end of the run. The degree of contagion was 45 percent, the amount of interspersion and juxtaposition was 20.15 percent. The contagion score was low relative to the abstract simulations, suggesting that the Midwestern model generated a more sprawl-like landscape. Interspersion and juxtaposition was much lower than that found in the abstract simulations, indicative of relatively lower homogeneity in the landscape, again an indicator of sprawl.

Analysis of the simulation run was performed across the lifetime of the simulation for the Midwestern simulation, to explore changes in the structure of the simulated city as it evolved

Table 21.3 Fractal dimensions for other cities

City	*Year*	*Fractal dimension*	*Source*
Albany, NY	1990	1.494	Batty and Langley (1994)
Beijing	1981	1.93	Frankhauser (1988)
Berlin	1980	1.73	Frankhauser (1988)
Boston	1981	1.69	Frankhauser (1988)
Budapest	1981	1.72	Frankhauser (1988)
Buffalo, NY	1990	1.729	Batty and Langley (1994)
Cardiff	1981	1.586	Batty and Langley (1994)
Cleveland	1990	1.732	Batty and Langley (1994)
Columbus	1990	1.808	Batty and Langley (1994)
Essen	1981	1.81	Frankhauser (1988)
Guatemala City	1990	1.702	Smith (1991)
London	1962	1.774	Doxiadis (1968)
London	1981	1.72	Frankhauser (1988)
Los Angeles	1981	1.93	Frankhauser (1988)
Melbourne	1981	1.85	Frankhauser (1988)
Mexico City	1981	1.76	Frankhauser (1988)
Moscow	1981	1.6	Frankhauser (1988)
New York	1960	1.71	Doxiadis (1968)
Paris	1960	1.862	Doxiadis (1968)
Paris	1981	1.66	Frankhauser (1988)
Pittsburgh	1981	1.59	Frankhauser (1988)
Pittsburgh	1990	1.775	Batty and Langley (1994)
Potsdam	1945	1.88	Frankhauser (1988)
Rome	1981	1.69	Frankhauser (1988)
Seoul	1981	1.682	Batty and Langley (1994)
Stuttgart	1981	1.41	Frankhauser (1988)
Sydney	1981	1.82	Frankhauser (1988)
Syracuse	1990	1.438	Batty and Longley (1994)
Taipei	1981	1.39	Frankhauser (1988)
Taunton	1981	1.636	Batty and Langley (1994)
Tokyo	1960	1.312	Doxiadis (1968)

Source: Adapted from Batty and Langley (1994).

within the simulation. Analyzed across time-steps, the results suggest that the simulated city-system developed in stages, with rapid changes in initial conditions, followed by a period of relative stability, and a sharp transition toward sprawl at the end of the model run. This is consistent with the life-cycle stages of an urban system, whereby cities go through periods of relative compaction, expansion, and decentralization (Hall 1983). The end result of that evolution, however, is the now-all-too-familiar sprawl that is readily apparent in much of the urbanized United States, and elsewhere.

The number of patches demonstrated a progressive increase from just a handful of seed sites to more than 5,000 by t = 138. After that point, the number of patches started to decline steadily, as fragmented areas began to coalesce (Figure 21.18). The fractal dimension fluctuated over the course of the simulation run, although it remained within a reasonable range, as compared to dimensions for other cities that have been mentioned in the literature (Table 21.3). This suggests

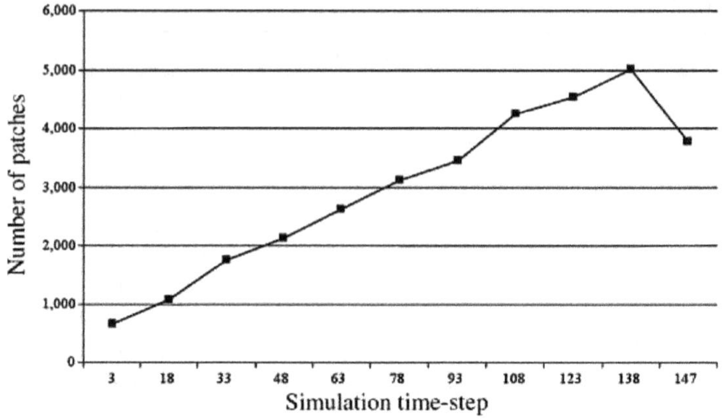

Figure 21.18 Change in the number of patches in the Midwestern simulation

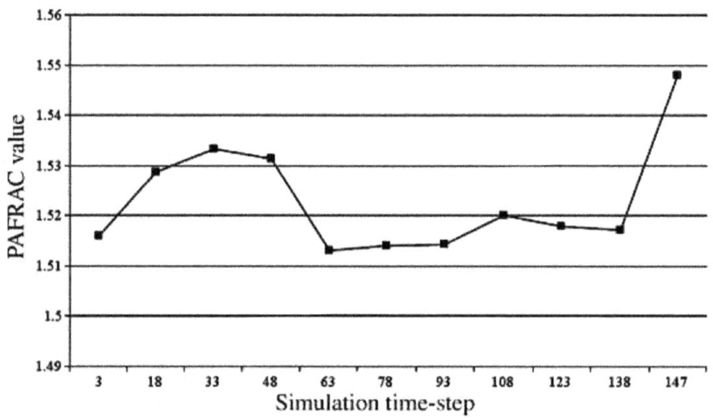

Figure 21.19 Perimeter-area fractal dimension change (PAFRAC) in the Midwestern simulation

that the simulated city went through stages of growth, with rapid space-filling at the beginning of its evolution followed by a period of relatively stable growth. The value climbed toward the end of the simulation run as the simulated city began to sprawl at a growing rate (Figure 21.19). Contagion and interspersion and juxtaposition demonstrated an almost inverse relationship over the simulation run. The degree of contagion in the landscape grew early in the simulation, declining thereafter before climbing rapidly toward the end of the model run (Figure 21.20). This is consistent with the results suggested by the other metrics—the city went through an early growth stage dominated by compaction. The decline in contagion thereafter is indicative of relative sprawl. The value of interspersion and juxtaposition in the simulation started off quite high, and subsequently declined quite rapidly before rising in value, mostly, throughout much of the simulation run (Figure 21.20). Once again, there was a sharp change at the end of the model run, where the value dipped to its lowest level. This suggests that the simulated city started off with relatively homogeneous conditions, losing homogeneity thereafter and entering into a sustained period in which there was poor interspersion. Toward the end of the simulation, there is a strong tendency for interspersion, with a growth in homogeneity, which we can associate with sprawl.

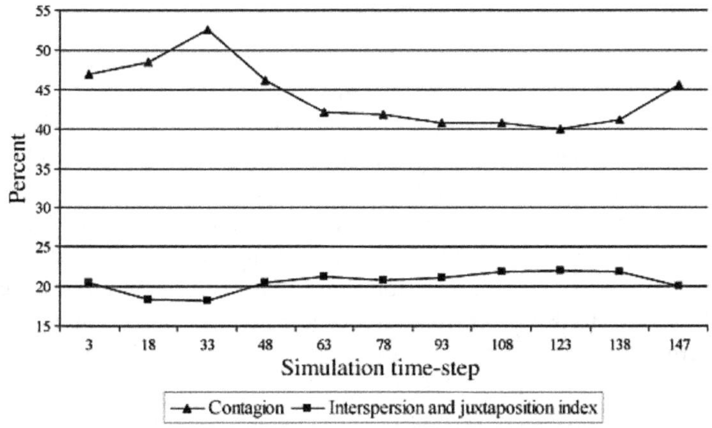

Figure 21.20 Contagion and interspersion change in the Midwestern simulation

Implications for Understanding Sprawl

It is evident from each of the simulations discussed in this chapter that sprawl is, to a certain extent, inevitable. It is the likely end-state in the natural evolution of a city-system. This is obvious in the context of most well-established cities in the United States. However, it is particularly important in the context of newly forming cities, such as those developing and growing rapidly in what were previously termed "Sun Belt" cities, predominantly situated in the southwestern region of the United States.

For these cities, there is a propensity for urban evolution to jump or skip the natural evolution process, fueled by higher-than-average growth rates and contemporary development regimes, and go straight to sprawl. However, there is also opportunity to plan cities in such a way that this situation does not occur. The simulations described in this chapter suggest a few—geographic—ways in which policies could be developed to mitigate circumstances.

Unchecked growth leads to low-density, blanket sprawl, with all the associated costs discussed in the literature implied. Essentially, controlling sprawl requires mechanisms to *manage* growth sustainably. A number of mechanisms are understood to drive sprawl, and several of these are represented in the simulations described here. We are most interested in geographic scenarios, and the results of the simulation exercises advocate some options.

Encouraging polycentric development appears to be one solution—allowing leap-frogging, but encouraging sustainable and compact independent clusters, in close proximity, rather than isolated patches. Edge cities (Garreau 1992) may be one way to achieve this; transit villages (Cervero 1998) are a more likely, sustainable, option; desakota-style clusters have been successful in Asia (Heikkila, Shen, and Kaizhong 2003). It is important to actually permit sprawl to occur locally on the periphery of these clusters, to facilitate infill and to avoid by-passing large areas of land. This idea is reminiscent of much older theories of urban development, notably the idea of central place theory. Road-like growth can also be used effectively to link isolated clusters. (Transit-oriented development may have even greater potential.) However, care must be taken to avoid isolated linear development—ribbon sprawl.

Conclusions

This article has demonstrated the application of a geographically derived automata methodology to the simulation of sprawl. The framework is particularly beneficial in modeling sprawl, allowing for the description of system dynamics as a function of spatial interactions between mobile, agent-like entities and a static, CA-like environment. Moreover, the framework allows for the generation of very realistic macroscale urban structures from these local-scale mechanisms.

The simulations described in this chapter were developed as artificial laboratories for exploring the relative—and geographic—impact of proposed causes of sprawl. The model generated sprawl-like cities in each of the simulation scenarios, and by varying the influence of rules within the model, facilitated exploration of the potential drivers of sprawl.

After measuring sprawl through the use of fractal analysis and metrics from landscape ecology, various potential options for managing sprawl were inferred. The results suggest that sprawl might best be tackled geographically, by encouraging compact and sustainable clusters of leapfrog development in close proximity. Sprawl on the periphery of these clusters then serves as an in-fill mechanism rather than continuing on the periphery of a larger urban mass in an unsustainable fashion. Moreover, it was determined that road-influenced growth could help to link isolated fragments of sprawl on the urban periphery under certain conditions.

The simulations discussed in this chapter were designed to explore geographic dimensions of sprawl, focusing on mimicking the spatial distribution of growth in dynamic contexts. In the literature on sprawl, however, it is clear that there are other important components to the phenomenon that these simulations have not addressed—namely, preference-based drivers at within-neighborhood geographies. Elsewhere, the authors have applied a similar methodology to the modeling of preference-based behavior in an artificial residential submarket, roughly equivalent to a single fixed-infrastructure automaton in the growth-based simulations described in this chapter (Torrens 2007). This remains a topic of ongoing research.

Acknowledgment

This chapter originally appeared in *Annals of the Association of American Geographers*, 96, 248–275 (2006).

References

American Farmland Trust. 1995. *Alternatives for Future Urban Growth in California's Central Valley: The Bottom Line for Agriculture and Taxpayers*. Washington, DC: American Farmland Trust.

Arai, Takeshi, and Tetsuya Akiyama. 2004. Empirical analysis of the land use transition potential in a CA based land use model: Application to the Tokyo Metropolitan Region. *Computers, Environment and Urban Systems* 28 (1/2): 65–84.

Audirac, I., A. H. Shermyen, and M. T. Smith. 1990. Ideal urban form and visions of the good life: Florida's growth management dilemma. *Journal of the American Planning Association* 56 (4): 470–82.

Bae, Chan-Hee C., and Harry W. Richardson. 1994. *Automobiles, the Environment and Metropolitan Spatial Structure*. Cambridge, MA: Lincoln Institute of Land Policy.

Bahl, Roy W. 1968. A land speculation model: The role of the property tax as a constraint to urban sprawl. *Journal of Regional Science* 8 (2): 199–208.

Barnett, J. 1995. *The Fractured Metropolis: Improving the New City, Restoring the Old City, Reshaping the Region*. New York: Icon-Editions.

Batty, Michael, and Paul Langley. 1994. *Fractal Cities*. London: Academic Press.

Batty, Michael, and Yichun Xie. 1994. From cells to cities. *Environment and Planning B* 21: s31–s48.

——. 1997. Possible urban automata. *Environment and Planning B* 24: 175–92.

Batty, Michael, Yichun Xie, and Zhanli Sun. 1999. Modeling urban dynamics through GIS-based cellular automata. *Computers, Environment and Urban Systems* 23 (3): 205–33.

Bell, Martin, Christopher Dean, and Marcus Blake. 1999. A model for forecasting the location of fringe urbanisation, with GIS and 3D visualisation. Paper presented at the Sixth International Conference on Computers in Urban Planning and Urban Management (CUPUM), September, at IUAV, Venice.

Benenson, Ithzak, and Paul M. Torrens. 2004a. *Geosimulation: Automata-based Modeling of Urban Phenomena.* London: Wiley.

——. 2004b. A minimal prototype for integrating GIS and geographic simulation through Geographic Automata Systems. In *GeoDynamics*, Ed. E. Wu, 347–69. Boca Raton, FL: CRC Press.

Benfield, E. K., M. D. Raimi, and D. D. T. Chen. 1999. *Once There Were Greenfields: How Urban Sprawl is Undermining America's Environment, Economy, and SocialFfabric.* Washington, DC: Natural Resources Defense Council.

Brown, Daniel G., Scott E. Page, Rick Riolo, and William Rand. 2002. Modeling the effects of greenbelts on the urban-rural fringe. Paper presented at iEMSs: Integrated Assessment and Decision Support symposium, June 24–27, at Lugano, Switzerland.

Burchell, Robert W., Naveed A. Shad, David Lisotkin, Hilary Phillips, Anthony Downs, Samuel Seskin, Judy S. Davis, Terry Moore, David Helton, and Michelle Gall. 1998. *The Costs of Sprawl Revisited.* Washington, DC: National Academy Press.

Cairncross, E. 1995. The death of distance. In *The Economist.* September 30.

Calthorpe, Peter, William Fulton, and Robert Fishman. 2001. *The Regional City: Planning for the End of Sprawl.* Washington, DC: Island Press.

Carruthers, John I. 2003. Growth at the fringe: The influence of political fragmentation in United States metropolitan areas. *Papers in Regional Science* 82: 472–99.

Cervero, Robert. 1998. *Transit Metropolis: A Global Inquiry.* Washington DC: Island Press.

Chapin, E. Stuart, and Shirley E. Weiss. 1962. Factors influencing land development. An Urban Studies Research Monograph. Chapel Hill, NC: Center for Urban and Regional Studies, Institute for Research in Social Science, University of North Carolina.

——. 1965. Some input refinements for a residential model. An Urban Studies Research Monograph. Chapel Hill, NC: Center for Urban and Regional Studies, Institute for Research in Social Science, University of North Carolina.

——. 1968. A probabilistic model for residential growth. *Transportation Research* 2: 375–90.

Clarke, Keith C., and L. Gaydos. 1998. Loose coupling a cellular automaton model and GIS: Long-term growth prediction for San Francisco and Washington/Baltimore. *International Journal of Geographical Information Science* 12 (7): 699–714.

Clarke, Keith C., S. Hoppen, and L. Gaydos. 1997. A self-modifying cellular automaton model of historical urbanization in the San Francisco Bay area. *Environment and Planning B* 24: 247–61.

Clawson, Marion. 1962. Urban sprawl and speculation in suburban land. *Land and Economics* 38 (2): 99–111.

Downs, Anthony. 1994. *New Visions for Metropolitan America.* Washington, DC: Brookings Institute.

Doxiadis. 1968. *Ekistics: An Introduction to the Science of Human Settlements.* London: Hutchinson.

Duany, Andres, Elizabeth Plater-Zyberk, and Jeff Speck. 2000. *Suburban Nation: The Rise of Sprawl and the Decline of the American Dream.* New York: North Point Press.

Duany, Andres, Jeff Speck, and Elizabeth Plater-Zyberk. 2001. *Smart Growth: New Urbanism in American Communities.* New York: McGraw-Hill.

Engelen, Guy, Roger White, and Inge Uljee. 2002. The MURBANDY and MOLAND models for Dublin. Maastricht: Research Institute for Knowledge Systems (RIKS) BV.

Engelen, Guy, Roger White, Inge Uljee, and Paul Drazan. 1995. Using cellular automata for integrated modelling of socioenvironmental systems. *Environmental Monitoring and Assessment* 30: 203–14.

Environmental Protection Agency. 1993. Costs of providing government services to alternative residential plans. Report prepared for the Chesapeake Bay Program's Subcommittee on Population and Growth. Washington, DC: EPA.

——. 2000. *National Air Pollutant Emission Trends: 1900–1998. EPA-454/R-00-002.* Research Triangle Park, NC: U.S. Environmental Protection Agency.

Epstein, Joshua M. 1999. Agent-based computational models and generative social science. *Complexity* 4 (5): 41–60.

Ewing, R. 1994. Causes, characteristics, and effects of sprawl: A literature review. *Environmental and Urban Issues* 21 (2): 1–15.

——. 1997. Is Los Angeles-style sprawl desirable? *Journal of the American Planning Association* 63 (1): 107–26.

Ewing, Reid, Rolf Pendall, and Donald D. T. Chen. 2002. Measuring sprawl and its impact. Washington DC: Smart Growth America.

Farley, R., H. Schuman, S. Binachi, D. Colasanto, and S. Hatchett. 1978. Chocolate city, vanilla suburbs: Will the trend toward racially separate communities continue? *Social Science Research* 7: 319–44.

Ferber, Jacques. 1999. *Multi-agent Systems: An Introduction to Distributed Artificial Intelligence*. Harlow, UK: Addison-Wesley.

Frank, James E. 1989. The costs of alternative development patterns: A review of the literature. Washington, DC: Urban Land Institute.

Frankhauser, Pierre. 1988. Fractal aspects of urban structures. International Symposium des Sonderforschungsbereich 230: Naturliche Konstruktionen-Leichtbau in Architektur und Natur: Teil I. Universities of Stuttgart and Tubingen, Germany.

Fujita, Masahisa, Paul Krugman, and Anthony J. Venables. 2001. *The Spatial Economy: Cities, Regions, and International Trade*. Cambridge, MA: MIT Press.

Galster, George. 1991. Black suburbanization: Has it changed the relative location of the races? *Urban Affairs Quarterly* 26: 621–28.

Garreau, Joel. 1992. *Edge City: Life on the New Frontier*. New York: Anchor Books/Doubleday.

Goldstein, Noah C., J. T. Candau, and Keith C. Clarke. 2004. Approaches to simulating the "March of Bricks And Mortar." *Computers, Environment and Urban Systems* 28 (1/2): 125–47.

Gordon, Peter, and Harry W. Richardson. 1997. Are compact cities a desirable planning goal? *Journal of the American Planning Association* 63 (1): 95–106.

Gottmann, Jean. 1967. *Megalopolis: The Urbanized Northeastern Seaboard of the United States*. Cambridge, MA: MIT Press.

Gottmann, Jean, and R. A. Harper. 1967. *Metropolis on the Move: Geographers Look at Urban Sprawl*. New York: Wiley.

Hall, Peter. 1983. Decentralization without end?, In *The Expanding City: Essays in Honor of Professor Jean Gottmann*. London: Academic Press.

Harvey, David. 1969. *Explanation in Geography*. London: Edward Arnold.

Harvey, R. O., and W. A. V. Clark. 1965. The nature and economics of sprawl. *Land Economics* 61 (1): 1–9.

Hasse, John. 2004. A geospatial approach to measuring new development tracts for characteristics of sprawl. *Landscape Journal* 23 (1): 52–67.

Hasse, John, and Richard G. Lathrop. 2003a. A housingunit-level approach to characterizing residential sprawl. *Photogrammetric Engineering and Remote Sensing* 69 (9): 1021–30.

——. 2003b. Land resource impact indicators of urban sprawl. *Applied Geography* 23: 159–75.

Heikkila, Eric J., Tiyan Shen, and Yang Kaizhong. 2003. Fuzzy urban sets: Theory and application to desakota regions in China. *Environment and Planning B* 30 (2): 239–54.

Herold, Martin, and Keith C. Clarke. 2002. The use of remote sensing and landscape metrics to describe structures and changes in urban land uses. *Environment and Planning A* 34: 1443–58.

Herold, Martin, Xiao Hang Liu, and Keith C. Clarke. 2003. Spatial metrics and image texture for mapping urban land use. *Photogrammetric Engineering and Remote Sensing* 69 (9): 991–1001.

James Duncan & Associates; Van Horn, Gray Associates; Ivey, Bennett, Harris, and Walls, Inc.; Wade-Trim, Inc. 1989. The search for efficient urban growth patterns: A study of the fiscal impacts of development in Florida. Report presented to the Governor's Task Force on Urban Growth Patterns and the Florida Department of Community Affairs.

Johnson, Michael. 2001. Environmental impacts of urban sprawl: A survey of the literature and proposed research agenda. *Environment and Planning A* 33: 717–35.

Klosterman, Richard E. 1999. The What if? Collaborative planning support system. *Environment and Planning B: Planning & Design* 26: 393–408.

——. 2001. The What if? Planning support system. In *Planning Support Systems: Integrating Geographic Information Systems, Models, and Visualization Tools*, Eds. R. K. Brail and R. E. Klosterman, 263–84. Redlands, CA: ESRI.

Knox, Paul. 1989. The vulnerable, the disadvantaged, and the victimized: Who they are and where they live. In *Social Problems and the City*, Eds. D. T. Herbert and D. M. Smith, 32–47. Oxford: Oxford University Press.

Krugman, Paul. 1996. *The Self-organizing Economy*. Malden, MA: Blackwell.

Krummel, J. R., R. H. Gardner, G. Sugihara, and R. V. O'Neill. 1981. Landscape patterns in a disturbed environment. *Oikos* 48: 321–24.

Kunstler, James H. 1993. The geography of nowhere: The rise and decline of America's man-made landscape. New York: Simon & Schuster.

Landis, John. 1994. The California urban futures model: A new generation of metropolitan simulation models. *Environment and Planning B: Planning & Design* 21: 399–420.

——. 1995. Imagining land use futures: Applying the California urban futures model. *Journal of the American Planning Association* 61 (4): 438–57.

——. 2001. CUF, CUF II, and CURBA: A family of spatially explicit urban growth and land-use policy simulation models. In *Planning Support Systems: Integrating Geographic Information Systems, Models, and Visualization Tools*, Eds. R. K. Brail and R. E. Klosterman, 157–201. Redlands, CA: ESRI.

Landis, John, and Ming Zhang. 1998a. The second generation of the California urban futures model: Part 1: Model logic and theory. *Environment and Planning B: Planning & Design* 25 (5): 657–66.

_1998b. The second generation of the California urban futures model. Part 2: Specification and calibration results of the land-use change submodel. *Environment and Planning B: Planning & Design* 25 (6): 795–824.

Lessinger, J. 1962. The cause for scatteration: Some reflections on the National Capitol Region Plan for the Year 2000. *Journal of the American Institute of Planners* 28 (3): 159–70.

Li, Xia, and Anthony Gar-On Yeh. 2000. Modelling sustainable urban development by the integration of constrained cellular automata and GIS. *International Journal of Geographical Information Science* 14 (2): 131–52.

——. 2002. Neural-network-based cellular automata for simulating multiple land use changes using GIS. *International Journal of Geographical Information Science* 16 (4): 323–43.

McGarigal, K., and B. Marks. 1995. FRAGSTATS: Spatial pattern analysis program for quantifying landscape structure. *PNW-GTR-351:* USDA Forest Service.

Meyer, Jean-Aready, and Agnès Guillot. 1994. From SAB90 to SAB9: Four years of animat research. In *From Animals to Animats 3. Proceedings of the Third International Conference on Simulation of Adaptive Behavior*, Eds. D. Cliff, P. Husbands, J.A. Meyer, and S. Wilson, 2–12. Cambridge, MA: MIT Press.

Morrill, Richard. 1991. Myths about metropolis. In *Our Changing Cities*, Eds. J. E. Hart, 1–11. Baltimore, MD: Johns Hopkins University Press.

Nakajima, Takashi. 1977. Application de la théorie de l'automate à la simulation de l'évolution de l'espace urbain (Application of automata theory to simulation of urban space evolution). In *Congrès sur la Méthodologie de l'Aménagement et du Development* (Congress for Land Development Management Methodology). Montreal: Association Canadienne-Française Pour L'Avancement des Sciences et Comité de Coordination des Centres de Recherches en Aménagement, Développement et Planification (CRADEP).

National Association of Home Builders. 1999. NAHB'S Smart Growth Report. Washington, DC: National Association of Home Builders.

OTA. 1995. The technological reshaping of metropolitan america. OTA-ETI-643. Washington, DC: U.S. Congress Office of Technology Assessment.

Parker, Dawn C., Steven M. Manson, Marco A. Janssen, Matthew J. Hoffmann, and Peter Deadman. 2003. Multi-agent system models for the simulation of land-use and land-cover change: A review. *Annals of the Association of American Geographers* 93: 314–37.

Peiser, Richard. 1989. Density and urban sprawl. *Land Economics* 65 (3): 193–204.

Pendall, Rolf. 1999. Do land-use controls cause sprawl? *Environment and Planning B* 26: 555–71.

Peterson, George E. 1980. Federal tax policy and the shaping of urban development. In *The Prospective City*, Ed. A. P. Solomon, 48–49. Cambridge, MA: MIT Press.

Power, Conrad, Alvin Simms, and Roger White. 2000. Hierarchical fuzzy pattern matching for the regional comparison of land use maps. *International Journal of Geographical Information Systems* 15 (1): 77–100.

Rand, William, Moira Zellner, Scott E. Page, Rick Riolo, Daniel G. Brown, and Luis Fernandez. 2002. The complex interaction of agents and environments: An example in urban sprawl. Paper presented at Agent 2002, Chicago.

Real Estate Research Corporation. 1974. The costs of sprawl: Environmental and economic costs of alternative residential patterns at the urban fringe. Washington, DC: U.S. Government Printing Office.

Reynolds, Craig. 1987. Flocks, herds, and schools: A distributed behavioral model. *Computer Graphics* 21 (4): 25–34.

Russell, Stuart, and Peter Norvig. 1995. *Artificial Intelligence: A Modern Approach.* Upper Saddle River, NJ: Prentice Hall.

Sanders, L., D. Pumain, H. Mathian, R. Guérin-Pace, and S. Bura. 1997. SIMPOP: A multiagent system for the study of urbanism. *Environment and Planning B* 24: 287–305.

Semboloni, Fernando. 1997. An urban and regional model based on cellular automata. *Environment and Planning B* 24: 589–612.

——. 2000. The growth of an urban cluster into a dynamic self-modifying spatial pattern. *Environment and Planning B: Planning & Design* 27 (4): 549–64.

Silva, Elizabeth A., and Keith C. Clarke. 2002. Calibration of the SLEUTH urban growth model for Lisbon and Porto, Portugal. *Computers, Environment and Urban Systems* 26 (6): 525–52.

Smith, B. 1991. *Morphological Similarities: Taunton, England and Guatemala City, Guatemala.* Buffalo, NY: Department of Geography, State University of New York at Buffalo.

Straatman, Bas, Roger White, and Guy Engelen. 2004. Towards an automatic calibration procedure for constrained cellular automata. *Computers, Environment and Urban Systems* 28 (1/2): 149–70.

Tobler, Waldo. 1970. A computer movie simulating urban growth in the Detroit region. *Economic Geography* 46 (2): 234–40.

——. 1979. Cellular geography. In *Philosophy in Geography,* Eds. S. Gale and G. Ollson, 379–86. Dordrecht: Kluwer.

Torrens, Paul M. 2007. A geographic automata model of residential mobility. *Environment & Planning B: Planning and Design*, 34, 200–222.

Torrens, Paul M., and Itzhak Benenson. 2005. Geographic automata systems. *International Journal of Geographical Information Science* 19 (4): 385–412.

Turner, M. G. 1989. Landscape ecology: The effect of pattern on process. *Annual Review of Ecological Systems* 20: 171–97.

Turner, M. G., and R. H. Gardner, Eds. 1991. *Quantitative Methods in Landscape Ecology, Ecological Studies.* New York: Springer.

Waddell, Paul A. 2000. A behavioural simulation model for metropolitan policy analysis and planning: Residential location and housing market components of UrbanSim. *Environment and Planning B* 27 (2): 247–63.

——. 2001. Between politics and planning: UrbanSim as a decision-support system for metropolitan planning. In *Planning Support Systems: Integrating Geographic Information Systems, Models, and Visualization Tools*, Eds. R. K. Brail and R. E. Klosterman, 201–28. Redlands, CA: ESRI.

——. 2002. UrbanSim: Modeling urban development for land use, transportation and environmental planning. *Journal of the American Planning Association* 68 (3): 297–314.

Waddell, Paul A., Alan Borning, Michael Noth, Nathan Freier, Michael Becke, and Gudmundur Ulfarsson. 2003. Micro-simulation of urban development and location choices: Design and implementation of UrbanSim. *Networks and Spatial Economics* 3 (1): 43–67.

Ward, Douglas P., Alan T. Murray, and Stuart R. Phinn. 2000. A stochastically constrained cellular model of urban growth. *Computers, Environment and Urban Systems* 24: 539–58.

Webster, Christopher J., and Fulong Wu. 1998. Simulations of urban growth with models of pollution property rights and subcentre formation. In Graphics, visualisation and the social sciences: Report from workshop held May 8–9. *Advisory Group on Computer Graphics (AGOCG) Technical Report Series No. 33*, Ed. A. Mumford, 113–22. Loughborough, UK: Loughborough University Joint Information Systems Committee (JISC).

Webster, Chris, Fulong Wu, and Sheng Zhou. 1998. An object-based urban simulation model for interactive visualisation. Paper presented at the Third International Conference on GeoComputation, September 17–19, University of Bristol.

White, Roger, and Guy Engelen. 1993. Cellular automata and fractal urban form. *Environment and Planning A* 25: 1175–99.

——. 1994. Urban systems dynamics and cellular automata: Fractal structures between order and chaos. *Chaos, Solitions, and Fractals* 4: 563–83.

——. 1997. Cellular automata as the basis of integrated dynamic regional modelling. *Environment and Planning B* 24: 235–46.

——. 2000. High-resolution integrated modelling of the spatial dynamics of urban and regional systems. *Computers, Environment and Urban Systems* 24: 383–400.

White, Roger, Guy Engelen, and Inge Uljee. 1997. The use of constrained cellular automata for high-resolution modelling of urban land use dynamics. *Environment and Planning B* 24: 323–43.

Wilson, Emily Hoffhine, James D. Hurd, Daniel L. Civco, Michael P. Prisloe, and Chester Arnold. 2003. Development of a geospatial model to quantify, describe and map urban growth. *Remote Sensing of Environment* 86: 275–85.

Wu, Fulong. 1996. A linguistic cellular automata simulation approach for sustainable land development in a fast growing region. *Computers, Environment and Urban Systems* 20: 367–87.

——. 1998a. An experiment on the generic polycentricity of urban growth in a cellular automatic city. *Environment and Planning B* 25: 731–52.

——. 1998b. SimLand: A prototype to simulate land conversion through the integrated GIS and CA with AHP-derived transition rules. *International Journal of Geographical Information Science* 12 (1): 63–82.

——. 1999. A simulation approach to urban changes: Experiments and observations on fluctuations in cellular automata. Paper presented at the Sixth International Conference on Computers in Urban Planning and Urban Management, September 7–10, Venice.

——. 2002. Calibration of stochastic cellular automata: The application to rural–urban land conversions. *International Journal of Geographical Information Science* 16 (8): 795–818.

Wu, Fulong, and Chris J. Webster. 1998. Simulation of land development through the integration of cellular automata and multicriteria evaluation. *Environment and Planning B* 25: 103–26.

——. 2000. Simulating artificial cities in a GIS environment: Urban growth under alternative regulation regimes. *International Journal of Geographical Information Science* 14 (7): 625–48.

Xie, Yichun. 1996. A generalized model for cellular urban dynamics. *Geographical Analysis* 28 (4): 350–73.

Yeh, Anthony Gar-On, and Xia Li. 1999. Measurement of urban sprawl in a rapid growing region using entropy. In Towards digital earth: Proceedings of the International Symposium on Digital Earth. Beijing: Science Press.

——. 2000. Simulation of compact cities based on the integration of cellular automata and GIS. In *Theoretical and Practical Issues on Cellular Automata*, Eds. S. Bandini and T. Worsch, 170–78. London: Springer-Verlag.

——. 2001. A constrained CA model for the simulation and planning of sustainable urban forms by using GIS. *Environment and Planning B: Planning & Design* 28 (5): 733–53.

——. 2002. A cellular automata model to simulate development density for urban planning. *Environment and Planning B: Planning & Design* 29 (3): 431–50.

22

AGENTS OF CHANGE

Mixed-Race Households and the Dynamics of Neighborhood Segregation in the United States

Mark Ellis, Steven R. Holloway, Richard Wright, and Christopher S. Fowler

Research Question: What role do mixed-race households play in the development of neighborhood segregation?
System Science Method(s): Agent-based models
Things to Notice:

- Combining simulation analysis with conventional census data analysis
- Adaptation of an existing agent-based model

This chapter explores the effects of mixed-race household formation on trends in neighborhood-scale racial segregation. Census data show that these effects are nontrivial in relation to the magnitude of decadal changes in residential segregation. An agent-based model illustrates the potential long-run impacts of rising numbers of mixed-race households on measures of neighborhood-scale segregation. It reveals that high rates of mixed-race household formation will reduce residential segregation considerably. This occurs even when preferences for own-group neighbors are high enough to maintain racial separation in residential space in a Schelling-type model. We uncover a disturbing trend, however: levels of neighborhood-scale segregation of single-race households can remain persistently high even while a growing number of mixed-race households drives down the overall rate of residential segregation. Thus, the chapter's main conclusion is that parsing neighborhood segregation levels by household type—single versus mixed race—is essential to interpret correctly trends in the spatial separation of racial groups, especially when the fraction of households that are mixed race is dynamic. More broadly, the chapter illustrates the importance of household-scale processes for urban outcomes and joins debates in geography about interscalar relationships.

Scholars continue to dispute the causes of residential racial segregation. In the United States, arguments that discrimination against minorities, especially Blacks, remains a potent force (e.g., Galster 1992; Massey and Denton 1993; Yinger 1997; Meyer 2000; Massey and Lundy 2001; Adelman 2005) contrast with claims that preference for own-group neighbors rather than covert discrimination undergirds today's levels of segregation (e.g., Clark 1991, 1992; Thernstrom and

DOI: 10.4324/9781315748771-25

Thernstrom 1997; Fossett 2006; Clark and Fossett 2008).[1] Scholars do agree on this: Declines in segregation occur when members of one racial group move into neighborhoods where other groups account for a larger fraction of the population. We add a twist to this account by considering the effects on residential racial segregation of changes in the fraction of households that are racially mixed. Whenever such households form they combine individuals from two or more racial groups in one neighborhood. Consequently, the more numerous these households, the lower the overall extent of neighborhood-scale segregation between racialized groups. This chapter investigates the sensitivity of neighborhood segregation levels to the presence of mixed-race households using two strategies—one based on trends observable in confidential 1990 and 2000 U.S. Census long-form data, the other deploying an agent-based simulation—to explore how increases in mixed-race household formation have affected neighborhood racial segregation in the recent past and how they might shape it in the long run.

Our project is situated within a well-established and diverse tradition of research on geographies of racial mixing. Segregation analysts investigate the extent to which people from different racialized groups share neighborhoods (e.g., Duncan and Duncan 1955; Taeuber and Taeuber 1964; Clark 1991; Massey and Denton 1993; Logan, Stults, and Farley 2004; Johnston, Poulsen, and Forrest 2007; Brown and Chung 2008). Research on geographies of multiracial people treats the body as the scale of interest, asking where multiracial people live and how their identities are contingent on these locations (e.g., Harris and Sim 2002; Mahtani 2002; Holloway et al. 2009). Studies linking these two scales together find that multiracial people are generally less segregated in residential space than their single-race counterparts (Johnston, Poulsen, and Forrest 2006; Clark and Maas 2009). Our contribution is to foreground the household scale, which, apart from research on the spatial determinants of mixed-race marriage (e.g., Peach 1980), has received scant attention as a scale of racial mixing in geographical inquiry (cf. Wright et al. 2003; Holloway et al. 2005).

Mixed-race households take several forms. Census 2000 revealed that 68 percent of these households were headed by opposite-sex couples, each person in the couple claiming a different race; the remainder included same-sex couples, households with adopted, step- or foster children, and unrelated roommates (Ellis et al. 2007). These contrasting living arrangements signify different degrees and types of intimacy; regardless of the form or depth of the attachment, they all unite people of two or more races in one household location and thereby affect segregation measured by neighborhood. The percentage of individuals in mixed-race households, especially for Whites and Blacks, is relatively small. U.S. Census 2000 microdata revealed that 94.9 percent of Whites, 92.0 percent of Blacks, 83.9 percent of Latinos, and 80.3 percent of Asians lived in single-race household—that is, households made up of people from the same racial group. Prior research shows how these living arrangements exerted a measureable downward effect on neighborhood segregation levels in 1990. Specifically, the residential segregation for those who live in single-race households was higher than the conventionally reported measures of segregation, which includes those who live in all racial household types—single- *and* mixed-race (Ellis et al. 2007). It stands to reason that changes in the frequency of mixed-race households over time could be partly responsible for changes in assessments of aggregate neighborhood segregation (i.e., measured conventionally using all members of racial groups regardless of living arrangement). In what follows we gauge how changing numbers of mixed-race households affect aggregate segregation trends and whether this development obscures what is happening to the spatial separation of the majority who live in single-race households.

We seek not only to devise appropriate empirical strategies to investigate and report on these issues but also to highlight the broader ramifications of our work for conceptualizing the role of mixed-race households, and household demographic change more generally, in the remaking of urban space (Buzar, Ogden, and Hall 2005). With few exceptions, investigations of the formation and location of mixed-race households—most notably how and where mixed-race couples form and where they subsequently live—do not speak to the implications of these developments for the transformation of spaces beyond the household (cf. Wong 1998; Ellis et al. 2007). Bringing the study of mixed-race households into investigations of residential racial segregation joins the study of interracial intimacy within homes to the literature on residential segregation and neighborhood change. This fusion illustrates how racialized processes at one scale—the private space of households—affect racialized dynamics at another scale—the public space of neighborhoods. As such, our work is mindful of arguments about public–private dichotomies and interdependencies featured prominently in scholarship about space and social relations (Marston 2000; Fincher 2004; England 2008; Staeheli and Mitchell 2008). Before reporting on our empirical work, we selectively survey key literatures on mixed-race households, residential racial segregation, and scale that inform our approach.

What Drives Changes in Neighborhood Segregation?

The research on residential racial segregation is voluminous. The general consensus is that between 1990 and 2000, U.S. segregation levels fell, slowly, for Whites and Blacks but increased for Asians and Latinos (Iceland 2004; Logan, Stults, and Farley 2004). The increase in neighborhood spatial isolation for Asians and Latinos resulted from immigration-led population growth in which newcomers located in close proximity to members of their own conational groups rather than behavioral shifts by these groups or others that promoted spatial separation (Iceland and Scopilliti 2008). Absent this immigration, the isolation of these groups would likely have fallen. Several papers disaggregate these and other trends, picking apart the details for specific groups by income, nativity, or some other variable, to gain additional insight. The bulk of this research shows that even as minority incomes or socioeconomic status rises, groups remain segregated, albeit at slightly lower levels than before (Clark and Blue 2004; Iceland and Wilkes 2006; Clark 2007). This persistent segregation leads to two interpretations (Charles 2003): Discrimination still operates, albeit "under the table" given its legal proscription (e.g., Yinger 1997), or people prefer own-group neighbors sufficiently to maintain racial divisions in urban space in the absence of discrimination (e.g., Clark 1991, 1992; Fossett 2006). The preference story has several variants. Some argue that neighborhood racial composition is not the direct motivation for neighborhood choice but a proxy for other neighborhood characteristics, such as crime and poverty, that people with resources seek to avoid (Harris 1999). Others note that Blacks often cite the chilly reception they expect to receive in majority White neighborhoods and so choose to live with similar others (Farley, Fielding, and Krysan 1997; Logan, Stults, and Farley 2004).

Whether one thinks that discrimination still shapes residential real estate markets or considers the evidence of preference studies sufficient to explain persistent segregation, residential mobility is the mechanism that produces changes in the spatial separation of groups. Simply put, people move into one neighborhood from another and the aggregate effect of this mobility on the geography of group population distributions drives residential racial segregation up or down.

Why do people move? Although most segregation research takes a limited view of this process by highlighting the role of own-group preference or racial discrimination (e.g., Farley, Fielding, and Krysan 1997; Bruch and Mare 2006; Clark and Fossett 2008; Krysan and Bader 2008), the residential mobility literature takes a broader perspective. It conceptualizes moves as the result of households searching for suitable housing and neighborhood environments (Rossi 1955; Clark and Dieleman 1996). Transitions in the life course, such as family formation, having children, or entry or exit from the workforce, condition the utility of housing and its surrounds, motivating when and where households move contingent on their income and neighborhood affordability (Brown and Moore 1970; Speare 1974; Clark, Deurloo, and Dieleman 1984, 2006; Clark and Huang 2003). Given geographical variation in the availability of housing types and neighborhood amenities, the aggregate outcome of these decisions is a patterning of urban residential spaces by household size, type, and age (Gober 1981). Families with children tend to cluster together in areas where single-family homes predominate; young single people are disproportionately found in areas where apartments prevail; and the elderly often gravitate to locales nearest particular sets of amenities and services. This geographic sorting will generate racial segregation if racial groups differ in their predominant household types. Thus, understanding segregation dynamics requires not only consideration of the dueling forces of discrimination and preference for own-group neighbors emphasized in the segregation literature but also an awareness of changes in household organization and the distribution of groups across household types.

Empirical work on how household organization and type affect segregation is relatively sparse but finds, for example, that families with children are more racially segregated than single-person households (Iceland et al. 2010). Thus, changes in the way groups organize into households, such as having relatively more single-person households and fewer families with children, could change levels of segregation. Average family size varies across U.S. metropolitan areas (Gober 1981), so this effect might partially explain the observed interurban variation in segregation. Innovative U.K. research shows that differences in life-cycle stage and housing size between non-White immigrants and Whites helps to account for rising rates of White–non-White segregation (Simpson, Gavalas, and Finney 2008).

We aspire to add to this literature by assessing the effects of changes in the frequency of one particular living arrangement—mixed-race households—on residential racial segregation dynamics. Mixed-race households necessarily mingle people from different racial groups in the same neighborhood. A greater propensity to mix within households should drive down (or restrain increases in) segregation at the neighborhood scale, and declines in household-scale mixing should increase (or limit decreases in) neighborhood segregation. The magnitude of these effects will depend additionally on where mixed-race households locate within residential space, specifically, on where they tend to live relative to those in single-race households.

Judging these mixed-race household effects requires some modifications of existing models of segregation dynamics. For example, consider the components of the Schelling schema from which most other models of segregation dynamics are derived. Schelling (1971) investigated how preferences for neighborhood racial composition affect the sorting of people by groups across residential space. The individual or household units who do the preferring, and who move if they do not like the neighborhood's racial mix, belong to one group or another; none of the households are racially mixed. The introduction of mixed-race households to such a model necessitates a distinction between the preferences of single- and mixed-race households because research shows that the latter, particularly households headed by Black–White couples, have a greater preference for racial diversity than the former (Dalmage 2000; Holloway et al. 2005; Wright, Holloway, and Ellis 2011).

Researchers must also decide how potential movers racialize members of mixed-race households when they are evaluating neighborhood environments. For example, do they view them as if they are an all-minority household (a household scale "one-drop rule"; Hollinger 2005), or do they assess neighboring households' populations based on the race of individuals within them (i.e., treating a Black–White couple as one Black neighbor and one White neighbor)? Such assessments will likely condition the effect of changes in the frequency of mixed-race households on residential segregation.

Mixed-Race Households and Neighborhood Segregation

The idea of a relationship between racial mixing in households and neighborhoods has a long history in research on mixed marriage (e.g., Bossard 1932). The relationship we posit—that mixed-race households affect residential segregation across neighborhoods—reverses the directionality of the effects presumed by that literature. The traditional view holds that residential segregation measured at the neighborhood scale—as a measure of social or spatial distance or both—determines mixed-marriage rates: The greater the level of segregation between groups, the lower the rate of marriage between them (e.g., Bossard 1932; Davie and Reeves 1939; Abrams 1943; Kennedy 1943; Clarke 1952; Peach 1980; Morgan 1981; Coleman and Haskey 1986; Lieberson and Waters 1988; Kalmijn and Flap 2001). By suggesting that the effect might work the other way we do not reject the empirical findings of existing mixed-race marriage research, but we are suggesting that they require modification. Much of this work is several decades old and from a time when it was much easier to imagine that neighborhoods circumscribed social worlds, limiting the possibilities for partner selection from further afield. In segregated cities, these circumstances necessarily curtailed partner choice to those in one's own group or to those from groups who were least segregated from it.

Fewer socio-spatial constraints on partner choice exist today. Antidiscrimination laws and affirmative action policies have made workplaces and colleges more racially diverse (Bowen and Bok 1998; Estlund 2003). This, plus the increased entry of women into the labor force, has expanded the range of possibilities for partner selection. At the same time, traditional constraints on partner choice, such as family, have weakened since the 1960s with the growth in independence of young adults (Rosenfeld 2007). More recently, the Internet and social networking tools of the last few years have added to the chances for meaningful contact beyond the boundaries of one's immediate segregated residential milieu (Houston et al. 2005).

We are not arguing that all of the barriers to intimate social interaction between groups have come down or that removal of the remaining obstacles to such contact will be smooth and inevitable. Race remains a salient force in U.S. society, and there are disturbing signs of rollbacks in public policies that promoted desegregation in schools and colleges (e.g., Bowen and Bok 1998; Boger and Orfield 2005). Persistent neighborhood-based segregation, however, is plainly not the constraint to social interaction across racial lines that it was even two decades ago. Increased contact in nonneighborhood spaces (Ellis, Wright, and Parks 2004) combined with relaxing attitudes to interracial intimacy (Romano 2001) and interaction help explain why rates of mixed-race household formation have accelerated in recent years while changes in residential segregation have been sluggish.

We know little about where people meet their spouses or intimate others (Rosenfeld and Thomas 2010). Historically, marital distance—the distance between the residences of spouses prior to marriage—was short (Morrill and Pitts 1967; Coleman 1977; Coleman and Haskey 1986). Most people married someone close by, typically from the neighborhood. This, though, is changing. Recent studies from France and the Netherlands suggest that the neighborhood has declined in importance as a locale for meeting romantic partners, with alternatives such as workplaces becoming more important (Bozon and Héran 1989; Kalmijn and Flap 2001). U.S. research shows that between 15 and 18 percent of people met their partners in the workplace (Laumann et al. 1994). The growth of gender-balanced workplaces appears to be partly responsible for this phenomenon (Svarer 2007). One can easily imagine how racially diverse and gender-balanced workplaces would elevate the odds that some of these romances are racially mixed.

The forces that extend the spatial range of potential romantic partners will also diversify the pool of roommate candidates, which raises the probability that this type of household living arrangement will be also of mixed race. Increased racial diversity in workplaces and colleges within educational and income classes similarly increases the odds that roommates are of different races. These populations—roughly equal on social and economic dimensions other than race—will probably search for housing in overlapping areas from the same workplace or college location.

The expanding contact space argument is not relevant in the case of mixed-race households that form through adoption or foster parenting. Relaxed attitudes are most likely responsible for the increased frequency of these arrangements. Although children from some groups remain heavily favored for transracial adoption over others, Black children, traditionally the least preferred for adoption, are now being adopted at slowly increasing rates in these arrangements, too (Clemetson and Nixon 2006).

The Household Scale and Segregation

Residential racial segregation varies with the spatial scale of measurement (Cowgill and Cowgill 1951; Cortese, Falk, and Cohen 1976; Wong 1997; Kaplan and Holloway 2001). In the United States, scholars typically use census tracts (which mark off areas usually containing between 4,000 and 8,000 people) for segregation analysis because they provide the greatest range of associated social and economic data (Massey and Denton 1988). A few scholars advocate for census blocks, usually the equivalent of a city block but sometimes smaller, to capture the microgeographies of residential racial segregation existing within tracts (Cowgill and Cowgill 1951). Block groups—clusters of blocks—offer a third alternative between larger tracts and smaller blocks (e.g., Frey and Farley 1996). Neighborhood racial segregation measured at the block scale is typically greater than in tracts; generally, the finer the scale of analysis, the higher the degree of segregation (Wong 1997, 2003, 2004). This scale effect is not just a residential phenomenon; segregation analyses of schools and workplaces disguise higher levels of racial separation within specific classrooms, lunchrooms, playgrounds, and workstations (Steinhorn and Diggs-Brown 1999; Tatum 2003).

Tracts and blocks delimited the range of scale possibilities for measuring residential segregation for many years. This has begun to change with the widespread availability of sophisticated geographic information systems and spatial statistics, especially interpolation and smoothing methods. These technologies make it relatively easy to infer segregation at multiple scales. For example, some have interpolated segregation from census data into 100-meter grids, which will correspond roughly to a single household in low-density suburban areas but is the equivalent of many households in more dense multiunit parts of metropolitan areas (Wu and Sui 2001; Sui and Wu 2006). Others propose to measure segregation at the individual scale, taking the view that types of interpersonal contact provide a useful assessment of racial diversity in social spaces (Schnell and Yoav 2001; Echenique and Fryer 2007). In a more integrative vein, some now advocate a continuous approach to scale effects on segregation, aiming to identify which scales matter the most for racial separation (Lee et al. 2008; Reardon et al. 2008, 2009; cf. Fischer et al. 2004).

As part of this growing interest in scale and segregation, some researchers have mused that, with the right data, one could treat the household as a microspatial container of segregation, in effect using the household as another spatial scale of measurement within a continuum that includes blocks and tracts (Omer and Benenson 2002; Reardon et al. 2008). Although this is potentially useful, the household has social characteristics that complicate treating it as just a microspatial version of administrative spatial units or as simply another scale of measurement within the domain of the modifiable areal unit problem. Households are the basic social and demographic organizational unit in society. They form and dissolve, adding and losing members through individual and joint decisions about who to live with (e.g., marriage, cohabitation, divorce) and because of life course events (e.g., births, deaths). These capacities occur within a context of intimacy where interactions have different meaning and significance from those occurring outside the home, regardless of their proximity. Feminists have emphasized the distinctive role of the household as the private locus of reproduction, consumption, and domesticity, and they have contrasted social relations at this scale from those in public scales of work, production, and civic life (Marston 2000; England 2008). This literature cautions, however, that we resist viewing this public–private divide as impermeable and fixed; such a position risks obscuring the substantial impact of changing social relations and demographic events within the domestic sphere on scales beyond the home (Fraser 1989; Marston 2000; Fincher 2004). What goes on within households, including changes in their size and composition, transforms the social geography of the city at large (Buzar, Ogden, and Hall 2005).

Our view of mixed-race households and residential racial segregation fits within this vision of the household as a distinctive private scale where interracial intimacy is qualitatively different in meaning from other forms of close contact between races in public spaces. Increased racial mixing within households signifies reduced social distance between racialized groups in the most intimate of social settings (Kennedy 2003; Moran 2003; Wright et al. 2003). This reduction in turn affects segregation at the public scale of neighborhoods; when people from different race groups live together, it necessitates a change in the distribution of race groups in blocks or tracts (Wong 1998). Such thinking builds on prior research on scale and segregation empirically and conceptually. We seek to measure the effect of changes in the rate of racial mixing within households on levels of racial mixing in neighborhoods. We do this by framing this linkage as a relation between interracial intimacy in the private spaces of mixed households and racial segregation in the public spaces of neighborhoods.

Analysis

The analysis has two goals: first, to show that mixed-race households affect residential segregation dynamics in the present; and, second, to explore how potential growth in the share of households that are racially mixed might drive change in neighborhood segregation levels in the future. Restricted-use census files from 1990 and 2000 provide the data for the first objective. For the second, we deploy a simulation experiment designed to anticipate the possible long-run effects of mixed-race households on residential segregation under a variety of scenarios.

Data

The residential geography of mixed-race households has received relatively scant attention to date partly because there is no publicly released census data on household racial composition at fine geographical scales. Only restricted-use U.S. Census decennial long-form data supply this information and thus we use it for our analysis of mixed-race household effects on residential segregation in the present. We accessed these files in a secure facility after gaining Census Bureau approval.

The analysis takes place at the tract scale for twelve large metropolitan areas[2]—Atlanta, Chicago, Dallas, Detroit, Houston, Los Angeles, Miami, New York, Philadelphia, San Diego, San Francisco, and Washington, DC—examined individually or averaged depending on the question at hand. These twelve locations are home to 39 percent of mixed-race households in the United States, but they are by no means uniform in the local prevalence of mixed-race

Table 22.1 Characteristics of the twelve metropolitan areas

	% of Households mixed race	*National share of mixed-race households (%)*	*% White*	*% Black*	*% Asian*	*% Latino*
United States	8.2	100.0	68.3	12.1	4.2	13.8
12 metros[a]	9.8	39.1	53.9	16.2	7.7	21.0
Atlanta	5.8	1.0	59.1	29.8	3.6	6.6
Chicago	6.9	2.7	59.1	18.7	4.6	16.7
Dallas	9.0	2.0	58.2	14.1	4.3	22.2
Detroit	6.6	1.5	70.6	22.3	2.7	3.0
Houston	9.0	1.8	48.0	16.8	5.3	28.9
Los Angeles	13.2	9.3	38.7	7.6	11.5	40.5
Miami	14.0	2.3	36.4	20.3	2.1	40.4
New York	8.3	7.5	56.2	16.7	7.4	18.3
Philadelphia	6.0	1.6	70.5	19.5	3.5	5.8
San Diego	17.2	2.1	54.8	6.0	10.6	26.8
San Francisco	15.4	4.7	50.3	7.5	20.7	19.7
Washington, DC	8.2	2.6	59.3	26.9	6.1	6.7

Note: (a) The percentages in this row are calculated by pooling the data from the twelve metropolitan areas.

households; some places have rates of household-scale racial mixing well above the U.S. average and some are below the U.S. average (see Table 22.1). They also vary substantially in racial population structure; some metropolitan areas are mostly populated by Whites and Blacks; others have larger percentages of Latinos and Asians. All in all, this small set of metropolitan areas captures a large range of experience in household-scale racial mixing and racial population diversity.

Mixed-race households in our study are those households that contain people who claim different races on the census, regardless of their relationships. Single-race households include only people who claimed the same race. As is typical in most census-based work on segregation, we limited our definitions of racial categories to those who self-identified as White, Black, Asian-Pacific Islander (Asian for short), American Indian, Other, and Latino, with all groups other than Latino restricted to non-Hispanics.[3] Latino, of course, is a multiracial ethnic category representing a wide range of backgrounds, but we follow the overwhelming majority of analysts who treat this group as a unitary "race" category for the purposes of segregation research. This approach allows us to assess the sensitivity of White–Latino segregation as typically reported to the presence of White–Latino households.

Census Results

Using 2000 census data we calculated sets of dissimilarity indexes for our sample metropolitan areas for two populations: the total (i.e., for those living in single-race and mixed-race households) and for those living in single-race households only (i.e., excluding those who live in mixed-race households from the calculation of the index of dissimilarity).[4] The total population measure is that typically reported in the segregation literature. The difference between single-race household and total population segregation measures is a gauge of how

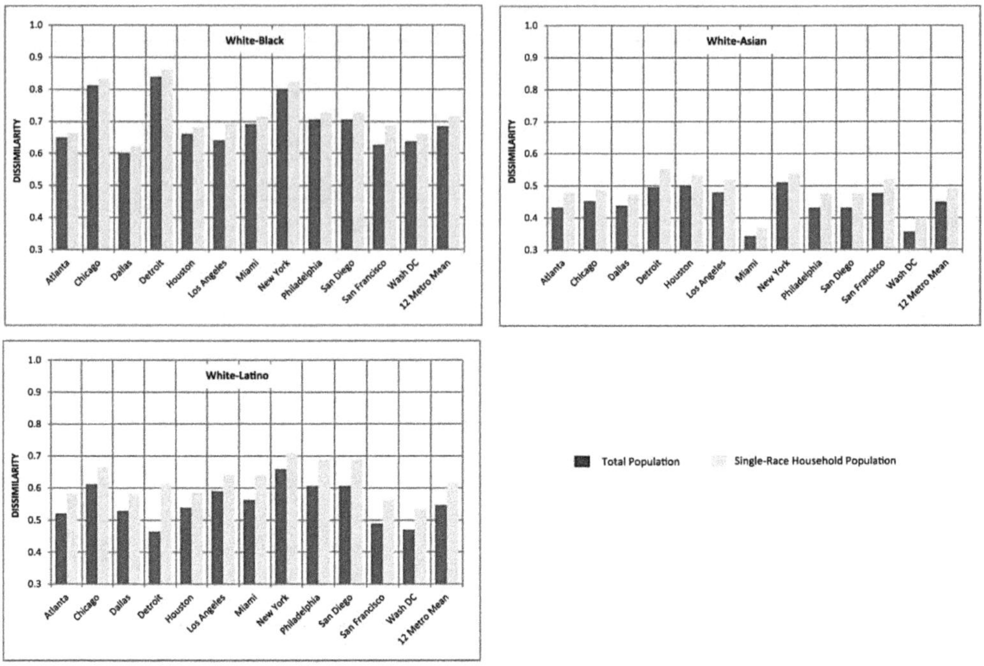

Figure 22.1 Dissimilarity for total population and single-race household population, 2000

much those who live in mixed-race households affect total neighborhood segregation. We restrict our presentation to the analysis of segregation between Whites and the three largest minority groups—Latinos, Blacks, and Asians. These pairings account for the bulk of the population and a very large majority of mixed-race households (Holloway et al. 2005).

Figure 22.1 plots White–Black, White–Asian, and White–Latino total population and single-race household only dissimilarity index values by metropolitan area. In line with previous findings, White–Black dissimilarity for the total population is consistently the highest in all metropolitan areas and White–Asian dissimilarity tends to be the lowest. In addition, single-race household populations are always more segregated than are total populations (lighter bars are higher than darker bars for all pairs in all metropolitan areas), confirming previous research based on 1990 census data (Ellis et al. 2007). Figure 22.2 renders the gaps between single-race household and total population segregation more visible. Each panel depicts the difference between these two values for a specific pair of groups by metropolitan area. These differences are generally smallest in the White–Black case, largest for White–Latino segregation, and somewhere between these extremes for White–Asian segregation. Without mixed-race households, White–Black segregation would be an average of 0.03 points higher (on a 0–1 scale); White–Asian segregation higher by 0.04 points; and White–Latino segregation 0.07 points higher. In percentage terms, White–Black segregation without mixed-race households would be about 5 percent greater than currently recorded; White–Asian and White–Latino segregation would be 10 percent greater.

Figure 22.3 expands the analysis another step. Consistent with the low rate of Black–White intermarriage (Qian and Lichter 2007), Whites and Blacks are the least likely to share households; this accords with the small difference between single-race household and total population segregation for this pairing. On average, about 12 percent of Asians and Latinos share households with Whites, which is three times the same percentage for Blacks. The average percentage of

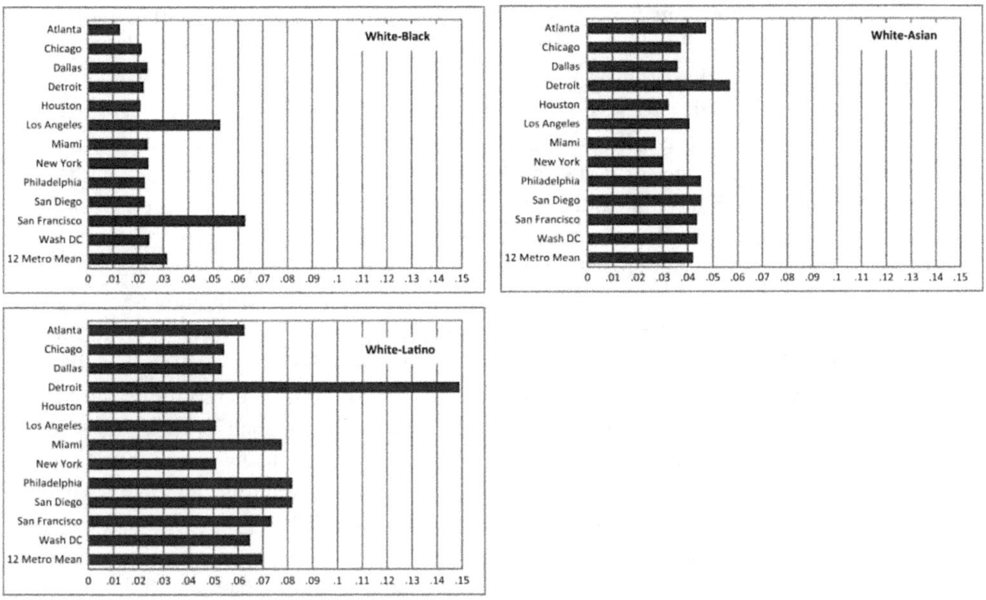

Figure 22.2 Difference between total population and single-race household population dissimilarity, 2000

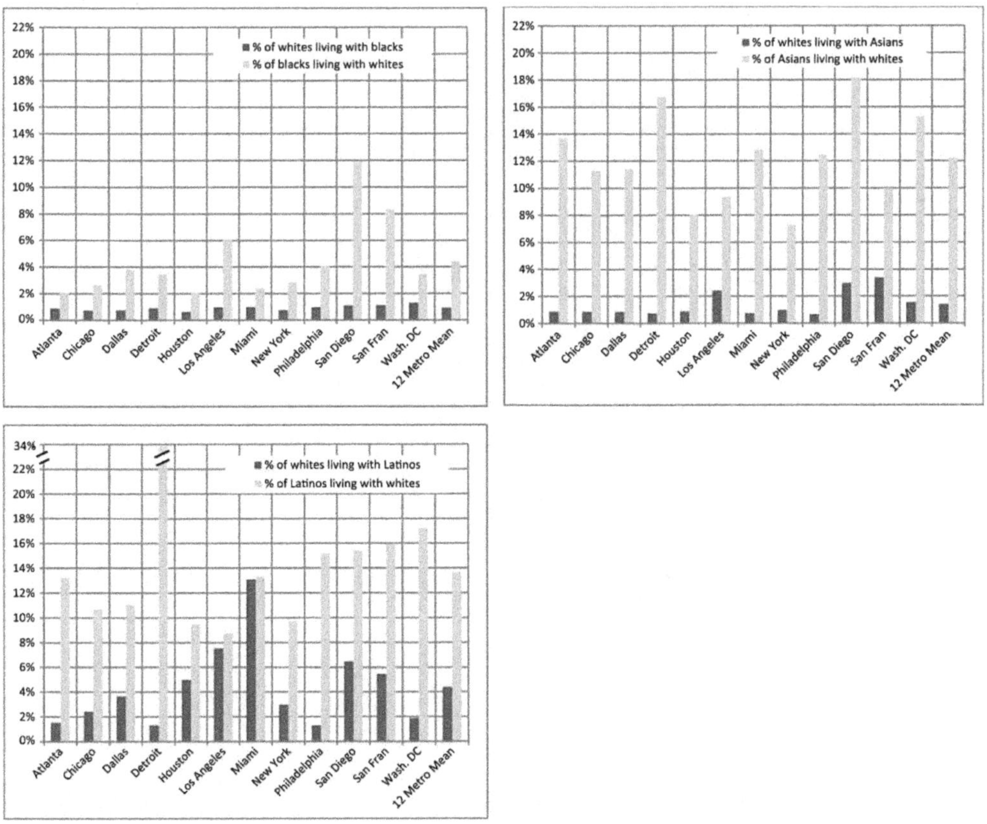

Figure 22.3 Percentage of group's population sharing households with other groups

Whites living with Latinos, however, is double that for Asians. Thus, it is unsurprising that the mixed-race household effect—the gap between single-race household and total population segregation—is greater in the White–Latino than the White–Asian case. As one would expect, metropolitan areas with high rates of household-scale mixing tend to have larger gaps between single-race household and total population segregation. For example, Los Angeles and San Francisco have relatively high percentages of Blacks living with Whites, and this corresponds with larger gaps between single-race household and total population segregation in these places. In contrast, San Diego, which also has an above-average percentage of Blacks living with Whites, does not fit this trend. Similar tendencies and occasional anomalies are evident in the White–Asian and White–Latino cases. With a couple of exceptions, metropolitan areas with above-average gaps between single-race household and total population segregation have above-average rates of household mixing (especially by minority populations with Whites). This suggests that although local rates of mixing are important, they are probably not the only factor determining the mixed-race household effect in specific metropolitan areas. Other forces, such as differences in the neighborhood geography of single-race and mixed-race households, likely matter, too.

This all funnels into the central question motivating this article: How did changes in rates of household mixing in the 1990s affect changes in segregation? The cross-sectional evidence suggests that spatial variation in mixing rates plays a substantial role in inter-metropolitan variations in the magnitude of the mixed-race household effect. Changes in those rates could thus leverage meaningful change in the intensity of neighborhood racial segregation in specific metropolitan areas over time. Rates of racial mixing in households certainly changed during the 1990s, with shifts varying in magnitude and direction by group and by metropolitan area (Figure 22.4).

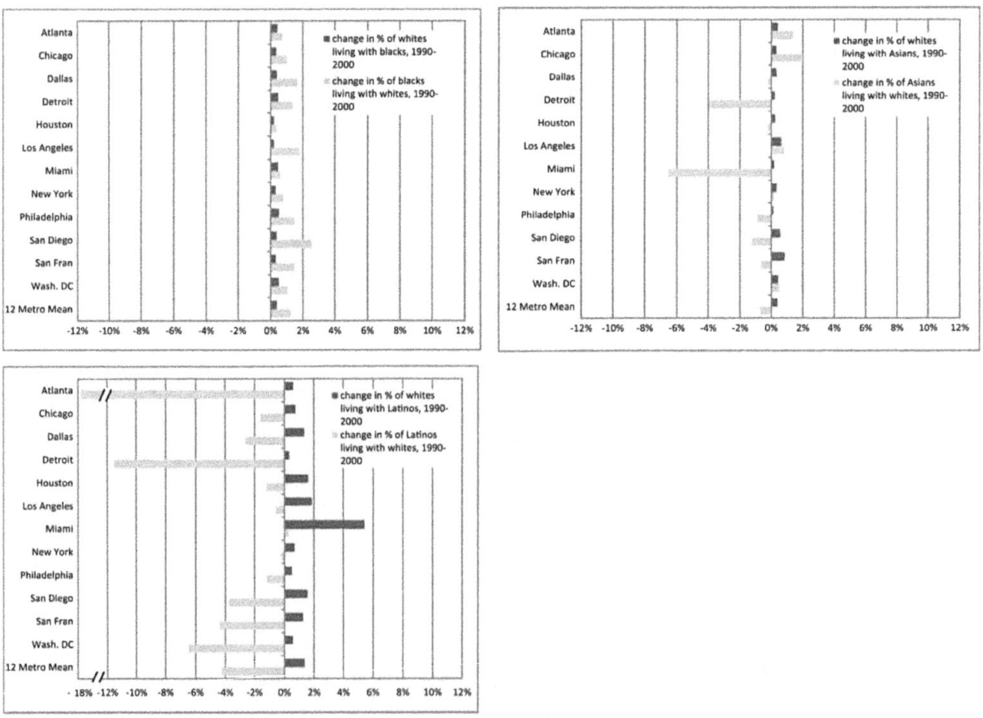

Figure 22.4 Change in percentage of group's population sharing households with other groups, 1990–2000

White–Black rates of household racial mixing increased within a modest range in all metropolitan areas, with Dallas, Detroit, Los Angeles, Philadelphia, San Diego, and San Francisco registering above-average increases in the percentages of Blacks living with Whites. White–Asian and White–Latino pairings demonstrate greater intermetropolitan variation. Whites were more likely to live in households with both Asians and Latinos everywhere at decade's end; Asians and Latinos were less likely to live with Whites in 2000 than in 1990 in the majority of metropolitan areas, a drop most likely due to the arrival of already partnered Asians and Latinos through immigration. This percentage decline in living with Whites was especially strong for Latinos, with only Miami bucking the downward trend. For Asians, the percentage decline in living with Whites is large only in metropolitan areas with small Asian populations (Detroit and Miami); averaged across metropolitan areas, this decline is quite small.

What if we could eliminate the changes shown in Figure 22.4 from our measurement of segregation in 2000? This would allow us to gauge White–Black segregation change in the 1990s absent the increase in White–Black cohabitation, and White–Asian and White–Latino segregation change without Asian and Latino declines in mixed-race living arrangements. A counterfactual experiment provides the means to do just this and involves estimating segregation levels in 2000 holding mixed-race household formation at 1990 rates. The counterfactual mimics the categories in earlier analysis by splitting a group's population into two components: those living in single-race households and those living in mixed-race households. We assume that the relative distribution of these subpopulations across tracts will be as observed in 2000. We alter the numbers who live in these two types of households in 2000, however, to reproduce the percentage living in mixed-race households in 1990.[5] In effect, the counterfactual assumes that neighborhood racial geography would evolve in accordance with observed changes from 1990 to 2000—the most reasonable assumption one could make about such geography without additional information—but that the frequency of racial mixing within households would stay constant at 1990 levels. This allows us to measure counterfactual segregation in 2000—the level of segregation that would exist in 2000 without the preceding decade's change in the prevalence of household-scale racial mixing.

Figure 22.5 charts actual total population segregation change between 1990 and 2000 alongside counterfactual segregation change (the difference between 1990 total population segregation and 2000 counterfactual segregation) for the three pairings of interest. The key feature to note in Figure 22.5 is whether and to what extent the counterfactual bars end to the right or left of the actual bars. The panel for White–Black segregation shows that in all twelve metropolitan areas the counterfactual bars end to the right of the actual bars, which indicates that household racial mixing had a consistent integrative effect—that is, the 1990s decline in segregation between these groups would have been smaller if the fraction of their populations in mixed-race households was fixed at the (lower) 1990 levels. For all twelve metropolitan areas, the average decline in total population segregation (the conventionally reported measure) between Blacks and Whites was a little over .03 points (on a scale of 0–1). Setting the percentage in mixed-race households to 1990 levels reduces this decline by .01 points (i.e., about one third), a substantial reduction in a slow declining index. Put differently, in the absence of increases in the fraction of the population who live in mixed-race households, overall White–Black segregation in these twelve metropolitan areas would have dropped by only about two-thirds of the actual recorded decline. In terms of specific places, the counterfactual shows that increased percentages in mixed-race household populations have been especially important in driving down the level of White–Black segregation in San Diego and San Francisco, two metropolitan areas with high percentages of Blacks living with Whites in 2000, and above-average increases in those percentages in the 1990s. In Houston, White–Black segregation would

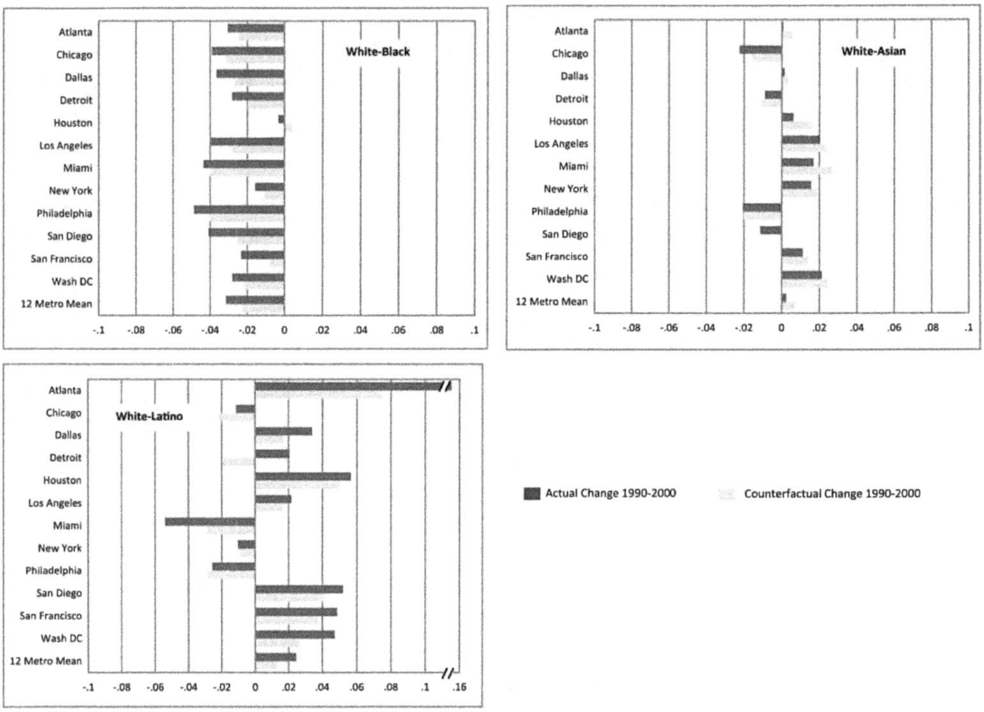

Figure 22.5 Actual versus counterfactual change in dissimilarity, 1990–2000

have increased in the 1990s without its very small increases in the percentages of Blacks and Whites living together.

In the White–Asian panel, the counterfactual bars end to the right of the actual bars in every metropolitan area except for Detroit. Overall, then, changes in the percentage of Whites and Asians living with each other again had an integrative effect. These changes either reduced the increase in White–Asian segregation (Atlanta, Dallas, Houston, Los Angeles, Miami, New York, San Francisco, and Washington, DC) or augmented the decline in White–Asian segregation (Chicago, Philadelphia, and San Diego). The percentage of Whites and Asians living together increased in five of these metropolitan areas (Atlanta, Chicago, Los Angeles, New York, and Washington, DC). In the remaining metropolitan areas, the percentage of Asians living with Whites declined. In all but one of these places—Detroit—the increased percentage of Whites living with Asians was apparently large enough to more than counteract this decline and yield, on balance, an aggregate tendency toward neighborhood integration.

In the White–Latino panel, the counterfactual bars end to the left of the actual bars in ten of the twelve metropolitan areas, which signals that compositional changes (i.e., the decrease in the percentage of Latinos living with Whites relative to the increase in the percentage of Whites living with Latinos) had a segregative effect by either increasing residential segregation or reducing the magnitude of neighborhood segregation decline. Counterfactually, if Latinos and Whites lived together at 1990 rates, segregation would have declined more than it did (Chicago and Philadelphia), declined instead of increased (Detroit), or increased less than it did (Atlanta, Dallas, Houston, Los Angeles, San Diego, San Francisco, and Washington, DC). Miami, and to a lesser degree New York, are outliers from this trend (i.e., 1990–2000 changes in the White–Latino propensity to share households had an integrative impact). Miami's status is easy to explain: It was the only metropolitan area in the 1990s where both the percentage of Whites living with Latinos and the percentage of Latinos living with Whites increased. In New York, the decline in the percentage of Latinos living with Whites was slight (–0.28 percent), which appears to have been insufficient to outweigh the integrative effect of the growing percentage of Whites living with Latinos.

The general finding of the counterfactual experiment is that changes in the propensity of people to live in mixed-race households during the 1990s, even when the direction of these changes diverged among race groups, was an important component of residential segregation change between 1990 and 2000 for all three pairs of groups in most of the twelve metropolitan areas studied. The direction of this effect was always integrative for White–Black segregation and almost always integrative for White–Asian segregation. In the latter case, increases in the percentage of Whites living with Asians generated integrative effects that were stronger than the segregative effect produced by the declines in the Asian percentage living with Whites. With the exception of Miami and New York, the opposite occurs in the White–Latino pairing in the majority of metropolitan areas; declines in the percentage of Latinos living with Whites in the 1990s more than offset the integrative tendency of increases in the fraction of Whites living with Latinos. Regardless of these specific results for each pairing, our analysis indicates that current interpretations of the causes of segregation change in the 1990s are incomplete unless they account for changes in the propensity of people to form mixed-race households.

Simulation Results

Census data provide glimpses of how the changing organization of populations in single- and mixed-race households can affect neighborhood segregation dynamics over a relatively short time span of ten years. What about the long run? What will happen to residential segregation at the neighborhood scale if the rates of living in mixed-race households dramatically increase? What will happen under these circumstances to the segregation of those who continue to live in single-race households? And how might these trends be affected by where people in mixed-race households prefer to live (i.e., by the sorts of neighbors they prefer to live among)?

Investigating these sorts of questions requires simulating residential change. These experimental approaches are a familiar option in residential segregation studies because they allow researchers to explore the marginal effect of changes in household preference for neighborhood racial diversity on long-run, system-wide segregation outcomes (e.g., Schelling 1971; Clark 1991; Zhang 2004; Fossett and Waren 2005; Bruch and Mare 2006; Clark and Fossett 2008). As far as we can tell, all prior studies that implemented this type of experimental analysis conceived of the "household" as single race. This means that the only way segregation can decline in these models is if preference structures encourage single-race households to live in diverse neighborhoods. Our variation on this approach lets households be single or mixed race, allowing each household type to have different racial preferences for neighbors and varying the proportion of mixed-race households. This structure allows mixed-race households to lower neighborhood residential segregation in two ways: through greater preference for sharing neighborhood space with other races and through increased proportions of the population living in households with other races.

Schelling's (1971) agent-based model of segregation is the foundation for all subsequent simulation investigations of residential segregation. The model is straightforward: "Agents" are divided into two groups and randomly scattered across a spatial network. They are then set in motion based on a simple preference rule: Move to locations where one's own group is at least 50 percent of the local population. Under this condition, individual agent preference would initially appear to favor a degree of group integration, but the aggregate outcome of individual agents moving to satisfy their apparently diversity-friendly preference rules is extreme segregation. Schelling's contribution was to show how neighborhood space could be highly segregated even when individual agents tolerate near complete integration.

Like others before us, we build on Schelling and adopt preference rules for single-race households that the literature has repeatedly confirmed will generate segregation (e.g., Clark 1991; Laurie and Jaggi 2003; Fossett and Waren 2005; Bruch and Mare 2006). Each single-race household prefers a neighborhood that has a population equal to or greater than 50 percent of the same race.[6] We opt for this well-understood segregation-generating preference structure because our aim is to see how increasing percentages of mixed-race households might change segregation outcomes in a preference environment that otherwise would produce high levels of segregation. In effect, we want to see whether household-scale racial mixing can negate or moderate the known segregating effects of neighborhood preferences in Schelling's agent-based model of segregation.

With these segregation-generating preferences of single-race households as a backdrop, we explore the effects of two preference structures for mixed-race households. We recognize that these two options restrict the possibilities for mixed-race household preference; where such households locate will probably depend on the racial and gender mixtures involved. The two options we selected, however, correspond to those for which the limited literature on mixed-race household residential location has found the clearest empirical support (Dalmage 2000; Wright, Holloway, and Ellis 2011). In the first of these, mixed-race households prefer neighborhoods that have a population equal to or greater than 50 percent of one race present in the household. In effect, this means that mixed-race households favor neighborhoods where one particular race is always in the majority—akin to Schelling's experiment. The difference between this trial and Schelling's is that we can vary the proportion of mixed-race households in the analysis. The second option gives mixed-race households a preference for neighborhood diversity, a possibility that aligns with empirical research that finds Black–White households disproportionately gravitate to racially diverse locales (Wright, Holloway, and Ellis 2011). Under this condition, mixed-race households seek neighborhoods where diversity—measured by entropy—is maximized.[7] Maximum diversity is realized when a neighborhood's population is split evenly between groups.

When the agents in our simulation make choices about where to live—whether to stay or move depending on their neighborhood's racial composition—they have to decide how to evaluate the race of those in their existing or future neighborhood who live in mixed-race households. The simple method is for them to count individuals by race, paying no heed to their living arrangements. But lingering social proscriptions against mixed-race marriage and other mixed-race living arrangements suggest that some might use an alternative rule based on who lives with whom. Specifically, agents might ignore the internal diversity in mixed-race households and identify all of its members as if they were from the same race. This is the household equivalent of the noxious notion of hypodescent or the so-called one-drop rule (Hollinger 2005) in which the racialization of one household member transfers to all others of a different race for the purposes of evaluating neighborhood suitability (Haslanger 2005; Houston 2009). Our simulation analyzes both ontologies—the simple count of individuals and the household one-drop rule—to see whether they modify the effect of the main variables of interest (the preference structures of single- and mixed-race households and the proportion of households that are racially mixed) on segregation.

To explore the effect of these preferences and evaluations, the simulations of two groups, arbitrarily labeled R and G, are run under three distribution conditions—when Gs comprise

10 percent, 25 percent, or 50 percent of the total population—to assess how relative population size affects the results. For each group, we let 20 percent live in one-person households and 80 percent in two-person households. We allow a fraction of households to be single-person units because their existence in real urban environments constrains the fraction of households that can be of mixed race and thus limits the effect household-scale racial mixing can have on neighborhood segregation. Our simulation enhances any effect of this constraint in actual cities because the 20 percent share is roughly double the percentage of persons in the United States who live in single-person households (9.7 percent in 2000). Two-person households can, of course, be single- or mixed-race households. We vary the fraction of households that are mixed race from 0 percent to 50 percent of all G's households in 5 percent increments. We determine the rate of household mixing based on G's population because R's potential mixing rate is constrained by the availability of partners from G when G's population share falls below 50 percent of the total.

The simulation is enacted on a standard 50 × 50 grid with 2,500 cells. Populating this space are 2,950 individuals aggregated into approximately 1,750 households. Each household occupies one cell, yielding a cell vacancy rate of approximately 30 percent to create opportunities for mobility (Fossett and Waren 2005). These households choose whether to move and where to move to on the basis of a neighborhood's racial composition. Households begin their decision process by evaluating the racial composition of their immediate surroundings. A key assumption within models of this type is the rule that defines the extent of these surroundings. The results presented here employ a Queen's second-order contiguity rule, meaning that the neighborhood size for a single cell ranges from eight cells for corners to as many as twenty-four cells for central cells that are not constrained by an edge.[8] Following Fossett and Waren (2005), we also introduce a rule that every household *must* move the first time they are selected by the simulation to evaluate the suitability of their neighborhood.

The preference and counting options we have described correspond to four scenarios listed in Table 22.2. All four employ the same preference structure for single-race households, varying only in the neighborhood preferences of mixed households and the rules used to count

Table 22.2 Simulation scenarios

Scenario	*Preferences*		*How those in RG households are counted in evaluations of neighborhood preferences*
	R and G households (single-race)	*RG households (mixed-race)*	
1.	Rs prefer neighborhoods > 50% R Gs prefer neighborhoods > 50% G	Prefer neighborhoods at maximum diversity/entropy (E_i)	Rs as Rs, Gs as Gs
2.	Rs prefer neighborhoods > 50% R Gs prefer neighborhoods > 50% G	Prefer neighborhoods at maximum diversity/entropy (E_i)	Rs and Gs counted as Gs
3.	Rs prefer neighborhoods > 50% R Gs prefer neighborhoods > 50% G	Prefer neighborhoods > 50% G	Rs as Rs, Gs as Gs
4.	Rs prefer neighborhoods > 50% R Gs prefer neighborhoods > 50% G	Prefer neighborhoods > 50% G	Rs and Gs counted as Gs

Note: Simulations of these four scenarios are run varying two conditions: (1) the percentage of G in the total population (10%, 25%, 50%); and (2) the fraction of G's households that are mixed race (i.e., RG households as a percentage of all households that include a G person) from 0% to 50% in 5% intervals. R and G are arbitrary group labels.

members of mixed households in the evaluation of neighborhood preferences. We produced 1,000 simulations of each of these scenarios at each permutation of G's percentage of the total population and by 5 percent increments in the percentage of households that are mixed race. Paralleling the census analysis, each run generated two segregation (dissimilarity) scores—one for the total population of R and G (i.e., including single- and mixed-race households) and the other for the single-race household population of R and G. To calculate "neighborhood" segregation, we partitioned the simulation grid into 100 five by five cell groups.

The four panels in Figure 22.6 correspond to the four preference scenarios; each one plots means of these 1,000 segregation scores against the percentage of G's households that are mixed race. All four panels chart three pairs of slopes; each of these pairs plots total population and single-race household population segregation for a specific percentage of G in the total population—10 percent, 25 percent, and 50 percent. Regardless of scenario, three results are conspicuous. First, as the percentage of G's households mixed with R's increases, total population segregation between G and R decreases—that is, growth in mixed-race households drives down segregation in a preference environment known to promote segregation between group members who live in single-race households. Second, single-race households remain highly segregated even as total population segregation declines with increases in the propensity to live in mixed-race households. In other words, growth in the fraction of the population in mixed-race households has little or no effect on the high level of residential segregation of those in single-race households. Scenario does affect this result—single-race household segregation declines slightly with increased percentages of mixed-race households in the household one-drop rule scenarios (2 and 4)—but not enough to challenge the general tendency of single-race households to remain segregated while total segregation drops. Third, as the minority population approaches a 50 percent share of the total population, the mixed-race household effect—the

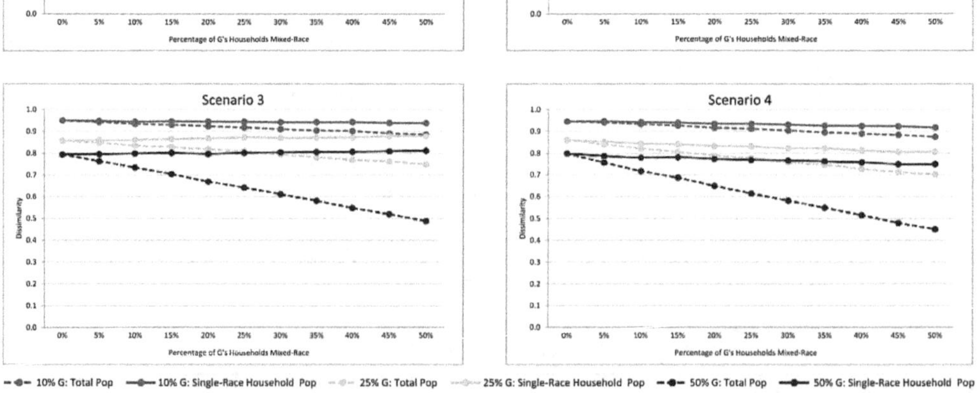

Figure 22.6 Four simulation scenarios showing the effect of rising percentages of mixed-race households on segregation

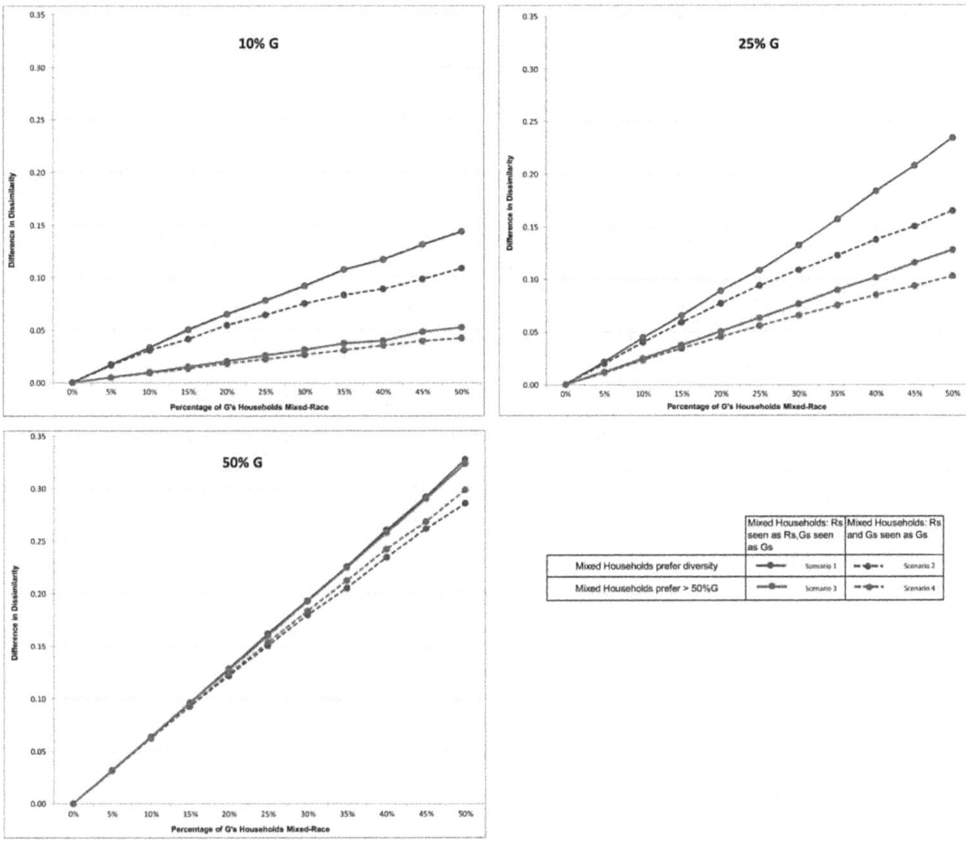

Figure 22.7 Difference in dissimilarity between single-race household and total population by scenario and G's percentage of the population

gap between single-race household and total population segregation—becomes more pronounced at any given percentage of mixed-race households. This is logical; as G's population approaches half of the total population, then the fraction of R's population that can live in mixed-race households necessarily increases. This renders a larger fraction of all households mixed race and therefore increases the size of the mixed-race household effect on total population segregation.

To explore mixed-race household preferences and counting rules some more, Figure 22.7 charts the difference between single-race household and total population segregation by scenario. The three panels depict this difference by the percentage of G's households that are of mixed race. This yields four plotted lines—one for each scenario—in each panel. In effect, Figure 22.7 is the simulation's version of the difference between single-race household and total population segregation plotted from census data in Figure 22.2. When Gs are in the minority—10 percent or 25 percent of the population—a mixed-race household preference for diverse residential space (scenarios 1 and 2) generates a much larger gap between the measurement of single-race household and total population neighborhood segregation than does a preference to be in minority neighborhoods (scenarios 3 and 4). This gap exists regardless of counting rules, but it is smaller when Rs who live in mixed-race households are counted or racialized as Gs (i.e., the household one-drop rule is in effect).

These findings imply, first, that the neighborhood preferences of mixed-race households condition the mixed-race household effect on total population segregation. If these households seek out diverse rather than minority-dominated (i.e., G) neighborhoods, then the mixed-race household effect on total population segregation will be more substantial than without this preference. If, though, mixed-race households gravitate to minority neighborhoods, then the segregation-reducing effects of an increase in their frequency will be attenuated. Second, this neighborhood preference effect interacts with how people in mixed-race households are racialized. Mixed-race household effects on measures of neighborhood segregation are smaller when all those living in a mixed-race household arrangement are racialized as minorities. In this complex dynamic, mixed RG households seek out diverse spaces; applying the notion of hypodescent, the resulting congregations of mixed households appear as a cluster of Gs, which attracts single-race household Gs but repels single-race household Rs. In effect, diverse neighborhoods are unstable, always trending toward domination by Gs. Thus, at any given percentage of mixed-race households, the one-drop counting rule generates greater segregation than the alternative individual counting rule.

When Gs comprise 50 percent of the population these patterns of results almost evaporate. Scenario hardly affects segregation when levels of household-scale racial mixing equal 50 percent. Under conditions of equal population share there is no constraint of minority group size on the majority's odds of finding a minority housemate; accordingly, the effect of the percentage of households that are racially mixed overwhelms any influence of variation in mixed-race household preference or counting rules.

Conclusion

Current understandings of neighborhood racial segregation largely rest on two forces: discrimination in housing market institutions and preferences for neighborhoods with particular racial compositions. Reductions in segregation occur when shifts in these forces change residential mobility patterns sufficiently to increase neighborhood mixing. The results in this chapter show that mixed-race households can also drive change in segregation. Alterations in the propensity to mix racially at the household scale can change residential segregation in neighborhoods; where mixed-race households prefer to live and how others perceive them also plays a role.

Our analysis of U.S. Census data shows that those who live in single-race households, who constitute the bulk of the population, are more segregated than indicated by measures of total population segregation. Without mixed-race households, dissimilarity measures of segregation between Whites and Blacks, and Whites and Latinos would be higher by nontrivial quantities. Counterfactual methods show that in the broadest terms, changes in the extent of household-scale racial mixing during the 1990s affected the magnitude of changes in neighborhood-scale segregation between 1990 and 2000. In this decade, Black–White neighborhood segregation would have declined by a considerably smaller magnitude than recorded—about a third less on average—if not for increases in the percentage of those groups living in Black–White households. Conversely, decreases in household-scale racial mixing had a segregative effect for the White–Latino pairing. Between 1990 and 2000, White–Latino neighborhood segregation increased overall and in most metropolitan areas; a nontrivial share of that increase was due to the increased percentage of Latinos living in Latino-only households. Miami and New York are the only metropolitan areas where changes in rates of White–Latino household mixing in the 1990s had an integrative impact, contributing to the decline in White–Latino residential segregation in those metropolitan areas. Changes in White–Asian mixed-race household percentages in the 1990s generally had an integrative effect, either accelerating declines or slowing increases in White–Asian dissimilarity. This integrative effect is present even in metropolitan areas where the percentage of Asians living with Whites declined; it appears that small increases in the percentage of Whites living with Asians were more than enough to cancel out the segregative impact of the larger percentage declines in Asians living with Whites. This result is possible because the White population is much larger than that of Asians in most places.

Our results also illustrate that the percentage living in mixed-race households does not always increase for non-White groups. For Latinos, ten out of twelve metropolitan areas sampled experienced a decline in the percentage living with Whites. We suspect that this reflects the impact of the considerable immigration of single-race households in the 1990s. The impacts of this demographic shift are legible in the neighborhood segregation statistics—when non-Whites become less likely to share households with Whites, White–non-White residential segregation either increases or it declines with smaller magnitude.

Our simulation experiment revises Schelling's classic conclusions. We show how growing shares of a population in mixed-race households lower segregation even under neighborhood preference conditions known to be segregation inducing. In other words, if mixing at the scale of the household is increasing, then segregation declines even when single-race households prefer to live in neighborhoods in which they are the majority. This trend is moderated by where mixed-race households prefer to live. If racially mixed households opt for diverse neighborhoods, segregation between minority and majority drops more than if they prefer to live in minority neighborhoods.

The significance of these results is in their demonstration that mixed-race households can lower overall segregation even while the segregation of single-race households remains constantly high. More specifically, researchers must be cautious about interpretations of declining segregation in the presence of growing fractions of the population living in mixed-race households. These declines might emanate in large part from increases in the portion of each racial group sharing households rather than from increasing preference for neighborhood diversity by single-race households or by reductions in housing market discrimination. Growth in the number of mixed-race households is unquestionably an expression of greater interracial tolerance and understanding. This outcome, however, might muddy the waters when it comes to making sense of change in neighborhood-scale segregation for the majority who live in single-race households.

This investigation should be seen as a sketch for a much broader research agenda. We need to know more about where mixed-race households locate; about how these locations depend on factors such as income; and about how key housing market agents, such as realtors and lenders, treat mixed-race versus single-race households. Ideally, any assessments of change in neighborhood racial segregation between 2000 and 2010 should take account of changes in the propensity to live in mixed-race households. More generally, the idea that households matter for neighborhood racial change needs greater empirical excavation. In this regard, our study underlines Buzar, Ogden, and Hall's (2005) call for greater attention to how the dynamics of household configurations and urban structure interrelate.

Acknowledgments

This chapter originally appeared in *Annals of the Association of American Geographers*, 102, 549–570 (2012). This research was made possible by grants from the National Science Foundation and the Russell Sage Foundation. Thanks to Rebecca Acosta, Angela Andrus, and Kevin McKinney at the California Census Research Data Center for assistance with the data. Warm thanks to Margaret East, who provided indispensible research support, and Serin Houston, who has been in on our conversations almost from the start. We are grateful to seminar participants at the University of St. Andrews, Dartmouth College, University of California Berkeley, and University of Connecticut for their insightful comments on earlier versions of this article; and to the anonymous referees at the Annals for their constructive suggestions on this article. This article reports the results of research and analysis undertaken while the authors were conducting research approved by the Center for Economic Studies at the U.S. Census Bureau. Any opinions and conclusions expressed herein are those of the author(s) and do not necessarily represent the views of the U.S. Census Bureau. All results have been reviewed to ensure that no confidential information is disclosed. Support for this research at the UCLA and Berkeley Research Data Centers from the National Science Foundation (ITR-0427889) is gratefully acknowledged. Additional support came from a Eunice Kennedy Shriver National Institute of Child Health and Human Development research infrastructure grant, 5R24HD042828, to the Center for Studies in Demography & Ecology at the University of Washington.

Notes

1. The current segregation debate does not engage much explicitly with issues of class, even though these issues have been of great concern in past work (Kaplan and Holloway 1998). Although some research continues to explore the persistence and decline of racial segregation by income strata (e.g., Darden and Kamel 2000), most current work tends to subsume such questions within the preferences versus discrimination rubric. We do not minimize class issues (whether understood in Marxian or Weberian terms) but rather situate our analysis in the context of the segregation debates that currently dominate the literature.
2. We define metropolitan areas in accordance with U.S. Census definitions as Consolidated Metropolitan Statistical Areas (CMSAs) that encompass groups of spatially contiguous and economically linked metropolitan regions (e.g., New York and Los Angeles) and Metropolitan Statistical Areas (MSAs) for smaller free-standing metropolitan regions.
3. These categories are consistent with those in use in the 1990 Census. Because we also use 2000 Census data to assess the effect of mixed-race households on decadal segregation trends, we shoehorn changes in racial categorizations from that census back into 1990 categories. Some of these recategorizations are simple (e.g., merging Pacific Islanders and Asians into an aggregate Asian-Pacific Islander category). Reassigning the 2.4 percent of those who chose more than one race in 2000 into a single-race category consistent with 1990 data is more complex. We opted for one of the deterministic whole-race assignment methods recommended by the Office of Management and Budget (OMB). This method—"Largest Group Other than White"— assigns responses that include White with some other racial group, to the other group, but responses with two or more racial groups other than White are assigned into the group with the highest single-race count" (U.S. Office of Management and Budget 2000, 88). Regardless of method, these assignments are at best awkward and no doubt offensive to many multiracial people. If one is to undertake any analysis of segregation over time, however, the use of such assignment methods is, unfortunately, a necessity.
4. We used the ubiquitous dissimilarity index (D) to capture unevenness in residential distributions between mixed-race and same-race households:

$$D = .5 * \sum_{j=1}^{J} \left| \left(\frac{w_j}{W} - \frac{x_j}{X} \right) \right|$$

where j indexes census tracts, and w and x index two racial groups. W and X are the total populations of groups w and x, respectively, across all tracts, and w_j and x_j are tract counts of the respective groups. Values of D range between 0 (no segregation) and 1 (maximum segregation).

5. The counterfactual procedure adjusts the 2000 census tract populations of each race group living in single-and mixed-race households to levels that, when summed across a metropolitan area, produce the same percentage of population living in mixed-race households—for each race group in each metropolitan area—that existed in 1990. The 2000 tract data are adjusted thus:

$$W^c_{s_j} = \left[(1 - M_{9w}) \sum_j W_{0j}\right] \cdot q0w_{s_j}$$

$$W^c_{m_j} = \left[M_{9w} \sum_j W_{0j}\right] \cdot q0w_{m_j}$$

where W^c_{sj} is the 2000 single-race household counterfactual population of group W in tract j; W^c_{mj} is the 2000 mixed-race household counterfactual population of W living in the same tract; M_{9w} is the fraction of the metro area population of W living in mixed-race households in 1990; W_{0j} is the population of W in tract j in 2000; q/w_{s_j} is the fraction of W's metro area population who live in single-race households in tract j in 2000; and q/w_{m_j} is the fraction of W's metro area population who live in mixed-race households in tract j in 2000.
6. There is ample evidence that although both Whites and non-Whites prefer neighborhoods in which their own racial group is in the compositional majority, there are distinct differences in their preferences. Blacks, for example, report that they prefer neighborhoods that are relatively evenly mixed (i.e., 50 percent White and 50 percent Black), whereas Whites prefer neighborhoods in which the White share never falls below 80 or 85 percent.
7. The entropy of neighborhood i's racial mix is calculated as:

$$E_i = \sum_{m=1}^{M} \pi_m \star \ln\left(\frac{1}{\pi_m}\right)$$

where π_m is the proportion of the population of group m in the neighborhood i. E_i is maximized when there is an equal probability in i of being from any of the M groups. $E_i = 0$ when only one race is found in a neighborhood.
8. We conducted sensitivity analyses employing Queen's first-, third-, and fourth-order contiguity rules. Although these assumptions do alter the results in terms of magnitude of the effects (larger vision equates to larger effect), they do not change the ordering or direction of the effects or change the arguments and conclusions we can draw from the results.

References

Abrams, R. H. 1943. Residential propinquity as a factor in marriage selection: Fifty year trends in Philadelphia. *American Sociological Review* 8 (3): 288–94.

Adelman, R. 2005. The roles of race, class, and residential preferences in the neighborhood racial composition of middle-class Blacks and Whites. *Social Science Quarterly* 86 (1): 209–28.

Boger, J. C., and G. Orfield, Eds. 2005. *School Resegregation: Must the South Turn Back?* Chapel Hill, NC: University of North Carolina Press.

Bossard, J. H. S. 1932. Residential propinquity as a factor in marriage selection. *The American Journal of Sociology* 38 (2): 219–24.

Bowen, W. G., and D. C. Bok. 1998. The shape of the river: Long-term consequences of considering race in college and university admissions. Princeton, NJ: Princeton University Press.

Bozon, M., and F. Héran. 1989. Finding a spouse: A survey of how French couples meet. *Population* 44 (1): 91–121.

Brown, L. A., and S. Y. Chung. 2008. Market-led pluralism: Rethinking our understanding of racial/ethnic spatial patterning in U.S. cities. *Annals of the Association of American Geographers* 98 (1): 180–212.

Brown, L. A., and E. G. Moore. 1970. The intra-urban migration process: A perspective. *Geografiska Annaler Series B, Human Geography* 52 (1): 1–13.

Bruch, E. E., and R. D. Mare. 2006. Neighborhood choice and neighborhood change. *American Journal of Sociology* 112 (3): 667–709.

Buzar, S., P. E. Ogden, and R. Hall. 2005. Households matter: The quiet demography of urban transformation. *Progress in Human Geography* 29 (4): 413.

Charles, C. Z. 2003. The dynamics of racial residential segregation. *Annual Review of Sociology* 29 (1): 167–207.

Clark, W. A. V. 1991. Residential preferences and neighborhood racial segregation: A test of the Schelling segregation model. *Demography* 28 (1): 1–19.

——. 1992. Residential preferences and residential choices in a multiethnic context. *Demography* 29 (3): 451–66.

——. 2007. Race, class, and place: Evaluating mobility outcomes for African Americans. *Urban Affairs Review* 42 (3): 295–314.

Clark, W. A. V., and S. A. Blue. 2004. Race, class, and segregation patterns in U.S. immigrant gateway cities. *Urban Affairs Review* 39 (6): 667–68.

Clark, W. A. V., M. C. Deurloo, and F. M. Dieleman. 1984. Housing consumption and residential mobility. *Annals of the Association of American Geographers* 74 (1): 29–43.

——. 2006. Residential mobility and neighbourhood outcomes. *Housing Studies* 21 (3): 323–42.

Clark, W. A. V., and F. M. Dieleman. 1996. Households and housing: Choice and outcomes in the housing market. New Brunswick, NJ: Center for Urban Policy Research.

Clark, W. A. V., and M. Fossett. 2008. Understanding the social context of the Schelling segregation model. *Proceedings of the National Academy of Sciences* 105 (11): 4109.

Clark, W. A., and Y. Huang. 2003. The life course and residential mobility in British housing markets. *Environment and Planning A* 35 (2): 323–40.

Clark, W. A. V., and R. Maas. 2009. The geography of a mixed-race society. *Growth and Change* 40 (4): 565–93.

Clarke, A. C. 1952. An examination of the operation of residential propinquity as a factor in mate selection. *American Sociological Review* 17: 17–22.

Clemetson, L., and R. Nixon. 2006. Overcoming adoption's racial barriers. *The New York Times*, August 17. Available at: www.nytimes.com/2006/08/17/us/17adopt.html (last accessed February 6, 2011).

Coleman, D. A. 1977. The geography of marriage in Britain, 1920–1960. *Annals of Human Biology* 4 (2): 101–32.

Coleman, D. A., and J. C. Haskey. 1986. Marital distance and its geographical orientation in England and Wales, 1979. *Transactions of the Institute of British Geographers* 11 (3): 337–55.

Cortese, C. F., R. F. Falk, and J. K. Cohen. 1976. Further considerations on the methodological analysis of segregation indices. *American Sociological Review* 41 (4): 630–37.

Cowgill, D. O., and M. Cowgill. 1951. An index of segregation based on block statistics. *American Sociological Review* 16 (6): 825–31.

Dalmage, H. M. 2000. *Tripping on the Color Line: Black–White Multiracial Families in a Racially Divided World.* New Brunswick, NJ: Rutgers University Press.

Darden, J. T., and S. M. Kamel. 2000. Black residential segregation in the city and suburbs of Detroit: Does socioeconomic status matter? *Journal of Urban Affairs* 22 (1): 1–13.

Davie, M. R., and R. J. Reeves. 1939. Propinquity of residence before marriage. *American Journal of Sociology* 44 (4): 510.

Duncan, O. D., and B. Duncan. 1955. A methodological analysis of segregation indexes. *American Sociological Review* 20 (2): 210–17.

Echenique, F., and R. G. Fryer. 2007. A measure of segregation based on social interactions. *Quarterly Journal of Economics* 122 (2): 441–85.

Ellis, M., R. Wright, and V. Parks. 2004. Work together, live apart? Geographies of racial and ethnic segregation at home and work. *Annals of the Association of American Geographers* 94 (3): 620–37.

Ellis, M., S. R. Holloway, R. Wright, and M. East. 2007. The effects of mixed-race households on residential segregation. *Urban Geography* 28 (6): 554–77.

England, K. 2008. Welfare provision, welfare reform, welfare mothers. In *Handbook of Political Geography*, Eds. K. Cox, M. Low, and J. Robinson, 141–53. Los Angeles, CA: Sage.

Estlund, C. 2003. *Working Together: How Workplace Bonds Strengthen a Diverse Democracy*. New York: Oxford University Press.

Farley, R., E. L. Fielding, and M. Krysan. 1997. The residential preferences of Blacks and Whites: A four-metropolis analysis. *Housing Policy Debate* 8: 763–800.

Fincher, R. 2004. From dualisms to multiplicities: Gendered political practices. In *Mapping Women, Making Politics: Feminist Perspectives on Political Geography*, Eds. L. Staeheli, E. Kofman, and L. Peake, 49–70. London and New York: Routledge.

Fischer, C. S., G. Stockmayer, J. Stiles, and M. Hout. 2004. Distinguishing the geographic levels and social dimensions of US metropolitan segregation, 1960–2000. *Demography* 41 (1): 37–59.

Fossett, M. 2006. Ethnic preferences, social distance dynamics, and residential segregation: Theoretical explorations using simulation analysis. *Journal of Mathematical Sociology* 30: 185–273.

Fossett, M., and W. Waren. 2005. Overlooked implications of ethnic preferences for residential segregation in agent-based models. *Urban Studies* 42 (11): 1893–1917.

Fraser, N. 1989. *Unruly Practices: Power, Discourse, and Gender in Contemporary Social Theory*. Minneapolis, MN: University of Minnesota Press.

Frey, W. H., and R. Farley. 1996. Latino, Asian, and Black segregation in U.S. metropolitan areas: Are multi-ethnic metros different? *Demography* 33 (1): 35–50.

Galster, G. C. 1992. Research on discrimination in housing and mortgage markets: Assessment and future directions. *Housing Policy Debate* 3 (2): 639–83.

Gober, P. 1981. Falling household size and its effect on metropolitan population growth and density. *The Annals of Regional Science* 15 (3): 1–10.

Harris, D. R. 1999. "Property values drop when Blacks move in, because . . . ": Racial and socioeconomic determinants of neighborhood desirability. *American Sociological Review* 64 (3): 461–79.

Harris, D. R., and J. J. Sim. 2002. Who is multiracial? Assessing the complexity of lived race. *American Sociological Review* 67 (4): 614–27.

Haslanger, S. A. 2005. You mixed? Racial identity without racial biology. In *Adoption Matters: Philosophical and Feminist Essays*, Eds. S. A. Haslanger and C. Witt, 265–90. Ithaca, NY: Cornell University Press.

Hollinger, D. A. 2005. The one-drop rule and the one-hate rule. *Daedalus* 134: 18–28.

Holloway, S. R., M. Ellis, R. Wright, and M. Hudson. 2005. Partnering "out" and fitting in: Residential segregation and the neighbourhood contexts of mixed-race households. *Population, Space and Place* 11 (4): 299–324.

Holloway, S. R., R. Wright, M. Ellis, and M. East. 2009. Place, scale and the racial claims made for multiracial children in the 1990 U.S. Census. *Ethnic and Racial Studies* 32 (3): 522–47.

Houston, S. 2009. Scales of whiteness and racial mixing: Challenging and confirming racial categories. *The Geographical Bulletin* 50 (2): 93–110.

Houston, S., R. Wright, M. Ellis, S. Holloway, and M. Hudson. 2005. Places of possibility: Where mixed-race partners meet. *Progress in Human Geography* 29 (6): 700–17.

Iceland, J. 2004. Beyond black and white metropolitan residential segregation in multi-ethnic America. *Social Science Research* 33 (2): 248–71.

Iceland, J., and R. Wilkes. 2006. Does socioeconomic status matter? Race, class, and residential segregation. *Social Problems* 53 (2): 248–73.

Iceland, J., and M. Scopilliti. 2008. Immigrant residential segregation in U.S. metropolitan areas, 1990–2000. *Demography* 45 (1): 79–94.

Iceland, J., K. A. Goyette, K. A. Nelson, and C. Chan. 2010. Racial and ethnic residential segregation and household structure: A research note. *Social Science Research* 39 (1): 39–47.

Johnston, R., M. Poulsen, and J. Forrest. 2006. Ethnic residential segregation and assimilation in British towns and cities: A comparison of those claiming single and dual ethnic identities. *Migration Letters* 3: 11–30.

——. 2007. The geography of ethnic residential segregation: A comparative study of five countries. *Annals of the Association of American Geographers* 97 (4): 713–38.

Kalmijn, M., and H. Flap. 2001. Assortative meeting and mating: Unintended consequences of organized settings for partner choices. *Social Forces* 79: 1289.

Kaplan, D. H., and S. R. Holloway. 1998. *Segregation in Cities*. Washington, DC: Association of American Geographers.

——. 2001. Scaling ethnic segregation: Causal processes and contingent outcomes in Chinese residential patterns. *GeoJournal* 53 (1): 59–70.

Kennedy, R. 2003. *Interracial Intimacies: Sex, Marriage, Identity, and Adoption*. New York: Random House.

Kennedy, R. J. 1943. Premarital residential propinquity and ethnic endogamy. *American Journal of Sociology* 48 (5): 580–84.

Krysan, M., and M. Bader. 2008. Perceiving the metropolis: Seeing the city through a prism of race. *Social Forces* 86 (2): 699–733.

Laumann, E. O., J. H. Gagnon, R. T. Michael, and S. Michaels. 1994. *The Social Organization of Sexuality: Sexual Practices in the United States*. Chicago: University of Chicago Press.

Laurie, A. J., and N. K. Jaggi. 2003. Role of "vision" in neighbourhood racial segregation: A variant of the Schelling segregation model. *Urban Studies* 40 (13): 2687–2704.

Lee, B. A., S. F. Reardon, G. Firebaugh, C. R. Farrell, S. A. Matthews, and D. O'Sullivan. 2008. Beyond the census tract: Patterns and determinants of racial residential segregation at multiple scales. *American Sociological Review* 73 (5): 766–91.

Lieberson, S., and M. C. Waters. 1988. *From Many Strands: Ethnic and Racial Groups in Contemporary America.* New York: Russell Sage Foundation.

Logan, J. R., B. J. Stults, and R. Farley. 2004. Segregation of minorities in the metropolis: Two decades of change. *Demography* 41 (1): 1–22.

Mahtani, M. 2002. Tricking the border guards: Performing race. *Environment and Planning D* 20 (4): 425–40.

Marston, S. A. 2000. The social construction of scale. *Progress in Human Geography* 24 (2): 219–42.

Massey, D. S., and N. A. Denton. 1988. The dimensions of residential segregation. *Social Forces* 67 (2): 281–315.

——. 1993. *American Apartheid: Segregation and the Making of the Underclass.* Cambridge, MA: Harvard University Press.

Massey, D. S., and G. Lundy. 2001. Use of black English and racial discrimination in urban housing markets: New methods and findings. *Urban Affairs Review* 36 (4): 452–69.

Meyer, S. G. 2000. *As Long as They Don't Move Next Door: Segregation and Racial Conflict in American Neighborhoods.* Lanham, MD: Rowman & Littlefield.

Moran, R. F. 2003. *Interracial Intimacy: The Regulation of Race and Romance.* Chicago: University of Chicago Press.

Morgan, B. S. 1981. A contribution to the debate on homogamy, propinquity, and segregation. *Journal of Marriage and the Family* 43 (4): 909–21.

Morrill, R. L., and F. R. Pitts. 1967. Marriage, migration, and the mean information field: A study in uniqueness and generality. *Annals of the Association of American Geographers* 57 (2): 401–22.

Omer, I., and I. Benenson. 2002. Investigating fine-scale residential segregation by means of local spatial statistics. *Geography Research Forum* 12: 41–60.

Peach, C. 1980. Ethnic segregation and intermarriage. *Annals of the Association of American Geographers* 70 (3): 371–81.

Qian, Z., and D. T. Lichter. 2007. Social boundaries and marital assimilation: Interpreting trends in racial and ethnic intermarriage. *American Sociological Review* 72 (1): 68–94.

Reardon, S. F., C. R. Farrell, S. A. Matthews, D. O'Sullivan, K. Bischoff, and G. Firebaugh. 2009. Race and space in the 1990s: Changes in the geographic scale of racial residential segregation, 1990–2000. *Social Science Research* 38 (1): 55–70.

Reardon, S. F., S. A. Matthews, D. O'Sullivan, B. A. Lee, G. Firebaugh, C. R. Farrell, and K. Bischoff. 2008. The geographic scale of metropolitan racial segregation. *Demography* 45 (3): 489–514.

Romano, R. 2001. *Race Mixing: Black–White Marriage in Post-War America.* Cambridge, MA: Harvard University Press.

Rosenfeld, M. J. 2007. *The Age of Independence.* Cambridge, MA: Harvard University Press.

Rosenfeld, M. J., and R. J. Thomas. 2010. Meeting online: The rise of the Internet as a social intermediary. Available at: www.stanford.edu/_mrosenfe/ (last accessed February 6, 2011).

Rossi, P. H. 1955. *Why Families Move: A Study in the Social Psychology of Urban Residential Mobility.* New York: Free Press.

Schelling, T. 1971. Dynamic models of segregation. *Journal of Mathematical Sociology* 1: 143–86.

Schnell, I., and B. Yoav. 2001. The sociospatial isolation of agents in everyday life spaces as an aspect of segregation. *Annals of the Association of American Geographers* 91 (4): 622–36.

Simpson, L., V. Gavalas, and N. Finney. 2008. Population dynamics in ethnically diverse towns: The longterm implications of immigration. *Urban Studies* 45 (1): 163–83.

Speare, A., Jr. 1974. Residential satisfaction as an intervening variable in residential mobility. *Demography* 11 (2): 173–88.

Staeheli, L., and D. Mitchell. 2008. *The People's Property? Power, Politics and the Public.* London and New York: Routledge.

Steinhorn, L., and B. Diggs-Brown. 1999. *By the Color of our Skin.* New York: Dutton Adult.

Sui, D. Z., and X. B. Wu. 2006. Changing patterns of residential segregation in a prismatic metropolis: A lacunaritybased study in Houston, 1980–2000. *Environment and Planning B* 33 (4): 559–79.

Svarer, M. 2007. Working late: Do workplace sex ratios affect partnership formation and dissolution? *Journal of Human Resources* 42 (3): 582–95.

Taeuber, K. E., and A. F. Taeuber. 1964. The Negro as an immigrant group: Recent trends in racial and ethnic segregation in Chicago. *American Journal of Sociology* 69 (4): 374–82.
Tatum, B. D. 2003. *"Why are all the Black kids sitting together in the cafeteria?": A psychologist explains the development of racial identity*. 5th ed. New York: Basic Books.
Thernstrom, S., and A. Thernstrom. 1997. *America in Black and White: One Nation, Indivisible: Race in Modern America*. 3rd ed. New York: Simon & Schuster.
U.S. Office of Management and Budget. 2000. *Provisional guidance on the implementation of the 1997 standards for federal data on race and ethnicity*. Executive Office of the President, December 15. Available at: www.whitehouse.gov/sites/default/files/omb/assets/information_and_regulatory_affairs/re_guidance2000 update.pdf (last accessed October 26. 2011).
Wong, D. W. 1997. Spatial dependency of segregation indices. *The Canadian Geographer* 41 (2): 128–36.
——. 1998. Spatial patterns of ethnic integration in the United States. *The Professional Geographer* 50 (1): 13–30.
——. 2003. Spatial decomposition of segregation indices: A framework toward measuring segregation at multiple levels. *Geographical Analysis* 35 (3): 179–95.
——. 2004. Comparing traditional and spatial segregation measures: A spatial scale perspective 1. *Urban Geography* 25 (1): 66–82.
Wright, R., S. Holloway, and M. Ellis. 2011. Where Black–White couples live. *Urban Geography* 32 (1): 1–22.
Wright, R., S. Houston, M. Ellis, S. Holloway, and M. Hudson. 2003. Crossing racial lines: Geographies of mixed-race partnering and multiraciality in the United States. *Progress in Human Geography* 27 (4): 457–74.
Wu, X. B., and D. Z. Sui. 2001. An initial exploration of a lacunarity-based segregation measure. *Environment and Planning B* 28 (3): 433–46.
Yinger, J. 1997. *Closed doors, opportunities lost: The continuing costs of housing discrimination*. New York: Russell Sage Foundation Publications.
Zhang, J. 2004. A dynamic model of residential segregation. *The Journal of Mathematical Sociology* 28 (3): 147–70.

23

THE (IN)COMPATIBILITY OF DIVERSITY AND SENSE OF COMMUNITY

Zachary P. Neal and Jennifer Watling Neal

Research Question: Why do residentially segregated communities tend to be more socially cohesive?
System Science Method(s): Agent-based models & Networks
Things to Notice:

- Adaptation of an existing agent-based model
- Uncovering micro-level mechanisms the produce an observed phenomenon

Community psychologists are interested in creating contexts that promote both respect for diversity and sense of community. However, recent theoretical and empirical work has uncovered a community-diversity dialectic wherein the contextual conditions that foster respect for diversity often run in opposition to those that foster sense of community. More specifically, within neighborhoods, residential integration provides opportunities for intergroup contact that are necessary to promote respect for diversity but may prevent the formation of dense interpersonal networks that are necessary to promote sense of community. Using agent-based modeling to simulate neighborhoods and neighborhood social network formation, we explore whether the community-diversity dialectic emerges from two principles of relationship formation: homophily and proximity. The model suggests that when people form relationships with similar and nearby others, the contexts that offer opportunities to develop a respect for diversity are different from the contexts that foster a sense of community. Based on these results, we conclude with a discussion of whether it is possible to create neighborhoods that simultaneously foster respect for diversity and sense of community.

Both respect for diversity and the promotion of a sense of community are longstanding, explicit values of the field of community psychology (e.g. Kelly 1971; Sarason 1974; Townley et al. 2011). Community psychologists view each of these phenomena as vital to thriving contexts noting that "respect and appreciation for diverse identities promotes personal and collective wellness" (Prilleltensky 2001, p. 754) and "the psychological sense of community is the overarching criterion by which one judges any community development" (Sarason 1974, p. 158). However, to simultaneously promote respect for diversity and sense of community in a particular context, it is necessary to understand the relationship between these two phenomena.

Recently, Townley et al. (2011) called attention to a potential "community-diversity dialectic," noting that the contextual conditions that foster respect for diversity often run in opposition to those that foster sense of community (p. 70) – that is, diversity and sense of community are negatively related, creating a paradox for community psychologists (Rappaport

DOI: 10.4324/9781315748771-26

1981). Townley et al. (2011) recommended changing the definition of sense of community, but this provides only a semantic, not a practical, solution to the paradox. Thus, the goal of this chapter is to understand why the community-diversity dialectic exists in an effort to determine whether and how community psychologists can address this paradox. To this end, we begin with a discussion of what fosters respect for diversity and sense of community in one important context: neighborhoods. Using agent-based modeling to simulate neighborhoods and neighborhood social network formation, we explore whether the community-diversity dialectic emerges from two principles of relationship formation: *homophily*, the tendency to associate with similar others, and *proximity*, the tendency to associate with nearby others. We conclude with a discussion of whether it is possible to create neighborhoods that simultaneously foster respect for diversity and sense of community.

Background

Fostering Respect for Diversity in Neighborhoods

Frameworks for diversity within community psychology eschew a deficit model in which differences from the dominant culture are viewed as inferior or deviant, and instead embrace a position of cultural relativity or pluralism where multiple cultures are valued (Harrell and Bond 2006; Rappaport 1977; Ryan 1976; Trickett et al. 1994). Fostering respect for diversity is important for community psychologists and is embedded in the mission statement of the Society for Community Research and Action (SCRA), Division 27 of the American Psychological Association. Specifically, one of the goals of SCRA is "to promote . . . greater inclusion for historically marginalized groups, and respecting all cultures" (SCRA 2010, p. 13). To this end, community psychologists aim to encourage contexts that facilitate respect for diversity, and view these contexts as promoting individual and collective wellbeing (Prilleltensky 2001).

Directly fostering respect for diversity can be quite challenging, but environmental modifications may provide an indirect route to the extent that some ecological contexts are more likely to promote a respect for diversity than others. In the particular context of neighborhoods, the vast literature on the contact hypothesis suggests that the opportunity for social contact between diverse groups can diminish animosities and stereotypes, and foster tolerance and ideally respect for one another (e.g. Allport 1954; Amir 1969; Hewstone and Brown 1986; Sigelman and Welch 1993; Dixon et al. 2005). To be sure, it would be naive to view social contact alone as sufficient for promoting a respect for diversity. Indeed, there is some evidence that superficial contact in the absence of more meaningful interactions can lead to intergroup tension (Townley et al. 2011), and others have argued that exposure to diversity may lead to social withdrawal or "hunkering down" (Putnam 2007). However, while contact is surely not a sufficient condition for promoting a respect for diversity, it is likely a necessary condition – that is, one must first have knowledge of and opportunities to interact with diverse others before one can develop a respect for their viewpoints and ways of life. In residentially integrated neighborhoods, people are more likely to come into contact with diverse others, increasing their opportunities for meaningful exposure to and acceptance of diverse perspectives. Therefore, residentially integrated neighborhoods are contexts that offer residents more opportunities to develop a respect for diversity than residentially segregated neighborhoods.

Fostering Sense of Community in Neighborhoods

In addition to valuing respect for diversity, community psychologists have also expressed a desire to foster a sense of community among individuals (e.g. Chavis and Pretty 1999; Riger 1993;

Sarason 1974). Here, sense of community is conceptualized as psychological and reflects individual perceptions rather than external states. Community psychologists have struggled to consistently define sense of community (Hill 1996), but have often cited four dimensions outlined by McMillian and Chavis (1986): membership, influence, integration and fulfillment of needs, and emotional connection (see also Long and Perkins 2003; Peterson et al. 2008). As a set, these dimensions speak to individuals' perceptions of belongingness, cohesion, and bond with a group. Neighborhoods are commonly viewed as one context that might foster psychological sense of community, and community psychologists have expressed an interest in understanding what features of neighborhoods facilitate sense of community among residents (see Chavis and Pretty 1999 for review).

Directly fostering a psychological sense of community can be quite challenging, but environmental modifications may provide an indirect route to the extent that some ecological contexts are more likely to promote a psychological sense of community than others. The feelings of belongingness and cohesion associated with a psychological sense of community are often found to be strongest for those with relatively dense personal social networks. This phenomenon has been describing using a range of terms: Coleman (1988) and Burt (2001) refer to the relational density as yielding social "closure," while Granovetter (1973) and Putnam (2001) view it as arising from "strong" or "bonding" ties, respectively. Despite minor differences, these theorists all point to a common mechanism whereby network density generates feelings of belongingness. When one's friends are also friends with one another, a relational feedback loop (what social network theorists call a "cycle") is established. For example, if A is friends with B and C, and B and C are also friends with each other, there is a closed loop or cycle A → B → C → A. In such cases, when A seeks social support from one friend (e.g. B), other friends (e.g. C) can also learn of her need and can provide assistance as well. In contrast, this sharing of social support cannot occur in *sparse* or *open* personal social networks where, when A seeks social support from one friend, her other friends would remain unaware of her need. Thus, when people have dense personal social networks, we would expect them to have a strong *psychological* sense of community, wherein they view themselves as a member of a strong community able to work together and support one another. By extension, in neighborhoods populated by such people, we would expect to see strong sense of community (see Granovetter 1973, p. 1373; Grannis 2009, p. 38) – that is, neighborhoods characterized by dense personal social networks are contexts that are likely to foster sense of community than neighborhoods characterized by sparse and fragmented personal social networks.

The Community-Diversity Dialectic

Community psychologists and others often seek to promote both respect for diversity and a strong sense of community, which begs the question: Are the ecological contexts that afford opportunities to develop a respect for diversity (i.e. residentially integrated neighborhoods) the same ecological contexts that foster a sense of community (i.e. dense personal social networks)? In search of an answer to this question, Townley et al. (2011) recently proposed a "community-diversity dialectic," noting that the ecological contexts that foster respect for diversity may be distinct from those that foster sense of community (p. 70). Supporting this proposition, they review several empirical studies have highlighted an inverse relationship between the integrated conditions that promote respect for diversity and sense of community. Separately, Portes and Vickstrom (2011) offer a similar review, finding that demographic homogeneity has often been linked with higher levels of trust, social cohesion, and belongingness typically thought to compose sense of community.

In the interest of space, we will not duplicate Townley et al.'s (2011) or Portes and Vickstrom's reviews, but do wish to highlight some additional studies that have indicated a negative relationship between the contextual conditions that promote respect for diversity and sense of community. In university settings, White freshman exhibited less racial prejudice but also less relationship satisfaction when they were randomly assigned an African American roommate rather than a White roommate (Shook and Fazio 2008). Similarly in neighborhood settings, diversity was an obstacle to the creation of neighborhood social ties by Italian adolescents (Lenzi et al. 2013), and of neighborhood collective efficacy by American homeowners and renters (Lindblad et al. 2013). Finally, in an ethnographic account, Berryhill and Linney (2006) highlighted the challenges inherent in bringing together a biethnic group of African American and Latino residents to work together on community issues. Of note, they described ethnic tensions associated with the group's diversity that may have dampened resident participation. Notably, these studies were conducted in a range of different settings (e.g. university residence halls, neighborhoods), using a range of methods (e.g. ethnography, controlled experiment), with participants ranging in age (e.g. adolescents, adult homeowners). Thus, taken together with Townley et al.'s (2011) review, they offer strong evidence that the goals of promoting respect for diversity and sense of community may not be compatible.

The community-diversity dialectic presents a paradox for community psychologists because it highlights the conflictual nature of two core values in the field. Townley et al. (2011) argue for an expansion in the definition of sense of community to realign it with the goals of promoting diversity. More specifically, they suggest that sense of community should be redefined to focus on bridging social capital (i.e. ties across diverse groups or communities that facilitate the flow of resources) rather than bonding social capital (i.e. trust, belongingness, social cohesion). This is a semantic solution that calls for a fundamental change in the conceptualization of sense of community. However, in this respect, it dodges rather than addresses the paradox. In this paper, we aim to increase our understanding of the community-diversity dialectic by examining why diversity and sense of community are negatively related. By understanding the mechanisms that place these two values of community psychology in conflict, we are better positioned to understand what, if anything, community psychologists can do about the community-diversity dialectic.

Methods

Agent-based models (ABM) are a powerful methodological tool for building theory by allowing researchers to explore the consequences of different behaviors in different contexts through simulation (Macy and Willer 2002; Hoffer et al. 2009). These models are, by definition, very simple, rooted in the notion that an agent's (e.g. a person's) behavior is driven by following a set of rules that dictate responses to environmental forces and reactions to other agents. Even when agents follow simple behavioral rules, complex phenomena often emerge from these models, highlighting that patterns that may be difficult to understand when viewed at a macroscopic scale (e.g. the community-diversity dialectic) can often be understood as the result of interactions occurring at the microscopic scale (e.g. relationship formation).

Most ABMs consist of two stages: an initial context-setup stage, and an agent-interaction stage. In the context-setup stage, a simulated world (e.g. a neighborhood) with specified characteristics (e.g. level of integration) is created and populated with agents (e.g. residents). In the agent-interaction stage, each agent simultaneously follows a common set of behavioral rules (e.g. homophily) that govern how they respond to their environment, which includes the other agents. After one or more periods of agent interaction, the researcher observes the macroscopic patterns that have emerged in the setting, then repeats the simulation with slightly different contextual characteristics and behavioral rules. By manipulating the characteristics of contexts and the way agents behave within them, which would be impossible in reality, the researcher develops an understanding of their relationships among the variables and their role in producing complex patterns. Thus, agent-based models are particularly promising for building theory in community psychology because they allow researchers to consider phenomena of interest not only in one or two contexts, but in all possible contexts.

In our ABM, developed using the NETLOGO software package (Wilensky 1999), the context-setup stage adapts Schelling's (1969) model of segregation to create simulated neighborhoods that are populated by two types of people. It is important to note that these "types" could represent any kind of socially consequential distinction made by those involved, including race/ethnicity, socioeconomic status, or religion; the models presented below should not be viewed narrowly as models of the effect of racial diversity, but as models of diversity on any socially consequential characteristic. Additionally, the "types" are not assumed to be different in any real sense, but only are assumed to be perceived or socially constructed as different by those involved. The neighborhoods are characterized by their level of residential integration among the two types, which we measure by the average percentage of one's neighbors who are dissimilar to oneself. This index ranges from 0% in a completely segregated neighborhood to 50% in a completely integrated neighborhood.[1] The top row of Figure 23.1 illustrates three simulated neighborhoods with varying levels of residential integration, which we contend is associated with the extent to which they promote a respect for diversity. In the highly integrated neighborhood, light and dark gray households are evenly mixed and the index of integration is 50%: half of one's neighbors are similar, and half are different, on average. Through exposure to difference, individuals living in this type of neighborhood have opportunities to develop a respect for diversity. In contrast, in the highly segregated neighborhood, households are tightly clustered with others of the same color and the index of integration is only 5%: on average, nearly all of one's neighbors are similar. Because they live in homogeneous clusters, individuals living in this type of neighborhood have few opportunities to develop a respect for diversity.

In the agent-interaction stage, individual people decide whether or not to form a relationship with one another. Although we intend to use the term "relationship" broadly to cover a range of positive affiliations, for simplicity and clarity we use the terms "friendship" and "friend" below. Many different factors play a role in determining whether two individuals become friends, but in the agent-interaction stage of this model, we focus only on two of the strongest and most widely documented: homophily and proximity. Homophily refers to the tendency for friendships "between similar people [to occur] at a higher rate than among dissimilar people" (McPherson et al. 2001, p. 416) and is the basis of the aphorism that "birds of a feather flock together." A tendency toward homophily is nearly always observed in human populations along such dimensions as race, ethnicity, age, education, social class, and attitudes and beliefs; notably it is also often observed in non-human animal populations also (Lazarsfeld and Merton 1964; McPherson et al. 2001; Fu et al. 2012).[2] In a given setting, the strength of homophily may be very strong (e.g. new immigrants in an ethnic enclave), or may be relatively weak (e.g. students on a university campus), but nonetheless is typically present to some degree. It is important to note that the existence of a tendency toward homophily does not necessarily imply feelings of prejudice or animosity. Instead, homophily can emerge from the simple fact that similar people tend to do similar things, and thus are more likely to have opportunities to form relationships.

The second force of friendship formation included in our model is proximity. Proximity refers to the tendency for friendships to occur between people who live nearby at a higher rate

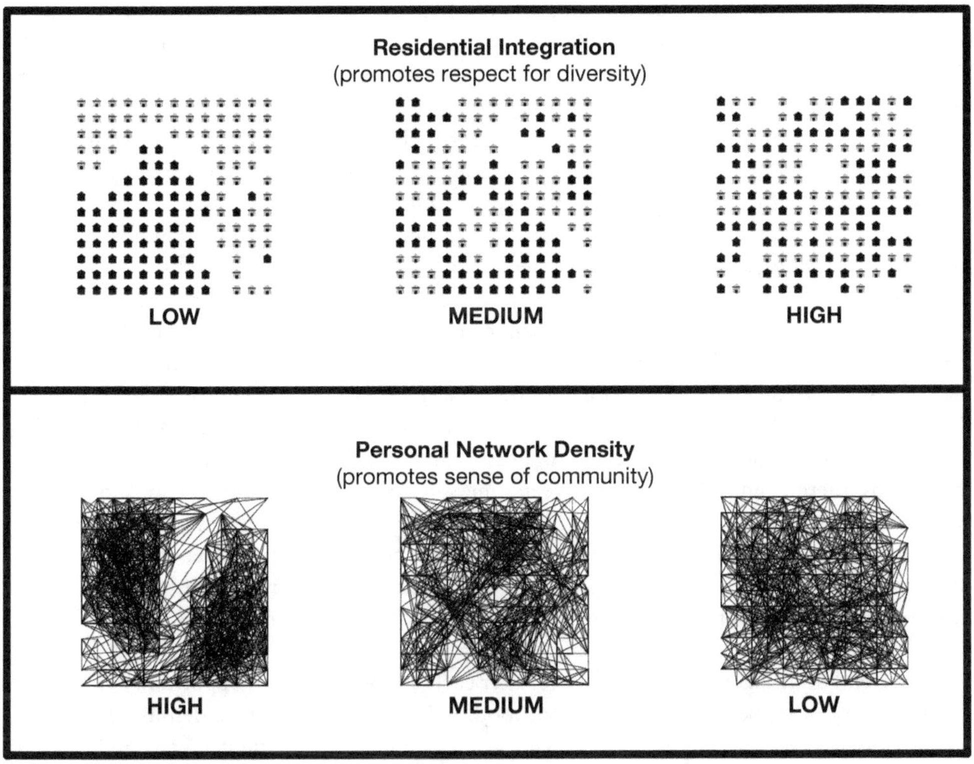

Figure 23.1 Examples of simulated neighborhoods

than between people who live far apart. As with homophily, a tendency toward proximity is nearly always observed in human populations (Moreno 1934; Festinger et al. 1950; Grannis 2009). The strength of a proximity tendency likely depends on the setting (e.g. stronger in a gated community, weaker in a large city) and on available technology (e.g. stronger before phones and cars, weaker with the advent of social media), but nonetheless is typically present to some degree. The existence of this tendency is not the result of an individual's explicit preference for nearby friends, but rather the result of the simple fact that one is more likely to have chance encounters and thus more opportunities to form friendships with those living nearby.

The probability that any two people, *i* and *j*, become friends in the model's agent-interaction stage is defined by a logistic selection function

$$\Pr(F_{ij}=1) = \frac{\exp(\beta_0 + \beta_H \delta_{ij} + \beta_P \text{Dist}'_{ij})}{1+\exp(\beta_0 + \beta_H \delta_{ij} + \beta_P \text{Dist}'_{ij})} \quad \text{where } \text{Dist}'_{ij} = \frac{1}{1+\exp\dfrac{\text{Dist}'_{ij}-5}{0.5}}$$

that depends on whether they are similar ($\delta_{ij} = 1$) or different ($\delta_{ij} = 0$) and the physical distance between them (Dist_{ij}). The β_H parameter controls the direction and strength of the tendency toward homophily in the setting: when it is positive, two people are more likely to be friends when they are similar (i.e. homophily), while when it is negative, they are more likely to be friends when they are different (i.e. heterophily). Likewise, the β_P parameter controls the direction and strength of the tendency toward proximity: when it is positive, two people are more likely to be friends when they live nearby, while when it is negative, they are more likely to be friends when they live far apart. Thus, by adjusting the values of these two parameters, this function allows friendship probabilities to be estimated under different combinations of behavioral tendencies. The intercept, β_0, determines the maximum probability of any relationship forming; throughout all simulations, $\beta_0 = -(\beta_H + \beta_P)$, which sets the maximum probability of a relationship forming at 50%. To capture the nonlinear effects of distance, we use a generalized logistic transformation of raw physical distance (i.e. Dist → Dist′).

A simple example serves to illustrate how this function is used. Imagine a world in which individuals are moderately more likely to become friends with others who are similar and who live nearby. This typical set of behavioral tendencies can be captured by setting $\beta_H = 2.5$ and $\beta_P = 2.5$ in the selection function, which can then be used to estimate probabilities of friendship. Figure 23.2 illustrates the estimated probability that two people will be friends in a world characterized by these behavioral tendencies. They are most likely to become friends if they are similar and live nearby, and least likely to become friends if they are different and live far apart. Note the nonlinear effect of distance: when it comes to opportunities for forming friendships, there is little difference between a person who lives one house away and a person who lives two houses away, and likewise little difference between a person who lives 10 blocks away and a person who lives 10 miles away. Although the behavioral tendencies described by these probability curves are fairly typical in human communities, friendship formation under different behavioral tendencies can be estimated by changing the values of β_H and β_P, which would yield probability curves with different shapes.

At the end of the agent-interaction stage, after each person has had an opportunity to befriend (or not) every other person in the neighborhood, we examine the average density of residents' personal social networks, or what is known as the clustering coefficient (Watts and Strogatz 1998). The clustering coefficient ranges from 0, when a person's friends are not friends with one another, to 1, when a person's friends are also friends with one another. The bottom row of Figre 23.1 illustrates whole neighborhood social networks with varying levels of average personal network density, as indexed by the clustering coefficient, which we contend promotes a sense of community. In the neighborhood where personal social networks are relatively dense on average (CC = 0.33), clusters of relationships around which a sense of community

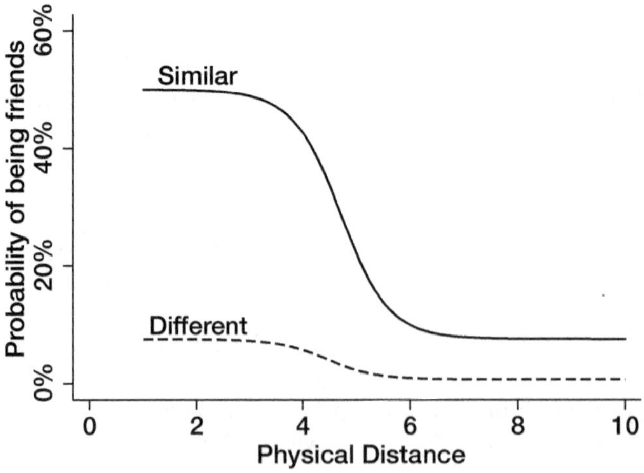

Figure 23.2 Friendship selection function when $\beta_H = 2.5$, $\beta_P = 2.5$

might develop are readily visible. Because dense personal social networks facilitate feelings of belongingness and social cohesion, individuals living in this type of neighborhood are most likely to enjoy a psychological sense of community. In contrast, in the neighborhood where personal social networks are relatively sparse on average (CC = 0.22), the fragmented and random neighborhood-level network provides few natural clusters to facilitate the formation of a sense of community. Because sparse personal social networks facilitate feelings of isolation and anomie, individuals living in this type of neighborhood are least likely to enjoy a psychological sense of community.

An interactive version of this model is available at www.msu.edu/~zpneal/research/nhood net.html, or on request from the author. The model allows users to replicate the results described below, and to investigate other patterns in the relationship between diversity and sense of community, in two ways. After setting the simulation's homophily, proximity, and integration parameters using the three sliders, the "Manual" button runs both the context-setup and agent-interaction stages of the simulation using the selected parameter, then plots the neighborhood's level of diversity and sense of community. Alternatively, the "Automated" button repeatedly runs the simulation using the selected homophily and proximity parameters and a randomly selected level of integration, each time plotting the neighborhood's level of diversity and sense of community, yielding a scatterplot like the one shown below in Figure 23.4.

Results

Figure 23.3 schematically illustrates the steps we follow to obtain the results we discuss below; the agent-based model itself appears in steps 2 and 3, while the other steps describe how we vary the model's parameters to explore diversity and sense of community in different contexts. We begin by examining the relationship between diversity and sense of community in a typical world where individuals are more likely to form relationships with similar than dissimilar others (i.e. homophily) and with nearby than distant others (i.e. proximity). Holding the intensity of the behavioral tendencies toward homophily and proximity constant at the moderate levels illustrated in Figure 23.2 (step 1), we simulated social network formation in 500 neighborhoods that varied in their level of integration, each time computing the resulting network's clustering coefficient (steps 2–5, the integration loop). Figure 23.4 plots each neighborhood's opportunity for residents to develop a respect for diversity (as measured by its level of residential integration) and its capacity to foster a sense of community (as measured by its residents' personal network density). A very clear, albeit somewhat non-linear, negative correlation between diversity and sense of community emerges ($r = -0.85$, $p < 0.001$; step 6). Neighborhoods with the greatest opportunity for residents to develop a respect for diversity (i.e. highly integrated neighborhoods) have the least capacity to foster a sense of community. Likewise, neighborhoods with the least opportunity for residents to develop a respect for diversity (i.e. highly segregated neighborhoods) have the greatest capacity to foster a sense of community. This finding suggests that, the values of community psychology notwithstanding, it is not possible to simultaneously promote respect for diversity and sense of community in a typical world where relationship formation is driven by homophily and proximity.

The results shown in Figure 23.4 represent the relationship between diversity and sense of community in a typical world relationship formation is driven by the particular levels of homophily and proximity described by setting $\beta_H = 2.5$ and $\beta_P = 2.5$. But, the relationship between these phenomena may be different when relationship formation is driven by different levels of these social forces. Perhaps it is possible to simultaneously promote diversity and sense of community in a slightly different worlds where behavioral tendencies toward homophily and/or proximity

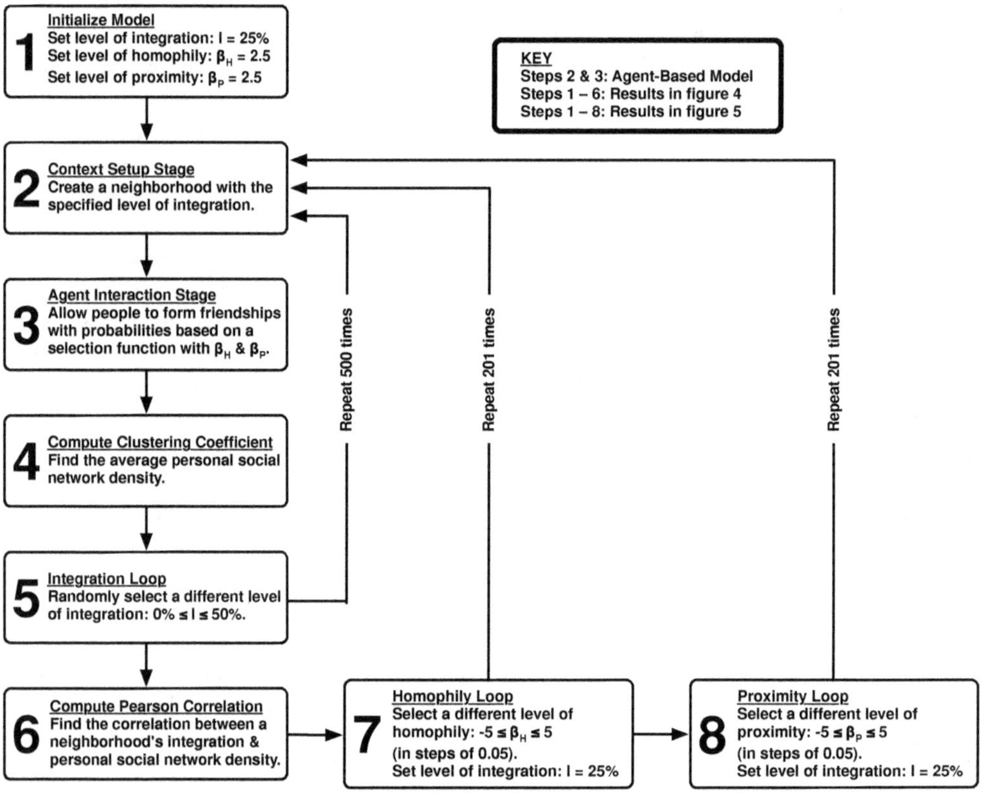

Figure 23.3 Schematic of model workflow

are weaker, or stronger, or even reversed. To consider this possibility, we repeated the analysis shown in Figure 23.4 using different levels of homophily and proximity. Specifically, we examined diversity and sense of community in 500 simulated neighborhoods varying in their level of integration (steps 2–6), for every level of homophily between –5 and 5 (in increments of 0.05; step 7) and every level of proximity between –5 and 5 (in increments of 0.05; step 8). This required slightly more that 20 million separate simulations (i.e. 500 neighborhoods × 201 levels of homophily × 201 levels of proximity).

The results of these simulations are shown in Figure 23.5. Each point in this heatmap plot represents a distinct world where relationship formation is governed by a specific level of homophily and proximity. In worlds toward the right, individuals exhibit progressively stronger behavioral tendencies toward forming relationships with similar others (i.e. homophily), while in worlds toward the left, they exhibit progressively stronger tendencies toward forming relationships with dissimilar others (i.e. heterophily). In worlds toward the top, individuals exhibit progressively stronger behavioral tendencies toward forming relationships with nearby others (i.e. proximity), while in worlds toward the bottom, they exhibit progressively stronger tendencies toward forming relationships with distant others. The shading of each point indicates the relationship between diversity and sense of community (as measured by the correlation between integration and network density) in the respective world. Darker points indicate worlds where the particular combination of homophily and proximity behavioral tendencies yields a negative relationship between diversity and sense of community (cf. Figure 23.4). Lighter points

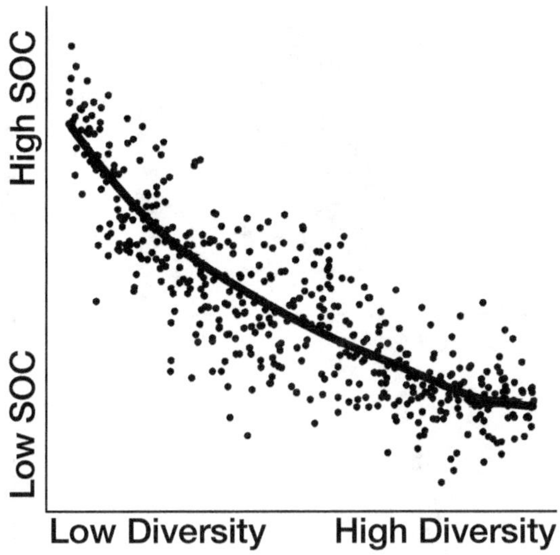

Figure 23.4 Relationship between diversity and SOC when $\beta_H = 2.5$, $\beta_P = 2.5$

indicate worlds where the particular combination of homophily and proximity behavioral tendencies yields a positive relationship between diversity and sense of community.

The findings illustrated in Figure 23.5 confirm that the negative relationship between diversity and sense of community observed in Figure 23.4 is not simply an artifact of the particular combination of behavioral tendencies toward homophily and proximity (i.e. $\beta_H = 2.5$ and $\beta_P = 2.5$) we initially examined. All points in the upper-right quadrant of Figure 23.5 are dark, indicating that all combinations of homophily and proximity yield a negative relationship between diversity and sense of community – that is, in any world where individuals exhibit at least some tendency to form relationships with similar others (i.e. $\beta_H > 0$) and at least some tendency to form relationships with nearby others (i.e. $\beta_P > 0$), diversity and sense of community are negatively related. It is important to note that all studies of human social networks have observed behavioral tendencies toward both homophily and proximity, while none have found worlds where one or both of these behavioral tendencies was missing. Thus, while the findings illustrated in Figure 23.4 suggest that diversity and sense of community are negatively related *in a typical world*, those illustrated in Figure 23.5 suggest this negative relationship would persist *in all reasonably likely worlds*.

Examining the other quadrants in Figure 23.5 allows us to ask: what would need to change about individuals' behavioral tendencies in order for diversity and sense of community to have a positive relationship? That is, in what kind of world could we simultaneously promote diversity and sense of community? Here, our focus shifts to the two light-colored regions of the heatmap. One such region appears in the upper-left quadrant, which corresponds to worlds in which individuals tend to form relationships with others who are nearby but dissimilar (i.e. heterophily). Although this type of behavior is conceptually possible, it is highly unlikely and has not been observed in empirical studies of people before. Importantly, a behavioral tendency toward heterophily does not describe what is sometimes called "multiculturalism" in which individuals express no particular tendency toward similar or dissimilar others, but rather describes a situation in which individuals actively avoid similar others and explicitly prefer dissimilar others. More colloquially, our findings in the upper-left quadrant suggest that diversity and sense of community could be simultaneously promoted in a world where birds of a feather avoid each other.

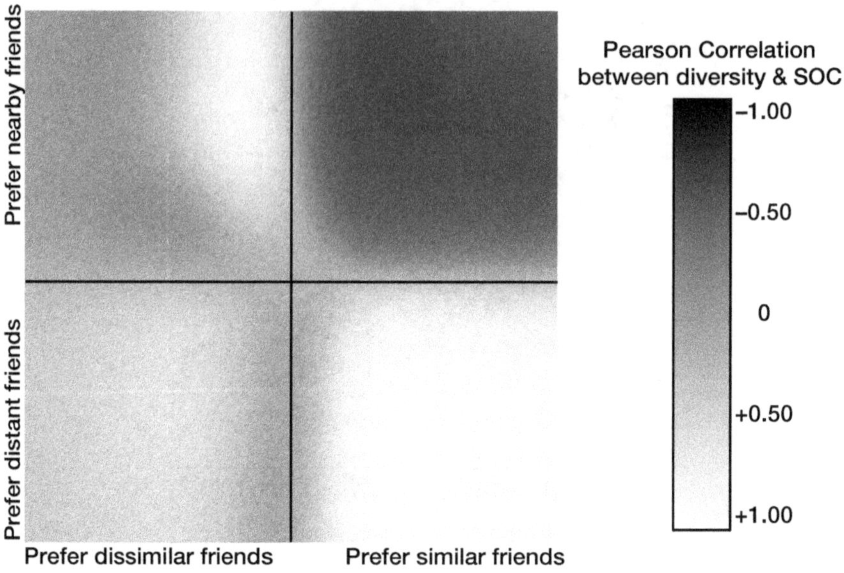

Figure 23.5 Relationship between diversity and SOC for all β_H and β_P

A second-light colored region appears in the lower right quadrant, which corresponds to worlds in which individuals tend to form relationships with others who are similar but live far away. Again, although this type of behavior is conceptually possible, it is highly unlikely and has not been observed in empirical studies of people before. Indeed, the physical laws of the universe essentially prohibit it. Such a behavioral tendency does not describe what might be called "cosmopolitanism" in which individuals express no particular tendency toward nearby or distant others, but rather describes a situation in which individuals actively avoid nearby others and explicitly prefers distant others. More colloquially, our findings in the lower-right quadrant suggest that diversity and sense of community could be simultaneously promoted in a world where neighbors avoid each other.

Discussion

Statistician George Box famously noted that all models are wrong, but some are useful (1976). This is certainly the case for agent-based models, which strive for parsimony in explaining complex phenomena. In the case of the model we present here, it is "useful" because it demonstrates how the frequently observed negative relationship between diversity and sense of community can emerge from two relatively simple behavioral tendencies, and is "wrong" in the sense that it omits certain complexities that exist in reality. For example, this model considers only a single, binary dimension of diversity: individuals are either light gray or dark gray. But, in reality, any given social distinction comes in many shades (e.g. race), and intersects with other distinctions (e.g. with ethnicity, with social class, etc.). Similarly, this model views relationships between people as either present or absent, when in reality relationships can be stronger or weaker (e.g. best friend vs. neighborly acquaintance). Finally, this simulated neighborhoods in these models do not include social spaces like schools or parks that some have hypothesized may mitigate the effects of homophily or distance by drawing people together (Lenzi et al. 2013). Thus, we view this model as a starting point that provides a baseline understanding of the community-diversity dialectic, and onto which additional complexities may be added by future studies.

Of these simplifications, the omission of individuals' multiple and potentially intersecting statuses and identities may be of greatest concern, and warrants additional comment. The consequences for our model and findings depend on precisely how these statuses intersect. One possibility is that multiple statuses are correlated, as is often the case for race and socioeconomic status in the United States, for example. To the extent that multiple statues are correlated, they collapse into a single status, reinforcing one another. Consistent with Blau's (1977) finding that "strongly correlated parameters consolidate status and group differences and thereby impede intergroup relations" (p. 45), we would expect that incorporating additional and correlated dimensions of difference would intensify our main finding. A second possibility is that multiple statuses are uncorrelated, as for example gender and race. In such cases, two individuals may differ on one dimension and thus have a diminished likelihood of interaction, and yet be similar on another dimension and thus have an increased likelihood of interaction. Consistent with Blau's (1977) finding that "intersecting parameters promote intergroup relations" (p. 45), we would expect that incorporating additional and uncorrelated dimensions of difference would mitigate our main finding.

Still a third possibility is the presence of an additional and nearly universally held status that draws individuals together, as, for example, when potentially different individuals are united by a common goal (e.g. cleaning up the neighborhood, cheering for the home team). Such unifying characteristics are the stuff that sense of community is made of, and indeed we would expect them to significantly increase the network densities observed in our simulations. However, it is also important to observe that the introduction of a widely held common status, such as a common goal, *ipso facto* reduces the diversity of the community. A community with some rich residents and some poor residents who all support the home team is less diverse than a community with some rich and some poor residents among whom only some support the home team. Thus, we would expect that incorporating additional and near-universal dimensions of difference would have no effect on our main finding.

A final possibility is that multiple statuses are not additive, but instead have unique effects when they are combined in different ways. Intersectionality theory suggests, for example, that the experience of a black woman is not simply the combination of the experience of women and the experience of blacks, but rather something completely unique (Crenshaw 1991). This is likely the most realistic perspective on how individuals' multiple statuses and identities function, but it is also the most complex and the most difficult to model. Here, the challenge is a purely computational not theoretical one: to incorporate intersectionality effects in our model would require a minimum of two additional parameters, one for the additional status and one for the interaction effect, and would require still more if additional statuses were included or if their interactions were non-linear. Together, these possibilities concerning the operation of multiple statuses and identities in the context of diversity and sense of community highlight some ways that our preliminary model might be extended.

Despite these limitations, these findings we present above help us to understand why the community-diversity dialectic may exist – that is, why it may be so challenging to simultaneously promote a respect for diversity and a sense of community in a single setting. The model demonstrates that this perennial challenge to community psychology praxis can emerge from two relatively simply, but universal behavioral tendencies: homophily and proximity. We find that when tendencies toward homophily and proximity in relationship formation exist, even if in a very weak form, the contexts that foster a respect for diversity are different from the contexts that foster a sense of community. When people behave as they usually do, community programs designed to shape the local ecology into one that fosters a respect for diversity are likely to have a problematic unintended consequence: also shaping the local ecology into one that diminishes a sense of community. Likewise, community programs designed to shape the local ecology into one that fosters a sense of community are likely to also shape it into one that diminishes respect for diversity.

How, then, might community psychologists approach the community-diversity dialectic? One possibility involves seeking to shift behavioral tendencies away from those responsible for the incompatibility between respect for diversity and sense of community. The location of the light regions in Figure 23.5 indicate that if behavioral tendencies toward homophily and/or proximity reversed – that is, if people were more likely to form relationships with dissimilar and/or distant others – then the dialectic would evaporate. At first glance, this may appear a promising avenue for future community-based work, but on closer inspection is quite problematic. First, no human population has been observed that did not exhibit at least some tendency toward both homophily and proximity, thus it is not clear how or even whether these tendencies can be reversed. Second, it is important to consider what it would mean "on the ground" to reverse these behavioral tendencies. The worlds in which our model suggests one can simultaneously promote respect for diversity and sense of community are those in which people (a) actively avoid similar others and (b) actively avoid their neighbors. This, it seems to us, is not the kind of world we would want to live in, even if it did allow us to achieve the goals of community psychology.

If we concede that behavioral tendencies toward homophily and proximity, and thus the community-diversity dialectic, are likely here to stay, then engaging with the dialectic requires a different approach. Our finding of a negative relationship between diversity and community suggests that, within each setting, community psychologists and community members must seek to find a contextually appropriate balance – that is, engaging the community-diversity dialectic involves the generation of multiple, context-dependent solutions, or what Rappaport (1981) calls "divergent reasoning." In some neighborhoods, it may be preferable to promote a respect for diversity, even at the expense of sense of community, while in others, a sense of community may be more beneficial than a respect for diversity. The point is that when it comes to pursuing the goals of community psychology, one is likely to encounter trade-offs and opportunity costs that invite difficult decisions. Ultimately, our model does not provide guidance on the optimum balance between diversity and sense of community in any given, real-world community; this is an important future direction for on-the-ground, non-simulation research. That is, our model suggests that community psychologists shift their focal question from "how can we promote diversity and sense of community in this setting" to "what is the right balance between diversity and sense of community in this setting?"

Although the right approach will be context-dependent and must take into account the needs and viewpoints of community members, we speculate that favoring a respect for diversity over a sense of community may often be preferable. There are few downsides (aside from diminished sense of community) to promoting respect for diversity. Beyond ensuring inclusiveness and reducing opportunities for oppression and marginalization, openness to diverse points of view is essential for creativity (Florida 2002) and lies at the heart of "bridging" or "weak tie" social capital (Granovetter 1973; Coleman 1988; Burt 2001; Putnam 2001). In contrast, there are potential downsides to promoting high levels of cohesion and sense of community beyond simply a diminished respect for diversity. The dense "closed" social networks that facilitate feelings of belongingness also isolate individuals from new ideas and other resources (Portes 1998). For example, residents in a poor inner-city neighborhood with a strong sense of community may benefit from the fact that neighbors provide one another assistance and social support, but are also likely walled off from access to key economic resources outside the community. Here, the dense networks facilitate working together, but also serve to concentrate and reinforce poverty.

Over 30 years ago, Rappaport (1981) noted that "the most important and interesting aspects of community life are by their very nature paradoxical" (p. 20). Such is the case for two of community psychology's core values: promoting contexts that are likely to increase respect diversity and promoting contexts that are likely to increase a sense of community (Townley et al. 2011). Results of our model suggest that this community-diversity dialectic can result from common behavioral tendencies toward homophily and proximity. Moreover, given the universality of these behavioral tendencies, it is unlikely that community psychologists can shift them sufficiently to simultaneously promote respect for diversity and sense of community. However, through divergent reasoning, community psychologists can seek a contextually appropriate balance between these two opposing goals that are near and dear to our field.

Acknowledgment

This chapter originally appeared in *American Journal of Community Psychology*, 53, 1–12 (2014).

Notes

1. Values greater than 50% are conceptually possible and describe what is known as dissortative mixing, where a person is surrounded primarily by dissimilar others. In a neighborhood context, this might occur for one or two households (e.g. a single minority household in a majority neighborhood). However, it can occur neighborhood-wide only if minority and majority households are arranged in a very precise "stripe" pattern, which guarantees that any given household has a maximum of only two similar neighbors (i.e. one in one direction, and another in the opposite direction). Such an arrangement does not seem realistic, so we have excluded it from our simulations. This exclusion does not affect our results, which we find remain the same even if we had also simulated such unrealistic hyper-integrated neighborhoods.
2. The only widespread example of heterophily, the opposite of homophily wherein relationships are more likely between dissimilar people, is along gender lines in the formation of romantic and sexual relationships among heterosexual individuals: heterosexual men form relationships primarily with women, and not with other men, and vice versa.

References

Allport, G. W. (1954). *The Nature of Prejudice*. Cambridge, MA: Perseus Books.

Amir, Y. (1969). Contact hypothesis in ethnic relations. *Psychological Bulletin*, *71*(5), 319–342.

Berryhill, J. C., & Linney, J. A. (2006). On the edge of diversity: Bringing African Americans and Latinos together in a neighborhood group. *American Journal of Community Psychology*, *37*(3/4), 247–255. doi:10.1007/s10464-006-9012-0.

Blau, P. M. (1977). A macrosociological theory of social structure. *American Journal of Sociology*, *83*(1), 26–54.

Box, G. E. P. (1976). Science and statistics. *Journal of the American Statistical Association*, *71*(356), 791–799.

Burt, R. S. (2001). Structural holes versus network closure as social capital. In N. Lin & R. S. Burt (Eds.), *Social capital: Theory and research* (pp. 31–56). New York: Aldine de Gruyter.

Chavis, D. M., & Pretty, G. M. H. (1999). Sense of community: Advances in measurement and application. *Journal of Community Psychology*, *27*(6), 635–642.

Coleman, J. S. (1988). Social capital in the creation of human capital. *American Journal of Sociology*, *94*(S), S95–S120.

Crenshaw, K. (1991). Mapping the margins: Intersectionality, identity politics, and violence against women of color. *Stanford Law Review*, *43*(6), 1241–1299.

Dixon, J., Durrheim, K., & Tredoux, C. (2005). Beyond the optimal contact strategy: A reality check for the contact hypothesis. *American Psychologist*, *60*(7), 697–711.

Festinger, L., Schachter, S., & Back, K. (1950). *Social Pressures in Informal Groups: A Study of Human Factors in Housing*. New York: Harper & Brothers.
Florida, R. (2002). *The Rise of the Creative Class: And How it's Transforming Work, Leisure, Community, and Everyday Life*. New York: Basic Books.
Fu, F., Nowak, M. A., Christakis, N. A., & Fowler, J. H. (2012). The evolution of homophily. *Scientific Reports, 2*, 845.
Grannis, R. (2009). *From the Ground Up: Translating Geography into Community Through Neighbor Networks*. Princeton, NJ: Princeton University Press.
Granovetter, M. S. (1973). The strength of weak ties. *American Journal of Sociology, 78*(6), 1360–1380.
Harrell, S., & Bond, M. A. (2006). Listening to diversity stories: Principles for practice in community research and action. *American Journal of Community Psychology, 37*, 365–376.
Hewstone, M., & Brown, R. (Eds.). (1986). *Contact and Conflict in Intergroup Encounters: Social Psychology and Society*. Cambridge, MA: Basil Blackwell.
Hill, J. L. (1996). Psychological sense of community: Suggestions for future research. *Journal of Community Psychology, 24*(4), 431–438.
Hoffer, L. D., Bobashev, G., & Morris, R. J. (2009). Researching a local heroine market as a complex adaptive system. *American Journal of Community Psychology, 44*, 273–286.
Kelly, J. G. (1971). Qualities for the community psychologist. *American Psychologist, 26*, 897–903.
Lazarsfeld, P. F., & Merton, R. K. (1964). Friendship as a social process: A substantive and methodological analysis. In M. Berger, T. Abel, & C. H. Page (Eds.), *Freedom and Control in Modern Society* (pp. 18–66). New York: Octagon Books.
Lenzi, M., Vieno, A., Santinello, M., & Perkins, D. D. (2013). How neighborhood structural and institutional features can shape neighborhood social connections: A multilevel study of adolescent perceptions. *American Journal of Community Psychology, 51*, 451–467.
Lindblad, M. R., Manturuk, K. R., & Quercia, R. G. (2013). Sense of community and informal social control among lower income households: The role of homeownership and collective efficacy in reducing subjective neighborhood crime and disorder. *American Journal of Community Psychology, 51*, 123–139.
Long, D. A., & Perkins, D. D. (2003). Confirmatory factor analysis of the sense of community index and development of a brief SCI. *Journal of Community Psychology, 31*(3), 279–296.
Macy, M. W., & Willer, R. (2002). From factors to actors: Computational sociology and agent-based models. *Annual Review of Sociology, 28*, 143–166.
McMillian, D. W., & Chavis, D. M. (1986). Sense of community: A definition and theory. *Journal of Community Psychology, 14*, 6–23.
McPherson, M., Smith-Lovin, L., & Cook, J. M. (2001). Birds of a feather: Homophily in social networks. *Annual Review of Sociology, 27*, 415–444.
Moreno, J. L. (1934). *Who Shall Survive? A New Approach to the Problem of Human Interrelations*. Washington, DC: Nervous and Mental Disease Publishing Co.
Peterson, N. A., Speer, P. W., & McMillian, D. W. (2008). Validation of a brief sense of community scale: Confirmation of the principal theory of sense of community. *Journal of Community Psychology, 36*, 61–73.
Portes, A. (1998). Social capital: Its origins and applications in modern sociology. *Annual Review of Sociology, 24*, 1–24.
Portes, A., & Vickstrom, E. (2011). Diversity, social capital, and cohesion. *Annual Review of Sociology*, 37, 461–479. doi:10.1146/annurev-soc-081309–150022.
Prilleltensky, I. (2001). Value-based praxis in community psychology: Moving toward social justice and social action. *American Journal of Community Psychology, 29*, 747–778.
Putnam, R. D. (2001). Bowling alone: The collapse and revival of American community. New York: Simon & Schuster.
Putnam, R. D. (2007). E Pluribus Unum: Diversity and community in the twenty-first century. *Scandinavian Political Studies, 30*, 137–174.
Rappaport, J. (1977). *Community Psychology: Values, Research, and Action*. New York: Holt, Rinehart, & Winston.
Rappaport, J. (1981). In praise of paradox: A social policy of empowerment over prevention. *American Journal of Community Psychology, 9*, 1–25.
Riger, S. (1993). What's wrong with empowerment. *American Journal of Community Psychology, 21*(3), 279–292.
Ryan, W. (1976). *Blaming the Victim*. New York: Random House.

Sarason, S. B. (1974). The psychological sense of community: Prospects for a community psychology. San Francisco, CA: Jossey-Bass.

Schelling, T. (1969). Models of segregation. *American Economic Review, 59*(2), 488–493.

Shook, N. J., & Fazio, R. H. (2008). Interracial roommate relationships: An experimental field test of the contact hypothesis. *Psychological Science, 19*(7), 717–723.

Sigelman, L., & Welch, S. (1993). The contact hypothesis revisited: Black-white interaction and positive racial attitudes. *Social Forces, 71*(3), 781–795.

Society for Community Research and Action (SCRA). (2010). *Policies and Procedures Manual.* Retrieved from: www.scra27.org/documents/scradocume/policiesandproceduresmanualrevisedjuly2010pdf.

Townley, G., Kloos, B., Green, E. P., & Franco, M. M. (2011). Reconcilable differences? Human diversity, cultural relativity, and sense of community. *American Journal of Community Psychology, 47*, 69–85. doi:10.1007/s10464-010-9379-9.

Trickett, E. J., Watts, R. J., & Birman, D. (1994). Toward an overarching framework for diversity. In E. J. Trickett, R. J. Watts, & D. Birman (Eds.), *Human Diversity: Perspectives on People in Context* (pp. 7–26). San Francisco, CA: Jossey Bass.

Watts, D. J., & Strogatz, S. H. (1998). Collective dynamics of small-world networks. *Nature, 393*, 440–442.

Wilensky, U. (1999). NetLogo: Center for connected learning and computer-based modeling. Evanston, IL: Northwestern University.

24

SPATIALIZING SOCIAL NETWORKS

Using Social Network Analysis to Investigate Geographies of Gang Rivalry, Territoriality, and Violence in Los Angeles

Steven M. Radil, Colin Flint, and George E. Tita

> **Research Question:** How is gang violence simultaneously organized by geography and relationships?
> **System Science Method(s):** Networks
> **Things to Notice:**
> - Incorporation of spatial details in network data
> - Use of a network of negative relations

Social network analysis is an increasingly prominent set of techniques used in a number of social sciences, but the use of the techniques of social network analysis in geography has been challenged because of a perceived lack of geographic nuance or consideration of spatialities of context in social networks. The concept of social position and the associated technique of structural equivalence in social network analysis are explored as a means to integrate two different kinds of embeddedness: relative location in geographic space and structural position in network space. Using spatialized network data, this chapter compares the geography of rivalry relations that connect territorially based criminal street gangs in a section of Los Angeles with a geography of the location of gang-related violence. The technique of structural equivalence uses the two different spatialities of embeddedness to identify gangs that are similarly embedded in the territorial geography and positioned in the rivalry network, which aids in understanding the overall context of gang violence. The technique demonstrated here has promise beyond this one study of gang crime as it operationalizes spatialities of embeddedness in a way that allows simultaneous systematic evaluation of the way in which social actors' positions in network relationships and spatial settings provide constraints on and possibilities for their behavior.

Social network analysis is an increasingly prominent technique in a number of social sciences and seemingly has obvious connections to geographies of networks and flows that have become popular in studies of globalization as well as identity politics (Murdoch and Marsden 1995; Dicken et al. 2001; Lantham 2002; Sheppard 2002). The compatibility of the techniques of social network analysis with geographic theories of networks has been challenged, however, because of a lack of geographic nuance or consideration of the spatialities of power and other social relations in

DOI: 10.4324/9781315748771-27

social networks (Allen 2003; Bosco 2006a). We recognize these shortcomings and explore the technique of structural equivalence in social networks as a means to incorporate theoretically informed geographies of situation or embeddedness into social network analysis, a specific step in the broader project of integrating social theories of geography and spatial analytical techniques (Goodchild et al. 2000).

Specifically, we explore how an actor's position in geographic space can be analyzed simultaneously with his or her position in social networks. The geographic premise that social behavior is context specific, and that space and society are mutually constituted, requires the incorporation of multiple spatialities into the analysis of social processes (Leitner, Sheppard, and Sziarto 2008). A typical social network analysis is a one-dimensional spatiality, identifying an actor's location in a social network. By spatializing social networks to include actors' simultaneous positions in networks of relations and places, we offer a technique to analyze the simultaneous embeddedness of actors in both network space and geographic space. The term *embeddedness* has become popular in discussing social networks to illustrate the many situations that social actors create and must negotiate in their behavior (Bosco 2006b). Embeddedness can be seen as a process of creating an increasing intensity of relationships (Bosco 2006b), but it is also a recognition that relationships, distance, and place-specific social relations are intertwined to situate actors (Sheppard 2002; Ettlinger 2003; Staeheli 2003; Leitner, Sheppard, and Sziarto 2008). Spatializing social networks facilitates the analysis of social behavior within the simultaneous and related contexts of network position and relative location in geographic space.

The ability of spatial analysis to incorporate the relative location of social actors, and the linkages between them, can, paradoxically, atomize actors being studied through a "spatial fetishism" that ignores or is unable to address the social relations that construct the spaces within which actors operate.[1] Simply put, spatial analysis is good at analyzing clusters of social behaviors and phenomena (such as crime or disease) but struggles to illustrate the underlying causal structures and relationships. When spatial analysis is overly dependent on reasoning from spatial form to social process, the risk of reducing people to the spaces they occupy grows while the likelihood for new insights shrinks. A spatial analysis that is grounded in theories of the social construction of space and that can model the spatial patterning of relevant social relationships would represent a meaningful advance.

We argue that such an outcome is possible through making use of the concept of embeddedness and the related social network analytic technique of structural equivalence. By performing a hybrid analysis that integrates a spatial analytic approach into the analysis of social networks, we believe that we make a first useful step toward spatializing social network analysis while reducing the possibility of privileging space over social process. The technique we outline combines relative position in geographic space with social network position in a manner that identifies similarly situated actors in network and geographic spaces simultaneously. This in turn allows hypothesis development and evaluations of how differences in position in multi-dimensional spatialities can be said to relate to material outcomes. The technique and its ability to inform are illustrated by an analysis of gang violence in Los Angeles.[2]

In the following section, a discussion of embeddedness and how it relates to network perspectives and methods is offered. We then describe the study area and introduce the data used: a social network of gang rivalries and the geographic distribution of gang-related violence in the Hollenbeck Policing Area in Los Angeles, California. We theorize territoriality, geographic embeddedness, and network position as the specific spatialities at work in the rivalry network and discuss how to consider these simultaneously. In the subsequent section, we present the results of a multirelational positional analysis using social network methods. We conclude the chapter with a discussion of the findings, which demonstrate that the spatialized social network is indeed a useful lens on gang violence in Hollenbeck. The geographic patterns in measures of violence in Hollenbeck are interpretable through and clarified by an understanding of both the network and spatial relationships of the rivalry relations between gangs in the area. The technique demonstrated here has promise beyond this one study of gang crime. It operationalizes relational data in a way that allows simultaneous systematic evaluation of the way in which social actors' positionality in network relationships and spatial settings provide constraints and possibilities on their behavior. In the Conclusion we briefly explore how this technique can be further developed to allow for the addition of other spatialities into a systematic analysis.

Embeddedness, Spatial Analysis, and Social Network Analysis

One particular way in which geographers have tried to understand the behavior of social actors (individuals, groups, organizations, or other social collectives) situated in specific contexts is through the concept of embeddedness. As noted by Ettlinger (2003), Hess (2004), and others, the use of embeddedness in geography largely arose from the work of sociologist Mark Granovetter. In a parallel to calls for more "spatialized" social science approaches (e.g., Goodchild et al. 2000), Granovetter (1985) argued for a more "socialized" understanding of "the extent to which economic action is embedded in structures of social relations," and described what he called the argument of embeddedness: "that the behavior and institutions to be analyzed are so constrained by ongoing social relations that to construe them as independent is a grievous misunderstanding" (481–82). This idea of embeddedness as a form of structural constraint on social action made use of a reading of the word *embed* that implies a state of being surrounded tightly, enveloped, or otherwise constrained. Another reading of the word suggests not just closeness, however, but the state of something becoming part of an integral whole. Later, Granovetter (1992) refined his arguments to differentiate between the kind of embeddedness that suggests closeness (which he called *relational embeddedness*) and the kind that suggests the importance of position in a larger whole (*structural embeddedness*).

Both perspectives on embeddedness have been important for geography. Embeddedness as closeness has resonance for geographers who explore the themes of geography, space, and place as contexts with implications for human behavior. Entrikin (1991) offered a helpful example with the claim that "my life is always embedded in the story of those communities from which I derive my identity" (9). For Entrikin, the "embeddedness" of being within a place-specific social milieu is inseparable from the particulars of human activity. Entrikin's sentiments also reflect Pred's (1984) theoretical understanding of place as a historically contingent process. Embeddedness as position in a larger social structure is also an emerging theme. For example, Flint (2002) argued that space is partly produced by the "connections [of a given place] to the rest of the world" (33) and that, presumably, different connections lie at the heart of the production of different kinds of spaces. In fact, as Staeheli (2003) pointed out, this perspective on space blurs with concepts of place in geography. Although space has historically been understood as referencing the general or universal in geography, as Staeheli noted, contemporary concepts of the social construction of space lead to a refocus from the universality of processes across space to the unique outcomes present in different spaces: Spaces become "social locations" embedded in "webs of cultural, social, economic, and political relationships" (Staeheli 2003, 160). From this perspective, the distinctiveness of spaces and places is due to the embeddedness of being differently located in larger social structures.

As Granovetter (1992) suggested, these different perspectives on embeddedness are not mutually exclusive or necessarily discrete categories, but rather are points that blend into the other. As such, both can be seen as important elements in geographic arguments about the social production of space and how geography mediates social behavior in places. For example, when Agnew (1987) described places as composed of three related elements (locale, location, and sense of place), embeddedness as closeness is similar to Agnew's description of locale as "the settings in which social relations are constituted," whereas embeddedness as position is similar to Agnew's notion of location as "the geographical area encompassing the settings for social interaction . . . at a wider scale" (28). Massey (1993, 66) provided another useful example: Her power-geometry concept is concerned with the position of individuals and social groups relative to the spatial flows of people, information, and capital between places, whereas her progressive sense of place concept incorporates both kinds of embeddedness, emphasizing the role of social relations that are both "in a situation of co-presence" and "stretched out over space." Drawing on these lines of thinking, social behavior can be understood as affected and produced not just by the specific embedded practices, conditions, institutions, or identities of a single place, social location, space, or geography, but also by the way these features are in turn affected and produced by the specific embedded practices, conditions, institutions, or identities of other places, social locations, spaces, or geographies to which they are connected.

Many kinds of spatial analysis make use of the concept of embeddedness as closeness. For example, spatial econometric models consider the influence of places on each other by formalizing the geographic connectivity between the units of analysis that is used to create new variables for inclusion into statistical models (Anselin 2002). More plainly, the degree to which a focal spatial unit is embedded in its closest neighboring geographies is modeled, usually either by identifying its areal contiguity with other spatial units or by selecting neighbors based on the smallest geographic distances between them. Beyond the issues of spatial fetishism, there are some meaningful limitations to this particular approach. All the processes that might contribute to the production of certain spatial patterns are modeled in the same fashion—by using the relative location of each unit of analysis. For example, in an analysis of U.S. crime patterns aggregated to the scale of counties, Baller, Anselin, and Messner (2001) accounted for all the processes of social interaction between counties by defining a set number of the nearest counties (measured by the geographic distance between the approximate center of each county) as influential neighbors for a given focal county. Then, crime data from the set of influential neighbors were used to create a "spatial lag" variable that was included with other explanatory variables for regression modeling (Baller, Anselin, and Messner 2001).

In this example, embeddedness is simply the degree to which each county exhibits similarity in crime and location. If the social relationships that are theorized to underlie a particular observed geographic form are one of embeddedness as closeness, then the model used by Baller and his colleagues might be an appropriate choice. What this approach does not allow is a consideration of any spatiality that cannot or should not be operationalized through an examination of the qualities of the closest units in geographic space. This limitation is meaningful because as observed by Leitner, Sheppard, and Sziarto (2008), multiple spatialities are bound up in issues of interest to geographers and these should be examined together where possible. Ettlinger (2003, 161) emphasized similar themes when describing what she refers to as overlapping networks: "the intersection of different networks in which individuals are engaged," where each network can be thought to represent different kinds of relationships and where only some relationships are "based on proximity." Kwan (2007) also argued that the complexity of human spatial behavior cannot be captured in spatial models with a single type of spatial measurement, such as distance. Quantitative analytic techniques that allow for the consideration of more than a single kind of embeddedness, including position within a social network, would be an important step in addressing the concerns previously noted.

Embeddedness that is based on occupying a particular position within a network is a central concept within sociology. Since the work of Simmel (1955; see also Breiger 1974; Grabher 2008) in the late nineteenth and early twentieth centuries to the very present, one of the primary goals of social network analysis has been to formalize and model the theoretical concepts of social position and to "reveal subsets of actors that occupy equivalent social positions" (Freeman 2005, 248). Social position refers to a collection of actors who are similar in social activity or interactions with respect to actors in other positions; in other words, a social position is "defined by a collection of actors who are similarly embedded in networks of relations" (Wasserman and Faust 1994, 348).

Investigating the effects and consequences of different social positions is a major theme in the social network literature. For example, Friedkin (1984) applied the relative contributions of positions to the study of social homogeneity, finding equivalence on multiple relations to be a useful indicator of group homogeneity. Other notable examples include Snyder and Kick's (1979) examination of the positions of states in international trade networks and Burt's (1987) look at the effect of positions in networks of professional relationships on the adoption of new drugs by physicians. The unifying theme across these different research topics and domains is the assumption that structural position in a social network is an important factor in understanding how actors behave and influence one another.

For social network analysis, similarly patterned actors are seen as occupying distinct social positions in network structures, which is to say that they are similarly embedded in the webs of relationships that constitute the social network in terms of links to other actors (Granovetter 1985; Wellman 1988; Wasserman and Faust 1994). As one of the primary goals of social network analysis is to formalize the theoretical concepts of social position (Freeman 2005), social network analysis is a useful way to explore the concepts of embeddedness in a quantitative fashion through highlighting different social positions as realized in networked data. As Bosco (2006a) and others have observed, however, such network analyses are largely devoid of any geographic specificity. Positions in network space are rarely considered in a way that attempts to incorporate the actual geography of the network although geographers tend to think of embeddedness in purely "territorial" ways.[3]

In network analytic terms, structural equivalence is one of the most common concepts and methods used to identify different social positions in a network of actors (Doreian, Batageli, and Ferligoj 2005). Actors in a network are said to be structurally equivalent if they have identical ties to and from the same other actors in the network (Lorrain and White 1971). Strict structural equivalence is a mathematical property of nodes in a network and typically unrealized in real data. For this reason the common approach in network-based analyses is to identify actors who are "approximately structurally equivalent" (Wasserman and Faust 1994, 366) or to employ variations on structural equivalence.[4] Identifying social positions as collections of actors with similar measures of equivalence allows theories of similar behaviors and outcomes for similar actors to be operationalized and tested. These sorts of questions are drawn from theories of social influence, which generally posit that identically positioned actors in a relational network use each other as a frame of reference for appropriate behavior even if the actors have no direct interaction with the other (Burt 1987). From this perspective, influence is directly tied to the perception of what constitutes proper actions for actors in specific network positions (Galaskiewicz and Burt 1991).

Network-based operationalizations of theories of social position offer potential insight into the spatiality of social behavior. We begin with the recognition that space is socially constructed and emphasize that part of the construction process relates to the simultaneous geographic embeddedness and network position of actors. Conceptualizing the geographic spaces within which actors are embedded as a kind of relational network and using the ability of network methods to identify similarly positioned actors can lead to a consideration of the spatiality of these networks. In other words, combining geographic and social networks in such a way that identifies differently structured spaces (beyond just a consideration of the relative location of these spaces) might offer insight into the relational nature of space and how measures of equivalence relate to specific behavioral outcomes in different spaces of social position.

We call this approach *spatializing social networks*. As an analytic framework, this approach allows influence to take place not just between geographically proximate neighbors (as with conventional spatial analysis) but also between actors who are close in terms of social network space. We see this as a first step toward addressing the complex nature of embeddedness (Bosco 2006b) and considering more than one spatiality of embeddedness (Leitner, Sheppard, and Sziarto 2008). Spatializing social networks allows the identification of *structurally equivalent geographies* along multiple relational spatialities; hypotheses of structured outcomes between and among spaces with similar social positions can then be tested empirically. The aim of this approach is a systematic consideration of the role of actors' embeddedness in space and network positionality as a partial explanation of their behavior.

Street Gangs, Rivalries, and Territoriality in Hollenbeck

This study focuses on violence involving urban street gangs in the Hollenbeck Community Policing Area in Los Angeles, California.[5] Located east of downtown Los Angeles, the Hollenbeck Policing Area "has a population of roughly 200,000 people and is 15.2 square miles in size. It encompasses the communities of El Sereno, Lincoln Heights, and Boyle Heights" (Los Angeles Police Department (LAPD) 2008). According to U.S. Census statistics, most of the population is Latino (84.5 percent) and nearly 40 percent (39.4 percent) of the total population was born in Mexico. A total of 30 percent of the population lives below the poverty line and of the total population that is at least 25 years old, 35 percent have less than a high school degree or equivalent (Tita et al. 2003).

According to Tita et al. (2003), homicide rates in Hollenbeck have been higher than both Los Angeles and U.S. national homicide rates since the early 1990s. Hollenbeck consistently ranks among the top three or four of the LAPD's eighteen policing areas in violent crime. LAPD crime statistics for 2007 show that violence in Hollenbeck remains high, as there were 799 violent crimes reported in the Hollenbeck area, which translates to 4.7 percent of the citywide totals for violent crimes in that year (LAPD 2008). Gangs and gang-related issues are central to violent crime in Hollenbeck: Gangs were involved in nearly 75 percent of all homicides in Hollenbeck from 1995 to 1998 (Tita et al. 2003) and in a 2008 report by the Los Angeles County District Attorney, the Hollenbeck Policing Area was classified as an area of "Very Heavy Gang Activity," the highest category of the classification scheme used in the report (Cooley 2008).

Tita et al. (2003) argued that the combination of physical barriers and political geographic boundaries that define the Hollenbeck area serves to limit interactions with gangs from neighboring areas. As seen in Figure 24.1, Hollenbeck is delimited in the west by the Los Angeles River and along the northwest by the Pasadena Freeway. The city of Vernon, California, which lies to the immediate south of Hollenbeck, is an industrial area with a total population of only 91 at the 2000 Census. Thus, there no are spatially proximate gangs in either of these directions. To the southeast, Hollenbeck is bordered by an unincorporated area of Los Angeles County (East Los Angeles). To the northeast, Hollenbeck shares a border with the city of Pasadena. Both of these areas do have urban street gangs, yet none of these gangs are rivals with any of

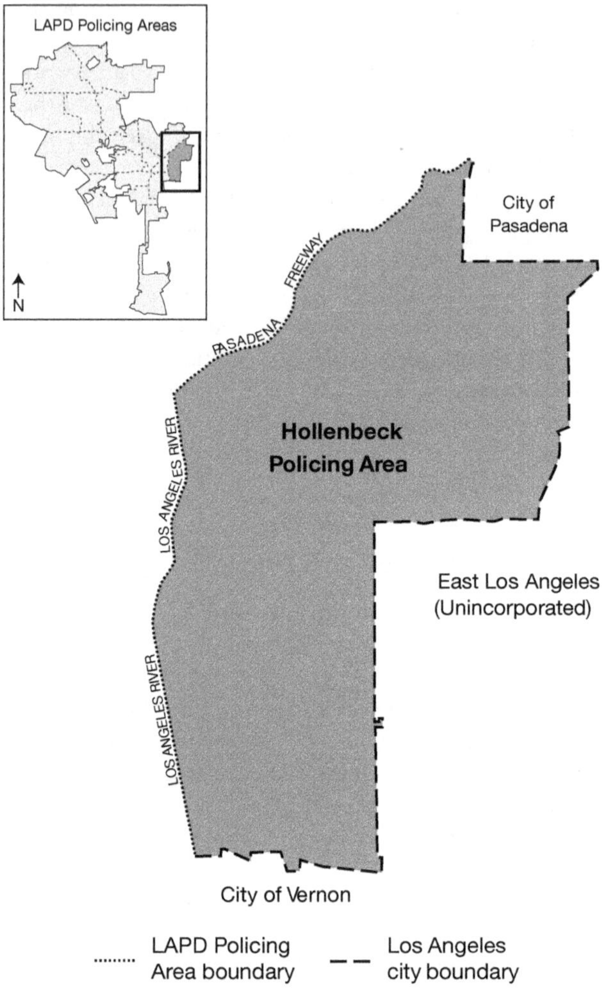

Figure 24.1 The Hollenbeck Policing Area is east of downtown Los Angeles. Numerous street gangs are active in Hollenbeck, but elements of the urban landscape, including the Los Angeles River and the Pasadena Freeway, serve to limit interactions with gangs from other areas (LAPD = Los Angeles Police Department)

the Hollenbeck gangs. There are several reasons for this. First, although no physical barrier serves to impede movement between Hollenbeck and either East Los Angeles or Pasadena, the fact that each is served by different public school districts greatly restricts across-place social interactions (Grannis 2009). Although every other gang in the region might be a potential rival, with no history of interaction among local youth the gangs outside of Hollenbeck remain outside of the awareness space of the Hollenbeck gangs. Second, there exists a simple propinquity effect, as none of the gangs found in either East Los Angeles or Pasadena occupy space on the border shared with Hollenbeck. The net effect of these border features, both physical and political, is to create a landscape within which the rivalries of the Hollenbeck gangs are wholly contained (Tita et al. 2003).

Hollenbeck is no stranger to gangs and gang activity. The history of urban street gangs in East Los Angeles, including Hollenbeck, is a long one, with some gangs documented back to the late 1940s (Moore 1991). From 2000 to 2002, twenty-nine active gangs were identified in the Hollenbeck area (Tita et al. 2003). Control over territory is a central theme for the gangs of Hollenbeck. The gangs in Hollenbeck are what Klein (1995) described as "traditional" in that they have a strong attachment to turf, or the territory under the direct control of a gang. Tita et al. (2003) made a similar argument and characterized the gang violence in Hollenbeck as expressly tied to the defense of turf and control over territory. Although they arise from different motivations, the antigang activities of the LAPD also revolve around control of territorial space. As described by Herbert (1997), "police (LAPD) strategies to create public order involve enacting boundaries and restricting access" (11). The key point here, made by Sack (1986) and others (e.g., Paasi 2003), is that territory is not the static result of social processes but is instead what Newman (2006) calls an "imperative" and an "essential component of human behavior" (88–89). The attempts by the various gangs to control the spaces of Hollenbeck result in violence between the different street gangs themselves. Understanding the spatial patterning of the relationships between the gangs, themselves wrapped up in issues of contesting and controlling space, is key to understanding the spatial patterning of gang violence in Hollenbeck.

The emphasis on territorial control by gangs in Hollenbeck relates to a key way in which spaces and places are socially constituted. In geography, territoriality is often seen as the "delimitation of boundaries" and the interrelated "behavior within those boundaries" (Kahler 2006, 2). Sack's (1986) influential work on territoriality defines it as the use of territory for political, social, and economic ends and it has been most often associated with the spatiality of the nation-state (Paasi 2003). We can say that territoriality is conventionally understood in geography as involving both a partitioning of space into distinct units and ongoing attempts to control the space to maintain the borders between the units (Kuus and Agnew 2008). These characteristics track well with how Tita, Cohen, and Enberg (2005) described gang turf: a well-defined geographic area of a city, such as a neighborhood, that is claimed by the gang as its "domain."

Territoriality as domain and partitioning behavior led Newman (2006, 91) to conclude that territorial behavior is quite meaningful at "local levels": "rivalries are played out through the daily life practices of segregated groups residing in their own distinct . . . neighborhood turfs." Both Newman's emphasis on spatial segregation and Cresswell's (1996) work on places as geographic expressions of cultural norms and transgression to those norms suggest a reason why the persistent territoriality of Hollenbeck's various gangs might result in violence between them. When a gang member enters the turf another gang a spatial transgression has occurred and the gang member is now "out of place." In these situations, the response to such spatial transgressions might involve violence. If presence in other turfs can be seen as transgressive, there is an expectation that local geographic embeddedness or the relative nearness to differently controlled spaces is an important element to certain kinds of outcomes, such as violence (Kahler 2006).

Mapping Gang Violence in Hollenbeck

As described in Tita (2006), from 2000 to 2002, Hollenbeck experienced 1,223 violent crimes by or against gang members. This kind of violence is defined by Tita as "gang related" and includes the protection of turf from an incursion by rival members as well as all other violence involving gang members. The list of crimes over this time period include the legal classifications of aggravated assaults, simple assaults, assault with a deadly weapon, attempted homicides, homicides, robberies, kidnappings, and firing a gun into an inhabited dwelling or vehicle.[6] When aggregated by U.S. Census block groups, mapping the violent crimes suggests two important features (see Figure 24.2). First, violence has penetrated all areas of Hollenbeck and, second, there might be some spatial clustering of gang-related violence.

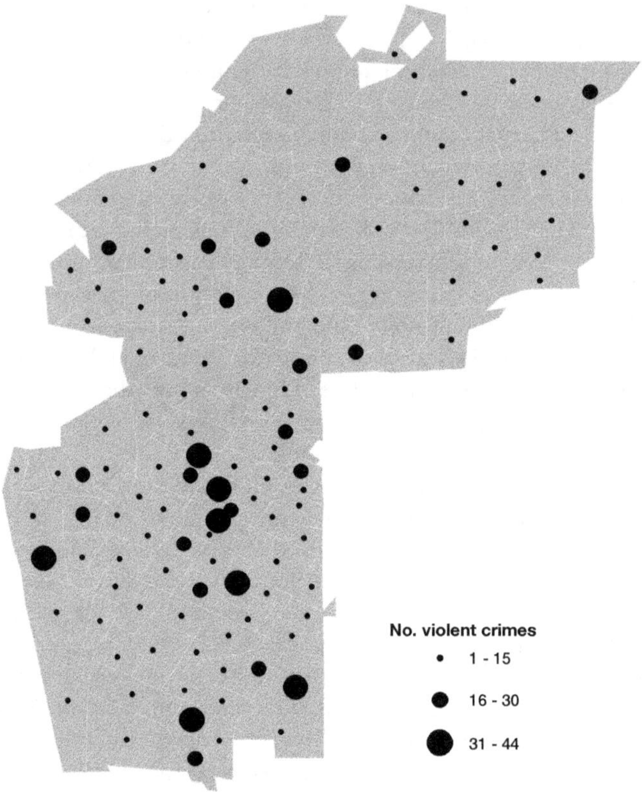

Figure 24.2 Gang and gang-related violence in Hollenbeck from May 2000 through December 2002 by U.S. Census block group

Table 24.1 Global Moran's *I* results of violence counts by census block

	First-order contiguous neighbors only	*First- and second-order contiguous neighbors*
Rook contiguity	0.09	–0.09
Queen contiguity	0.08	–0.04

Note: $N = 120$.

The first point is straightforward, as violent crimes were present in every block group. Incident counts by block group range from a low of 1 (occurring 6 times) to a high of 44 (occurring once), with a mean across block groups of 10.19. To evaluate the second observation, a global Moran's *I* test of spatial autocorrelation was performed by identifying block groups as neighbors on the basis of either sharing a common length of border ("rook" contiguity) or sharing a single border point ("queen" contiguity).[7] Despite the presentation in Figure 24.2, neither rook nor queen configurations resulted in strong statistical measures of dependence (see Table 24.1). The Moran's *I* ratios of 0.09 for rook and 0.08 for queen are interpreted as very weak positive dependence (positive dependence means neighboring values are similar). In fact, repeating the test by increasing the contiguity from first-order neighbors (those that share border lengths or

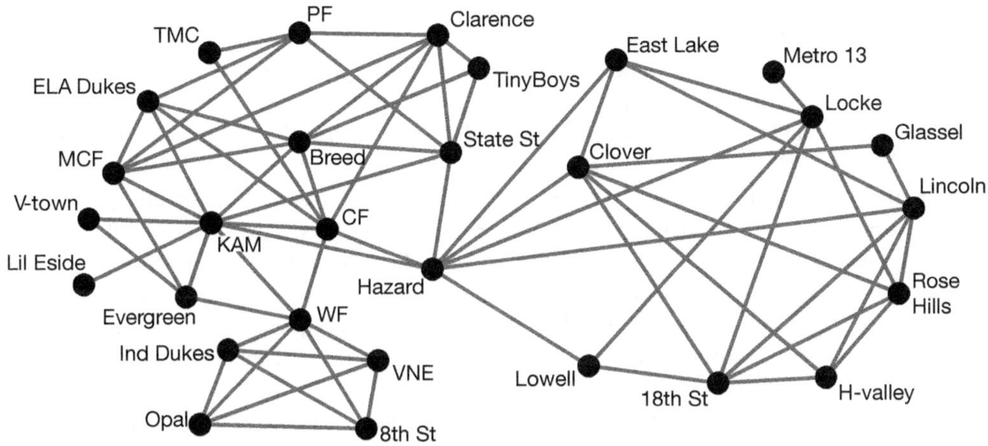

Figure 24.3 A network diagram of the rivalry network among gangs in Hollenbeck. Each of the 29 gangs is shown as a node and the presence of a rivalry between two gangs results in a link between them. Although all gangs are connected to the network by at least one rivalry, all but two gangs are involved in multiple rivalries.

points) to also include second-order neighbors (neighbors of the first set of neighbors using the same criteria of shared border length or point) results in measures that are interpreted as very weak negative spatial dependence (neighboring values are dissimilar; Moran's *I* ratios of –0.09 for second-order rook and –0.04 for second-order queen). Although the absence of evidence for robust global spatial dependence might be due to the level of aggregation in the data, it also does not recommend a conventional spatial analytic approach.

Gang Rivalry and Territorial Networks

Tita et al.'s (2003) analysis of violence pertaining to the 29 different gangs active in Hollenbeck from 2000 to 2002 identified a social network of gang rivalries. Rivalry is a meaningful relation in this circumstance, as urban street gangs are committed to the defense of their turf and have negative relationships (i.e., rivalries) that explicitly tie them to other gangs (Tita et al. 2003). This interpretation is related to implications of rivalries as a key social relation in the attempt to understand other forms of violence (see Diehl and Goertz 2000; Flint et al. 2009, for examples). The rivalry relations were identified through the use of a survey by Tita et al. (2003) that asked informants from both the LAPD and some of the gangs to identify the rivalries between each of the gangs.[8] A network diagram of these rivalries is shown in Figure 24.3. The nodes in this network represent the gangs and the connections between them represent the rivalries. The measurement of rivalry was binary where the presence of a rivalry resulted in a link and the absence of a rivalry resulted in no link. Every gang was connected to the network structure and the number of rivalries ranged from a minimum of one to a maximum of ten. As is common in network analyses, the network can also be described as a matrix. In this case, a 29 × 29 binary matrix was produced where the presence of a rivalry between the gangs was coded as one and the absence of a rivalry was coded as zero (Tita et al. 2003).

In addition to identifying gang rivalries, the location and boundaries of the turf of each of the 29 gangs was mapped through the participation of the LAPD (Tita et al. 2003). Tita (2006) later used U.S. Census block group areal units to establish the presence or absence of gang turf at a geographic level that possessed data of interest to ecological studies of crime. The Hollenbeck area includes 120 such block groups and gang turf was present in 103 of these units. In seven instances, more than one gang claimed turf in the same block group, but close spatial proximity was not always a predictor of rivalry (Tita 2006).

After identifying both a social network of rivalries and the relative geographic locations of gang turf with a disaggregated territorial map, Tita (2006) made a unique methodological choice by attempting to blend these structures together. This was done by first reimagining the geography of gang turf as a network of neighbors, a conventional way to produce a spatial weights matrix for use in spatial econometric models (Anselin 2002). Each areal unit of Hollenbeck (census block groups) became a single node in a new network and the links between the nodes were based on geographic contiguity. The coding scheme was again binary where block groups that shared borders were formally connected and those that did not remained unconnected. After distributing the gang rivalries to the disaggregated territorial map, the two separate network matrices (29 × 29 gang rivalries and 120 × 120 census block groups) were then associated using matrix multiplication to produce a single 120 × 120 matrix for use in a spatial econometric model. This weights matrix was then used to create a new explanatory variable, which was included in a regression along with a variety of other common measures in criminology, such as education and income (Tita 2006).

Tita's (2006) use of the hybrid weights matrix as the connectivity input for a spatial econometric regression model was an innovative approach in the spatial econometric modeling of crime, but it also kept at arm's length one of the central claims of relational social science—that relationships between actors have more explanatory power than do attribute-based categories (Emirbayer and Goodwin 1994; Emirbayer 1997). Despite using relational data (the relative locations of the gangs and the rivalry relationships), the initial analytic product was another attribute-based explanatory variable and the analytic focus remained on the perceived causal power of attributes. Additionally, and most important for this chapter, this approach forced network position to be modeled as local geographic embeddedness. In other words, the theoretically distinct spatialities of embeddedness were operationalized as a single form of embeddedness and all interactions between the geographic units of analysis became overly territorialized (Hess 2004).

The lack of fidelity to the explanatory power of relationships along with the fact that both the spatialities of geographic embeddedness and network positionality must be operationalized in the same way leaves something to be desired. On this basis and by drawing on Anselin's (2002) observation that the construction of the spatial weights matrix in spatial econometric models is actually based on social network concepts, the authors began a series of conversations that extended Tita's previous analysis while engaging theoretical discussions of embeddedness. In this chapter, the particular outcome is a consideration of whether structural similarity based on both kinds of embeddedness contributes to similar outcomes for actors.[9] We examine this question in a way that fits the spatialities under consideration. The hypothesis here is whether or not similarly embedded and positioned territorial spaces experience similar amounts of violence. In other words, can structurally equivalent geographies be identified from both the embedded gang turf and the rivalry network that connects them and do differently structured geographies exhibit different violence patterns?

Spatializing the Social Networks of Hollenbeck's Gang Rivalries

Positional analyses involve identifying social positions in a relational network based on similar patterns of links between individual nodes. The process is well developed in social network analysis and most positional analyses "focus on identifying subsets of equivalent actors" (Wasserman and Faust 1994, 354). Identifying subsets in a complex network involves simplifying the networked data. Whether presented as a network diagram or a matrix, it is difficult to visually identify meaningful patterns that might exist. For example, the spatialized rivalry network is shown in Figure 24.4. The complexity of the ties precludes visual interpretation, but a matrix of spatialized rivalry relations can be reorganized on the basis of similar ties between actors in the network. This activity is referred to in network terms as *matrix permutation* and allows similar actors to be grouped together (Wasserman and Faust 1994).

As previously discussed, relationships among the 29 different gangs of Hollenbeck can be represented in matrix form, but the rows and columns of the resulting 29 × 29 matrix represent both a social unit (gangs) and a territorial unit (turf). As such, there are two distinct spatialities of embeddedness that must be considered: local geographic embeddedness and position in the overall network structure. Geographic embeddedness, especially among territorial units with demarked boundaries and that are mutually exclusive of other units (such as gang turf), can be represented in the conventional spatial econometric way through a consideration of relative location. Neighboring gang turf (those that comprise the geography in which a given gang's turf is embedded) can be identified through shared borders. The rivalry network can be represented in a similar way, with rivalries between gangs coded in the matrix. The result is two 29 × 29 matrices, one for each spatiality.

Although the 29 × 29 matrices capture all the social and territorial units identified by Tita et al. (2003) in the Hollenbeck Policing Area, the matrices do not capture the entire geography of Hollenbeck. There are areas of Hollenbeck that are not claimed as turf by any of the gangs but that do experience gang-related violence. Although these areas of Hollenbeck do not represent the territorial claims of the gangs, we include these areas in both the geographic embeddedness matrix and the network positionality matrix. An important reason for this choice is that some of the gangs do not share borders with other gangs and are instead bounded by unclaimed areas. These unclaimed areas still must be negotiated in some sense to reach the turf of other gangs, though, and therefore can be understood as a kind of connective tissue to the overall gang geography (see Tita and Cohen 2004, 199). Further, as an empirical reality, gang-related violence is not wholly contained within claimed turf and excluding unclaimed areas would also involve excluding information that could affect the outcome of the analysis. More plainly, a consideration of local geographic embeddedness that did not include unclaimed spaces would be partial and incomplete. For these reasons, the unclaimed areas were included by simply adding another row and column to each matrix. In the case of the geographic embeddedness matrix, the unclaimed areas were treated as another possible neighbor in the embedded geography of the gang turf. For the network positionality matrix, the unclaimed areas never resulted in a link, as they did not represent a social unit that another gang could have a rivalry with. The final result was two binary 30 × 30 matrices.

The identification of equivalent geographies on the patterning of ties of both geographic embeddedness and network positionality was accomplished using the network analysis program UCINET (Borgatti, Everett, and Freeman 2005). The particular technique used for this study is called *convergence of iterated correlations* (CONCOR; Breiger, Boorman, and Arabie 1975; see also Wasserman and Faust 1994, 376–81). This procedure computes Pearson product–moment correlation coefficients among the rows and columns of the input matrices by comparing the value of a given cell to the mean value of both the row and column in which it occurs.[10] The result is a new single matrix where cell values are the calculated correlation coefficients, which is meant to represent the structural similarities between pairs of actors based on similarities in the patterns of the ties between them. Actors who are perfectly equivalent on all relations will

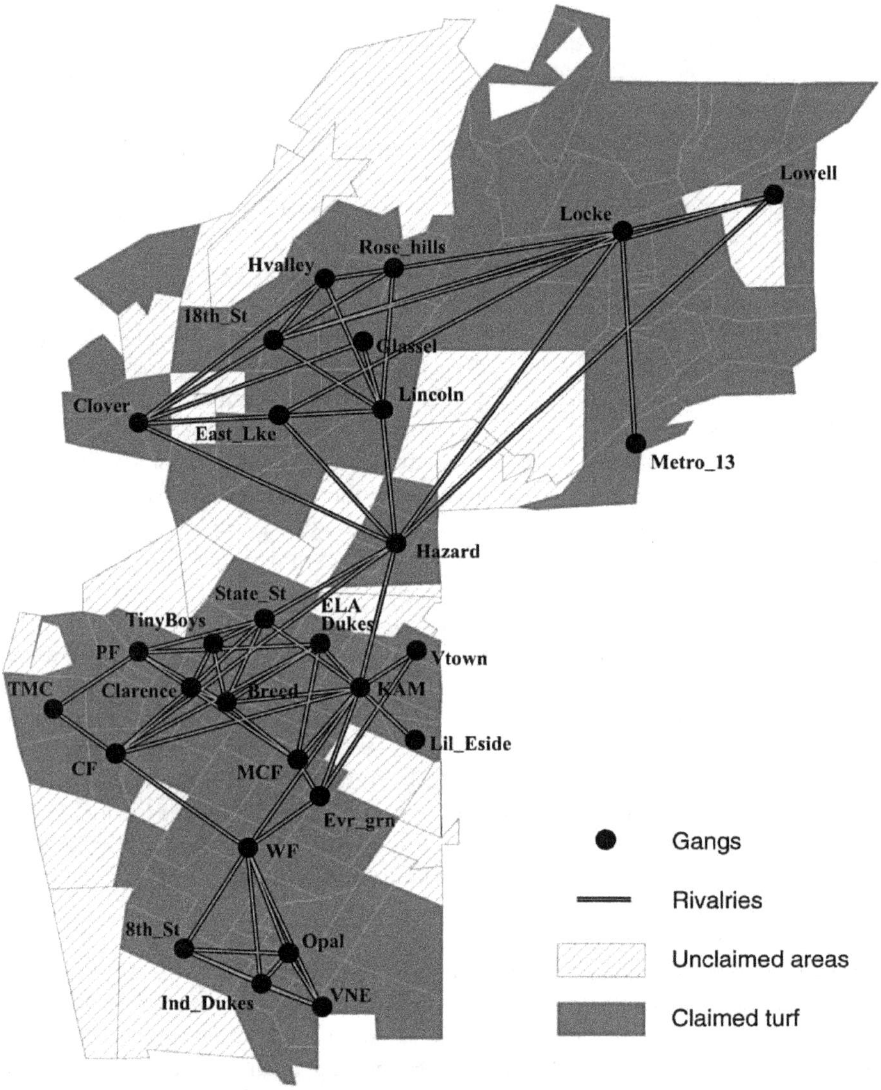

Figure 24.4 Placing the gang rivalry network (based on turf locations) into the geographic space of Hollenbeck shows both the complexity of the social relations and how some relations "stretch" long distances to link gangs, whereas others link only immediate neighbors

have correlation coefficients of +1 between their rows and columns of the original input matrices. As noted earlier, however, perfect equivalence rarely occurs in most actual network data. For this reason, the CONCOR process uses the correlation matrices as input for a new round of correlation computations. The output from this calculation is used as input for yet another round of correlations, and the process continues in this fashion. As noted by Wasserman and Faust (1994, 377), "after several iterations of this procedure, the values of all correlations in the matrix are equal to either +1 or −1."[11] The final correlation matrices are dichotomized to allow all actors to be grouped into one of two categories. In network terms, actors in the same category are similarly structured in the network, which is the functional definition of equivalence.

The CONCOR process also allows analysts to define the number of positions to identify. For example, the process will identify exactly two social positions for all the actors in a given multirelational network. This might be a useful generalization, but it could also be an oversimplification and the analyst might desire to identify more detailed patterning. This can be done by performing another round of CONCOR to submatrices of the original data, which comprise the actors of each of the two positions identified in the first round as described earlier. In other words, all the actors from each category are reorganized into new and separate matrices (one for each relation) and the CONCOR process begins again to each of these new sets of matrices. Each set of matrices is split into two positions again; two becomes four, four becomes eight, and so on.[12]

For this study, the CONCOR process was applied three times, which produced eight positions. As previously mentioned, more positions could be identified, up to a maximum of thirty.[13] In keeping with the overall goal of a positional analysis to simplify patterns, however, it was concluded that identifying more than eight positions could be counterproductive. A common way to represent the product of the CONCOR process is with a dendrogram. The dendrogram that resulted from applying the CONCOR process three times to the spatialized rivalry data is shown in Figure 24.5. Each of the 30 units in the two relational matrices was classified into one of 8 positions, many of which comprised multiple gangs. Only two positions were made up of a single gang or unit of analysis: After the third split, the unclaimed areas of Hollenbeck were identified as a unique geography, as was the turf of one of the gangs (see Figure 24.5).

Each time the CONCOR process was applied, two further analytic steps were taken to interpret the results. First, the resulting positions were mapped using a geographic information system (GIS) to aid in understanding and describing the overall geography of the equivalence categories. Second, analysis of variance (ANOVA) tests were performed on violence counts to determine statistically significant differences between the geographies identified by each CONCOR iteration.[14] Statistically significant results suggest that the observations on violence are drawn from different populations. We interpret this not just as empirical evidence of differences in the amount of gang-related violence between differently embedded and positioned geographies but also as evidence of different social processes at work. The results of each split and of the associated ANOVA tests are described next and presented in Table 24.2.

First Split

The first application of CONCOR resulted in the identification of two differently structured positions based on patterns of geographic embeddedness and network positionality. As seen in Figure 24.6A, these positions result in a clear north–south division in the gang geography in Hollenbeck. The northern position, labeled as Position 1 in Figure 24.6A, comprises the turf of 11 different gangs plus the unclaimed areas (which accounts for the presence of this position in some of the census block groups in the southern half of Hollenbeck). The southern position, labeled as Position 2 in Figure 24.6A, is made up of the turf of 18 different gangs. This north–south gang geography corresponds to an observation in Tita et al. (2003) of a strong north–south division in the rivalry network based on a landscape feature: The San Bernadino Freeway (Interstate 10) bisects Hollenbeck and might constrain gang interaction in the same way that Tita et al. (2003) argued that built landscape and political boundary features do. Interestingly, CONCOR grouped the unclaimed areas of Hollenbeck together with the turf of the gangs in the northern half of Hollenbeck. This suggests two insights: (1) the spatiality of geographic

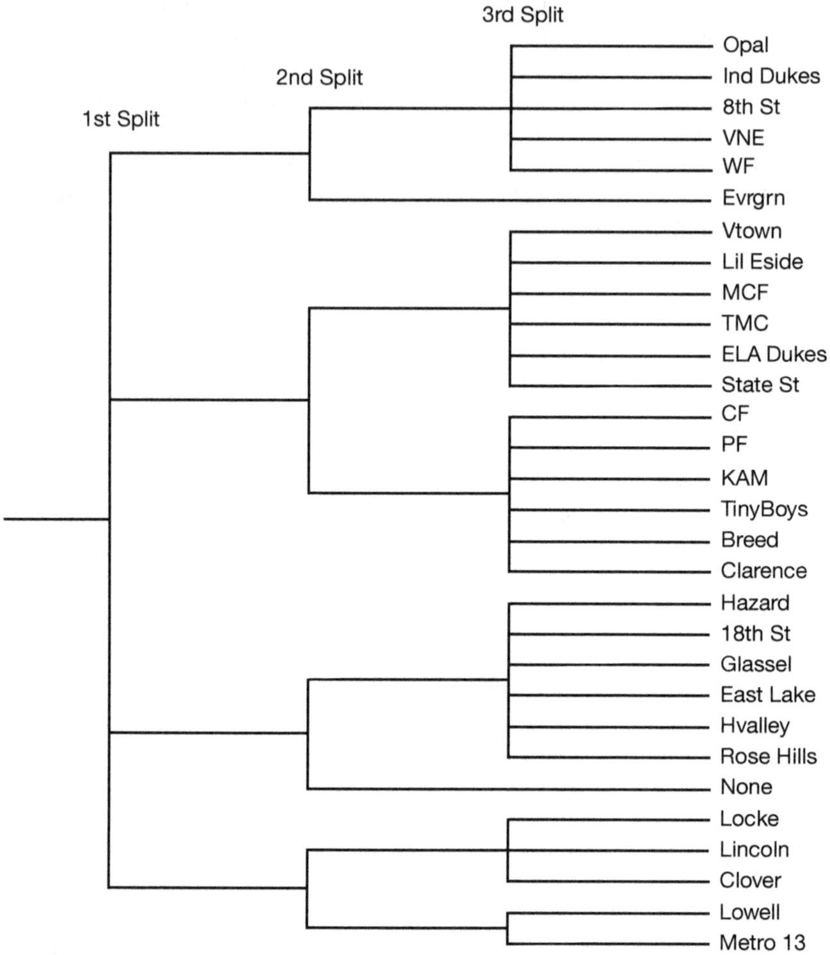

Figure 24.5 A dendrogram of the CONCOR (convergence of iterated correlations) positional analysis process through three splits. Each of the thirty units (29 gangs plus one unit for unclaimed turf) in Hollenbeck is classified into distinct positions at each stage of the process.

embeddedness is mediating the rivalry links between gangs in Hollenbeck as the rivalries are clearly geographically organized at this level; and (2) that the material geography of violence in claimed turf in the north might be more similar to the patterns in the unclaimed areas than to the claimed turf in the southern position. We return to the first point in the concluding discussion of this article, but the second point is easily evaluated using gang-related violence as a metric. The southern position in Hollenbeck clearly experienced more violence as the mean amount of gang violence in the southern position was over 30 percent higher (33.76 percent) than that in the north. The ANOVA tests clearly pick up on the fact that the differences between the variations in violence between the two positions were larger than the variations in violence among each position.

Table 24.2 Analysis of variance results for each iteration of the CONCOR process

	Position	No. of gangs	Mean no. of violent incidents	n (block groups)	Parametric ANOVA (F)	Nonparametric ANOVA (χ^2)
1st split	1	11[a]	8.00	52	9.586[b]*** (p = 0.002)	6.518[b]** (p = 0.011)
	2	18	13.06	68		
2nd split	1.1	6[a]	8.34	41	3.294[c]** (p = 0.023)	7.595* (p = 0.055)
	1.2	5	7.48	27		
	2.1	12	12.48	29		
	2.2	6	13.78	23		
3rd split	1.1.1	0[a]	7.00	25	2.404[d]** (p=0.025)	11.738[d] (p = 0.110)
	1.1.2	6	10.44	16		
	1.2.1	2	6.00	4		
	1.2.2	3	7.74	23		
	2.1.1	6	14.30	10		
	2.1.2	6	11.53	19		
	2.2.1	1	32.00	1		
	2.2.2	5	12.95	22		

Notes: CONCOR = convergence of iterated correlations; ANOVA = analysis of variance. a Includes unclaimed areas. b df = 1. c df = 3. d df = 7. *p < 0.1. **p < 0.05. ***p < 0.01.

Second Split

The second application of CONCOR to the spatialized social network subdivided the first two categories, resulting in a total of four new differently structured categories as shown by Figure 24.6B. The northern position was partitioned into two new geographies, which suggest a center–periphery arrangement: Position 1.1 describes a geography comprising the turf of six different gangs that roughly occupy the center of the original northern position plus the unclaimed areas of Hollenbeck, whereas Position 1.2 describes a geography of the turf of five gangs to both the east and west of the first new position. Interestingly, the CONCOR correlations identify the unclaimed areas as most similar to the gang turf in the center of the northern position (1.1), which might suggest lower violence levels than in the east–west position (1.2). The southern position was also partitioned, but a different geographic pattern was evident. This position was subdivided into two new north–south oriented geographies: Position 2.1 describes the northernmost geography and comprises the turf of 12 different gangs, whereas Position 2.2 describes the southernmost geography of 6 gangs. The number of gangs in Position 2.1 is at least double that of any other position, suggesting the possibility of more violence.

Despite the identification of four distinctly structured geographies based on patterns of geographic embeddedness and network positionality, violence data for the four new positions suggest fewer material differences, at least between each of the subgroups of Positions 1 and 2. Mean violence among the four positions was lowest in the northern periphery group (Position 1.2, M = 7.48) but at very similar levels in the northern core group (Position 1.1, M = 8.34), which contained the unclaimed areas. Mean violence counts were highest in the southernmost group (Position 2.1, M = 13.78) and violence counts for the final group were slightly lower (Position 2.2, M = 12.48). Despite different patterns of geographic embeddedness and network positionality, there were not significant material differences in violence patterns between these two pairs of geographies. We believe this is an important matter and return to discuss it in more detail in the Conclusion.

Figure 24.6 (A) The first split of the convergence of iterated correlations (CONCOR) process reveals two positions in the gang network, one in the north of Hollenbeck and one in the south. (B) The second split of the CONCOR process subdivides each of the first two positions. (C) The third split of the CONCOR process continues to subdivide positions. In both the north and south areas of Hollenbeck, core–periphery positions continue to be suggested.

Third Split

The third application of CONCOR to the spatialized rivalry relations again subdivided the previously identified positions into eight differently structured positions, which are mapped in Figure 24.6C. At this stage the geographies are considerably more complex but we wish to highlight the following features. First, only at this stage does the CONCOR process identify the unclaimed areas as a unique and separate position (Position 1.1.1 in Figure 24.6C), which is meaningful as it shows that these areas are indeed implicated in the relationships that constitute violence in Hollenbeck. Second, of the northernmost positions derived from Position 1 (Positions 1.1.1, 1.1.2, 1.2.1, and 1.2.2) the violence levels are similarly low except for the gang geography that essentially lies between those where violence is highest (Position 1.1.2, $M = 10.44$). For example, Position 1.1.2, composed of the turf of six different gangs, lies roughly geographically between the all the gangs of the southernmost positions and the gangs of Positions 1.2.1 and 1.2.2 but also geographically between the turf of the three different gangs that make up Position 1.2.2. This finding points to a new spatiality, that of geographical betweenness, which might be implicated in producing a higher level of violence. Finally, of the southernmost positions (Positions 2.1.1, 2.1.2, 2.2.1, and 2.2.2) violence is also similar for three of the positions and noticeably higher for the other position. Position 2.2.1, composed of the turf of a single gang (Evr grn; see Figure 24.4), experienced 32 incidents of gang-related violence, one of the highest census block group counts for the entire study area. This outcome might also be attributable to a relational betweenness, as although the gang's turf is not obviously geographically situated between many other gangs or positions as in the previous example, it is situated between two important gangs, WF and KAM (see Figures 24.3 and 24.4), which both have rivalries with Evr grn and with each other. This relational betweenness might play a significant role in producing a high level of violence. The ability of the technique to highlight new and previously unobserved spatialities is an important outcome that we expand on in the Conclusion.

Conclusion

The spatialities of social relations among and between gangs in Hollenbeck undoubtedly have implications and manifestations that cannot be completely captured and interpreted by the hybrid spatial analytic and social analytic methods used here. For example, dynamism is not present in our data and, as such, our analysis presents a presumably unchanging rivalry network and associated turf map. Social relations are inherently dynamic and, although changes occur within existing structures, today's rivalries might in fact be tomorrow's alliances. Our approach does not prevent the consideration of different network structures over time and disaggregating networks temporally (in additional to distributing them geographically) would be a positive step toward introducing dynamism in a way that we were not able to with these data. Further, our example was quite streamlined in that only rivalries and relative location in space were considered for network position. This limitation undoubtedly overlooks some factors in the production of gang violence and therefore is not a perfect example of an analysis of multiple spatialities.

Nonetheless, accepting multiple networks of relations as analytic inputs is a hallmark of the techniques we presented and including other kinds of relationships that would better reflect the concerns of the overlapping networks of Ettlinger (2003) or the multiple spatialities of Leitner, Sheppard, and Sziarto (2008) is certainly possible. Even when considering the limitations of our example, it is evident that using territoriality as a lens through which to focus on the rivalry relations in Hollenbeck leads to several important conclusions. Although the specific findings of the gang rivalry networks in Hollenbeck are certainly dependent on the geographic and historical context of gangs in Hollenbeck, we interpret these findings in a way that emphasizes the utility of the demonstrated concepts and methods for other topics.

The specific findings of this study offer meaningful evidence that, in Hollenbeck, the overall geographic pattern of violence is interpretable by spatializing the rivalry relationships and that new spatialities can emerge from the complex web of relationships in geographic and network space. To the first point, the spatialized positional analysis clearly reveals that the rivalry network has produced distinct spatial patterns. These distinct geographies, easily verified by the material geography of gang-related violence, are formed by the rivalry network but also clearly mediated by the relative location of each gang in geographic space. This geographic mediation of the rivalry network is more than just an argument of space as mechanism of integration of social processes (e.g., Goodchild et al. 2000) and one would be hard pressed to reach this observation through either a conventional spatial analysis that did not consider social relations or a conventional social network analysis that did not consider geography.

To the second point, social network analytic techniques can simplify complex and multi-dimensional network structures, which might also highlight new kinds of spatialities that emerge from the interplay of different spatialized networks. For example, spatialities of betweenness might be important to the overall geography of violence in Hollenbeck. We use betweenness here in its social network sense, as a type of network centrality that has to do with being located between and connecting different actors in a network structure (Freeman 1977; Friedkin 1991). In the social network literature, the betweenness concept has typically been used to evaluate the importance of particular actors in gatekeeping roles (also known as *brokerage*) or to explain the benefits for actors that bring together otherwise disconnected networks ("structural holes"), both major themes in social network analysis when dealing with flows of information, capital, or anything else that can move through a network.[15] Our concept of spatialities of betweenness is related to these ideas in that being situated between other actors either in social space (relational betweenness), geographic space (geographical betweenness), or both might be important elements in understanding the overall geographic patterns of gang violence. Given that part of what is flowing in our example of the territorialized network of rivalries is violence, betweenness is not necessarily a positive and bridging a structural hole (such as connecting different spatial domains) might mean being targeted for more violence.

Although the research presented in this chapter concerns a specific kind of violence between specific kinds of social units in a specific locality, there are general implications for research into other contexts that are framed here in terms of methodology. First, and most important, the patterning of ties in the social network, or the structure of social relations, can and should be understood geographically. Despite arguments that geographic perspectives on social networks (especially while utilizing social network analytic methodologies) are difficult to achieve (Leenders 2002; Bosco 2006a), this article demonstrates a relatively simple yet powerful technique to link relational networks with geography. Second, because many relations have a material geographic nature, positional analyses can yield insights into how social relations might be implicated in the production of differentiated spaces and places. These insights can be particularly valuable because they are arrived at in a novel fashion. By and large, investigations into spatiality in geography are almost exclusively qualitative affairs. In as much as the conventional approach to spatiality might not easily penetrate other social science disciplines where quantitative methodologies remain central, the addition of a quantitative and empirically minded set of techniques that engages with issues of spatiality is significant.[16] Third and finally, although the methods demonstrated in this article are capable of standing alone, they seem particularly valuable when used in concert with other ways of knowing. As noted in the discussion of specific findings about the spaces of gang rivalry in Hollenbeck, the analytic methods used in this chapter answer some questions and suggest others, and are not offered as an absolute substitute for granular, situated, and ethnographic knowledge.

This chapter focused on introducing a set of techniques that we feel can be compatible with issues in geography concerned with the patterning of social relations. Because we focus on methodology here, however, we are not making an explicit argument about theories of social relations and, as such, the issues of precisely how territorially defined gang rivalry networks produce different patterns of violence in Hollenbeck are not exhausted in our study. Many paths for future inquiry are open. Although not intended as a comprehensive list, we wish to use our example to draw attention to a few possibilities that we also feel suggest the flexibility and applicability of social network analysis methods and techniques to other kinds and scales of research in geography. As we have already mentioned, issues associated with changing patterns of relationships over time provide a potentially fruitful line of inquiry. Beyond this, perhaps the most important frontier is the investigation of gang rivalries in other geographic contexts. Relational data on other gang rivalry networks at similar geographic scales would allow comparisons that might reveal domain-specific laws applicable to other contexts (e.g., O'Loughlin 2000) or suggest other kinds of relationships that should also be considered. Even if the focus remains with the gangs of Hollenbeck, network methods offer potential lines of inquiry.

Positional analyses, such as the type performed in this chapter, are holistic and global in that they are concerned with the structural properties of an entire network. They are also but one possible approach to applying social network analysis methods to issues of the social construction of space. For example, an egocentric approach, which focuses on properties of the network from the perspective of different actors situated in particular locations, might tell a great deal more about the differences and similarities between places (social locations) and also refocus inquiry from networks of gangs to networks of individuals.[17] It is our hope that through the example of Hollenbeck, the issues and methods presented in this study encourage future research into how social networks are involved in the social production of space and how social network analysis methods can be utilized to understand if and how the structure of social networks has implications for material geographic outcomes.

Acknowledgment

This chapter originally appeared in *Annals of the Association of American Geographers*, 100, 307–326 (2010).

Notes

1. This critique can be traced to Wolpert's (1964) behaviorist approach but saw a more recent manifestation in Harley's (1989) critiques of the cultural norms within cartographic representations and Pickles's (1995) concerns of the unacknowledged representations in GIS. More recently Bosco (2006a) and Ettlinger and Bosco (2004) have noted the need to consider power relations in social network analysis. See also Collinge (2005, 191–201) for a contemporary discussion of spatial fetishism.
2. This chapter reports on initial findings in a collaborative project between geographers and a criminologist and extends the results of an exploratory spatial analysis of gang violence by the criminologist (Tita et al. 2003).
3. The data used in this article are the same as used in Tita et al. (2003). In this sense, territorial means an emphasis on geographically local relations (see Ettlinger 2003; Hess 2004). In a related critique, Bosco's (2006b) discussion of embeddedness warns against focusing on simple measures, such as whether ties in a network are weak or strong, or the actual geographic distance separating actors. In his qualitative analysis, the situation of actors in places that are themselves situated in broader networks is identified

as a key feature of the situation of actors. A quantitative analysis of social networks will struggle to uncover the precise mechanisms by which a few actors developed their position in a network as well as the role of emotions in social behavior, both important features of Bosco's (2006b) work. The conclusion that "network processes are affected by, and cannot be divorced from, the conditions governing the context in which they are produced and in which they operate" (Bosco 2006b, 360), however, is one that indicates a role for quantitative analysis in identifying the interaction between geographic and network spaces in contextualizing social behavior. Hence, and in a complementary fashion, a quantitative analysis of social networks can attempt to integrate the insights gained from contemporary social theory regarding space and geography (Goodchild et al., 2000).

4. The two most common approaches to equivalence are structural equivalence and regular equivalence. The most important difference between the two is that structural equivalence requires that equivalent actors have the same connection to the same neighbors while regular equivalent actors have the same or similar patterns to potentially different neighbors (Doreian, Batageli, and Ferligoj 2005).
5. The LAPD has a geographic structure that organizes policing activities. The LAPD divides Los Angeles into eighteen different geographic regions, called Community Policing Areas. These Community Areas are organized into one of four Bureaus. Hollenbeck is one of five Community Areas in the LAPD's Central Bureau. See www.lapdonline.org/hollenbeck community police station (last accessed January 20, 2010) and Tita et al. (2003).
6. Rape and domestic violence reports were not available and not included in violence counts (Tita et al. 2003).
7. Contiguity matrices and Moran's *I* tests were constructed and performed using the GeoDa spatial econometric software package (Anselin, Syabri, and Kho 2006).
8. Rivalry relationships between the gangs were identified through the use of a survey of local police and former gang members by Tita et al. (2003). Each informant was provided with a survey that included one page for each gang. At the top of the page, a particular gang was identified and the respondent was asked to "Please Identify All of the Gangs that are an *Enemy* of the <insert gang name>." Law enforcement experts and several current and former gang members completed the survey and there was perfect agreement across the gang members' and law enforcement experts' surveys.
9. See Leenders (2002) for a discussion of the nature of the spatial weights matrices in spatial econometrics and similarities with social network analysis.
10. Correlations are not the only method to identify equivalences in a network. It was chosen for this study as it is identified by Wasserman and Faust (1994) as the preferred measure for pattern similarities. The formula for calculating correlations on a single relation between an actor *i* and an actor *j* is

$$r_{ij} = \frac{\sum (x_{ki} - \bar{x}_{\bullet i})(x_{kj} - \bar{x}_{\bullet j}) + \sum (x_{ik} - \bar{x}_{i\bullet})(x_{jk} - \bar{x}_{j\bullet})}{\sqrt{\sum (x_{ki} - \bar{x}_{\bullet i})^2 (x_{kj} - \bar{x}_{\bullet j})^2}\ \sqrt{\sum (x_{ik} - \bar{x}_{i\bullet})^2 (x_{jk} - \bar{x}_{j\bullet})^2}}$$

where $\bar{x}_{i\bullet}$ is the mean of the values in row *i* of the matrix (excluding the diagonal) and $\bar{x}_{\bullet i}$ is the mean of the values in column *i* of the matrix (excluding the diagonal). This formula can accept matrices coded as 0 or 1 as it subtracts mean row and column values from the focal unit cell values. Calculating correlations on multirelational data is done using a generalized version of the formula for single relations:

$$r_{ij} = \frac{\sum_{r=1}^{2R} \sum_{k=1}^{g} (x_{ikr} - \bar{x}_{\bullet i})(x_{jkr} - \bar{x}_{\bullet j})}{\sqrt{\sum_{r=1}^{2R} (x_{ikr} - \bar{x}_{\bullet j})^2}\ \sqrt{\sum_{r=1}^{2R} (x_{jkr} - \bar{x}_{j\bullet})^2}}$$

where there are $r = 1, 2, \ldots, R$ relations (Wasserman and Faust 1994,368–69). The result for both formulas is a single matrix where the nondiagonal cells are correlation coefficients. Our study uses two matrices as inputs—one for geographical embeddedness and one for network position.
11. The default number of iterations in UCINET is 25 and this was used for this analysis.
12. The CONCOR procedure always splits a set of actors into exactly two subsets and repeating the process results in a series of binary splits. This has been critiqued in the social network literature as imposing a form on the identification of social positions that might not connect well with theory (Wasserman and Faust 1994). This critique can be seen in this study as imposing a hierarchical structure to the gang turf geography. Even in an exploratory study such as this one, this issue is worth noting and

future efforts along these lines would benefit from comparing the positions identified by CONCOR with those identified by other commonly used techniques. See Borgatti and Everett (1992) for a helpful discussion of the challenges in matching the methods used to identify social positions with the different theoretical concepts of equivalence in social network analysis.

13. Because each matrix contained only thirty rows and columns, the maximum number of possible equivalence categories is thirty, which would occur when each row or column in the matrix becomes its own unique category.
14. Parametric and nonparametric ANOVA tests were performed and results of each are reported in Table 24.2. The nonparametric test used is the Kruskal–Wallis test, which uses ranks in place of actual data. The Kruskal–Wallis test, like all rank-based statistics, lacks resolving power (as seen by the second and third split results in Table 24.2) but is generally considered robust against non-normal data, such as the count data we used. The census block group geography used by Tita et al. (2003) and Tita (2006) was retained for the ANOVA tests. Counts were aggregated by census block group (n = 120) and each block group was coded to one of the twenty-nine gangs. Block groups that were outside any gang turf were coded as part of the single unclaimed area unit previously described.
15. Influential examples of these concepts related to betweenness in the social network literature include Granovetter's (1973) investigation of how information about job opportunities is transmitted in professional social networks and Burt's (1992) analysis of structural position on entrepreneurial success. Various methods to operationalize and evaluate measures of betweenness have also been developed, mostly inspired by the work of Freeman (1977, 1979). See also Wasserman and Faust (1994) and Everett and Borgatti (2005).
16. See O'Loughlin (2000) for an example of how different methodologies contribute to keeping ideas from conventional political geography apart from the mainstream of thought and practice in political science.
17. A focus on individuals is a persistent concern for geographers as a way to avoid spatial fetishism. Adapting these arguments to social networks, Ettlinger (2003, 146) argued that the unit of analysis (or node) in a network should be at least partly composed of individuals as a way to avoid "a reification of firms or other organizations or networks themselves." In our example, the units of analysis are both social collectives (gangs) and the spaces they control (turf). Although we feel that in our example the gangs themselves reify turf with their abiding emphasis on territorial control, we acknowledge the importance of Ettlinger's (2003) argument to certain research objectives and advocate careful consideration of her argument when selecting the unit(s) of analysis.

References

Agnew, J. A. 1987. *Place and Politics: The Geographical Mediation of State and Society*. Boston, MA: Allen & Unwin.

Allen, J. 2003. *Lost Geographies of Power*. Malden, MA: Blackwell.

Anselin, L. 2002. Under the hood: Issues in the specification and interpretation of spatial regression models. *Agricultural Economics* 27: 247–67.

Anselin, L., I. Syabri, and Y. Kho. 2006. GeoDa: An introduction to spatial data analysis. *Geographical Analysis* 38: 5–22.

Baller, R. D., L. Anselin, and S. F. Messner. 2001. Structural covariates of U.S. county homicide rates: Incorporating spatial effects. *Criminology* 39: 561–90.

Borgatti, S. P., and M. G. Everett. 1992. Notions of position in social network analysis. *Sociological Methodology* 22: 1–35.

Borgatti, S. P., M. G. Everett, and L. C. Freeman. 2005. *UCINET 6 for Windows: Software for Social Network Analysis*. Boston, MA: Analytic Technologies.

Bosco, F. J. 2006a. Actor-network theory, networks, and relational approaches in human geography. In *Approaches to Human Geography*, Eds. S. Aitken and G. Valentine, 136–46. London: Sage.

——. 2006b. The madres de plaza de mayo and three decades of human rights' activism: Embeddedness, emotions, and social movements. *Annals of the Association of American Geographers* 96: 342–65.

Breiger, R. L. 1974. The duality of persons and groups. *Social Forces* 53: 181–90.

Breiger, R. L., S. A. Boorman, and P. Arabie. 1975. An algorithm for clustering relational data with applications to social network analysis and comparison with multidimensional scaling. *Journal of Mathematical Psychology* 12: 328–83.

Burt, R. S. 1987. Social contagion and innovation: Cohesion versus structural equivalence. *American Journal of Sociology* 92: 1287–1335.

——. 1992. *Structural Holes: The Social Structure of Competition*. Cambridge, MA: Harvard University Press.

Collinge, C. 2005. The différance between society and space: Nested scales and the returns of spatial fetishism. *Environment and Planning D: Society and Space* 23 (2): 189–206.

Cooley, S. 2008. Gang crime and violence in Los Angeles County: Findings and proposals from the District Attorney's Office. Available at: http://da.co.la.ca.us/pdf/LADA Gang Crime & Violence APR 2008.pdf (last accessed May 29, 2009).

Cresswell, T. 1996. *In Place/Out of Place: Geography, Ideology and Transgression*. Minneapolis, MN: University of Minnesota Press.

Dicken, P., P. F. Kelly, K. Olds, and H. W.-C. Yeung. 2001. Chains and networks, territories and scales: Towards a relational framework for analysing the global economy. *Global Networks* 1 (2): 89–112.

Diehl, P. F., and G. Goertz. 2000. *War and Peace in International Rivalry*. Ann Arbor, MI: University of Michigan Press.

Doreian, P., V. Batageli, and A. Ferligoj. 2005. Positional analyses of sociometric data. In *Models and Methods in Social Network Analysis*, Eds. P. J. Carrington, J. Scott, and S. Wasserman, 77–97. New York: Cambridge University Press.

Emirbayer, M. 1997. Manifesto for relational sociology. *American Journal of Sociology* 103: 281–317.

Emirbayer, M., and J. Goodwin. 1994. Network analysis, culture, and the problem of agency. *American Journal of Sociology* 99: 1411–54.

Entrikin, J. 1991. *The Betweenness of Place: Toward a Geography of Modernity*. Baltimore, MD: Johns Hopkins University Press.

Ettlinger, N. 2003. Cultural economic geography and a relational and microspace approach to trusts, rationalities, networks, and change in collaborative workplaces. *Journal of Economic Geography* 3 (2): 145–71.

Ettlinger, N., and F. J. Bosco. 2004. Thinking through networks and their spatiality: A critique of the US (public) war on terrorism and its geographic discourse. *Antipode* 36: 249–71.

Everett, M., and S. P. Borgatti. 2005. Extending centrality. In *Models and Methods in Social Network Analysis*, Eds. P. J. Carrington, J. Scott, and S. Wasserman, 57–76. New York: Cambridge University Press.

Flint, C. 2002. The theoretical and methodological utility of space and spatial statistics for historical studies: The Nazi Party in geographic context. *Historical Methods* 35 (1): 32–42.

Flint, C., P. Diehl, J. Scheffran, J. Vasquez, and S.-H. Chi. 2009. Conceptualizing ConflictSpace: Toward a geography of relational power and embeddedness in the analysis of interstate conflict. *Annals of the Association of American Geographers* 99 (5): 827–35.

Freeman, L. C. 1977. A set of measures of centrality based on betweenness. *Sociometry* 40 (1): 35–41.

——. 1979. Centrality in social networks: Conceptual clarification. *Social Networks* 1 (3): 215–39.

——. 2005. Graphic techniques for exploring social network data. In *Models and Methods in Social Network Analysis*, Eds. P. J. Carrington, J. Scott, and S. Wasserman, 248–69. New York: Cambridge University Press.

Friedkin, N. E. 1984. Structural cohesion and equivalence explanations of social homogeneity. *Sociological Methods and Research* 12 (3): 235–61.

——. 1991. Theoretical foundations for centrality measures. *American Journal of Sociology* 96 (6): 1478–1504.

Galaskiewicz, J., and R. S. Burt. 1991. Interorganizational contagion in corporate philanthropy. *Administrative Science Quarterly* 36: 88–105.

Goodchild, M., L. Anselin, R. P. Appelbaum, and B. H. Harthorn. 2000. Toward spatially integrated social science. *International Regional Science Review* 23: 139–59.

Grabher, G. 2008. Trading routes, bypasses, and risky intersections: Mapping the travels of "networks" between economic sociology and economic geography. *Progress in Human Geography* 30 (2): 163–89.

Grannis, R. 2009. *From the ground up: Translating geography into community through neighbor networks*. Princeton, NJ: Princeton University Press.

Granovetter, M. 1973. The strength of weak ties. *American Journal of Sociology* 78 (6): 1360–81.

——. 1985. Economic action and economic structure: The problem of embeddedness. *American Journal of Sociology* 91 (3): 481–510.

——. 1992. Problems of explanation in economic sociology. In *Networks and Organizations: Structure, Form, and Action*, Eds. R. G. Eccles and N. Nohria, 25–26. Boston, MA: Harvard Business School Press.

Harley, J. B. 1989. Deconstructing the map. *Cartographica* 26 (1): 1–20.

Herbert, S. K. 1997. *Policing Space: Territoriality and the Los Angeles Police Department*. Minneapolis, MN: University of Minnesota Press.

Hess, M. 2004. "Spatial" relationships? Towards a reconceptualization of embeddedness. *Progress in Human Geography* 28 (2): 165–86.

Kahler, M. 2006. Territoriality and conflict in an era of globalization. In *Territoriality and Conflict in an Era of Globalization*, Ed. M. Kahler, 1–24. Cambridge, UK: Cambridge University Press.

Klein, M. 1995. The American street gang: Its nature, prevalence and control. New York: Oxford University Press.

Kuus, M., and J. A. Agnew. 2008. Theorizing the state geographically: Sovereignty, subjectivity, territoriality. In *Handbook of Political Geography*, Eds. K. Cox, M. Low, and J. Robinson, 95–106. Thousand Oaks, CA: Sage.

Kwan, M.-P. 2007. Mobile communications, social networks, and urban travel: Hypertext as a new metaphor for conceptualizing spatial interaction. *The Professional Geographer* 59 (4): 434–46.

Lantham, A. 2002. Retheorizing the scale of globalization: Topologies, actor-networks, and cosmopolitanism. In *Geographies of Power: Placing Scale*, Eds. A. Herod and M. W. Wright, 115–44. Oxford, UK: Blackwell.

Leenders, R. T. A. J. 2002. Modeling social influence through network autocorrelation: Constructing the weight matrix. *Social Networks* 24: 21–47.

Leitner, H., E. Sheppard, and K. M. Sziarto. 2008. The spatialities of contentious politics. *Transactions of the Institute of British Geographers* 33 (2): 157–72.

Lorrain, F., and H. C. White. 1971. Structural equivalence of individuals in social networks. *Journal of Mathematical Sociology* 1: 49–80.

Los Angeles Police Department. 2008. Hollenbeck crime statistics. Available at: www.lapdonline.org/assets/pdf/hbkprof.pdf (last accessed May 29, 2009).

Massey, D. 1993. Power-geometry and a progressive sense of place. In *Mapping the Future: Local Cultures, Global Change*, Eds. J. Bird, B. Curtis, T. Putnam, G. Robertson, and L. Tickner, 59–69. London and New York: Routledge.

Moore, J. 1991. *Going Down to the Barrio: Homeboys and Home-girls in Change*. Philadelphia, PA: Temple University Press.

Murdoch, J., and T. Marsden. 1995. The spatialization of politics: Local and national actor-spaces in environmental conflict. *Transactions of the Institute of British Geographers* 20 (3): 368–80.

Newman, D. 2006. The resilience of territorial conflict in an era of globalization. In *Territoriality and Conflict in an Era of Globalization*, Ed. M. Kahler, 85–110. Cambridge, UK: Cambridge University Press.

O'Loughlin, J. 2000. Geography as space and geography as place: The divide between political science and political geography continues. *Geopolitics* 5: 126–37.

Paasi, A. 2003. Territory. In *A Companion to Political Geography*, Eds. J. A. Agnew, K. Mitchell, and G. Ó Tuathail,109–22. Malden, MA: Blackwell.

Pickles, J., Ed. 1995. *Ground Truth: The Social Implications of Geographical Information Systems*. London and New York: Routledge.

Pred, A. 1984. Place as historically contingent process: Structuration and the time-geography of becoming places. *Annals of the Association of American Geographers* 74: 279–97.

Sack, R. D. 1986. *Human Territoriality: Its Theory and History*. Cambridge, UK: Cambridge University Press.

Sheppard, E. 2002. The spaces and times of globalization: Place, scale, networks, and positionality. *Economic Geography* 78 (3): 307–30.

Simmel, G. 1955. *Conflict and the Web of Group Affiliations*. New York: Free Press.

Snyder, D., and E. L. Kick. 1979. Structural position in the world system and economic growth, 1955–1970: A multiple-network analysis of transnational interactions. *The American Journal of Sociology* 84 (5): 1096–1126.

Staeheli, L. A. 2003. Place. In *A Companion to Political Geography*, Eds. J. A. Agnew, K. Mitchell, and G. Ó Tuathail, 158–70. Malden, MA: Blackwell.

Tita, G. 2006. Neighborhoods as nodes: Combining social network analysis with spatial analysis to explore the spatial distribution of gang violence. Paper presented at the Workshop on Space, Networks, and Social Influence conference, University of California, Irvine, February 9–11, 2006.

Tita, G., and J. Cohen. 2004. Measuring spatial diffusion of shots fired activity across city neighborhoods. In *Spatially Integrated Social Science*, Eds. M. F. Goodchild and D. G. Janelle, 171–204. Oxford, UK: Oxford University Press.

Tita, G., J. Cohen, and J. Enberg. 2005. An ecological study of the location of gang "set space." *Social Problems* 52 (2): 272–99.

Tita, G., K. J. Riley, G. Ridgeway, C. Grammich, A. F. Abrahamse, and P. W. Greenwood. 2003. *Reducing Gun Violence: Results from an Intervention in East Los Angeles*. Santa Monica, CA: Rand.

Wasserman, S., and K. Faust. 1994. *Social Network Analysis: Methods and Applications.* Cambridge, UK: Cambridge University Press.

Wellman, B. 1988. Structural analysis: From methods and metaphor to theory and substance. In *Social Structures: A Network Approach*, Eds. B. Wellman and S. D. Berkowitz, 19–61. New York: Cambridge University Press.

Wolpert, J. 1964. The decision process in spatial context. *Annals of the Association of American Geographers* 54 (4): 337–58.

25

SOCIAL NETWORKS AND THE STUDY OF LANGUAGE VARIATION AND CHANGE

Suzanne Evans Wagner and
Maya Ravindranath Abtahian

Research Question: How do social networks influence how we talk?
System Science Method(s): Networks
Things to Notice:

- Using networks to explain long-term and subtle changes in behavior
- Networks affect behavior in every century and every country

By the age of four children have, with seemingly little effort and arguably insufficient input, acquired the complexities of their native language(s) such that their language is hardly different from that of their caregivers. Understanding how children do this, and to what extent the blueprint for language may be innate, is the premise of much of the enterprise of modern linguistics. At the same time we have evidence for the fact that language is not perfectly transmitted from generation to generation. In the long-term, evidence for disruption in the transmission process comes from the fact that languages change over time. Modern readers of Shakespeare, for example, encounter many difficulties of comprehension due to the cumulative effects of changes to the word order, sound system, and lexicon of English. Even in the short term we can observe differences in the way that different generations speak: English speakers under the age of fifty today are unlikely to ever use the adjective groovy, for example, even while older speakers may still use it. How do these innovations arise? How are they propagated across the generations, across geographic space, and across social groups? These questions are at the root of the study of language variation and change, often known as sociolinguistics. This chapter will begin with a brief overview of the advances that have been made in understanding language variation and change ("variationist sociolinguistics"). The majority of the chapter will constitute a survey of studies of language change that have applied social network theories to their analyses of linguistic data.

Both language change and language variation can occur at any level of linguistic structure. This includes variation in:

- pronunciation (as in the pronunciation of *time* as "taam" in the southern USA; *laugh* as "laff"and "lahf" in northern and southern Britain respectively);
- word form (as in *I've got* in British English versus *I've gotten* in US English);

 DOI: 10.4324/9781315748771-28

- word meaning (as in the different meanings of *boot*—trunk of a car, footwear, to vomit, etc.—in different parts of the English-speaking world);
- variation in sentence structure (as in *Turn off the light* versus *Turn the light off*).

Although in some cases multiple variants of a particular linguistic variable will survive for centuries (as in the case of the Old English variants *ascian* and *axian* of the verb "to ask" that persist today in the variable pronunciations of *ask* as "ask" or "ax"), some variation will lead to long-term variation. Shakespeare, for example, uses the irregular forms *durst* for the past tense of *dare* and *holp* for the past tense of *help* where regular forms *dared* and *helped* are always used today, as well as the lexical items *afeard* (scared) and *befall* (happen) that have fallen almost completely out of use. English speakers did not wake up one day deciding to use the regular *dared* instead of *durst*; rather a period of variation would have preceded this change and been characterized by both inter speaker and intra speaker variation —that is, some speakers or communities using more of the innovative form, as well as individual speakers using more of the innovative form in casual or "vernacular" speech versus more formal speech.

One of the guiding principles of the sociolinguistic research paradigm founded in the 1960s and known as "variationist sociolinguistics" has been the "use of the present to explain the past" (Labov, 1975). Language change in the present can be studied in greater depth, with more attention to social detail and with more completeness than is possible in post-hoc historical study. If all language changes are preceded by a period of variation (two or more ways of saying the same thing), therefore at least some of the linguistic variation that can be observed today is evidence of change in progress. Insights from contemporary studies of change can then be extrapolated backwards to explain historical language changes. Sociolinguists seek to observe and quantify linguistic variation such as that described here, to determine whether any particular instance of synchronic language variation is evidence of diachronic change, and to understand how changes spread. In general, the first step is to identify the linguistic variable and its variants. In the case of the *ask* example above, for instance, one would identify the variants "ask" and "ax" of the abstract variable *ask*. The second step is to find a way to observe natural language use that is likely to contain the variable, select speakers from relevant social groups and across the age spectrum, and design a data collection method. The final step is to analyze the distribution of the linguistic variants with respect to relevant social factors in the community.

Sociolinguists are attracted to social network theory in particular for the insights it can offer into two of five "problems" that must be explained by the study of language change according to seminal work by Weinreich, Labov, and Herzog (1968). The five problems identified by Weinreich et al. are as follows: *constraints* (What linguistic changes are possible?), *transition* (How does the change progress?), *evaluation* (How is the new form socially evaluated by the community, and what effect does it have on communicative efficacy?), *actuation* (Why do changes happen in a particular place and not another?) and *embedding* (How does a language change diffuse throughout the language community?). The effect of interpersonal relationships on language choices is a part of most variationist research for the insight it can offer into the spread of change. The subset of these studies that have explicitly focused on the network concept are generally those that are most concerned with the problems of *actuation* and *embedding* – that is, why changes happen in a particular place (i.e. with particular individuals and not others) and how linguistic changes become embedded in the language community (i.e. what types of community structures promote the diffusion of change). The adoption of social network analysis to answer these problems started with James and Lesley Milroy's interest in Granovetter's (1973) work on so-called "weak ties" in network structure, which they employed in their study of eight linguistic variables in Belfast, Northern Ireland in the 1970s (Milroy and Milroy, 1978). We describe their study in the next section.

The application of network theory to language change by the Milroys was made possible by several decades of established work on the actuation and embedding of other types of social changes. This is outlined in the classic work on the diffusion of innovations by Rogers (2003 [1962]) as well as more recent analysis of the same by Valente (1995). Much of the early work in this field focused on the diffusion of innovative technologies and practices, such as agricultural innovations (Ryan and Gross, 1943), driver training among schools (Mort, 1953, in Rogers, 2003), and the spread of drugs like the antibiotic tetracycline (Coleman, Menzel and Katz, 1966). In the classic hybrid corn study, for instance, Ryan and Gross (1943) sought to explain why some farmers adopted hybrid seed corn faster than others. The diffusion of any innovative technology such as these requires adopters first, to acquire knowledge about the innovation and second, make the decision to adopt. Ryan and Gross conclude that both mass communication and interpersonal communication networks played a role in the dissemination of knowledge about the technology and the decision to adopt (with the former playing a larger role in disseminating information and the latter playing a larger role in the decision to adopt). The fact that the adoption of so many social changes follows a similar S-shaped curve over time led Rogers to the conclusion that the diffusion of innovations was a "universal process of social change" (2003: xvi).

Moreover, in addition to the conscious adoption of new technologies, a few studies have shown that the unconscious adoption of behaviors and habits may follow the same process. Christakis and Fowler (2007), for instance, show that the spread of not only infectious diseases but also non-communicable diseases such as obesity may be dependent on social network. In their study individuals who had strong ties to obese individuals were more likely to become obese themselves.

Linguistic changes may fall into both categories—those that people are more consciously aware of (e.g. vocabulary such as *to google something, hashtag, fracking, to crowdsource*) and those that are mostly unconscious, such as subtle pronunciation changes or the initial proportional rise in frequency (the base of the S-curve) of the intensifier *so* at the expense of *really* and *very*.[1] Labov (2002) cites the work of Stanley Lieberson (2001) on trends in children's names as being particularly relevant to the discussion of linguistic change and diffusion, since "individuals select their children's names as a matter of . . . personal choice, without being aware of the social factors that determine that choice."

The majority of the cases that are of interest to linguists and that we describe in this chapter are of unconscious linguistic actuation and embedding. Since the spread of linguistic features (especially unconscious ones) typically cannot be observed directly, social network theory has been of special utility in allowing linguists to make ever more sophisticated inferences about this process.

The Spread of Language Change

Studies of social networks have demonstrated that cities are loci of innovation (Florida, 2003). As such, they have been the focus of most variationist sociolinguistic research to date. In large cities, members of different social class and ethnic groups do not always use the same words, nor do they sound exactly alike in their pronunciation or sentence structure. Yet members of urban speech communities do share many features of the local dialect, because language changes do not only diffuse from place to place, but from group to group and speaker to speaker (Labov, 1966). For instance, Labov (2001) proposed that most of Philadelphia's twentieth-century vowel changes first emerged in the working-class neighborhoods of the inner city before spreading to middle-class neighborhoods. Small differences in pronunciation between these neighborhoods were an artifact of middle-class speakers still lagging behind their working-class counterparts.

Generalizing about the pathway of language change in terms of social class or ethnicity, however, is to rely upon social constructs that may or may not have universal applicability. In the 1970s, Lesley and James Milroy introduced social network theory to sociolinguistics as a means to avoid such heavy reliance on categories,[2] an approach in line with that of network analysis in sociology (see, e.g., Wellman 1983). The Milroys were influenced by Granovetter's (1973) ideas about "weak ties." "Weak" ties (roughly, "acquaintanceships"), he argued, were more likely to serve as bridges between networks. Granovetter initially defines relatively "strong" and "weak" ties on an intuitive basis: Tie strength between individuals increases with "the amount of time, the emotional intensity, the intimacy (mutual confiding), and the reciprocal services which characterize the tie" (Granovetter, 1973: 1361). Thus, especially strong ties could include those that exist between family members, or best friends; weak ties could include acquaintances and co-workers with whom one only intermittently interacts. Since there is a strong motivation for one's good friends to know one another, strong ties typically form dense networks in which many of the possible connections between individuals exist. Conversely, one's acquaintances typically comprise a loose, low-density network in which many of the possible connections between individuals are absent. These acquaintances, in turn, are embedded within their own dense networks characterized by strong ties. Weak ties between acquaintances therefore create "bridges" between dense networks. Without weak bridging ties, members of such networks "will be deprived of information from distant parts of the social system and will be confined to the provincial news and views of their close friends" (Granovetter, 1983: 202).[3]

To illustrate these concepts, Figure 25.1 shows, for a single individual, "A," a hypothetical social network. A has a dense network constituted of strong ties, but also has several weak

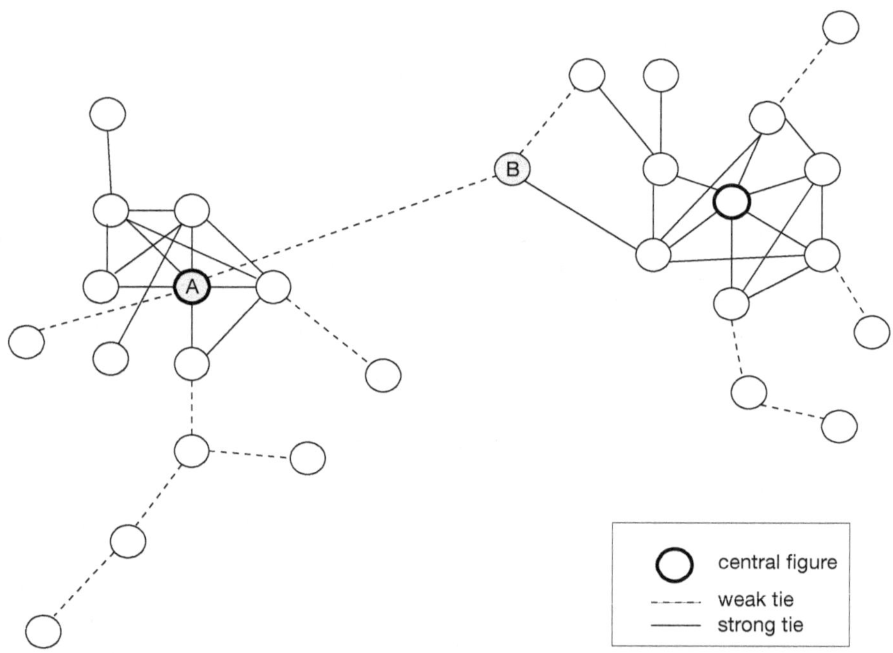

Figure 25.1 Hypothetical model of two networks, each comprising dense and loose network clusters, bridged by a weak tie between A and B

ties to acquaintances who form a looser cluster within A's network. The weak tie between A and his/her acquaintance "B" is a bridge between A and B's respective local dense networks. Since A and B have no mutual friends, information and influence can flow from one network to the other only across the weak bridging tie between A and B. In a scenario in which A transmits something to B, whether consciously (e.g. information about a job opportunity, as in Granovetter, 1973) or unconsciously (e.g. obsesity, as in Christakis and Fowler, 2007, or a linguistic innovation), the likelihood of B acting on A's information or adopting A's innovation is related to B's position in his/her network. Citing Becker (1970), Granovetter uses the example of public health innovations. Central figures (such as A) in a public health network are less likely to adopt innovations such as new drug programs than are marginal figures (such as B), since central figures have a more prominent professional reputation that they wish to protect from risk. Only if the innovation is low risk and uncontroversial will central figures lead in adopting it. The behavior of central figures is germane to our discussion, later in this chapter, of their role in conservatively maintaining existing linguistic norms.

For the Milroys, marginal figures were of greater relevance to their study. The Milroys applied Granovetter's theory to language changes spreading between densely networked neighborhoods in Belfast, Northern Ireland. Given Granovetter's proposals outlined above, they expected that "mobile individuals who are rich in weak ties but (as a consequence of their mobility) relatively marginal to any given cohesive group are . . . in a particularly strong position to diffuse innovation" (Milroy and Milroy, 1985: 366). Participant-observation over many months was used to record the sampled speakers' network structures, including both strong and weak ties, and to assign sociometric scores. Scores of one or zero were assigned to each informant for five indicators on a Network Strength Scale. The indicators captured whether the informant was a member of a high-density local group, such as a soccer team; had kinship ties to more than two households in the neighborhood; worked at the same place as two or more neighbors; worked at the same place as two or more neighbors of the same gender; and chose to spend free time with co-workers. The higher an informant's total score, the stronger his or her ties to their neighborhood. The Milroys found a positive correlation between network score and strength of local accent.

One of the sound changes the Milroys examined was the backing of the vowel /a/. In Belfast this vowel can be pronounced in a variety of ways. From front to back of the mouth, these include, e.g., "beg" for *bag*, "bot" (like the US English vowel in *hot*) for *bat*, and "bawd" (like New York City stereotyped "cawfee" for *coffee*) for *bad*. The Milroys reported that middle-class speakers typically did not produce the "aw" pronunciation; it was most widespread in the speech of working-class male speakers, with those in East Belfast ahead of all men in the city. But surprisingly, it was working-class young *female* speakers who most frequently used this pronunciation in West Belfast. The Milroys argued that the "aw" pronunciation had diffused from East to West Belfast via the weak ties contracted between young West Belfast women in service jobs and their East Belfast customers. In contrast, men in the same West Belfast neighborhood were mostly unemployed, and were embedded in dense, multiplex neighborhood networks with few ties to external groups. Older women were similarly unlikely to have weak ties beyond the neighborhood, since they were stay-at-home parents and housewives.

The weak ties model can also explain the diffusion of language changes over much greater geographic distances. Labov (2003) traces the spread of the Philadelphia dialect word *hoagie* ("sandwich") to Pittsburgh restaurants via the sale of pizza ovens manufactured in Philadelphia. He posits that informal contact between businesses provided the necessary weak ties between food-related industries in the two cities. A series of changes in vowel pronunciation known as the Northern Cities Shift (discussed further below) has been observed to have "hopped" or "cascaded" from large northern US city to large northern city over the last two centuries via trading along the Erie Canal (Labov, 2010) and commuting along major interstate highways (Labov, Ash and Boberg, 2006).

Locating the Leaders of Language Change

The leaders of linguistic change are those people who are ahead of their neighborhood or group with respect to some innovative way of speaking. It happens that these people are usually also the individuals in a network who have many ties within a network but who also have many ties to other networks—that is, position in a network is a crucial factor in the spread of language change. To acquire innovations, speakers must have ties that extend beyond their immediate network, but it is speakers who also have a high degree of network centrality who are in a position to rapidly pass on innovations to the rest of the network. Among the young West Belfast women studied by Milroy and Milroy, the most advanced speakers with respect to the backing of /a/ were not only connected to people from East Belfast via weak ties, but were also central figures in their neighborhood networks, with lots of local ties.

The prototypical example of such a speaker was identified by William Labov in a landmark study of the neighborhoods of Philadelphia, Pennsylvania (Labov, 2001). Labov and his team undertook ethnographic studies of the residents of selected city blocks in a subset of neighborhoods that represented a variety of socioeconomic groups. Over several years in the early 1970s, fieldworkers recorded informal interviews with residents, and also collected information about their social network ties both on and off their block. "Celeste S.," a middle-aged, lower middle-class Italian-American woman, emerged as linguistically more progressive than anyone else on her block in South Philadelphia. For several ongoing sound changes (mostly changes in pronunciation of vowels), Celeste was ahead of her friends and neighbors. A colorful, popular and forthright character, Celeste was well connected to everyone on her block and in block network terms was highly central, but more than 80% of her named friends lived elsewhere. Labov (2001: 351) suggests that linguistic leaders like Celeste pick up advanced forms of ongoing language changes in the wider community, and bring them back to their neighborhood, where "people look to her as a point of reference, and are likely to be influenced by her actions, behavior and opinions."

Penelope Eckert (2000: 216) showed how speakers like Celeste S. might become linguistically influential through her analysis of some pre-teen girls who were both social and linguistic leaders among their peers. "Rachel," for example, was a student at an elementary school in Northern California that Eckert studied in the 1990s. Eckert followed students in a cohort from fifth through sixth grades (ages 10–12) as they made the transition from childhood to pre-adolescence. Rachel exhibited advanced pronunciations of the vowel /o/ (in, e.g., *go*, *phone*), which in California is becoming "fronter" over time, i.e. the tongue advances toward the same space in which /i/ (in, e.g., *see*, *feed*) is pronounced. Eckert (2011: 93) described Rachel as "the main drama queen of the popular crowd" who employed fronted /o/ when performing a "cool teenage persona" and a more retracted /o/ when performing a "'poor little me' persona." Leaders of sound changes such as /o/-fronting may thus emerge through an association of /o/-fronting with a socially desirable performance (in this case "teenager"), particularly when that performance is produced by a central network figure such as Rachel.

In keeping with the notion of applying the present to the past, Nevalainen, Raumolin-Brunberg and Mannila (2011) looked at leaders of language change in fifteenth- to seventeenth-century England. They examined six language changes that were occurring in English during this time, such as the loss of final [n] in possessive *mine, thine* → *my, thy*; the change in the third person suffix, e.g. *he goeth* → *he goes*; and multiple negation. From a large corpus of personal letters,[4] 778 people from a range of social backgrounds were selected, including clergy, nobility, gentry and merchants, among others. Even royalty was represented in the corpus. In a section of the study in which 48 writers between 1500 and 1619 were compared, Queen Elizabeth I was found to be the most linguistically progressive writer of all. Her proportional use of innovative variants such as *my*, *goes* and single negation was above average for five of the six language changes. As the authors note, "It might be surprising that a person of the topmost social stratum should lead such a large number of linguistic changes, as present-day sociolinguistic studies tend to claim that the topmost layers are rarely active in promoting linguistic change" (Nevalainen et al., 2011: 28).

Their work bears similarities to Padgett and Ansell's (1993) study of Florentine social networks and the rise of the Medici family in the fifteenth century. The authors conclude that the Medicis' political and socioeconomic ascendance could not be attributed to their relative wealth or other sociodemographic characteristic, but instead was due to their network centrality. Families who supported the Medicis were almost all tied directly to them, and in many cases were connected to other supporters only through the Medicis, usually because of factional distinctions between neighborhood groups and lineages (patrician or "new money") within the network. This made supporters dependent on the Medicis for personal, economic and political power, and consolidated the Medicis' position as leaders, ultimately leading to the Medicis' conquest of their rivals. The Medicis' opponents, on the other hand, formed a dense, multiplex network. This created "too many status equals, each with plausible network claims to leadership" (Padgett and Ansell, 1993: 1277–1278), making the opposition ineffective in its resistance to the Medicis.

Likewise, Nevalainen and colleagues attribute Queen Elizabeth I's role as a leader in language change to several factors, including her central position in her social networks. In fact, they compare Elizabeth directly to the leaders of sound change such as Celeste in Philadelphia. Like Celeste, and like Cosimo Medici, Elizabeth was at the center of a dense network (her Court) but tied through her administrative role to other political and social networks. Elizabeth's structural network position facilitated her adoption of changes below the level of public comment, such as *my* for *mine*, while her education and power facilitated her access to and adoption of changes that were overtly promoted among elites, such as single negation. Members of the Royal Court network also adopted these language changes, reinforcing the new norms established by Elizabeth, and providing a model for the rest of the country.

Dense Networks: Maintaining Linguistic Norms

Although language variation is a precondition for language change, not all variation is evidence of ongoing change. Sometimes, two or more ways of saying the same thing ("sociolinguistic variants") may persist as alternatives in the language for generations, such as the "ax" and "ask" pronunciations of *ask* mentioned above. This occurs even though usually at least one of the variants is viewed negatively by the public as non-standard, non-mainstream and uneducated-sounding, as is often the case for "ax." They remain in the language because they also have positive social value, such as making the speaker sound friendly and casual. It is hard to imagine a classic rock 'n' roll lyric such as *I can't get no satisfaction* having popular impact without multiple negation (*I can't get any satisfaction*), or fast-food restaurants conveying their casual dining ethos without "g-dropping" (*Dunkin' Donuts; I'm lovin' it; Finger-lickin' good*). Likewise, non-standard regional or ethnic dialects (e.g. Appalachian English, African-American English), or non-official languages (e.g. Jamaican Creole) are maintained in informal contexts because they are valued for the ways in which they can convey intimacy, solidarity, and pride in one's identity.

Sociolinguists have drawn upon social network theory to better understand the persistence of such non-standard linguistic features. One of the earliest studies in this vein was conducted by Jenny Cheshire in Reading, England (1982). At the level of word composition (morphology) and word order (syntax), the Reading dialect exhibits numerous non-standard features. Cheshire looked at 14 of these features in the speech of adolescents, such as third person plural *-s* (1); second person *has* and *was* (2a and 2b); multiple negation and *ain't* (3); and relative pronoun *what* (4).

(1) They calls me all the names under the sun.
(2a) You just has to do what the teachers tell you.
(2b) You was with me, wasn't you?
(3) It ain't got no pedigree or nothing.
(4) Are you the little bastards what hit my son over the head?

Cheshire collected her data by regularly visiting two parks where teenagers typically hung out when they were skipping school. As such, they represented speakers who were oriented away from institutional and standard norms, and oriented instead toward local and non-standard norms of behavior, including linguistic behavior. Cheshire calculated a social network status for each participant using two methods. First, she asked the teenagers to tell her with whom they spent the most time. Each teenager's centrality was calculated by counting the number of others who reciprocally reported spending time with them. Second, Cheshire used her observations and the information she gathered in conversations to assign each participant a score on a "vernacular culture index" to determine their integration into the playground network. The components of the index included indicators such as swearing, fighting, style of clothing, whether the participant was involved in crime, and the kinds of jobs he or she aspired to. The overall network score was used to group participants into "core," "secondary," and "peripheral" members of their respective playground networks. Core members were the most frequently mentioned in the naming task, and had the highest vernacular index scores. They were also the most frequent users of non-standard language features such as those listed in (1) to (4). Secondary members used non-standard features less frequently, and peripheral members used them least often.

Cheshire's study indicates that linguistic behavior is regulated by social norms, and that these norms are most strongly maintained by individuals with a high degree of network centrality within dense social networks. Central figures ("core members" in Cheshire's study) are simultaneously leaders in the use of vernacular language variants to whom other members accommodate, and also victims, in a sense, of social pressure to maintain this high frequency of vernacular speech. The core of the network is a linguistic echo chamber, reinforcing the norms of linguistic behavior that characterize the group, and simultaneously being constrained by them.

Labov (1972a, 1973) makes the same observations, in this case about the differential linguistic behavior and social practices of core gang members versus peripheral "lames" in a participant-observation study conducted in an African-American community in Harlem, New York City. He hypothesizes: "If peer group pressures are important in maintaining the vernacular in its present uniform state, and in resisting the pressures of other dialects, then those who are most bound by the norms of the group should show the most consistent form of the vernacular" (Labov, 1973: 98). This was tested by calculating network centrality for each gang member from responses to a question asking the individual to list all of the people he "hung out" with, and who he perceived to be the "leader" of the group. Members of the most central clusters (such as a core cluster in one gang described by other members as "the six best fighters"), did indeed exhibit the most structurally systematic use of African-American English vernacular, and the most frequent use of vernacular features such as the reduction of word-final consonant clusters such as [st], e.g. *past me* > *pas' me* and contraction and deletion of *is* e.g. *He is crazy* > *He's crazy* > *He crazy*. Some vernacular features appeared at especially high frequency within particular clusters as symbolic markers of affiliation, such as the pronunciation of the tag *and shit* as *an' shi-it* ("Eh well, they learned me about reefer an' shi-it."—i.e. *They taught me about marijuana and things like that*; Labov, 1973: 106). In contrast, lames were the most marginal individuals in network terms, with the fewest "hang out" mentions and the least convergence with other group members regarding perceived identity of the gang leader. Lames were also the least familiar with or adept at negotiating the vernacular street culture of the gangs –fighting, "sounding," "playing the dozens," smoking, taking drugs etc.—as their negative epithet suggested. Lames were furthermore the most likely to speak a vernacular African-American English that was "tainted" Labov, 1972a: 269) with features of mainstream standard English. According to Labov (1972a: 180):

> because [the lames] are not constantly subject to the supervision and control of a peer group, they lack any social mechanism whereby a highly focused set of vernacular norms can be consistently maintained against the constant pressure of a competing set of institutionally legitimized norms, and so they drift away from that vernacular.

Labov argues that network centrality, and not just familiarity with the local gang culture, is critical to the adoption and maintenance of vernacular linguistic norms. By way of illustrating this point, he describes Vaughn, a recent transplant from another neighborhood. As an outsider, Vaughn had been unaware of local gang practices ("These men have taught me everything I know about all this bullshit, because I'm uptown, that like a different world an' shi-it"; Labov, 1973: 106), but quickly integrated himself into the center of the network, becoming one of the aforementioned "six best fighters." Thus, in contradistinction to the lames, he was under greater pressure to conform to the norms of the group, and had already adopted the *an' shi-it* pronunciation that was absent from the lames' speech. Vaughn's case is also illustrative of the importance of interaction frequency within the core cluster's reciprocal network ties: although he quickly acquired superficial features such as *an' shi-it*, more systematic alignment with the cluster (e.g. in terms of the frequency of consonant clusters reduction or *is* deletion, and the linguistic conditions under which these phenomena occur) had not yet occurred. Labov (1973: 107) concludes that "The remarkably consistent grammar of the [core members] is the result of ten years of their continuous interaction with each other."

Thus, because of their function as norm-maintaining systems, dense networks can act as a buffer against linguistic innovations. When dense networks do not have many ties to one another, linguistic differentiation occurs: between social classes, for example, or between ethnic groups, or between geographic localities. The lack of ties between networks can come about through social avoidance, geographic hindrance (mountains, a river), or lack of habitual mobility between groups or places—such as lack of commuting—that are dictated by economics rather than active social choice. Speakers within these networks sound more and more like one another, and less and less like speakers from other networks. For example, Black Americans in early twentieth-century recordings sound hardly different from their White counterparts, but after the Great Migration, African-American Vernacular English diverged more and more from mainstream US English. This can be understood in social network terms, as Black and White residential integration decreased during this period (Van Herk, 2008).

Dense Networks: Maintaining Minority Languages and Dialects

Further understanding of the similarities of the effect of geographic and social isolation comes from studies of social networks in immigrant and migrant communities, where researchers have shown that denser social networks tend to inhibit the acquisition of local linguistic features: either dialect features, in the case of migration within a monolingual community, or new languages, in the case of immigration. In general, dense, multiplex networks tend to foster longer maintenance of immigrant and minority languages in the face of language shift to the dominant language.

Evans (2004), for instance, used social network analysis to examine dialect contact and the extent to which migrants to an area might adopt local linguistic norms. In her study of a migrant community in Ypsilanti, Michigan, she interviewed people who had migrated to or were children of those who had migrated to Ypsilanti from Appalachia, in the southern United States, mostly seeking employment in the Ford Motor Company in the 1940s–60s. Most of the migrants settled in the same part of the city, which came to be known as "Ypsitucky" (a combination of **Ypsi**lanti and the Appalachian region state of Ken**tucky**), reflecting the density of migrant settlement. The linguistic norm that Evans examined in her analysis was the Northern Cities Shift (NCS), a shift in the pronunciation of five short vowels of English including the vowel /ae/, as in the words *cat* and *ham*, which in this dialect are pronounced more like "kee-at" and "hee-am" (i.e. the vowels are "raised"). The NCS is not a part of the dialect of the southern part of the United States; thus, by looking at their production of the vowel /ae/ (a raised /ae/ is considered the first step of the NCS), Evans could measure the degree to which her respondents had adapted to the local norm. Evans asked each of her respondents to read a word list and assigned them an "Appalachian integration score" that measured the density and multiplexity of their Appalachian social network. She measured their production of the vowel [ae] and assigned each speaker a score from 1–5 based on their production of this vowel: speakers whose [ae] vowels were significantly raised and fronted as they would be if they were participating in the NCS got a score of 5, while speakers whose /ae/ vowels were not significantly raised and fronted (and thus sounded no different than they would for a Southern speaker), got a score of 1. The most significant predictor of participation in the Northern Cities Shift as measured by the raised /ae/ vowel was the speaker's social network, with Appalachian-dominant networks reinforcing Southern speech norms and preventing acquisition of the local Michigan dialect. Similarly, in studies of migrants to a Philadelphia suburb and to Milton Keynes, UK, respectively, Payne (1980) and Kerswill and Williams (2000) found that children of migrants with few local social ties were least likely to adopt features of the local dialect.

Dialect acquisition like that examined by Evans, Payne, and others is not so different from language acquisition in immigrant and minority language communities, where the motivation to acquire the politically and socially dominant language is often in competition with the desire to maintain the home language. Although degree of language shift to the dominant language is closely tied to the generation to which an immigrant belongs, with first generation immigrants more likely to continue to use the home language and the third generation more likely to be monolingual in the new language, the use of social network analysis in studies of language shift has offered great insight into how immigrant or minority languages can be maintained, particularly within the second and to some extent the third generations, for whom language choice shows more individual variation. Li (1994) showed how language choice in multilingual communities can be closely tied to social network in his studies of code-switching among Chinese–English bilinguals in Chinese immigrant community in Tyneside, England. As expected, the strongest ethnic networks were associated with the first-generation immigrants and the most use of Chinese, while the weakest ethnic networks were associated with third-generation immigrants and the most use of English. However, he also found that membership in a particular church community (which made up a dense network of speakers of multiple generations) led to more maintenance of Chinese among third- generation speakers, as evidenced by more code-switching in the speech of church-network third-generation speakers than that seen in other third- generation speakers, who primarily used monolingual English discourse. Gal (1978) similarly found that makeup of an individual's network contributed to his or her degree of language shift or maintenance. In her study of language shift from Hungarian to German in a previously Hungarian village in Austria, she found that the makeup of the individual's social network (made up more of "peasants" vs. "workers") was the strongest predictor of whether they continued to use Hungarian or had shifted to more use of German.

Zentella (1997) used a less quantitative approach to social networks but concluded that social network analysis is crucial in the examination of such language shift scenarios. She showed that focusing on the linguistic input of parents in a community like the bilingual Puerto Rican community she studied in East Harlem (*el bloque*) would not have adequately described the language experience of the community's children, who were members of dense and multiplex networks that were characterized by variable bilingualism. Just because an individual child's parents only spoke English to the child, for instance, did not mean that that child did not have other adults who regularly spoke Spanish with the child. Dense, multiplex networks contributed to the maintenance of Spanish–English bilingualism. Most tellingly, Zentella's longitudinal work showed the break-up of these dense networks in the ensuing decade due to a variety of social and political changes in East Harlem that affected *el bloque*. She demonstrates the negative effect that this had on the bilingual proficiency of the children. Once strong network ties in *el bloque* had eroded, children's proficiency in Spanish or English changed due to their individual life trajectories, but the bilingual competence that was so closely tied to being a member of *el bloque* no longer remained. Zentella's findings on the erosion of networks with strong ties demonstrate the corollary relationship between strong ties and the enforcement of linguistic norms and weak network ties and language diffusion and change.

The Transfer of Social Meaning

Most of the language changes that we have so far described originated as unconscious innovations in the speech of low status groups, subsequently spread to high status groups, and in some cases became part of the dialect of all speakers in the community. How do non-prestigious innovations spread upwards through the social hierarchy? Even if the social network conditions are structurally amenable to the diffusion of a working-class innovation to the middle class, it does not necessarily follow that the middle class will want to adopt it. Yet somehow this occurs in the majority of language changes. Starting with the Milroys in the 1970s, sociolinguists have posited that during diffusion via weak ties, the social meanings associated with linguistic innovations may be altered or even lost, and that this may help to make innovations (consciously or unconsciously) desirable to new audiences. For example, the "aw" pronunciation in Belfast was a feature of the speech of working-class men, and presumably indexed additional secondary attributes associated with working-class men, such as toughness. It is not obvious why such attributes would be attractive to the young West Belfast women who adopted "aw." Yet in the fleeting, transactional encounters these women had with residents of East Belfast in the stores where they worked, they somehow adopted *aw*. The Milroys (1985) suggested that first, the very high frequency of interactions with East Belfast speakers promoted the adoption of "aw" by the store workers and second, that in its transmission along these weak ties, "aw" became detached from its male, working-class, tough social meanings. Indeed, for the young women it may have come to be associated with the relative prestigiousness of their jobs and with a general sense of upward mobility.

Modaressi (1978) observed that a working-class sound change in Tehran, Iran—the raising of /a/ in, e.g., *nan* (bread) to "nun"—was adopted in the provincial city of Ghazvin by the highest social class. The original social meanings associated with /a/-raising were transformed during their transmission along loose network ties with Tehran, from something like "sounds working-class" to "sounds sophisticated and urban." On Martha's Vineyard, MA, a return to traditional pronunciations of *right, wife, house, about* as "roight," "woife," "hoose," "aboot" was spearheaded by islanders who most strongly resisted rising tourism (Labov, 1972b). Their pronunciations of these vowels were sometimes more extreme even than those of the older fishermen who had always used the traditional pronunciations. In their transmission through the island's social networks from fishermen to pro-island non-fishermen, the social meanings ("fisherman," "Yankee," "old") of the traditional pronunciations became diluted to mean simply "positive orientation to Martha's Vineyard" (Labov, 1972b: 38). In a study of language shift in a Cajun community in Louisiana, Dubois and Horvath (1998) found a similarly surprising pattern with respect to age, gender, and social network and the pronunciation of words such as *think* and *that* with stops instead of fricatives ("tink" and "dat"). The use of these variants was characteristic of older Cajun English speakers but decreased significantly with the middle generations, along with increased access to other varieties of English and stigmatization of the Cajun accent. In the youngest generation, use of these variants increased again, and the authors attribute this to a Cajun cultural renaissance that increased cultural pride and was accompanied by the use of stereotypically Cajun variants. Social networks also played a role, however, in that the young women with open networks (i.e. outside the Cajun community) continued to shift away from the use of "t" and "d," while young men with open networks were *more* likely to use the "d" variant. Dubois and Horvath attribute this to the ties that young men with open networks have to the tourist industry and the cultural capital associated with the use of Cajun variants in that context. In other words, the social meaning associated with the use of Cajun variants had undergone a generational shift.

Eckert (1989, 2000) mapped the friendship networks of an entire senior class at a high school in a Detroit suburb. The network included the students' out-of-school friends. Within the school, students self-identified along a continuum of peer-group affiliation that had two opposing poles: the pro-school, middle-class-oriented Jocks at one end, and the anti-school, working-class-oriented Burnouts at the other. Most students identified themselves as "In-Betweens," but the continuous nature of the cohort's network structures meant that even In-Betweens were connected to a greater or lesser extent to students near the two poles. All students were participating in the Northern Cities Shift vowel changes, but Burnouts had more advanced pronunciations than their peers. Accounting for the Burnouts' linguistic behavior was not difficult: language changes generally originate in urban centers, and the Burnouts were likelier than other students to spend time cruising or hanging out in Detroit, and to have friends in the city. They would therefore have had plenty of exposure to advanced Northern Cities Shift pronunciations, and would have been positively oriented to the social meanings of urban-ness, toughness, danger etc. that were associated with those advanced pronunciations. Accounting for the diffusion of advanced NCS pronunciations outwards from the Burnouts to the In-Betweens and then the Jocks was less straightforward. Why would a Jock want to sound like a Burnout? Eckert (2000) suggested that the social meanings of advanced NCS vowels were imperfectly transferred from individual to individual in the cohort network. Some meanings, such as toughness, were sufficiently attractive to some members of the Jock clusters (notably the athlete Jock boys) to facilitate adoption of the advanced NCS, from whence the sound changes became progressively less associated with urban toughness and more with toughness more generally, or perhaps just edginess. In this way, the social meanings of linguistic innovations shift and change as they are passed from group to group, shedding some meanings and gaining others.

New Applications

Linguists are using network analysis not only to better understand the past, but to keep pace with language change in the rapidly changing present. How is language affected by new forms of communication such as e-mail, Twitter, Facebook, text messaging, instant messaging, and video conversation? And how can modern computing technology, such as complex modeling software, help us to refine our understanding of interactions between speakers?

Turning to the first question, studying new online forms of interaction allows for larger scale and less labor-intensive projects than have been usual in sociolinguistics, where ethnographic techniques have been common. Danescu-Niculescu-Mizil, West, Jurafsky and Potts (2013) examined the acquisition and maintenance of linguistic norms in two large online communities: BeerAdvocate and RateBeer. Contributors to the sites come to rate and discuss types of beer with one another asynchronously through posted reviews and user forums. The authors found that acquisition and retention of within-site norms (such as the use of *aroma* versus *smell*) depended on a user's "lifestage" on the site. Rapid alignment with community norms occurred early in the user's contributions to the site. But as community norms shifted (e.g. *smell* increased in overall frequency over 10 years, at the expense of *aroma*), established users did not shift too, but maintained their use of earlier linguistic norms. This finding may have wider applicability for an understanding of social network effects on language use, insofar as longstanding interactions with other actors in a network may have less effect on language than do novel interactions.

Bamman, Eisenstein and Schnoebelen (2014) examined the language of Twitter users. As in many other sociolinguistic studies (Labov, 1990), they found that gendered patterns of language use emerged. However, Twitter users whose language was not strongly gender-marked were most likely to have non-homophilous networks, i.e. networks composed of many users of the opposite sex. Similar findings have been reported before, such as in Cameron's (2005) face-to-face study of speakers in San Juan, Puerto Rico, in which he showed that gendered language use waxes and wanes over the lifespan as children and adolescents go through stages of preferences for same-gender and mixed-gender peer groups. A "big data" approach using Twitter or other large-scale online databases allows for a finer-grained perspective on the ways in which the language patterns associated with gender, race, age and other social categories are not fixed, but constituted through network interactions.

With respect to new computer technology, agent-based modeling (see, e.g., Gilbert, 2008) is a method for simulating interactions between individual "agents" to assess the effect of those interactions on the entire system or community being modeled. In sociolinguistics, agent-based modeling has been employed as a means of understanding how language change increments over generations, and how it diffuses from place to place. Stanford and Kenny (2013) used this technique to model an imaginary Chicago and an imaginary St. Louis, both populated by agents modeling "children" and "adults." They tested a hypothesis of Labov's (2007), under which complex language changes are perfectly transmitted across the generations when learned by children, but imperfectly diffused across space by adults, because adults are not such good language learners as children. For the Northern Cities Shift, for which (real) Chicago is an epicenter, local children appear to learn all components of the change. When diffused to (real) St. Louis via adult-to-adult interactions, not all of these components appear in the local dialect, because they arose through adult-to-adult interaction. Yet in their simulation, Stanford and Kenny found that there is no need to propose, as Labov did, "two different kinds of language learning" (Labov, 2007: 349). There is more economical possibility—namely, that language changes are passed faithfully from adults to children in Chicago largely because of the density and frequency of their interactions with one another, but the same changes are imperfectly diffused between adults in Chicago and adults in St. Louis because they have fewer interactions with one another and have looser social network ties. This underlines the nature of linguistic diffusion: weak ties are needed for the spread of language change, but such ties necessarily alter the linguistic feature in some way, either structurally or in terms of their social meanings.

Conclusion

One of the major advantages of a social network approach is that it has allowed analysts to move beyond fixed social categories in their explanations of language change. Stanford and Kenny's (2013) model of the Northern Cities Shift, for example, suggests that network interactions, rather than speaker age (child vs adult), can account for the imperfect diffusion of the NCS from Chicago to St. Louis. Additionally, sociolinguistic studies that incorporate social networks usually employ ethnographically oriented data collection techniques, rather than aiming to select a random sample stratified by age, social class, gender, and ethnicity, which had been the general approach of early, larger urban sociolinguistic studies (e.g. Labov, 1966). A particular advantage of this approach is that it avoids preconceived social categories and instead uses locally meaningful categories once the researcher is familiar with the community. As a result, these kinds of studies have also been helpful in illuminating the interaction between social network structures and macro categories such as gender, social class and ethnicity, as well as community-specific categories such as "Jocks."

Many studies have shown, for example, that the leaders of language change tend to be middle-class women (Labov, 1990), but studies that incorporate a social network approach have illuminated this generalization further by showing that in some societies being a woman and being middle class simply makes it more likely that one will have a higher number of weak ties to outside networks—in itself a better predictor of language diffusion than gender or class. The fact that network effects outweigh any "inherent" gender effects can be seen if we return to Zentella (1997), who describes the fact that girls on *el bloque*, who generally spent more time at home and on the block, as well as more time with monolingual Spanish-speaking mothers and female caretakers, were more likely to be more proficient in Spanish than boys, who had looser networks, spent more time away from home and off the block, and interacted with more African American playmates. Boys or girls whose language proficiency was dissimilar to their "gender-mates" were those whose networks were also different from the norm—e.g. the teen male who shared more interests with the girls and spent more time at home was more proficient in Spanish than his "male-bonded, baseball- and football- playing brother" (Zentella, 1997: 52). The gender differences found between peasants in Oberwart, Austria (Gal, 1978) were attributed to aspirational differences between young peasant men, for whom the independent, outdoor, peasant lifestyle was appealing, and peasant women, for whom the hardworking life of a peasant wife was less appealing than that of a German-speaking "worker" wife. As well, gender differences between young Cajun men and women (Dubois and Horvath, 1998) were better attributed to differences in the makeup of their "open" social networks.

Indeed, network effects may be the only relevant explanation for language change when the speech community is relatively socially homogeneous. Evans's (2004) Ypsitucky speakers did not have a great deal of variability in their socioeconomic status, mostly because there is not a great deal of socioeconomic diversity in the community itself. To the extent that her speakers did have socioeconomic differences, social network was nonetheless a much stronger predictor of whether speakers participated in local dialect norms or not. Milroy and Milroy (1992) have argued that the many linguistic changes in which the middle socioeconomic groups are shown to lead are explicable in terms of network structure if we consider that upper and lower classes tend to have denser and more multiplex networks; in a summary of different approaches to social class Dodsworth (2009) argues that the social network model is an underutilized approach to considering class differences in language variation and change.

Ethnographic studies of social networks can also help sociolinguists understand the nature of linguistic influence, which appears to be largely a subconscious process of alignment with a speaker or group of speakers. Since the adoption of most linguistic innovations (except for lexical innovations, i.e. new words and phrases) occurs largely without explicit comment until the innovations are very widespread, it is the actuation and early adoption of innovations that remains most mysterious to analysts. As the Milroys (1985: 370) observed, "for the very reason that persons who actuate linguistic change may do so in the course of fleeting, insignificant encounters with others occupying a similarly marginal position in their social groups, direct observation of the actuation process may be difficult, if not impossible." Given the methodological difficulties in observing the actuation of language change and its transmission across weak ties, the use of large-scale Internet corpora and of modeling technology is likely to be more suited to solving this issue than ethnographic approaches. In addition, ethnographic studies are typically conducted on a small scale and may generate hypotheses about the nature of language and social networks that are not generalizable to a larger scale. However, in taking advantage of large corpora and computer modeling, sociolinguists must take care to keep the human agents of language change in view. Reducing language change to a set of mechanistic processes and outcomes may not prove to be the most accurate reflection of social reality. Achieving a balance between micro-level and macro-level studies of networks, and between modeling and direct observation, will be challenges for sociolinguists going forward.

We have shown how the use of social network theory in the study of language variation and change has offered insights into how linguistic norms are maintained, how innovative linguistic forms diffuse within or between communities, what type of individual leads language change, and how social meaning is transferred when a linguistic innovation spreads in a community. To summarize the major findings: dense, multiplex networks composed of strong ties maintain local norms, even of stereotyped features like g-dropping and the use of *ain't*, and work to preserve linguistic differences between communities of migrants/immigrants and dominant cultures. Conversely, the weakening of dense network ties can help to explain the process of linguistic assimilation to the dominant language in both types of communities, as networks of weak ties promote the spread of linguistic change. Leaders of change, like "Celeste," "Rachel," and Elizabeth I, have certain characteristics: they are key figures in dense networks but have numerous weak ties to outside networks, and these leaders play a key role in the diffusion of linguistic change from network to network. Finally, linguistic features are always transmitted along with social meanings, and it is these meanings that make linguistic innovations unconsciously desirable or undesirable to those who are exposed to them. All of these general principles can be successfully applied to studies of the past, and will no doubt be refined in the studies of the future.

Acknowledgments

We are grateful to Ashley Hesson for help with bibliographic research, and Richard Cameron for reading an earlier version of this chapter. Any remaining errors or oversights are our own.

Notes

1. Typically, but not always, language changes of any kind that advance past the middle of the S-curve of diffusion will attract public attention, even if they were initially spread quite unconsciously.
2. This is not to say that social categories are not linked to social network structure, however, and we return to this in our concluding section.
3. Although strong ties can also be bridges, this occurrence is rare, because Granovetter adopts a definition of "bridges" as being the shortest (usually *only*) path between two individuals (Hararay, Norman and Cartwright 1965).
4. The Corpus of Early English Correspondence, www.helsinki.fi/varieng/domains/CEEC.html

References

Bamman, D., Eisenstein, J. and Schnoebelen, T. (2014). Gender identity and lexical variation in social media. *Journal of Sociolinguistics*, *18*, 135–160.

Becker, M. (1970). Sociometric location and innovativeness. *American Sociological Review*, *35*, 267–282.

Cameron, R. (2005). Aging and gendering. *Language in Society*, *34*, 23–61.

Cheshire, J. (1982). *Variation in an English Dialect: A Sociolinguistic Study*. Cambridge: Cambridge University Press.

Christakis, N.A. and Fowler, J.H. (2007). The spread of obesity in a large social network over 32 years. *New England Journal of Medicine*, *357*, 370–379.

Coleman, J.S., Katz, E. and Menzel, H. (1966). *Medical Innovation: A Diffusion Study*. Indianapolis, IN: The Bobbs-Merrill Co.

Danescu-Niculescu-Mizil, C., West, R., Jurafsky, D., and Potts, C. (2013). No country for old members: User lifecycle and linguistic change in online communities. In *Proceedings of the 22nd World Wide Web Conference* (pp. 307–317). Rio de Janeiro, Brazil: ACM.

Dodsworth, R. (2009). Modeling socioeconomic class in variationist sociolinguistics. *Language and Linguistics Compass*, *3*, 1314–1327.

Dubois, S. and Horvath, B. (1998). Lets think about dat: Interdental fricatives in Cajun English. *Language Variation and Change*, *10*, 245–261.

Eckert, P. (1989). *Jocks and Burnouts*. New York: Teachers College Press.

Eckert, P. (2000). *Linguistic Variation as Social Practice*. Oxford: Blackwell.

Eckert, P. (2011). Language and power in the preadolescent heterosexual market. *American Speech*, *86*, 85–97.

Evans, B. (2004). The role of social network in the acquisition of local dialect norms by Appalachian migrants in Michigan. *Language Variation and Change*, *16*, 153–167.

Florida, R. (2003). Cities and the creative class. *City and Community*, *2*, 3–19.

Gal, S. (1978). Peasant men can't get wives: Language change and sex roles in a bilingual community. *Language in Society*, *7*, 1–16.

Gilbert, N. (2008). *Agent-based Models*. New York: Sage Publications.

Granovetter, M. S. (1973). The strength of weak ties. *American Journal of Sociology*, *78*, 1360–1380.

Granovetter, M. S. (1983). The strength of weak ties: A network theory revisited. *Sociological Theory*, *1*, 201–233.

Hararay, F., Norman, R. and Cartwright, D. (1965). *Structural Models*. New York: Wiley.

Kerswill, P. and Williams, A. (2000). Creating a new town koiné: Children and language change in Milton Keynes. *Language in Society*, *29*, 65–115.

Labov, W. (1966). The social stratification of English in New York City. Ph.D. dissertation, Columbia University. Published 2006 by Cambridge University Press.

Labov, W. (1972a). *Language in the Inner City*. Philadelphia, PA: University of Pennsylvania Press.

Labov, W. (1972b). *Sociolinguistic Patterns*. Philadelphia, PA: University of Pennsylvania Press.

Labov, W. (1973). The linguistic consequences of being a lame. *Language in Society*, *2*, 81–115.
Labov, W. (1975). On the use of the present to explain the past. In L. Heilmann (Ed.), *Proceedings of the 11th International Congress of Linguists* (pp. 825–851). Bologna: Il Mulino.
Labov, W. (1990). The intersection of sex and social class in the course of linguistic change, *Language Variation and Change*, *2*, 205–254.
Labov, W. (2001). *Principles of Linguistic Change: Social Factors*. Oxford: Blackwell.
Labov, W. (2002). Driving Forces in Linguistic Change. 2002 International Conference on Korean Linguistics. Seoul National University, August 2, 2002.
Labov, W. (2003). Pursuing the cascade model. In D. Britain and J. Cheshire (Eds.), *Social Dialectology: In Honour of Peter Trudgill* (pp. 9–22). Amsterdam and Philadelphia, PA: John Benjamins.
Labov, W. (2007). Transmission and diffusion. *Language*, *83*, 344–387.
Labov, W., Ash, S. and Boberg, C. (2006). *The Atlas of North American English: Phonetics, Phonology and Sound Change*. Berlin and New York: Mouton de Gruyter.
Li, W. (1994). *Three Generations, Two Languages, One Family*. Clevedon: Multilingual Matters.
Lieberson, S. (2001). *A Matter of Taste*. Cambridge, MA: Harvard University Press
Milroy, J.and Milroy, L. (1978). Belfast: Change and variation in an urban vernacular. In P. Trudgill (Ed.), *Sociolinguistic Patterns in British English* (pp. 19–36). London: Arnold.
Milroy, J.and Milroy, L. (1985). Linguistic change, social network and speaker. *Journal of Linguistics*, *21*, 339–384.
Milroy, J.and Milroy, L. (1992). Social network and social class: Toward an integrated sociolinguistic model. *Language in Society*, *21*, 1–26.
Modaressi, Y. (1978). A sociolinguistic investigation of modern Persian. Ph.D. dissertation, University of Kansas.
Nevalainen, T., Raumolin-Brunberg, H. and Mannila, H. (2011). The diffusion of language change in real time: Progressive and conservative individuals and the time depth of change. *Language Variation and Change*, *23*, 1–43.
Padgett, J.F. and Ansell, C.K. (1993). Robust action and the rise of the Medici, 1400–1434. *American Journal of Sociology*, *98*, 1259–1319.
Payne, A. (1980). Factors controlling the acquisition of the Philadelphia dialect by out-of-state children. In W. Labov (Ed.), *Locating Language in Time and Space* (pp. 143–178). New York: Academic Press.
Rogers, E. (2003). *Diffusion of Innovations*, 5th edition. New York: Free Press.
Ryan, B., Gross, N.C. (1943). The diffusion of hybrid seed corn in two Iowa communities. *Rural Sociology*, *8*, 15.
Stanford, J. N.and Kenny, L.A. (2013). Revisiting transmission and diffusion: An agent-based model of vowel chain shifts across large communities. *Language Variation and Change*, *25*, 119–153.
Valente, T.W. (1995). *Network Models of the Diffusion of Innovations*.Cresskill, NJ: Hampton Press.
Van Herk, G. (2008). Fear of a Black phonology: The northern cities shift as linguistic White flight. *University of Pennsylvania Working Papers in Linguistics*, *14*(2), 157–161.
Weinreich, U., Labov, W., and Herzog, M. (1968). *Empiral Foundations for a Theory of Language Change*, Austin, TX: University of Texas Press.
Wellman, B. (1983). Network analysis: Some basic principles. *Sociological Theory*, *1*, 155–200.
Zentella, A. C. (1997). *Growing up Bilingual*. Oxford: Blackwell.

26

COMMUNITY AS METHOD, COMMUNITY AS NET

Social Network Analysis as a Tool for Studying Mutual Aid between Therapeutic Community Residents

Keith Warren and Nathan Doogan

Research Question: In self-help and theraputic communities, what motivates participants to support one another?
System Science Method(s): Networks
Things to Notice:

- Networks unfold over time, with future relationships depending on past relationships
- Ability of network analysis to reveal findings that individual-focused analyses could not

Therapeutic communities (TCs) are residential programs in which the primary method of treatment is mutual aid between peers. In the phrase of De Leon (2000), in TCs the community is the method of treatment. One aspect of mutual aid-based treatment in TCs is a system of resident peer feedback, in which residents affirm each other for prosocial actions and correct each other for actions that contravene TC norms. Several aspects of this system are incompletely understood. It is not clear what motivates residents to offer peer feedback. It is not clear whether residents are more likely to give feedback to peers who are more similar to them. It is not clear which demographic characteristics predict that residents are more or less likely to offer feedback. In this chapter we analyze data drawn from a large clinical database of peer affirmations and corrections kept at six TC units, treating peer interactions as a dynamic directed social network. In our analysis ego is treated as level 2 of a multilevel event history analysis (shared frailty) model, while opportunities for affirming peers are treated as level 1. We find that TC residents show both direct and generalized reciprocity in affirming peers, consistent with experimental results on cooperation in groups. Residents are more likely to affirm peers of the same race and who entered the facility at about the same time, consistent with sociological findings on homophily. Residents who have given more peer affirmations up to a given day are less likely to affirm a peer on that day. Female residents are more likely to affirm peers and residents at higher risk of reincarceration are less likely. We find no evidence that residents are more likely to affirm peers after receiving a correction. We find no evidence that the time residents have spent in the program or resident age influence the likelihood of affirming a peer. These results suggest that dynamic social network analysis, with its ability to treat interpersonal interactions as predictor variables, can give insights into TC processes that are not available through either qualitative research or statistical methods based on the personal characteristics of residents.

DOI: 10.4324/9781315748771-29

Therapeutic communities (TCs) are residential programs for the treatment of substance abuse and criminal behavior in which mutual aid between residents constitutes the primary clinical method (De Leon, 2000; Perfas, 2012; Vandevelde, Broekaert, Yates & Kooyman, 2004). Mutual aid between residents is seen as growing out of responsible concern for peers; residents are expected to show responsible concern by participating in community meetings, contributing to the daily operations of the program through work crew assignments and being willing to encounter peers who are flagging in their commitment to the program. The goal is a milieu in which residents cooperate and work together to learn responsible and caring behavior from peers (De Leon, 2000). Both TC researchers and TC clinicians have repeatedly reiterated that the process of mutual aid growing out of responsible concern for peers is the defining element of the TC (De Leon, 2010; Vandevelde, Broekaert, Yates & Kooyman, 2004). For TCs, the community is the method of treatment (De Leon, 2000).

Systematic reviews and meta-analyses indicate that TCs overall are effective in reducing criminal recidivism and substance abuse. There is, however, considerable variability in the effect sizes reported (De Leon, 2010; Lees, Manning & Rawlings, 2004; Pearson & Lipton, 1999; Mitchell, Wilson & McKenzie, 2006). Moreover, many TCs graduate a low proportion of their residents, a significant problem since program graduation is the most important predictor of outcomes such as recidivism (De Leon & Schwartz, 1984; De Leon, Wexler & Jainchill, 1982).

This suggests that research aimed at understanding and evaluating TC processes might play an important role in the future development of these programs (De Leon & Wexler, 2009). There is, in fact, a rich clinical and empirical theory of TC processes which has formed the basis of several books and numerous articles (De Leon, 2000; Goethals, Soyez, Melnick, De Leon, Broekaert, 2011; Hawkins & Wacker, 1986; Loat, 2006; Mandell, Edelin, Wenzel, Dahl & Ebener, 2008; Miller, Sees & Brown, 2006; Neville, Miller & Fritzon, 2007; Perfas, 2012). Drawing on this literature, Pearce and Pickard (2012) have argued that TCs foster change in part through developing a sense of "responsible agency" (p. 636) in residents. This sense is to a great extent developed by a system of mutual peer monitoring, which encourages residents to take responsibility both for their own actions and for those of peers (Pearce & Pickard, 2012). Peer monitoring occurs within a hierarchical structure of jobs that are necessary for running the TC and in which senior residents hold more senior positions and are expected to mentor junior residents (De Leon, 2000; Perfas, 2010).

The reaction of residents to this regime of peer feedback is critical to successful TC treatment. Ravndal and Vaglum (1994) followed a sample of thirteen female TC residents after graduation and found that positive peer relationships were strongly correlated with outcome. A positive first week response to the TC environment predicts program retention at one month, while increase in residents' perception of peer support, peer enthusiasm and comfort with peer interactions over the course of the first month of treatment predicts program retention at three, six and nine months (Mandell, Edelin, Wenzel, Dahl & Ebener, 2008). Residents who perceive the TC environment as being orderly were more likely to complete the program (Carr & Ball, 2014).

However, qualitative studies have repeatedly shown that it is often difficult for residents to accept the process of peer monitoring and feedback that forms the core of TC clinical methodology. In a focus group study of prisoners in a British corrections-based therapeutic community, Miller, Sees and Brown (2006) found that they valued peer feedback even though they sometimes found it difficult to receive. In a similar focus group study of prisoners in an American corrections-based therapeutic community, Patenaude (2005) found that residents and graduates expressed distrust of the motives of the senior peers in the TC, those who are expected to provide feedback to more junior members. Hawkins & Wacker (1986) noted that TC residents would go to elaborate lengths to avoid either giving or receiving peer corrections that might endanger their own or a peer's stay. In a study of residents' perceptions of their own stays, Burnett (2001) found that residents valued the insights that peers provided into their own behavior, but did take several months to learn to take full advantage of the mutual aid available in the TC.

TC treatment therefore depends on the exercise of responsible agency through the practice of mutual peer feedback. The practice is of demonstrable value but residents often find it difficult to accept. This raises several questions that are germane to an understanding of TC clinical process. Who is most likely to show responsible agency by offering feedback to others? Why do residents participate in a system of mutual feedback? With whom do they (and should they) participate? A quantitative approach to these questions requires a statistical methodology that focuses on interactions between individuals in addition to the traditional analysis of individual characteristics.

Social Network Analysis, Social Science Theory and Therapeutic Communities

It is possible to visualize an interaction between two individuals as an arrow between the two. If Dawn says hello to Michelle, there is an arrow going from Dawn to Michelle. If Michelle then says hello back, an arrow goes in the other direction—from Michelle to Dawn. The arrows could represent brief interactions such as saying hello or more long-term relationships such as liking a person. These particular relationships are represented as arrows because there is no guarantee that they will be reciprocated; Dawn can say hello to Michelle while Michelle ignores Dawn, and Dawn can like Michelle without Michelle liking Dawn. The arrows are technically known as directed ties, edges or arcs (Prell, 2012; Wasserman & Faust, 1994). (In the case of a tie that is always mutual, such as the tie between siblings, we don't need arrows and we can just use a straight line.)

If Michelle says hello to Alicia, and we add Lisa, Ginny, Dawn, Jacqueline, Holly, Andrea and Audrey to the mix and draw arrows from each person who says hello to each person who receives a greeting, we now have a network with nine individuals (in network terminology, nodes) and some number of directed ties between them. An example is given in Figure 26.1. If we decide to focus on a particular node, we refer to that node as ego and to other nodes as alters. We can study the network at one point in time or we can study the network as it changes over time. We call such a network of directed ties between people a directed social network, and the statistical techniques used to analyze it are known as social network analysis (Borgatti & Everett, 2012; Prell, 2012; Wasserman & Faust, 1994).

Social network analysis, because of its flexibility as a general framework for studying interpersonal interactions, allows us to systematically integrate and apply multiple theories in a TC context. Peer feedback in TCs occurs in particular patterned forms. Residents are expected to affirm peers when they show prosocial behavior. These affirmations are known as push-ups. They are also expected to correct peers for behavior contravening TC norms. These corrections are known as pull-ups (De Leon, 2000). In this study we begin with the assumption that affirming a peer in a TC constitutes an act of feedback to a peer and therefore one of responsible agency (Pearce & Pickard, 2012). If we are willing to assume this, we can make our questions about TC participation specific to the act of sending an affirmation to a peer—why do residents participate in the feedback system by affirming peers, and whom do they affirm?

Evolutionary game theorists and experimental economists have developed a substantial body of theory on the maintenance of cooperative interactions in social settings, a core issue for TCs

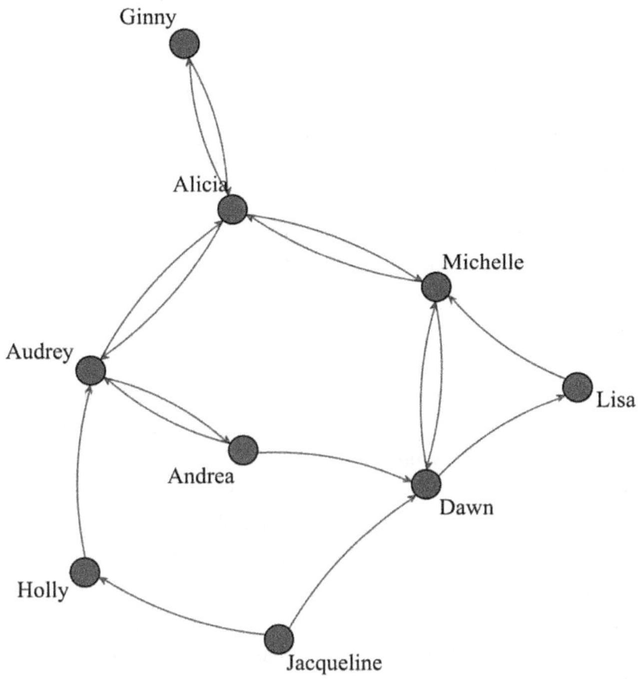

Figure 26.1 A network of nine individuals greeting each other—some return greetings, others do not

Nowak (2006) gives an article-length summation of several of the core ideas, while Nowak & Highfield (2012) offer a book-length introduction aimed at popular audiences. Several of the core mechanisms are testable within a social network analysis framework. There is both theory and experimental evidence suggesting that direct reciprocity—"you scratch my back, I'll scratch yours—"is a common mode of cooperation in social systems (Axelrod, 1984; Axelrod, 1997; Stanca, 2009). This principle has parallels in the TC clinical literature, in which residents are seen as helping each other and learning from each other in a reciprocal manner (De Leon, 2000; Miller, Sees & Brown, 2006; Perfas, 2012). Direct reciprocity would imply that if ego affirms alter for a positive TC behavior today, then alter will be more likely to affirm ego tomorrow.

Individuals who receive help are also more likely to aid third parties, a phenomenon known as generalized reciprocity or upstream indirect reciprocity. This has been observed repeatedly in experiments (Bird, 1996; Isen & Levin, 1972; Stanca, 2009). Evidence suggests that generalized reciprocity arises from positive feelings after receiving help (Isen & Levin, 1972). Moreover, it doesn't take a lot of help—when confederates hold the door for an experimental participant, the participant is more likely to hold the door for someone else (Bird, 1996) and students who receive a cookie from a confederate are more likely to offer to help peers with homework (Isen & Levin, 1972). It's not surprising, therefore, that De Leon (2000) notes that receiving an affirmation can lead to an increase in TC resident effort. Miller, Sees & Brown (2006) cite a TC resident as stating, "positive feedback makes such a difference, makes you feel good and keep on going" (p. 122). In the case of generalized reciprocity, if ego affirms alter today, alter becomes more likely to affirm someone other than ego tomorrow.

There is an important but subtle difference between generalized reciprocity and positive reinforcement as it is typically understood in the clinical literature (Spiegler & Guevremont,

2010). If affirmations come about through positive reinforcement of previous affirmations, the sequence of actions would be that ego affirms one alter, a second alter rewards ego by affirming him or her, and ego responds to the reward by becoming more likely to affirm someone else. In the case of generalized reciprocity, ego will become more likely to affirm alter after receiving an affirmation, regardless of any previous affirmations that he or she has sent. In order to establish generalized reciprocity it is therefore necessary to control for the previous affirmations that ego has sent.

TCs also incorporate ongoing and systematic negative peer feedback—in TC terminology these are known as pull-ups or corrective reminders (De Leon, 2000; Perfas, 2012). These corrections are seen as showing responsible concern for peers by reminding them of behaviors that contravene TC norms and, by extension, the norms of the society outside the TC (De Leon, 2000; Perfas, 2012). TCs put a great deal of emphasis on the idea that a correction is not a purely negative event. At one TC the first author of this chapter frequently observes residents accept a correction with the words, "Saving my life." The peer who has sent the correction will then often state, "Life worth saving."

That having been said, peer corrections clearly operate as a mild sanction. There is experimental evidence that peer sanctions can act to stabilize cooperation in groups (Bowles & Gintis, 2011; Fehr & Gachter, 2000), even when those sanctions are as mild as light peer teasing (Barr, 2001; Masclet, Noussair, Tucker & Villeval, 2003). The literature on group cooperation therefore supports the use of verbal peer corrections in TCs. This would appear to contradict received wisdom in correctional studies, which emphasizes the importance of positive reinforcement (Gendreau, 1996), which occurs in TCs when residents affirm peers. But the focus of the cooperation literature lies in maintaining cooperation in groups rather than on outcomes following graduation. The contradiction is therefore more illusory than real—regardless of their effect on outcomes, it is entirely possible that corrections act to increase the willingness of an individual to offer peer feedback in the TC.

The experimental literature on cooperation in groups sheds light on TCs, which depend on the cooperation of residents to achieve clinical change. However, there are two important differences between the TC setting and the laboratory and field settings in which cooperation studies have been done. First, while experimental studies of cooperation have been conducted in a number of cultures (Poteete, Janssen & Ostrom, 2010), they have not been conducted in TCs. Second, experiments are nearly always short, lasting from a few seconds to a few hours. It is therefore unclear whether studies can replicate experimental findings within a time frame of several days.

Just as the experimental literature on evolution and cooperation offers hypotheses on TC residents' motivations for participation, social network theory and empirical literature offer hypotheses on which peers residents are likely to interact with. One of the most common findings in studies of social networks is homophily, the tendency of people who are alike in some way to associate with each other (McPherson, Smith-Lovin & Cook, 2001). Any form of homophily would imply that affirmations occur disproportionately within groups—the arrows disproportionately connect individuals who are similar in some way. At least two forms of homophily are of potential concern in TCs. Racial homophily is endemic in the United States (McPherson, Smith-Lovin & Cook, 2001) and is particularly problematic in the American prison system (Trammell, 2009). Homophily by entry time could also be problematic for TCs, since they depend on senior residents to mentor junior residents (De Leon, 2000; Perfas, 2012). If TC residents preferentially affirm peers who arrived at roughly the same time they did, this could erode the quality of the mentoring system.

Social network analysts have also extensively documented the phenomenon of transitivity. In a transitive relationship, if Dawn is a friend of Michelle, and Michelle is a friend of Lisa, then Dawn is more likely to be a friend of Lisa (Wasserman & Faust, 1994). The reason is simple and intuitive—if Dawn is a friend of Michelle and Michelle is a friend of Lisa, but Dawn is not a friend of Lisa, the situation is likely to be uncomfortable for all three. This reasoning does not directly apply to affirmations in TCs, unless residents are simply affirming their friends. However, we could extend the argument by arguing that if Dawn affirms Michelle and Michelle affirms Lisa, then Dawn is more likely to affirm Lisa as if taking a cue from Michelle, for whom Dawn has previously demonstrated respect. In other words, a transitive relationship in affirmations could represent a mutual judgment of an individual's behavior.

Finally, several individual variables might influence the rate at which residents affirm their peers. It seems likely that more senior residents might affirm peers at a higher rate than junior residents (De Leon, 2000). Relational theorists emphasize the importance of connection to women; this raises the possibility that women might affirm peers at a higher rate than men do (Covington, 2008). It is possible that older residents affirm peers at a higher rate than younger residents out of a desire to mentor peers (Erikson, 1974), and it is possible that those TC residents at higher risk of future criminal behavior affirm peers at a lower rate than residents who are at lower risk do.

Methodology

Data

The data for this study are based on clinical records of 282,604 written peer affirmations received, 290,539 written affirmations sent and 152,797 written peer corrections received, exchanged between 5,105 residents of three different community corrections-based TCs in the Midwestern United States. Each affirmation and correction included the date on which it was sent, the name of the sender and the name of the receiver. The difference in the number of affirmations sent and received in the dataset comes from a combination of affirmations that were sent but not sent to peers—for instance, it was possible to affirm staff members—and clerical errors which sometimes made it impossible to determine to whom a given affirmation had been sent. Two of the TCs drew on predominantly rural catchment areas. Of these two, one was a 60-bed entirely male TC while the other had a 90-bed male unit and a 17-bed female unit. The third TC drew on a mixed urban/rural catchment area. This TC included two 80-bed male units and one 85-bed female unit. These facilities also kept data on gender, ethnicity, age, time since entry and residents' scores on the Level of Services Inventory-Revised (LSI-R), a standardized measure of risk of recidivism (Andrews & Bonta, 1995). The LSI-R includes information on substance abuse, education level, previous offenses, employment level and social support, among other known risk factors for criminal behavior.

Analysis

Since the data on affirmations and corrections included information on sender, receiver and date it was possible to analyze the dataset as a longitudinal social network. Several methods of longitudinal social network analysis exist. Certainly, the best known, most thoroughly researched and most used method of longitudinal network analysis is the RSiena package (Ripley, Snijders, Boda, Voros, Preciado, 2014; Snijders, 2005; Snijders, Steglich & Schweinberger, 2006).

However, RSiena is based on a stochastic actor oriented model (SAOM), which assumes that actors create, maintain, and terminate ties (Ripley, Snijders, Boda, Voros, Preciado, 2014). This makes the package inappropriate for the data in this project, in which connections represent single events that are not maintained and are automatically terminated.

Another framework for modeling longitudinal change in social networks is multilevel modeling, in which ego is treated as the Level 2 unit and alters are treated as Level 1 units (De Nooy, 2011; Kossinets & Watts, 2006; Warren, Doogan, De Leon, Phillips, Moody & Hodge, 2013). Warren et al. (2013) used a multilevel negative binomial regression to model the influence of affirmations and corrections received on the weekly number of affirmations that residents sent. The authors found that the number of affirmations that residents sent in any given week was associated with the number of corrections and the affirmations that residents received in the week prior. However, because both affirmations and corrections were aggregated at a weekly level, the analysis could not differentiate between direct reciprocity, generalized reciprocity and transitivity and could not address homophily effects.

An analysis of daily additions to the event network of affirmations and corrections allows us to distinguish between these different sources of network structure. An appropriate model for such an analysis is a multilevel event history model, in which the sending of an affirmation from ego to alter is the event of interest (Kosinetts & Watts, 2006; Kleinbaum & Klein, 2005). In the case of this dataset, it is possible to choose a random selection of days and choose one resident who was living in a given facility during that day. That resident, ego, becomes one of the level 2 units of the multilevel event history analysis model. Level 1 then constitutes all potential affirmations that ego might send to alters (fellow TC residents) over a series of days. While social network data is by definition not independent, taking a comparatively small sample from a large network addresses this problem (Kosinetts & Watts, 2006). Such an analysis can be done using a Cox proportional hazards model (Kleinbaum & Klein, 2005) that allows for the addition of random effects (also known as a shared frailty model) to capture correlation of observations on the same level 2 ego unit. This allows us to treat dynamic peer interactions as predictors in a multilevel statistical model.

Several of the predictors in our model are included for control, as is typical of data analysis of naturalistic observational studies. Our primary interest is to identify mechanisms that are theorized to promote cooperation in groups. First, we would like to know whether a tendency to engage in the community by giving push-ups to others results from receiving push-ups or pull-ups from peers. If residents are more likely to send a push-up after receiving one, this would be an example of generalized reciprocity. If they are more likely to send a push-up after receiving a pull-up, this would be a positive response to a mild sanction. Direct reciprocity (i.e., returning a push-up to a specific other), in addition to being a mechanism that might promote cooperation, is a common phenomenon that will inflate an estimate of generalized reciprocity if not controlled.

Figure 26.2 visualizes generalized reciprocity and three other phenomena that are included in the model—reciprocity, transitive closure, and homophilous selection, or homophily. In the table, solid lines between nodes represent ties that have occurred while dotted lines represent potential ties. We include a reciprocity predictor to control for whether a potential tie to actor *j* could be due to a recently received tie from actor *j*. We include a transitive closure predictor to control for whether some two-path of interactions connecting *i* to *j* through another actor *h* might have an effect on whether the potential tie directly from *i* to *j* will form. Transitive closure is an extremely common feature of social networks because humans tend to connect with their connections' connections, thus supporting the formation of what we often call network clustering. We also want to isolate generalized reciprocity from an exogenous force known as

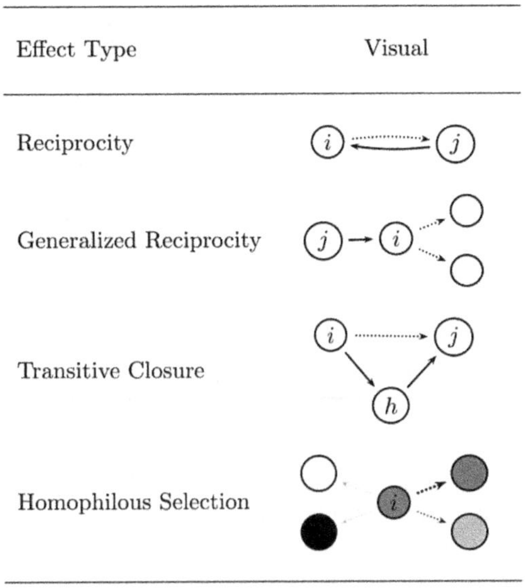

Figure 26.2 Social network relationships included in the current analysis

homophily—the tendency for like individuals (with respect to some covariate) to connect. In the table, it can be seen that the darkest dotted tie—intended to identify the most probable tie—is the one between *i* and the alter that is shaded most similarly. Homophily model terms can be either continuous indicating similarity on some trait like time spent in treatment, or dichotomous indicating sameness on a categorical trait like race.

One additional model term type not visualized is a main effect of some nodal characteristic on a node's tendency to engage in the community through push-up interactions. An example is the effect of *age* on an individual's tendency to interact; we call these terms "ego" effects. We also designate the generalized reciprocity terms with "ego" because they similarly represent something about the focal actor that may alter their push-up giving tendencies. The only difference is that the generalized reciprocity predictors change dynamically with the network through time (i.e., they are endogenous), while *age* is fixed. While *time in program* is a covariate that changes, it is not dependent on the changing network and so is time varying but exogenous.

Sample

For this analysis a clustered random sample of single days and egos was chosen. First, a TC unit was selected randomly with weight proportional to the number of beds in the unit. Second, a starting date was selected uniformly at random from the unit. Third, an ego was selected from all those present in the unit during that starting date. Fourth, all other residents present with ego over the twelve-day period beginning with the starting date were selected as level 1 alters with whom the level 2 ego could potentially establish a network tie by sending an affirmation.

A total of 310 ego/days were randomly sampled from a sampling frame of 552,203 resident days in the three facilities. The first five days of each twelve-day period were used to model the history of the sixth day, which was the first day during which the affirmations that ego sent were modeled. Ego's possibility of forming a tie by affirming each alter was therefore modeled

Table 26.1 Descriptive statistics for continuous predictor variables

	Mean	*SD*	*Median*	*Minimum*	*Maximum*
Ego Age	27.942	8.540	25.000	18.000	60.000
Ego LSI-R	27.390	7.177	27.000	7.000	47.000
Absolute Time Difference	7.300	5.277	6.143	0.000	25.000
Ego Time In Program	72.934	43.537	70.000	4.000	173.000
Ego Affirmations Sent	3.742	6.103	1.000	0.000	97.000
Ego Affirmations Received	4.282	4.256	3.000	0.000	28.000
Ego Corrections Received	2.617	3.345	1.000	0.000	27.000
Reciprocity Opportunities	0.056	0.255	0.000	0.000	4.000
Transitivity Opportunities	0.194	0.927	0.000	0.000	28.000
Outcome Tie	0.006	0.077	0.000	0.000	1.000

for a seven-day (one week) period. Once ego affirmed a given alter further affirmations of that alter were not modeled, but further affirmations of other alters continued to be. All network variables were cumulative over the twelve days. Thus, if ego received corrections on days one and four, the cumulative total of corrections on day six would be two, and if ego received a further correction on day eight the cumulative total of corrections on day nine would be three.

Descriptive statistics for continuous variables can be found in Table 26.1. Unsurprisingly, the egos were on average young adults (mean ego age of 27.94), and age was positively skewed. Mean ego LSI-R scores were 27.39, within the range that researchers have typically seen as high-risk (Arnold, 2007). Egos had on average resided on the facility for 72.93 days. The absolute value of the mean difference of time in the facility between ego and alter was 7.3 weeks, with a minimum of 0 and a maximum of 25.

The mean cumulative number of affirmations per day that ego sent was 3.72, the mean cumulative number of affirmations that ego received from all alters was 4.28, and the mean cumulative number of corrections that ego received from all alters was 2.62. These are all treated as level 2 variables, applying to ego. Reciprocity and transitivity are level 1 variables, applying to the relationship between ego and alter. Thus, the mean cumulative number of reciprocity opportunities for ego per day is .056, reflecting the fact that in a unit of as many as ninety residents the vast majority of peers have not affirmed ego, resulting in relatively few reciprocity opportunities. Transitivity constituted a count of two paths linking ego to alter—in any given day the mean value is .194, suggesting that ego has the opportunity to form transitive relationships with roughly one out of five alters. The outcome tie variable is the tie being modeled and captures whether ego has affirmed a specific alter—and, at a mean of .006, implies that the average ego took advantage of a bit less than 1% of all available affirmation opportunities. While zero is the minimum value for all of these variables, the maximum values range widely for all except the outcome variable, which by definition is capped at one.

Categorical variables were not included in the table. The TCs from which the sample is drawn were gender segregated and the great majority of the residents were male. The sampling scheme yielded a gender proportional sample in which 8.9% of the included participants are female. On average any given ego occupies a unit in which 68.8% of peers are of the same race.

Results

Results of the model are given in Table 26.2. As hypothesized, residents who had received more affirmations affirmed peers at a higher rate (B = .074, SE = .030, HR = 1.077). Contrary

Table 26.2 Results of multilevel event history analysis model

	Est	*SE*	*z*	*p*	*HR*
Fixed Effects					
Ego Affirmations Received	0.074	(0.030)	2.488	.006	1.077
Ego Corrections Received	0.007	(0.032)	0.202	.420	1.007
Reciprocity Opportunities	0.588	(0.093)	6.341	<.001	1.801
Transitivity Opportunities	–0.037	(0.044)	–0.835	.202	0.964
Ego Affirmations Sent	–0.201	(0.017)	–11.953	<.001	0.818
Female	2.626	(0.627)	4.189	<.001	13.814
Ego Age	–0.002	(0.005)	–0.507	.306	0.998
Ego LSI	–0.063	(0.028)	–2.229	.013	0.939
Ego Time In Program	0.002	(0.005)	0.360	.359	1.002
Absolute Time Difference	–0.040	(0.008)	–5.001	<.001	0.960
Same Race	0.266	(0.100)	2.663	.004	1.305
Variance Components					
Ego (random intercept)	2.844				
Date (random intercept)	0.890				

to our expectations, there was no evidence that residents who had received more corrections affirmed peers at a higher rate (B = .007, SE = .032, HR = 1.007). As hypothesized, residents showed a tendency to reciprocate affirmations (B = .588, SE = .093, HR = 1.801). Residents showed no particular tendency toward transitive closure of triads (B = – .037, SE = .044, HR = .964). Ego affirmations sent, the relationship between the affirmations that residents had sent to peers in previous days to the likelihood of affirming a peer on the current day, was negative (B = – .201, SE = .017, HR = .818).

Female residents sent more affirmations than male residents (B = 2.626, SE = .627, HR = 13.814). Residents with higher LSI-R scores sent fewer affirmations (B = – .063, SE = .028, HR = .939). Age was not a statistically significant predictor of the number of affirmations that residents gave (B = – .002, SE = .005, HR = .998), nor was length of time in the program (B = .002, SE = .005, HR = 1.002).

Residents were more likely to affirm a peer of the same race (B = .266, SE = .100, HR = 1.305) and less likely to affirm a peer who had entered the program at a different time (B = – .040, SE = .008, HR = .960). Both homophily hypotheses were therefore confirmed.

Rerunning the model after the removal of two outlier individuals did not alter the statistical significance of any findings.

In thinking about these results it is important to realize that the hazard ratio as reported is the exponentiated change in hazard of sending an affirmation per unit of the predictor variable (Kleinbaum & Klein, 2005). This is straightforward when the predictor variable has only two values; thus, an HR of 1.305 for the same race variable implies that TC residents a peers of the same race at a roughly 30% higher rate. It's more difficult to interpret the hazard ratio when the variable is continuous. For instance, the hazard ratio affirmations received is 1.077, which is simply the exponentiated value of the estimated coefficient of .074. This represents the change in hazard of sending an affirmation if a resident receives one affirmation rather than none. However, if the resident receives ten affirmations, the HR equals exp (.074 × 10), or 2.10, a doubling of the hazard of sending an affirmation. Similarly, the HR for absolute difference is .96, meaning that residents are 96% as likely to affirm a peer who arrived a week before or after they did. When the difference is 20 weeks, the HR equals exp(–.04 × 20), or .45. Residents in these TCs affirm peers who arrived 20 weeks before or 20 weeks after they did at about 45% the rate they affirm peers who arrived during the same week.

Discussion

In this chapter we have treated the hazard of a resident affirming a peer as a measure of resident involvement in the process of peer feedback that underlies TC clinical practice (De Leon, 2000; Pearce & Pickard, 2012). The study used longitudinal social network analysis to answer three questions. First, which interactions make it more or less likely that a resident will affirm a peer? Second, which peers do residents affirm? Third, which resident characteristics predict affirmations?

The most obvious limitation of this analysis is that the feedback that residents give peers by affirming them is only one type of activity in a TC (De Leon, 2000; Perfas, 2012). However, written affirmations are both an important and a common activity in the TC, and peer feedback in general lies at the core of TC clinical practice (Pearce & Pickard, 2012). As long as we remember that it is entirely possible that a variable that predicts whether and to whom a resident will send an affirmation might not predict other TC activities, written affirmations are legitimate objects of analysis.

Both TC clinical literature and laboratory experiments on cooperation would have predicted that residents would respond to peer affirmations with both direct and generalized reciprocity (De Leon, 2000; Stanca, 2009). The analysis supported both of these predictions, finding both that the number of affirmations that residents received and the number of opportunities to reciprocate affirmations predicted affirmations sent. Since the analysis controls for the affirmations that ego sent, positive reinforcement does not explain the relationship.

The predictions of direct and generalized reciprocity were supported within a week-long time frame. To put this in perspective, even a relatively long laboratory experiment on cooperation in groups lasts for about 90 minutes (Stanca, 2009). The time frame for this study is roughly two orders of magnitude longer. As the literature on cooperation would predict, both direct and generalized reciprocity appear to play a significant role in maintaining the TC system of peer feedback, and they do so over a length of time that is pragmatically significant for program management. This also replicates and adds detail to the finding of Warren et al. (2013) that the affirmations that residents receive in one week predict the affirmations that they will send in the next.

Peer corrections did not appreciably influence the number of affirmations that residents sent. This fails to replicate the finding of Warren et al. (2013) that the corrections that residents receive in one week predict the affirmations that they will send in the next. But the dataset used in Warren et al. (2013) was substantially larger than the dataset used in this study, and the relationship between the corrections residents received and the affirmations residents sent was substantially weaker than that between the affirmations residents received and the affirmations they sent. It is possible that the current dataset was too small to detect the weak relationship between the corrections that a resident received and the affirmations that he or she sent.

Both the current and previous analyses support an emphasis on affirmations in TCs, since direct and generalized reciprocity appear to have more influence over residents' prosocial participation, as measured by the affirmations they send to peers, than peer sanctions. However, it is important to note that the primary purpose of peer corrections is obviously to discourage antisocial behavior. It is entirely possible that peer corrections reduce antisocial behavior while having little or no influence on prosocial behavior. Further research on the role of peer corrections in TC clinical process is warranted.

The analysis found no statistically significant tendency toward transitivity in this data—when ego affirms an alter who affirms a second alter, ego is no more likely to affirm the second alter. Residents do not seem to take cues from peers whom they have affirmed. Future research could employ an alternative view of transitive closure to test whether residents take cues from those who have affirmed them. For example, if Dawn affirms Keith and Audrey, perhaps Keith is more likely to affirm Audrey. It is possible, though, that written affirmations are both too brief and too common for residents to keep effective track of who affirms whom, beyond the comparatively simple task of remembering who has affirmed oneself.

The negative hazard for ego's previous affirmations has the clinical implication that one can expect the number of affirmations that any given resident gives to oscillate—more today means on average fewer tomorrow, fewer today means on average more tomorrow.

The finding that both direct and generalized reciprocity influence resident interactions does not mean that residents are equally likely to interact with all peers. Homophily in these TCs arose from at least two sources. One was race, unsurprising given the ubiquity of racial homophily both in American life and in the American correctional system (McPherson, Smith-Lovin & Cook, 2001; Trammel, 2009). This should be seen in the context of an earlier study that drew on data from the second rural facility and found that on average African American and European American residents sent and received nearly identical numbers of affirmations (Linley, Warren & Davis, 2010). This suggests that, at least as far as the amount of peer mentorship that residents receive, TCs are resilient in the face of racial homophily as long as residents of different ethnic backgrounds give similar numbers of affirmations to peers. This might not be true in a more staff-centered model; if, for instance, positive reinforcement primarily comes from staff, and staff members give less positive reinforcement to African American residents, there is no one to make up the deficit. On the other hand, it is possible that racial homophily in affirmations indicates underlying tension that needs to be addressed in the TC.

Since TCs depend on senior residents to mentor junior residents, cohort homophily is likely to be problematic. The finding that more senior residents are no more likely to affirm peers than more junior residents is also problematic. These findings suggest that it may be challenging for TC staff to establish an effective mentoring system. There are possible upsides, though—for instance, staff could potentially use cohort homophily to balance racial homophily by consciously selecting mixed-race entrance groups. It is also possible that new residents will adjust to the TC environment more effectively if they trade affirmations with a friend who is also new. Cohort homophily could act to facilitate early engagement in the TC. Any upsides, however, appear to come at the cost of decreased mentorship.

Cohort homophily could indicate multiple possible structures in the network of affirmations. It is possible that senior residents affirm each other and junior residents also affirm each other. It is also possible that senior residents affirm each other while affirming junior residents at somewhat lower rates, while the junior residents do not affirm each other much at all. This would correspond to a core/periphery structure (Borgatti & Everett, 1999), with the implication that as residents gain seniority in a TC they tend to move into a more intense level of peer interaction. The current analysis could not distinguish between these possibilities, and further research on this finding is therefore needed.

As expected, female residents affirmed peers at a higher rate than male residents. While there is a feminist literature that is critical of TCs (Beck, 2006; Eliason, 2006), in this study female residents were more active in providing positive feedback to peers. High-risk residents, on the other hand, were less likely to affirm peers. Clinicians may wish to focus more effort on engaging high-risk residents in TC practices such as peer feedback, particularly since there is evidence that severe criminality correlates with lower retention in TCs (De Leon & Wexler, 2009). This is a particularly salient point since the risk principle in criminology, which has considerable empirical support, states that treatment should focus on high-risk individuals, since without it they recidivate at very high rates (Andrews & Dowden, 2006). Age was unrelated to the hazard of affirming a peer.

Conclusion

TCs are founded on the principle that mutual aid and peer feedback in a residential setting under professional supervision are effective in treating substance abuse (De Leon, 2000). Any residential treatment setting is a complex system, but TCs are unusual in their effort to leverage the complexity of interpersonal interactions for clinical treatment. Process research in such a milieu-based setting is a challenge, because peer interactions are the therapy. In such a situation the structure of most social science, which emphasizes the influence of personal characteristics such as gender or motivation on behavior, fights the research goal. Even very good qualitative and quantitative research, when based on the characteristics of individuals rather than interactions between individuals, at best elicits data on the overall history of peer relationships. Social network analysis allows us to consider the dynamic behavior of an individual as a function of peer interactions.

Social network theory and analysis allow us to quantify the idea that the community itself is the method of treatment, because the links between individuals, which together form the structure of the community, become the units of fundamental interest. This lets us frame and test hypotheses that cannot otherwise be considered. To use this study as an example, it would be possible to simply sum the affirmations that residents send over their careers and use the individual level variables (time in program, LSI-R, gender, age and summed corrections) as predictors of the summed affirmations in a multilevel regression model in which level 1 is the individual and level 2 is the TC unit (Pinheiro & Bates, 2009). But this approach could not adequately replicate any of the other results—reciprocity, transitivity and homophily are irreducibly network concepts (Wasserman & Faust, 1994; Prell, 2012), while summing the affirmations and corrections that residents receive would destroy the longitudinal nature of the dataset. By treating the TC as a social network we see more, we see differently, and we can test the insights that we gain.

Acknowledgments

The authors would like to acknowledge James Moody. This research was funded by NIH grant R21DA023474.

References

Andrews, D.A. & Bonta, J. (1995). *The level of service inventory–Revised.* Toronto, Canada: Multi-health Systems.

Andrews, D.A. & Dowden, C. (2006). Risk principle of case classification in correctional treatment: A meta-analytic investigation. *International Journal of Offender Therapy and Comparative Criminology, 50,* 88–100.

Arnold, T.K. (2007). Dynamic changes in the Level of Service Inventory Revised (LSI-R) scores and the effects on prediction accuracy. Unpublished Master's thesis.

Axelrod, R.M. (1984). *The Evolution of Cooperation.* New York: Basic Books.

Axelrod, R.M. (1997). *The Complexity of Cooperation: Agent-based Models of Competition and Collaboration.* Princeton, NJ: Princeton University Press.

Barr, A. (2001). Social dilemmas and shame-based sanctions: Experimental results from rural Zimbabwe. The Centre for the Study of African Economies Working Paper Series, paper 149, hosted by the Berkeley Electronic Press.

Beck, J. A. (2006). Recovering selves: Women and the governance of conduct in a residential drug treatment program. *Women and Therapy, 29*(3/4), 239–259.

Bird, R.J. (1996). Chaos and social reality: An emerging perspective. In Eve, R.A., Horsfall, S.T. and Lee, M. (Eds.), *Chaos, Complexity and Sociology: Myths, Models and Theories*, pp. 141–152. London: Sage Publications.

Borgatti, S.P., Everett, M.G. and Johnson, S.C. (2013). *Analyzing Social Networks.* London: Sage Publications.

Bowles, S. & Gintis, H. (2011). *A Cooperative Species: Human Reciprocity and its Evolution.* Princeton, NJ: Princeton University Press.

Burnett, K. (2001). Self-help or sink-or-swim? The experience of residents in a UK concept-based therapeutic community. In Rawlings, B. & Yates, R. (Eds.), *Therapeutic Communities for the Treatment of Drug Users.* London: Jessica Kingsley.

Carr, W. A. & Ball, S. A. (2014). Predictors and treatment outcomes of perceived ward atmosphere among therapeutic community residents. *Journal of Substance Abuse Treatment, 46*, 567–573.

Covington, S.S. (2008). Women and addiction: A trauma-informed approach. *Journal of Psychoactive Drugs*, SARC Supplement 5, pp. 377–385.

De Leon, G. (2000). *The Therapeutic Community: Theory, Model and Method.* New York: Springer.

De Leon, G. (2010). Is the therapeutic community an evidence-based treatment? What the evidence says. *Therapeutic Communities, 31*(2), 104–128.

De Leon, G. & Schwartz, S. (1984). Therapeutic communities: What are the retention rates? *American Journal of Drug and Alcohol Abuse, 10*(2), 267–284.

De Leon, G. & Wexler, H. (2009). The therapeutic community for addictions: An evolving knowledge base. *Journal of Drug Issues*, 39, 167–178.

De Leon, G., Wexler, H. K. & Jainchill, N. (1982). The therapeutic community: Success and improvement rates 5 years after treatment. *International Journal of the Addictions, 17*(4), 703–747.

de Nooy, W. (2011). Networks of action and events over time: A multilevel discrete-time event history model for longitudinal network data. *Social Networks*, 33(1), 31–40.

Eliason, M. (2006). Are therapeutic communities therapeutic for women? *Substance Abuse Treatment, Prevention, and Policy, 1*(3).

Erikson, E. H. (1974). *Dimensions of a New Identity.* New York: Norton.

Fehr, E. & Gachter, S. (2000). Cooperation in public goods experiments. *The American Economic Review, 90*(4), 980–994.

Gendreau, P. (1996). The principles of effective intervention with offenders. In Harland, A.T. (Ed), *Choosing Correctional Options that Work: Defining Demand and Evaluating the Supply.* Thousand Oaks, CA: Sage.

Goethals, I., Soyez, V., Melnick, G., De Leon, G. & Broeckaert, E. (2011). Essential elements of treatment: A comparative study between European and American therapeutic communities for addiction. *Substance Use & Misuse, 46*(8), 1023–1031.

Hawkins, J. D. & Wacker, N. (1986). Side bets and secondary adjustments in therapeutic communities. In G. De Leon & J. T. Ziegenfuss (Eds.), *Therapeutic Communities for Addictions: Readings in Theory, Research and Practice*, pp. 141–155. Springfield, IL: Charles C. Thomas.

Kleinbaum, D. G, & Klein, M. (2005). *Survival Analysis: A Self-learning Text.* New York: Springer.

Isen, A.M. & Levin, P.F. (1972). Effect of feeling good on helping: Cookies and kindness. *Journal of Personality and Social Psychology, 21*(3), 384–388.

Kossinetts, G. & Watts, D. (2006). Empirical analysis of an evolving social network. *Science, 311*(5757), 88–90.

Lees, J., Manning, N. & Rawlings, B. (2004). A culture of enquiry: Research evidence and the therapeutic community. *Psychiatric Quarterly, 75*(3), 279–294.

Linley, J., Warren, K. & Davis, T. (2010). A comparison of affirmations given and received by minority and non-minority status men in a correctional therapeutic community. *Therapeutic Communities, 31*(1), 37–49.

Loat, M. (2006). Sharing the struggle: An exploration of mutual support processes in a therapeutic community. *Therapeutic Communities: International Journal for Therapeutic and Supportive Organizations, 27*(2), 211–228.

Mandell, W., Edelen, M.O., Wenzel, S.L., Dahl, J. & Ebener P. (2008). Do dimensions of therapeutic community treatment predict retention and outcomes? *Journal of Substance Abuse Treatment, 35*(3), 223–231.

Mascelet, D., Noussair, C., Tucker, S., & Villeval, M. (2003). Monetary and non-monetary punishment in the voluntary contributions mechanism. *The American Economic Review, 93*(1), 366–380.

McPherson, M., Smith-Lovin, L. and Cook, J. M. (2001). Birds of a feather: Homophily in social relationships. *Annual Review of Sociology, 27*, 415–444.

Miller, S., Sees, C., & Brown, J (2006). Key aspects of psychological change in residents of prison therapeutic community. *The Howard Journal, 45*, 116–128.

Mitchell, O., Wilson, D.B. & MacKenzie, D.L. (2006). The effectiveness of incarceration-based drug treatment on criminal behavior. *Campbell Collaboration Systematic Review*. Available at: www.campbell collaboration.org/

Neville, L., Miller, S. & Fritzon, K. (2007). Understanding change in a therapeutic community: An action systems approach. *The Journal of Forensic Psychiatry & Psychology, 18*(2), 181–203.

Nowak, M. A. (2006). Five rules for the evolution of cooperation. *Science, 314*(5805), 1560–1563.

Nowak, M. A. & Highfield, R. (2011). *Supercooperators: Altruism, Evolution and Why We Need Each Other to Succeed*. New York: Free Press.

Patenaude, A. L. (2005). A qualitative exploration into a prison substance abuse treatment program: "I tell them what they want to hear." In Sims, B. (Ed.), *Substance Abuse Treatment with Correctional Clients: Practical Implications for Institutional and Community Settings*. New York: The Haworth Press.

Pearce, S. & Pickard, H. (2012). How therapeutic communities work: Specific factors related to positive outcome. *International Journal of Social Psychiatry, 59*(7), 636–645.

Pearson, F. S. & Lipton, D. S. (1999). A meta-analytic review of corrections-based treatments for drug abuse. *The Prison Journal, 79*(4), 384–410.

Perfas, F.B. (2012). *Deconstructing the Therapeutic Community: A Practice Guide for Professionals*. North Charleston, SC: Create Space.

Pinheiro, J.C. & Bates, D. (2000). *Mixed Effects Models in S and S-Plus*. New York: Springer.

Poteete, A. R., Janssen, M. A. & Ostrom, E. (2010). *Working Together: Collective Action, the Commons and Multiple Methods in Practice*. Princeton, NJ: Princeton University Press.

Prell, C. (2012). *Social Network Analysis: History, Theory, Methodology*. Thousand Oaks, CA: Sage.

Ravndal, E. and Vagnum, P. (1994). Treatment of female addicts: The importance of relationships to parents, partners and peers for the outcome. *International Journal of the Addictions, 29*(1), 115–125.

Ripley, R., Snijders, T., Boda, Z., Voros, A. and Preciado, P. (2014). Manual for RSiena. Available at: www.stats.ox.ac.uk/~snijders/siena/RSiena_Manual.pdf

Snijders, T.A.B. (2005). Models for longitudinal network data. In Carrington, P.J., Scott, J. & Wasserman, S. (Eds) *Models and Methods in Social Network Analysis*, pp. 215–247. New York: Cambridge University Press.

Snijders, T., Steglich, C. & Schweinberger, M. (2007). Modeling the coevolution of networks and behavior. In Montfort, K. v., Oud, J. and Satorra, A. (Eds) *Longitudinal Models in the Behavioral and Related Science*, pp. 41–71. European Association of Methodology series. Mahwah, NJ: Lawrence Erlbaum Associates.

Spiegler, M. D. and Guevremont, D. C. (2010). *Contemporary Behavior Therapy*. Belmont, CA: Wadsworth.

Stanca, L. (2009). Measuring theories of indirect reciprocity: Whose back do we scratch? *Journal of Economic Psychology, 30*(2), 190–202.

Trammell (2009). Values, rules and keeping the peace: How men describe order and the inmate code in California prisons. *Deviant Behavior, 30*(8), 746–771.

Vandevelde, S., Brokaert, E., Yates, R. and Kooyman, M. (2004). The development of the therapeutic community in correctional establishments: A comparative retrospective account of the 'democratic' Maxwell Jones TC and the hierarchical concept-based TC in prison. *International Journal of Social Psychiatry, 50*(1), 66–79.

Warren, K., Doogan, N., De Leon, G., Phillips, G., Moody, J. & Hodge, A. (2013). Short-run prosocial behavior in response to receiving corrections and affirmations in three therapeutic communities. *Journal of Offender Rehabilitation, 52*(4), 270–286.

Wasserman, S., and Faust, K. (1994). *Social Network Analysis: Methods and Applications*. New York: Cambridge University Press.

27

CIRCLES OF ASSOCIATION

The Connections of Community-Based Food Systems

David S. Connor and Ralph Levine

Research Question: How can community-based food systems help overcome health, economic, and environmental problems?
System Science Method(s): System dynamics
Things to Notice:

- Use of system dynamics to inform interventions
- Building a complete model from multiple submodels

Michigan, like many states, is experiencing a number of deep and persistent problems, including high rates of obesity, unemployment, fiscal deficits and farmland loss. This chapter develops a conceptual framework, showing how these problems are related to options available for food and agricultural systems. Community-based food systems (CFS) lie at the center of a number of causal loops which may reinforce a series of potentially positive or negative outcomes. We suggest a number of interventions that would lead to a more community-based food system: (1) strategies to make fresh food and local food more accessible; (2) policy interventions to reverse bias against CFS and promote locally grown good; (3) nutrition and food system education; (4) training for entrepreneurial agriculture. We also discuss countervailing forces such as the influence of those benefiting from the current system: agribusiness and food corporations such as input suppliers, manufacturers, realties and fast food chains. The unintended consequence of gentrification is also discussed. We conclude with a discussion of future research directions, utilizing a system dynamics model to identify levers of and barriers to change and ultimately the most effective pathway toward realizing CFS's benefits.

Michigan, like many other states of the United States, is dealing with a number of deep and persistent problems: high rates of chronic disease and obesity, high unemployment, fiscal deficits, sprawl; and loss of farms and farmland. Although these problems may seem unconnected and are indeed often dealt with in isolation, they are bound together by a common thread: each has a profound connection to and is impacted by the type of food system we have and the choices we make in how to feed ourselves.

This chapter will argue that a community-based food system (CFS) is a critical component of solutions to these problems, and is able to support positive loops or virtuous cycles within Michigan's communities. The chapter's purpose is to raise awareness among decision makers and contribute to the dialogue concerning the role of community-based food systems in achieving broad community goals, by outlining the connections among food and agriculture, land use,

 DOI: 10.4324/9781315748771-30

economic growth and public health. By informing a broad array of citizens, such as policymakers and government officials (federal, state, and local), academics, economic development officers and planners, representatives of nongovernmental organizations and others, and encouraging multidisciplinary and crosscutting collaborations, this chapter will attempt to guide decisions that will reinforce the positive loops and avoid the negative which would inhibit the processes. It will emphasize how community-based food systems can contribute positively to each of these arenas, and how gains in these arenas can mutually reinforce each other.

It is important to note that the associations between arenas discussed in this chapter vary in their certainty and concreteness. Those not supported in the literature are suggested in a way that we hope stimulates discussion and draws further attention and interest to the topic of community-based food systems. Although Michigan is the focus of this chapter, we believe these associations exist in many other areas too.

Conceptual Framework and Supporting Literature

The way people feed themselves—the food system of a given place or population—lies at the intersection of a number of loops that can contribute positively or negatively to community health and economic development. The ties between the food and agricultural system, on one hand, and economy, public health and land use, on the other are fairly straightforward. For example, Michigan's food and agriculture sector has been estimated as accounting for $60 billion annually and 1.05 million jobs.[1] Janet Olszewski, Director of the Michigan Department of Community Health lists poor diet as a leading cause of preventable death in the state.[2] Michigan has 10,142,958 acres of farmland, accounting for about 28 percent of its land area.[3,4] What is not clear is the way that positive outcomes in one arena could reinforce those in other arenas and in turn, perpetuate the original outcome. Although not providing precise quantitative measures of impacts, our framework draws attention to qualitative changes that can arise from the presence or absence of a healthy, community-based food system.

The conceptual framework, in the form of a causal loop diagram in Figure 27.1, depicts the circles of association. Causal loop diagrams have been found useful in representing relationships among variables that lead to potential actions of dynamic feedback processes. An introduction to causal loop diagramming can be found in Eden, Jones, and Sims,[5] Wolstenholme[6] and Sterman.[7] This figure shows the virtuous cycle aspect, the mutually reinforcing positive outcomes.

A community-based food system can be defined as a collaborative effort to build more locally based food systems and economies.[8] CFSs prioritize local resources and local markets, emphasize social equity and environmental sustainability, and rely on relationships among growers and eaters, retailers and distributors, processors and producers of food within the community (paraphrased from Heller[9]). "When local agriculture and food production are integrated in community, food becomes part of a community's problem-solving capacity rather than just a commodity that is bought and sold."[9] CFSs link efforts of two grassroots movements—sustainable agriculture and community food security—in a way that addresses each group's core concerns: viability of ecologically friendly and socially just farms, and access to healthy foods for all people. The food is generally purchased through short supply chains, either directly from farmers through face-to-face channels such as farmers' markets, community supported agriculture and farm stands, or through local businesses (e.g., cooperative health-food stores, buying clubs, specialty shops, restaurants) committed to building relationships and placing shared values at the forefront. It implies a cohort of conscious and conscientious consumers who place value on *how, where and by whom food is produced and distributed* as well as on traditional drivers of food demand, such as price and convenience.

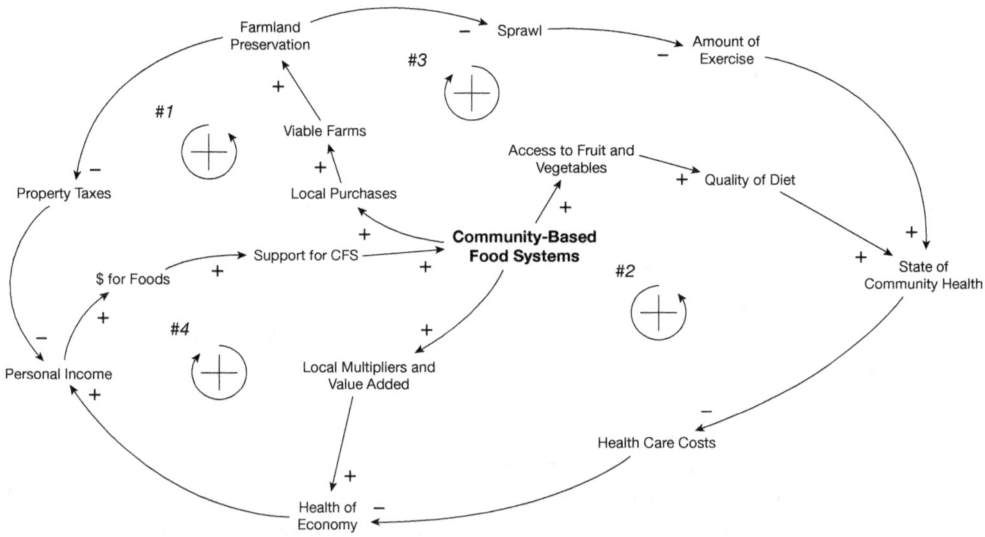

Figure 27.1 Circle of associations

There are four loops in Figure 27.1, each with a CFS in the middle. The following paragraphs will outline each loop—in no particular order—and provide supporting literature where available.

Loop 1. CFS → (+) Local Purchases → (+) Viable Farms → (+) Farmland Preservation → (–) Property Taxes → (+) Personal Income, → (+) $ for Food → (+) Support for CFS → CFS

Farm viability is a major goal of a CFS. Linking farmers and consumers in relationship-based market transactions builds customer loyalty. These links increase farm revenue by increasing the market share of local farms, allowing price-making behavior through product differentiation, and directing a larger percentage of the food dollar in the farmers' pockets compared with commodity markets.[10–13] Higher and more stable revenues help farms to remain economically viable in the face of rising land costs and development pressures.

Farmland conservation contributes positively to local fiscal revenues. Farmland has significantly higher revenue-to-expenditure ratios than residential, commercial or even industrial land uses,[14] promoting fiscal balance and/or lower property taxes. In addition, Michigan state taxes are distributed to local governments in revenue sharing agreements to address shortcomings in local budgets, so local fiscal balance can decrease state taxes as well. Therefore, farmland preservation may mitigate the "crowding out" effect of taxation and fiscal deficits, leading to more investment in local business. (Crowding out refers to government deficit spending competing with and limiting private business investment, in part through higher interest rates. It is not a universally accepted macroeconomic concept.) Lower taxes increase consumer spending power which leads to higher personal income. Higher income has been shown to impact likelihood of food purchases from local farms positively.[15,16]

Loop 2. CFS → (+) Access to Fruits and Vegetables → (+) Quality of Diets → (+) State of Community Health → (–) Health Care Costs (+) Health of Economy → (+) Personal Income → (+) $ for Food → (+) Support for CFS → CFS

Another hallmark of CFSs is the increasing access and availability of healthy foods such as fruits and vegetables to areas often lacking food choices (e.g., the "food deserts" of inner cities). Improving access to high quality, healthy foods will help to increase their consumption, and improve the health of community members. Efforts in Detroit, Rochester, New York, Philadelphia and Chicago are helping to bring fresh produce to underserved areas.[17–20] Diet-related diseases are a major public health problem in Michigan, whereas "eat better" is one of the steps in the state Surgeon General's Healthy Michigan plan. *The Detroit News* states that obesity costs Michigan residents $3 billion per year in increased health care expenses.[21] Michigan State University Extension states that the cost of treating chronic disease has increased by about 8 percent annually over recent years.[22]

Healthcare costs are creating difficulties for Michigan businesses; decreasing their costs would improve the business climate for firms of various sizes. A survey conducted for the Small Business Association of Michigan finds that rising health care costs have caused businesses to hire fewer employees.[23] A total of 96 percent of surveyed businesses were strongly concerned about the high costs of health care premiums; 24 percent say the high costs threaten their ability to stay in business. According to the U.S. Small Business Administration, small businesses employ about half of all private sector workers, pay about 44 percent of total payroll and generate 60–80 percent of net new jobs annually. The difficulties caused by high healthcare costs are not limited to small businesses: they negatively impact the automobile industry so critical to Michigan, as well as other corporations.[24]

Decreasing healthcare costs would help businesses of all types, spur employment and increase income, while providing a better business environment to recruit new employers. All this will improve the economy, increase personal income and spawn purchases from CFSs. Finally, obesity and poverty are closely linked: in the United States, incidence of obesity and type 2 diabetes falls disproportionately on low-income people.[25] So, conversely, with no intervention, poor diets lead to higher healthcare costs, decreasing income and employment, thus making purchases of healthy food more difficult.

Loop 3. CFS → (+) Local Purchases → (+) Viable Farms → (+) Farmland Preservation → (–) Sprawl → (+) Exercise → (+) State of Community Health → (–) Health Care Costs (+) Health of Economy → (+) Personal Income → (+) $ for Food → (+) Support for CFS → CFS

Farmland preservation is a component in efforts to limit sprawl. According to the American Farmland Trust between 1992 and 1997, Michigan lost farmland at a rate 67 percent higher than over the previous five years.[26] Much of the land being lost is the best farmland. Furthermore, it is an uncontrolled growth that causes this loss:

> From 1982 to 1997, U.S. population grew by 17 percent, whereas urbanized land grew by 47 percent. Over the past 20 years, the acreage per person for new housing almost doubled and since 1994, 10+ acre housing lots have accounted for 55 percent of the land developed.[27]

Sprawl, in turn, is associated with unhealthy lifestyles. Residents living in sprawling counties are likely to walk less, weigh more and have greater prevalence of hypertension than those living in compact counties.[28] Farmland is seen as a key to livable communities for many Michiganders. According to a study by the University of Michigan's Institute for Social Research, 65 percent of surveyed residents believed that preserving farmland would improve the quality of life for future generations. Livable communities are important in their own right, and contribute to attracting and retaining bright, educated and capable people needed to own and run businesses.[29] Farmland brings a variety of other benefits (ecosystems services such as water filtration, carbon sequestration and biodiversity habitat; aesthetic beauty of open space) which are important but whose discussion is beyond the scope of this chapter.

Sprawl and the disappearance of farmland are complex issues, with no single or simple solutions. The manner in which our society and markets undervalue farmland relative to housing or business development is a fundamental driver of farmland loss. It can be seen as a market failure that the pubic goods supplied by a farm as opposed to other uses (e.g., ecosystem services) are not reflected in the price a retiring farmer faces when deciding to sell the land for development versus to a farmer.[30] There are many strategies to combat sprawl, with farmland preservation being just one part. To imply that CFSs can single-handedly save farms and end sprawl is grossly oversimplified. However, we believe the connection between food systems, farmland, sprawl and public health is meriting of further attention; illustrating the broader benefits of farmland conservation can be a key motivation for better land use policy.

Loop 4. CFS → (+) Local Multipliers and Value-added → (+) Health of Economy → (+) Personal Income → (+) $ for Food → (+) Support for CFS → CFS.

CFS can benefit local economies directly and indirectly. For example, farmers' markets, a key market channel of CFSs, have been found to be important business incubators, where farmers can try new ideas in a low-risk manner.[31] Farmers' markets draw additional consumers to downtown areas and spur patronage of other downtown shops and restaurants.[32] Food purchases from local businesses also increases the economic multiplier effect of these purchases and decrease municipal service costs such as police and road maintenance, compared, for example, with purchases from "Big Box" grocery retailers and fast food restaurants.[33] Local businesses can provide stable, well paying jobs and income.[33] Finally, smaller scale farms, which are commonly engaged in direct markets, benefit their local economies: several studies suggest that smaller scale farms tend to buy more from local businesses than do larger farms.[34–37]

Interventions

The overarching intervention is to change the mindset of consumers, citizens and government, away from the industrialized or commodity-based "cheap food" paradigm, toward the community-based one: in other words, to change our culture about and relationship with food. One good place to start is to apply a popular framework in the social sciences that helps gain insight into the change process, called the Stages of Change Theory.[38] This theory has been applied successfully to changing behaviors such as food habits and preference, and might be instrumental in the process of changing our association with food in a more healthy direction.[39,40] A list of interventions and initiatives, from the practical to the fantastic, which can begin to facilitate this transition follows.

Make Fresh, Locally Grown Food Accessible in Every Part of the Country

Many areas, especially some inner cities, are "food deserts," places with no easy access to grocery stores, where most food comes from liquor stores, pharmacies, etc. The "redlining" of these areas by supermarkets has been the topic of some research and discussion.[41,42] Redlining, originally associated with lenders, realtors and insurance providers discriminating racial and ethnic minorities, is now also used to describe food retailers' unwillingness to locate in poor inner city neighborhoods (see Heany and Hayes[42]).

Although bringing fresh foods to underserved areas has no single magic solution, examples of success stories here in Michigan and elsewhere include farmers' markets and buying clubs established by community members in poor urban neighborhoods that accept Electronic Benefit Transfer cards (Food Stamps) and WIC Farmers Market coupons. Tax incentives, tied to incentives to buy locally grown and processed foods, could be used to attract supermarkets to poor neighborhoods.

Make Local Food Outlets Accessible and Authentic

Listings of farmers' markets U-pick operations and farm stands, including locations and hours of operations, can be made widely available on Websites and in public buildings, through public service announcements and other such means. Farmers' markets can be incorporated into public spaces; private businesses and civic and religious organizations can encourage and host local food markets, buying clubs, etc. Information on how to participate in Community-Supported Agriculture programs, youth farm stands, and community gardens can also be listed.

Farmers' markets could become more one-stop shopping locations by allowing sales of meats, wine, processed foods, food vendors, etc. Anecdotes from farmers' markets in Michigan imply that market managers and farmers have difficulty understanding and complying with safety regulations regarding meat sales and/or giving samples to customers. Planck[43] suggests samples as a way to get customers to try new items. Research is needed to balance easing burdensome regulations while guaranteeing public safety; health inspectors can make a priority of helping vendors comply with regulations. Markets can include music and other entertainment to add to the social appeal.

Get Government Involved in a Positive Way

Government has played a huge role in creating the current industrialized, commodity-based food system, in numerous direct and indirect ways.[44] The Federal Government directly subsidizes feed grains and inputs to processed foods; it subsidizes highways to transport food over vast distances. It exempts agriculture from many occupational safety laws. Many observers believe concentration in the food industry damages competition, increases consumer prices and decreases farm income, yet anti-trust legislation has not corrected this problem.[45–47] Correcting these distortions would be an excellent start, although politically difficult. Two access points are submitting comments to USDA proposed rules and pressuring legislators to pass more CFS-friendly Farm Bill regulations.

State and local governments can also make supporting local food systems a priority. Increased funding to Buy Local campaigns like "Select Michigan" would raise the visibility of local foods. Numerous other states fund efforts to promote locally grown foods.[48] Leaders could use the bully pulpit to encourage support of local food systems. Government-funded institutions, from schools to jails to office cafeterias, should make buying local foods a priority. Many government and private institutions have committed to buying local food, including hospitals, colleges and universities, county governments, corporate cafeterias, and K-12 schools.[49–52]

Food Policy Councils (FPCs) are a good way to catalyze food system development efforts by gathering a broad array of stakeholders to focus on policy initiatives that will enhance the viability and performance of the food system. They are often organized on the state or city level. State examples include those in Michigan, Iowa and Oklahoma; cities such as Toronto and Hartford, CT also have FPCs.

Implement Comprehensive Food System and Nutrition Education

Concerned organizations such as Center for Science in the Public Interest have claimed that junk food advertising targeted at children has fueled the childhood obesity epidemic.[53] Certainly, companies selling these foods have more money and resources to garner children's attention than do nutrition educators serving up the counter-message. Nonetheless, there are success stories: anecdotal evidence implies farm to school, school gardens, field trips to farms, etc., can increase students' consumption of healthy foods.[54]

Our society has recognized the threat of tobacco smoking; we would never think of allowing advertisements aimed at children or cigarette machines in schools. We do not allow schools to be a venue for guns, pornography, gambling or alcohol, yet we often stand by while junk food is sold in schools, advertised on children's television shows, and so forth. However, owing to increased citizen awareness and activism, the tide may be turning: California passed a law limiting the fat and sugar content of food items sold in school vending machines or stores.[55] The soda industry has imposed restrictions on sales of sugary drinks in schools, in response to the threat of lawsuits and state legislation.[56,57]

Although schools require civics lessons to foster good citizenship, very little is done to create responsible consumers. Much is made of consumer freedom, but little is made of the associated responsibility. The social, economic, health, environmental and other ramifications of our purchasing decisions should be researched, taught and brought to the forefront of public dialogue.

Build Skills and Assets for Entrepreneurial Agriculture

University/extension services and government could play a key role in CFS development by placing higher priority on smaller scale, locally based farms and farmers. Specifically, efforts can be focused toward improving access to training, credit and land. Extension educators could prioritize practical education (including, for example, pasture-based livestock production, season extension for vegetables, business management and marketing) for farmers wishing to transition to community-based food markets. Capital access for beginning and transitioning entrepreneurial farmers can be increased by low interest loans, as well as Individual Development Accounts and other innovative credit programs (for description of Individual Development Accounts, see www.alternatives. org/ida.html).

Several states (e.g., CA, ME, MI, NJ, NY, PA, VT) have farm link programs that connect people seeking farmland with those looking to sell, lease or transfer land. Establishing farm link programs in all states, with high visibility, commitment and funding from public and private entities (state departments of agriculture, Farm Bureau, cooperative extension) would help expand their scope and effectiveness, maintaining communities' capacity to grow food for their people (see www. smallfarm.org/nell/index.htm for a list of programs in the northeast).

Leverage Points and Key Alliances

The purpose of this chapter is to highlight the connections between diverse sectors. Accordingly, building community-based food systems cannot go forward without broad support and alliances between diverse sectors.

State government agencies, such as public health, economic development, environmental protection and education departments are all vital stakeholders, as are universities and extension. Key motivations for action include diet-related illness, sprawl and other land use issues, environmental degradation, unemployment and poverty, declining farm numbers and rural community decline.

The industrial commodity-based food system is driven by survival of the high volume and/or low-cost producer. Leaving alone the socioeconomic, ecological and public health damage this system has brought, it is a poor long-term strategy:

1. It drives out of business the truly indispensable link: farmers.[58]
2. It relies on cheap petroleum to grow, store and transport food, a trend many believe has an increasingly tenuous future.[44,59–61]
3. It is difficult to imagine Michigan producers competing with those in nations like China and Brazil entering the market with lower land and labor costs, and longer seasons.

On the other hand, promoting Michigan products as something truly different, coming from a place with high standards of ecological, animal, labor and community stewardship and creating brand loyalty on that basis mitigates the race to the bottom mentality and lays the foundation for long-term prosperity for agriculture.

Counterarguments

No Money on the Table

There is an adage in economics that there is "no money on the table." If there were economic incentives and opportunity for community-based food systems, some smart entrepreneur would have already snapped it up. Some supermarkets redline poor areas for good reason: they don't make money there.

Certainly, our food system is set up so that food follows money, not hunger or need. Government and private charitable interests have roles in filling gaps left by the private sector. The alternative, allowing hunger and malnutrition to go unabated, likely costs far more in external costs than would measures to kick-start the local food system. Furthermore, poor people must and do eat; and to capture these dollars (often in the form of food stamps) within the local community, rather than having many of them leak out, would provide a boost to struggling economies. Finally, the current system is heavily dependent on government subsidies, which may not persist in the face of rising fiscal deficit and trade agreements.

Consumers Have Spoken Already

One might argue that our food system exists in its present form because this is what people want. The sum total of people voting with their dollars has created the system people want. There is some truth here. However, this argument ignores the huge subsidies that support the growth of inputs for processed foods, grain-fed meats, etc. A recent editorial in the *Los Angeles*

Times spelled out how federal subsidies tend to support inputs for unhealthy processed foods, such as corn and soybeans, instead of making fruits and vegetables more accessible and affordable.[62] A report by the Institute for Agriculture and Trade Policy supports this view.[63] The environmental costs of chemical intensive production and transportation are not reflected in the food's price. In short, while the food system is the result of decisions consumers have made in the market place, it is a market place skewed in countless ways to favor the industrialized commodity-based status quo.[44] Part of the challenge of creating community-based food systems will be leveling the playing field.

People Don't Want to Change; Will Keep Eating Unhealthy Foods and Buy Them Elsewhere; Banning Junk Food from Schools Will Decrease School Revenues

Certainly, people's food choices are guided by innumerable factors, and simply making healthy food more available will not change people's behaviors overnight. But public health initiatives have succeeded in decreasing consumption of tobacco in the United States. A recent study by the Centers for Disease Control and prevention and U.S. Departments of Agriculture and Education finds that a majority of schools experienced increased school revenues as a result of initiatives to make healthy food choices more available to students.[64] It would be interesting to model the cost-benefit ratio of school revenue versus, for example, heathcare costs. Michigan's obesity epidemic has been estimated to cost every citizen, on average, $300 per year.[20] Although it is difficult to attribute what percentage of that figure is caused, directly and indirectly, by a single policy decision such as allowing soda machines in schools, the costs of America's epidemic of diet-related disease are staggering in simple economic terms and demonstrate that quick fixes like junk food revenues to schools are no bargain in the long run.

Certainly, people will continue to eat unhealthy foods in some quantity. However, raising the consciousness about diet's role in our health care crisis and limiting children's exposure to advertising and access may prevent bad eating habits from forming at young and impressionable ages.

Countervailing Forces

Change always involves winners and losers. Forces opposed to a transition to a community-based food system are those who benefit from the current system: agribusiness input suppliers, large-scale farmers, food manufacturers, distributors, supermarkets, and the fast food industry, among others; all have a stake in the status quo. Others who stand to lose include land developers, advertisers, and convenience/liquor/drug store operators who sell food in poor neighborhoods. Poverty, substance abuse, crime and apathy in poor neighborhoods also make lasting changes difficult.

It is important to acknowledge the daunting nature and enormity of the task of creating CFSs. The lags between stages in the positive loops (involving improved diets, improved health, decreased healthcare costs, improved economies and increased personal income, etc.) require perseverance when results are slow to manifest. Finally, the negative forces that oppose or dampen the above-discussed positive loops are present and, most likely, are currently better funded than community groups, local governments and other agents advocating CFS. These inhibiting forces that counter growth of CFSs can be represented as negative (balancing) loop structures (see Senge).[65]

Unintended Consequences

An important unintended consequence of food-based community development through CFSs is a form of "gentrification." As CFSs become established in low-income urban neighborhoods, they may become hubs of local economic activity, drawing in people from outside the neighborhood and crowding locals out of the food, job and housing markets. Effort must be made to increase local residents' ownership in community assets, to ensure they benefit from progress they help create. In general, communities must guard against the danger of co-option of success by stakeholders and beneficiaries of the industrial commodity-based food systems.

Future Directions of Research: A Systems Approach

Finding the dynamic interactions among such components as community health, farmland preservation, and economic well-being demands a systems approach. System dynamics (SD) and a soft systems approach[66–68] have been utilized to gain insight and understanding of complex issues at a variety of levels and settings, such as regional and community levels[69,70] and at the organizational level.[71,72]

Indeed, one viable future direction for this framework is to develop an SD simulation model[7,73] that would embrace the four positive loops and would, in addition, include some of the processes described in the section on counterarguments. These processes might greatly inhibit the growth of CFS. Forces that would inhibit the growth can be represented by a set of negative loop processes that attempt to maintain the status quo. The advantages of developing a full-blown: SD analysis are: (1) The model can help determine which specific loops dominate the system over time, and (2) insight can be gained by getting the stakeholders together and testing the policy options.[74] In addition, the relationships between variables do not have to be straight lines. Indeed, in SD models, relationships between variables can be specified as curves (curvilinear) to conform to reality.

The model can be quite helpful in finding a pathway to CFS that might include a "better" mixture of CFS and traditional industrialized commodity food systems. Finally, the model would attempt to address the thorny problem of gentrification, by finding levers of change around local residents' ownership in community assets.

Conclusions

This chapter creates a conceptual framework that demonstrates the interconnectedness of food, farming, land use, economics and health. Its purpose is to stimulate further research and discussion into interventions and initiatives that have great potential to address many of the greatest ills facing states such as Michigan—unemployment and lack of economic opportunity, diet-related illness and obesity, farmland loss and sprawl, access and availability of healthy foods—by creating community-based food systems. The appeal in this framework is the virtuous cycle, or positive feedback loop, aspect: progress in one area can be self-reinforcing in ways discussed in this chapter. Currently, much of our food system is constrained by negative loops, where the industrialized food system is at the center of cycles marked by unemployment, diet-related illness, farmland loss and other conditions.

The list of interventions, leverage points, counterarguments, countervailing forces and unintended consequences is preliminary and far from being exhaustive. Furthermore, we acknowledge that much of the evidence we cite is anecdotal, informed opinion and/or not from peer-reviewed journal sources. We hope that this chapter will encourage compilations of success and "lessons learned" stories from all areas of the United States, spurring research that will guide the development and operation of CFSs that capitalize on opportunity and minimize unintended and negative consequences. It leaves more questions unasked and unanswered than it addresses. However, we hope it is a rallying point for academics, policymakers, economic developers, community members, agricultural professionals and others to form alliances and create the political will to put food at the center of community development, public health and economic decisions.

Acknowledgment

This chapter originally appeared in *Journal of Hunger and Environmental Nutrition*, 1, 5–25 (2006).

References

1. Peterson HC, Knudson WA, Abate G. The Economic Impact and Potential of Michigan's Agri-Food System. Strategic Marketing Institute Working Paper, 1–1606. Michigan State University. 2006. Available at: http://aec.msu.edu/productdocuments/Working/Economic%20Impact%20of%20Michigan%20Agri-Food%20Final%20010906. pdf. Accessed July 19, 2006.
2. Wilkerson R. ACES Goes Island Hopping for A Fit Michigan. Michigan Department of Community Health. 2006. Available at: www.michigan.gov/mdch/0,1607,7–132–2939–141309–,00.html. Accessed July 19, 2006.
3. Land: 2002 and 1997. USDA National Agricultural Statistics Service: Washington DC. 2002 (Table 8). Available at: www.nass.usda.gov/census/census02/volume1/mi/st26_1_008_008.pdf. Accessed July 19, 2006.
4. Michigan Quick Facts from the US Census Bureau. US Census Bureau: Washington DC. 2005 Available at: http://quickfacts.census.gov/qfd/states/26000.html. Accessed July 19, 2006.
5. Eden C, Jones S, Sims D. *Messing About in Problems*. New York: Pergamon; 1983.
6. Wolstenholme, EF. *System Enquiry*. West Sussex, UK: John Wiley & Son; 1990.
7. Sterman JD. *Business Dynamics: Systems Thinking and Modeling for a Complex World*. Boston, MA: McGraw-Hill; 2000.
8. Peters J. Community Food Systems: Working Toward a Sustainable Future. *Journal of the American Dietetic Association*. 1997; 97(9): 955–957.
9. Heller M. (2005). Food Connection: Capitol Area Community Food Profile. C.S. Mott Group for Sustainable Food Systems at Michigan State University. Available at: www.mottgroup.msu.edu/downloads/CACfoodprofile.pdf. Accessed July 19, 2006.
10. Stevenson G, Lev L. Common Support for Local Agriculture in Two Contrasting Oregon Communities. *Renewable Agriculture and Food Systems*. 2004; 19(4): 210–217.
11. Kambara K, Shelley C. The California Agricultural Direct Marketing Study, USDA AMS and California Institute of Rural Studies. 2002. Available at: www.ams.usda.gov/directmarketing/CIRSDM Report.pdf. Accessed July 19, 2006.
12. McCauley MA. Buying Local Food: The Smart Alternative. Oxfam America. Available at: www.oxfamamerica.org/whatwedo/where_we_work/united_states/news_publications/food_farm/art 2564.html. Accessed July 19, 2006.
13. Adelaja S. Preserving Working Landscapes Through Farm and Forest Viability. Michigan State University Land Policy Program. Available at: www.landpolicy.msu.edu/ongoing/ag_viability.ppt#335. Accessed July 19, 2006.
14. Fact Sheet: Cost of Community Services Studies. American Farmland Trust Farm Information Center. 2004. Available at: www.farmlandinfo.org/documents/27757/FS_COCS_8–04.pdf. Accessed July 19, 2006.
15. Brown C. Consumers' Preferences for Locally Produced Food: A Study in Southeast Missouri. *American Journal of Alternative Agriculture*, December 2003; 18(4): 213–224.
16. Jekanowski M, Williams D, Schiek W. Consumers' Willingness to Purchase Locally Produced Agricultural Products: An Analysis of an Indiana Survey. *Agricultural and Resource Economics Review*. 2000; 29: 43–53.

17. Score M. Process and Success: Establishing an Inner City Food Buying Club. Welcome to the Feast. *Food System Economic Partnership Newsletter.* 2006; 1(1, Winter/Spring 2006). Available at: www.fsepmichigan.org/newsletter/v1_i1.pdf. Accessed July 19, 2006.
18. Aboelata M. The Built Environment and Health. Prevention Institute. Prevention Institute. 2004. Available at: www.preventioninstitute.org/pdf/BE_Rochester_NY.pdf. Accessed July 19, 2006.
19. Supermarket Campaign. The Food Trust. Available at: www.thefoodtrust.org/php/programs/super.market.campaign.php#2. Accessed July 19, 2006.
20. Redmond, L. Creating Local Food Options in an Urban Setting. *The New Farm* [serial online]. November 2004. Available at: www.newfarm.org/features/1104/urban_farm/. Accessed July 19, 2006.
21. Upton J. Obesity Surge Costs Michiganians $3 billion. *The Detroit News*, March 21, 2004. Available at: www.detnews.com/2004/specialreport/0403/23/a01–98368.htm. Accessed July 19, 2006.
22. Outcomes for Michigan's Future: Reducing Health Care Costs for At Risk Populations. Michigan State University Extension. 2005. Available at: http://web1.msue.msu.edu/home/outcomes/omf papers/HealthCare11.pdf. Accessed July 19, 2006.
23. Hamilton D. Don't Get Sick. *Future Magazine* [serial online]. Winter, 2003. Available at: www.futuremagonline.com/archive/2003/winter/4.html. Accessed July 19, 2006.
24. Bailey A. Ford says Rising Health Care Costs Trouble for Automaker. *The Detroit News*, May 31, 2003.
25. Drewnowsi A, Specter SE. Poverty and Obesity: The Role of Energy Density and Energy Costs. *American Journal of Clinical Nutrition* 2004; 79: 6–16.
26. Farming on the Edge Report: Listing of Loss by State. American Farmland Trust. Available at: www.farmland.org/resources/fote/states/allStates.asp. Accessed July 19, 2006.
27. Farming on the Edge Report: What's Happening to Our Best Farmland? American Farmland Trust. Available at: www.farmland.org/resources/fote/default.asp. Accessed July 19, 2006.
28. Ewing R, Schmid T, Killingsworth R, Zlot A, Raudenbush S. Relationship Between Urban Sprawl and Physical Activity, Obesity, and Morbidity. *American Journal of Health Promotion* [serial online]. 2003; 18(1): 4757. Available at: www.activelivingbydesign.org/fileadmin/template/documents/killingsworth pubs/Ewing_AJHP_2003.pdf. Accessed July 19, 2006.
29. Some Stress, much Satisfaction in Quality of Life in the Detroit area. University of Michigan News Service. 2002. Available at: http://ipumich.temppublish.com/cgi-bin/print.cgi?Releases/2002/Mar02/chr032002a. Accessed July 19, 2006.
30. Antle JM, Valdivia RO. Modelling the Supply of Ecosystem Services from Agriculture: A Minimum-Data Approach. *Australian Journal of Agricultural and Resource Economics.* 2006; 50(1): 1–15.
31. Lyson TA, Gillespie GW, Hilchley D. Farmers' Markets and the Local Community: Bridging the Formal and Informal Economy. *American Journal of Alternative Agriculture.* 1995; 10(3): 108–113.
32. Lev L, Stephenson G. Analyzing Three Farmers Markets in Corvallis and Albany, Oregon. Oregon State University Extension Service, Small Farm Working Group, October 1998. Available at: http://smallfarms.oregonstate.edu/publications/techreports/TechReport3.pdf#search=%22Analyzing%20Three%20Farmers%20Markets%20in% 20Corvallis%20and%20Albany%2C%20Oregon%22. Accessed July 19, 2006.
33. Rockne J, Milchen J. Local Ownership Pays Off for Communities. Reclaim Democracy. Available at: 2003. www.reclaimdemocracy.org/independent_business/local_ownership_pays.html. Accessed July 19, 2006.
34. Chism J, Levins R. Farm Spending and Local Selling: How Do They Match Up? *Minnesota Agricultural Economist.* 1994; 676: 1–435.
35. Foltz, J, Jackson-Smith, D, Chen, L. Do Purchasing Patterns Differ Between Large and Small Dairy Farms?: Econometric Evidence From Three Wisconsin Communities. *Agricultural and Resource Economic Review.* 2002; 31: 28–38.
36. Lawrence J, Otto D, Meyer S. Purchasing Patterns of Hog Producers: Implications for Rural Agribusiness. *Journal of Agribusiness.* 1997; 15(1): 1–18.
37. Marousek G. Farm Size and Rural Communities: Some Economic Relationships. *Southern Journal of Agricultural Economics.* 1979; 11(2): 57–61.
38. Prochaska JO, Norcross JC, DiClimente CC. *Changing for Good.* New York: William Harrow & Company; 1994.
39. Greene GW, Horacek TM, White A, Ma J. Use of a Diet Interview Method to Define Stages of Change in Young Adults for Fruit, Vegetable and Grain Intake. *Topics in Clinical Nutrition.* 2003; 18: 32–41.

40. Glanz K, Patterson RE, Kristal AR, DiClemente CC, Heimendinger J, Linnan L, McLerran DF. Stages of Change in Adopting Healthy Diets: Fat, Fiber, and Correlates of Nutrient Intake. *Health Education Quarterly*. 1994; 21(4): 499–519.
41. Perry, D. Food for Every Child: The Need for More Supermarkets in Philadelphia. The Food Trust. 2005. Available at: *www.thefoodtrust.org/pdf/supermar.pdf* Accessed July 19, 2006.
42. Heaney J, Hayes, J. Redlining Food: How to Ensure Community Food Security. Food First. Available at: *www.foodfirst.org/node/1452* Accessed July 19, 2006.
43. Planck, N. Some Thoughts on Selling at Farmers' Markets. The New Farm [serial online]. May 2004. Available at: www.newfarm.org/features/0504/farmmarkets/index.shtml Accessed July 19. 2006.
44. Conner DS. Expressing Values in Agricultural Markets: An Economic Policy Perspective. *Agriculture and Human Values*. 2004; 21(1): 27–35.
45. Harl NE. Antitrust Issues in the New Food System. Presented at American Agricultural Economics Association Workshop, "Policy Issues in the Changing Structure of the Food System," Tampa, Florida, July 29, 2000. Available at: www.farmfoundation.org/tampa/harl.pdf. Accessed July 19, 2006.
46. Johnson PD. Our Agriculture Needs a Public Threshing. The Land Institute. 2003. Available at: www.landinstitute.org/vnews/display.v/ART/3edd1a7a65539 Accessed July 19, 2006.
47. Rosset P. Food Sovereignty: Global Rallying Cry of Global Movements. Back-grounder [serial online]. 2003; 9(4): 1–4. Food First Institute. Available at: www.foodfirst.org/pubs/backgrdrs/2003/f03v9n4.html. Accessed July 19, 2006
48. Patterson PM. State Grown Promotion Programs: Fresher, Better? Choices [serial online]. 2006; 21(1). Available at: www.choicesmagazine.org/2006–1/grabbag/2006–1-08.pdf Accessed July 19, 2006.
49. Kulick M. Healthy Food, Healthy Hospitals, Healthy Communities. Institute for Agriculture and Trade Policy, Food and Health Program. 2005. Available at: www.environmentalobservatory.org/library.cfm?refid=72927 Accessed July 19, 2006.
50. Pirog R. Institutional Buying Models and Local Food Markets: The Iowa Experience. Presented to Farm to Cafeteria: Healthy Farms, Healthy Students Conference Seattle, Washington October 5, 2002. Leopold Center. 2002. Available at: www.leopold.iastate.edu/pubs/speech/files/100502_cafeteria.pdf Accessed July 19, 2006.
51. Farm to School. Available from: www.farmtoschool.org/. Accessed July 19, 2006.
52. Marqusee R. Iowa County Adopts Organic/Local Food Purchase Policy. Organic Consumers Association. 2006, Organic Consumers Association: Finland, MN. Available at: www.organicconsumers.org/organic/iowa011006.cfm. Accessed July 19, 2006.
53. CSPI Hits Marketing Junk Food to Kids. Center for Science in the Public Interest. 2003. Available at: http://cspinet.org/new/200311101.html. Accessed July 19, 2006.
54. Vallianatos M, Gottlieb D, Haase M. Farm-to-School Strategies for Urban Health, Combating Sprawl, and Establishing a Community Food Systems Approach. *Journal of Planning Education and Research*. 2004; 23: 414–423.
55. BBC News (2005). California Bans School Junk Food. September 15, 2005. Available at: http://news.bbc.co.uk/2/hi/americas/4251928.stm. Accessed July 19, 2006.
56. Burros M, Warner M. (2006). Bottlers Agree to a School Ban on Sweet Drinks. *The New York Times*, May 4, 2006. Available at: www.nytimes.com/2006/05/04/health/04soda.html?ex=1304395200&en=80149e91b4b733bb&ei=5088&partner=rssnyt&emc=rss. Accessed July 19, 2006.
57. Otto M, Aratani L. Soda Ban Means Change at Schools. *Washington Post*. May 4, 2006; Page B03. Available at: www.washingtonpost.com/wp-dyn/content/article/2006/05/03/AR2006050302399.html. Accessed July 19, 2006.
58. Olson RD. Below-Cost Feed Crops Indirect Subsidy for Industrial Animal Factories. Institute for Agriculture and Trade Policy Trade and Global Governance Program. 2006. Available at: www.agobservatory.org/library.cfm?refid=88122. Accessed July 19, 2006.
59. Duffy M, Smith D. Rising Energy Prices and Iowa Farmers. Leopold Center, Iowa State University. 2005. Available at: www.leopold.iastate.edu/pubs/staff/files/energy_impact0405.pdf. Accessed July 19, 2006.
60. Kunstler JH. End of the Binge. The American Conservative [serial online]. September 12, 2005. Available at: www.amconmag.com/2005/2005_09_12/cover.html. Accessed July 19, 2006.
61. Murray D. Rising Oil Prices Will Impact Food Supplies. People and Planet. 2005. Available at: www.patp3.webbler.co.uk/doc.php?id=2532&PHPSESSID=36cd6c4b35809dc23c9e8871b391ee05. Accessed July 19, 2006.

62. Quaid L. Feds Subsidizing Obesity, Not Fruits and Vegetables. Organic Consumers Association. 2005. Available at: www.organicconsumers.org/ofgu/obesity081105.cfm. Accessed July 19, 2006.
63. Schoonover H, Muller M. Food Without Thought: How US Farm Policy Contributes to Obesity. Institute for Agricultural and Trade Policy. 2006. Available at: www.iatp.org/iatp/publications.cfm?accountID=421&refID=80627. Accessed July 19, 2006.
64. Making It Happen, School Nutrition Success Stories. Food and Nutrition Service, U.S. Department of Agriculture; Centers for Disease Control and Prevention, U.S. Department of Health and Human Services; and U.S. Department of Education. FNS-374, Alexandria, VA, January 2005. Available at: www.cdc.gov/healthyyouth/nutrition/Making-It-Happen. Accessed July 19, 2006.
65. Senge PM. *The Fifth Discipline: The Art and Practice of the Learning Organization.* New York: Doubleday; 1990.
66. Lane D. With a Little Help from Our Friends: How System Dynamics and Soft OR Can Learn from Each Other. *System Dynamics Review.* 1994; 10: 101–132.
67. Checkland P, Scholes J. Eds. *Soft Systems Methodology in Action.* 1999; Chichester: Wiley.
68. Habron G, Kaplowitz M, Levine RL. A Soft Systems Approach to Watershed *Management: A Road Salt Case Study. Environmental Management. 2004;* 33(6): 776–787.
69. Ford A. *Modeling the Environment: An Introduction to System Dynamics Modeling of Environmental Systems.* Washington, DC: Island Press; 1999.
70. Van den Belt M. *Mediated Modeling: A System Dynamics Approach to Environmental Consensus Building.* Washington, DC: Island Press; 2004.
71. Levine, RL. The Dynamics of the Transition from Being a Supervisor to a Team Leader. Presented to the International System Dynamics conference, Atlanta, GA. July, 2001 (on CD).
72. Garcia R, Calantone R, Levine RL. The Role of Knowledge in Resource Allocation to Exploration vs. Exploitation in Technologically Oriented Organizations. *Decision Sciences.* 2003; 34: 323–349.
73. Coyle RG. *System Dynamics Modelling: A Practical Approach.* London: Chapman & Hall; 1996.
74. Vennix JA. *Group Model Building: Facilitating Team Learning Using System Dynamics.* New York: John Wiley; 1996.

28

USING SYSTEM DYNAMICS MODELING TO UNDERSTAND THE IMPACT OF SOCIAL CHANGE INITIATIVES

Gary B. Hirsch, Ralph Levine, and Robin Lin Miller

Research Question: What factors promote or hinder curriculum innovation and the adoption of new curricula?
System Science Method(s): System dynamics
Things to Notice:

- Use of system dynamics to inform interventions
- Model building as a stepwise process

Community psychologists have a long history of interest in understanding social systems and how to bring about enduring positive change in these systems. However, the methods that community psychologists use to anticipate and evaluate the changes that result from system change efforts are less well developed. In the current chapter, we introduce readers to system dynamics modeling, an action research approach to studying complex systems and the consequences of system change. We illustrate this approach by describing a system dynamics model of educational reform. We provide readers with an introduction to system dynamics modeling, as well as describe the strengths and limitations of the approach for application to community psychology.

From its earliest days, community psychology put complex community systems at the front and center of its psychological science and action (Sarason, 2000). Salient themes in our field highlight our closely held commitment to understanding how social systems affect individual lives and how to bring about enduring positive change in these systems to benefit individuals. Located throughout our theory and practice are the core tenets of an orientation toward system change. Our theoretical frameworks (e.g., Kelly, 1968, 1987; Seidman, 1988; Trickett, Kelly, & Vincent, 1985) provide blueprints for thinking about how social systems structure individual lives, how community processes are sustained, and where levers of change might reside. Our action-oriented projects illustrate principles and tactics for advancing change among individuals by changing elements of families, schools, work places, neighborhoods, and communities.

The methods that community psychologists use to anticipate and evaluate the changes that result from system change efforts are less well developed than our interest in stimulating systems change might imply (see, for example, Luke, 2005). We continue to rely on methods that assume

 DOI: 10.4324/9781315748771-31

a very different kind of world than the one that is reflected in the settings in which we work. The unidirectional models we use to try to draw links between a set of variables and an outcome are not consistent with what we know about the complexity of the phenomena we hope to study. Rather than develop small-scale complex models of what we believe actually happens over time, we willingly suspend our disbelief that an uncomplicated, linear, and unidirectional snapshot fairly represents those processes of interest as they actually seem to unfold. In this way, we permit methods to obscure complexity by seeking to reduce a phenomenon to a simple set of linear pathways. We also limit our opportunity to understand paradoxical and counter-intuitive behavior, relegating these system behaviors to the black box of poor model fit. In choosing our tools, we too often elect to ignore the interdependencies among real world processes that might provide us insight into how best to approach the task of creating and evaluating system change.

In the current chapter, we introduce readers to an alternative approach to model development and assessment that community psychologists might use to gain insight about system change. The approach is called system dynamics modeling and is part of a larger class of systems approaches to understanding and solving complex problems. We first provide readers with an overview of the system dynamic modeling approach and describe its origins and potential applications. We describe how system dynamics modelers understand systems and change. We then describe how models are built and the potential for including stakeholders in the process. Next we illustrate the approach by presenting an example derived from the area of school reform. We examine how the system dynamics model was developed to provide insight on how to plan for a school-based system change effort around curricular innovation. We conclude by highlighting the strengths and limitations of the approach for community psychologists.

Systems, Systems Change, and System Dynamics

To help readers understand what system dynamics methods bring to the task of designing and evaluating system change, we first define the key terms, "system" and "system change," from the vantage point of system dynamics. We then highlight how a system dynamics expert thinks differently about modeling problems from a traditionally trained psychologist.

Defining the System in System Dynamics

A system is a functional whole, composed of a set of components, coupled together to function in a way that might not be apparent from the functioning of the separate component parts (Levine & Fitzgerald, 1992). Systems change deals with changing the root causes of a problem through actions, policies, and new infrastructure. Systems change is qualitatively different from changing the intensity of a few system components to make minor corrections when a system gets a bit out of line. Systems change occurs when there are substantial changes in the structural, relational, and institutional makeup of a system or its subsystems. From this standpoint, system change is akin to redesigning some or all of the major systems in a car; it is not akin to a tune-up to improve how well the car runs. From a system dynamics point of view, efforts to change systems carefully consider how components of a system are coupled together, and not just what the components are, for the coupling of components allows one to locate the root causes of problems. Also, by considering the coupling of system components, one can understand why at times systems change efforts produce no apparent change and at other times change occurs in directions that are counter to what we desire and intend.

System Dynamics Modeling and Thinking

System dynamics modeling is an action research approach to studying complex systems, such as families, organizations, and communities. System dynamics grew out of work by Forrester and others at the Massachusetts Institute of Technology who were trying to understand complex organizational behavior (see, for example, Forrester, 1961). Historically, system dynamics has focused on comparatively tangible processes that have discrete boundaries such as sales and the production of goods. As the field of system dynamics has developed, however, it has expanded in focus from internal organizational and intra-organizational dynamics to encompass the complex dynamic phenomenon of the kind that interests community psychologists. It has been used to explore problems such as highway congestion (Goodman, 1974; Sterman, 2000), the dynamics of urban growth and decline (Alfeld & Graham, 1976; Forrester, 1969), implementation of innovations (Repenning, 2002), community health status (Hirsch & Immediato, 1998, 1999; Homer, Hirsch, Minniti & Pierson, 2004), and human service delivery (Hirsch, 1976; Miller, Levine, Khamarko, Valenti, & McNall, 2006), among many other issues. In this sense, system dynamics has become a broadly applicable school of systems science that emphasizes both a system's behavior and the feedback mechanisms that are assumed to underlie a system's behavioral patterns.

System dynamics builds on diverse sources of data and on group process techniques to develop computer simulations that allow for virtual experimentation with system change policies. On the face of it, the process of using system dynamics to explore a problem appears quite simple. A modeler or model-building team develops a visual representation of a problem, specifying what processes they hypothesize give rise to problematic behavior in a system (e.g., high rates of student failure). The modeler or team then develops a set of mathematical equations to represent the model. Next, the team explores the model's behavior and various policy actions via computer simulation. By policy actions, system dynamics modelers refer to the operational policies and actions that individuals and groups use to attain goals. Finally, the team, having derived insights into the consequences of various policy actions, initiates real-world change efforts. This seemingly simple effort reflects a highly refined set of methodological procedures and very different way of thinking about model building than most psychologists are trained to do.

Basic Elements of System Dynamics Models

Before we describe how models are built, we highlight critical elements of system dynamic models to introduce readers to its basic vocabulary and principles. A comprehensive introduction is beyond the scope of this chapter, but many excellent texts (e.g., Sterman, 2000) and training programs exist to provide interested readers with a more thorough introduction.

System dynamics modelers think in terms of feedback processes to account for problematic behavior patterns. The notion of causation flowing in one direction is pervasive among the social sciences. A vast majority of conceptual and statistical tools, from logic models to structural equation models, almost always go through casual chains that move in one direction. On the other hand, systems thinking stresses chains of reciprocal, causal relations among the variables. To system dynamics modelers, loops that form a nexus of closed relationships become important units of analysis per se, having a purpose and differing in importance or dominance over time. The loop becomes a higher conceptual unit than the variables that make up the circular chain. To the system dynamics modeler, a given variable may be in more than one feedback loop. Consider a variable that is common to two loop processes. The first loop may come around to affect the common variable in a positive way, while at the same time a second loop may

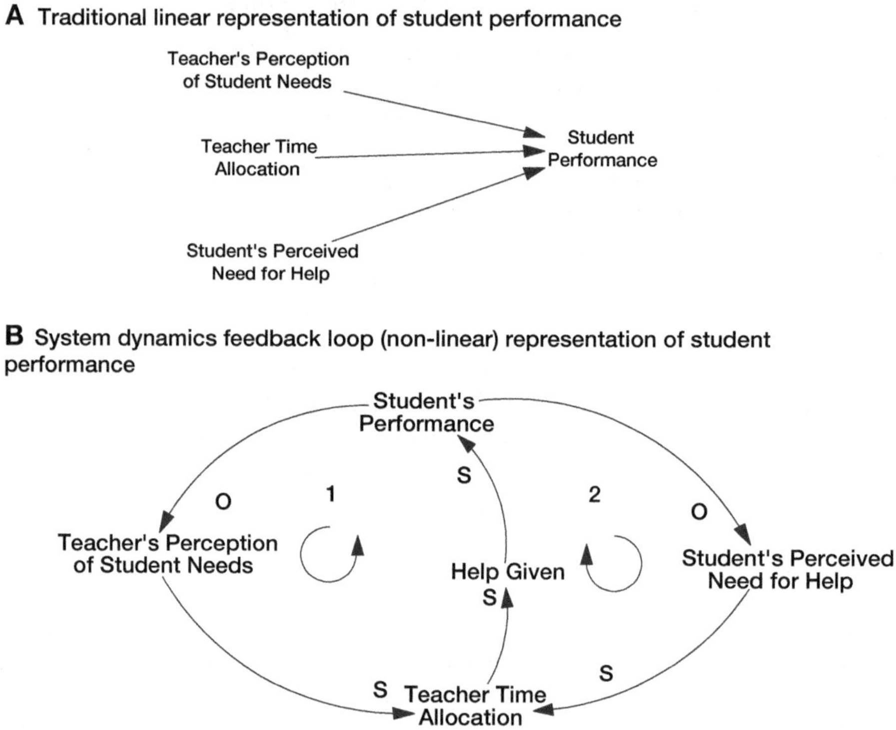

Figure 28.1 Comparison of traditional and system dynamics representations of student performance. (A) Traditional linear representation of student performance, and (B) system dynamics feedback loop (nonlinear) representation of student performance.

affect the common variable in a negative direction. With experience, students who learn system dynamics eventually think in terms of both the variables that are involved in pushing the system around and the loop processes in which these variables might be embedded.

To illustrate, consider Figure 28.1A, which is an adaptation of a model of the dynamics of the classroom developed by Nancy Roberts (Roberts, 1976a, 1976b). An educational researcher who may be interested in student performance in the classroom could develop a traditional multiple regression model of how well both teacher and students function in a classroom setting. A regression analysis could be performed by generating a list of variables that affect the dependent variable; in this case, *Student Performance* (see Figure 28.1). The regression model assumes that *Student Performance,* designated as the dependent variable, does not affect the antecedent variables. However, rarely does one find a real world process characterized by non-recursive, reciprocal relationships among the variables. Figure 28.1A is an example of what we might call, "unidirectional thinking."

A system dynamics modeler thinks about dynamic processes through closed loop or feedback thinking. Figure 28.1B, based on Robert's work, shows an example of closed loop thinking applied to the variables shown in Fig. 28.1A. In this case, *Student Performance* is affected by the help given by the teacher, which in turn is a function of the teacher putting more time and resources toward helping the student. The allocation of resources itself is affected by both the *Students' Perceived Need for Help* and the *Teacher's Perception of Student Needs*. These two variables are in turn determined by the level of *Student Performance.* In this configuration, the direction of causation is partly reversed from the model represented in Figure 28.1A. *Student Performance* is affected by the other variables, but *Student Performance* also directly and indirectly affects them. Figure 28.1B is composed of two closed loop processes, which in this case appear to move *Student Performance* in the same direction. In this network of loops, there are no pure dependent variables. Every variable plays the role of a mediator. To a system dynamics modeler, the two loops in Figure 28.1B, each composed of variables, are core elements in understanding the potential for controlling student performance.

Representing Dynamic Structures

Causal loops. Figure 28.1B is what system dynamics modelers call a causal loop diagram or a CLD. System dynamics modelers indicate the nature of the change relationship between two variables using a simple sign system. Consider the variables in Figure 28.1B. If a change in one variable, such as Help Given, generates a change in the same direction in another variable, such as Student Performance, then one would put an "S" for same, or in some system dynamics circles a "+" at the head of the arrow going from the first variable to the second variable. On the other hand, if a change in a variable, such as Student Performance, moves a variable in the opposite direction, such as Teacher's Perception of Student Needs, then one would put an "O" (for opposite) or an "–" at the head of the arrow drawn between those variables. In Figure 28.1B, the "S" on the end of the arrow drawn between Help Given and Student Performance indicates that as the student receives increased help from the teacher, the student's level of performance increases. The "O" on the end of the arrow drawn between Student Performance and Teacher's Perception of Student Needs indicates that as the student's levels of performance increases, the teacher begins to pay less attention to assessing the student's needs. Again, "S" means going in the same direction and an "O" means going in the opposite direction.

Reinforcing and balancing loops. Rather than focus on the potential causal relations between two variables at a time (the trees), system dynamics modelers first focus on larger units of causation (the forest) by trying to ask "what is the purpose of this process?" The answer is frequently a causal loop mechanism. In the situation depicted in Figure 28.1B, the teacher has a policy of monitoring student performance to determine the appropriate allocation of time to each student. The teacher tries to deal with a decrease in student performance by allocating more time to helping the student improve performance. In turn, improvements in student performance lead to changes in allocation of time to the student. In a sense, that is the big picture. The system dynamics modeler has identified at least one causal loop that attempts to accomplish the system's task, in this case generating adequate levels of student performance.

Causal loops fulfill one of two principal functions. Balancing (B) loops attempt to counteract change. If one of the variables in the loop increases or decreases, this type of loop works to bring the system back to the status quo, just as a thermostat functions to maintain a steady temperature. On the other hand, reinforcing (R) loops amplify the directional change in any of the variables in the loop by prompting either growth or decay (e.g., accelerating achievement scores, declining school enrollments). In Figure 28.1B, the model has two balancing (B) loops, indicating that the system modeled here acts to keep student performance in a state of equilibrium relative to whatever the desired goals are for student performance.

System dynamics modelers are particularly interested in identifying what happens when loops go into "collapse mode," or what some people call a vicious cycle. An example of a reinforcing loop that drives the system into collapse mode would be what might happen with the growth of alternative charter schools that receive public funds and are located alongside public schools in public school districts. As the funds are taken away from the district's traditional public schools, quality of the existing programs in the traditional schools may decrease, causing more pressure to charter more alternative schools and divert more funding to the existing charter schools. In turn, more charter schools and increases to their funds leads to cutting the resources of the public school system's traditional schools even more. Ultimately, this pattern resembles a vicious cycle. A system dynamics modeler would attempt to understand why other loops that should prevent a vicious cycle from occurring, such as loops that ought to encourage gains in quality in public schools, fail to operate in such a circumstance.

Stocks and flows. System dynamics modelers distinguish between stocks (accumulations) and flows (decisions and actions). Psychologists, on the other hand, generally do not distinguish between stocks and flows. In system dynamics, the stocks represent the accumulation of materials, energy, or information over time. A typical stock would be the number of teachers assigned to a school, the number of youth at risk of low academic attainment in a setting, the number of patients being served by a new clinic, or the level of experience with a new curriculum. In the psycho-

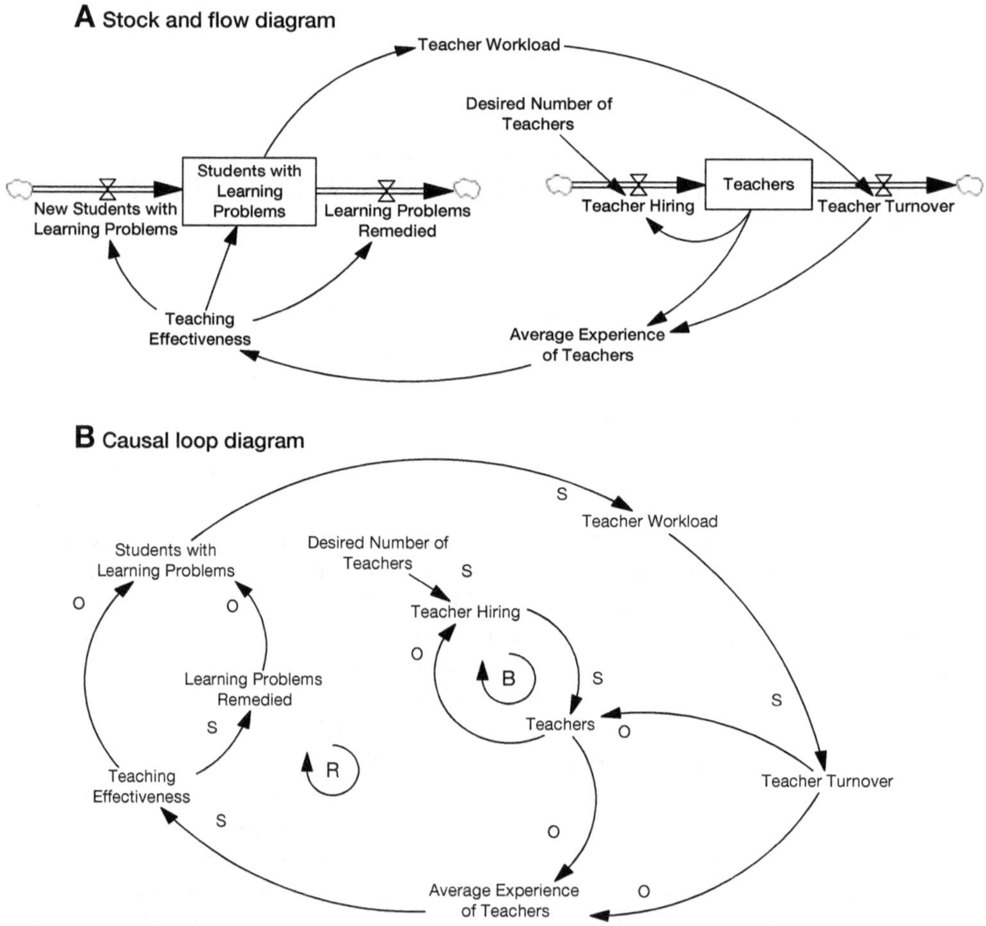

Figure 28.2 Comparison of stock and flow and causal loop diagrams of same problem–student learning problems and teacher turnover. (A) Stock and flow diagram, and (B) Causal loop diagram.

logical realm, stocks are processes that also take time to accumulate, such as aspiration, depression, prejudice, trust, power, attitudes, and beliefs. Flow variables represent the information processes and actions that change the value of the stocks. It is important to note that the only way a stock can change its level or intensity occurs through the action of one or more flow variables. Usually, stocks are what people follow or monitor over time. For instance, a state Department of Education may monitor scores on an achievement test in a given school district rather than monitor other potential indicators of student performance and school quality. For this reason, stocks are an essential feature of system dynamics models.

There are two major diagramming schemes used in system dynamics to convey the relationships among variables. The CLD, like the one found in Figure 28.1B, makes it easy to locate the loop structure representing the root causes of the problem being studied. CLDs do not differentiate between the accumulating stocks and the flows that change those accumulations. Stock and flow diagrams provide a second way of representing the dynamics of the problem and are the basis for developing the model's equations. Figure 28.2 shows an example of stocks and flows as well as their CLD counterparts. In Figure 28.2A, the boxes are the stocks, in this case the Number of Students with Learning Problems and the (number) of Teachers. The four flow variables that change the stocks are represented by arrows that come in and out of the boxes and have spigots on them to signify the potential for controlling the rates of input or output. For instance, teachers flow out of the stock Teachers as a result of retirement, termination, quitting, or reassignment.

Now consider the differences between a stock and flow diagram and a CLD. Note that it is easier to tell where the loops are using a CLD representation. However, in Figure 28.2A, one finds it easier to tell the difference between a stock variable and a flow variable. It takes time to hire and accumulate teachers, especially if there is turnover. Thus, one can see that a stock and flow representation points out the existence of lags in the system. The explicit modeling of lags or time delays in a system points out another critical difference between system dynamics and approaches that are more conventional within community psychology: our models typically do not explicitly acknowledge time delays.

Loop Processes and Changing the System

Why is this emphasis on specifying the underlying core structure via stocks, flows, and causal loops so important in system dynamics? If the loop structure helps to understand the problematic behavior patterns of a system, then the loop structure might help in the design and analysis of new policies and interventions that lead to desirable, long term solutions. System dynamics modelers use their models to explore how changes to the system might resolve the problem of interest.

A system dynamics modeler may suggest changes in a system by modifying its loop structure or creating a new loop process. Then the modeler can assess, via computer simulations, the impact of this new loop structure on how the system behaves. System dynamics modelers may also change the system by finding ways to block or slow the action of a loop that significantly contributes to the problem or by strengthening a loop that can counterbalance problem loops but that has not been able to operate effectively in the past.

An example of changing the loop structure and proposing one or more interventions can be illustrated by returning to the problem of helping students with learning problems and the problem of maintaining a supply of effective teachers in Figure 28.2. The CLD version of the proposed feedback structure has been reproduced and augmented in Figure 28.3. In this figure, the bolded arrows represent possible interventions that might be explored. First, one might attempt

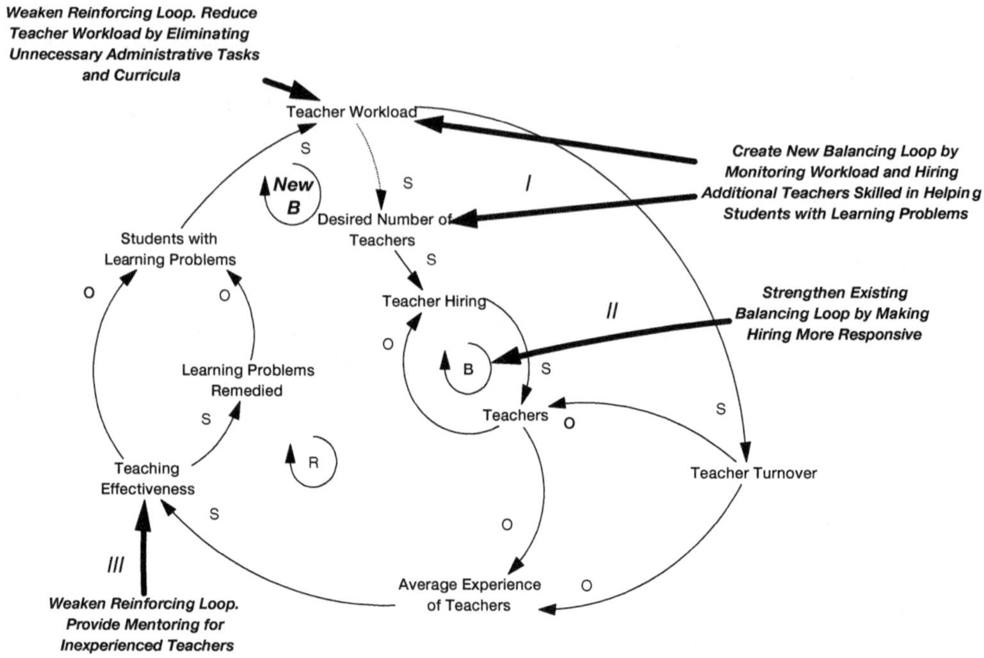

Figure 28.3 Use of loop structure for policy design to remedy student learning problems and teacher turnover

to cope with an increase in *Teacher Workload* by hiring more teachers (*I*). This, in effect, adds a connection between two variables in the model, *Teacher Workload* and the *Desired Number of Teachers*, creating a new model loop. The new loop is a balancing loop that is designed to deal with an increase in *Teacher Workload* and return workloads to manageable levels. It implies a new policy action in which the school monitors teacher workload and responds to increases in teacher workload through hiring.

Another possible intervention (*II*) might be to strengthen an existing balancing loop, one that serves to maintain the system in status quo. In the middle of the figure, there is a loop between hiring efforts and the number of teachers. Making the hiring process more efficient and responsive to a need to hire more teachers would strengthen the existing negative loop by speeding up how rapidly teacher hiring resulted in more teachers. An example of weakening a reinforcing loop that plays a key role in the dynamics of the problem would be to lighten the *Teacher Workload* by cutting down on non-essential tasks (*III*). As one can see from the figure, reducing the workload lowers *Teacher Turnover.* The *Teacher Workload* loop can become less dominant by finding ways to increase *Teaching Effectiveness*, such as developing a good mentoring system for inexperienced teachers.

Model Building and Exploration

We now turn to how these concepts are put together in the process of building and exploring the dynamic behavior of a model. In this section, we review how modelers collect and use data, and how models are operationalized. We also discuss model building as a participatory process. For clarity, we present the process as if it proceeds in stages, when in fact model building typically proceeds iteratively. We do not discuss the technical details of developing model parameters, procedures for assessing model fit, and performing analyses. We refer readers to Graham (1980) and Sterman (2000) for an introduction to these topics.

Data Sources

System dynamics uses a variety of techniques to develop and explore a model. Model building may rely on in-depth review of existing empirical and theoretical literature, collection of new qualitative or quantitative data, secondary data analyses, or on the experiences and opinions of people who are close to the process of interest. For instance, Homer and Milstein (2002) used multiple methods of inquiry to develop a model to look at the impacts of outside assistance, such as might be given by the CDC, on communities that might be afflicted by multiple diseases with multiple causes. Repenning (2002) used literature reviews and theory to examine why some companies that attempt to replicate new innovations fail and some are very successful, building his model exclusively from descriptions of variables in the literature. Lounsbury (2002) used epidemiological data and expert opinion to model the Michigan AIDS epidemic and changes in emphasis on care and treatment. Thus, system dynamics models may be built primarily on theory, extant knowledge, or data, or any combination of these.

A key characteristic of system dynamics is that one can combine quantitative and qualitative methods (Luna-Reyes & Andersen, 2003). System dynamics modelers, when given quantitative time series data, use fairly sophisticated estimation procedures to fit the system dynamics model to the data (see, for example, Peterson, 1980; Sterman, 2000). On the other hand, modelers may also include psychological and unmeasured processes for which there are not any time-ordered quantitative data. Modelers may use qualitative data to estimate trends in and behaviors of variables over time. For instance, modelers may have stakeholders plot curves to portray how variables qualitatively relate to one another over time. These plots can then be used to approximate functions.[1]

Operationalization of Constructs and Use of Data

Assuming a model has been specified using CLDs or stocks and flows, system dynamics modelers face the challenge of operationalizing a model into equations in order to conduct simulations. Available simulation software allows modelers to develop the CLD and stock and flow diagrams directly on the screen, as well as the underlying equations of the model. Powersim (Bergen, Norway), STELLA (ISEE Systems, Lebanon, NH), and Vensim (Ventana Systems, Harvard, MA) are the three main simulation packages used by system dynamics modelers. Each is capable of allowing the modeler to sketch out non-linear relationships between variables, develop simultaneous model equations, and run model simulations.

To see how a modeler might apply available data and measures to specify model variables, consider Figure 28.2. The dynamic picture suggests a mixture of variables that are not all equally easy to operationalize and quantify. There are two stocks in this figure, the number of teachers presently in the system and the number of students with learning problems. It may be trivial to obtain the size of the teaching staff over time from official personnel records. *Students with Learning Problems* might be a bit more difficult to measure, but once the construct is rigorously defined, such as the number of students who score below a defined level of performance, one can obtain records to estimate the number of students who teachers have found to have learning problems over time. *Teacher Turnover*, a flow or action variable, might be measured in terms of the number of teachers leaving per year or over some other unit of time. The hardest construct to measure in Figure 28.2 may be *Teaching Effectiveness*, but measurement of this construct could be accomplished by developing a meaningful scale to represent the idea of effective teaching[2] —that is, a model may not require an exact measure of how effective teaching in any particular school may be but require instead a scale of teaching effectiveness that has meaning logically and empirically.[3] Finally, the *Desired Number of Teachers* could be assessed by talking to the school's administrator or from formal documents suggesting a goal of having X number of teachers to teach in Y number of classrooms.

As the prior example highlights, system dynamics modelers operate under a different paradigm than that of many social scientists developing models. Interest in discerning patterns over time, such as growth, overshoot, and collapse are a hallmark of the system dynamics method as opposed to a more intense interest in determining the statistical significance of a set of predictors or how much variance in the dependent variable any one predictor may explain. System dynamics focuses on explaining the structures that underlie the patterns in data over time. A valid, adequately detailed, and properly parameterized model will always fit the real world data fairly well. The true test of the model is that it can both reproduce the general pattern of behavior that the system exhibits in the real world and help the modeler and model users to understand what features of the system produce the pattern of behavior. Once they understand the source of that behavior, they can determine how to deal with the problems it creates.

Working with Stakeholders: Group Modeling Efforts

System dynamics is intended to promote in-depth learning and insight about dynamically complex problems, so the methodology works especially well when individuals who initially hold diverse perspectives on an issue come to have a stake in the model by actively engaging in the model building process. For example, Van den Belt, a system dynamics modeler with a background in economics, has written an excellent text on what she calls "mediated modeling" to help build consensus among stakeholders who may not even care to meet others around the table (Van den Belt, 2004).

Modelers use a variety of techniques to build models in a participatory manner, such as creating model-building teams composed of both modelers and substantive experts representing diverse points of view and using group process techniques such as scenario planning. Ideally, stakeholders should help formulate the problem behavior patterns to be modeled, evaluate the model's assumptions and versions of the model runs, and also help discern the meaning and implications of computer simulations. In participatory model building, the stakeholders are typically trained and encouraged to do simulation runs on their own to find out for themselves how the model responds to ideas they might have about a new policy. The advantages to participatory model building cannot be overstated. As the constructs are refined, the modelers may obtain a better grasp on what the model should address by including stakeholders on the model-building team. Also, being asked to help in developing a model can be a very useful way to have people express and then challenge their cognitive maps or mental models of how the system works. It also makes explicit other participants' cognitive maps, highlighting alternative sets of beliefs and values. Finally, as the model is generated from existing data, both researchers and participants can begin to understand what data are missing from the picture. This can lead to the design of new studies to obtain relevant data that were previously considered unimportant.

Once a model is up and running, having the participants generate and evaluate scenarios based on conditions that they have observed can lead to insight. Sometimes the output of the model matches the participants' expectations. In this case it may help to validate the model in their minds. On the other hand, many models generate results that are counterintuitive. Perhaps a policy a participant favors works well at first and then becomes worse over time. An advantage of system dynamics models is that one can follow the logic of the model over time to see clearly why the expected results did not happen. The model may produce one or more unintended consequence that even those participants who know the system well could not have anticipated. This leads to better knowledge of the problem and impacts of policies.

Using System Dynamics to Understand System Change in Schools

To this point, we have discussed system dynamics in rather abstract terms. We now turn to a concrete example in which we describe how the first author worked with stakeholders to build a model of system change in schools. We briefly present the model and describe the process by which it was developed. The model we present concerns the process of changing schools by introducing innovations. Getting new curricula and other innovations adopted in schools is a notoriously difficult process.

The system dynamics modeling effort reported in this section arose from a conversation between Ted Sizer, a well-known educator and former Dean of the Harvard Graduate School of Education, and Jay Forrester, founder of the field of system dynamics who has devoted many years to its application to K-12 education. Sizer and Forrester pondered why adoption of innovation by schools has been so difficult and felt that there was a dynamic explanation in how innovations interact with all of the other things going on in schools. They felt that a better understanding of these dynamic phenomena could help to improve schools' acceptance of new ideas.

The Model-Building Team and its Goals

A working group was formed by Forrester and Sizer to examine the innovation process from a dynamic standpoint and included a system dynamics modeler, Gary Hirsch, several faculty from schools of education including one with extensive experience in system dynamics and K-12 education, the principal of a middle school organized around the principles of systems thinking, the director of a clearinghouse that provides systems curricula for schools, and a science teacher who has used system dynamics in his classroom.

The group first articulated the following goals for the curriculum innovation modeling: to identify factors that promote or hinder curriculum innovation and the adoption of new curricula; to develop an understanding of the dynamics of curriculum innovation and explain how the interaction of these factors over time can lead to successful innovations or innovations that fail to be adopted; to learn what characteristics of schools and school systems encourage or resist successful adoption of innovations; to identify policies and programs that serve as leverage points for increasing the likelihood of successful innovation; and to simulate the potential impact of those interventions and determine which combinations of policies and programs are likely to be most effective.

Defining the Model

The work of the group and their consultant followed a classic sequence for system dynamics modeling efforts alluded to earlier. The first meeting was an open-ended session in which these goals and key variables were identified. This was initially a straightforward process of eliciting and explaining important factors. The discussion included some basic questions such as "What is an innovation?" and "What is successful adoption?" It was agreed that an innovation is a significant change in how a subject is taught rather than incremental change in content.

The next task of the group was to identify the factors that enable curriculum innovations to be successfully adopted in schools or to fail. Some of the factors identified were relatively straightforward such as the amount of time teachers have available for learning about and working with the new curriculum. Adopting a new curriculum competes with teachers' many other responsibilities and lack of time may doom a curriculum innovation. Another factor was teacher motivation. Teachers' motivation, in turn, depends on their past experience with innovation, the need for innovation that they perceive, an awareness of innovation going on elsewhere, and the stress experienced by teachers created by the combination of their day-to-day responsibilities and adopting new ways of doing things. The group also began talking about causal relationships between these variables.

The group also identified the mode of student evaluation as an important variable. Having the appropriate mode of student evaluation enables a school to both identify student needs that might be met by new curricula and the impact of innovations that are adopted. For example, the impact of an innovation designed to improve students higher-level thinking skills may not be adequately measured by multiple choice tests and may appear to offer little advantage over learning by rote. More elaborate methods of evaluation such as the preparation and presentation of student portfolios may be necessary to reflect higher-level skills. The group described structural flexibility as another co-requisite of innovation. Certain innovations, for example, may require different arrangements such as double-periods for certain types of exploratory exercises. A scheduling system that makes these changes easier will increase the likelihood that an innovation will succeed. Another important variable that came out of the group's discussions was the trust between the schools and the community. Past successes will allow the community to give the school the necessary resources and to be patient while awaiting the results of the innovation. Past failures, on the other hand, may make the community members resistant to curriculum innovation because they do not trust the schools to implement it properly.

The working group was the principal source of this information. As suggested earlier, other modeling efforts may draw on a wider range of people and input. We certainly would have done so if we were working with a particular set of stakeholders in a real-world school system that needed to be informed about and invested in the process, as well as providing qualitative data.

As these factors were identified in a series of meetings, the modeler sketched CLDs that related the factors to each other. Figure 28.4 shows a sample of the work at this stage. In Figure 28.4, the relationships among experience with innovation, teacher motivation, and curriculum innovations adopted are presented. The relationships in Figure 28.4 form a reinforcing loop, explained earlier, in which a high level of teacher motivation will improve the chances of innovations being adopted, and lead to better experience with innovation and a continued high level of motivation and receptivity to innovations in the future. Conversely, poor experience with innovation will reduce motivation and make it difficult to sustain effort on those innovations or have future innovations adopted.

Identifying the Reference Modes

The next step was to identify "reference modes" or alternative patterns of behavior typically observed in the real world systems being modeled. The group discussed how innovations typically proceed. When innovations are successful, does adoption proceed at a steady pace or in fits and starts? Is there sometimes an interim success as part of the curriculum is adopted followed ultimately by failure when the implementation process "runs out of steam" before the innovation

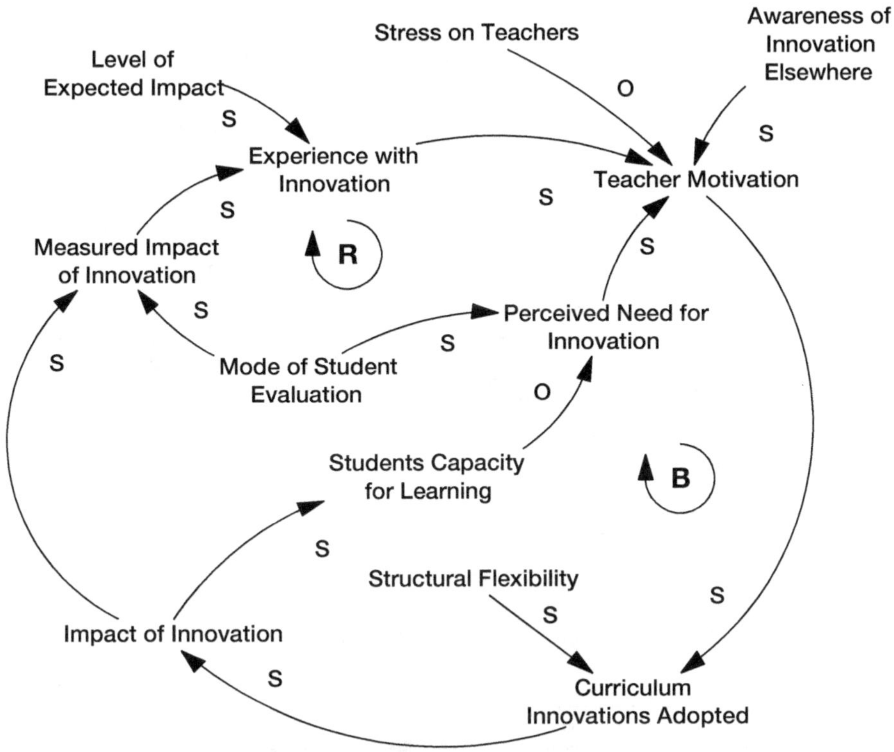

Figure 28.4 Factors affecting teacher motivation

is fully installed? Identifying the reference modes allowed us another check on the model to see that there were the necessary factors to explain the different trajectories that an innovation might take.

Figure 28.5 provides an overview of the complete model and shows how trust between schools and the community interacts with the other variables. Additional reinforcing loops through this trust variable enable success to build on success and cause failure to initiate a downward spiral that becomes an impediment to future innovation.

Developing Stock and Flow Diagrams

The structure shown in Figure 28.5 was then converted to a stock and flow diagram prior to creating the simulation model. A portion of the stock and flow model derived from this causal structure is shown in Figure 28.6. Adding this level of specificity provoked more discussion about the exact nature of the relationships.

Figure 28.6 shows a central part of the stock and flow structure for the curriculum innovation model. *Curriculum Innovations Initiated* depends on school system's policy and the system's capacity to deal with change. The fraction of *Curriculum Innovations in Process* that is adopted depends on the levels of *Teacher Motivation* and *Structural Flexibility* (e.g., ability of scheduling system to accommodate different-sized blocks of time). Innovations not adopted are discarded. *Curriculum Innovations Adopted* can remain in place for a considerable period of time until they become obsolete.

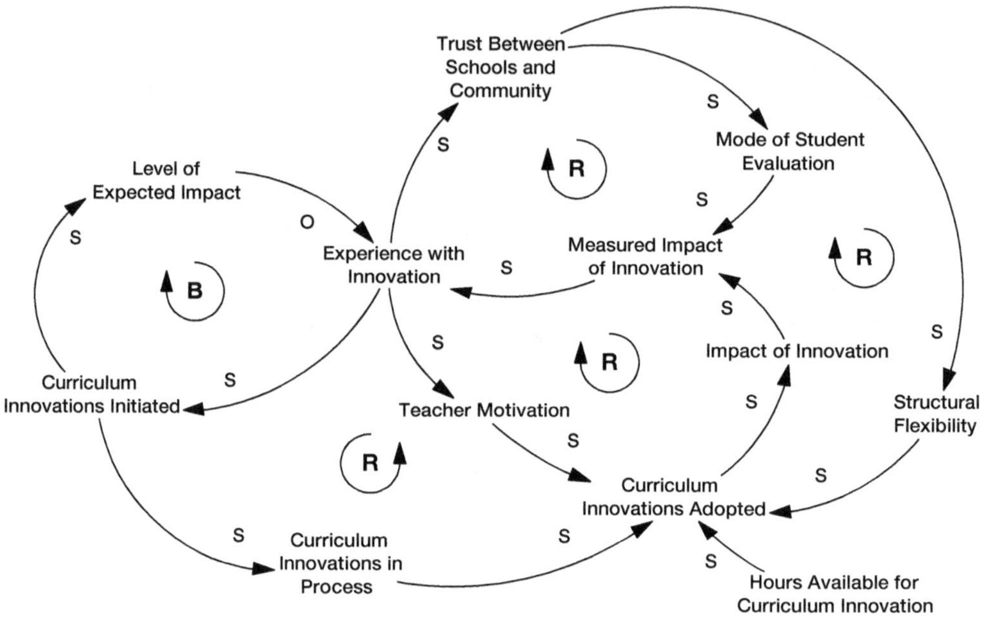

Figure 28.5 Overview of factors affecting curriculum innovation

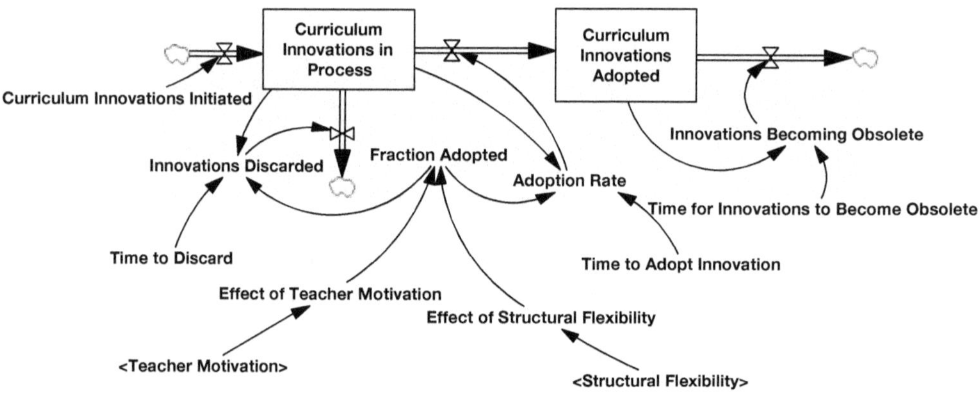

Figure 28.6 Stock and flow structure related to curriculum innovation

Simulating Policy Impacts

These stock and flow diagrams were then converted to a mathematical simulation model that could be used to pose "what if?" questions about different strategies for promoting the adoption of innovations. A software package called Vensim, mentioned earlier, was used to develop the simulation model. Vensim enables modelers to sketch out diagrams such as those shown in Figures 28.4–28.6, develop an equation for each relationship represented by arrows in the diagrams, and then run simulations. Equations can include simple algebraic relationships or complex ones that include logical statements (IFTHENELSE) and non-linear relationships that are drawn on graphs rather than being expressed algebraically. Simulations are done by starting

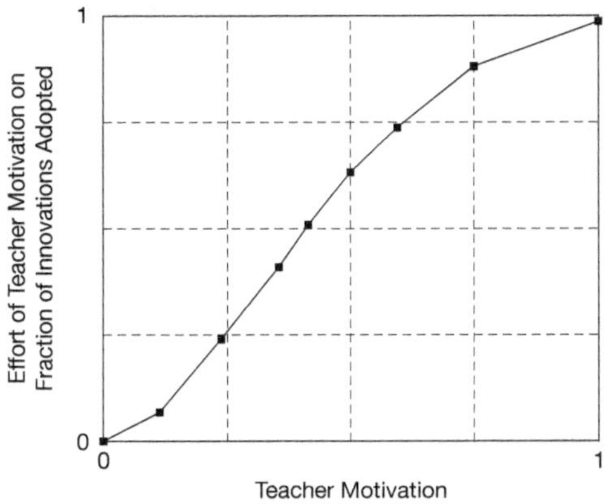

Figure 28.7 Relationship for determining the effect of teacher motivation on fraction of innovations adopted

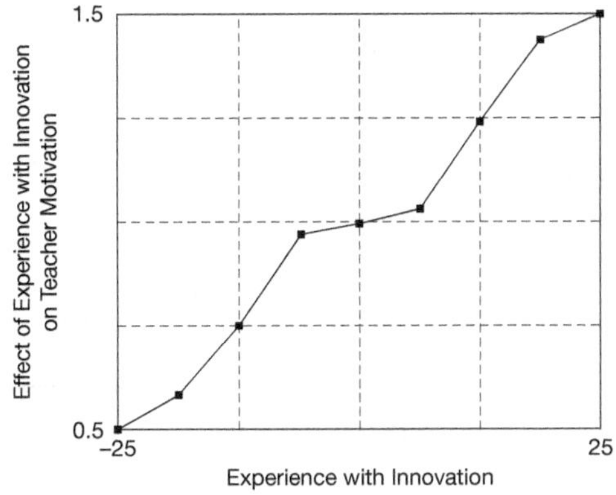

Figure 28.8 Relationship for determining the effect of experience with innovation on teacher motivation

with a specified set of initial conditions, using the equations to calculate changes that occur over the first time interval, updating the state of the system to reflect those changes, and then using the new system state to calculate the changes over the next time period.

As indicated above, many of the variables the group identified were deemed important, but there were no data from studies that could help with the model's quantification. As we previously noted, it is not uncommon that the data available for variables that are central in understanding the dynamics of the problem have not been collected over time. System dynamics modelers will often use qualitative data instead because the variables in the model are essentially latent variables and need only be quantified enough to reproduce the pattern of behavior that has been observed by actors in the system. Figure 28.7 shows an example of one of the relationships developed with the working group's help. This one relates the level of *Teacher Motivation* to the *Fraction of Innovations Adopted.* Motivation is scaled from zero to one. A high level of motivation will result in the maximum *Fraction of Innovation Adopted* while a low level of *Teacher Motivation* may keep any innovations from being adopted at all.

How is *Teacher Motivation* assumed to be affected over time? One determinant is the relationship between *Experience with Innovation* and its *Effect on Teacher Motivation.* Poor past experience with innovation will leave teachers relatively unmotivated to adopt new curricula. On the other hand, positive experiences may make them eager to do so. The model's equations are written so that results of innovation over time falling below the expected impact will cause *Experience with Innovation* to take on a negative value *Experience with Innovation* can vary from values of –50 to +50. Better than expected results will produce a positive value. The working group helped develop the relationship shown below in Figure 28.8 in which very negative *Experience with Innovation* will cause *Teacher Motivation* relative to adopting innovation to be 50% lower than if *Experience with Innovation* had been neutral. Very positive past *Experience with Innovation* (the cumulative effect of exceeding expectations over time) will result in *Teacher Motivation* that is 50% higher, all other things being equal.

Once the model was quantified, a series of simulations provided several insights, only a few of which space allows us to highlight here (see Hirsch, 1998 for a more complete discussion of the model and simulation results). For these simulations, starting values of community characteristics such as *Trust between School and Community* were chosen to be neutral and typical of the "average" community. The working group had a critical role in terms of reality-testing the results and helping to assess whether the model was responding appropriately or might still be lacking some elements.

Figure 28.9 shows the typical output from a simulation, a set of graphs that display the behavior of variables of interest over time. The graphs show the result of implementing a curriculum innovation by itself without doing anything to change the methods used for student evaluation or increasing structural flexibility.

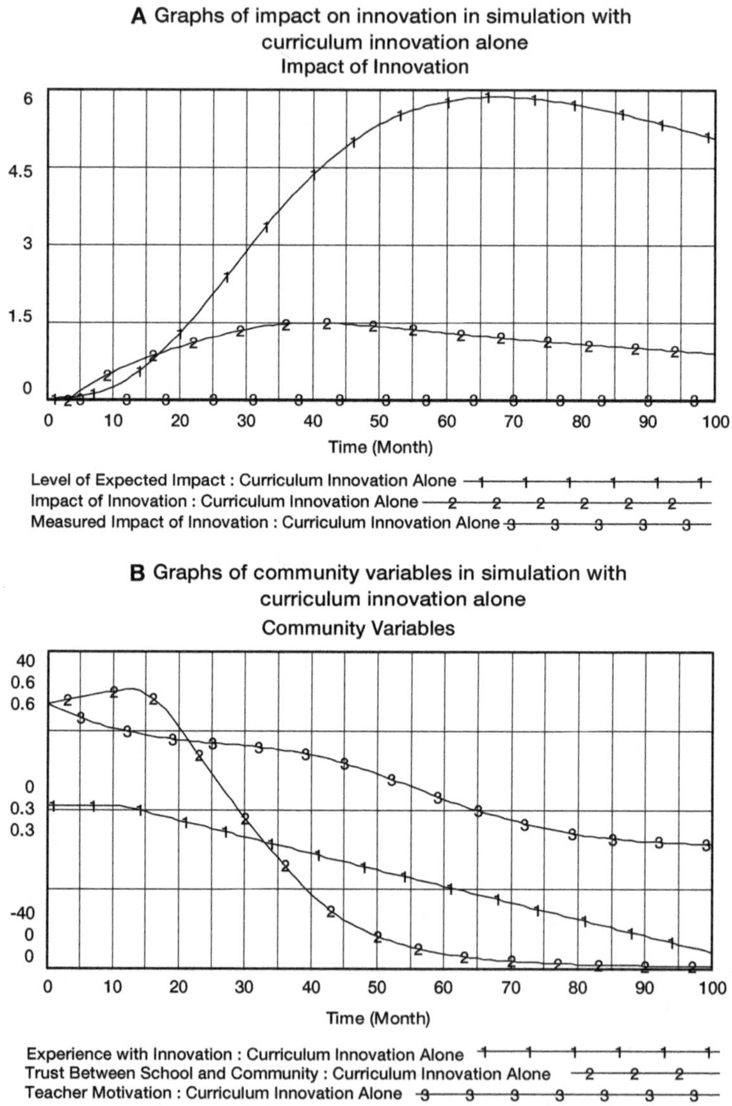

Figure 28.9 Results of simulation with curriculum innovation alone (Curriculum innovation introduced at time = 0). (A) Graphs of impact on innovation in simulation with curriculum innovation alone, and (B) Graphs of community variables in simulation with curriculum innovation alone. Note that scales on vertical axes of all of these graphs correspond to the variables listed below the graphs in the same order. Thus in graph B, the first variable, Experience with Innovation, is plotted on a scale of –40 to +40. Trust Between School and Community and Teacher Motivation are plotted on scales of 0–0.6. The *x*-axis on all graphs is time in months.

Not surprisingly, as the graphs in Figure 28.9 demonstrate, implementation of curriculum innovation on its own is likely to fail. The *Impact of Innovation* reaches a level that is significantly below the *Level of Expected Impact. Expected Impact* could reflect results that other school districts have achieved with that curriculum innovation or results promised by its proponents. Structural changes would have been needed to facilitate adoption. The *Measured Impact of Innovation*, reflecting both the *Impact of Innovation* and *Mode of Student Evaluation*, is essentially zero since the innovation is being evaluated with traditional methods in this simulation. Changes in the mode of student evaluation are necessary for the impact of innovation to be properly measured. The poor *Experience with Innovation* sets off a downward spiral of *Trust between School and Community* and *Teacher Motivation* that makes adoption impossible. Clearly, the results of this first simulation suggest that curriculum innovation needs to be coordinated with how students are evaluated and other structural changes. However, as the group found when they suggested a more comprehensive strategy, trying to do everything at once—implement a curriculum innovation, change the mode of student evaluation, and increase structural flexibility—also produces poor results. The simulation results revealed that doing everything at once leaves too little time for any one task and causes curriculum innovation to stretch out over an unacceptably long time period. As shown in Figure 28.10, temporary success gives way to failure as teachers succumb to the stress of trying to do too much at once and *Teacher Motivation* falls. As a result, the innovation fails to meet expectations and is assumed to be a failure, setting off a downward spiral in the loops through *Teacher Motivation* and *Trust Between Schools and the Community*.

This result suggested another experiment. What if the school initiated changes in student evaluation and structural flexibility, but gave those changes two years to have their effect before undertaking the bulk of the curriculum innovation? The results shown in Figure 28.11 suggest that delaying curriculum innovation for two years until the new mode of student evaluation and structural flexibility are in place helps to overcome this problem and allows enough time to then be concentrated on adopting the innovation. Successful Experience with Innovation helps keep Teacher Motivation and Trust Between Schools and the Community at a high level.

Explorations of this sort carried out with the model, often referred to as sensitivity analyses, provided some additional insights of where there might be sources of leverage. For example, changes in the amount of time devoted to the "traditional curriculum" can have a significant effect on the success of innovations. Introducing a policy that reduces the time devoted to the traditional curriculum, in addition to delaying the introduction of a new curriculum, produces even better results. With this policy addition, successful curriculum innovation will set another reinforcing loop in motion in which increases in students' ability to learn will reduce the time required for the traditional curriculum even further and leave more time for the new curriculum. On the other hand, increases in the traditional curriculum, possibly required by "teaching to the test" in response to statewide high-stakes testing, will leave too little time for adopting innovation and cause attempts at innovation to fail at a time when they are badly needed. Other activities that seem like a good idea, such as professional development, can add to teachers' time burden and indirectly serve as impediments to innovation.

Our illustration highlights how system dynamics creates opportunities for gaining insight about changing systems. For example, our group learned that curriculum innovations, especially in the form of large projects, are likely to fail unless they are undertaken in the context of the larger set of changes needed for them to succeed. Strategies work best when they focus on leverage points in a system. These leverage points are often found on reinforcing loops where they can help to promote growth processes or prevent downward spirals. In this example, leverage was achieved through coordinated and properly sequenced introduction of related innovations in structural flexibility and student evaluation that led to favorable conditions for curriculum

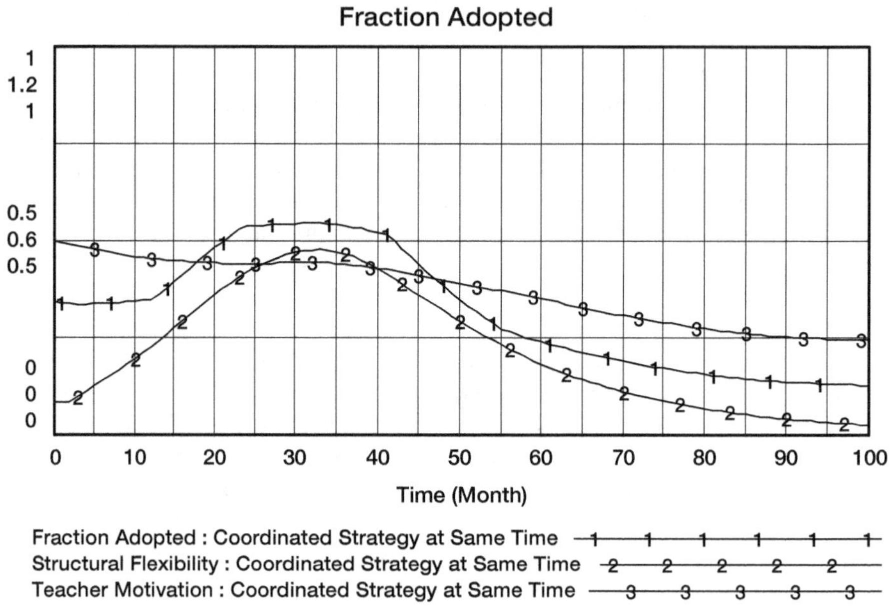

Figure 28.10 Fraction adopted and other results of simulation with coordinated strategy that combines curriculum innovation with new mode of student evaluation and increase in structural flexibility, all initiated at time = 0

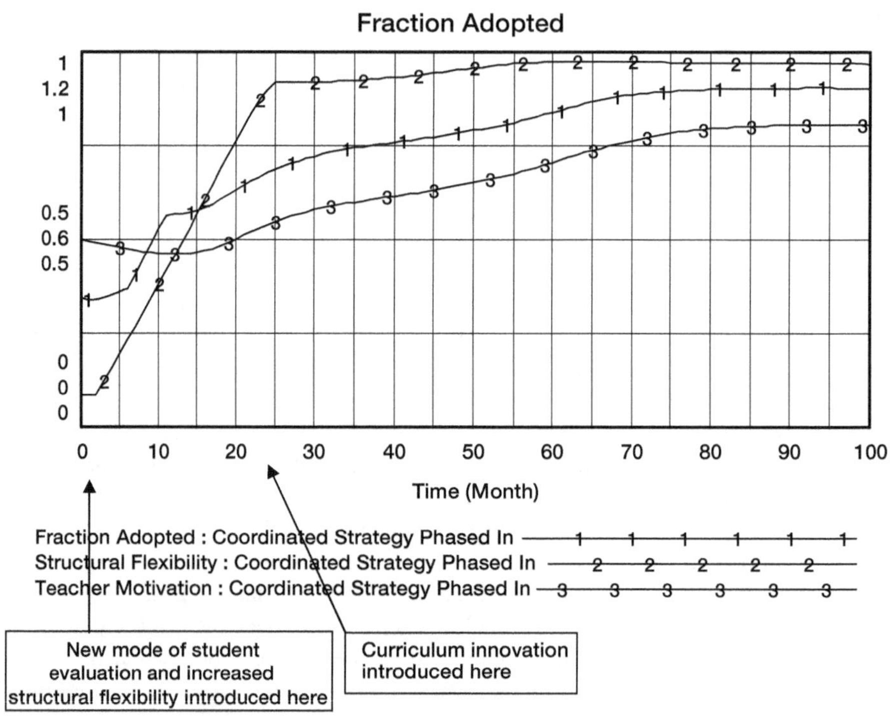

Figure 28.11 Fraction adopted and other results of simulation with coordinated strategy that combines curriculum innovation with new mode of student evaluation and increase in structural flexibility, introduced in phases as shown rather than all at once

innovation, successful innovation, and sustained conditions supporting further innovation. Working with the simulator to explore multiple "what if?" scenarios highlighted why it is essential to understand the systemic effects of changes, including possible unintended consequences, rather than simply focusing on their direct effects.

Conclusion

System dynamics provides community psychologists with a useful approach to modeling a system's complexity and to understanding why it generates particular patterns of behavior. By better understanding what processes produce a system's behavior, we gain insight into why particular system change efforts produce the effects that they do. As we have shown, system dynamics modeling also allows us to experiment with alternative change efforts and learn about the probable consequences of these efforts as a basis for policy and intervention planning.

System dynamics enables community psychologists to work with a range of variables including those that have been the subject of rigorous empirical research and others that people know from experience are important, but have not been as rigorously studied. By incorporating both types of variables into a model, system dynamics allows us to treat community members' insights as equally valid as empirical research and submit these ideas to logical scrutiny. Further, we can test out community members' ideas and beliefs about the causes of social problems and examine the conditions under which these might hold. Simulation is a valuable tool for testing the potential impact of interventions. It is too difficult for the unaided mind to understand how the complex interactions in systems will affect the efficacy of interventions and the possibility of unintended consequences that can worsen problems. Group model building techniques let real-world practitioners get involved in model-building and enhance their own understanding while keeping the work grounded in practical concerns.

System dynamics allows us to represent problems in terms of underlying causal structures amenable to intervention rather than treating them simply as a string of seemingly isolated variables. System dynamics provides a tool for understanding structures responsible for multiple community problems such as poverty, crime, and drug use rather than seeing them as separate entities (see, for example, syndemics work of Homer and Milstein, 2002). Models also enable a problem focus that crosses disciplinary boundaries and incorporates clinical, epidemiological, economic, political, and other variables in a framework that makes sense to professionals and community members tackling the problems in the real world. The ability of system dynamics to integrate diversity of perspectives into a single model is among its unique strengths.

System dynamics, like any approach, is not without its limitations, four of which we highlight here. First, although the models system dynamics produces may be more realistic representations of complex processes than are path diagrammatic models and though system dynamics can handle more complexity than traditional social science modeling approaches, system dynamics cannot capture the full complexity of a process. At some point, modelers must choose which system elements and interactions can produce useful insight about a problem and are feasible to handle in a single model. System dynamics can bring us closer to examining complex systems, though it cannot help us to do so perfectly.

Second, system dynamics models are also only as good as the thinking and assumptions that underlie them. A principal reason for participatory, multisectoral, multidisciplinary model building and reliance on multiple forms of data is to make the assumptions that underlie models as credible as possible to as diverse an audience as possible. Yet, no model can fully satisfy all of its critics and system dynamics models are no exception. Third, model building requires skill and experience. Learning how to build and examine models can take many years of training, so community psychologists may not be able to apply these techniques without the assistance of a system dynamics collaborator. Fourth, system dynamics model results are not predictions. Rather, models allow us to play out a set of scenarios for change in ways that challenge the simple mental model of problems that our current methods of study force us to adopt. Yet, it is in these experiments that we can challenge our own thinking about how to create and evaluate planned efforts at system change.

Acknowledgment

This chapter originally appeared in *American Journal of Community Psychology*, 39, 239–253 (2007).

Notes

1. These patterns may also be used as reference modes. Reference modes are typical patterns of behavior that the system has been observed to produce.
2. One must be careful if the scales used are interval, because the time series patterns of Likert-type scales may differ from patterns generated by using a ratio scale technique. This problem has been addressed by Levine and Fitzgerald (1992).
3. In a recent project on healthcare, for example, we used meta-analytic data to quantify the average effectiveness of a particular type of program and the range of effectiveness.

References

Alfeld, L., & Graham, A. (1976). *Introduction to Urban Dynamics*. Cambridge, MA: The MIT Press.

Forrester, J. (1961). *Industrial Dynamics*. Cambridge, MA: The MIT Press.

Forrester, J. (1969). *Urban Dynamics*. Cambridge, MA: The MIT Press.

Goodman, M. R. (1974). *Study Notes in System Dynamics*. Cambridge, MA: The MIT Press.

Graham, A. K. (1980). Parameter estimation in system dynamics modeling. In J. Randers (Ed.), *Elements of the System Dynamics Method* (pp. 224–245). Cambridge, MA: Productivity Press.

Hirsch, G. (1976). The causes of patient drop-out from mental health services. In G. Levin & E. B. Roberts (Eds.), *The Dynamics of Human Service Delivery* (pp. 41–66). Cambridge, MA: Ballenger Publishing Company.

Hirsch, G. B. (1998) Innovation in schools: A model to help structure the discussion and guide the search for strategies. Acton, MA: The Creative Learning Exchange, available at www.clexchange.org.

Hirsch, G. B., & Immediato, C. S. (1998). Design of simulators to enhance learning: Examples from a health care microworld. Proceedings of the 16th International System Dynamics Society Conference. Quebec City, Canada.

Hirsch, G. B., & Immediato, C. S. (1999). Microworlds and generic structures as resources for integrating care and improving health. *System Dynamics Review, 15*, 315–330.

Homer, J., Hirsch, G., Minniti, M., & Pierson, M. (2004). Models for collaboration: How system dynamics helped a community organize cost-effective care for chronic illness. *System Dynamics Review, 20*, 199–222.

Homer, J., & Milstein, B. (2002). Communities with multiple afflictions: A system dynamics approach to the study and prevention of syndemics. Proceedings of the 20th International System Dynamics Conference. Palermo, Italy. Available at: www.systemdynamics.org/conf2002/papers/Homerl.pdf

Kelly, J. G. (1968). Toward an ecological conception of preventive interventions. In J. W. Carter (Ed.), *Research Contributions from Psychology to Community Health* (pp. 76–100). New York: Behavioral Publications.

Kelly, J. G. (1987). An ecological paradigm: Defining mental health consultation as a preventive service. In J. G. Kelly & R. E. Hess (Eds.), *The Ecology of Prevention: Illustrating Mental Health Consultation* (pp. 1–35). New York: Haworth Press.

Levine, R. L., & Fitzgerald, H. (1992). *Analysis of Dynamic Psychological Systems: Basic Approaches to General Systems, Dynamic Systems, and Cybernetics* (Vol. 1). New York: Plenum Press.

Lounsbury, D. W. (2002). Understanding the dynamics of prevention, care, and empowerment: A systems approach to HIV/AIDS policy innovation. Unpublished doctoral dissertation. Michigan State University, East Lansing.

Luke, D. A. (2005). Getting the big picture in community science: Methods that capture context. *American Journal of Community Psychology, 35*, 185–200.
Luna-Reyes, L. F., & Andersen, D. L. (2003). Collecting and analyzing qualitative data for system dynamics: Methods and models. *System Dynamics Review, 19*, 271–290.
Miller, R. L, Levine, R., Khamarko, K., Valenti, M., & McNall, M. (2006). Recruiting clients to a community-based HIV-prevention program: A dynamic model system. Proceedings of the 24th International Conference of the Dynamics Society Conference. Nijmegen, Netherlands.
Peterson, D. (1980). Statistical tools for system dynamics. In J. Randers (Ed.), *Elements of the System Dynamics Method* (pp. 224–245). Cambridge, MA: Productivity Press.
Repenning, N. A. (2002). A simulation-based approach to understanding the dynamics of innovation implementation. *Organization Science, 13*, 109–127.
Roberts, N. (1976a). The structure of student–teacher interaction in the elementary classroom. In G. Levin & E. B. Roberts (Eds.), *The Dynamics of Human Service Delivery* (pp. 93–103). Cambridge, MA: Ballinger Publishing Company.
Roberts N. (1976b). A computer analysis of student performance. In G. Levin & E. B. Roberts (Eds.), *The Dynamics of Human Service Delivery* (pp. 105–145). Cambridge, MA: Ballinger Publishing Company.
Sarason, S. B. (2000). Barometers of community change. In J. Rappaport & E. Seidman (Eds.), *Handbook of Community Psychology* (pp. 919–929). New York: Kluwer Academic/Plenum Publishers.
Seidman, E. (1988). Back to the future, community psychology: Unfolding a theory of social intervention. *American Journal of Community Psychology, 16*, 3–24.
Sterman J. D. (2000). *Business Dynamics: Systems Thinking and Modeling for a Complex World.* Boston, MA: McGraw-Hill.
Trickett, E. J., Kelly, J. G., & Vincent, T. (1985). The spirit of ecological inquiry in community psychology. In E. C. Susskind & D. C. Klein (Eds.), *Community Research* (pp. 283–333). New York: Praeger.
Van den Belt, M. (2004). *Mediated Modeling: A System Dynamics Approach to Environmental Consensus Building.* Washington, DC: Island Press.

ABOUT THE CONTRIBUTORS

(Only contributors of original chapters are listed below; email addresses are provided only for corresponding authors.)

Tom Andersen is Professor of Aquatic Biology and Toxicology in the Department of Biosciences at the University of Oslo.

Patrick d'Aquino (patrick.d'aquino@cirad.fr) is a Researcher at the Centre de coopération internationale en recherche agronomique pour le développement (CIRAD).

Alassane Bah is Associate Professor at Université Cheikh Anta Diop de Dakar.

David N. Barton (david.barton@nina.no) is a Senior Research Scientist at the Norwegian Institute for Nature Research.

Abigail W. Batchelder (abatchelder@mgh.harvard.edu) is a Clinical Research Fellow of Behavioral Medicine, Department of Psychiatry, Massachusetts General Hospital/Harvard Medical School.

Olvar Bergland is an Associate Professor in the School of Economics and Business at the Norwegian University of Life Sciences.

Robert Boutilier (info@stakeholder360.com) is president of Boutilier and Associates and an Associate at the Centre for Sustainable Community Development at Simon Fraser University, Vancouver.

Brian Bush is the Principal Strategic Analyst in the Strategic Energy Analysis Center at the National Renewable Energy Laboratory.

Nathan Doogan is a Postdoctoral Researcher in the Center of Excellence in Regulatory Tobacco Science Health Behavior and Health Promotion at The Ohio State University.

Alexander Engebretsen is a Ph.D. candidate in the Department of Chemistry at the University of Oslo.

Tatiana Foroud is the P. Michael Conneally Professor of Medical and Molecular Genetics at Indiana University.

Sue Grady is Associate Professor of Geography, Environment, and Spatial Scences at Michigan State University.

Tony H. Grubesic is a Professor in the College of Computing & Informatics at Drexel University and Director of the Center for Spatial Analytics and Geocomputation.

Cornelia Guell is a Career Development Fellow at the MRC Epidemiology Unit and the UKCRC Centre for Diet and Activity Research (CEDAR), at the University of Cambridge.

Witold Henisz is the Deloitte & Touche Professor of Management at The Wharton School, The University of Pennsylvania.

Daniel Inman (Daniel.Inman@nrel.gov) is a Research Scientist in the Strategic Energy Analysis Center at the National Renewable Energy Laboratory.

Laura M. Koehly is a senior investigator in the Social and Behavioral Research Branch, National Human Genome Research Institute (NHGRI) at the National Institutes of Health.

Mariah Kornbluh is an Assistant Professor of Psychology at California State University, Chico.

Arika Ligmann-Zielinska (ligmannz@msu.edu) is Associate Professor of Geography at Michigan State University.

David W. Lounsbury is Assistant Professor in the Department of Epidemiology & Population Health and the Department of Family and Social Medicine at the Albert Einstein College of Medicine, Montefiore Medical Center.

Jeremy McWhorter is a Geographic Field Analyst with TomTom in the Detroit metropolitan area.

Liang Mao (liangmao@ufl.edu) is Assistant Professor of Geography at the University of Florida.

Jack K. Martin was Director of Research for the Institute of Social Research at Indiana University. Professor Martin sadly passed away on 22 March 2015.

Eric M. Meslin is founding Director of the Indiana University Center for Bioethics, Associate Dean for Bioethics in the Indiana University School of Medicine, and Professor of Medicine; of Medical & Molecular Genetics; of Bioethics and Law; and of Philosophy.

S. Jannicke Moe is a Research Scientist at the Norwegian Institute for Water Research.

Jennifer Watling Neal (jneal@msu.edu) is Associate Professor of Psychology at Michigan State University.

Zachary P. Neal (zpneal@msu.edu) is Associate Professor of Psychology and Global Urban Studies at Michigan State University.

Emily Newes is an Energy Analyst in the Strategic Energy Analysis Center at the National Renewable Energy Laboratory.

Anders Nielsen is a Senior Lecturer in Medicine at the University of the West Indies (UWI), Cave Hill Campus.

Sigrun Olafsdottir is Professor of Sociology at the University of Iceland.

Geir I. Orderud is a Senior Researcher at the Norwegian Institute for Urban and Regional Research.

Corey Peck is a Principal Owner at Lexidyne, LLC.

Brea L. Perry is Associate Professor of Sociology at Indiana University.

Bernice A. Pescosolido (pescosol@indiana.edu) is Distinguished Professor of Sociology at Indiana Consortium for Mental Health Services Research, and Co-Director, Indiana University Network Science Institute.

Steve Peterson is a Principal at Lexidyne, LLC.

William Pridemore is a Distinguished University Professor of Criminal Justice and Criminology in the Andrew Young School of Policy Studies and a member of the GSU Second Century Initiative's cluster on Evidence-Based Policy.

Maya Ravindranath Abtahian is Assistant Professor of Linguistics at the University of Rochester.

Eirik Romstad is an Associate Professor in the School of Economics and Business at the Norwegian University of Life Sciences.

Anantha Shekhar is the Associate Dean for Translational Research, Raymond E. Houk Professor of Psychiatry, and Professor of Pharmacology and Neurobiology at Indiana University.

Natasha Sobers-Grannum is Lecturer in Public Health at the University of the West Indies (UWI), Cave Hill Campus.

Olaf Sporns is Provost Professor in Psychological and Brain Sciences at Indiana University, and Co-Director, Indiana University Network Science Institute.

Dana Stright a Senior Data Analyst at Lexidyne, LLC.

Koji Tominaga is a Researcher in the Centre for Ecological and Evolutionary Synthesis at the University of Oslo.

Nigel Unwin (nigel.unwin@cavehill.uwi.edu) is Visiting Professor of Population Health Sciences, Chronic Disease Research Centre, The University of the West Indies.

Alessandro Vespignani is Sternberg Family Distinguished University Professor of Theoretical Condensed Matter and Biological Physics at Northeastern University.

Laura Vimmerstedt is a Senior Environmental Analyst in the Strategic Energy Analysis Center at the National Renewable Energy Laboratory.

Michael Vitevitch (mvitevit@ku.edu) is Professor of Psychology at the University of Kansas.

Rolf D. Vogt is Professor of Environmental Analysis in the Department of Chemistry at the University of Oslo.

Suzanne Evans Wagner (wagnersu@msu.edu) is Associate Professor of Linguistics at Michigan State University.

Keith Warren (warren.193@osu.edu) is Associate Professor of Social Work at The Ohio State University.

Bennet Zelner is Associate Professor of Logistics, Business & Public Policy at the Robert H. Smith School of Business at the University of Maryland.

INDEX

abatement decisions, lake eutrophication 208, 297–320
Abtahian, Maya Ravindranath 421, 553–569
adaptation: coupled human and natural systems 272, 279; Dryland Societies 209, 384–402
Add Health *see* National Longitudinal Study of Adolescent Health
adolescents: cognitive social structures 23–24, 147–163; commitment development 24, 191–206; peer relationships 23–24, 147–163; romantic and sexual networks 24, 164–190
aerobics 51, 57
affection 24, 164–190
affirming peers 421, 570–584
African American students 155–156, 158–160
African Dryland rainfall 209, 384–402
African mining 209–210, 403–417
AFV *see* alternative fuel vehicles
age, obesity and weight loss 50, 53, 56, 57–59
agent-based models 2–7; African Dryland rainfall 209, 384–402; African mining 209–210, 403–417; alternative fuel vehicles 209, 341–366; communities and social change 419–421, 423–527; comparative analysis 207–208, 267–296; computational experimentation 49–65; coupled human and natural systems 207–208, 229–296; cross-site analysis 207–208, 267–296; diet 49–52, 54–56, 64; endangered species habitats 208, 229–292; environment and sustainability 207–210, 229–296, 341–366, 384–417; health 22, 49–65, 68–81; household/residential mobility 419, 423–447; human development 22, 49–65, 68–81; influenza epidemics 22, 68–81; language change/variation 565–566; limitations 6–7; modeling processes 4–6; multidisciplinary/multiscale integration 208, 229–266; obesity and weight loss 22, 49–65; rules 3; segregation 419–421, 423–447, 486–527; social change 419–421, 423–527; sprawl 419–420, 423–485; strengths 6–7; sustainable mobility simulations 209, 341–366; transtheoretical model combinations 49, 54–64; urban dynamics 419–421, 423–527; weight loss 22, 49–65
aggregation, peer relationships 148, 150–161
agribusiness 585, 593
agriculture: Vanemfjord lake, Norway 208, 297–319; *see also* farmers/farming
AIDS risks 86–87, 90, 170
Ailuropoda melanoleuca 233, 269
alcohol: chronic disease prevention 23, 107, 109, 112–115, 118–120, 125; dependence 25, 27, 34–40
algal blooms 297, 304–305, 307, 309–314, 317
alliances, community food systems 592
alternative fuels 208–209, 321–340
alternative fuel vehicles (AFV) 209, 341–366
ANalysis Grid for Environments Linked to Obesity 50
analysis of variance (ANOVA) 541–543
Andersen, Tom 208, 297–320
An, Li 207–208, 229–296
ANOVA *see* analysis of variance
anxiety disorders 141–143
Asgeirsson, Eyjolfur Ingi 209, 341–366
asset building 585, 591
assortative mixing 137–138
Atlanta, US 492
attribute-based partnership preferences 178
auditory cortex/scenes 135
automata models 230–231, 424, 448–452, 455–463, 470, 480

average path lengths 13
Axinn, William 207–208, 267–296

Bah, Alassane 209, 384–402
Barandabar forest 270, 276
Barton, David N. 208, 297–320
basal metabolic rates (BMR) 53–54
Bass diffusion 343–344
Batchelder, Abigail W. 22–23, 84–103
bathymetry 303
battery electric vehicles (BEV) 357, 359–360
Bayesian belief networks 208, 297–320
BCAP *see* Biomass Crop Assistance Program
BDM model 407
Bearman, Peter S. 24, 164–190
behavior: adolescents and children's peer relationships 23–24, 147–163; cognitive social structures 23–24, 147–163; illegal logging 207–208, 211–228; influenza epidemics 22, 68–81; language change/variations 421, 553–569; network analysis 130–131, 133, 140–141; obesity and weight loss 22, 49–65; peer relationships 23–24, 147–163; psychology 130–131, 133, 140–141; women living with HIV 85, 100
Behavioral Risk Factor Surveillance Systems (BRFSS) 50, 52, 56–59, 64
benefit-cost analysis 297–298, 311–314, 318–319
Bengal tigers 270
Bergland, Olvar 208, 297–320
betweenness centrality 371, 374, 376–379
BEV *see* battery electric vehicles
bias aspects, adolescent's romantic and sexual networks 165–166
biodiesel 357, 360
biofuels industry 208–209, 321–340, 357–360
biogas 357–360
biological embedding 33
biological networks 21–22, 25–40
biological psychology 130–131, 133–135, 138, 143
Biomass Crop Assistance Program (BCAP) 325–327
Biomass Scenario Model (BSM) 208–209, 321–340
biophysical aspects 278, 281
biorefinery investors 329–331
birth rates 239–243, 250, 252, 259
blood pressure problems 36, 57, 107
BMI *see* body mass index
BMR *see* basal metabolic rates
body mass index (BMI) 49–64
bottom-up models *see* agent-based models
Boutilier, Robert 209–210, 403–417
brain and psychology 130–131, 133–135, 138, 143
BRFSS *see* Behavioral Risk Factor Surveillance Systems
bridge symptoms 141–142
bridging process models 167, 176
Bronx, NY, USA 86–90, 100
BSM *see* Biomass Scenario Model
Buffalo, NY, USA 22, 68–81
building system dynamic models 24, 191–206, 606–617
built environment, obesity and weight loss 22, 49–65
Bush, Brian 208–209, 321–340
"buy one and get others free" strategy 22, 76
bystander effect 131

CA *see* cellular automata
calibrating models 191, 200–203
calories 53–59, 64
cancers 35, 38–39, 107, 126, 185
car dealers 209, 341, 346–349, 351–359
cardiovascular diseases (CVD) 23, 107, 112–115, 123–126
car manufacturers 209, 341–344, 346–356
CAS *see* complex adaptive systems
catchment management/run-off 297, 300, 302, 304, 306–307, 312, 317
causal loops/causal loop diagrams: community food systems 421, 585–587; education 603–607, 610; illegal logging 211, 213, 219; system dynamics 7–10, 421, 585–587, 603–607, 610; women living with HIV 85–90, 95, 100, 102
causal relationships, illegal logging 207–208, 211–228
causation, sprawl 452–453
CD *see* chronic diseases
CDC *see* Center for Disease Control and Prevention
cellular automata (CA) 230–231, 424, 451, 456–458, 463, 480
census data 420, 425, 434–435, 486–487, 491–506
Center for Disease Control and Prevention (CDC) 84–87, 112–115, 121–123, 125
centrality 371, 374, 376–379
centralization, natural resource management 369–371, 374
CHANS *see* coupled human and natural systems
characteristics: coupled human and natural systems 284–286; sprawl 449–451
charging stations, vehicles 341, 343–346, 350–352, 355–356, 358
"cheap food" paradigm 589
Cheshire, Jenny 560
Chicago, US 492
childbirth 239–243, 250, 252, 259
childhood sex abuse 86–88, 90, 93–95, 98, 101

children: cognitive social structures 23–24, 147–163; household/residential mobility 423, 428, 430–431, 434–435, 438, 443–445; language: cognitive psychology 130, 139; peer relationships: perception and behavior 23–24, 147–163
China: coupled human and natural systems 208, 229–266; endangered species habitats 208, 229–292; panda habitats 208, 229–266, 269, 271, 273, 276, 281–285, 288–290; rural population growth 208, 229–266; Wolong Nature Reserve 208, 229–292
Chitwan National Park, Nepal 208, 267–292
choice factors, residential mobility 419, 423–447
chronic diseases (CD) 23, 107–126, 585, 588
climate, African Dryland rainfall 209, 384–402
clinical interventions, women and HIV 84, 90, 97, 102–103
clinical psychology 130, 140–143
closed systems, influenza epidemics 70
closeness: adolescent's romantic and sexual networks 182–184; gang rivalry 531
clustering 13, 436, 439–440
coalitions, African mining 406, 411–413
cognitive psychology 130–131, 133–143
cognitive social structures (CSS) 23–24, 147–163
cognitive therapy 140–141
cohesive communities 420, 512–527
cohort data analysis 90, 97
cohort homophily 421, 581–582
collection, biofuel feedstocks 326–327
combination tests 249–257
commitment development, adolescents 24, 191–206
communities: African Dryland rainfall 386–387; African mining 404–406, 410, 413–414; agent-based models 419–421, 423–527; biological psychology 134; census data 420, 425, 434–435, 486–487, 491–506; Chitwan National Park, Nepal 275–276, 279–280, 283, 286–288, 292; curricula adoption/innovation 422, 599–619; diversity 420, 487–493, 499–505, 512–527; economics 419, 423–447; education 422, 585, 591–593, 599–619; food systems 421–422, 585–598; gang-related rivalry/violence 420, 528–552; household/residential mobility 419–420, 423–485, 488, 501, 504; illegal logging 211, 213–227; K-12 curriculum 422, 609–619; language variation 421, 553–569; migration 419–420, 423–485; mixed-race households 419–420, 486–511; network analysis 419–421, 423–447, 511–569, 575; pattern-oriented modeling 419–420, 423–447; psychology 512–516, 519, 523–525; race 419–420, 423–424, 428, 431, 434–435, 438, 441, 443–445, 486–511, 570, 574, 577–579, 581; residential mobility 419–420, 423–527; schools: K-12 curriculum 422, 609–619; segregation 419–421, 423–447, 486–527; sense of community 420, 512–527; simulations 419–420, 423–485, 612–617; social change 419–619; social cohesion 420, 512–527; social networks 419–421, 423–447, 511–584; socioeconomic segregation 419, 423–447; speech variations 421, 553–569; sprawl 419–420, 423–485; system dynamics 421–422, 585–595, 599–619; tie distributions 419, 423–447; urban dynamics 419–619; Wolong Nature Reserve, China 275–276, 279–280, 283, 286–288, 292; women and HIV 84, 86–87, 89–90, 96–101, 103; written word variations 421, 553–569
community psychology 512–516, 519, 523–525, 599, 605
comorbidity 141–143
comparative analysis 207–208, 267–296
complex adaptive systems (CAS) 168, 230, 345
complexity: coupled human and natural systems 208, 229–271, 274, 281, 284–291; endangered species habitats 208, 229–266; Network Science 27–29; obesity and weight loss 50, 61, 63–64; population growth 208, 229–266
composition measurements, sprawl 473–478
comprehension, language change/variation 421, 553–569
computer models: agent-based models 49–64; community-diversity 516; education/curricula impacts 612; household/residential mobility 449, 455–479; language change/variation 565–566; obesity and weight loss 49–64; women living with HIV 86, 90–91, 96–97
conceptual models: adolescent's commitment development 191–196; African Dryland rainfall 384–389, 391, 399; alternative fuel vehicles 209, 341–366; community food systems 586–592; household/residential mobility 449, 452–455; illegal logging 212–213; population growth 235–236
CONCOR *see* convergence of iterated correlations
configuration measurements, sprawl 473–478
Connor, David S. 421–422, 585–595
consciousness 135
consensus aggregation 148, 150–161
consequences: community food systems 585, 594; sprawl 449–451
conservation, community food systems 587–589
construct operationalization 607–608
consumers: alternative fuel vehicles 209, 341–363; biofuels industry 208, 335–338;

choice behavior 356–360; community food systems 585–595
contagion, sprawl 475, 478–479
controlling illegal logging 211–212, 214–218, 220, 223–225, 227
convergence of iterated correlations (CONCOR) 539–545
cooperation, therapeutic community residents 574
core models 166–169, 174, 176–177, 180, 183–185
corporations, community food systems 585, 593
corruption, illegal logging 212, 215
cost-effectiveness, lake eutrophication 297–300, 305, 311–318
Counties Manukau, New Zealand 113–115, 123–124
counting rules 503–504
coupled human and natural systems (CHANS): agent-based models 207–208, 229–296; Chinese rural population growth 208, 229–266; comparative analysis 207–208, 267–296; complexity 208, 229–271, 274, 281, 284–291; cross-site analysis 207–208, 267–296; endangered species habitats 208, 229–292; environment and sustainability 207–208, 211–296; illegal logging 207–208, 211–228; key features 207–208, 267–296; panda habitats 208, 229–266; rural population growth 208, 229–266; socioeconomic factors 208, 229–263, 268, 271–279, 281, 286; Wolong Nature Reserve, China 208, 229–292
credibility, African mining 405
crime: gang rivalry 528–552; mixed-race households 488; obesity and weight loss 52–53, 56–57, 60–61, 63–65; therapeutic community residents 571–582
crops, biofuels industry 322, 324–327, 332
cross-level interconnections 27, 38–39
cross-site analysis 207–208, 267–296
curricula adoption/innovation 422, 599–619
CVD *see* cardiovascular diseases
cyanobacteria blooms 297, 304, 317

Dallas, US 492
Danville, Illinois 419, 423–447
d'Aquino, Patrick 209, 384–402
data analysis: acquisition/processing 58–60; African mining 404, 409–415; children and adolescent's peer relationships 155–159; education 607–608; mixed-race households 492–504; obesity 49–50, 58–63; population growth 237–239; residential mobility 434–435
databases 49–50, 60
Davidsdottir, Brynhildur 209, 341–366
dealers, vehicles 209, 341, 346–349, 351–359
deaths: chronic disease prevention 23, 107, 113–114, 117–118, 122–123; population growth 239
decision making: African mining 209–210, 403–417; biofuels industry 321–329; chronic disease prevention 23, 107–126; lake eutrophication 208, 297–320; natural resource management 209, 367–383; residential mobility 433–434
degree centrality 371, 374, 376–379
degree distribution 138
demographics: children and adolescent's peer relationships 155, 159–160; influenza epidemics 69, 71–72, 80; mixed-race households 419–420, 486–511; obesity and weight loss 51, 58, 64; population growth 208, 229–263
dense networks, language change/variation 559–563
density, natural resource management 374
density at maximum reach 176
Department of Energy 208–209, 321–340
depression/depressive disorders 23, 100–101, 140–143
design: behavior promotional strategies 76; coupled human and natural systems 272, 277–280
Detroit, US 492
developers, sprawl 454
development, language 140–141, 421, 553–569
diabetes 23, 107, 109, 112–115, 121–123, 125, 588
dialects, language change/variation 555–563, 566–567
diet 49–52, 54–56, 64, 107–109, 120–123
diffusion: influenza epidemics 68, 70–80; sprawl simulations 459–460
Diffusion Tensor Imaging (DTI) 134
direct reciprocity 573, 576
discrete choice methods 343–344
discrete event uncertainty management 96–97
disease: adolescent's romantic and sexual networks 164–169, 172, 175–177, 184–185; chronic disease prevention 23, 107–126, 585, 588; community food systems 585, 588, 593; dynamic pathways 26, 29–30, 32, 34, 36, 40; influenza epidemics 22, 68–81
dissimilarity indexes 493–504
diversity: mixed-race households 487–493, 499–505; sense of community 420, 512–527
Doogan, Nathan 421, 570–584
DPSIR *see* driver-pressure-state-impact-response

driver-pressure-state-impact-response (DPSIR) chains 297, 299, 305, 308–319
drug use 84, 86–103
Dryland Societies 209, 384–402
dual-diffusion frameworks 70, 75–77
Dudley, Richard G. 207–208, 211–228
dyads 12, 15–16, 172, 176–185
dynamic pathways, health, disease and health care 26, 29–30, 32, 34, 36, 40
dynamics of neighborhood segregation 419–420, 486–511
dynamic systems models *see* system dynamics

E15/E85 transitions 336–337
EBM *see* energy balance model
economics: alternative fuel vehicles 341, 344–345, 349–351, 355–356, 360; food communities 421–422, 585–595; household mobility 419, 423–447; lake eutrophication 297–300, 303–305, 311–318; panda habitats: population growth impacts 229–263
ecosystems: African Dryland rainfall 209, 384–402; Chitwan National Park, Nepal 208, 267–292; Indonesian villagers and illegal logging 207–208, 211–228; lake eutrophication 208, 297–320; Wolong Nature Reserve, China 208, 229–292
edges, network analysis 11–17
education: community food systems 585, 591–593; curricula adoption/innovation 422, 599–619
ego, therapeutic community residents 570, 572–574, 576–581
electricity 234–239, 244–246, 250, 252–254, 259, 261–263, 269, 272, 275, 278–279, 281, 287
electrophysiological techniques 133–134
elementary schools 158–160
Ellis, Mark 419–420, 486–511
embeddedness, gang rivalry 420, 528–552
Emerald Hills, San Diego, California 51–52, 54, 60, 63–64
emissions, vehicles 49, 54–60, 63–64
emotional distress 86–92, 95–103
empirical observations 22, 49–63, 425–430, 449
employment: community food systems 585, 588, 594; household/residential mobility 424–425; illegal logging 216–217; mixed-race households 490–491; sprawl 453–454
endangered species habitats 208, 229–292
energy, biofuels 208–209, 321–340
energy balance model (EBM) 52–58, 63
energy supply systems, vehicles 341, 346–347, 349–350, 355–356, 358, 360
enforcement, illegal logging 211–212, 214–218, 220, 223–225, 227
Engebretsen, Alexander 208, 297–320
entrepreneurship 585, 591–592
environment: African Dryland rainfall 209, 384–402; African mining 209–210, 403–417; agent-based models 207–210, 229–296, 341–366, 384–417; alternative fuel vehicles 209, 341–366; biofuels industry 208–209, 321–340; Chinese rural population growth 208, 229–266; Chitwan National Park, Nepal 208, 267–292; coupled human and natural systems 207–208, 211–296; endangered species habitats 208, 229–292; eutrophication 208, 297–320; food communities 421–422, 585–595; Indonesian villagers and illegal logging 207–208, 211–228; integrating social drivers 208, 229–266; lake eutrophication 208, 297–320; logging practices 207–417; mining companies 209–210, 403–417; natural resource management 209, 367–401; network analysis 208–210, 297–320, 367–383, 403–417; obesity and weight loss 22, 49–65; panda habitats 208, 229–266, 269, 271, 273, 276, 281–285, 288–290; policy 208–209, 297–366; population growth impacts 208, 229–266; public policy 208–209, 297–366; rainfall 209, 384–402; resource management 209, 367–401; rural population growth 208, 229–266; social driver integration 208, 229–266; social network analysis 209, 367–383; stakeholder engagement 209–214, 367–417; sustainability 207–417; system dynamics 207–209, 211–228, 297–366; Wolong Nature Reserve, China 208, 229–292
epidemic-behavior models 22, 68–81
epidemiological burdens 86–87
equality factors 424, 440–445
error measurements 436–437
ethanol biofuels 322, 328–339
ethical issues 39
ethnicity 274
ethnographics 558, 565–567
European Union Water Framework Directive 208, 297–320
eutrophication 208, 297–320
evaluations: adolescent's commitment development 203–204; syndemic risks: women and HIV 98–100
evolutionary game theorists 572–573
exercise, obesity and weight loss 22, 51–65
ex-offenders' behavior 421, 571–582
explanatory pathways, obesity and weight loss 49–50, 57
exploration: in breadth 195–199, 202; commitment development 192–205; in depth 195, 198, 202
external change, sprawl simulations 458–459

external organization relationships, African mining 403–410, 412
extreme combination/value tests 249–257

factor analysis 405
family planning 239, 248, 250, 252, 254, 259, 261–262, 274
family ties 424
farmers/farming: African Dryland rainfall 209, 384–402; biofuels industry 208; community food systems 421, 585–595; lake eutrophication 297, 303–306, 309, 312, 315–319; population growth: panda habitat impacts 236, 243; Vanemfjord lake, Norway 297, 303–306, 309, 312, 315–319
fast food chains 585, 593
FCV *see* fuel cell vehicles
feedback methods: biofuels industry 322–325, 328, 339; chronic disease prevention 108, 111, 116–118; coupled human and natural systems 268, 284–285, 290; illegal logging 214–217; therapeutic community residents 574
feedstocks, biofuels industry 322–339
fertility 239, 248, 250, 259, 262, 274, 278, 282
FFV *see* flex-fuel vehicles
field reports, illegal logging 211–213
filtered Monte Carlo experiments 439–440
financial hardship, women and HIV 86, 88–103
fiscal deficits, community food systems 585, 587, 592
flex-fuel vehicles (FFV) 336–338
Flint, Colin 420, 528–552
flows: biofuels industry 323–324; chronic disease prevention 108, 111–115, 117–118, 121–124; education 604–605, 607–608, 611–617; syndemic risks: women and HIV 85, 90–91, 95–96; system dynamics 9–10, 604–605, 607–608, 611–617
flu *see* influenza epidemics
food: community systems 421–422, 585–598; diet 49–52, 54–56, 64, 107–109, 120–123; distribution 35; obesity and weight loss 49–50, 64
Food Policy Councils (FPC) 591
forests: Chitwan National Park, Nepal 208, 267–292; endangered species habitats 208, 229–266; illegal logging 207–208, 211–228; panda habitats: population growth impacts 208, 229–266; Vanemfjord lake, Norway 208, 297–319; Wolong Nature Reserve, China 208, 229–292
forming adolescent commitments 192–195, 203
Foroud, Tatiana 25–40
Forrester, Jay 7, 207, 609
Fowler, Christopher S. 419–420, 486–511
FPC *see* Food Policy Councils
fractal dimension measures 474–478, 480
fragility, adolescent romantic and sexual networks 173–174
free hand-sanitizer programs 68–69
fresh food 585, 588, 590
freshwater eutrophication 208, 297–320
friendships: community-diversity 514, 516–522; household/residential mobility 425, 428, 430; language change/variation 564–565
fuel cell vehicles (FCV) 344–347, 359–360
fuels: alternative fuel vehicles 209, 341–366; biofuels industry 208–209, 321–340; fuel stations 341, 343–346, 350–352, 355–356, 358; fuelwood consumption 229, 234–263, 269, 272–285, 289
functional networks, brain 130–131, 133–135, 138, 143
future research, community food systems 585, 594

GA *see* geographic automata
games: game theorists 572–573; number guessing games 132–133; role-playing 384, 386–390, 393–400
gang members, language change/variation 560–561
gang-related rivalry/violence 420, 528–552
gender, mixed-race households 490
generalized reciprocity 421, 570, 573–574, 576–577, 580–581
genes, network links 21–22, 25–40
gentrification consequences 585, 594
geodesic lengths 176–177
geographic aspects: gang-related rivalry/violence 420, 528–552; household/residential mobility 419–420, 423–485; language change/variation 421, 553–566; obesity and weight loss 49–51, 53, 64; racial segregation 487–492, 496–497; Social Symbiome frameworks 25, 28, 32–35, 37–39; sprawl 419–420, 423–485
geographic automata (GA) 455–462
geographic inertia 459
geographic information systems (GIS) 49–50, 53, 230–232, 237–238, 243, 259, 261–263
Geopolitical Influence and Strategy Tool (GIST) 409, 411, 414
Germany, alternative fuel vehicles 343–344
Gini coefficient 440–442
GIS *see* geographic information systems
GIST *see* Geopolitical Influence and Strategy Tool
global network structures, adolescent romantic and sexual networks 164
governing norms, adolescent romantic and sexual networks 180–185

government: alternative fuel vehicles 341, 343–344, 346, 348–349, 352–355; community food systems 590–591, 593
Grady, Sue C. 49–64
graph theory 39; *see also* network analysis
greenhouse gases 208–209, 321–340
group innovation, social psychology 131–133
group modeling efforts, system dynamics 608–609
group relationships, African mining 403–410, 412
Grubesic, Tony H. 25–40
Guell, Cornelia 23, 107–126
guns 535
gyms 22, 52–53, 56–57, 59, 61–64

H1NI 35
Hamming distances 136
hand-sanitizer programs 68–69
hand washing 22, 68–69, 73
harvesting: biofuel feedstocks 326–327; illegal logging 211, 213–214, 216–220, 223–224, 227
health 21–206; adolescent's peer perception and behavior 23–24, 147–163; adolescent's romantic and sexual networks 24, 164–190; African mining 413–414; agent-based models 22, 49–65, 68–81; alcohol 23, 25, 27, 34–40, 107, 109, 112–115, 118–120, 125; cardiovascular diseases 23, 107, 112–115, 123–126; care 22, 26, 29–30, 32, 34, 36, 40, 68–81; children's peer perception and behavior 23–24, 147–163; chronic disease prevention 23, 107–126; community-based food systems 421–422, 585–595; databases 49–50, 60; diabetes 23, 107, 109, 112–115, 121–123, 125; dynamic pathways 26, 29–30, 32, 34, 36, 40; food communities 421–422, 585–595; influenza epidemics 22, 68–81; networks 21–40, 130–190; obesity 22, 35, 49–65, 107–109, 111–112, 114, 120–123, 125–126; psychology 22–23, 84–103, 130–146; smoking control 23, 107–109, 112–118, 122–126; Social Symbiome framework 21–22, 25–40; syndemic risks 22–23, 84–103; system dynamics 22–24, 84–103, 107–126, 191–206; tobacco control 23, 107–109, 112–118, 122–126; weight loss 22, 49–65, 120, 123; women and HIV 22–23, 84–103
healthy food 421–422, 585–598
height, dynamic systems model 50, 53, 56, 58
Henisz, Witold 209–210, 403–417
heterogeneity: coupled human and natural systems 232, 268, 284–285, 289–290; lake eutrophication valuation 300, 306, 311; obesity and weight loss 49–52, 60
high schools 168–176
Himalayas 270
Hirsch, Gary B. 422, 599–619
HIV: adolescent romantic and sexual networks 164, 179, 185; Network Science 35; women 22–23, 84–103
hog farming 236
Hollenbeck Policing Area, Los Angeles, California 529, 533–547
Holloway, Steven R. 419–420, 486–511
home ownership 427–430, 433–435, 443–445
homicide 533, 535
homophily: adolescent's romantic and sexual networks 165–166, 168, 178–180, 182–184; diversity and community 512–525; household/residential mobility 419–420, 424–425, 430–431, 438, 444–445; natural resource management 371; therapeutic community residents 570, 574–577, 579, 581–582
households: Chitwan National Park, Nepal 269, 271–287; communities and social change 419–420, 423–485, 488, 501, 504; Danville, Illinois 419, 423–447; geographic aspects 419–420, 423–485; housing status 427–430, 433–435, 443–445; influenza epidemics 68–81; population growth: panda habitat impacts 208, 229–266; residential mobility 419–420, 423–485, 488, 501, 504; segregation 420, 423–447, 488, 501, 504; sprawl 419–420, 423–485; urban dynamics 419–420, 423–485, 488, 501, 504; Wolong Nature Reserve, China 208, 229–266, 269, 271–287
Hubacek, Klaus 209, 367–383
Human Connectome Project 134
human development 21–206; adolescent's commitments 24, 191–206; adolescent's peer perception 23–24, 147–163; adolescent's romantic and sexual networks 24, 164–190; agent-based models 22, 49–65, 68–81; children and adolescent's peer perception 23–24, 147–163; Chitwan National Park, Nepal 208, 267–292; chronic disease prevention 23, 107–126; commitment development 24, 191–206; endangered species habitats 208, 229–292; influenza epidemics 22, 68–81; networks 21–40, 130–190; obesity 22, 35, 49–65, 107–109, 111–112, 114, 120–123, 125–126; psychology 22–23, 84–103, 130–146; Social Symbiome framework 21–22, 25–40; system dynamics 22–24, 84–103, 107–126, 191–206; weight loss 22, 49–65, 120, 123; Wolong Nature Reserve, China 208, 229–292; women and HIV 22–23, 84–103; *see also* health

human–environment interactions 207–417; African Dryland rainfall 209, 384–402; African mining 209–210, 403–417; agent-based models 207–208, 229–296; alternative fuel development 208–209, 321–340; alternative fuel vehicles 209, 341–366; biofuels industry 208–209, 321–340; Chitwan National Park, Nepal 208, 267–292; Indonesian villagers and illegal logging 207–208, 211–228; lake eutrophication 208, 297–320; mining companies 209–210, 403–417; natural resource management 209, 367–402; panda habitats 208, 229–266; population growth 208, 229–266; public policy 208–209, 297–366; stakeholder engagement 209–210, 367–417; Wolong Nature Reserve, China 208, 229–292; *see also* coupled human and natural systems; environment and sustainability
Human Genome Project 27
hybrid modeling 209, 341–383
hydrogen, alternative fuel vehicles 344–347, 359–360
hypothetical intervention 49, 68–69

ICE *see* internal combustion engines
Iceland, alternative fuel vehicles 344, 355–362
identity development 192–196, 204–205, 373
IJU *see* Interspersion and Juxtaposition Index
illegal logging 207–208, 211–228
illegal mining 414–415
illegal workers 214–215
imaging tools, biophysical psychology 133–134
immigrants 562–563; *see also* migration
IMSHED *see* Integrative Model for Simulating Household and Ecosystem Dynamics
incentive strategies: biofuels industry 208–209, 321–340; influenza epidemics 68, 76–81
inclusionary methods, African Dryland rainfall 384, 386, 393, 397
income: household/residential mobility 423–424, 429–435, 438, 440–445; illegal logging 211–227; mixed-race households 488–489, 491, 505
indigenous knowledge/principles 398–400
indirect reciprocity 421, 570, 573–574, 576–577, 580–581
individual persons: adolescent romantic and sexual networks 164–169, 172, 175–177, 184–185; ex-offender's behavior 421, 571–582; household/residential mobility 419, 423–447; obesity and weight loss 22, 46–65; population growth 208, 229–263
Indonesian villagers and illegal logging 207–208, 211–228
industry development: biofuels 208–209, 321–340
inequality, urban dynamics 424, 440–445
infectious diseases 22, 68–81, 107
influenza epidemics 22, 68–81
information delay representation 95–96
information types, lake eutrophication valuation 301, 312–313
initializing coupled human and natural systems 280–281
Inman, Daniel 208–209, 321–340
in-migration 239
innovation: K-12 curricula 422, 609–619; social psychology 131–133
institutional trust 405
integrated agent-based–system dynamics modeling 209, 341–366
integrated lake eutrophication valuations 208, 297–320
integration, social–environmental drivers 208, 229–266
Integrative Model for Simulating Household and Ecosystem Dynamics (IMSHED) 235–236, 238–241, 245–246, 261
interdisciplinarity: lake eutrophication valuation 300, 311; network analysis 25, 130
internal combustion engines (ICE) 342–345, 357–360
internal system dynamics, alternative fuel vehicles 209
International Obesity Taskforce 50
inter-organizational networks 36
Interspersion and Juxtaposition Index (IJI) 475–476, 478–479
interviews: African mining 410, 414; children and adolescent's peer perception and behavior 147–148
intra-organizational networks 36
intraurban household mobility 419, 423–447
inverse core models 166–168, 177, 184–185
investors, biofuel feedstocks 329–331
irrational behaviors, illegal logging 207–208, 211–228
irregular movement, sprawl 461–462
I tests 536

Java-Swarm programming 232, 236–238, 244, 246, 262–263
Jefferson High School 168–176
junk food 591, 593
Juxtaposition Index 475–476, 478–479

K-12 curriculum 422, 609–619
Kentucky–Tennessee–Delaware (KTD) neighborhood association 427–428
knowledge/knowledge systems: adolescent commitment development 191, 194,

204–205; African Dryland rainfall 384–386, 392, 397–401; lake eutrophication valuation 301, 312
Koehly, Laura M. 25–40
Kornbluh, Mariah 23–24, 147–163
Krackhardt, D. 148–151, 155, 160–161
KTD *see* Kentucky–Tennessee–Delaware neighborhood association
Kunnen, Saskia 24, 191–206

Labov, William 558, 560–562, 564–566
La Jolla, San Diego, California 51–52, 54, 60–62, 63–64
lake eutrophication 208, 297–320
Lake Michigan, US 470–474, 476
land tenure 397, 399
land use: African Dryland rainfall 209, 384–402; Peak District National Park 370, 372; Vanemfjord lake, Norway 297, 303–306, 309, 312, 315–319
language: change/variation 421, 553–569; cognitive psychology 130, 135–140; network analysis: psychology 130–131, 135–140, 143–144
LAPD *see* Los Angeles Police Department
laws: biofuel feedstocks 326–327; illegal logging 211–212, 214–218, 223–225, 227
LDC *see* light-duty vehicle demand
leaders, language change/variation 558–559
leap-frog movement 462–463
learning, coupled human and natural systems 272, 279
legacy effects 268, 284–285
legitimacy factors, African mining 405
Levenshtein distances 136
Levine, Ralph 421–422, 585–619
lexicons, cognitive psychology 136–139
licenses, African mining 209–210, 403–416
life histories, population growth 208, 229–263
lifestyles: influenza epidemics 80; obesity and weight loss 22, 49–65
light-duty vehicle (LDV) demand 336, 357
Ligmann-Zielinska, Arika 22, 49–65
Linderman, Marc 208, 229–266
linguistics 421, 553–569
literature reviews/searches: chronic disease prevention 109–112, 125; community food systems 586–592; syndemic risks: women and HIV 89–90
Liu, Jianguo 207–208, 229–296
local food systems 585–594
locally aggregated structures 148, 150–152, 154–156, 160
local ties, household/residential mobility 423, 431–432, 438, 441, 443–445
Logan Heights, San Diego, California 51–52, 54, 60–64
logging practices 207–208, 211–228
logistic issues, Social Symbiome frameworks 39
loop processes *see* causal loops/causal loop diagrams
Los Angeles Police Department (LAPD) 529, 533–538
Los Angeles, US: gang-related rivalry/violence 420, 528–552; mixed-race households 492
Lounsbury, David W. 22–23, 84–103
low-income urban schools 157–160
low-resource urban regions 22–23, 84–103

macro-economic models 344
McWhorter, Jeremy 49–65
magnetic resonance imaging (MRI) 130–131, 133–134
major depressive disorders (MDD) 140–143
male therapeutic community residents 575, 578–579, 581
management: African mining 209–210, 403–417; lake eutrophication valuation 300, 304–305, 311
manufacturers: alternative fuel vehicles 209, 341–344, 346–356; community food systems 585, 593
Mao, Liang 22, 68–81
mapping gang rivalry/violence 535–538
Marcia's identity status model 192
markets: alternative fuel vehicles 343–347, 360–361; biofuel feedstocks 331–334; community food systems 590
marriages: coupled human and natural systems 274, 278, 280; mixed-race households 487, 490, 492, 494, 500; population growth 239–242, 247–252, 254, 259
Martin, Jack K. 25–40
mathematical modeling, alternative fuel vehicles 347–355
maximum/mean geodesic lengths 176–177
MDD *see* major depressive disorders
meaning, language change/variation 554, 563–567
measuring sprawl 473–480
mental illness/disorders 23, 130
mental processes, cognitive psychology 130–131, 133–143
mentoring 574
Meslin, Eric M. 25–40
Mesquita, Bruce Bueno de 407–408
meta-models 208
Metcalf, Sara S. 419, 423–447
metropolitan areas, mixed-race households 493–504
Miami, US 492
Michigan, US 470–474, 476, 585–595
micro-simulations 344

Midwestern United States 24, 164–190, 470–474, 476–479
migration: coupled human and natural systems 274, 282; household mobility 419, 423–447; language change/variation 562–563; population growth 239–241, 250, 252, 254, 261
Miller, Robin Lin 422, 599–619
Milroy, James and Lesley 554–558, 563, 567
mining companies 209–210, 403–417
Ministry of Health (MOH) 117
minority languages 562–563
mixed-race households 419–420, 486–511
mobility: alternative fuel vehicles 209, 341–366; residential mobility 419–420, 423–527
modeling/models: adolescent commitment development 24, 191–206; adolescent romantic and sexual networks 166–169, 174, 176–177, 180, 183–185; alternative fuel vehicles 209, 341–366; biofuels industry 208–209, 321–340; building system dynamic models 24, 191–206, 606–617; curricula adoption/innovation 422, 599–619; illegal logging 207–208, 211–228; processes in 4–6; step-wise model building 84–85, 91, 93, 96–97, 103; sustainable mobility 209, 341–366; *see also* agent-based models
model parsimony 101
model testing 246–261
Moe, S. Jannicke 208, 297–320
MOH *see* Ministry of Health
molecular networks 21–22, 25–40
Monte Carlo experiments 436, 439–440, 443–444
Moody, James 24, 164–190
Moran's I test 536
Moreno, Jacob 419
morphology, language 136
Morsa catchment, Norway 208, 297–319
movement *see* migration
MRI *see* magnetic resonance imaging
multiple criteria analysis, lake eutrophication 297, 313, 315
multiple data source triangulations 84
multiple disciplinary integration 208, 229–266
multiple health problems, women and HIV 22–23, 84–103
multiple level approaches: African Dryland rainfall 209, 384–402; Social Symbiome frameworks 34; therapeutic community residents 576
multiple risk factors, women and HIV 22–23, 84–103
multiple scale focus: African Dryland rainfall 386, 392–394; agent-based models 208, 229–266; Social Symbiome frameworks 25
multiple submodels, illegal logging 207–208, 211–228
mutual aid-based treatments 421, 570–584

national health policies 23, 114
National Longitudinal Study of Adolescent Health (Add Health) 168–171
National Renewable Energy Laboratory 208–209, 321–340, 343
natural resource management 209, 367–401; African Dryland rainfall 384–401; social network analysis 367–383; stakeholder analysis 367–383
Neal, Jennifer Watling 23–24, 147–163, 420, 511–527
Neal, Zachary P. 1–19, 420, 511–527
nearby movement, sprawl simulations 461
neighborhood associations 426–428
neighborhood environments: mixed-race households 419–420, 486–511; obesity and weight loss 22, 49–53, 56–58, 60, 63–65; residential mobility 419, 423–445; safety 51, 57–58, 59, 63–65; segregation 419–420, 486–527; social cohesion 420, 512–527
NEM *see* Network-Episode Model
Nepal, Chitwan National Park 208, 267–292
nested networks 21–22
NETLOGO software package 516
network analysis 11–17; African mining 209–210, 403–417; brain/biological psychology 130–131, 133–135, 138, 143; clinical psychology 130, 140–143; cognitive psychology 130–131, 133–143; dyad level analysis 12, 15–16; edges 11–17; environment and sustainability 208–210, 297–320, 367–383, 403–417; household/residential mobility 419, 423–447; interdisciplinarity 25, 130; lake eutrophication 208, 297–320; language: psychology 130–131, 135–140, 143–144; language: variation 421, 553–569; mutual aid 421, 570–584; natural resource management 209, 367–383; nodes 11–17; psychology 23, 130–146; segregation 419–420, 423–447, 511–527; social psychology 130–133; Social Symbiome frameworks 21–22, 25–40; stakeholder engagement 209–210, 403–417; therapeutic community residents 421, 570–584; triad level analysis 16; urban dynamics 419–421, 423–447, 511–569, 575; *see also* networks; social network analysis
Network Embedded Symbiome *see* Social Symbiome frameworks
Network-Episode Model (NEM) 27, 29–30
networks: adolescent behavior and peer relationships 23–24, 147–163; adolescent romantic and sexual networks 24, 164–190;

centralization 176; children's behavior and peer relationships 23–24, 147–163; cognitive social structures 23–24, 147–163; cohesive communities 420, 512–527; communities and social change 421; data types 147–161; health and human development 21–40, 130–190; influenza epidemics 69–74; language 130; multiple scales of 25; peer relationships 23–24, 147–163; romantic and sexual networks 24, 164–190; *see also* network...
Network Science 21–22, 25–40
network theory 554–556, 560, 567
neuroimaging tools 133–134
neuroscience 130–131, 133–135, 138, 143
Newes, Emily 208–209, 321–340
New York, US: mixed-race households: segregation 492; syndemic risks; women and HIV 86–87
New Zealand 23, 112–118, 123–126
Nielsen, Anders 23, 107–126
nodes, network analysis 11–17
"no money on the table" 592
non-communicable disease prevention 23, 107–126
non-invasive tools 133–134
nonlinear coupled human and natural systems 268, 285
nonromantic sexual relationship 169
non-villager logging workers 211, 214, 219–225, 227
norm-maintenance, language change/variation 559–561
Norway, lake eutrophication 208, 297–320
number guessing games 132–133
nutrient abatement 208, 297–320
nutrition, community food systems 585, 591–592
Nutritional Health and Nutrition Examination Survey 50

obesity: agent-based models 22, 49–65; behavior modifications 22, 49–65; chronic disease prevention 107–109, 111–112, 114, 120–123, 125–126; community food systems 585, 588, 591, 593–594; computational experimentation 49–65; data analysis 49–50, 58–60; databases 49–50, 60; energy balance model 52–59, 63; environments 22, 49–60, 63–65; exercise 22, 51–65; geographic aspects 49–51, 53, 64; health databases 49–50, 60; intervention strategies 22, 49–65; neighborhood environments 22, 49–53, 56–57, 60, 63–65; neighborhood safety 51, 56–57, 59, 63–64; Network Science 35; physical activity 22, 49–65, 107–109, 120–123; policy interventions 22, 49, 51, 54–56, 63–64; preventative tools 22, 49–65; safety aspects 51, 56–58, 60, 63–65; safe walking/neighborhoods 51, 56–58, 60, 63–64; San Diego, California, USA 22, 49–64; simulations 22, 49–60; transtheoretical models 49, 54–60, 63–64; weight loss 22, 49–65
obesogenic systems 50, 64–65
objectives, coupled human and natural systems 272, 279
Object-Oriented Bayesian Networks (OOBN) 297, 299, 305, 308–319
object-oriented programming (OOP) 231–232, 242, 261, 458
observational networks 147–148, 150, 153, 156–157, 171–174
ODD *see* overview, design, details frameworks
Olafsdottir, Sigrun 25–40
one-horned rhinoceros 270
online search engines 109–110
OOBN *see* Object-Oriented Bayesian Networks
OOP *see* object-oriented programming
operation licenses 209–210, 403–416
opportunity management 209–210, 403–416
Orderud, Geir I. 208, 297–320
organization-based networks 28, 31–33, 35–36, 38
Østfold County, Norway 302
out-migration 239–240, 261
output-focused biofuel incentives 329–331
outside workers 211, 214, 219–225, 227
overview, design, details (ODD) frameworks 208, 267, 270–271, 275, 279, 289, 291
overweight problems 50–51, 60, 64, 107; *see also* obesity
ownership: African Dryland rainfall 394; household/residential mobility 427–430, 433–435, 443–445

PAFRAC *see* perimeter-area fractal dimension measures
panda habitats 208, 229–266, 269, 271, 273, 276, 281–285, 288–290
Panthera tigris tigris 270
participatory modeling 209, 384–402
partnership patterns 24, 164–190
patches, sprawl 473–478
path finding, population growth 243–244
path lengths, network analysis 13
pathway diversity-focused biofuel incentives 329–331
pattern-oriented modeling 419–420, 423–447
Peak District National Park, United Kingdom 209, 367–380
Pearson product–moment correlation coefficients 539–545
Peck, Corey 208–209, 321–340
peer-inferred relationships 152, 154–155

peer relationships: African mining 407–409, 412; behavior and 23–24, 147–163; children and adolescents 23–24, 147–163; language change/variation 560–561, 564–565; perception of 23–24, 147–163; therapeutic community residents 421, 570–584
perceptions: peer relationships 23–24, 147–163; women and HIV financial hardships 86, 88–103
perimeter-area fractal dimension (PAFRAC) measures 474–478, 480
Perry, Brea L. 25–40
personal social networks 31–33, 36–39
persons: Chitwan National Park, Nepal 267–292; population growth 208, 229–263; Wolong Nature Reserve, China 208, 229–263, 267–292
Pescosolido, Bernice A. 21–22, 25–40
Peterson, Steve 208–209, 321–340
PHEV *see* plug-in hybrid electric vehicles
Philadelphia, US 492
phonology 130, 136–138
physical activity 22, 49–65, 107–109, 120–123
physical networks, brain/biological psychology 130–131, 133–135, 138, 143
pig farming 236
pixels (forests) 234, 236–239, 243–245, 257, 260–263
planners, sprawl 454–455
plug-in hybrid electric vehicles (PHEV) 344, 357, 359
plural values, lake eutrophication valuation 300–301, 311, 314–315
point-of-production-focused biofuel incentives 330–331
poles, Social Symbiome frameworks 30–31, 39
police, gang rivalry 529, 533–547
policy: African Dryland rainfall 209, 384–402; alternative fuel vehicles 209, 341–366; biofuels industry 208–209, 321–340; chronic disease prevention 23, 107–126; community food systems 585–595; education/curricula impacts 612–617; environment and sustainability 208–209, 297–366; lake eutrophication 208, 297–320; obesity and weight loss 22, 49, 51, 54–56, 63–64; population growth: panda habitat impacts 232, 234, 236, 243–244, 248–251, 259, 261–262; sprawl 454–455
political dynamics simulations, African mining 407–409
polycentric processes, sprawl 452, 468–470, 472, 476
population growth 208, 229–266, 419–420, 423–485
positional analysis 420, 528–552
positive reinforcement 573–574, 580–581
power: African mining 404, 407–408, 410–412, 415; gang rivalry 420, 528–552
preferential attachment 24, 139, 164–190
Prell, Christina 209, 367–383
prevention: chronic disease prevention 23, 107–126; influenza epidemics 22, 68–81; obesity and weight loss 22, 49–65
pricing biofuels 334–335
PRIZM data 80
probability of ties 431–432
producers/production, biofuel feedstocks 325–328
product–moment correlation coefficients 539–545
prohibitions 181–182
projectomics 134
promoting influenza epidemic prevention 22, 68–81
pronunciation, language change/variation 553–564
prophylactic medicines 68–69, 73–77, 79
prostitutes 166
protein networks 21, 32, 37–38
proximity: diversity and community 512–513, 516–525; gang rivalry 531, 533, 538; household/residential mobility 420, 423–426, 430, 434, 444
psychology: communities and social change 512–516, 519, 523–525; education: curricula adoption/innovation 599, 605; health 22–23, 84–103, 130–146; network analysis 23, 130–146; women and HIV 22–23, 84–103
psychopathology 130, 140–143
public elementary schools 158–160
public health: chronic disease prevention 23, 107–126; Social Symbiome frameworks 21–22, 25–40; system dynamics modeling 23, 107–126; women and HIV 84–103
public high schools 168–176
public policy: alternative fuel vehicles 209, 341–366; biofuels industry 208–209, 321–340; environment and sustainability 208–209, 297–366; lake eutrophication 208, 297–320
push-up interactions 572, 576–577

Qi, Jiaguo 208, 229–266
qualitative studies: African Dryland rainfall 384, 386, 395; syndemic risks: women and HIV 85, 89–90, 97, 100–102; system dynamics: education 607; therapeutic community residents 571–572
quantitative system dynamics: adolescent commitment development 24, 191–206; alternative fuel vehicles 342; education 607; illegal logging 211, 213; syndemic risks: women and HIV 85, 101

Raberto, Marco 209, 341–366
race: household/residential mobility 419, 423–424, 428, 431, 434–435, 438, 441, 443–445; mixed-race households 419–420, 486–511; therapeutic community residents 570, 574, 577–579, 581
Radil, Steven M. 420, 528–552
rainfall, African Drylands 209, 384–402
random mixing 165, 168
random rewiring 180–185
Ravindranath, Maya Abtahian 421, 553–569
real networks, peer relationships 147
real spatial environment simulations 22, 49–60
reciprocity: coupled human and natural systems 268–269, 271, 284–286; therapeutic community residents 421, 570, 573–574, 576–582
recreational use, Vanemfjord lake, Norway 297, 299, 304–307, 309–314
Reed, Mark 209, 367–383
reference mode identification 610–612
reforms, education/curricula 422, 599–619
regression analysis 437–439, 444
regulations, biofuels industry 321, 331
relational embeddedness 530
relationships: adolescent's romantic and sexual networks 24, 164–190; children and adolescent's peer perception and behavior 23–24, 147–163; gang rivalry 420, 528–552
renewable energy: alternative fuel vehicles 209, 341–366; biofuels industry 208–209, 321–340
renewable inventory numbers (RIN) 331–334, 337–338
renewable super premium (RSP) 336–337
rented properties 427–430, 433–435, 443–445
resident affirming 421, 570–584
residential mobility 419–420, 423–527; Danville, Illinois 419, 423–447; geographic aspects 419–420, 423–485; mixed-race households 488, 501, 504; segregation 419–420, 423–447, 486–527; sprawl 419–420, 423–485; urban dynamics 419–420, 423–485, 488, 501, 504
residential segregation 419–420, 423–447, 486–527
residential therapeutic communities 421, 570–584
resilience, coupled human and natural systems 268, 284–285
resistance training 51, 57, 63
resource management 209, 367–401; African Dryland rainfall 384–401; social network analysis 367–383; stakeholder analysis 367–383
respiratory disease 107
retail outlets, biofuels industry 334–335
rewiring networks 180–185
Rhinoceros unicornis 270
RIN *see* renewable inventory numbers
risk/risks: adolescent romantic and sexual networks 164–169, 172, 175–177, 184–185; African Dryland rainfall 209, 384–402; African mining 209–210, 403–416; chronic disease 23, 107–126; psycho-social issues HIV 22–23, 84–103; women and HIV 22–23, 84–103
rivalries, gang-related 420, 528–552
road-like movement, sprawl 462
role-model strategies 68, 70, 76, 79–81
role-playing games 384, 386–390, 393–400
romantic networks 24, 164–190
romantic partners 487, 490–492, 494, 500
Romstad, Eirik 208, 297–320
RSP *see* renewable super premium
run-off 297, 302, 306–307, 312, 317
rural population growth 208, 229–266

safety nets 30, 32–33
safe walking/neighborhoods 51, 56–58, 60, 63–64
sanctions 408–409, 574, 580
San Diego, California, USA 22, 49–64, 492
San Francisco, US 492
SARS, Network Science 35
satellite imaging 237–238, 470–474
scaling aspects: African Dryland rainfall 386, 392–394; agent-based models 208, 229–266; coupled human and natural systems 271–276; lake eutrophication valuation 301, 314; mixed-race households 491–492; Social Symbiome frameworks 25
scenario analysis, population growth 248–249, 258
scheduling, coupled human and natural systems 272, 276–277, 283–284, 288–289
Schelling, Thomas 419, 486, 489, 499–500, 505
schools 158–160, 422, 593, 609–619
scientific knowledge 384–386, 392, 397, 399–401
search engines 109–110
second-hand smoke 23, 117
segregation: agent-based models 419–421, 423–447, 486–527; household/residential mobility 419, 423–447; mixed-race households 419–420, 486–511; network analysis 419–420, 423–447, 511–527; residential mobility 419–420, 423–447, 486–527; single-race households 486–505; social cohesion 420, 512–527; social networks 419–420, 423–447, 512–527
self-design models 384–402
self-eliciting workshops 387
self-reported measures 147–148, 150–156, 159–161
self-worth 22–23, 86, 88–95, 98–103
semantics 130–131, 136, 139

semi-urbanized catchments 208, 297–319
Senegalese farmers 209, 384–402
sense of community 420, 512–527
sensing, coupled human and natural systems 272, 279–280
sensitivity analysis 68, 70, 76, 79–81
sensitivity testing 249, 252, 254
sentence structure 136, 554–555
service places, influenza epidemics 70–73, 75
sewage 298, 304–306, 308, 310, 313–314
sex abuse 86–88, 90, 93–95, 98, 101
sexually transmitted diseases (STD) 164–166, 168, 172, 175–177, 184–185
sexual networks/relationships 24, 86–88, 90–103, 164–190
sexual self-care 86, 88, 90–92, 95–100, 103
sex workers 166
Shafiei, Ehsan 209, 341–366
Shekhar, Anantha 25–40
short cycle adolescent romantic and sexual networks 177
Shortridge, Ashton 208, 229–266
sign tests 169
simulations: adolescent commitment development 191–192, 200–205; adolescent romantic and sexual networks 24, 164, 168, 173–183; African Dryland rainfall 384, 386–387, 389–390, 392–400; African mining 404, 407–409; alternative fuel vehicles 209, 341–366; biofuels industry 322–332, 336; communities 419–420, 423–485, 612–617; education/curricula impacts 612–617; household/residential mobility 419, 423–447; influenza epidemics 22, 68–81; mixed-race households 419–420, 486–511; obesity and weight loss 22, 49–60; population growth 229–232, 235–240, 243–249, 252, 256–259, 262; social change 419–420, 423–485, 612–617; sprawl 419–420, 448–485; syndemic risk 22–23, 84–103; urban dynamics 419–420, 423–485, 612–617; women and HIV 22–23, 84–103
single-race households, segregation 486–505
SIR *see* susceptible-infectious-recovered model
Sizer, Ted 609
skew of the reach distribution 177
skill building 585, 591
slices, CSS: peer relationships 148, 150, 152, 154, 157–158, 160–161
SLO *see* social license to operate
small-world networks 131–133, 137
smoking control 23, 107–109, 112–118, 122–126
SNA *see* social network analysis
Sobers-Grannum, Natasha 23, 107–126
social and environmental drivers 208, 229–266
social capital 405
social change 419–619; agent-based models 419–421, 423–527; census data 420, 425, 434–435, 486–487, 491–506; community-based food systems 421–422, 585–598; curricula adoption/innovation 422, 599–619; diversity 420, 487–493, 499–505, 512–527; economics 419, 423–447; education 422, 585, 591–593, 599–619; food systems 421–422, 585–598; gang-related rivalry/violence 420, 528–552; household/residential mobility 419–420, 423–485, 488, 501, 504; K-12 curriculum 422, 609–619; language variation 421, 553–569; mixed-race households 419–420, 486–511; mutual aid 421, 570–584; network analysis 419–421, 423–447, 511–569, 575; pattern-oriented modeling 419–420, 423–447; psychology 512–516, 519, 523–525; race 419–420, 423–424, 428, 431, 434–435, 438, 441, 443–445, 486–511, 570, 574, 577–579, 581; residential mobility 419–420, 423–527; schools: K-12 curriculum 422, 609–619; segregation 419–421, 423–447, 486–527; sense of community 420, 512–527; simulations 419–420, 423–485, 612–617; social cohesion 420, 512–527; social networks 419–421, 423–447, 511–584; socioeconomic segregation 419, 423–447; sprawl 419–420, 423–485; system dynamics 421–422, 585–595, 599–619; therapeutic community residents 421, 570–584; tie distributions 419, 423–447
social cognitive mapping 147–148
social cohesion 420, 512–527
social contagions 22, 68–81
social contract 405
social health, women and HIV 22–23, 84–103
social license to operate (SLO) 209–210, 403–416
social meaning, language change/variation 554, 563–567
social network analysis (SNA): African mining 209–210, 403–416; gang-related rivalry/violence 420, 528–552; natural resource management 367–383; psychology 131–132; therapeutic community residents 421, 570–584
social networks: adolescent and children's behavior and peer relationships 23–24, 147–163; cognitive social structures 23–24, 147–163; gang-related rivalry/violence 420, 528–552; influenza epidemics 69–74; language variation 421, 553–569; mutual aid 421, 570–584; peer relationships 23–24, 147–163; residential mobility 419, 423–445; segregation 419–420, 423–447, 512–527; therapeutic community residents 421, 570–584
social position, gang rivalry 528–552
social preferences/structures 403–416

social psychology 130–133
social science theory 572–575
social support 28–29
Social Symbiome frameworks: alcohol dependence 25, 27, 34–40; background 29–30; biological networks 21–22, 25–40; cross-level interconnections 27, 38–39; ethical issues 39; foundations 30–33; geographical networks 25, 28, 32–35, 37–39; individual network systems 32, 37; interdisciplinarity of network analysis 25; level illustrations 33–39; logistic issues 39; molecular networks 21–22, 25–40; multiple scale networks 25; network analysis 21–22, 25–40; organization-based networks 28, 31–33, 35–36, 38; personal social networks 31–33, 36–39; public health 21–22, 25–40
society organization levels, lake eutrophication valuation 300, 313–314
socioeconomics: coupled human and natural systems 208, 229–263, 268, 271–279, 281, 286; household/residential mobility 419, 423–447; mixed-race households 488; population growth 208, 229–263; segregation 419, 423–447, 488
sociolinguistics 421, 553–569
sociopolitical risks 404, 411–413
software packages: community-diversity 516; education/curricula impacts 612; illegal logging 211; population growth 232, 236–238, 244, 246, 262–263; syndemic risks: women and HIV 86, 90–91, 96–97
Soil and Water Assessment Tool (SWAT) 305–307, 309–310, 312, 314–315
South-Eastern Norway 208, 229–266, 297–319
spanning tree models 167–168, 172–173, 176–177, 180–185
spatial analysis: agent-based models 22, 49–60, 68–81, 208, 229–266, 419–420, 423–485; endangered species habitats 208, 229–266; gang rivalry 420, 528–552; household/residential mobility 419–420, 423–485; inequality 424, 440–445; influenza epidemics 22, 68–81; obesity and weight loss 22, 49–60; population growth 237–239
speech variations 421, 553–569
speed of convergence 132
Sporns, Olaf 25–40
sprawl: agent-based models 419–420, 423–485; community food systems 589; household/residential mobility 419–420, 423–485
Stages of Change Theory 589
stakeholders: African Dryland rainfall 209, 384–402; African mining 209–210, 403–417; alternative fuel vehicles 341–343; curricula adoption/innovation 600, 607–610; environment and sustainability 209–214, 367–417; illegal logging 211–214; natural resource management 209, 367–383; ownership aspects 394; social network analysis 209, 367–383
state variables, coupled human and natural systems 271–276
status dislocation 182–184
status quo, African mining 411–413
STD *see* sexually transmitted diseases
Stefansson, Hlynur 209, 341–366
step-wise model building 84–85, 91, 93, 96–97, 103, 191–205
stochastic coupled human and natural systems 272, 278, 280
stocks-and-flow: chronic disease prevention 108, 111–115, 117–118, 121–124; education 604–605, 607–608, 611–617; syndemic risks: women and HIV 85, 90–91, 95–96; system dynamics 9–10, 604–605, 607–608, 611–617
storage, biofuel feedstocks 326–327
Stovel, Katherine 24, 164–190
street gang rivalries 420, 528–552
strength, commitment development 192–200–5
Stright, Dana 208–209, 321–340
structural embeddedness, gang rivalry 530
structural equivalence 528–529, 532
structural validity 98–100
student performance 602–618
submodels: Chitwan National Park, Nepal 273, 281–284, 287–288; illegal logging 207–208, 211–228; population growth 239–246; Wolong Nature Reserve, China 239–246, 273, 281–284, 287–288
substance abuse 84, 86–103, 571, 575, 582
suburban sprawl 448–450
suicide 31–32
supermarkets 590
supply chains, biofuels 208–209, 321–340
surprise aspects, coupled human and natural systems 268, 284–286, 289–292
susceptible-infectious-recovered (SIR) model 72
sustainability 207–417; African Dryland rainfall 209, 384–402; African mining 209–210, 403–417; agent-based models 207–210, 229–296, 341–366, 384–417; alternative fuel vehicles 209, 341–366; biofuels industry 208–209, 321–340; Chinese rural population growth 208, 229–266; Chitwan National Park, Nepal 208, 267–292; coupled human and natural systems 207–208, 211–296; endangered species habitats 208, 229–292; eutrophication 208, 297–320; Indonesian villagers and illegal logging 207–208, 211–228; lake eutrophication 208, 297–320;

logging practices 207–417; natural resource management 209, 367–401; network analysis 208–210, 297–320, 367–383, 403–417; panda habitats 208, 229–266, 269, 271, 273, 276, 281–285, 288–290; policy 208–209, 297–366; population growth impacts 208, 229–266; public policy 208–209, 297–366; rainfall 209, 384–402; resource management 209, 367–401; rural population growth 208, 229–266; social network analysis 209, 367–383; stakeholder engagement 209–214, 367–417; system dynamics 207–209, 211–228, 297–366; Wolong Nature Reserve, China 208, 229–292
Swarm programming 232, 236–238, 244, 246, 262–263
SWAT *see* Soil and Water Assessment Tool
Switzerland, alternative fuel vehicles 343
Syndemic Prevention Network 85, 87
syndemic risk, women and HIV 22–23, 84–103
system boundaries, lake eutrophication valuation 301–309, 316–317
system conceptualization, syndemic risks: women and HIV 87–90
system dynamics: adolescent commitment development 24, 191–206; alcohol consumption 23, 107, 109, 112–115, 118–120, 125; alternative fuel vehicles 209, 341–366; biofuels industry 208–209, 321–340; building models 24, 191–206, 606–617; cardiovascular disease 23, 107, 112–115, 123–126; causal loops 7–10, 421, 585–587, 603–607, 610; chronic disease prevention 23, 107–126; commitment development 24, 191–206; communities and social change 421–422, 585–595, 599–619; construct operationalization 607–608; definitions 111; diabetes 23, 107, 109, 112–115, 121–123, 125; diet 107–109, 120–123; environment and sustainability 207–209, 211–228, 297–366; eutrophication 208, 297–320; flows 9–10, 604–605, 607–608, 611–617; food communities 421–422, 585–595; group modeling efforts 608–609; health and human development 22–24, 84–103, 107–126, 191–206; illegal logging 207–208, 211–228; lake eutrophication 208, 297–320; limitations 10–11; model building 24, 191–206, 606–617; obesity 107–109, 111–112, 114, 120–123, 125–126; psycho-social health issues 22–23, 84–103; public health 23, 107–126; reference mode identification 610–612; smoking control 23, 107–109, 112–118, 122–126; stocks-and-flows 9–10, 604–605, 607–608, 611–617; strengths 10–11; sustainable mobility simulations 209, 341–366; syndemic risks: women and HIV 22–23, 84–103; tobacco control 23, 107–109, 112–118, 122–126; top-down models 7–11; women and HIV 22–23, 84–103
systems thinking, chronic diseases 108–109, 111, 123, 125–126

tax credits 331–334
teachers 148, 152, 154, 156–160, 602–618
teenagers 419, 558, 560, 566
temporal ties 172–174
territoriality 420, 528–552
testing, population growth 209, 246–261
theory–model linkages, commitment development 191–195, 200–205
therapeutic community residents 421, 570–584
thresholds, coupled human and natural systems 268, 285
tie distributions: adolescent romantic and sexual networks 164, 166, 173–177, 180–181; African mining 408–409, 412; household/residential mobility 419, 423–447; language changes/variation 554–560; natural resource management 369–371, 374–379; therapeutic community residents 572, 576
timber: fuelwood consumption 229, 234–263, 269, 272–285, 289; illegal logging 207–208, 211–228; panda habitats 229, 234–263
time lags, coupled human and natural systems 230, 232, 235, 249, 262, 268, 284–285
time representation: Chitwan National Park, Nepal 268, 275–276, 284–285; Wolong Nature Reserve, China 230, 232, 235, 249, 262, 268, 275–276, 284–285
Tita, George E. 420, 528–552
tobacco control 23, 107–109, 112–118, 122–126
Tominaga, Koji 208, 297–320
top-down models *see* system dynamics
Torrens, Paul 420, 448–485
training aspects, community food systems 585, 591
transitivity, therapeutic community residents 575–582
translational science 25–26, 28, 40
transmission, influenza epidemics 68, 70–73, 77
transportation: alternative fuel vehicles 209, 341–366; biofuel feedstocks 326–327; sprawl 453
transtheoretical obesity and weight loss models 49, 54–60, 63–64
triads 12, 16, 172, 185
triangulation 84
trust 405
type 2 diabetes 23, 107, 109, 112–115, 121–123, 125, 588

Udry, J. Richard 185
UK *see* United Kingdom

uncertainty: African Dryland rainfall 209, 384–402; lake eutrophication 299–301, 305–308, 311, 317; population growth 247–248; syndemic risks: women and HIV 96–97
UNHLM *see* United Nations High Level Meeting
United Kingdom (UK): chronic diseases 121, 126; natural resource management 209, 367–380
United Nations High Level Meeting (UNHLM) 107
Unwin, Nigel 23, 107–126
upstream indirect reciprocity 421, 570, 573–574, 576–577, 580–581
urban regions: agent-based models 419–421, 423–527; census data 420, 425, 434–435, 486–487, 491–506; diversity 420, 487–493, 499–505, 512–527; dynamics 419–619; economics 419, 423–447; food systems 421–422, 588–590; gang-related rivalry/violence 420, 528–552; household/residential mobility 419–420, 423–485; influenza epidemics 22, 68–81; language variation 421, 553–569; migration 419–420, 423–485; mixed-race households 419–420, 486–511; network analysis 419–421, 423–447, 511–569, 575; obesity and weight loss 22, 49–65; pattern-oriented modeling 419–420, 423–447; public elementary schools 158–160; race 419–420, 423–424, 428, 431, 434–435, 438, 441, 443–445, 486–511, 570, 574, 577–579, 581; residential mobility 419–420, 423–527; schools 157–160; segregation 419–421, 423–447, 486–527; sense of community 420, 512–527; simulations 419–420, 423–485, 612–617; social cohesion 420, 512–527; social networks 419–421, 423–447, 511–584; socioeconomic segregation 419, 423–447; sprawl 419–420, 423–485; therapeutic community residents 575; tie distributions 419, 423–447; women and HIV 22–23, 84–103
use suitability models, lake eutrophication 307–308

vaccinations 68–69
validating models: adolescent commitment development 203–204; Chitwan National Park, Nepal 284–285, 289–291; population growth 246–261; syndemic risks: women and HIV 98–100; Wolong Nature Reserve, China 246–261, 284–285, 289–291
value plurality, lake eutrophication 300
value tests, population growth 249–257
Vanemfjord lake, South-Eastern Norway 208, 297–319
Vansjø *see* Vanemfjord lake, South-Eastern Norway
Vansjø-Hobøl basin *see* Morsa catchment
variationist sociolinguistics 421, 553–569
vehicle-kilometre traveled (VKT) 354–355, 358
vehicles: sustainable mobility 209, 341–366
Vensim software 211, 612
verifying models: Chitwan National Park, Nepal 284–285, 289–291; natural resource management 378–379; population growth 246–261; Wolong Nature Reserve, China 246–261, 284–285, 289–291
Vespignani, Alessandro 25–40
villagers and illegal logging 207–208, 211–228
Vimmerstedt, Laura 208–209, 321–340
violence: gang-related rivalry 420, 528–552; women and HIV 86–103
Vitevitch, Michael 23, 130–146
VKT *see* vehicle-kilometre traveled
Vogt, Rolf D. 208, 297–320

Wagner, Suzanne Evans 421, 553–569
walking, obesity and weight loss 22, 50–53, 56–58, 59–61, 63–65
Warren, Keith 421, 570–584
Washington, DC 492
Water Framework Directive (WFD) 208, 297–320
water quality 208, 297–320
water-treatment programs 414–415
weight loss 22, 49–65, 120, 123; *see also* obesity
western Vansjø *see* Vanemfjord lake, South-Eastern Norway
WFD *see* Water Framework Directive
willingness, illegal logging 207–208, 211–228
willingness-to-pay, lake eutrophication valuation 304, 306, 308, 310–311, 314
Wolong Nature Reserve, China 208, 229–292
women: HIV 22–23, 84–103; therapeutic community residents 570–581
word form, language change/variation 553
workers, illegal logging 207–208, 211–228
workplace influenza scenarios 68, 70, 76, 79–81
Wright, Richard 419–420, 486–511
written word variations 421, 553–569

Zelner, Bennet 209–210, 403–417
Zvoleff, Alex 207–208, 267–296

www.ingramcontent.com/pod-product-compliance
Lightning Source LLC
LaVergne TN
LVHW080111011026
849160LV00013B/1070
* 9 7 8 0 4 1 5 8 4 3 3 4 8 *